Ambrish Maurya · Anmesh Kumar Srivastava ·
Pradeep Kumar Jha · Shailesh Mani Pandey
Editors

Recent Trends in Mechanical Engineering

Select Proceedings of PRIME 2021

Springer

Editors
Ambrish Maurya
National Institute of Technology
Patna, Bihar, India

Anmesh Kumar Srivastava
National Institute of Technology
Patna, Bihar, India

Pradeep Kumar Jha
Mechanical and Industrial Engineering
Indian Institute of Technology
Roorkee, India

Shailesh Mani Pandey
National Institute of Technology
Patna, Bihar, India

ISSN 2195-4356 ISSN 2195-4364 (electronic)
Lecture Notes in Mechanical Engineering
ISBN 978-981-19-7708-4 ISBN 978-981-19-7709-1 (eBook)
https://doi.org/10.1007/978-981-19-7709-1

This Springer imprint is published by the registered company Springer Nature Singapore Pte Ltd.
The registered company address is: 152 Beach Road, #21-01/04 Gateway East, Singapore 189721, Singapore

Lecture Notes in Mechanical Engineering

Lecture Notes in Mechanical Engineering (LNME) publishes the latest developments in Mechanical Engineering—quickly, informally and with high quality. Original research reported in proceedings and post-proceedings represents the core of LNME. Volumes published in LNME embrace all aspects, subfields and new challenges of mechanical engineering.

To submit a proposal or request further information, please contact the Springer Editor of your location:

Europe, USA, Africa: Leontina Di Cecco at Leontina.dicecco@springer.com
China: Ella Zhang at ella.zhang@springer.com
India: Priya Vyas at priya.vyas@springer.com
Rest of Asia, Australia, New Zealand: Swati Meherishi at swati.meherishi@springer.com

Topics in the series include:

- Engineering Design
- Machinery and Machine Elements
- Automotive Engineering
- Engine Technology
- Aerospace Technology and Astronautics
- Nanotechnology and Microengineering
- MEMS
- Theoretical and Applied Mechanics
- Dynamical Systems, Control
- Fluid Mechanics
- Engineering Thermodynamics, Heat and Mass Transfer
- Manufacturing
- Precision Engineering, Instrumentation, Measurement
- Tribology and Surface Technology

Indexed by SCOPUS and EI Compendex. All books published in the series are submitted for consideration in Web of Science.

To submit a proposal for a monograph, please check our Springer Tracts in Mechanical Engineering at https://link.springer.com/bookseries/11693

Preface

The first International Conference on Mechanical Engineering (PRIME-2021) was successfully held at the National Institute of Technology, Patna, a beautiful city of Patna sahib, from August 5 to 7, 2021. The sole theme of PRIME-2021 is "Progressive Research in Industrial and Mechanical Engineering" which focuses on the recent development from advanced technologies to digital technologies, surface engineering, environment safety, robotics, structural engineering, automobiles, and another interdisciplinary and incipient fields.

This conference provided an excellent platform to scientific researchers, industries, scholars, academicians, and engineers to exchange their areas of expertise, ideas, and the latest innovation in the area of their interest. Approximately, 100 abstracts were selected from more than 500 submitted abstracts, for the presentation purpose.

This book contains the selected proceedings of the International Conference on "Progressive Research in Industrial and Mechanical" (PRIME-2021). This volume targets the distinguished advanced topics of mechanical and industrial engineering. It covers current trends in nanotechnology, nano-mechanics, thermal, blockchain, ANN, IoT, material science, and many more. This book would be highly beneficial to researchers, academicians, students, and professionals who are working in the area of mechanical, allied science, and interdisciplinary fields.

We believe that advanced technologies of mechanical and industrial engineering plays a crucial role in energy saving, time saving, and environmental protection in the future and definitely would contribute to the economy of the country.

We are very much obliged for the valuable financial support of the National Institute of Technology, Patna, for the highly operative preparation and laborious work of the organizers.

At last, we thank to all the authors and members of PRIME-2021 committees and the publishers who made the successful publication of this lecture series possible through their endless effort and kind support.

Patna, India — Ambrish Maurya
Patna, India — Anmesh Kumar Srivastava
Roorkee, India — Pradeep Kumar Jha
Patna, India — Shailesh Mani Pandey

Contents

About the Editors

Dr. Ambrish Maurya is working as an Assistant Professor in the Department of Mechanical Engineering at National Institute of Technology (NIT) Patna, India. He has a teaching experience of more than 5 years and during his academic career, he has guided several Master's thesis and B.Tech. projects. He is presently also guiding two Ph.D. scholars. Dr. Maurya has completed his Master of Technology in Production and Industrial Engineering and Ph.D. in physical and mathematical modeling of continuous casting process of steelmaking from the Indian Institute of Technology (IIT) Roorkee in 2010 and 2017 respectively. He also has research expertise in the area of processing and fabrication of metal matrix composite, polymer composites, advanced casting technologies, computational fluid dynamics, and solar energy. Dr. Maurya has published more than 20 research articles and book chapters in international journals and conferences. He is also a reviewer for journals of national and international repute.

Dr. Anmesh Kumar Srivastava is currently working as Assistant Professor in the Department of Mechanical Engineering, National Institute of Technology (NIT) Patna, India. He obtained his B.Tech. in Mechanical Engineering in 2010 from Gautam Buddh Technical University, Lucknow. He received his M.Tech. in Energy Engineering from Malaviya National Institute of Technology (MNIT) Jaipur in 2012 and Ph.D. from MNIT Jaipur in 2018. His areas of interest in research include I.C. engines, alternative fuels for I.C. engines, emissions and renewable energy. He is the author of a number of research publications that have been published in reputed journals and conferences. He is a member of various professional societies.

Dr. Pradeep Kumar Jha is currently working as Professor in the Department of Mechanical and Industrial Engineering. After receiving Bachelor's and Master's degree from M.I.T. Muzaffarpur, Bihar and National Institute of Foundry and Forge Technology, India in 1995 and 1999 respectively, Dr. Jha obtained his Ph.D. from Indian Institute of Technology (IIT) Kharagpur in 2004. Prior to joining as Assistant Professor in IIT Roorkee in 2007, he also taught at IIT Guwahati and IIT (ISM) Dhanbad. Dr. Jha specializes in the area of manufacturing engineering, with special emphasis on casting operations, modelling and simulation of continuous casting operations and metal matrix composites. He has to his credit more than 100 publications in international and national journals, conferences and book chapters. In his broad area of expertise, he has supervised 10 doctoral, 40 M.Tech. theses and 25 B.Tech. theses. He has dealt many sponsored research projects from government funding agencies. Dr. Jha has been involved with teaching of subjects with manufacturing and thermal streams. He has been an active contributor to the online education platform, NPTEL, sponsored by Government of India, by developing and running eight different courses in the domain of his teaching and research.

Dr. Shailesh Mani Pandey is an Assistant Professor in the Department of Mechanical Engineering at National Institute of Technology (NIT) Patna, India. He has over 13 years of teaching/research and industry experience. Dr. Pandey completed his M.Tech and Ph.D. from Delhi Technological University. His research interest lies in tribology, surface modification, coating, friction, wear, SEM, XRD, composite materials, etc. Dr. Pandey has published various research articles in journals of national and international repute and guided several master's theses.

Development of a Combined Evaporative Cooling System with Eco-Friendly Nozzles and Encapsulated Phase Change Material

Shubham S. Gosavi, Aditya P. Magdum, Ranjit K. Kharade, Harshad A. Deshmukh, and Mandar A. Jadhav

Abstract Power outages affect the functionality of commercial and industrial air conditioners. The processes followed by AC units and emissions impacts on global climate. Recently, Indian government banned imports of AC and refrigeration products and components which were supplied from China. But Indian Government understands people's need for cooling and realizes that an exponential rise in demand for cooling is inevitable. Considering such critical situations, we have designed an eco-friendly air-conditioning module which contains PCM-based heat exchanger for indirect evaporative cooling and eco-friendly designs for direct evaporative cooling. Eco-friendly designs of convergent shape are manufactured by considering software analysis results and by following the same steps required for manufacturing of soil pot. Software analysis results of the whole experimental setup showed a temperature reduction in the range of 7–10 °C. Required humidity is achievable by controlling some parameters. Material used for this project is completely economical and easily available. Poisonous and harmful refrigerants and greenhouse gases are not used in this project. It is smaller, compact, economical and efficient. Also, there are no noticeable direct and indirect emissions from this experimental setup. This research paper includes software analysis for studying cooling effect and humidity control, design and performance analysis of PCM-based heat exchanger and related research work to achieve desired temperature reduction.

Keywords Eco-friendly air-conditioning · Direct evaporative cooling · Indirect evaporative cooling · Economical design

1 Introduction

Energy conservation is energy independence in the true sense. Due to the increased rate of global warming, environmental planning became important to achieve sustainable goals. Although industrialization plays an important role in achieving sustainable

S. S. Gosavi (✉) · A. P. Magdum · R. K. Kharade · H. A. Deshmukh · M. A. Jadhav
Sharad Institute of Technology College of Engineering, Yadrav, India
e-mail: shubhamgosavi5055@gmail.com

A. Maurya et al. (eds.), *Recent Trends in Mechanical Engineering*, Lecture Notes in Mechanical Engineering, https://doi.org/10.1007/978-981-19-7709-1_1

goals, energy generation devices like diesel generators used in industries pollute the air and heat rejected from it becomes the reason for global warming.

Cooling is not only important for enhancing the productivity, health and wellbeing of people but also in some cases, be necessary for survival. The operation of air-conditioning units will also have an impact on our climate. The emissions associated with their electricity consumption and refrigerants will only exacerbate the global temperature rise. The direct and indirect emissions from room ACs alone could contribute to as much as a 0.5 °C increase in global warming by 2100. Currently, India contributes only 5% of the global annual emissions from room ACs. India is predicted to account for over 25% of annual emissions globally by 2050 due to the unprecedented rise in comfort cooling demand, particularly in the residential sector.

In this project, combined evaporative techniques [1, 2] have been implemented. Phase change material (PCM) has been used for indirect evaporative cooling [3] and water and eco-friendly designs for direct evaporative cooling [4] using evaporative techniques. Mud pots are an important part of India's tradition and art. This inspired us to solve the above-mentioned problems. It will help us to save energy and reduce pollution and global warming problems. In this project, encapsulation method for phase change materials (PCMs) is effectively utilized.

2 Experimental Setup Designing and Investigation

Figure 1 is a proposed model of our project. It is designed and rendered by using Autodesk's Fusion 360 software. This shows an actual design plan that is manufactured. Above design contains nozzles, heat exchanger, motor, water tank and support stand.

2.1 *Aqueous Salt Solutions for Cooling Application*

Phase change material is used in cold storage absorb heat from targeted body and liquefies. There are some feasible thermo-physical and chemical properties for a material to be used as a phase change material (PCM) [5].

Aqueous salt solutions of sodium chloride (NaCl) and potassium chloride (KCl) are found to be suitable phase change materials (PCMs) for the cooling applications and cold storage systems. These salts were selected because of their easy availability at low cost, and melting-freezing temperature ranges of these mixtures are feasible for cold storage. NaCl and KCl possess all mentioned properties. Hence, both materials are used as PCM in the aqueous solution [5, 6]. This aqueous salt solution is poured in the cooling module shown in Fig. 1 and is placed at the bottom of the experimental setup which helps to give additional cooling effect to flowing water in the system and air flowing over it. It also has space to keep water bottles for cooling. Phase change material-based cooling module has 5 L of volume and contains a mixture of water,

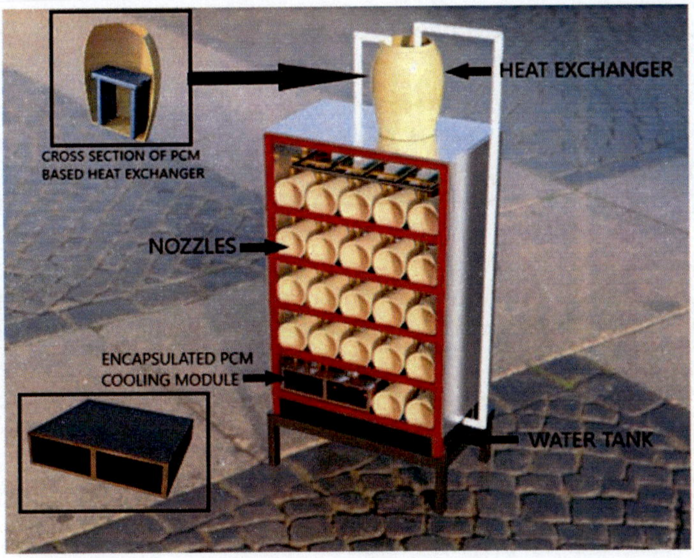

Fig. 1 Prototype of experimental setup

NaCl, KCl and glycerin. Percentage of NaCl in this mixture is 15%, percentage of KCl is 2.5% and glycerin is added to avoid volumetric expansion of KCl [5, 6].

2.2 *Phase Change Material-Based Water Cooler*

Encapsulation of phase change material (PCM) is a method to increase the heat transfer and it avoids phase change material (PCM) from mixing with the heat transfer fluid. [7, 8] The phase change material (PCM) container used for encapsulation should possess qualities such as strength, flexibility, corrosion resistance and thermal stability. While developing a thermal storage system, it is necessary to be focused on the corrosion of the construction material when it is used with different PCMs at low temperature [8, 9]. As shown in Fig. 1, three charged encapsulated silica gel pouches are used for this water cooler module (PCM-based heat exchanger). This is a type of macro-encapsulation. The temperature of cooled water reaches close to 5°C due to silica gel encapsulated pouches. Then, cooled water is sprinkled on the whole setup for further evaporation cooling.

2.3 *Eco-Friendly Convergent Nozzles*

A nozzle is a device which increases the velocity of fluid by decreasing the pressure in doing so the temperature of fluid also decreases [10, 11]. The reason behind the increase in velocity is to eliminate the use of fans at output to throw the air out. By the use of these clay nozzles, temperature decreases due to nozzle shape and temperature decreases due to evaporative cooling technique [12]. For nozzles, red soil is used because it has very good water holding capacity, which is beneficial for evaporative cooling technique. Evaporative cooling method is the method of reduction in temperature from the evaporation of liquid which removes latent heat from the surface from which evaporation takes place [12]. With all this comprehensive research work, experimental setup has been manufactured as shown in Fig. 2.

3 Design Characteristics, Transient Flow and Thermal Analysis of Convergent Nozzles

Smaller and compact cooling module is preferred for room cooling. Considering the aesthetics factor, two nozzles have been designed having a length of 30 cm so, it will occupy small space in the room and can be fit partially into the fill level below the window. Also, the outer diameter is kept 6 cm to compress the air stream and maintain linear flow. Inner diameters of those nozzles are 10.8 inlet diameter and 12 cm diameter, respectively.

Fig. 2 Manufactured experimental setup

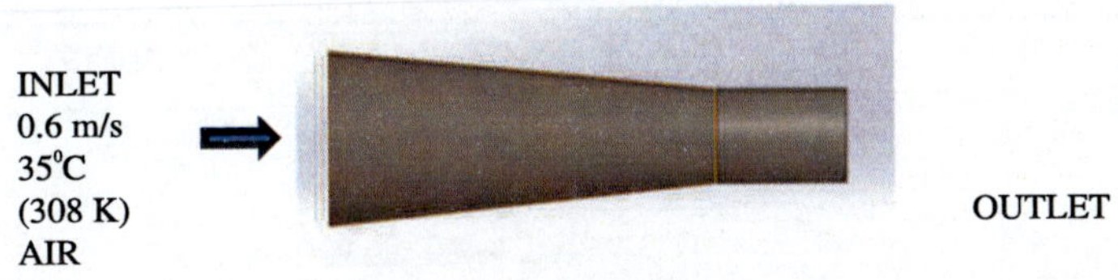

Fig. 3 Air flow direction

3.1 *CFD Analysis and Selection of Suitable Nozzle*

ANSYS Fluent v18.2 software is used to perform CFD analysis. The nozzles design and geometry were modeled in Fusion 360 software. For meshing of both the nozzle models, hexahedral and wedge-shaped elements are used with 3 mm size.

The analysis was performed to determine the velocity of air and temperature drop of air. Based on the CFD results, the velocity for the 10.8 cm nozzle is 1.76 m/s, and for the 12 cm nozzle, it is 2.49 m/s. Average reduced temperature is 306 K for both nozzles as given in Table 2.

3.1.1 Flow Conditions

Figure 3 is software design of convergent nozzle. As shown in Fig. 3, air flows through the major diameter of convergent nozzle hence due to compression effect, temperature of air decreases and velocity decreases.

3.1.2 Boundary Conditions and Other Considerations (For Model Shown in Fig. 3)

1. Inlet air velocity is taken as 0.6 m/s
2. Outlet pressure is taken as 0 pa
3. Temperature considered is room temperature (300 K)
4. Heat flow through soil is set as 'CONVECTION'
5. Bulk (Initial Temp) is taken as 5 °C (278 K)
6. Thermal heat coefficient is taken as 11.25 W/m^2K for convective heat transfer.

3.1.3 Meshing

Element size selected for this meshing is 3 mm. Hexahedral and wedge-shaped elements are used for the same.

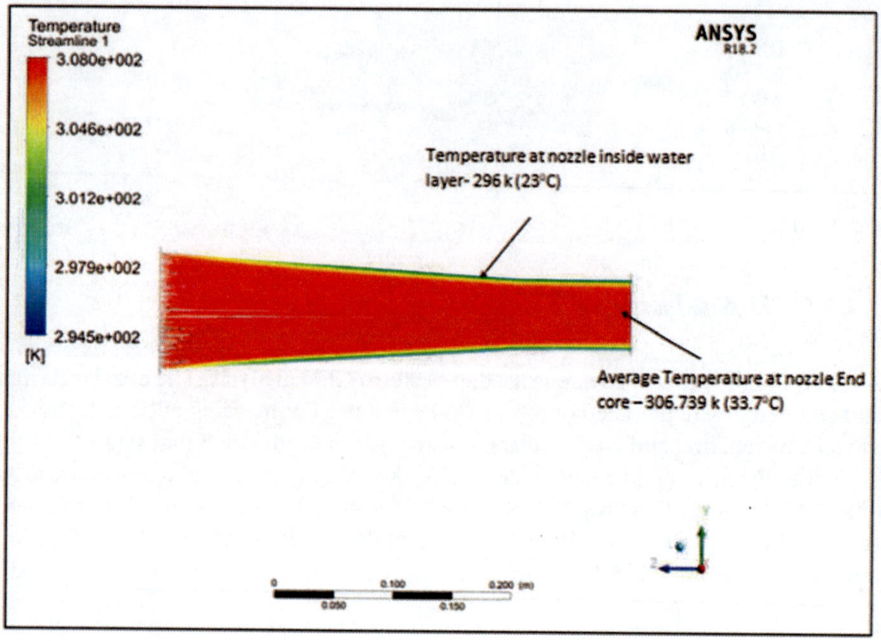

Fig. 4 Temperature analysis of convergent nozzle with 10.8 cm inlet diameter

4 Results of CFD Analysis and Discussion

4.1 CFD Analysis Results of Nozzle Having 10.8 cm Inlet Diameter

See Figs. 4 and 5.

4.2 CFD Analysis Results of Nozzle Having 12 cm Inlet Diameter

Considering Figs. 5 and 7, convergent nozzle having 12 cm inlet diameter show higher compression of air resulting in higher velocity output which is higher than velocity output of convergent nozzle having 10.8 cm inlet diameter by 0.74 m/s. Higher compression also helped to reduce temperature as shown in Figs. 4 and 6 (Table 1).

Table 1 Summary from Figs. 4, 5, 6 and Fig. 7

S. no.	Nozzle model	Average temp (reduced) (K)	Average velocity (increased) (m/s)
1	10.8 convergent nozzle	306.739	1.75924
2	12.0 convergent nozzle	306.867	2.49408

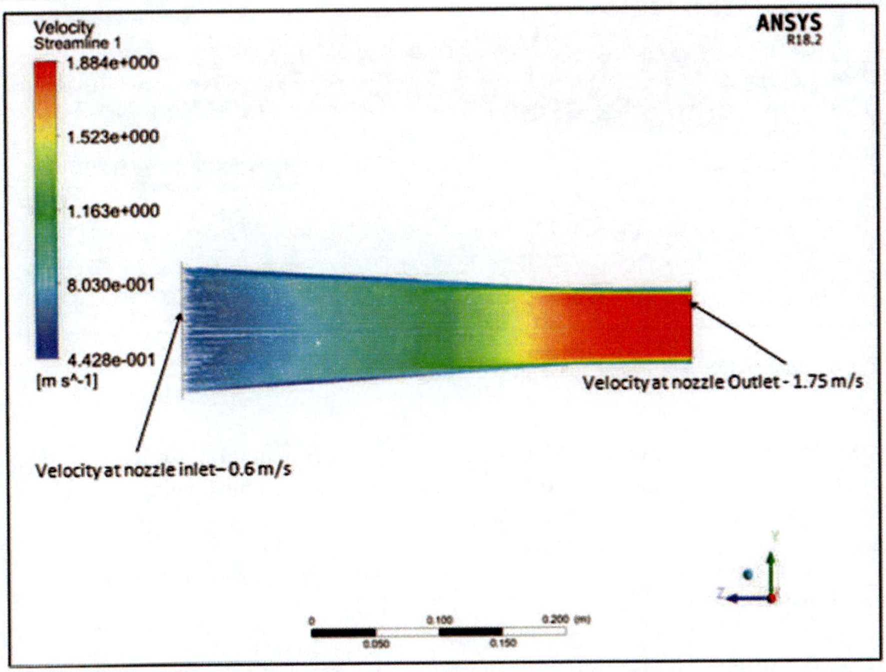

Fig. 5 Velocity analysis of convergent nozzle with 10.8 cm inlet diameter

5 Experimental Analysis

Following readings in Table 2 have been taken under natural atmospheric conditions on 19/3/2021.

6 Findings

As shown in Fig. 8, temperature reduction is achieved by combining the effect of direct and indirect cooling methods performed during experimentation. This cooling effect depends on air flow rate inside the nozzles, temperature of sprinkled water on nozzles and atmospheric temperature. So macro-encapsulation of PCM helped to

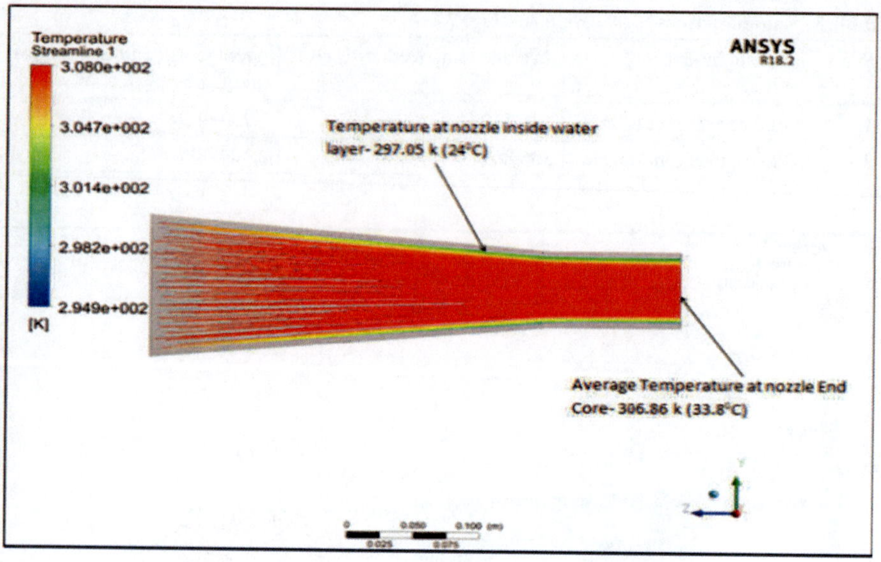

Fig. 6 Temperature analysis of convergent nozzle with 12 cm inlet diameter

lower the temperature of sprinkled water during experimentation. Convergent shape of nozzles plays a huge role in reducing the temperature of air flowing through it.

7 Conclusion

Comprehensive experimental work has been performed with this experimental module without affecting aspects and prospects of the window; it showed that phase change materials help to reduce temperature fluctuations in cooling applications by absorbing heat from the system. Proper water dripping method implemented in this project is useful for filtering air and direct evaporative cooling helps for controlling humidity and hence good air quality is achieved. Eco-friendly convergent nozzles made up of pot soil are cost effective, easy to manufacture, smaller in size and its material is reusable. Convergent nozzles having inlet diameter of 12 cm give best velocity and temperature output. Input velocity was 0.6 m/s considering diesel generator output, and after velocity analysis using ANSYS software, it increased more than 2 m/s. Arrangement of eco-friendly nozzles in a rectangular array and cooling capacity of this experimental setup increases with respect to time and no direct or indirect emissions have been detected during experimentation.

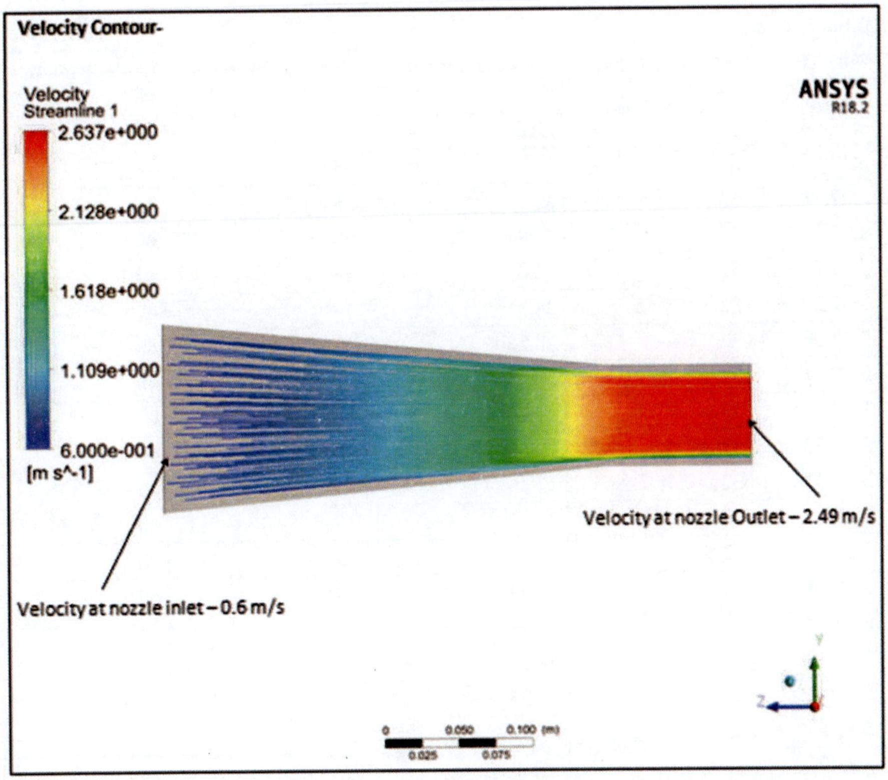

Fig. 7 Velocity analysis of convergent nozzle with 12 cm inlet diameter

Table 2 Observation table

Time (PM)	Inlet temperature (°C)	Outlet temperature (°C)	Inlet humidity (%)	Outlet humidity (%)
3:36:36	32	31.77	35	36
3:36:40	32	31.60	35	37
3:36:44	32	31.50	35	38
3:36:48	32	31.28	35	39
3:36:52	32	30.79	35	39
3:36:56	32	30.79	35	41
3:37:00	32	30.30	35	42

(continued)

Table 2 (continued)

Time (PM)	Inlet temperature (°C)	Outlet temperature (°C)	Inlet humidity (%)	Outlet humidity (%)
3:37:04	32	29.81	35	43
3:37:09	32	29.33	35	44
3:37:13	32	29.33	35	45
3:37:17	32	28.84	35	46
3:37:21	32	28.84	35	47
3:37:25	32	28.35	35	47
3:37:29	32	28.35	35	48
3:37:31	32	28.35	35	48
3:37:33	32	27.86	35	49
3:37:35	32	27.86	35	49
3:37:37	32	27.86	35	49
3:37:39	32	27.54	35	50
3:37:41	32	27.54	35	50
3:37:43	32	27.54	35*	50*
3:37:45	32	27.54		
3:37:49	32	27.37		
3:37:53	31	27.37		
3:37:57	31	27.37		
3:38:02	31	26.88		
3:38:06	31	26.88		
3:38:10	31	26.39		

*Humidity kept constant by controlling water flow on experimental setup. It is measured by using DHT 22 sensor

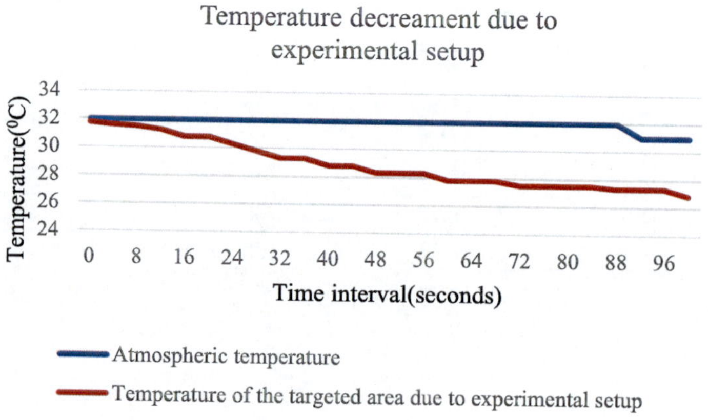

Fig. 8 Temperature decrement due to experimental setup

References

1. Amer O (2015) A review of evaporative cooling technologies. Int J Environ Sci Dev 6(2):111–117. https://doi.org/10.7763/IJESD.2015.v6.571
2. Cui X, Chua KJ, Yang WM (2014) Use of indirect evaporative cooling as a pre-cooling unit in humid tropical climate: an energy saving technique. Energy Proc 61:176–179
3. Porumb B, Ungureşan P, Tutunaru LF, Şerban A, Bălan M (2016) A review of indirect evaporative cooling technology. Energy Proc 85:461–471
4. Amer O, Boukhanouf R, Ibrahim HG (2015) A review of evaporative cooling technologies. Int J Environ Sci Dev 6(2):111–117. https://doi.org/10.7763/IJESD.2015.v6.571
5. Yilmaz S, Sheth Karathia F, Martorell I, Paksoy HO, Cabeza LF. Salt-water solutions as PCM for cooling applications
6. Oró a E, de Gracia A (2012) Review on phase change materials (PCMs) for cold thermal energy storage applications. Appl Energy 99:513–533. Elsevier
7. Oróa E, de Graciaa A, Castell A, Farid MM, Cabeza LF. Review on phase change materials (PCMs) for cold thermal energy storage applications. A GREA Innovació Concurrent, Universitat de Lleida, Edifici CREA, Pere de Cabrera s/n, 25001 Lleida, Spain Department of Chemical and Materials Engineering, The University of Auckland, Auckland, New Zealand
8. Surendra Reddy M, Venkatesh G et al (2016) Performance improvement of a domestic refrigerator by using phase change material. Int J Adv Res Sci Eng 5(7)
9. Hossen Khan Md I, Afroz HMM (2014) Diminution of temperature fluctuation inside the cabin of a household refrigerator using phase change material. Int J Recent Adv Mech Eng (IJMECH) 3(1)
10. Maddu YR, Saidulu S, Md, Azeem JS (April 2018) Design and fluid flow analysis of convergent-divergent nozzle. Int J Eng Technol Sci Res IJETSR, ISSN 5(4):2394–3386
11. Arif U, Arif M (2017) Feasibility study of using nozzles for air cooling in air cooling system. Int J Innov Res Sci Eng Technol 6(7):4
12. Inamdar SJ, Junghare A, Kale P (2019) Performance enhancement of evaporative cooling by using bamboo. Int J Eng Adv Technol Spec Issue 8(6S):856–860. https://doi.org/10.35940/ijeat.f1161.0886s19

Multi-objective Optimization of Green Drilling Parameters on HcHcr Steel Using GRA-TOPSIS with PCA Method

Sandeep Kumar and Abhishek Singh

Abstract Green CNC drilling is a machining process where deionised water, tap water and dry air are used as coolant to increase the efficiency, quality of product and reduce the harmful industrial waste (solid and gaseous). In this research paper, GRA and TOPSIS with PCA method are used to optimized the multiple response parameters of green CNC drilling on HcHcr using HSS twist drill bit. The work considered point angle, spindle speeds, drill diameters and feed rates as input variables, while tool vibration, burr length and thrust force as output parameters. Initially, experimentation is performed based on Taguchi's L_{27}-OA and recorded the response parameters. Additionally, ANOVA is carried out to determine the influential parameter at 95% confidence level. The results show that optimum conditions of drilling parameters using GRA-PCA are point angle 136°; spindle speed 1800 rpm; feed rate 1.2 mm/rev and drill diameters 8 mm and using TOPSIS-PCA point angle 136°; Spindle speed 900 rpm; feed rate 0.8 mm/rev and drill diameters 8 mm. Further, analysis of variance is performed on grey grade value (in case of GRA-PCA) and observed that the drill diameter is highly influential process parameters with the highest 49.63% contribution. Similarly, analysis of variance is performed on preference value (in case of TOPSIS-PCA) and it is found spindle speed is most significant process parameters with the highest 86.71% contribution. The confirmatory tests are performed on obtained optimal values and results show that the final gains are 0.007006 and 0.005528 in grey grade value and preference value respectably.

Keywords Green drilling · Taguchi's method · GRA · TOPSIS · PCA · Optimization · ANOVA

S. Kumar (✉) · A. Singh
Department of Mechanical Engineering, NIT Patna, Patna, India
e-mail: sandeepk.ph21me@nitp.ac.in

A. Maurya et al. (eds.), *Recent Trends in Mechanical Engineering*, Lecture Notes in Mechanical Engineering, https://doi.org/10.1007/978-981-19-7709-1_2

1 Introduction

The high-carbon high-chromium steel (HcHcr) is extensively used in tools, dies and moulds manufacturing because it exhibits outstanding abrasion, excellent dimensional stability, wear resistance and hot hardness. Among all machining process in manufacturing, drilling is popular method to producing cylindrical holes and enlarging the holes in solid material for riveting and fastening [1]. In the drilling process, shear deformation, material fracture, impact forces, multi-points contact and friction are involved which induced inaccuracy like delamination, roundness error, interlaminar crack propagation, surface roughness, splintering, swelling and burrs to affect the performance and quality characteristics [2]. Hence, the response parameters like MRR, TWR, delamination, thrust force, torque, crack propagation, surface roughness, burrs height, burr length, thermal damage, power consumption and hole diameter error are significantly affected by input process parameters like SS, FD, DD, temperature, tool geometry and other parameters. The existing manufacturing and production industry produce large amount of poisonous industrial waste disposal (solid, liquid and gaseous) which creates hazardous and serious health and environment problems. Also, manufacturing process is energy exhaustive activity which directly or indirectly influences the environment. Thus, the green manufacturing is used to enhance the efficiency of CNC drilling operation and decrease the health and environmental affects during drilling process and there are two major components of CNC green drilling, manufacturing components (process time, tool wear rate, MRR, and surface quality) and environmental components (process energy, dielectric consumption and exposure of aerosol). The amount of industrial waste disposal and response parameters of drilling process is extremely affected by drilling process parameters. Hence, the optimization of drilling parameters is important work to achieve the green CNC drilling [3]. Green CNC drilling is machining process where deionised water, tap water and dry air are used as coolant and it also includes efficient uses of resources, minimization of energy consumption and environmental impact. On other hand, a number of researcher worked on MRR, surface characterization, thrust force, torque, thermal damage and hole diameter error as response parameters with drills geometry, FD, SS, temperature as process parameters such as the effect of machining conditions, and temperature on quality holes and tools life [4]; the effects of process parameters on drilling damage of CFRP [5]; the analysis of effect of CNC drilling process parameter on delamination and surface quality of different composite materials[6]; the investigation of drilling parameters on AISI B_{1113} with drill bit of M_2 HSS[7]; the analysis of drilling parameters on Al-Sic composite of metals matrix and Al-Sic-Gr hydride composite[8]; the estimation of drilling input variables on response variable on machining of composites (Al-2219+8%B4C), hybrid composites (Al22198%B4C3%Gr) and pure base alloy (Al2219) [9]; the investigation and optimization of CNC drilling input parameters on response parameters like MRR and SR of AISI D_3 steel [10]. Additionally, to improve the performance and drill holes quality characteristics, the selection of optimal drilling parametric setting combination of process parameters is essential that affects directly or indirectly performance, quality and efficiency of CNC green drilling, while many researchers have been used different optimization technique to find out the optimal combination of process parameters like multi-objective optimization of CNC drilling parameter like PA, SS, FR and DD on response variable such as SR, DE and TF for HcHcr steel with HSS drill bit using Taguchi's-based utility concept and GRA-PCA methods [11]; Taguchi's technique is used to optimize the drilling parameter on H21 steel [12]. Hence, from past literature, no works have been found on multi-objective optimization of green drilling parameters on HcHcr steel using GRA-TOPSIS with PCA.

The aim of present research work is multi-objective optimization to determine and compare the optimal parametric setting and their importance on green CNC drilling process to minimize the tool vibration (TV), burr length (BL) and thrust force (TF) using GRA-TOPSIS with PCA on HcHcr by DD, SS, FD and PA as input process variable to attain GM.

2 Material and Methods

2.1 *Work Material*

In the present work, three rectangular plate of HcHcr of dimension 20 mm x 30 mm × 10 mm is used as a work material and with their chemical compositions are %C:1.83; %Si:0.25; %Mn:0.6; %Cr:12.45 and properties are: thermal expansion(/°C): 10.4×10^{-6}; Melting point (^{0}C):1421; Poisson's ratio: 0.27–0.3, elastic modulus (GPa): 190–210, hardness (BHN): 62; density (kg/m^3): 7.7×10^3.

2.2 *Design of Experiment*

In this research work, Taguchi's fractional factorial design of experiment is used to estimate the effects of green CNC drilling process parameters on response parameters. The four process parameters like DD, SS, FR and PA are varied with their levels to estimate the affects on response parameters such as TV and TF and considering process parameters and their levels L_{27} orthogonal array used for experimentation as revealed in Table 1.

2.3 *Experimental Procedure and Measurement*

In this research works, the experiments are performed on the KMC-11VC Vertical CNC machine using HSS twist drill as tool and deionized water as coolant. The L_{27}-OA of Taguchi's design of experiment is used for experimentation and all experiments

Table 1 Process parameters and their levels

Inputs parameters	Symbols	Units	Level 1	Level 2	Level 3
Point angle	PA	Degree	118	127	136
Speed of spindle	SS	rpm	900	1800	2700
Feed rate	FD	mm/min	80	120	160
Drill bit dia	DD	mm	8	10	12
Coolant	Deionised water				

repeated two time and mean of response value used for investigation to achieve the high accuracy. The vibrometer is used to determine the vibration signal and burr length measure using tool makers microscope. The thrust force (TF) is determined directly by dynamometer and recorded in Table 2.

2.4 *Principal Component Analysis (PCA)*

PCA is a dimensionality reduction method use to calculate weight of response variables in multi-attribute optimization problems. This method is based on matrix structure of variance and covariance by means of linear combination of each response parameters. The following steps involved to determine the weight using PCA [12].

Step-1: The response parameters orthogonal array is constructed as

$$Y_{ij} = \begin{bmatrix} y_{11} & y_{21} & \cdots\cdots\cdots & y_{1n} \\ y_{21} & y_{22} & \cdots\cdots\cdots & y_{2n} \\ \cdots\cdots & \cdots\cdots & \cdots\cdots\cdots & \cdots\cdots \\ y_{m1} & y_{m1} & \cdots\cdots\cdots & y_{mn} \end{bmatrix}, \ \mathrm{i} = 1, 2, 3 \ldots \mathrm{m} \ \text{ and } \mathrm{j} = 1, 2, 3 \ldots \mathrm{n} \qquad (1)$$

Y_{ij} is decision matrix, *m* is experimental trails n is numbers of response variables, y is response value on experimental trails and m = 27, n = 3.

Step-2: The estimation of correlation coefficient array is determined by using

$$C_{jl} = \left[\frac{\mathrm{cov}(y_i(j), y_i(l)}{\sigma_{y_{i(j)}} \times \sigma_{y_{i(l)}}}\right], \ \mathrm{j} = 1, 2, 3, \ldots, \ \mathrm{n} \ \text{ and } \ \mathrm{l} = 1, 2, 3, \ldots, \mathrm{m} \qquad (2)$$

$\mathrm{cov}(y_i(j), y_i(l)$ is the covariance of trails sequence yi(*j*), yi(*l*) and $\sigma_{y_{i(j)}} \times \sigma_{y_{i(l)}}$ are standard deviation(σ_x) of covariance sequences yi(*j*) and yi(*l*), respectively.

Step-3: The calculation of Eigen's value and Eigen's vector using correlation coefficient array as Eq. (3).

Table 2 L_{27} orthogonal array and response variable

Exp. No.	Inputs drilling parameters				Output drilling parameters			GRG value	Preference value
	Drill diameter	Feed rate mm/rev	Speed of spindle	Point angle (deg.)	Vibration (μm)	Burr length (μm)	TF (N)		
1	8	0.8	900	118	0.02665	0.202	46.95	0.216	0.116
2	8	1.2	1800	118	0.02165	0.107	57.33	0.231	0.062
3	8	1.6	2700	118	0.02665	0.483	65.79	0.141	0.042
4	10	1.6	900	118	0.03265	0.137	49.79	0.207	0.116
5	10	1.2	1800	118	0.02865	0.110	69.81	0.188	0.062
6	10	0.8	2700	118	0.03465	0.301	55.19	0.158	0.042
7	12	1.6	900	118	0.03565	0.054	75.19	0.194	0.116
8	12	0.8	1800	118	0.03165	0.429	59.08	0.143	0.062
9	12	1.2	2700	118	0.03765	0.132	68.79	0.167	0.042
10	8	0.8	900	127	0.02765	0.113	44.95	0.246	0.124
11	8	1.2	1800	127	0.02165	0.163	51.79	0.228	0.066
12	8	1.6	2700	127	0.02365	0.035	64.83	0.242	0.045
13	10	1.6	900	127	0.03365	0.093	55.79	0.204	0.124
14	10	1.2	1800	127	0.02565	0.178	66.61	0.181	0.066
15	10	0.8	2700	127	0.03065	0.437	49.85	0.166	0.045
16	12	1.6	900	127	0.03465	0.182	67.79	0.160	0.124
17	12	0.8	1800	127	0.02865	0.159	51.79	0.203	0.066
18	12	1.2	2700	127	0.03265	0.328	62.95	0.145	0.045
19	8	0.8	900	136	0.01965	0.211	43.79	0.266	0.131
20	8	1.2	1800	136	0.01865	0.120	50.79	0.266	0.070

(continued)

Table 2 (continued)

Exp. No.	Inputs drilling parameters				Output drilling parameters			GRG value	Preference value
	Drill diameter	Feed rate mm/rev	Speed of spindle	Point angle (deg.)	Vibration (μm)	Burr length (μm)	TF (N)		
21	8	1.6	2700	136	0.02265	0.379	64.85	0.166	0.048
22	10	1.6	900	136	0.02565	0.504	51.79	0.168	0.131
23	10	1.2	1800	136	0.02565	0.406	62.93	0.154	0.070
24	10	0.8	2700	136	0.02965	0.174	47.85	0.210	0.048
25	12	1.6	900	136	0.02865	0.322	63.45	0.153	0.131
26	12	0.8	1800	136	0.02665	0.420	49.79	0.177	0.070
27	12	1.2	2700	136	0.03165	0.192	61.79	0.169	0.048

$$(C - \lambda_k..I_m)U_{ik=0} \tag{3}$$

where λ_k is the Eigen values $\sum_{k=1}^{n} \lambda_k = n, k = 1, 2, 3, n$. $U_{ik} = [ak_1, ak, ... ak_m]^T$ are Eigen's vector consequent to the Eigen value λ_k.

Step-4: Principal component calculated as

$$P_{mk} = \sum_{i=1}^{n} Y_m(i).U_{ik} \tag{4}$$

where Pm_1 Pm_2 Pm_3 called first, second and third principal components, respectively, and Pm_1 is responsible for more variance in response parameters. The weight factor of each response parameters like TV, BL, and TF are 0.287, 0.409 and 0.304, respectively.

2.5 *Analysis of Variance (ANOVA)*

ANOVA is the statistical tool for analysis of effects of process variable on response parameters. In the present study, ANOVA performed on responses parameters to find out the most influential green CNC machining process parameters and estimate the comparative significance of process parameters in terms of their percentage contributions. Also in ANOVA, Fisher's ratio test and probability test are performed [13]. Tables 3 and 4 illustrate the result of ANOVA of GRG and preference value, respectively.

Table 3 Results ANOVA for grey grade value

Source	*Dof*	*Square SS*	*Adj.SS*	*Adj.MS*	F_{value}	P_{Value}	*%Contributions*
Regression	4	0.019279	0.019279	0.004820	6.18	0.002	52.90
Drill diameter	1	0.018211	0.013411	0.013411	17.19	0.000	49.63
Feed rate	1	0.005248	0.002088	0.002088	2.68	0.116	14.30
Spindle speed	1	0.008144	0.004244	0.004244	5.44	0.029	22.19
Point angle	1	0.000677	0.000377	0.000377	0.48	0.494	1.03
Error	22	0.003811	0.017163	0.000780			10.50
Total	26	0.036091					100.00

Table 4 Results of ANOVA for preference value

Source	*Dof*	*SquareSS*	*Adj.SS*	*Adj.MS*	F_{value}	P_{Value}	*% Contributions*
Regression	4	0.028441	0.028441	0.007110	81.84	0.000	93.70
Drill diameter	1	0.000000	0.000000	0.000000	0.00	1.000	0.00
Feed rate	1	0.001663	0.000173	0.000173	1.99	0.172	5.48
Spindle speed	1	0.026358	0.026358	0.026358	303.39	0.000	86.84
Point angle	1	0.000421	0.000421	0.000421	4.84	0.039	1.39
Error	22	0.001911	0.001911	0.000087			6.30
Total	26	0.030353					100.00

3 Results and Discussion

3.1 *Multi-objective Optimization Using GRA-PCA*

GRA is an optimization technique performed by multiple attributes transferred into an equivalent response characteristics which is known as grey grade value (GRG) [14]. The multi-response optimization of performance characteristics such TV, BL, and TF is carried out using GRA-PCA. The GRG value of all experimental trails of L_{27}-OA is used for optimization of CNC green drilling parameters as given in Table 2. The GRG of all parametric combination is ranked according GRG value, and highest GRG is considered as rank one. Thus obtained result of the experimental trail 20 shows the best experimental parametric combination for desirable performance characteristics which has highest GRG value of 0.266 and rank one. The optimal combination of process parameters considering highest GRG value is 8 mm DD, 1.2 mm FD, 1800 rpm SS and 136° PA.

3.2 *Confirmatory Experiments for GRA-PCA*

After the estimation of optimal process parameters, the determination and confirmation of the improvement in response parameters using optimal parametric combination of process parameters is performed and calculated GRG($\overline{\gamma}$) value using optimal set of design parameters are as follows:

$$\overline{\gamma} = \gamma_\eta + \sum_{j=1}^{n} (\overline{\gamma_j} - \gamma_\eta) \tag{5}$$

where γ_η = total GRG mean, γ_j = GRG mean at optimal level, n = number of variable influencing the response characteristics. Hence, estimated GRG value using Eq. (5) is 0.273006 and improvement in GRG value is 0.007006, i.e. 2.633%.

3.3 ANOVA for GRA-PCA

The effects of process parameters on various response variables are estimated by analysis of variance and results for grey relational gades value with factor response larger is better at 95% level using minitab17 are illustrated in Table 3 and observed that DD and SS are significant parameter influencing the response parameters with the highest contribution 49.63% and 22.19% respectively while FD, and PA are insignificant parameters.

3.4 Multi-objective Optimization Using TOPSIS-PCA

TOPSIS is a simple ranking multi-criteria decision-making method that helps to determine the most suitable alternative from a finite set by the solution of multi-objective optimization problem. The multi-response optimization of performance characteristics such as tool vibration, burr length and thrust force is carried out using TOPSIS-PCA [15]. The preference value of all experimental trails of L27-OA is used for optimization of CNC green drilling parameters, preference value and rank of alternatives which are described in Table 2. The preference value of all parametric combination is ranked according preference value and highest preference value which is considered as rank one. Thus, obtained results of the experimental trail 19 show the best experimental parametric combination for desirable performance characteristics which has highest preference value 0.131 and rank one. The optimal combination of process parameters considering highest preference value is 8 mm DD, 0.8 mm FR, 900 rpm SS and 136° PA.

3.5 Confirmatory Experiments for TOPSIS-PCA

The confirmation of improvement in response parameters using optimal parametric setting of process parameters is performed and using Eq. (5) and estimated preference value 0.1365228 and improvement in preference value is 0.005528, i.e.4.22%.

3.6 ANOVA for TOPSIS-PCA

The influence of process parameters on response parameters is estimated using analysis of variance and results for preference value with factor response larger are better at 95% level using minitab17 software as illustrated in Table 4 and observed that SS and FD are process parameter highly influencing the response parameters with the highest contribution 86.84% and 5.48%, respectively, and all process parameters significantly influence the response parameters.

4 Conclusion

In the present research work, analysis and multi-objective optimization are performed using GRA-PCA and TOPSIS-PCA. The conclusions of this research work are as follows:

- PCA is used for estimation of weight of each response parameters and the obtained weights of TB, BL and TF are 0.287, 0.409, and 0.304.
- The optimal green drilling process parameters are 8 mm DD, 1.2 mm FD, 1800 rpm SS and 136° PA using GRA-PCA and 8 mm DD, 0.8 mm FD, 900 rpm SS and 136° PA using TOSIS-PCA.
- ANOVA is performed on GRG value and found that DD and SS are more influencing on the response parameters with the highest contribution 49.63 and 22.19%, respectively, while FD and PA are insignificant process parameters.
- ANOVA is performed on preference value and found that SS and FR are more influencing on the response parameters with the highest contribution 86.84% and 5.48%, respectively, while FD and PA are insignificant process parameters.
- The confirmation experiments reveal that the enhancement of GRG value using GRA-PCA and preference value using TOPSIS-PCA is 0.007006, i.e. 2.633% and 0.005528, i.e. 4.22%, respectively, and
- TOPSIS-PCA reveals the better performance than GRA-PCA.

References

1. Benes J (2000) Whole making trends run deep, fast and dry. Am Mach 144(5):97–104
2. Melngk SA, Smith RT (1999) Green manufacturing. SME, Dearborn
3. Tsao CC, Hocheng H (2008) Evaluation of thrust force and surface roughness in drilling composite material using Taguchi analysis and neural network. J Mater Process Technol 2031(3):342–348
4. Lin SC, Chen IK (1996) Drilling carbon fiber-reinforced composite material at high-speed wear. J Compos Mater 194:156–162
5. Hocheng H, Puw HY (1992) On drilling characteristics of fiber-reinforced thermoset and thermoplastics. Int J Mach Tool Manuf 32(4):583–592
6. Kalita B, Nath T (2016) An experimental investigation and optimization of cutting parameter in drilling AISI B1113 using M2HSS drill bit. In: International conference on explorations and innovation in engg. &tech. (ICEIET)
7. Senthil Babu S, Vinayagam BK (2015) Modeling and analysis of surface roughness and thrust force in drilling of Al-based metal matrix/hybrid composites. Int Rev Model Simul (IREMOS) 8(4)

8. Ravindranath VM, Shiva Shankar GS, Basavarajappa S, Suresh R (2016) Effect of cutting parameters on thrust force and surface roughness in the drilling of Al-2219/B4C/Gr metal matrix composite. IOP Conf Ser Mater Sci Eng 149
9. Sharma K, Jatav A (2015) Optimization of machining parameter in drilling of AISI D2 Steel. Int J Sci Res Eng Technol (IJSRET)
10. Kumar S, Jagadish Singh AK, Kumar N (2020) Multi-objective optimization of CNC drilling parameters on HcHcr steel using Taguchi's based utility concept & GRA-PCA methods. Int J Adv Eng Sci Technol Res (IJAESTR-special issue ICAME-2020) 2394–9627:55–63
11. Kumar S, Singh I (2016) The influence of process parameters on cutting speed of WEDM using Taguchi's technique. Int J Emerg Technol (Special Issue on ICRIET-2016) 7(2):332–337
12. Hostelling H (1993) Analysis of a complex of statistical variables into principal components. J Educ Psychol 24:417–441
13. Kumar S, Ramola IC, Kumar R (2015) Analysis of surface roughness and material removal rate for high carbon high chromium steel on die sinking EDM using Taguchi technique. SSRG Int J Mech Eng (SSRG-IJME)-EFES 2348–8360
14. Kumar S, Jagadish Ray A (2021) Multi objective optimization CNC drilling Parameters on machining of HcHcr steel using Taguchi's technique and grey relational analysis. Adv Thermal Eng Manuf Prod Manage Proc ICTEMA2020 353–368. Springer
15. Zhongyou X (2012) Study on the application of TOPSIS method to the introduction of foreign players in CBA games. Phys Proc 33:2034–2039

Numerical Analysis of Supersonic Axisymmetric Inlet with Single Ramp

V. Phaninder Reddy, D. Govardhan, A. Rathan Babu, V. Varun, and V. Raghavender

Abstract A computational fluid dynamic analysis on axisymmetric inlet has been performed, to investigate the flow properties of a ram jet engine with a conical inlet. This paper describes about the analysis of different conical inlets with varying semi-vertex angles, angle of attack (AOA) and Mach numbers. Results have shown that as the angle of attack increases, the static pressure ratio decreases, for the constant semi-vertex angle, the static pressure ratio increases with increase in Mach number, as the semi-vertex angle increases, the static pressure ratio increases. It is observed that a pressure recovery of 96% can be achieved for a free-stream Mach number of 1.5 with angle of attack of 0° for 10° cone angle and 98.5% can be achieved for 15° cone angle.

Keywords Semi-vertex angle (cone angle) · Angle of attack (AOA) · Mach number · Static pressure ratio · Total pressure ratio

1 Introduction

Most of the air-breathing engines operate at moderate subsonic speed, at the order of M = 0.3. A supersonic aircraft has to undergo adverse gradients of flow properties during its flight. An aircraft moving at supersonic speeds, aircraft inlet plays a crucial role since it reduces the flow from supersonic to subsonic. This is due to the formation of shock waves. In external compression inlets, the oblique shock waves are formed outside the inlet followed by a terminal shock at the cowl lip of the inlet. The aircraft engine needs to be fuel-efficient with maximum total pressure. In a supersonic conical inlet, a conical oblique shock is formed which is weaker than a normal shock which facilitates the external compression to occur before the internal compression. The external compression is performed followed by a series of oblique shocks from compression corners and terminates by a normal shock. In this regard, this paper analyzes different conical inlets with varying semi-vertex angles, angles of attack

V. P. Reddy (✉) · D. Govardhan · A. R. Babu · V. Varun · V. Raghavender
Institute of Aeronautical Engineering, Dundigal, Hyderabad 500043, India
e-mail: phaninder280990@gmail.com

A. Maurya et al. (eds.), *Recent Trends in Mechanical Engineering*, Lecture Notes in Mechanical Engineering, https://doi.org/10.1007/978-981-19-7709-1_3

(AOA) and Mach numbers. In our design, external compression occurs with a conical shock followed by a terminal normal shock. The main objective is to analyze the inlet properties for all the above-mentioned cases and choose a best-optimized inlet at supersonic flight conditions.

2 Geometry Specifications

The conical inlet geometry of an aircraft engine was created by calculating the coordinates from a reference axisymmetric inlet. The geometry has designed according to the semi-vertex angle of the conical inlet angle (θ_e). Two conical inlets of semi-vertex angle 10° and 15° were designed to study the performance. The coordinates for the inlet designs of 10° and 15° cone angles are given in Table 1. The boundary coordinates of the x-y plane ((50, 70), (−50, 70), (30, 70), (30, −70)) are imported into ICEM CFD software, and these points are projected on the x-z plane with the same y-coordinate. Thus, a 3D structure is generated. These points were joined together with smooth lines to obtain curves. The surface geometry is created with curves. Each part of the geometry, the properties that need to be calculated, is named separately as cone, cowl and diffuser. A cylindrical boundary is designed along with an engine inlet of 140 mm diameter and a width of 80 mm. The large boundary domain is selected to obtain the properties accurately. The geometry is imported and meshed (Fig. 1c).

Table 1 Coordinates of cone with different angle

10° cone angle		15° cone angle	
X-coordinate	Y-coordinate	X-coordinate	Y-coordinate
7.51899	1.3258	7.51899	2.014707
7.51899	2.26286	7.51899	3.4386
10.8608	3.15554	10.8608	4.7952
15.4937	3.73208	15.4937	5.6713
20.3165	4.08954	20.3165	6.2145
30	4.28739	30	6.5152
30	3.3347	30	5.0675
9.11392	1.69281	9.11392	2.5724
15.8734	1.98577	15.8734	3.0176
30	1.58788	30	2.4130
0.15713	0.0277	0.15713	0.04210

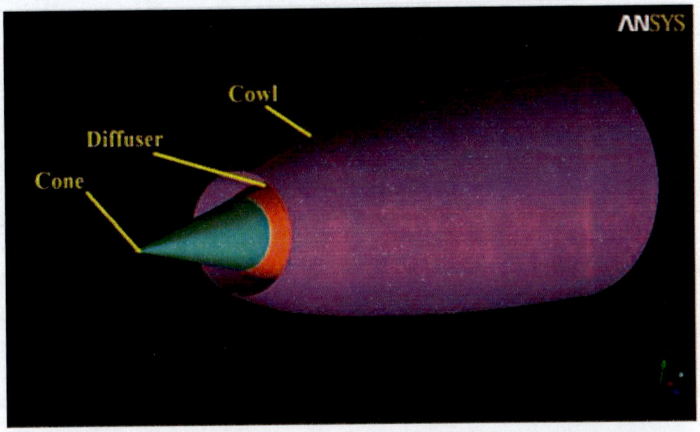

(a)10°Cone angle geometry

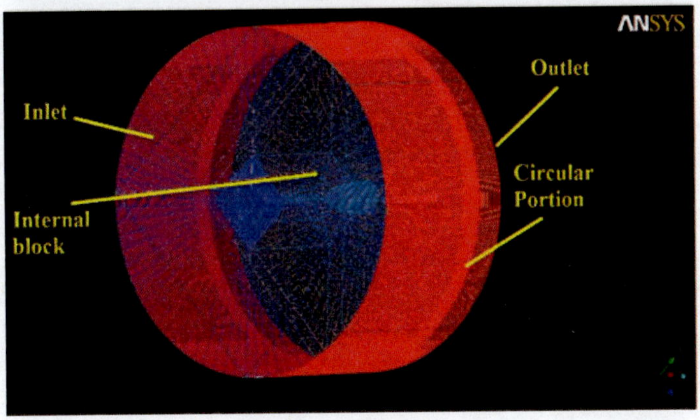

(c) Domain Mesh

Fig. 1 Geometry of axisymmetric inlet

3 Mesh

The grid generation around the engine inlet involves the paramount features that need to be taken care of. Any geometry subjected to supersonic flow conditions experiences variable effects such as shock waves and boundary layer interactions. These properties need to be captured accurately to analyze the flow on the geometry, and hence, the performance of the engine can be estimated. A large number of CFD software such as Techplot@, ANSYS ICEM CFD@, Numeca@ and Turbogrid@ enhanced grid generation. Some provide structural meshing techniques, and others ease in unstructured meshing. A structural hexahedral mesh is performed on the engine inlet and the surrounding boundary. The structural mesh includes the creation of blocks, O-grid blocks, splitting, deleting the unwanted blocks, etc. The O-grid

Table 2 Mesh parameters of domain

Properties	10°	15°
Mesh type	Structural hexahedral	Structural hexahedral
Number of elements	4,181,236	3,833,184
Quality: angle	45°	45°
Quality: determinant 2*2*2	0.7	0.7

Table 3 Boundary conditions of domain

Name of the entity	Type
Cone	Wall
Diffuser	Wall
Cowl	Wall
Inlet	Pressure far field
Circular portion	Pressure far field
Outlet	Pressure outlet
Internal block	Fluid

generation enables the generation of orthogonal grid lines on the object boundary. A structured mesh is used because it provides maximum control of the elements in the mesh and ease in capturing the required portions of the geometry, and Table 2 provides information regarding the number of grid points used, this is obtained from mesh independent studies, non-orthogonality and aspect ratio of mesh.

3.1 *Boundary Conditions*

After achieving the required mesh quality, the mesh is converted to unstructured mesh because the solver-ANYSYS Fluent@ reads only an unstructured mesh. The ANSYS Fluent solver is selected in the output menu, and the mesh is exported to the solver. The necessary boundary conditions and initial conditions used in simulations are given in Table 3. The ambient temperature is considered as 303.15 K, and gauge pressure and operating pressures are 100500 Pa and 0 Pa.

3.2 *Solver*

The mesh file is imported into ANSYS Fluent solver and sorted for errors or negative volumes. Viscous solver, Spalart–Allmaras, one equation solver in ANSYS Fluent is used in simulations to capture the wall boundary flows. The governing equations

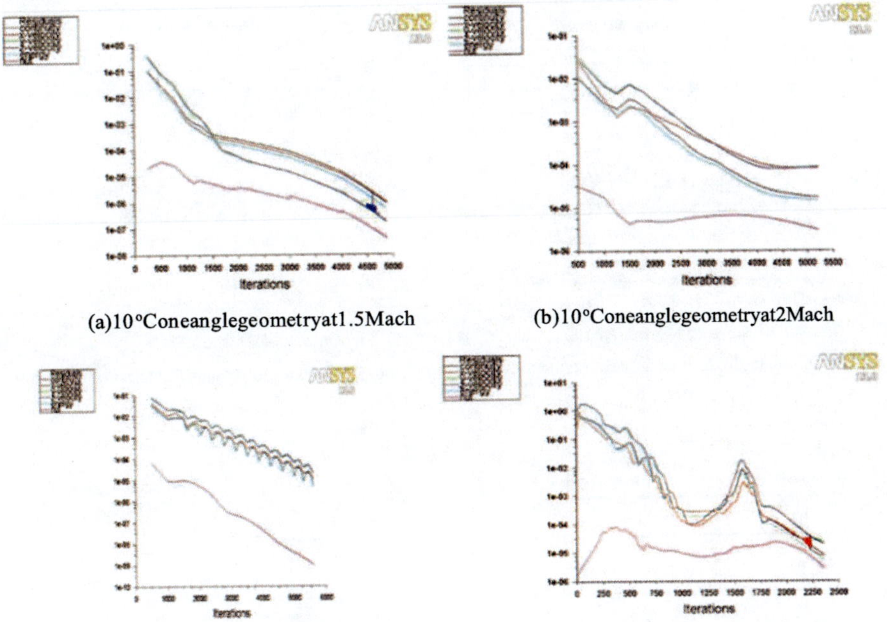

(a)10°Coneanglegeometryat1.5Mach (b)10°Coneanglegeometryat2Mach

(c)15° Cone anglegeometryat1.5Mach (d)15°Coneanglegeometryat2Mach

Fig. 2 Residual plot for 10° and 15° cone angles at 0° AOA

such as continuity, momentum and energy are solved for each case to obtain the flow properties. A density-based, absolute velocity formulated, steady simulations performed in a rotating reference frame. The energy equation is solved as it is a density-based solver. Spalart–Allmaras, a one equation viscous solver, is chosen to capture the viscous effects and chosen for its simplicity and accurate results. An ideal gas equation is used in pressure far-field boundary conditions. Mach numbers are 1.5 and 2 chosen based on the case. The angle of attack values is considered as cosine and sine values for every case. Simulations are carried out till the convergence criteria are satisfied, it is observed that around 5100 odd iterations, the solution has converged as shown in Fig. 2.

4 Results and Discussion

The static pressure across a shock wave will always increase and in this aspect, a conical shockwave is formed originating from the apex of the cone due to supersonic flow stream which is shown in Fig. 3. For every case, the initial free-stream pressure was chosen as $P_1 = 100500\ Pa$. For free-stream Mach number $M_0 = 1.5$, the compression in the inlet is terminated by a normal shock after which the pressure has increased to $P_3 = 244335\ Pa$. Hence, the static pressure ratio $\frac{P_3}{P_1} = 2.4312$, whereas

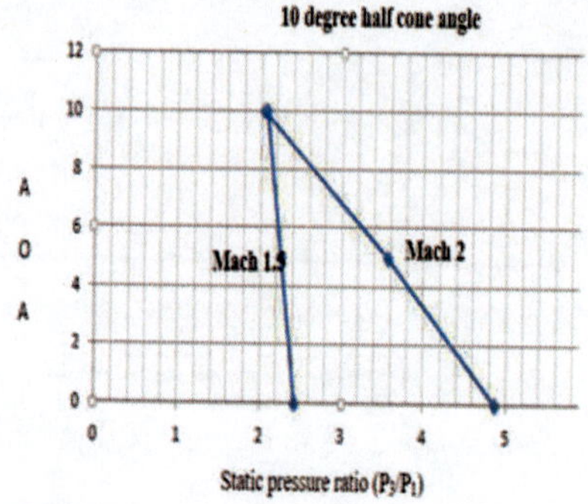

(a)10 °Coneangle:Angleofattackversusstaticp ressureratio

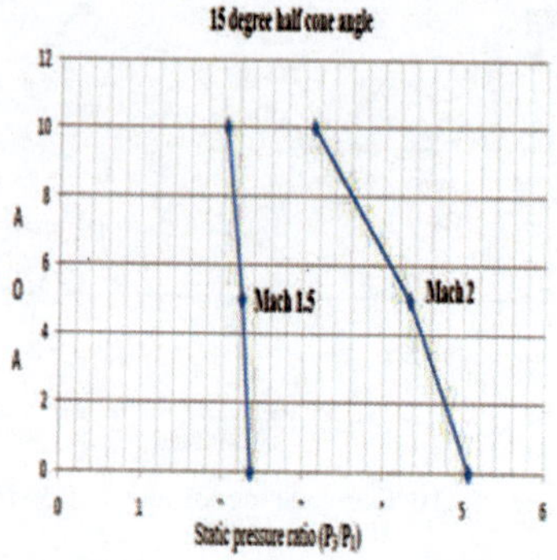

(b)For15 °Coneangle:Angleofattackversusstaticpre ssureratio

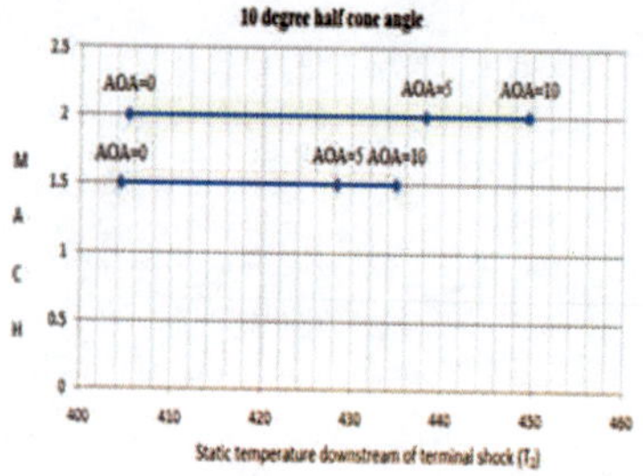

(c) For 10 °Cone angle: Static temperature versus Mach number

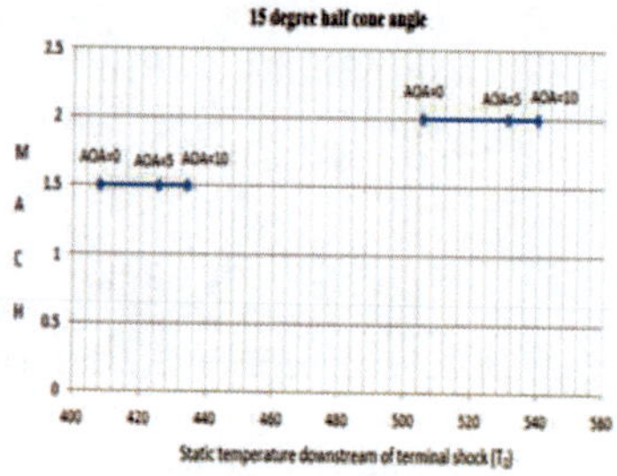

(d) For15 °Coneangle: Static temperature versus Mach number

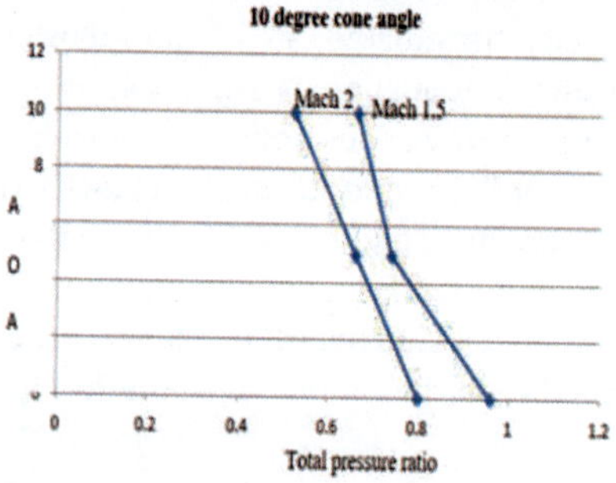

(e)For10 °Coneangle:AngleofattackversusTo talpressureratio

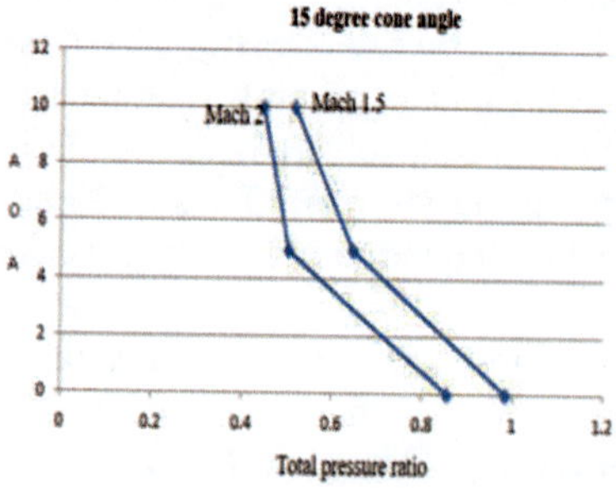

(f)For15 °Coneangle:AngleofattackversusTotal pressureratio

Fig. 3 Pressure ratio's and temperature plots for 10° and 15° cone angles

for $M_o = 2$, the pressure after the normal shock P_3 is 489354 Pa. Hence, the local pressure ratio, in this case, will be $\frac{P_3}{P_1} = 4.8692$. As the flow gets compressed, the Mach number across the shock wave will decrease, therefore, for a free-stream Mach number $M_0 = 1.5$, the Mach number after the conical shock is decreased to $M_1 = 1.372$, and further, it is decreased to $M_3 = 0.656$ after the terminal shock. When M_0

= 2, the Mach number after the conical shock is reduced to $M_1 = 1.83$, and further, it is reduced to $M_3 = 0.75$ after the terminal shock.

The total pressure is an important aspect of the performance of an engine. The optimization of the total pressure ratio is of great importance. Stagnation or total pressure decreases after the shock wave. For an efficient engine, the decrements of both Stagnation and total pressure have to be minimized to obtain maximum total pressure recovery and proper combustion. For $M_0 = 1.5$, the total pressure ratio $\frac{P_{03}}{P_{01}} = 0.96$, and for $M_0 = 2$, $\frac{P_{03}}{P_{01}} = 0.8019$, which is given in Table 4.

Table 4 provides the numerical results for various AOA's, Mach numbers and cone angles. As the angle of attack increases, the cross-flow velocity increases. Hence, the leeward (upward) portion of the cone experiences flow separation and embedded shocks is formed. They appear in most cases when the angle of attack is greater than the half cone angle (SV). Hence, for angles of attack greater or equal to θc, the pressure gradient on the leeward side becomes unfavorable. The total pressure ratio decreases as the angle of attack increases as shown in the graph of Fig. 3e, f. As the angle of attack is increased, embedded shocks were formed on the cone surface which increases flow separation in the leeward (upward) side. This portion experiences unfavorable pressure gradients due to which the total pressure losses increase. But the windward portion has favorable pressure gradient. For the same

Table 4 Results

AOA	Term	Value at 10° cone angle		Value at 15° cone angle	
		Mach 1.5	Mach 2	Mach 1.5	Mach2
0°	P_1	100500 Pa	100500 Pa	100500 Pa	100500 Pa
	P_3	244335 Pa	489354 Pa	236979 Pa	509595 Pa
	P_3/P_1	2.4312	4.8692	2.358	5.0706
	P_{03}/P_{01}	0.96	0.8019	0.9856	0.855
	M_1	1.372	1.39	1.27	1.72
	M_3	0.656	0.75	0.8070	0.6415
5°	P_1	100500 Pa	100500 Pa	100500 Pa	100500 Pa
	P_3	227331 Pa	357991 Pa	225220 Pa	434702 Pa
	P_3/P_1	2.262	3.5621	2.241	4.3254
	P_{03}/P_{01}	0.74	0.661	0.645	0.501
	M_1	1.382	1.89	1.32	1.63
	M_3	0.674	0.8	0.574	0.694
10°	P_1	100500 Pa	100500 Pa	100500 Pa	100500 Pa
	P_3	208286 Pa	204226 Pa	205170 Pa	314102 Pa
	P_3/P_1	2.0725	2.0321	2.0415	1.37
	P_{03}/P_{01}	0.6627	0.521	0.514	0.444
	M_1	1.27	1.67	1.37	1.55
	M_3	0.794	0.82	0.728	0.79

semi-vertex angle, as the Mach number is increased, the stagnation pressure losses are high. The total pressure ratio increased with the increase of the semi-vertex angle of the cone at constant Mach number as plotted in Fig. 3e, f. It is evident from the Fig. 3a, b that as

1. The angle of attack increases, the static pressure ratio decreases.
2. For the same semi-vertex angle, the static pressure ratio increases with an increase in Mach number.
3. As the semi-vertex angle increases, static pressure ratio increases.

The results of pressure ratio were acceptable and it was compared with experimental results conducted by Antonio Ferri (A. Ferri 1952) and James F. Connors (J. F. Connors 1955).

From Fig. 3a, it is observed that for Mach 1.5 and 10° cone angle, there is no much drop in static pressure with an increase in the angle of attack. At Mach 2 (10° cone angle), static pressure ratio decreases with an increase in the angle of attack. Approximately, at 10° angle of attack, Mach 1.5 and Mach 2 have the same static pressure ratio for 10° cone angle. A similar type of phenomena is found for 15° cone angle (Fig. 3b) but static pressure ratio varies.

5 Conclusion

Fluid flow analysis of axisymmetric inlet was carried out for different Mach numbers, angle of attack and semi-vertex angles. Numerical solutions have shown that pressure recovery of 96% can be achieved for a Mach number of 1.5 with an angle of attack of 00 for 10° cone angle and 98.5% can be achieved for 15° cone angle. As the angle of attack increases, the cross-flow velocity increases which reduces the mass flow into the engine, one need to look into the efficient engine design. In the present simulation, a single ramp inlet was used and in the future simulations of a multi-ramp, the inlet will be considered such that total pressure recovery will be efficient for higher Mach numbers, and additionally, shockwave boundary layer effect will be seen.

Bibliography

1. Ferri A, Nucci LM (1952) Preliminary investigation of new type of Supersonic inlet. Langley Aeronautical Laboratory Langley Field
2. Bangert LH, Santman DM, Horie G, Miller LD (1982) Some effects of cruise speed and engine matching on supersonic inlet design. J Aircr 19:58–64
3. Chima R (2011) Computational Analysis of a low-boom supersonic inlet. In: 29th AIAA applied aerodynamics conference. Honolulu, Hawaii
4. Connors JF, Meyer RC (1955) Design and analysis tool for external-compression supersonic inlets. Technical report. Lewis Flight Propulsion Laboratory, Cleveland Ohio
5. Loth E, Titchener N, Babinsky H, Povinelli L (2013) Canonical normal shock wave/boundary-layer interaction flows relevant to external compression inlets. AIAA J 51:2208–2217

6. Mavris HR, Dimitri (nd) Preliminary Design of a 2D Supersonic Inlet to Maximize Total Pressure Recovery." By AIAA 5th ATIO and 16th Lighter-Than-Air Sys Tech. and Balloon Systems Conferences. n.d.
7. Askari R, Soltani MR (2017) Two- and three-dimensional numerical simulations of supersonic ramped inlet. Sci Iranica 2198–2207
8. Slater JW (2012) Design and Analysis tool for external-compression supersonic inlets.In: Aerospace sciences meeting. AIAA 20120016
9. Slater J (2005) CFD methods for computing the performance of supersonic inlets. In: 40th AIAA/ASME/SAE/ASEE joint propulsion conference and exhibit. Arlington, Virginia
10. Soltani M Abedi, Askari R, Sepahi-Younsi J, Soltani MR (2020) Axisymmetric and three-dimensional flow simulation of a mixed compression supersonic air inlet. Propul Power Res 9:51–61
11. Zha G-C, Smith D, Schwabacher M, Rasheed K, Gelsey A, Knight D, Haas M (1997) High-performance supersonic missile inlet design using automated optimization. J Aircr 34:697–705

Numerical and Experimental Analysis of Friction Stir Welding of Aluminum Alloy AA2024-T4

Deepak Kumar, Md. Parwez Alam, and A. N. Sinha

Abstract The paper deals with the influence of heat input on metallurgical and mechanical properties that have been investigated through experimental and numerical investigations. Two process parameters were taken to conduct the experiments. Thermocouples were used to measure temperature distributions. The heat generated during the friction stir welding (FSW) affected the grain refinements strongly, which were revealed with optical microscope (OM) micrographs. A three-dimensional model was developed to obtain the simulated temperature during the FSW of AA 2024-T4. The simulated temperature data was very close to that of the experimental data. It was observed that the heat generation depends on both transverse and rotational speed, and the peak temperature obtained was about 80% of the melting point of base metal. Additionally, tensile tests were performed for joint strength.

Keywords Friction stir welding · Optical microscope · AA 2024-T4 · Tensile test

1 Introduction

Friction stir welding (FSW) is a solid-state thermomechanical joining technique. It is an environmental-friendly, energy-efficient, and ideal welding technique that produces high-performing joints at a low cost. A rotating cylindrical tool having two parts, shoulder and pin, is used. The pin is fed in the joint line of two clamped sheets until the shoulder touches its surface and subsequently traversed along the joint line [1]. Frictional and plastic deformational heat is generated during the process, which softens the material. During the forwarding movement of the tool along the weld line, it extruded behind the pin, and high pressure assists in forged consolidation of the weld [2]. A schematic diagram illustrates the FSW process, shown in Fig. 1. In the FSW process, heat generation has crucial effects on the weld joints' mechanical properties and metallurgical characteristics. A full understanding of heat generation is essential for controlling the joints' mechanical properties and metallurgical

D. Kumar · Md. P. Alam (✉) · A. N. Sinha
Mechanical Engineering Department, NIT Patna, Ashok Rajpath Patna 800005, India
e-mail: parwez.me13@nitp.ac.in

A. Maurya et al. (eds.), *Recent Trends in Mechanical Engineering*, Lecture Notes in Mechanical Engineering, https://doi.org/10.1007/978-981-19-7709-1_4

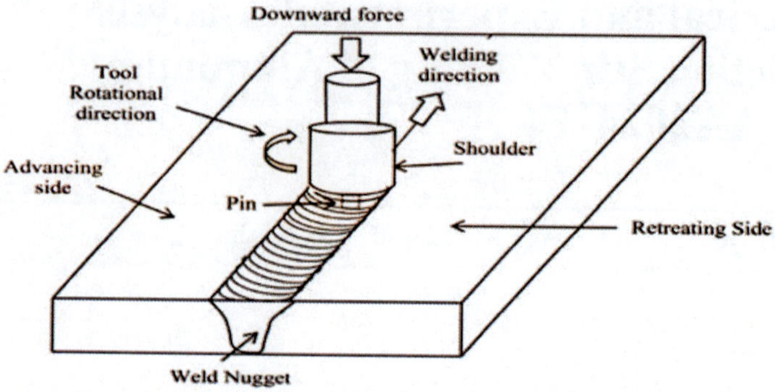

Fig. 1 Schematic representation of the FSW process [3]

characteristics. Numerical modeling offers an excellent prospect for understanding different phenomena of the FSW. For FSW, several heat generation models have been developed which describe friction and plastic deformation as the primary heat source.

Abdul-Sattar et al. [4] analyzed the temperature distribution through experimental work and modeling and observed higher temperature at the advancing side. Cartgueen et al. [5] concluded that the rate of heating and peak temperature highly depends on the transverse and rotational speed of the tool, respectively. Gadakh et al. [6] constructed a model for comparing heat generation by FSW using a taper and straight cylindrical pin and observed that the taper pin generated less heat than the straight pin. Zhang et al. [7] found that the contact area between shoulder-workpiece contributes more in heat generation than pin-workpiece. Bastier et al. [8] developed a heat transfer model for the FSW process. They observed that the plastic and frictional heat generation was equal to 4.4% and 95.6% of the total heat generation, respectively. Rajamanickam et al. [9] developed a transient model in which heat input from the tool shoulder alone was considered. The maximum temperature gradients were observed to be located just beyond the shoulder edge in both longitudinal and lateral directions. Many studies show that the moving heat technique is a reliable method to simulate heat generation during the FSW process. The use of ANSYS software can obtain the FSW process's temperature outputs at each required time step. In this study, two different process parameters had been used to perceive its sound effect on FSW temperature distribution. With the help of ANSYS software, a thermal model was developed to simulate the thermal history of AA2024-T4 during FSW. The model was validated by experimental data. Tensile tests were carried out to find out the joint strength. Besides, the microstructure of friction stir welded AA2024-T4 was also evaluated.

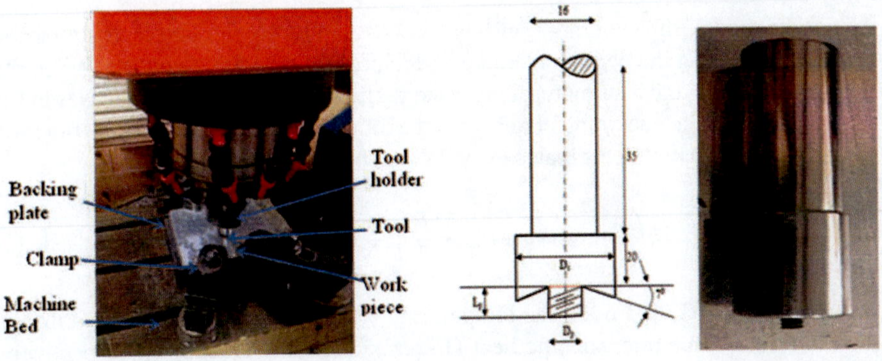

Fig. 2 Experimental setup and FSW tool with its geometry

2 Experimental Investigation

AA2024-T4 sheets (120 × 80 × 3 mm) were used in the present study. A vertical milling machine was used for the joining of these sheets. Figure 2 shows the experimental setup and FSW tool with its geometry. The tool was made of AISI H13 tool steel material. Tool shoulder was curved, and its diameter (D_S) was 18 mm. The pin was straight cylindrical threaded, and its diameter (D_P) and length (L_P) were 4.5 mm and 2.7 mm, respectively. The input parameters, 700, 850, and 1000 rpm as rotational speed, and 20, 30, and 40 mm/min as transverse speed, were taken for machine process parameters. 0.1 mm plunge depth was given in all experiments.

For temperature measurement, thermocouples were inserted in two holes, which were drilled at 15 mm and 25 mm away from the joint line. ASTM standards were followed for tensile samples. For metallurgical characterization, metallographic images were captured by an optical microscope (OM) after polishing and etching the samples.

3 Numerical Investigation

3.1 Heat Transfer Model

Developing a thermal profile model for the FSW process is an ideal beginning for simulating the process using simulation software. The model helps calculate the approximate heat generation value by the FSW process, which can be used as the input for evaluating the temperature distribution using ANSYS 17.2. Understanding the FSW process is essential to develop a sound mathematical model. Many researchers over the years have tried to develop mathematical models to predict heat generation. Hamilton et al. [10] developed the most suitable model for calculating the heat flux.

As per the model, the generation of heat is constant during the intermediate period of welding. During the beginning and end, the process is highly dynamic. When the welding is started, the heat increases exponentially, and at the end, it decreases in the same manner. In the heat transfer analysis, the transient temperature (T), the function of time (t), and spatial coordinates (x, y, z) are obtained by Eq. (1).

$$k\left(\frac{\partial^2 T}{\partial x^2} + \frac{\partial^2 T}{\partial x^2} + \frac{\partial^2 T}{\partial x^2}\right) + Q_{\text{int}} = cp\frac{\partial T}{\partial t} \quad (1)$$

where T, k, Q_{int}, C, and ρ are the temperature (K), thermal conductivity (W/m.K), internal heat source rate, specific heat (J/kg.K), and material density, respectively. Heat loss from the workpiece surfaces to the ambient due to convection and radiation heat transfer, but heat loss from the workpiece's bottom surface is due to conduction. But the evaluation of this heat loss is not a simple task; this can be evaluated by the simple convection Eq. (2).

$$q_b = \beta_b(T - T_0) \quad (2)$$

where β_b is the convection coefficient of heat transfer, in W/(m^2.K), between the backing plate and the workpiece. 200 W/m^2 K used for β_b value. In FSW, heat is generated by friction as well as plastic deformation. But in this present work, plastic deformational heat is not considered. Hence, only frictional heat is used as the heat source in this numerical investigation. It is assumed that the heat flux (rate of heat generation) is uniformly distributed at the pin tool shoulder Chao et al. [11]. The equation is used to obtain a heat generation (q_r) rate.

$$q_r = \frac{3Qr}{2\pi\left(r_o^3 - r_i^3\right)} \quad (3)$$

$$Q = \frac{\omega\pi\mu F\left(r_0^2 + r_o r_i + r_i^2\right)}{45(r_o + r_i)} \quad (4)$$

where F is a downward force (15 KN), ω is the tool rotational speed, μ is the coefficient of friction (0.3), Q is the total heat input, r_o and r_i are the shoulder radius and pin radius, respectively.

3.2 Finite Element Model

In this study, ANSYS 17.2 was used to carry out the transient heat transfer analysis. By using the ANSYS software, the workpiece of size 125 × 80 × 3 mm had been modeled. Table 1 gives the temperature-dependent properties of AA 2024-T4. Two different meshes were required for conducting the simulation. One of the elements selected was SOLID 70, a solid 3D element to predict the nodal solutions. Another

Table 1 Temperature-dependent properties of AA2024-T4 [12]

Temperature (K)	290	373	473	573	673
Specific heat capacity (Jkg^{-1} K^{-1})	864	921	1047	1130	1172
Thermal conductivity (Wm^{-1} K^{-1})	120	134.4	151.2	172.2	176.4
Yield strength (MPa)	350	331	140	45	21

one was SURF 55, a surface element used for allocating convection properties to the model. 293 K temperature was taken for initial as well as ambient temperature. 30 W/ (m^2.K) was taken for convective coefficient at the top surfaces of the plates for natural convection between air and aluminum alloy surfaces.

4 Results and Discussion

4.1 Temperature Distribution

Figure 3a–c shows the temperature distribution of temperature contour at 30, 150, and 270 s when the pin is translated along the joint line. The transient temperature is affected by both rotational speed and transverse speed. Also, it is affected by conduction convection and radiation. The peak temperature of friction stir welding of aluminum alloy 2024-T4 using the thermal model at different input process parameters is shown in Fig. 3d–h. Figure 3d–f shows that at constant welding speed (20 mm/min), the peak temperature increases when increasing the rotating speed. When the rotating speed increases, it increases the strain rate and plastic dissipation in the stir zone, which increases the peak temperature. Increased rotating speed from 700 to 1000 rpm leads to significant changes in the peak temperature. At 1000 rpm–20 mm/min, pick temperature 732.67 K is observed. Similar trends were observed by Cartigueyen et al. [5]

Figure 3e–h shows the effect of transverse speed on the peak temperature at 850 rpm. As the transverse increases from 20 mm/min to 40 mm/min, it decreases the peak temperature from 712.67 K to 652.128 K in the stir zone at a given point. With an increase in transverse speed, the time for which the tool is in contact with the workpiece decreases; hence, frictional heat decreases, which leads to the temperature drop at the welding position. At transverse speed 40 mm/min and constant rotational speed 850 rpm, peak temperature 652.128 K is observed, as shown in Fig. 3h.

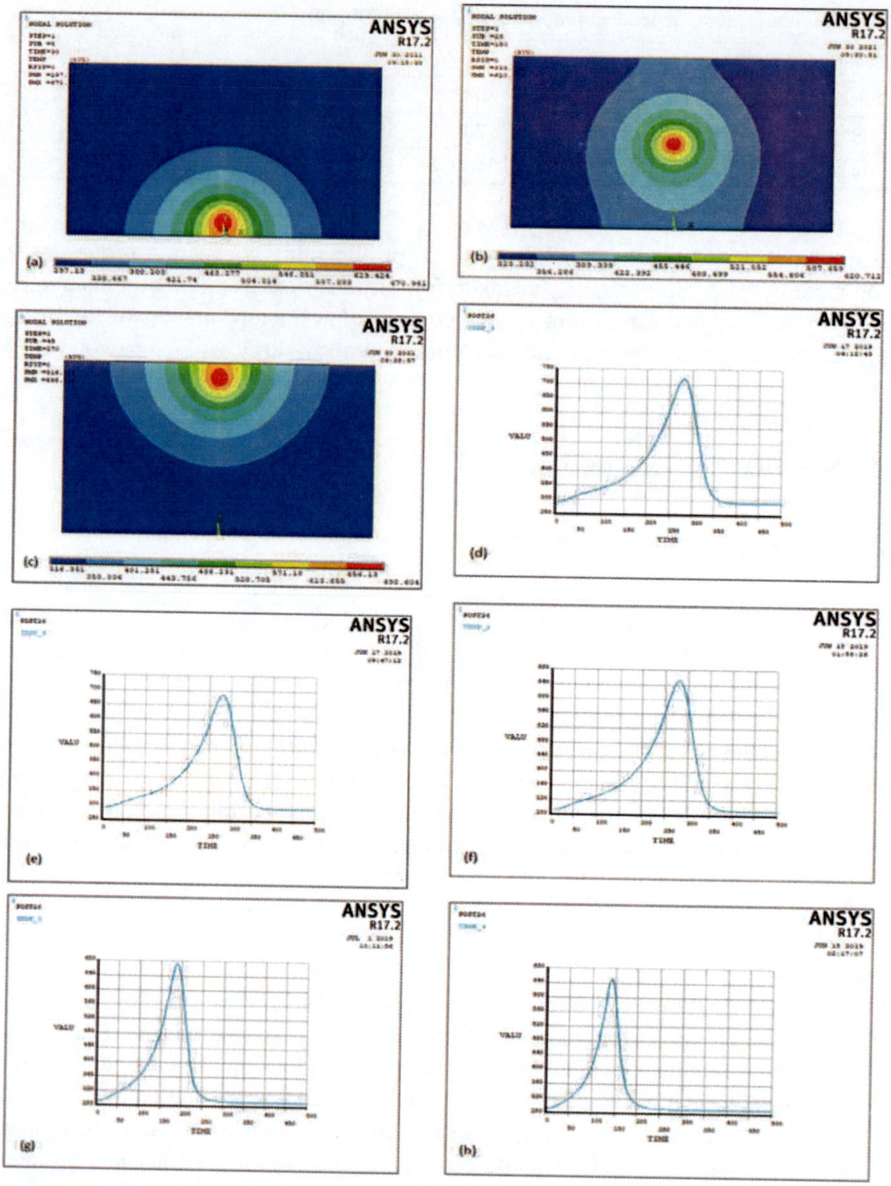

Fig. 3 Simulated temperature distribution of temperature contour at **a** 30 s, **b** 150 s, **c** 270 s and process parameters, **d** 1000 rpm and 20 mm/min, **e** 850 rpm and 20 mm/min, **f** 700 rpm and 20 mm/min, **g** 850 rpm and 30 mm/min, **h** 850 rpm and 40 mm/min

Table 2 Comparison between experimental temperature and simulated temperature

Sample no.	Parameters		Experimental temp. (K)		Simulated temp. (K)		% Error	
	Rotational speed (rpm)	Transverse speed	Temp. at P1 (max)	Temp. at P2 (max)	Temp. at P1 (max)	Temp. at P2 (max)	P1	P2
1	700	20	446	371	428	361	4	2.7
2	850	20	456	377	441	368	3.3	2.4
3	1000	20	465	385	449	372	3.4	3.4
4	700	30	424	356	408	350	3.8	1.7
5	850	30	436	361	424	358	2.8	0.83
6	1000	30	453	379	438	366	3.3	3.4
7	700	40	411	353	395	344	3.8	2.5
8	850	40	430	361	415	353	3.5	2.2
9	1000	40	443	369	427	360	3.6	2.4

4.2 *Validation of the Thermal Model*

Table 2 gives the comparison between experimental temperatures obtained using the thermocouples placed at a distance (P1) 15 mm and (P2) 25 mm away from the welding line and simulated temperature. The table gives that the experimentally measured temperature and the simulated temperature values lie close to each other, and the % error varies from 2.8% to 4% at location P1 and 0.83% to 3.4% at location P2. The variation between these two temperatures is due to ignoring the plastic deformational heat and heat generated by the pin of the tool.

4.3 *Tensile Test Study*

The tensile tests were carried out by an electromechanical universal testing machine. 1 mm/min feed rate was used for the testing. Nine welded samples and a base material sample were tested. For each condition, three specimens were tested. After performing the experimentation and testing, it was observed that the samples had failed due to localized softening at the stir zone / HAZ boundary.

Figure 4a shows the standard force (MPa) vs. strain relation of the tensile specimens, which were prepared from the plate's joint by friction stir welding at constant welding speed but varying rotational speed. 247 MPa ultimate tensile is observed at 700 rpm, but at 1000 rpm, it is increased to 343 MPa, which is approximately 80% of the base material's ultimate tensile strength. In this study, it is observed that for a given set of rotational parameters, the joint strength increased as the rotational speed had increased when welding speed was kept constant. Figure 4b shows the joint's strength for different welding speeds when the rotational speed is kept constant.

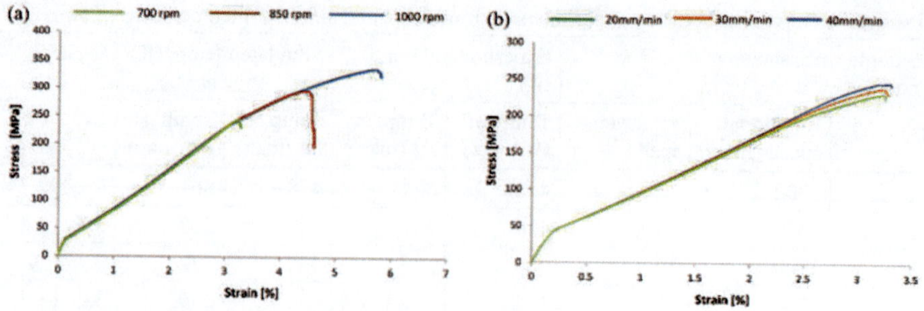

Fig. 4 Stress-strain diagram at **a** constant transverse speed, **b** constant rotational speed

It is observed that the strength of the joint increased when the welding speed had increased for the given set of welding speed parameters.

4.4 Microstructural Analysis

As the weld zone is affected by the thermal cycle during the process, it leads to a significant microstructure change. Three different zones were observed, namely stir zone (SZ), thermomechanical affected zone (TMAZ), and heat-affected zone (HAZ), as shown in Fig. 5. In the stir zone (SZ), there is direct contact between the pin and the material, due to which the material experiences deformation, heating, and friction. The thermomechanical affected zone (TMAZ) and heat-affected zone (HAZ) are not in contact with the tool. TMAZ is subjected to high temperature and deformation, and the heat emitted from the contact zone between the material and the tool affects the HAZ. It is clear from Fig. 5 that the SZ has the finest grain size as it undergoes severe plastic deformation; hence, the weldment's strength could be enhanced.

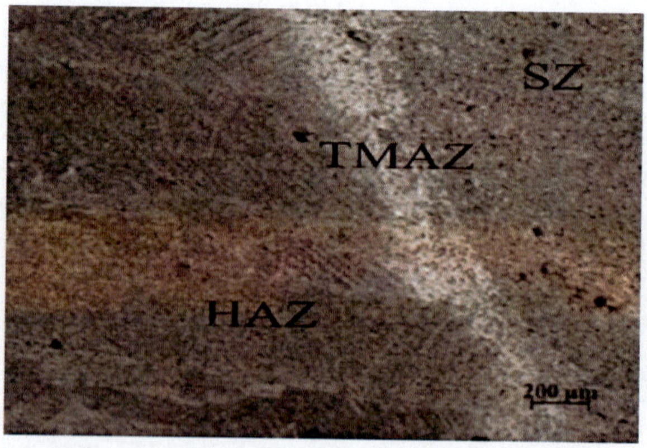

Fig. 5 Optical micrograph of friction stir welding region

Figure 6 shows the microstructure of base material and stir zone at different rotation and welding speeds. It was observed that the decrease of grain size from base material at constant welding speed 20 mm/min but decreasing rotational speed from 700 to 1000 rpm. It was also found that the constant rotational speed, 700 rpm, as the welding speed increases, the grain size is decreasing from the parent material grain size.

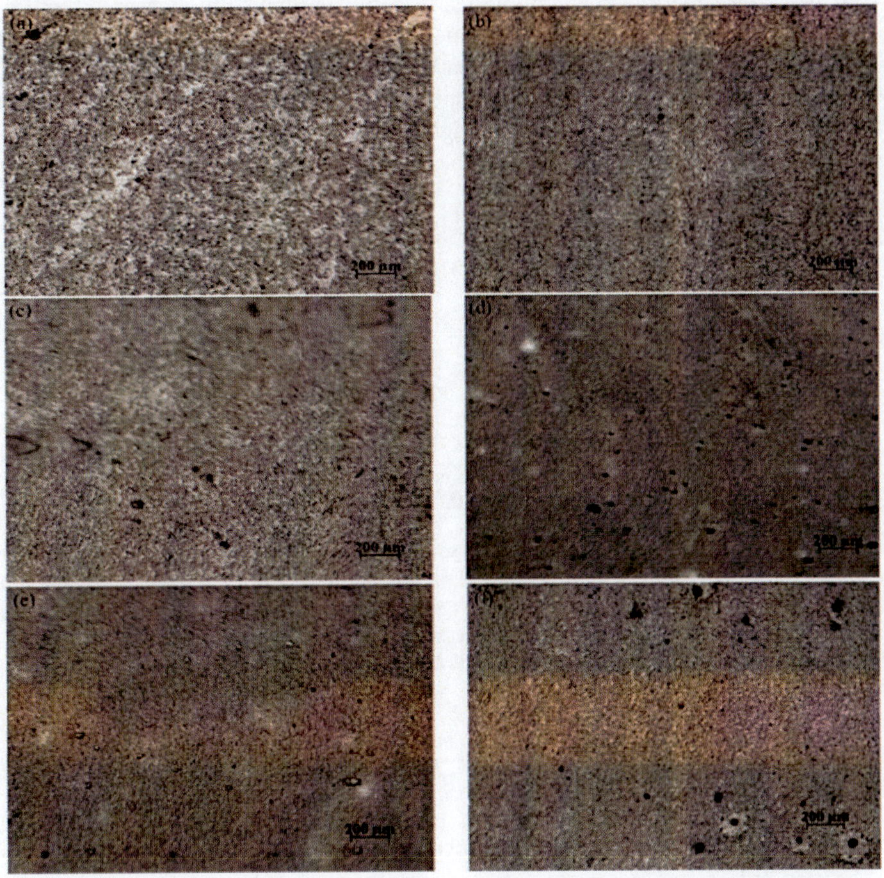

Fig. 6 Optical microstructure of **a** base material and stir zone of processed AA 2024-T4, **b** 1000 rpm-20 mm/min, **c** 850 rpm-20 mm/min, **d** 700 rpm-20 mm/min, **e** 700 rpm-30 mm/min, **f** 700 rpm-40 mm/min

5 Conclusions

Experimental and numerical investigations of FSW have provided many conclusions and silent features, which are summarized as follows:

- It is found that the tensile strength increases as both rotational and transverse speeds increase for a considered range of parameters.
- A good agreement between numerically and experimentally measured temperatures has been obtained at different locations. The maximum difference in temperature is observed to be 4% near the weld zone.
- The results show that the peak temperature increases as the rotational speed increases and decreases as the welding/ tool transverse speed increases.
- The increase in temperature affects the grain size, which further affects the joint properties.

References

1. Mishra RS, Ma ZY (2005) Friction stir welding and processing. Mater Sci Eng R Rep 50(1–2):1–78.41
2. Bozzi S, Etter AL, Baudin T, Robineau A, Goussain JC (2008) Mechanical behaviour and microstructure of aluminum-steel sheets joined by FSSW. Texture Stress Microstruct
3. Alam MP, Sinha AN (2019) Fabrication of third generation Al–Li alloy by friction stir welding: a review. Sādhanā 44(6):1–3
4. Abdul-Sattar M, Tolephih MH, Jweeg MJ (2012) Theoretical and experimental investigation of transient temperature distribution in friction stir welding of AA 7020–T53. J Eng 18(6):693–709
5. Cartigueyen S, Sukesh OP, Mahadevan K (2014) Numerical and experimental investigations of heat generation during friction stir processing of copper. Proc Eng 1(97):1069–1078
6. Gadakh VS, Adepu K (2013) Heat generation model for taper cylindrical pin profile in FSW. J Market Res 2(4):370–375
7. Zhang Z, Qi WU, Zhang HW (2014) Numerical studies of effect of tool sizes and pin shapes on friction stir welding of AA2024-T3 alloy. Trans Nonferrous Metals Soc China 24(10):3293–301
8. Bastier A, Maitournam MH, Van Dang K, Roger F (2006) Steady state thermomechanical modelling of friction stir welding. Sci Technol Weld Joining 11(3):278–288
9. Rajamanickam N, Balusamy V, Thyla PR, Hari Vignesh G (2009) Numerical simulation of thermal history and residual stresses in friction stir welding of Al 2014–T6. J Sci Ind Res 68:192–198
10. Hamilton C, Dymek S, Sommers A (2008) A thermal model of friction stir welding in alumi num alloys. Int J Mach Tools Manuf 48(10):1120–1130
11. Chao YJ, Qi X, Tang W (2003) Heat transfer in friction stir welding—experimental and numerical studies. J Manuf Sci Eng 125(1):138–145
12. Shi L, Wu CS (2017) Transient model of heat transfer and material flow at different stages of friction stir welding process. J Manuf Process 25:323–39

An Artificial Neural Network-Based Expert System for Bi-directional Prediction of the Parameters of Single-Layer Deposition by Wire and Arc Additive Manufacturing

Ananda Rabi Dhar, Dhrubajyoti Gupta, Shibendu Shekhar Roy, and Nilrudra Mandal

Abstract Wire and arc additive manufacturing (WAAM), especially its new variant cold metal transfer (CMT) process is regarded as one of the most potential and advanced additive manufacturing processes. The process parameters play an extremely influential role in determining the dimensional accuracy of the part manufactured, stability of the process and obtainment of other important process outcomes of interest. Hence, subtle determination of a suitable combination of the process parameters stands extremely crucial, as a result of which, process modelling and parameter optimization of WAAM has been an intriguing subject of research over a past few decades. In this paper, a reliable predictive system has been attempted using computational intelligence to estimate the input conditions so as to achieve the desired nominal responses qualifying the key performances. Artificial neural network models have been developed, envisaging Levenberg–Marquardt training algorithm in order to achieve a bi-directional predictive capability for a set of 3 inputs and 12 responses. Optimal network parameters like initial weight and hidden layer neurons have been determined by validation performance while training the same samples in multiple trials. R-squared values of the training samples and mean absolute percentage errors of the test samples for each response have been found quite satisfactory suggesting fairly adequate predictive models. With this approach, both forward and backward mappings have been successfully achieved.

Keywords Wire and arc additive manufacturing · Cold metal transfer · ANN · Predictive system · Optimization · Forward and backward mappings

---

A. R. Dhar (✉) · D. Gupta · S. S. Roy
Department of Mechanical Engineering, National Institute of Technology, Durgapur, West Bengal, India
e-mail: ard.16me1104@phd.nitdgp.ac.in

N. Mandal
Material Processing and Microsystems Laboratory, CSIR-CMERI, Durgapur, West Bengal, India

A. Maurya et al. (eds.), *Recent Trends in Mechanical Engineering*, Lecture Notes in Mechanical Engineering, https://doi.org/10.1007/978-981-19-7709-1_5

1 Introduction

Additive manufacturing technology (AMT) refers to all the techniques used for manufacturing parts especially of complex geometry by adding materials layer by layer directly from 3D CAD data. This technology saves materials wastes and reduces the lead time characterized by production lines in conventional manufacturing processes. It offers higher design flexibility with a penalty in the form of higher build time, compromised material property and in some cases poor dimensional accuracy. One of the most potential techniques of this class of manufacturing is wire and arc additive manufacturing (WAAM) processes, which use an electric arc, either the gas tungsten arc (GTA), gas metal arc (GMA), or the pulse arc (PA) as heat inputs and filler wire as feedstock [1]. Compared to the powder-feed process, the wire-feed process is cleaner and environment-friendly. The arc heat source enables nearly 90% energy efficient, which improves the deposition rate way beyond the normal capacity of additive manufacturing. The electric arc either continuously or in pulsated mode melts the filler wire to form droplets, which are transferred into the molten pool to get deposited on solidification in the form of additive layers. The continuous heat input leading to instability in thermal and mechanical behaviour of the deposited layers is mitigated by the Fronius developed cold metal transfer (CMT) process. In this technology, a reciprocating wire-feed mechanism is adopted to forcibly short-circuit the wire tip contact with the substrate material.

Any change in process parameters and working conditions in this complex process may have great impact on the dimensional tolerances and material properties of the parts produced. The most crucial factors that require optimal settings and ideally real-time monitoring along with enhanced decision support capabilities are heat dissipation, interlayer temperature, and quality of deposited previous layers. The geometry of the parts produced solely depends on the process parameters in a nonlinear fashion. Various research activities have already contributed in this direction to model the advanced welding, especially the WAAM process so as to understand the complex relationship between inputs and outputs.

A predictive system which can estimate the deposited layer geometry and other dependent process responses in function of the CMT input process parameters and the environmental conditions should be clearly recommended for improved efficiency. This kind of system can potentially reduce the time in selecting the process parameter values and also the trial phase wastes of costly materials often used in additive mode of manufacturing, thereby reducing the overall production cost. Many mathematical models and computational intelligence (CI)-based models were used [3–11] in the past addressing such typical advanced welding problems. The finite element method (FEM) was also adopted to predict the penetration and width of weld bead [12]. The modelling work of complex welding process so far has been restricted to only simple geometries and required oversimplifying assumptions to reduce computation demands conforming to the added complexity. It may be rewarding to future researchers to apply machine learning techniques to aid the development of predictive models even involving numerical simulation.

A literature review reveals in particular that all studies and simulations focused only on mapping input-output parameters in forward direction. The responses of practical importance in WAAM, especially CMT process, namely the geometry of weld bead often ask for nominal values unlike maximum or minimum ones, which necessitates the development of a predictive system which can estimate the process parameters to obtain a desired quality of responses. The backward mapping of input-output parameters has not been studied in this field. Due to multi-collinearity of responses, a backwardly mapped model cannot be precisely determined through regression. This has necessitated the usage of CI-based approaches like ANN.

The present study has attempted to carry out both forward and backward mappings of an exemplary single deposition CMT process reported in the past literature [2]. The model has been developed using MTALAB ANN toolbox. Here, wire diameter, wire-feed speed, and ratio of wire-feed speed to travel speed have been regarded as the input process parameters and 12 responses starting from bead width (W) to heat source length (HSL) as depicted in Table 1 have been considered as the dependent process outputs.

In this study, initially, a feed-forward ANN has been used to develop a forwardly mapped model, which predicts twelve responses given values of three input parameters. A large number of artificial data have been generated by regression equations. This data plus the actual experimental data have been together used as data samples for the model, out of which a definite portion has been used to train and the rest to validate and test the trained model. The model training has been accomplished envisaging

Table 1 Input parameters and responses

Process parameters and responses	Symbols	Input/response	Units	Ranges considered	
				Max	Min
Wire diameter	WD	Input	mm	1.2	0.8
Wire-feed speed	WFS	Input	m/min	1.68	13.35
Wire-feed speed/travel speed	(WFS/TS)	Input		6.62	19.34
Bead width	W	Response	mm	8.83	2.52
Bead height	H	Response	mm	3.62	1.61
Remelting ratio	RR	Response		0.48	0.02
Penetration	P	Response	mm	2.13	0.27
Dilution	D	Response		32.62	2.4
Deposition rate	DR	Response	Kg/hr	3.1	0.6
Contact angle	CA	Response	°	132.4	42.8
Reinforcement area	A_2	Response	mm^2	18.35	4.16
Penetration area	A_1	Response	mm^2	7.31	0.26
Aspect ratio	AR	Response		1.147	0.289
Heat input	HI	Response	J/mm	635.35	115.8
Heat source length	HSL	Response	mm	21.7	4.9

Levenberg–Marquardt algorithm. The backward mapping model has been developed using the similar procedure for the estimation of three process parameters from the twelve responses, and testing has been conducted to validate the model.

2 Methodology

2.1 Experimental Setup

A modified GMA-welding variant, which is an advanced technology utilizing controlled dip transfer mode, developed and coined as cold metal transfer (CMT) by Fronius™ was envisaged in this study. The CMT control panel allowed the selection of the pure CMT process, material type, shielding gas mixture, wire diameter, and contact-to-work-distance. A comprehensive investigation was carried out [2] to reveal the impact of three important factors on process responses related to bead geometry and plate fusion characteristics (Table 1).

Experimental investigations concerned with single deposition were conducted on substrates made of as-rolled S355J2+N grade structural steel. A solid mild steel wire made of G3Si1 was utilized throughout the study applied in three variants of diameters—0.8, 1.0, and 1.2 mm. The other inputs also had three designated levels encompassing the bounds as indicated in Table 1 designed for experimentations.

The dimensions of the base plates used in this study are 250 mm × 200 mm × 6 mm. A shielding gas flow rate of 15 l/min was regularized throughout this study for all experimental trials. Interested readers may please refer to the original article [2] to get acquainted with a detailed description of the experimental methodology.

2.2 Statistical Modelling and Data Generation

In order to conduct experimental studies on the CMT process, an IV-optimal design has been adopted. Minitab has been used to develop a statistical regression model by feeding the experimental data [2] as metadata for defining a custom design for the response surface methodology (RSM). The input variables have been codified using the respective maxima as $+1$ and respective minima as -1. By applying RSM to the 70 experimental data sets, regression equations of the coded type have been generated. 95% confidence intervals have been used and stepwise backward elimination methods considering all available terms (full quadratic) have been applied. Using the highest and lowest levels of input parameters, 9 hypothetical levels have been created. 729 samples have been spawned utilizing all possible combinations of different input parameter levels, as displayed in Table 1. Essentially, this has been done to provide maximum coverage. This is worth mentioning that the goodness of fit for each response has been found close and above 90% indicating adequate models.

2.3 ANN Modelling for Forward and Backward Mappings

ANN has been successfully utilized as classifier, function approximator, or predictor in many useful engineering applications including advanced welding processes. Neural network is simplified central nervous system model which is motivated by the enormous computations performed by human brains. The key advantages of using this technology include faster, robust, and parallel computing often on noisy data, and mapping capabilities.

In this work, the feed-forward architecture which is commonly used in welding models has been adopted. The first layer deals with the input information, followed by multiple hidden layers where the information can travel to reach the last output layer that produces the response information. MATLAB toolbox function NEWFF [13] has been aptly utilized along with other functions related to training and evaluation for this purpose. Three neurons contained in the input layer representing the input process parameters, two successive hidden layers each having 12 or more neurons to be optimized, and 12 neurons in the output layer have been configured for building the network topology. The basic ANN architecture in forward mapping can be understood by the illustration in Fig. 1a. The general delta rule has been adopted in the form of Levenberg–Marquardt training algorithm. For evaluating the performance of the network, mean square error (MSE) has been selected. The learning rate has been initially kept as low as 0.001 which gets reduced as the algorithm goes through the iterative cycles or epochs. The maximum number of epochs has been kept as 1000 for terminating the algorithm and the training has been accomplished in an incremental mode for each training sample.

The entire data set has been split into three subsets—the first one which is training sets comprises of 70% of the data, while the second and third one are for validation and testing each having 15% share of data. The data selection for training, validation, and testing have been truly random. The physical process model's inputs and outputs have been swapped to achieve a backward mapping of the process parameters. The new ANN model has been fitted with 12 process responses as input parameters, while the model's outputs have been represented by three process parameters (Fig. 1b).

3 Results and Discussions

The optimal network configuration in terms of initial weights and number of neurons in two consecutive hidden layers has been determined by running the validation trials after the training with a specific topology. The best setting has been saved as a pre-trained network for evaluating the test samples. The R-squared values for the

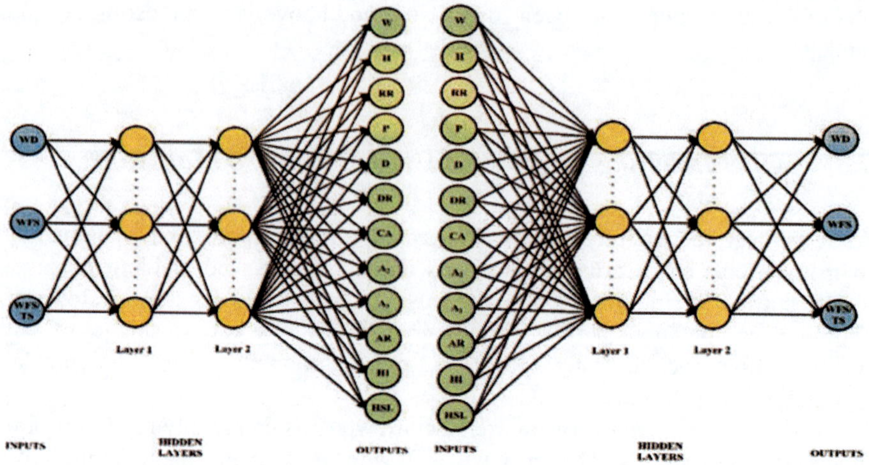

Fig. 1 ANN architecture of the forward (**a**) and backward (**b**) model

training samples and mean absolute percentage error for both validation and test data have been illustrated graphically in Fig. 2a. Larger prediction errors in case of HSL (32.38%) and AR (29.15%) have been witnessed, which has been mutually agreed by lower R-squared values (80.50% and 74.53%, respectively) observed for the responses in the trained model. The prediction accuracy has been encouraging in case of basic weld geometry parameters like W and H. The prediction accuracy and model fitness for D and DR have been also satisfactory. Both penetration depth and penetration area have been adequately modelled by forward mapping.

For the backwardly mapped model, the optimal network configuration has been determined similarly to that for the forwardly mapped model. The goodness of fit of the backwardly trained network model has been found considerably high and close to 99% for each output. Though the prediction accuracies of process parameters have been found adequate, there is still some scope for improvement especially for

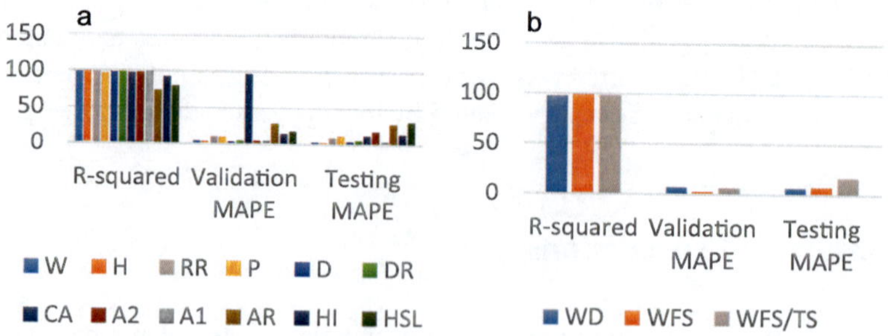

Fig. 2 Summary of performance of forward (**a**) and backward (**b**) mapped ANN model

the ratio terms WFS/TFS. The performance of the backward mapping is depicted in Fig. 2b.

4 Conclusions

1. A CMT process has been modelled with CI tool using three input variables and 12 responses conforming geometry and process stability.
2. Statistical regression model has been developed using the experimental data to generate a large volume of data for training a neural network.
3. The regression model has been used to replace the missing/erroneous values in the experimental data.
4. The forward and backward mappings of input and output parameters using a feed-forward ANN paradigm have been successfully built.
5. The network has been programmatically optimized for initial weight and number of neurons in the two hidden layers using the validation set of data.
6. The average error using the test data has been found satisfactorily low for each response in the predictive systems indicating the reliability.
7. The models can be hybridized with many state-of-the-art meta-heuristic optimization algorithms for the training purpose in order to improve their accuracy.

References

1. Furukawa K (2006) New CMT arc welding process—welding of steel to aluminium dissimilar metals and welding of super-thin aluminium sheets. Weld Int 20:440–445
2. Sequeira Almeida PM (2012) Process control and development in wire and arc additive manufacturing. Doctoral dissertation, Cranfield University
3. Meco S, Pardal G, Eder A, Quintino L (2013) Software development for prediction of the weld bead in CMT and pulsed-MAG processes. Int J Adv Manuf Technol 64(1–4):171–178
4. Palani PK, Murugan N (2006) Development of mathematical models for prediction of weld bead geometry in cladding by flux cored arc welding. Int J Adv Manuf Technol 30:669–676
5. Dey V, Pratihar DK, Datta GL, Jha MN, Saha TK, Bapat AV (2009) Optimization and prediction of weldment profile in Beadon-plate welding of Al-1100 plates using electron beam. Int J Adv Manuf Technol 48:513–528
6. Andersen K, Cook GE, Karsai G, Ramaswamy K (1990) Artificial neural networks applied to arc welding process modeling and control. IEEE Trans Ind Appl 26:824–830
7. Manikya Kanti K, Srinivasa Rao P (2008) Prediction of bead geometry in pulsed GMA welding using back propagation neural network. J Mater Process Technol 200:300–305
8. Nagesh DS, Datta GL (2002) Prediction of weld bead geometry and penetration in shielded metal-arc welding using artificial neural networks. J Mater Process Technol 123:303–312
9. Zhao Y, tao, Li W gang, Liu A, (2020) Optimization of geometry quality model for wire and arc additive manufacture based on adaptive multi-objective grey wolf algorithm. Soft Comput 24(22):17401–17416
10. Foorginejad A, Azargoman M, Mollayi N, Taheri M (2020) Modeling of weld bead geometry using adaptive neuro-fuzzy inference system (ANFIS) in additive manufacturing. J Appl Comput Mech 6(1):160–170

11. Xia C, Pan Z, Polden J, Li H, Xu Y, Chen S (2021) Modelling and prediction of surface roughness in wire arc additive manufacturing using machine learning. J Intell Manuf
12. Sathiya P, Jinu GR, Singh N (2009) Simulation of weld bead geometry in GTA welded duplex stainless steel (DSS). Sch Res Exch 2009:1–9
13. Demuth H (2009) Neural network toolbox TM 6 user's guide

Effect of Intermixing on Thermal Performance of Converged-Diverged Microchannel Heat Sinks for High Heat Flux Applications

Kshitij Bajpai and Abhishek Kundu

Abstract In the recent time, miniature type electronic chips are useful in aerospace technology, micro-electromechanical systems, hybrid data centres, and microfluidics. Usefulness of these chips increases if the generated heat can be eliminated for stable and reliable operation. Microchannel heat sinks are one of the possible ways to dissipate heat for larger heat flux generation. Over a last couple of decades, many researches have been performed on microchannel heat sink to measure the pressure drop and thermal resistance. With the help of these, one can determine the efficiency of a microchannel. To estimate the pressure drop and thermal resistance on microchannel, many researches have been concentrated on design of it. In the present study, converged diverged (CD) microchannel is used to compare its thermal performance with straight microchannel. Main aim of this study is to see the effect of intermixing on thermal performance of a convergent-divergent microchannel. A bar chart of average inlet and outlet temperatures has also been studied to see the effect of intermixing on thermal performances of a CD microchannel. The average inlet taken was 298 K, and average outlet temperature was found to be 313.47 K.

Keywords CFD · Microchannel · Simulation · Thermal performance · Intermixing · Vortices

1 Introduction

The miniature size electronic chips in such applications are the need of the century, thanks to improvements in aircraft technology, micro-electromechanical systems, hybrid data centres, and microfluidics. Moore's law asserts that when the operating

K. Bajpai (✉) · A. Kundu
Department of Applied Mechanics, Motilal Nehru National Institute of Technology Allahabad, Prayagraj 211004, Uttar Pradesh, India
e-mail: kshitijbajpai2181996@gmail.com

A. Kundu
e-mail: abhishekkunduamd@mnnit.ac.in

A. Maurya et al. (eds.), *Recent Trends in Mechanical Engineering*, Lecture Notes in Mechanical Engineering, https://doi.org/10.1007/978-981-19-7709-1_6

temperature of electronic boards rises, the component's durability decreases. Furthermore, the International Technology Roadmap for Semiconductors (ITRS, 2009) warned that temperature management in high-performance chip packages will be critical for the parent device's intermittent and efficient operation. Growing technological breakthroughs need the development of innovative and effective thermal solutions. Microchannel heat sinks are extremely promising in terms of high heat flux dissipation capacity and compatibility with the electronics system, despite the fact that research on diverse thermal heat dissipation has been undertaken all over the world. However, the performance of microchannel heat sinks is relatively poor, necessitating a great deal of study to improve the microchannel system.

Tuckerman et al. [1] were the first to investigate microchannels. Several studies have been conducted to improve heat transfer coefficients in microchannels with varying aspect ratios using single phase flow. Microchannel cooling has previously been shown to be effective in the mitigation of high heat flux transients in devices such as p-n diodes, metaloxide semiconductor field-effect transistors (MOSFETs), and resistance temperature detectors [2–5] (RTDs). The issue of flow boiling, flow reversal, reduced critical heat flux, thermal stresses, pressure, and temperature fluctuation in linear microchannels necessitated a new topology. Diverging microchannels were employed by Lee et al. [6] to reduce the impact of flow instabilities that produce nucleate boiling. Agarwal et al. [7] observed the presence of a critical angle that can lower the pressure drop in a single phase microchannel. Convective heat transport in diverging/converging microchannels was previously examined by certain scientists [8–13].

A mixture of converging and diverging microchannel heat sinks was investigated in this study. Solidworks 16 was used to build convergent-divergent (CD) microchannels, and ANSYS 16 Fluent was used to conduct a three-dimensional numerical analysis to assess the effect of intermixing on the thermal performance of a convergent-divergent microchannel. The CD microchannel heat sink's heat dispersion is studied using CFD software. Throughout this study, a constant heat flux of 120 W/cm^2 was applied on the surface of the heat sink, and de-ionized water is considered as the working fluid. On comparing to the straight microchannel, the CD microchannel proved better results in terms of overall thermal performance.

The objective is to analyze the reason for better thermal performance in CD microchannels through the middle line temperature plots.

2 Problem Definition

2.1 Geometry Description

The configuration of converging diverging microchannel used in the present study is as shown in Fig. 1a. The thermal performance of the proposed microchannel is compared with two of the basic straight microchannel which is also shown in the

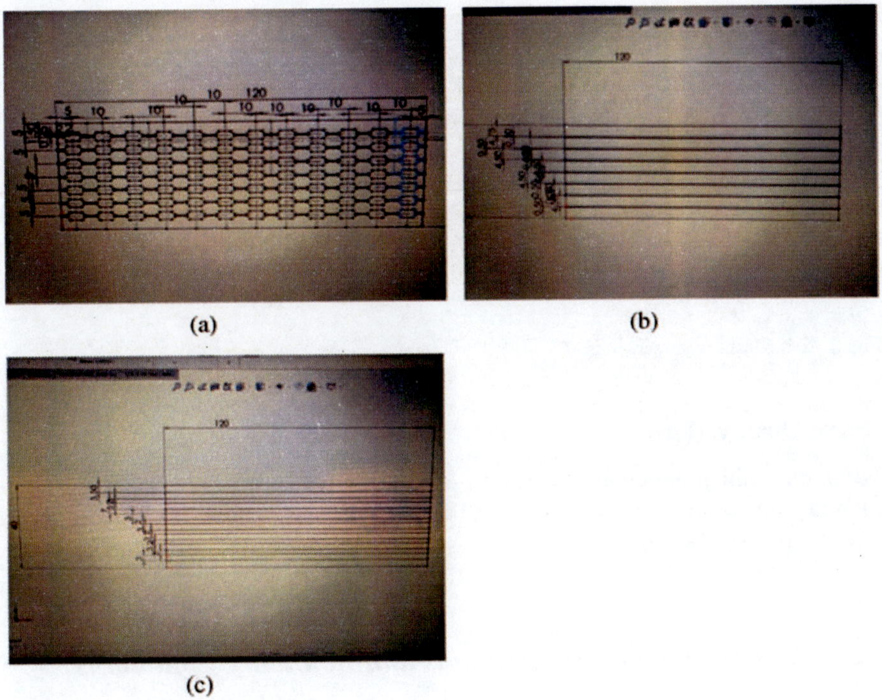

Fig. 1 **a** Converging diverging 3–0.5 mm microchannel sketch, **b** 0.5 mm straight microchannel sketch, **c** 3 mm straight microchannel sketch

Fig. 1b, c. The geometry considered in the computational domain is subjected to uniform wall conditions and boundary conditions.

The geometry is prepared in Solidworks 16 CAD software and then imported to ANSYS Workbench 16 for further simulation work.

The length, width, and thickness of all the microchannel heat sink are 120 mm, 40 mm, and 1.2 mm, respectively. The dimensions of the CD microchannel are to be optimized for higher heat dissipation. The optimization domain range for microchannel dimensions is 500 μm (minimum width) to 3000 μm (maximum width) with constant 51.34° converge-diverge angle.

2.2 Meshing Description

The cooling fluid zone has been meshed finely, and the heated zone is meshed coarsely because our zone of study is the cooling fluid (water) zone.

Method used for meshing: tetrahedrons; algorithm: patch conforming; behaviour: soft; element size: 0.4 mm; nodes: 68,328; elements: 266,875 (Fig. 2).

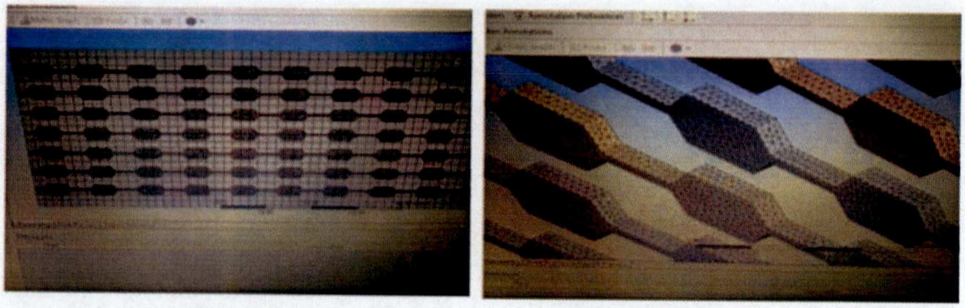

(a) (b)

Fig. 2 **a, b** Meshing screenshots

***Mesh Quality**: (Fine).

Minimum orthogonal quality = 1.51572e-01 (orthogonal quality ranges from 0 to 1); maximum ortho skew = 7.66533e-01 (ortho skew ranges from 0 to 1); maximum aspect ratio = 1.88731e + 01.

2.3 *CFD Simulation Assumptions and Boundary Conditions*

ANSYS Fluent 16 was used to run a three-dimensional steady-state simulation. For the simulation, the following assumptions and boundary conditions are used: 1. In the computational domain, fluid flow is laminar; 2. all of the surfaces were non-radiant; 3. the heat transfer process was in a steady-state condition; 4. inlet wall is "velocity inlet" at 25 °C, whilst the outlet wall is "environmental pressure"; 5. heat flux is constant which is 120 W/cm^2 and applied on the surface of the microchannel heat sink; 6. material assumed was homogeneous and isotropic; 7. 3D, steady state, incompressible flow; 8. effect of gravity is in normal direction.

Inlet velocity = 0.1 m/s; outlet pressure = 0 Pa (gauge pressure); inlet temperature = 25 °C; working fluid-de-ionized water; material of microchannel-aluminium.

2.4 *Conjugate Heat Transfer Module*

The classical Navier stokes equations: Continuity equation, momentum equation, and energy equation are solved by solver in the software.

For incompressible flow:

$$Continuity:\quad \frac{\partial u}{\partial x} + \frac{\partial v}{\partial y} = 0$$

$$NS\, equation: \quad \rho\frac{D\vec{V}}{Dt} = \rho\vec{g} - \nabla P + \mu\nabla^2\vec{V}$$

where

$$\frac{D}{Dt}(\vec{V}) = \frac{\partial}{\partial t}(\vec{V}) + u\frac{\partial}{\partial x}(\vec{V}) + v\frac{\partial}{\partial y}(\vec{V})$$

$$\nabla^2(\vec{V}) = \frac{\partial^2}{\partial^2 x}(\vec{V}) + \frac{\partial^2}{\partial^2 y}(\vec{V})$$

2.5 *Optimization Module*

Here in the model section, energy was set to on, and laminar flow has been considered. The materials considered for the cooling fluid domain and heated domain were water and aluminium, respectively. As for the boundary conditions, the inlet was set as velocity inlet and outlet as pressure outlet. The left and right sides were set as wall, and the top and bottom surfaces were treated as symmetry. Solution method considered was pressure-velocity coupling (SIMPLE), and momentum equation was solved using second order upwind scheme. Standard initialization from inlet has been done. And the residuals were as follows: 1.convergence criteria = 10^{-6}; 2. number of iterations = 1000.

3 Results and Discussion

3.1 *Validation*

Qualitative Analysis

The viscous forces along the walls of solid-fluid boundary reduce due to formation of flow recirculation vortices as shown in Fig. 3. At the exit furrows of the channels, there are symmetric vortices formed despite high mainstream velocity flow. These vortices induce the hot fluid from the main stream flow to mix and recirculate with the cold fluid. Hence, the viscous forces on the walls reduce maintaining low pressure drop in the CD microchannel.

Almost similar velocity contour is observed in present study when compared Chakravati et al.

Quantitative Analysis (Fig. 4)

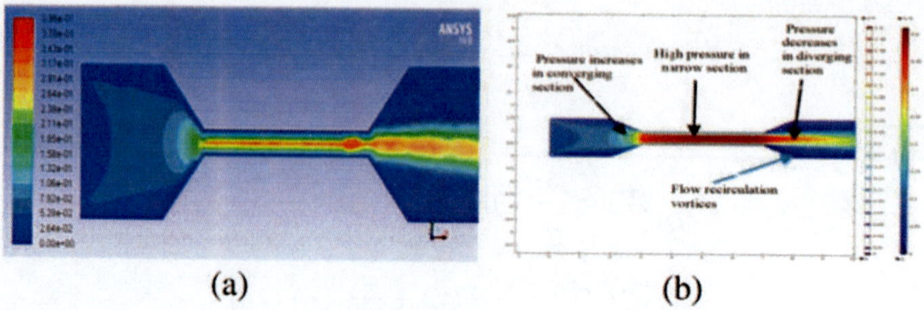

Fig. 3 **a** Velocity contour of present studies; **b** velocity contour of Chakravarti et al.

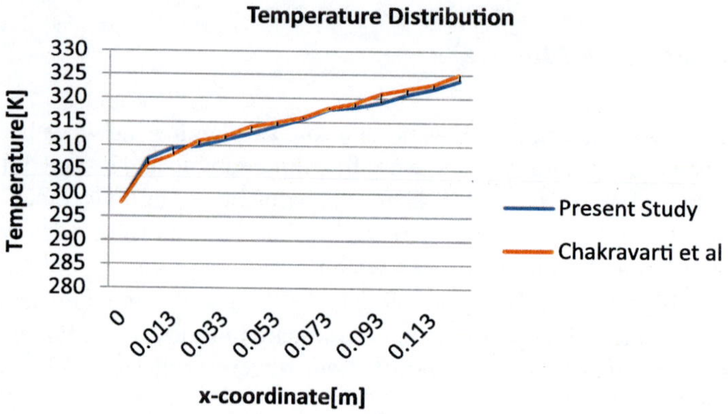

Fig. 4 Temperature distribution in the microchannel

3.2 *Grid Independence Test*

The grid independence test was performed using three different grids. The elements present in those three grids were Grid 1—183,436; Grid 2—205,855; Grid 3—266,875 (Fig. 5).

3.3 *Enhanced Observational Study*

From the Figs. 6, 7, 8, it can be observed as follows:

- In converging diverging microchannel, peaks and valleys are observed, and the curve is rough.
- In 0.5 mm straight microchannel, smoother curve is observed with very less peaks or valleys.

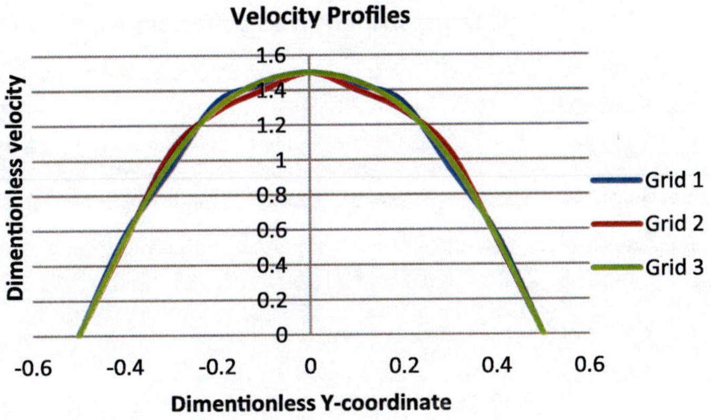

Fig. 5 Velocity profiles for different grids

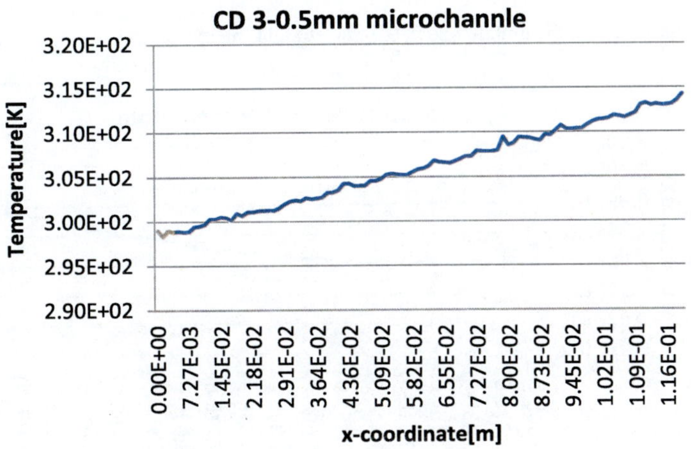

Fig. 6 Temperature on the middle line in CD microchannel

- In 3 mm straight microchannel, almost constant value is observed with a smooth curve.

The above observations clearly signify the formation of vortices inside the microchannel tubes which causes the intermixing of fluid and hence providing better thermal performance.

***Temperature Rise**

Values of avg. outlet temperatures from Fig. 9.

- CD microchannel = 313.47 K
- 0.5 mm microchannel = 311.799 K

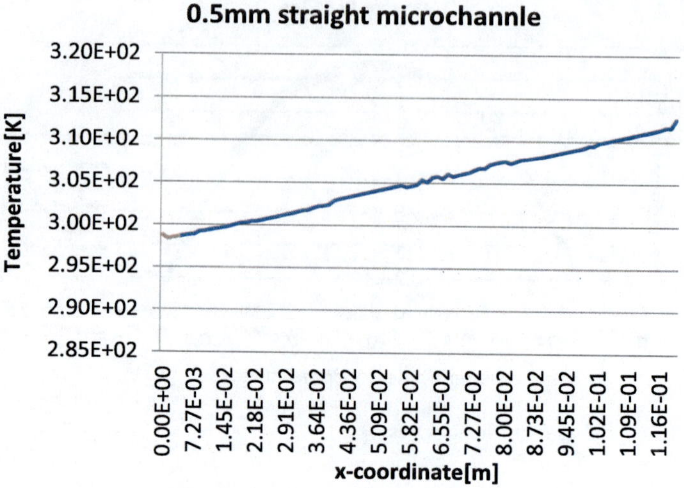

Fig. 7 Temperature on the middle line in 0.5 mm straight microchannel

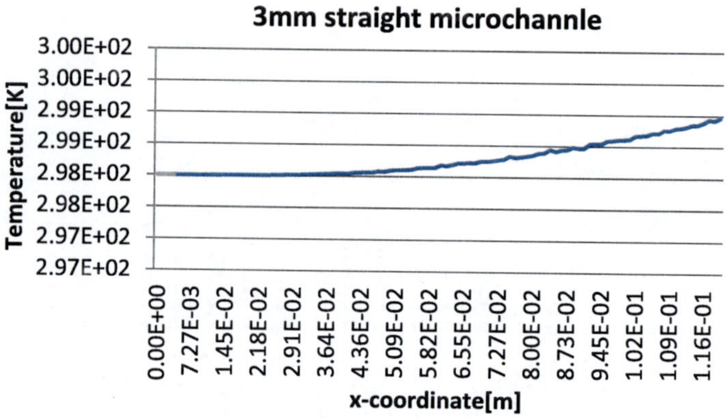

Fig. 8 Temperature on the middle line in 3 mm straight microchannel

- 3 mm microchannel = 301.11 K

It is clear from data above that higher temperature rise is observed in case of CD microchannel and hence a better thermal performance when compared to straight microchannels.

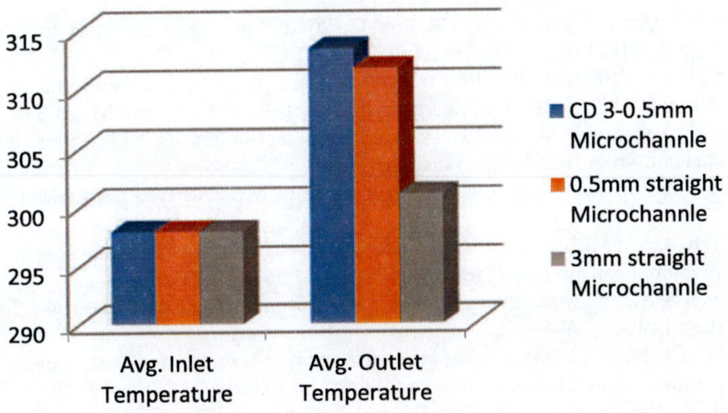

Fig. 9 Average inlet and outlet temperatures

4 Conclusion

The influence of converging diverging microchannel on heat transfer is proven for superior thermal performance. The temperature rise was found to be high for the microchannel dimension of 3000–500 μm, and this is considered as the optimal dimension for the CD microchannel heat sink. For the proposed microchannel, the pressure drop is calculated to be 49% lower than higher than the 0.5 mm straight microchannel. Converging diverging section in the CD microchannel prompts for non-nucleated flow and periodic velocity rise which contributes for increased heat transfer. And also the formation of vortices and intermixing of flow increases heat transfer rate. This can be proven by observing the middle line temperature charts. Hence, the proposed CD microchannel heat sink illustrates superior thermal performance.

Acknowledgements The author gratefully acknowledges Motilal Nehru National Institute of Technology and Asst. Professor Abhishek Kundu as my Guide in this study.

References

1. Tuckerman DB, Pease FW (1981) High performance heat sinking for VLSI. IEEE Electron Device Lett EDL-2 126–129
2. Singh SG, Agrawal A, Duttagupta SP (2011) Reliable MOSFET operation using twophase microfluidics in the presence of high heat flux transients. J Micromech Micro Eng 21:105002
3. Singh SG, Agrawal A, Duttagupta SP (2009) In-situ impact analysis of very high heat flux transients on non-linear p–n diode characteristics and mitigation using on-chip single-phase and two-phase microfluidic. J Microelectromech Syst 18(6):1208–1219
4. IV. Rosaguti N R, Fletcher D F, Haynes B S, (2006) Laminar flow and heat transfer in a periodic serpentine channel with semicircular cross section. Int J Heat Mass Transf 49:2912–2923

5. Sui Y, Teo CJ, Lee PS, Chew YT, Shu C (2010) Fluid flow and heat transfer in wavy microchannels. Int J Heat Mass Transf 53:2760–2772
6. Lee PC, Pan C (2008) Boiling heat transfer and two-phase flow of water in a single shallow microchannel with a uniform or diverging cross section. J Micromech Micro Eng 18:025005
7. Agrawal DVS, Singh SG (2012) Pressure drop measurements with boiling in diverging microchannel. Front Heat Mass Transf 3:013005
8. Kosar PY (2006) Thermal-hydraulic performance of MEMS-based pin fin heat sink. J Heat Transf 128:121–131
9. Peles Y, Kosar A, Mishra C, Kuo C, Schneider C (2005) Forced convective heat transfer across a pin fin micro heat sink. Int J Heat Mass Transf 48:3615–3627
10. Rahman MM (2000) Measurements of heat transfer in microchannel heat sinks. Int Commun Heat Mass Transf 27:495–506
11. Wu HY, Cheng P (2003) An experimental study of convective heat transfer in silicon microchannels with different surface conditions. Int J Heat Mass Transf 46:2547–2556
12. Bariyar VV, Prabhansu KR, Chandra P, April-June KMK (2016) CFD analysis of loop seal in the circulating fluidized bed system system. J Basic Appl Eng Res 3(8):697–700
13. Kumar R, Chandra P, Prabhansu, (2020) Innovative method for heat transfer enhancement through shell side and coil side fluid flow in SHCHE. Arch Thermodyn Polish Acad Sci 41(2):239–256

Static Analysis for Stress Concentration Due to Elliptical Notches of Laminated Plates

Rahul Kumar, Achchhe Lal, and B. M. Sutaria

Abstract In the present paper, stress concentration factors (SCF) of sandwich plates with elliptical notch at corner subjected to out plane load are studied. The mathematical modeling of the present problem is done using secant function-based shear deformation theory (SFSDT) with geometrical nonlinearity. Minimum potential energy method is used to derive the bending governing equation. Influence of cutout dimensions, fiber orientations, plate span to thickness ratios, and boundary conditions are investigated using a MATLAB program. The present outcomes are compared with the outcomes of already published work.

Keywords Laminated composite sandwich · Cutout dimensions · Stress concentration factor

1 Introduction

Sometimes cutouts are made into plates to fulfill many requirements as to access internal parts, to fastening one part to another one, and to avoid resonance. These geometrical discontinuities (notches) in the structural members cause strength degradation and therefore must be analyzed to avoid undesired deformation and any damage.

Huge number of works on bending behavior study of LCSPs with hole and notches are available. Paul and Rao [1, 2] determine SCF around the circular cutout of laminates. Jain et al. [3–5] investigated the significance of ratio of hole diameter to plate length (D/A) on SCF of laminated plates under bending. Firstly, they analyzed orthotropic and isotropic plates with hole and then parametric study to investigate effect of reinforcement angle on SCF under transverse loading. Xiwu et al. [6] done parametric study for SCF and examined the effect of various parameters on the same for laminated composite plate with hole.

R. Kumar · A. Lal (✉) · B. M. Sutaria
Department of Mechanical Engineering SVNIT, Surat, Gujarat, India
e-mail: lalachchhe@yahoo.co.in

A. Maurya et al. (eds.), *Recent Trends in Mechanical Engineering*, Lecture Notes in Mechanical Engineering, https://doi.org/10.1007/978-981-19-7709-1_7

Rzayying [7] investigated hole position along plate dimensions on SCF, ultimate stress, and deformation of plates. Ukadgaonker and Rao [8] examined the stress distribution around various shapes of hole of plate subjected to in-plane loadings. SCF values are first determined experimentally and then compared with Lekhnitskii theoretical model by Toubbal et al. [9]. Patel and Sharma [10] studied the influences of various parameters on failure strength and distribution of moments in laminate.

As per the above review, it is observed that sufficient research works are available to investigate the stress concentration factor due to geometrical discontinuities, but extremely limited work is there to investigate the influence of cutout dimension on the nonuniform stress distribution around the notches area.

Therefore, in the present paper, investigation for influence of cutout dimension on SCF of laminated composite sandwich plates has been done.

2 Problem Formulation

The LCSP with dimension parameters $1 \times 1 \times 0.1$ as shown in Fig. 1 is taken for present investigation. The out plane dimension of the considered plate is distributed into top and bottom face sheet thickness and core thickness. In the present analysis, sandwich plate having elliptical notch with major and minor axis radii m_a and m_i is taken.

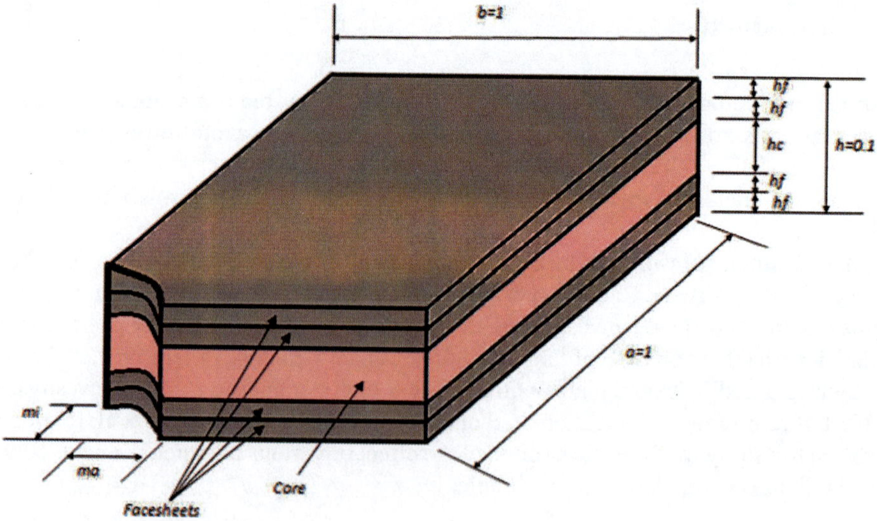

Fig. 1 LCSP with cutout (notch) at corner

2.1 Displacement Fields

The displacement field for the present analysis can be given as, [11]

$$u = u_0 - z\frac{\partial w_0}{\partial x} + (f(z) + z\mathrm{H}_0)\theta_x;$$

$$v = v_0 - z\frac{\partial w_0}{\partial y} + (f(z) + z\mathrm{H}_0)\theta_y;$$

$$w = w_0 \tag{1}$$

The above displacement field provides top and bottom surface of the plate free from out plane shear strain and C^0 continuity. The distribution of out plane shear stresses along the thickness of the plates is parabolic.

Where $f(z) = z\sec(\frac{rz}{h})$ and $\mathrm{H}_0 = -\frac{\sec(\frac{r}{2})}{1+\frac{r}{2}\tan(\frac{r}{2})}$.

The displacement vector is taken as

$$\{\vee\} = \begin{pmatrix} u_0 & v_0 & w_0 & \phi_y & \phi_x & \theta_y & \theta_x \end{pmatrix}^T \tag{2}$$

2.2 Strains Derived from Displacement Field

The strain vectors by differentiating displacement field can be given as, [11]

$$\{\varepsilon\} = \{\varepsilon_0\} + \{\varepsilon_{nl}\} \tag{3}$$

where $\{\varepsilon_0\}$ and $\{\varepsilon_{nl}\}$ are linear and nonlinear strains, respectively.

2.3 Stress Strain Relations

The stress strain relation for kth lamina can be given as, [12]

$$[\sigma] = \left[\overline{Q_k}\right]\{\varepsilon\} \tag{4}$$

where [Q] is reduced constitutive matrix for different layers and k = 3 for core layer and k = 1, 2, 4, and 5 for face sheets material of LCSP.

2.4 *Stress Concentration Factor (SCF)*

Laminates having cutouts are often used as structural member. Due to these cutouts, there is uneven stress distribution around these geometrical discontinuities. SCF can be expressed as

$$\text{SCF} = \frac{\text{Stress at a point on the circumference of cutout}}{\text{stress at the same point without cutout}}$$

2.5 *Minimum Potential Energy Approach*

The total strain energy of LCSP can be expressed as

$$U = \frac{1}{2}\sum_{k=1}^{NL}(U_l + U_{nl}) \tag{5}$$

where U_l and U_{nl} are linear and nonlinear strain energies, and *NL* is number of layers of LCSP. *k* refers to specific layer of LCSP.

Potential energy (Ω) due to work done may be written as

$$\Omega = \sum_{i=1}^{NE}\{\vee\}^T q \tag{6}$$

where q is load parameter and NE number of elements.

Here for transverse loading, q is taken as

$$q = (0 \;\; 0 \;\; q_0 \;\; 0 \;\; 0)^T \tag{7}$$

T stands for transpose of vector.

2.6 *Governing Equation*

Using principle of virtual displacement, the equation for bending can be given as [11],

$$\frac{\partial \prod}{\partial \vee^T} = 0 \tag{8}$$

where total potential energy of the system, $\prod = \text{U} + \Omega$.

On substituting corresponding values from (5) and (6) in (8), the governing equation for static deflection can be given as

$$[K]\{\vee\} = [q] \tag{9}$$

where $[K]$ is overall stiffness matrix.

2.7 Solution Approach

The governing equation for bending given in Eq. (9) is solved using Newton Raphson method, and maximum transverse deflection is determined.

After evaluating deflection elemental Gauss point, stresses are determined at desired nodes (*i*) of the element as [11],

$$[\sigma_i]^e = \left[\overline{Q_k}\right][B_i]\{\vee_i\}^e \tag{10}$$

where $[B_i]$ = displacement matrix consists of derivatives of interpolation function. Corresponding extrapolated stresses can be written as

$$\{\sigma_{i,ep}\}^e = \left[N_i'\right]\{\sigma_i\}^e \tag{11}$$

where [N] is interpolation function.

3 Results and Discussion

A MATLAB program based on FEM using mathematical formulation and solution technique is developed for present analysis. A nine noded iso-parametric element with 63 unknowns per element is considered.

The following boundary conditions are used for present work.

All edges are simply supported.

$u_0 = w_0 = \phi_y = \theta_y = 0$, at Y = 0 and Y = b ; $v_0 = w_0 = \phi_x = \theta_x = 0$, at X = 0 and X = a

All edges are clamped.

$u_0 = v_0 = w_0 = \phi_y = \phi_x = \theta_y = \theta_x = 0$, at Y = 0 and Y = b
$u_0 = v_0 = w_0 = \phi_y = \phi_x = \theta_y = \theta_x = 0$, at X = 0 and X = a

While applying these boundary conditions to the plate with cutout in the present work, the arc of cutout is considered as constraint free.

The normalizations used for uniformly distributed load (UDL), deflection, and stress can be given as

$$q = q0E_2(h/a)^4 \tag{12}$$

$$W_0 = w_{\max}/h \tag{13}$$

and

$$\left[\sigma_x, \sigma_y, \tau_{xy}, \tau_{yz}, \tau_{zx}\right] = \frac{1}{q0}[\sigma_{0x}, \sigma_{0y}, \sigma_{0xy}, \tau_{0yz}, \tau_{0xz}] \tag{14}$$

where $q0$ is input load parameter, W_0 is normalized maximum transverse deflection, h and a are plate thickness and length of plate, respectively.

Material-I (for face sheets)

$$E_{f1} = 220 \times 10^9, \quad E_{f2} = 13.79 \times 10^9, \quad \nu_{f12} = 0.2,$$
$$\nu_{f12} = 0.25, \quad E_m = 3.45 \times 10^9, \quad \nu_m = 0.35,$$

Material-II (for core)

$$[Q_c] = 10^9 \times \begin{bmatrix} 0.999781 & 0.231192 & 0 & 0 & 0 \\ 0.231192 & 0.524886 & 0 & 0 & 0 \\ 0 & 0 & 0.26810 & 0 & 0 \\ 0 & 0 & 0 & 0.266810 & 0 \\ 0 & 0 & 0 & 0 & 0.15991 \end{bmatrix}$$

Validation of present study is shown by Fig. 2. It is clear from the Fig. 2 that the current results have good agreement. Figure 3 shows variation of SCF with various cutout condition. The cutout cases taken on abscissa refer as 1 to ma = 0.2 and mi = 0.1, 2 to ma = 0.4 and mi = 0.1, 3 to ma = 0.6 and mi = 0.1, 4 to ma = 0.1 and mi = 0.2, 5 to ma = 0.1 and mi = 0.4, 6 to ma = 0.1 and mi = 0.6, 7 to ma = 0.1 and mi = 0.2, for simply supported condition under UDL of q0 = 100. SCF for is determined at corner point of cutout arc. It is clear from the figure that maximum SCF occurs for σ_y is maximum in case all edges of plate are simply supported. Figure 4 shows SCF for clamped conditions of LCSP. It is clear from the figure that maximum SCF occurs for σ_y is maximum in case all edges of plate are clamped. The SCF for case 6 ($m_a = 0.1$ and $m_i = 0.6$) is maximum as we are determining SCF on intersection of cutout arc and y-axis, and there is maximum cutout along the same axis. Figure 4

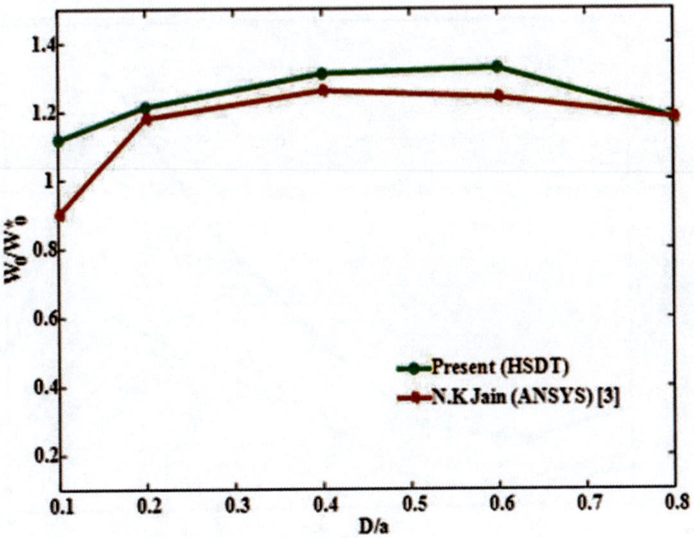

Fig. 2 Validation of ratio of transverse deflection with cutout to without cutout with diameter of hole to the plate length ratio (D/a)

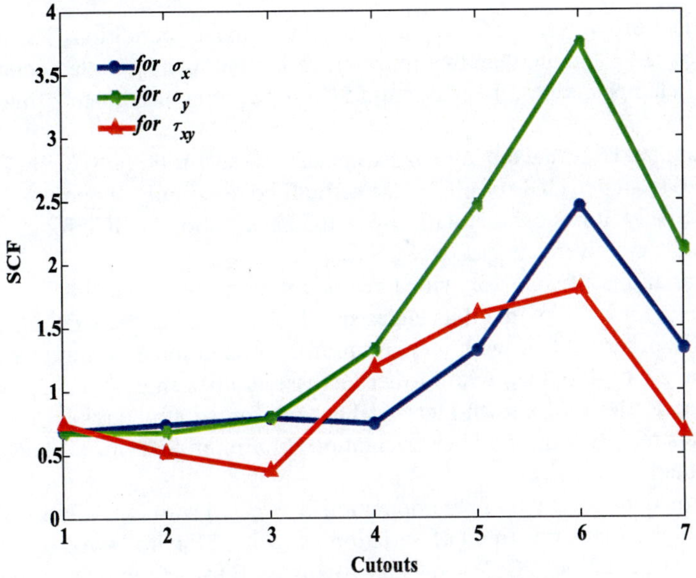

Fig. 3 Effect of cutouts on SCF for LCSP (0/90/C/90/0) in simply supported

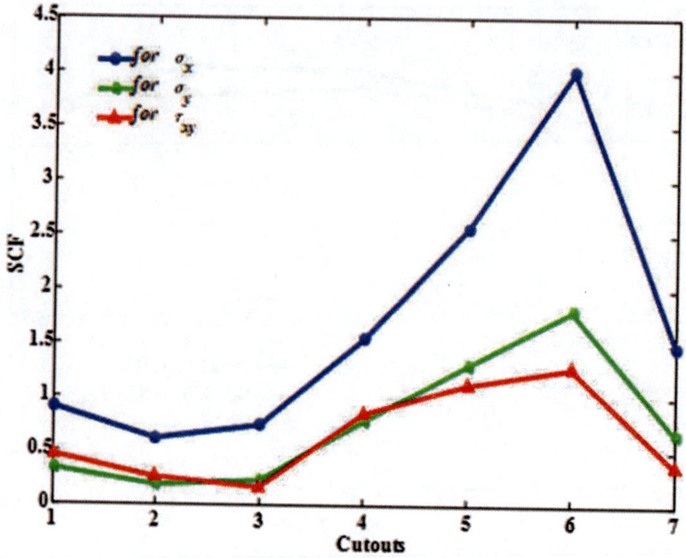

Fig. 4 Consequence of cutouts on SCF for LCSP (0/90/C/90/0) in clamped conditions

shows effect of cutout on SCF under clamped boundary condition. The maximum SCF for σ_y is 52% more than that for σ_x in case of simply supported condition and in case of clamped edges, the maximum SCF for σ_x is approximate 2 times that for σ_y.

Figures 5 and 6 reveal the influence of plate thickness ratio (a/h) on SCF of LCSP (0/90/C/90/0) under UDL of q0 = 100 with all edges simply supported for σ_x and σ_y, respectively. It is clear that with rise in thickness ratio, the SCF for σ_x decreases while results are reverse in case of SCF for σ_y.

The variations of SCF for with fiber orientation are shown in Figs. 7 and 8, respectively, for LCSP with all its edges simply supported under UDL of q0 = 100. It is clearly observed that, with increase in fiber orientation SCF initially increases up to an angle of 450, then with further increase in fiber angle, the SCF decreases. This is due to the well-known fact that fiber-reinforced structures have maximum transverse strength with 450 fiber orientation. Minimum SCF for is shown for 900 fiber orientation.

Variation of SCF for with fiber orientation is revealed with Fig. 9. It is clearly seen from the Fig. 9 that same trend of variation as with SCF is followed by. With fiber having 450 orientation, LCSP shows maximum SCF for and with an orientation of 900, the plate has least SCF.

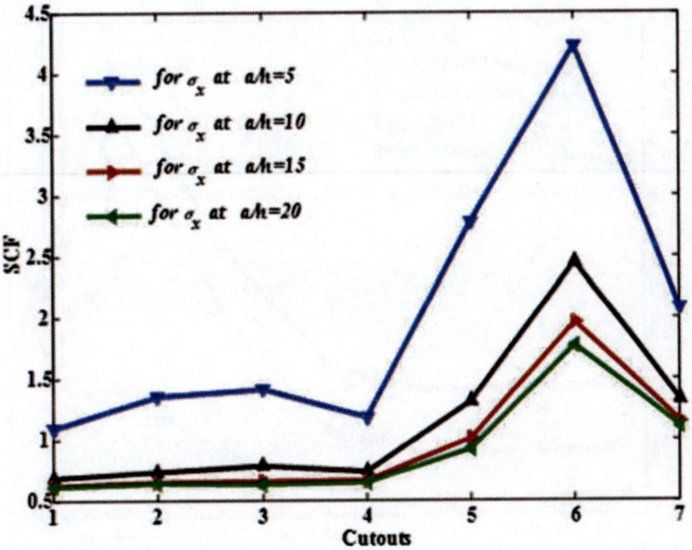

Fig. 5 Effect of cutouts and plate thickness ratio (a/h) on SCF for LCSP (0/90/C/90/0) in simply supported conditions for σ_x

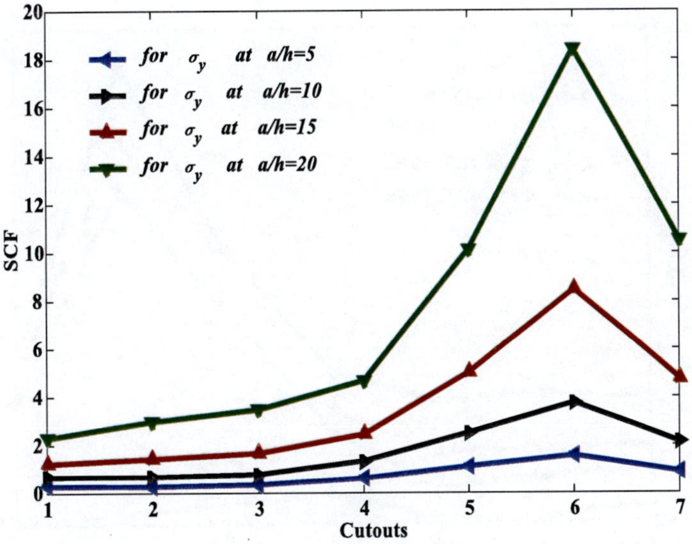

Fig. 6 Effect of cutouts and plate thickness ratio (a/h) on SCF for LCSP (0/90/C/90/0) in simply supported conditions for σ_y

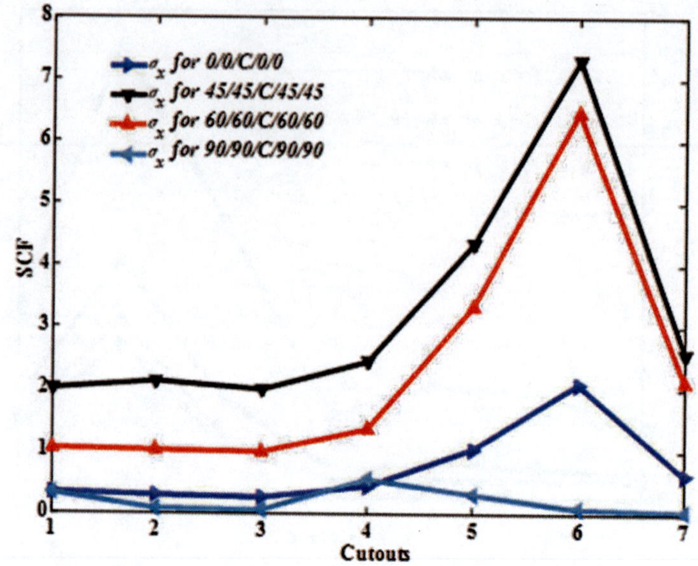

Fig. 7 Effect of fiber orientation with cutout on SCF for LCSP in simply supported conditions for σ_x

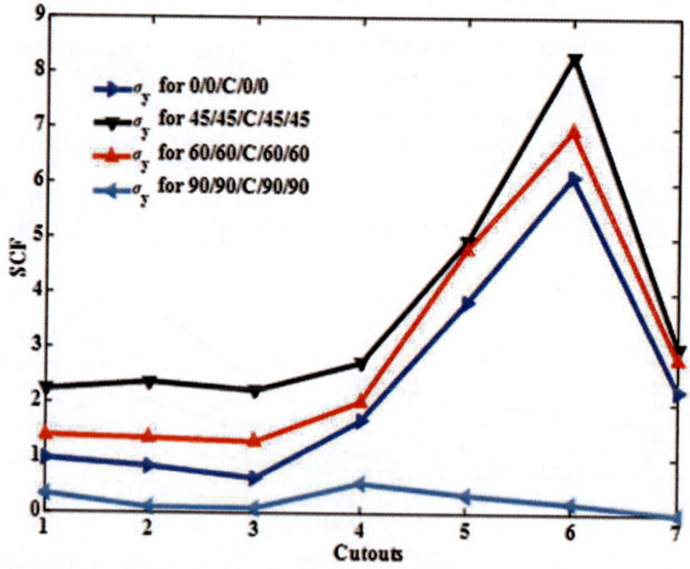

Fig. 8 Consequence of fiber orientation with cutout on SCF for σ_y

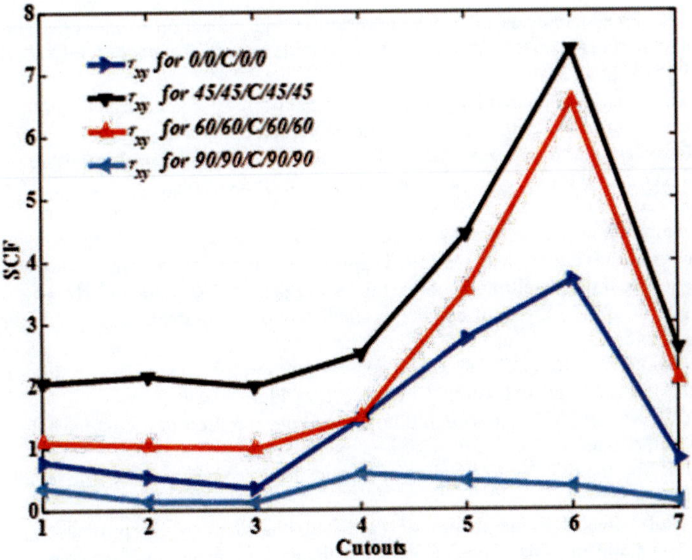

Fig. 9 Consequence of fiber angle with cutout on SCF for τ_{xy}

4 Conclusions

On the basis of the present work, the following conclusions can be given:

- The SCF increases with the increase in cutout dimensions and has maximum value at the intersection of edges of plate with cutout arc.
- The SCF for σ_x *and* σ_y significantly depends upon support conditions, while SCF for τ_{xy} shows less variation with the boundary conditions.
- With increase in plate span to thickness ratio, SCF for σ_x *and* σ_y decreases and increases, respectively.
- The SCF varies significantly with the reinforcement orientation of LCSP. The SCF increases with increase in fiber orientation till 45° after that it decreases with further increment in fiber orientation and has maximum value at 45° fiber orientation.

References

1. Paul TK, Rao KM (1989) Stress analysis around circular holes in FRP laminates under transverse load. Comput Struct 33(4):923–937
2. Paul TK, Rao KM (1993) Finite element evaluation of stress concentration factor of thick laminated plates under transverse loading. Comput Struct 48(2):311–317

3. Jain NK (2009) Analysis of stress concertation and deflection in isotropic and orthotropic rectangular plates with maximum circular hole under transverse loading. Int J Mech Mechatron Eng 3:1513–1519
4. Jain NK, Banerjee M, Sanyal S (2012) Three dimensional parametric analyses on effect of fibre orientation for stress concentration factor in fibrous composite cantilever plate with central circular hole under transverse loading. IIUM Eng J 13(2), ISSN no 2289-7860
5. Mittal ND, Jain NK (2012) Parametric analyses on effect of fibre orientation for stress concentration factor in two edges fixed and two edges simply supported fibrous composite plate with central circular hole under transverse loading. Int J Mech Eng Ser J 5, ISSN 0974-5823
6. Xiwu X, Liandxing S, Xuqi F (1995) Stress concentration of finite composite laminates weakened by multiple elliptical holes. Int J Solid Struct 32(20):3001–3014
7. Rzayyig AY (2011) Effect of cutouts on the behavior of clamped rectangular plates. Anbar J Eng Sci 2:45–59
8. Ukadgaonker VG, Rao DKN (2000) A general solution for stresses around holes in symmetric laminates under in-plane loading. Compos Struct 49:339–354
9. Toubal L, Karama M, Lorrain B (2005) Stress concentration in a circular hole in composite plate. Compos Struct 68:31–36
10. Patel NP, Sharma DS (2015) Bending of composite plate weakened by square hole. Int J Mech Sci 94:131–139
11. Lal A, Kulkarni NM, Siddaramaiah VH (2016) Stochastic hygro-thermo-mechanically induced non-linear static analysis of piezoelectric elastically support sandwich plate using secant function based shear deformation theory (SFSDT). Int J Comput Mater Sci Eng 6:16500201–16500246
12. Lal A, Kulkarni N, Singh BN (2015) Stochastic thermal post-buckling response of elastically supported laminated piezoelectric composite plate using micromechanical approach. Curved Layered Struct 2:331–350

Design of Subsonic Axial Flow Compressor Rotor Blade

Anand P. Darji and Beena D. Baloni

Abstract Axial flow compressors/fans are widely used machines in industrial as well as aircraft gas turbine engines. The performance improvement of such machines has improved significantly due to key efforts of researchers and engineers. The hard work and research inputs require for the development of a single compressor which includes years of careful study and investigations. Rotor is a primary element of any axial compressor/fan which imparts kinetic energy to fluid, and so, the designers have emphasized more on the rotor design. Present study incorporates an effort for the development of a low-speed axial flow compressor/fan rotor blade based on NACA 65-series airfoil. The limiting design parameters are selected from the grounded axial fan test rig available in the department lab. The study initiates with one-dimensional flow parameters calculation based on velocity triangle at mean location and is extended for total 11-radial locations from hub to tip. The blade angles are obtained from correlations available in literatures, and the corrected angles are used to set up selected airfoil profile along spanwise locations using commercial modeling tool. Once the airfoils are stacked on one another, a solid blade model is developed. This is further extended to develop entire rotor with calculated number of blades. The data obtained from the present study are validated with the available literature plots and are in good trend. The obtained blade has high pressure variation and total pressure loss close to tip. The blade has higher camber close to hub and lower camber at higher spanwise positions.

Keywords Axial flow compressor · Subsonic · Rotor blade · NACA 65-airfoil · Blade plots

A. P. Darji (✉) · B. D. Baloni
Mechanical Engineering Department, SardarVallabhabhai National Institute of Technology, Surat, Gujarat, India
e-mail: ananddarji99@gmail.com

A. Maurya et al. (eds.), *Recent Trends in Mechanical Engineering*, Lecture Notes in Mechanical Engineering, https://doi.org/10.1007/978-981-19-7709-1_8

1 Introduction

The widely known theory for design of such machines is blade elementary theory. Compared to the slipstream theory, blade element momentum theory accounts for the angular momentum of the rotor or rotational components of the fluid. In this method, the rotor area is segmented into number of annular sections of infinitesimally small thickness. This is done to assume that the axial flow parameters and tangential flow parameters are constant throughout the annular sections. An assumption of this approach is that annular sections are independent of one another, i.e., there is no interaction between the fluids of neighboring annular section. The flow across blade row is always subject to adverse pressure gradient in axial compressor. Thus, with increase in pressure, ratio makes it difficult to design rotor.

The conventional blade profiles like NACA 65-series and C4 type profiles are widely used and implemented in various gas turbine engines. As noted by Casey [1] and Casey et al. [2], one of the reasons to use such profile forthe present study is the availability of exhaustive cascade studies and documentation with correlations. Another series of airfoil popularly known as controlled diffusion airfoil (CDA) is also implemented in jet engines and ground-based gas turbine power plants. Eisenberg [3] observed that such airfoils have tendency to achieve higher peak efficiency and considerable larger flow rate. The advantages of such profiles over conventional blade profiles have been studied by Hobbs and Weingold [4] and Cumpsty [5]. Koller et al. [6] and Kusters et al. [7] have explained improved design approach of CDA in their respective studies. Falck N [8] has explained meanline design methodology to design axial flow compressor. The present study has followed similar methodology to initiate the design procedure and is extended to hub and tip sections. Keskin [9] has proposed a new method based on multi-objective optimization technique to design compressor. The method of optimized compressor/turbine blade design is well known nowadays. The proposed method is based on reducing overall compressor losses and to enhance the performance.

Present study includes the design procedure of developing three-dimensional low-speed axial compressor/fan rotor blades for optimum performance. Based on few initial fixed parameters, inlet and outlet flow velocities are determined using one-dimensional elementary blade theory. The design method is based on correlations available in literatures by Aungier [10], McKenzie [11, 12], Cumpsty [5], Dixon [13], Saravanamuttoo et al. [14], Mattingly et al. [15], Joseph Herrig et al. [16] and Roy and Pradeep [17]. The fixed parameters are taken from the grounded low-speed axial fan test rig facility available at the laboratory. Flow angles from calculated velocities are obtained from correlations available in literatures for airfoil stacking.

2 Design Methodology

Specification

Proposed design in the study is reconstructed based on nonoperational subsonic axial fan test facility available at Mechanical Engineering Department, Sardar Vallabhabhai National Institute of Technology–Surat. Initial design specifications obtained from the test facility for the present design are as mentioned in Table 1. The low-speed axial fan rotor contains eight blades with low blade twist as shown in Fig. 1. Few limiting parameters for the present study are obtained from the existing rotor. The proposed rotor has higher aspect ratio which also suggests having higher blade loading to achieve the desired pressure ratio. Authors would like to clarify that neither inlet guide vanes nor the downstream stator are considered while the design of rotor. Considered geometric limiting parameters and desired rotor efficiency and pressure ratio are mentioned in Table 1. Various design laws and assumptions are considered for the selected subsonic blade profile.

Table 1 Initial design specifications for rotor

Specification	Dimension	Specification	Dimension
Tip diameter, d_t	0.390 m	Blade chord, C	0.1 m
Hub diameter, d_h	0.150 m	Mass flow rate, $\dot{m}$	6 kg/s
Blade span, h	0.120 m	Operating speed, N	2500 RPM
Efficiency, $[\eta_{is}]_{Targeted}$	90%	P.R, $[P_{02}/P_{01}]_{Targeted}$	1.15

Fig. 1 Existing rotor of low-speed axial fan

The design requires certain assumptions to initialize with available physical parameters. The limiting operational parameters need to follow available design criteria, and so, the followings are the design assumptions considered while the design calculations.

- Flow studies at mean radial location of blade
- No inlet guide vanes available at upstream to the rotor
- Use of De-Heller's criteria for maximum allowable flow deflection
- Use of free vortex law ($C_w \times r = \text{Constant}$)
- Use of circular arc type airfoil (NACA 65-series)

Procedure

The design procedure initiates with the initial specifications as mentioned in Table 1. Entire design procedure is carried out in three steps. Step-1 includes the discretization of entire blade into eleven radial stations by considering initial design specifications. Each station will have local inlet–outlet flow conditions which are calculated in step-2 by deriving local velocity triangle. Step-3 includes the selection of subsonic CD type blade profile based on NACA 65-series due to availability of open literatures. Blade parameters are obtained using various limiting criteria and available correlations from literatures. Few correlations which are considered in the design are mentioned in prior. All the calculated data are mentioned in tabular form. Blade angles are obtained from available flow conditions from step-2 and corrected using correction factor. These are obtained for each radial station as the desire pressure ratio generates twisted blade profile. Each station has separate blade setting angles. The iterative procedure between the flow parameters and blade angles calculation ends once the smooth blade profile is obtained. The last step includes the angle definition, arrangement of selected airfoil profiles at local station, and generation of three-dimensional blade (Fig. 2).

Methodology

As mentioned earlier, one-dimensional blade elementary theory-based calculations are done to obtain inlet and outlet flow conditions. The design specifications and assumptions have given inflow and outflow conditions. Once the inflow and outflow conditions are known from velocity triangle, the blade parameters are calculated.

The correlation developed by Saravanamuttoo et al. [14] for degree of reaction ($\bigwedge$) with non-dimensional form of radii is used in present study.

$$\bigwedge = 1 - \frac{(Cw_2 \times r) + (Cw_1 \times r)}{2U_m[\frac{r^2}{r_m}]} \tag{1}$$

According to free vortex law,

$$C_w \times r = \text{Constant}$$

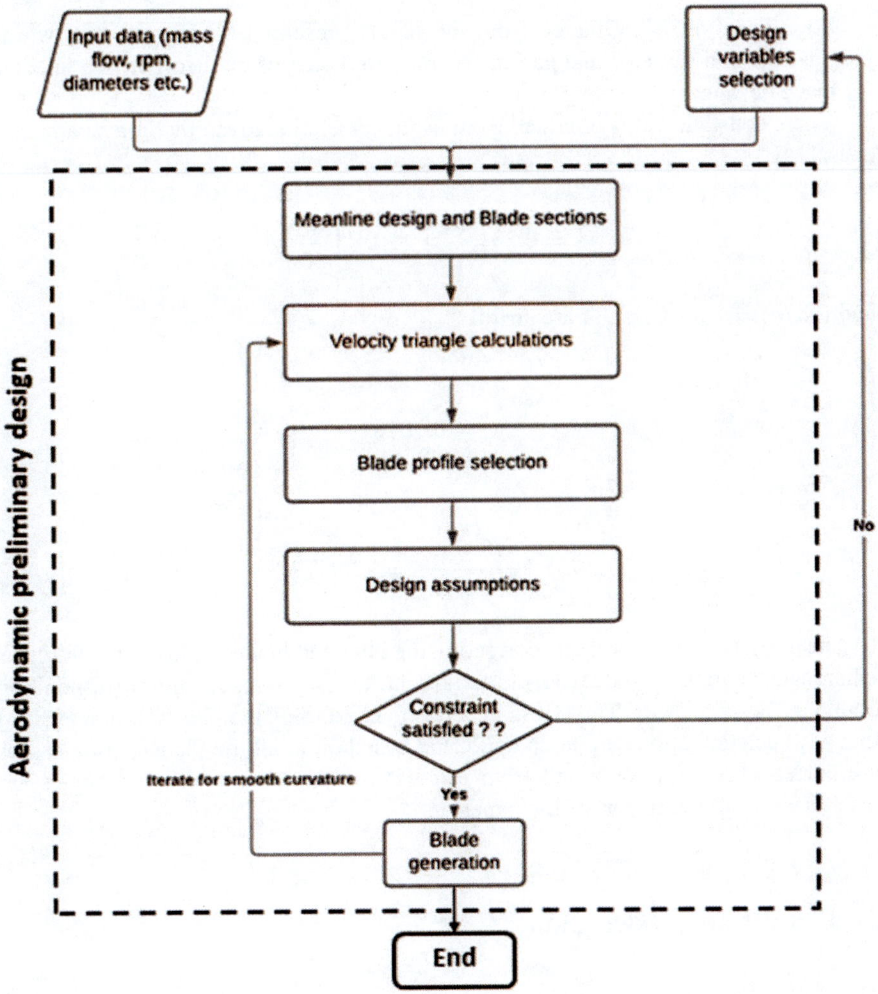

Fig. 2 Design procedure

Lieblein et al. [18] diffusion factor (D) depends on blade pitch to chord ratio (s/c). Pitch to chord (s/c) ratio is obtained from the relative plot presented by Sayed A [19] for 65-series profiles.

Diffusion factor NACA (D),

$$D = 1 - \frac{V_2}{V_1} + \frac{(\Delta C_w)}{(2V_1)}(\frac{s}{c}) = 0.3829 \tag{2}$$

Flow deviation is estimated from the report presented by Howell [20] which includes the non-dimensional parameter 'm' based on type of airfoil profile selected for blade design.

Where constant 'm' is obtained from using correlation given by Saravanamuttoo et al. [14],

$$m = 0.23\left(\frac{2a}{c}\right)^2 + 0.1\left(\frac{a_2}{50}\right), \tag{3}$$

and $(2a/c) = 1$ for Circular arc airfoil

so,

$$m = 0.30182 \tag{4}$$

Camber angle θ is given

$$\theta = \frac{\Delta\beta - i_r}{1 + m_r\sqrt{\frac{s}{c}}} = 22.9477 \tag{5}$$

The design practice suggests correcting the obtained blade angles as the fluid has inherent behavior to deviate from the surface due to viscous nature and corresponding boundary layer growth. Thus, it is general practice implemented in literature [21] that any such deviation can be up to 3°, and based on it, all the blade setting angles are corrected. Results obtained from calculated blade parameters and angles are discussed in following part of the paper.

3 Results and Discussion

The study of blade elementary theory-based calculations of inflow–outflow conditions and blade setting angles for the development of subsonic axial flow compressor/fan is done and is presented in Table 2.

The followings are the plots representing various parameters of present rotor design. The presented plots are compared with available literatures to meet the design criteria.

Figure 3a shows that exit air angle (α_2) is higher at hub region than other two flow angles, whereas relative flow angle at inlet (β_1) is dominant near tip region. The blade is designed with higher loading near tip region (as shown in Fig. 3b).

Radial distribution of degree of reaction (DR) at each station is plotted in Fig. 4. It can be seen that the blade has higher DR at meanline station. This shows the rotor contribution to achieve desired pressure ratio. The trend is compared with available cascade test data from Saravanamuttoo et al. [14].

Table 2 Calculated blade parameters

Parameter	Value	Parameter	Value
Degree of reaction, Λ_m	0.6672	Stagger angle, ?	34.4424°
Diffusion factor, D	0.3829	Deviation with WF, δ_w	7.2262°
Constant, m	0.30182	Pitch, s	0.042 m
Deviation, δ_1	7.4886° °	Number of blades, n	29
Camber angle, θ_1	25.9477°	Rotor efficiency, η_{is}	89.7967%

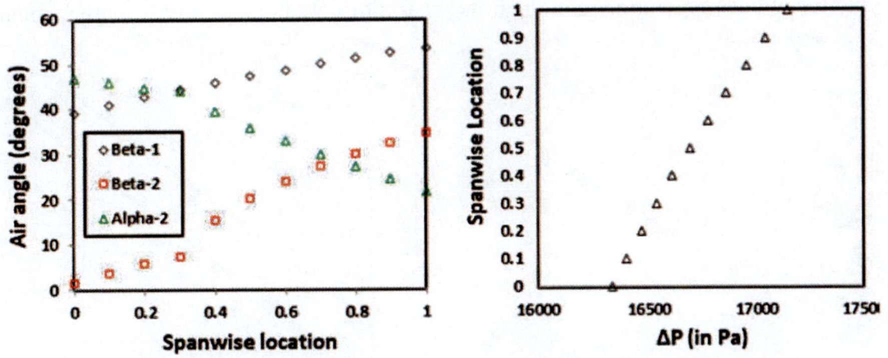

(A) Air angle distributions (B) Pressure variation distributions

Fig. 3 Spanwise distribution of air angles and pressure variation

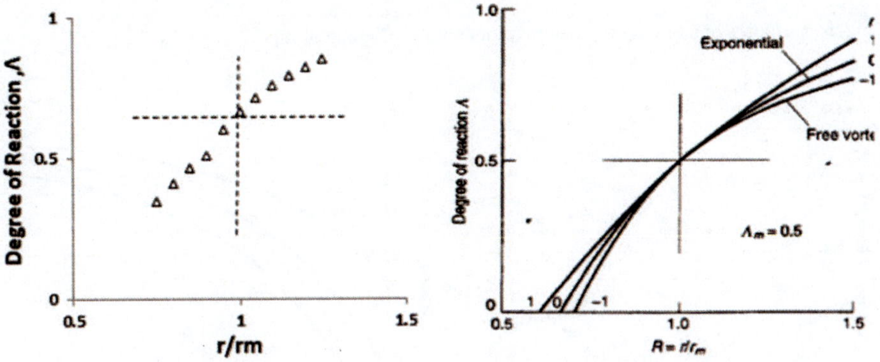

(A) Present study (B) Saravanamuttoo et al [2]

Fig. 4 Comparison of degree of reaction for NACA 65-series data

Variation of degree of reaction along radial location with respect to tip radius is presented in Fig. 6. Meanline degree of reaction for present design is 0.66, and r_m/r_t ratio is 0.8. The results are compared with Saravanamuttoo et al. [14], and the curve follows the ideal design behavior.

Deviation with and without workdone factor (WF) is plotted in Fig. 5.

The deviation factor with WF is higher than deviation without WF due to inclusion of blockage factor within blade rows. A linear regression curve shows the ideal deviation close to 5°. The trend is in good match with McKenzie plots [11] for the profile.

As per design criteria in Fig. 7a, blade meets the rule of higher camber angle and incidence angle for the near hub blade profiles. Higher the camber higher is the chances of flow separation, and the flow separations are not accepted at higher spans

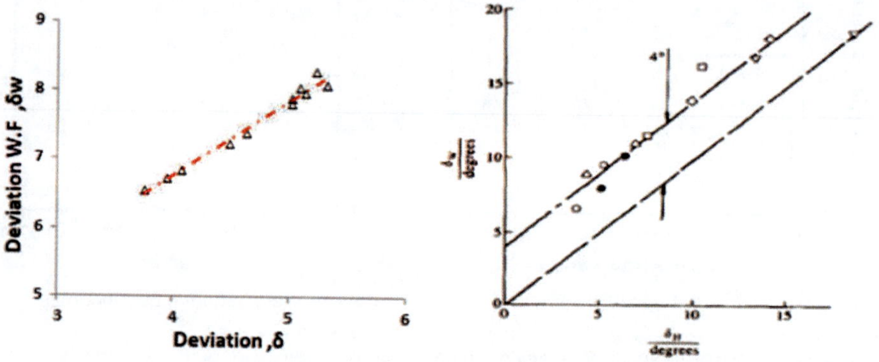

(A) Present study

(B) McKenzie plot [9]

Fig. 5 Comparisons of deviation angle plots with McKenzie plot [11]

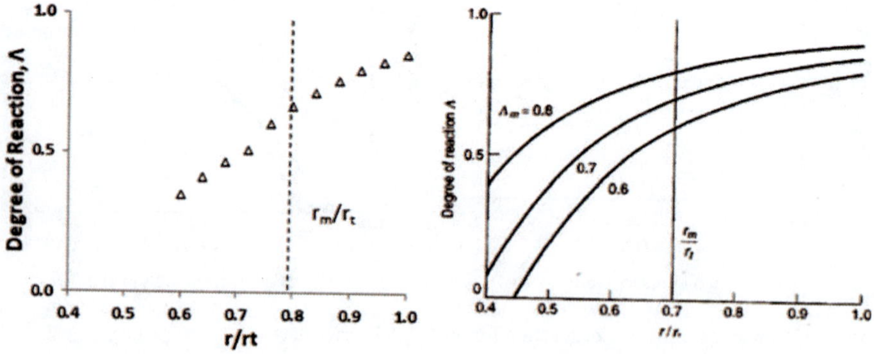

(A) Present study

(B) Saravanamuttoo et al [2]

Fig. 6 Comparison of degree of reaction distribution

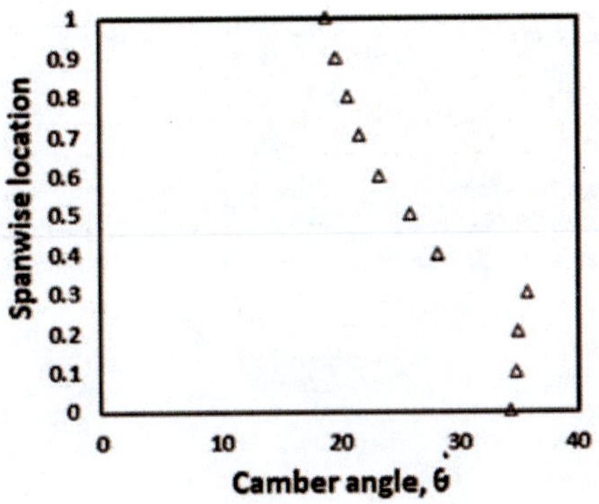

(A) Camber distribution

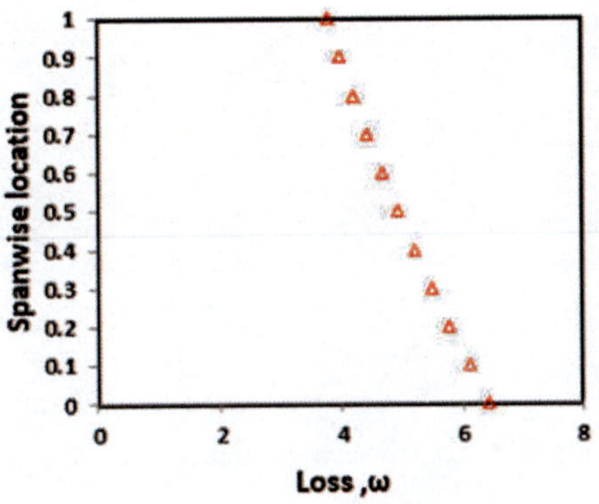

(B) Loss distribution

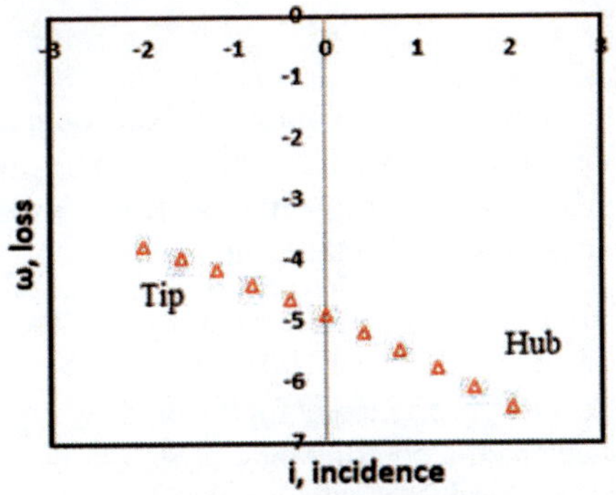

(C) Loss Vs incidence angle

Fig. 7 Camber, loss, and loss versus incidence angle distribution for the design rotor

as it leads to blockage near tip endwall. Figure 7b represents the distribution of loss along spanwise location, and higher losses are included for higher radial stations.

In Fig. 7c, loss vs incidence angle shows that the losses are lower near the hub at higher incidence, but it increases as the incidence reduces. This has been reported in all the literatures as the blade is designed to with stand at higher loadings and flow complexities occurring at the tip region.

For present study, NACA 65-410 blade profile has selected. The profile is designed for subsonic compressor/fan rotor. The 2D profile is stacked on one above from hub to tip at eleven radial locations. The commercially available CAD tool has been used to define blade positioning and angle definition as per calculated data. It is required to have smooth and clean blade profile without any extended surfaces or patches on it. The final geometry is as shown in Fig. 8.

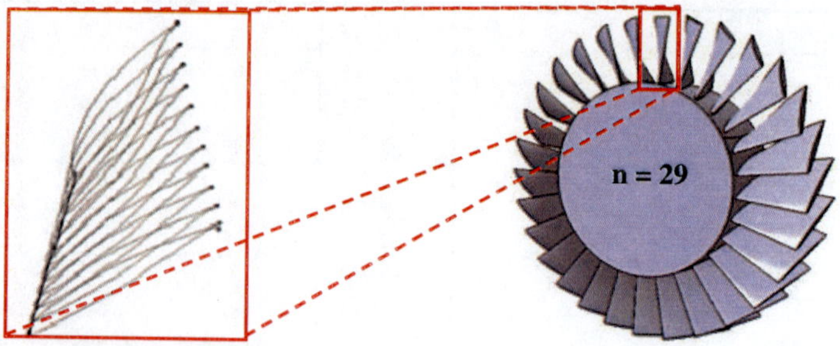

Fig. 8 CAD geometry generation

4 Conclusions

The study presented the steps and development of subsonic axial compressor blade based on one-dimensional blade element theory. The successful aerodynamic design is based on close match between the design rules and developed blade. The followings are the major conclusions drawn from the study:

- The low-speed axial fan/compressor rotor designed in the present study contains 29 numbers of blades having span of 100 mm.
- The assumed rotor isentropic efficiency for the blade was 90% which is close with the rotor performance calculated efficiency of 89.7967%.
- The designed degree of reaction for the blade at mean is close to 0.6 which tends to get higher pressure rise contribution by rotor, and this is advantageous for the axial compressor designs.
- Designed blade in the present study has higher camber near hub region and lower camber near tip region. This is suggested as per the blade design rule in order to reduce losses occur by three-dimensional flow phenomena.
- The designed blade has lower pressure gradient across the passage inlet to outlet at lower span and higher-pressure gradient close to tip region.

References

1. Casey MV (1987) A mean line prediction method for estimating the performance characteristic of an axial compressor stage. Turbomachinery—efficiency prediction and improvement. Robinson College, Cambridge, UK, ImechE, London, Paper C264/87
2. Casey MV, Hugentobler O (1988) The prediction of the performance of an axial compressor stage with variable stagger stator vanes. VDI-Ber. 706:213–227
3. Eisenberg B (1994) Development of a new front stage for an industrial axial flow compressor. J Turbomach 116:597–604
4. Hobbs D, Weingold H (1984) Development of controlled diffusion airfoil for multistage compressor application. ASME J Eng Gas Turbines Power 106:271–278
5. Cumpsty NA (1989) Compressor aerodynamics. Longman, London

6. Ko¨ller U, Mo¨ning R, Ku¨sters B, Schreiber H-A (1999) Development of advanced compressor airfoils for heavy-duty gas turbines, part I: design and optimization. ASME Paper No 99-GT-95
7. Ku¨sters B, Schreiber H-A, Ko¨ller U, Mo¨ning R (1999) Development of advanced compressor airfoils for heavy-duty gas turbines, part II: experimental and theoretical analysis. ASME Paper No 99-GT-96
8. Falck N (2008) Axial flow compressor mean line design. Master thesis. Lund University, Sweden
9. Keskin A (2007) Process integration and automated multi-objective optimization supporting aerodynamic compressor design. Phd Thesis, Brandenburg Technical University Cottbus
10. Aungier RH (2003) Axial flow compressor, 2nd edn
11. McKenzie AB (1988) The selection of fan blade geometry for optimum efficiency. Proc Inst Mech Eng Part A Power Process Eng 202:39–44
12. McKenzie AB (1997) Axial flow fans and compressors. Ashgate Publishing Limited
13. Dixon SL (1998) Fluid mechanics, thermodynamics of turbo machinery, 4th edn
14. Herbert IH Saravanamuttoo, Rogers GFC, Cohen H (2001) Gas turbine theory. Pearson Education
15. Mattingly JD, Heiser WH, Daley DH (1987) Aircraft engine design. AIAA, NewYork
16. Joseph Herrig L, Emery JC, Erwin JR (1957) Systematic two-dimensional cascade tests of NACA 65-series compressor blades at low speeds
17. Roy B, Pradeep AM (2014) Turbomachinery aerodynamics. Lect-8, Department of Aerospace Engineering, IIT Bombay. http://nptel.ac.in/courses/101101058/downloads/Lec-08ppts.pdf
18. Lieblein S, Schwenk FC, Broderick RL (1953) Diffusion factor for estimating losses and limiting blade loadings in axial-flow-compressor blade elements. National Advisory Committee for Aeronautics Cleveland Oh Lewis Flight, Technical report
19. Ahmed F (2016) El-Sayed. Springer, Fundamentals of aircraft and rocket propulsion.
20. Howell AR (1945) Fluid dynamics of axial compressors. Proc Inst Mech Eng 153:441–452
21. Sieverding F, Ribi B, Casey M, Meyer M (2001) Design of industrial axial compressor blade sections for optimal range and performance. J Turbomach 126:323–331

Fuzzy Logic-Based PID Controller Design for Car Suspension System with Magneto-Rheological Damper

Manav Kumar and Sharifuddin Mondal

Abstract To fulfill the high requirement of safety and comfort in automobiles, many new technologies are applied in suspension system. Damper acts an important role in suspension system for eliminating vibration. Bingham model of magneto-rheological (MR) damper is proposed for obtaining the control forces. In this work, proportional integral derivative (PID) controller gain parameters are adapted by the output parameters of fuzzy logic control to regulate the damper in the suspension system. The vehicle body deflections for two different road profiles are chosen for the performance analysis of fuzzy tuned PID controller and only PID controller. On the basis of simulation results, it is observed that fuzzy logic-based PID controller displays more improvement in efficiency and comfort level of riders by decreasing the amplitude of the vibration as compared to PID controller.

Keywords Bingham model · Fuzzy logic · MR damper · PID controller · Suspension system · Simulation

1 Introduction

There are mainly two major tasks needed to perform by a vehicle suspension system. The first one is to enhance the passenger comfort level by insulating the vehicle body from road disturbance, and the second one is to maintain the vehicle body displacement and make continuous close interaction in between the path of automobile and tire to give direction along the path. A conventional suspension system is also termed as passive suspension system, which comprises of an energy dissipating device known as shock absorber and a spring for energy storing [1]. The conventional suspension system is not able to maintain both safety factor and comfort in the ride simultaneously [2]. Semi-active and active systems are attracting a lot of attention due to better road holding and rider comfort. In semi-active suspension system, it uses semi-active damper whose force, i.e., damping force is controlled by controller

M. Kumar (✉) · S. Mondal
Department of Mechanical Engineering, National Institute of Technology, Patna 800005, India
e-mail: manavk.phd17.me@nitp.ac.in

A. Maurya et al. (eds.), *Recent Trends in Mechanical Engineering*, Lecture Notes in Mechanical Engineering, https://doi.org/10.1007/978-981-19-7709-1_9

using a changing damping property [3]. A parallel device which is known as MR damper is installed with conventional system. Sundar and Gangadharan [4] have designed and conducted experiments on the twin tube MR damper and analyzed the damping characteristic. It is concluded that as compared to silicon oil and air, MR damper fluid has better damping property and characteristics. Abdolvahab et al. [5] investigated and developed a methodology to design controller for active suspension system, which improves execution of system controller and gives satisfactory result in comparison of passive system. Liu et al. [6] have studied about performance of magneto-rheological fluids flowing through metal foams and their effect on shape of metal foams. Porosity and the penetrability relationship of metal foams and the variation in behavior of MR fluid passing throughout metal foam are measured. Liu et al. [7] have developed a high-level controller known as proportional integral derivative (PID) neural network controller. Talib and Darus [8] have studied iterative learning algorithm with PID controller for hydraulic actuator system and magneto-rheological (MR) damper suspension system. The result shows that hydraulic actuator performs better compared to semi-active suspension system incorporated with MR damper and conventional system. Patil et al. [9] have designed controller for active suspension in traveler automobile and observed improvement in the response to speed as compared to passive system. Several control strategies have been presented to reduce the vibration of vehicle suspension resulted in rider comfort in literature. Some controllers have certain limitations to perform the required task due to uncertainties in system parameters. The vehicle suspension system equipped with PID controller reduces the vibration to some extent and reaches steady state quicker than conventional vehicle suspension system incorporated with MR damper. The gains of PID controllers need to be tuned optimally according to the road profile (disturbance) for the improved level of rider's comfort. For this, a new fuzzy logic-based PID (FLPID) controller is designed and presented to minimize the sprung mass displacement, thereby increasing the efficiency and comfort level of riders.

In this paper, an active suspension of quarter car model equipped with MR damper is considered. The work is carried out for two different road profiles like step input and random input disturbances. The amplitudes of vibration are compared between by PID controller only and FLPID controller with respect to active suspension system. In FLPID controller, the gain parameters of PID controller are adjusted by output parameters of fuzzy logic control (FLC). The models are implemented in MATLAB/SIMULINK, and the results display that an active suspension system with MR damper fitted with FLPID controller is best to increase the rider comfort as compared to other cases.

The rest part of paper is arranged in following order: In Sect. 2, the mathematical models of the quarter car suspension and modified Bingham model of MR damper are presented. Section 3 describes the FLPID controller representing the structure and components of PID controller and basic of FLC approach. Section 4 describes model parameters and MATLAB/SIMULINK-based simulation results. Finally, conclusions and future scope of work are presented in the last Sect. 5.

2 Mathematical Model of Quarter Car Suspension System

The model of quarter car suspension system is displayed in Fig. 1. The governing equations of the model are defined by the Newton second law of motion and presented as [10]:

$$m_b\ddot{x}_1 = -c_1(\dot{x}_1 - \dot{x}_2) - k_1(x_1 - x_2) + f \tag{1}$$

$$m_u\ddot{x}_2 = c_1(\dot{x}_1 - \dot{x}_2) + c_2(\dot{x}_1 - \dot{x}_2) + k_1(x_1 - x_2) + k_2(w - x_2) - f \tag{2}$$

where m_u is the unsprung mass of the vehicle; m_b: the sprung mass; x_1: sprung mass displacement; x_2: wheel displacement; k_1: spring constant; k_2: tire spring constant; c_1: damping coefficient of the suspension system; c_2: damping coefficient for the tire; f: damping force in MR damper.

Here, two degrees of freedom for the suspension system are considered to analyze the system. Force f is the damping force for which controller is designed to control by the effect of magnetic field developed by current.

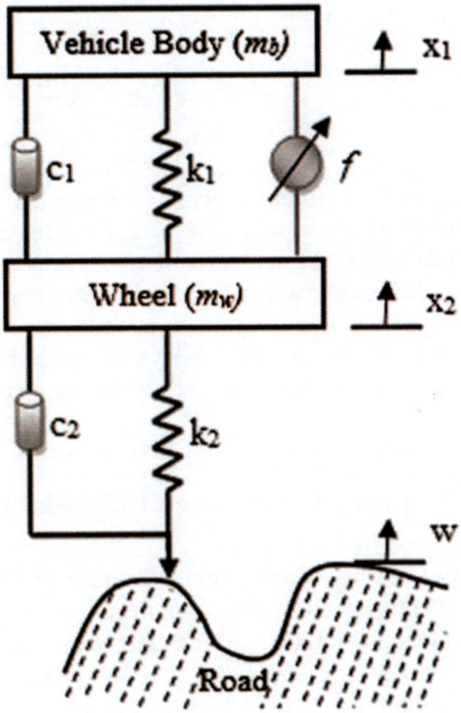

Fig. 1 Quarter car suspension model

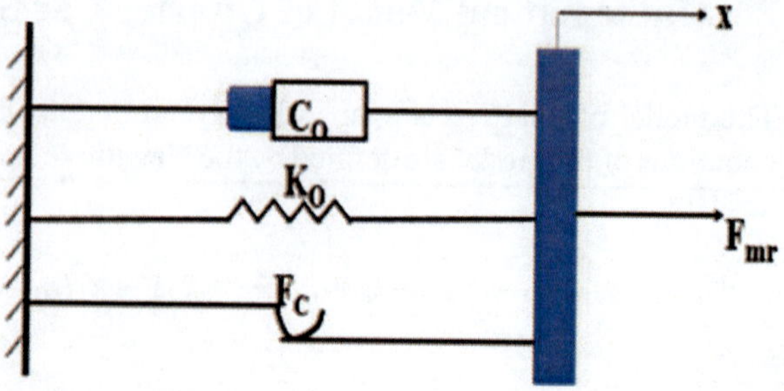

Fig. 2 Modified Bingham model of MR damper [7]

2.1 MR Damper of Suspension System

There are different types of mathematical models of MR damper to represent the dynamic behavior in the literature. Here, modified Bingham models of MR damper are selected [11].

MR Damper- Modified Bingham Model

The modified Bingham model of MR damper is presented in Fig. 2. The mathematical model is expressed by the following equation [13]:

$$F_{mr} = \frac{2F_c \tan^{-1}(d.x)}{\pi} + C_0\dot{x} + K_0 x + F_0 \tag{3}$$

where C_0: damping constant; F_c: frictional control force; K_0: stiffness of an elastic element; d: form factor; F_0 indicates the offset force (value of constant force); x: piston displacement corresponding to the displacement x_1 of the suspension system mass (m_b). The derivative of x with respect to time is the piston velocity.

MR fluid damper based upon the Bingham model is linked with PID controller in order to minimize the error (difference between the set point and instantaneous displacement of vehicle body). The proposed FLPID controller regulates the current according to the disturbance of the car chassis (suspension system).

3 Fuzzy Logic-Based PID (FLPID) Controller

FLC has been getting more attention of the researchers, especially in applications of complex nonlinear dynamic system in automatic controls. The FLPID controller offers the simplicity of PID and adjusts the control action to actual (real) operating state, thereby it providing the controller system with a sort of decision-making skill. In the fuzzy logic, linguistic variables are used to describe the complex dynamic system, that makes possible in decision-making in a analogous manner to human intellectual and later change it into an automatic controller based on the knowledge gained by the system.

The PID controller is useful due to its simplicity but found well suited only in solution of linear system and defined mathematical models. There are several approaches in literature to adjust the PID gains. Here, MATLAB control system toolbox is used for tuning of PID gains for the suspension system with MR damper. For this purpose, well-known and experiment-based Zeigler and Nichols method for tuning is considered.

3.1 Proportional Integral Derivative (PID) Controller

A PID controller tries to reduce the error between the set point and the measured value of the process parameters. The important objective of its algorithms is to minimize the error in a very short duration of time to make rapid and more sensitive process. For this, parallel structure of PID controller is used and mathematical definition of PID controller expressed by the equation as [8]:

$$u(t)_{PID} = K_P e(t) + K_I \int_0^t e(t)dt + K_D \frac{de(t)}{dt} \tag{4}$$

where $e(t)$ is error $(y_s(t) - y(t))$ or actuating error signal, $u(t)$ is the output signal from the controller. K_P, K_I, and K_D are the proportional gain, integral gain, and derivative gain, respectively.

3.2 Fuzzy Logic Controller

One of the promising solutions for dealing with the problem of uncertainty and imprecise is fuzzy set theory. It is an excellent and effective mathematical tool to tackle the uncertainty arising due to ambiguity or doubt. The structure of fuzzy logic is similar to human brain, and it operates with the help of linguistic variables. Hence, for the complex systems diagnosis and obtaining solution, fuzzy set theory has been getting more attention in recent years.

There are four fundamental components of the FLC process, which are shown in Fig. 3. In the fuzzification step, all the input and output measured data are converted into precise linguistic term. In the second step, two term, namely database and rule base, come into action. Membership functions are decided by the experts and/or operators in this step. The third step is fuzzy inference system. Fuzzy inference also referred to as approximate reasoning refers to evaluating linguistic descriptions. In this step, results of value (process parameter) coming from fuzzification are generated. The complete system outputs are converted into mathematical values with the help of the centroid method in the last step termed as defuzzification.

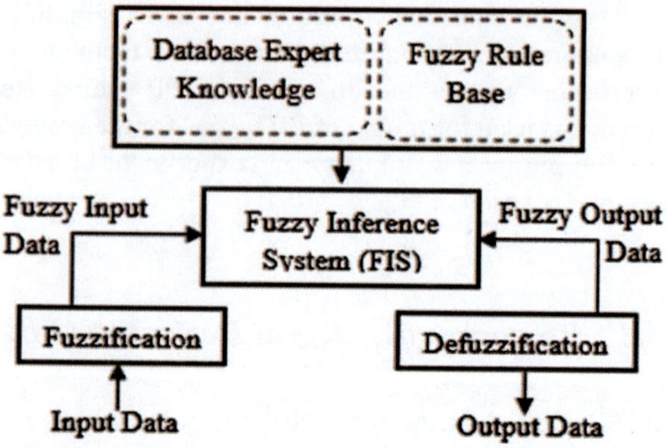

Fig. 3 Flowchart of components and structure of fuzzy logic system

The FLPID controller is modification of the conventional PID controller. There are two inputs and three outputs of FLC. The relationship between input parameters (error E and error change rate EC) and output parameters (PID gain parameters) can be established on the basis of fuzzy control theory. The output parameters (KP, KI, and KD) of FLC are self-adjusted with the variation of the E and EC on the basis of fuzzy rule. These self-adjusted parameters enable good dynamic and static performance of the controlled system.

In the proposed FLPID as shown in Fig. 4, a fuzzy set of seven triangular membership functions for two inputs (E and EC) and three outputs (K_P, K_I, and K_D) are considered with total 49 × 3 rules to obtain appropriate PID gain parameters. The membership functions for different inputs and outputs are shown in Figs. 5, 6, 7, 8 and 9.

The standard interval (range) for the input and output depends on the nature of system dynamic and the knowledge of experts and/or operators. Here, the range of input and output are normalized in the interval of [−11], and appropriate normalized

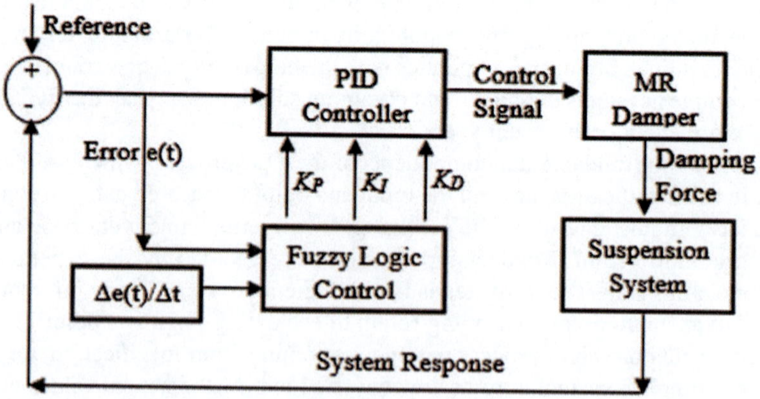

Fig. 4 Fuzzy-based PID controller structure

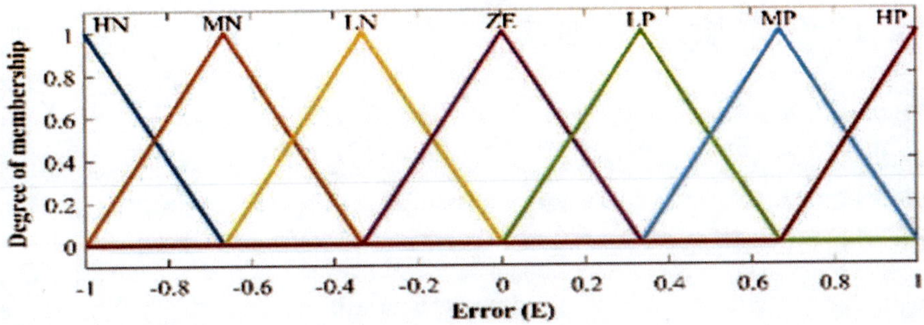

Fig. 5 Triangular membership function for input 1 of FLC (error (E))

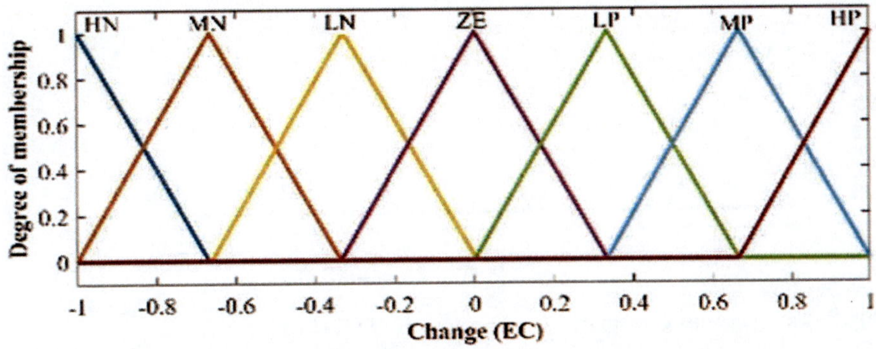

Fig. 6 Triangular membership function for input 2 of FLC (error change (EC))

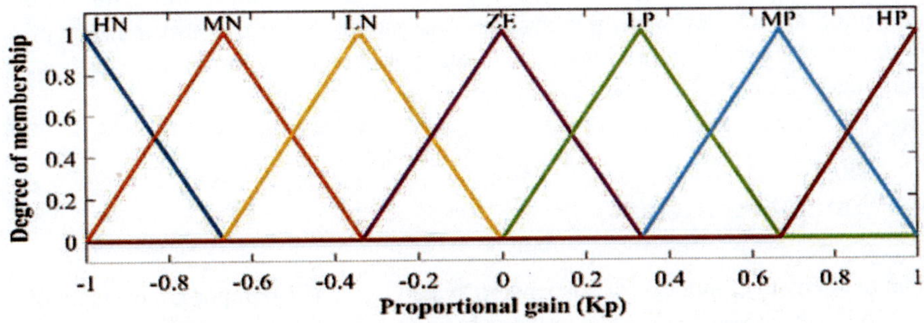

Fig. 7 Triangular membership function for output 1 of FLC (K_P)

gains are considered for the respective parameters. The linguistic labels and their descriptions are represented as high negative (HN), medium negative (MN), low negative (LN), zero (ZE), low positive (LP), medium positive (MP), and high positive (HP).

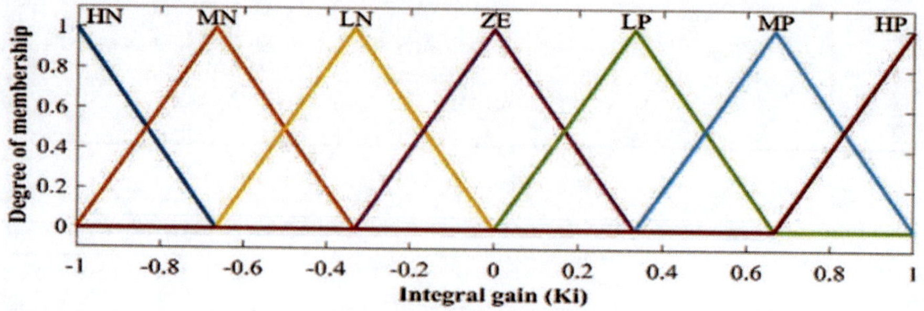

Fig. 8 Triangular membership function for output 2 of FLC (K_I)

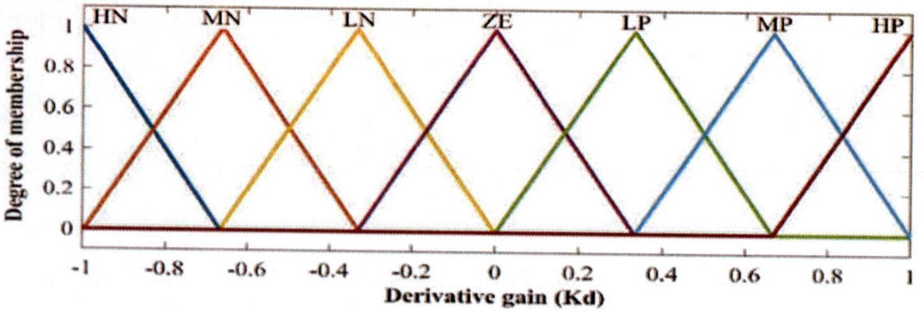

Fig. 9 Triangular membership function for output 3 of FLC (K_D)

Fuzzy rules play a vital role in the accuracy of fuzzy logic control. Appropriate fuzzy rules are established to adjust PID parameters to control the displacement (vibration amplitude) of vehicle suspension system. According to the fuzzy principle, the rule (decision rule) table for output parameters K_P, K_I, and K_D with input parameters E and EC is obtained and presented in the Table 1.

4 Simulation Results

The mathematical models for suspension system with MR damper are implemented in MATLAB/SIMULINK and simulated to compare the displacement response. The following numerical values of different parameters of car suspension system (as given in Table 2) are considered. The simulation works are performed for two different road profiles with only damper, damper-PID, and damper-fuzzy logic-PID controllers.

Table 1 Fuzzy rule for K_P, K_I, and K_D

E	EC							
		HN	MN	LN	ZE	LP	MP	HP
HN	K_P	HP	HP	MP	MP	LP	ZE	ZE
	K_I	HN	HN	MN	MN	LN	ZE	ZE
	K_D	LP	LN	HN	HN	HN	MN	LP
MN	K_P	HP	HB	MP	LP	LP	ZE	LN
	K_I	HN	HN	MN	LN	LN	ZE	ZE
	K_D	LP	LP	HN	MN	MN	LN	ZE
LN	K_P	MP	MP	MP	LP	ZE	LN	LN
	K_I	MN	MN	LN	LN	ZE	LP	LP
	K_D	ZE	LN	MN	MN	LN	LN	ZE
ZE	K_P	MP	MP	LP	ZE	LN	MN	MN
	K_I	MN	MN	LN	ZE	LP	MP	MP
	K_D	ZE	LN	LN	LN	LS	LS	ZE
LP	K_P	LP	LP	ZE	LN	LN	MN	MN
	K_I	MN	LN	ZE	LP	LP	MP	HP
	K_D	ZE	ZE	ZE	ZE	ZE	ZE	ZE
MP	K_P	LP	ZE	LN	MN	MN	MN	HN
	K_I	ZE	ZE	LP	LP	MP	HP	HP
	K_D	HP	LN	LP	LP	LP	LP	HP
HP	K_P	ZE	ZE	MN	MN	MN	HN	HN
	K_I	ZE	ZE	LP	MP	MP	HP	HP
	K_D	HP	MP	MP	MP	LP	LP	HP

Table 2 Quarter car model and modified MR Bingham model parameters [12]

Parameters	Value
Unsprung mass of the vehicle (m_b)	241.5 kg
Sprung mass (m_u)	41.5 kg
Spring constant of suspension (k_1)	80,000 N/m
Tire spring constant (unsprung) (k_2)	500,000 N/m
Damping coefficient of the suspension (c_1)	350 Ns/m
Damping coefficient of the tire (c_2)	15,020 Ns/m
Frictional force (F_c)	210 N
Damping constant (C_0)	650 Ns/m
Stiffness of an elastic element (K_0)	300 N/m
Form factor (d)	5
Offset force (F_0)	111 N

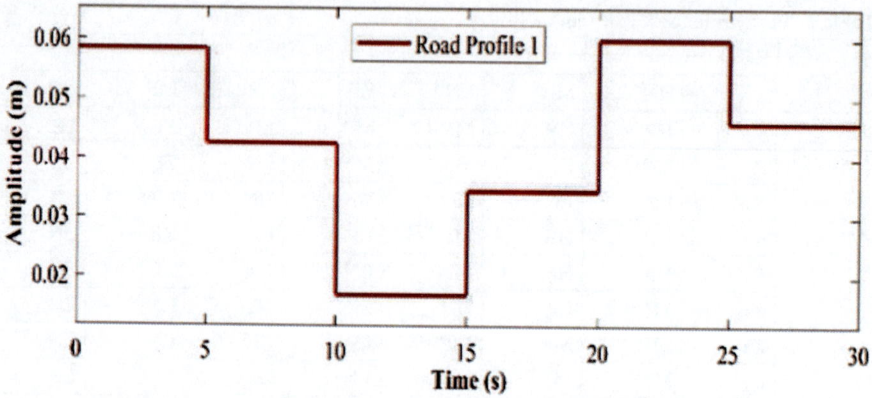

Fig. 10 Random input road profile 1

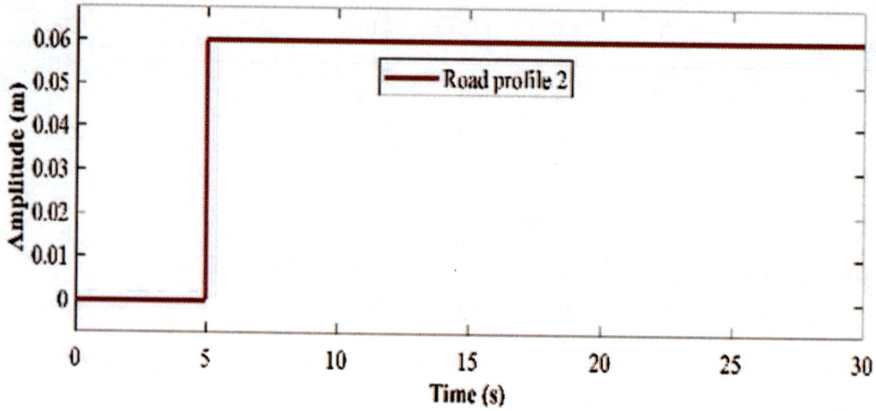

Fig. 11 Step input road profile 2

4.1 Road Profiles

Two different road surfaces are considered as inputs of the system to excite the semi-active suspension system. First road profile or disturbance is random input with the highest and lowest value 0.06 m and 0.01 m, respectively, as shown in Fig. 10. The second road profile or disturbance is step input of 0.06 m at 5 s shown in Fig. 11.

4.2 Analysis of Road Input Disturbance

Displacement of the vehicle suspension for different cases with damper, PID, and fuzzy-PID (FLPID) is compared with displacement of active suspension system and

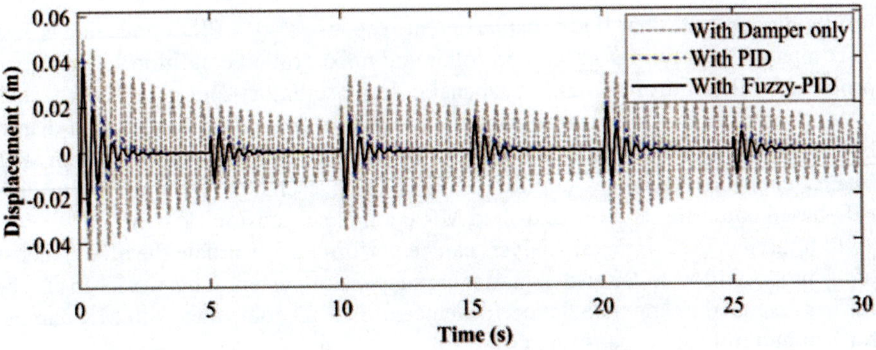

Fig. 12 Response of displacement for road profile 1 (random input)

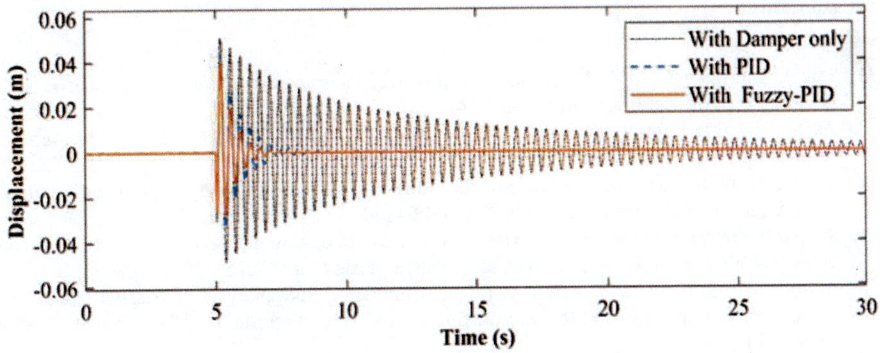

Fig. 13 Response of displacement for road profile 2 (step input)

shown in Figs. 12 and 13. For these, step input disturbances are introduced at 5 s. Also, random input of highest and lowest value 0.06 m and 0.01 m, respectively, is introduced.

It is observed that the response of vehicle body displacement of FLPID controller with MR damper system is improved as compared to PID controlled and uncontrolled system. Also, simulation results show that MR damper system with the FLPID controller becomes stable quickly as compared to MR damper with PID control. It is also concluded that settling time of the response reduces for FLPID controlled system as compared to PID controlled and uncontrolled system.

5 Conclusions

For analysis of quarter car suspension system, modified Bingham model-based MR damper is considered and simulated in MATLAB/SIMULINK. Two road profiles, namely step road profile and random road profile, are considered. Simulation results demonstrate that body displacement response for FLPID controller is less as compared to only PID-based controller and uncontrolled condition in each road profile. FLPID controller makes suspension less oscillatory and gains steady state faster as it takes 2–3 s less time compared to others depending on the road disturbances. Thus, simulated results confirm that MR damper suspension system with FLPID controller enhances comfort level of the passengers as compared to only PID-based controller and uncontrolled MR damper suspension system.

In future, an experimental analysis can be performed to validate the effectiveness of the proposed controller on the quarter car suspension system. This work can also be further extended to appraise the performance of FLPID controller with MR damper for half and full car suspension system.

References

1. Tan H, Bradshaw T (1997) Model identification of an automotive hydraulic active suspension system. Proc Am Control Conf 5:290–924
2. Zhang P (2011) Energy harvesting, ride comfort, and road handling of regenerative vehicle suspensions. Dyn Syst Control Conf 2:01–08
3. Rao K, Pavani K (2010) Modelling and vibration control of suspension system for automobiles using LQR and PID controllers. IJLTEMAS 4:64–68
4. Sundar B, Gangadharan K (2014) Experimental study of damping characteristics of air, silicon oi, magneto rheological fluid on twin-tube damper. Proc Mater Sci 5:2258–2262
5. Abdolvahab A, Ghobad S, Armin B (2012) Simulation and analysis of passive and active suspension system using quarter car model for different road profile. Int J Eng Trends Technol 3:636–644
6. Liu X, Fu Z, Yao X, Li F (2011) Performance of magnetorheological fluids flowing through metal foams. Measur Sci Rev 11:144–148
7. Liu W, Shi W, Liu D, Yan T (2010) Experimental modeling of magneto-rheological damper and PID neural network controller design. Sixth Int Conf Nat Comput 10:1674–1678
8. Talib M, Darus I (2013) Self-tuning PID controller with MR damper and hydraulic actuator for suspension system. Fifth Int Conf Comput Intell 13:119–124
9. Patil K, Jagtap S, Jadhav V, Bhosale A, Kedar B (2017) Performance evaluation of active suspension for passenger cars using MATLAB. IOSR J Mech Civil Eng 334:06–14
10. Pekgokgoz R, Gurel M, Bilgehan M, Kisa M (2010) Active suspension of cars using fuzzy logic controller optimized by genetic algorithm. Int J Eng Appl Sci 2:27–37
11. Syed K, Kumar D (2018) Response and control of jacket structure with magneto- rheological damper at multiple locations/combinations. Ocean Syst Eng 8:201–221
12. Ferdaus M, Rashid M, Bhuiyan M (2014) Development of an advanced semi-active damper using smart fluid. Adv Mater Res 939:615–622
13. Sulaymon L (2016) Modeling and simulation of non-linear and hysteresis behavior of magneto-rheological dampers in the example of quarter-car model. Eng Math 1:19–38

Multi-frequency Approximation for a Hysteretically Damped Tuned Mass System

Khogesh Kumar Rathore and Saurabh Biswas

Abstract In this paper, we study a hysterically damped tuned mass system. We note that tuned mass dampers are extensively used in practical systems in order to reduce the dynamic response of the primary structures. The theory of tuned mass systems with viscous damping is well established. Simple tuned mass systems, with linear viscous damping, are analytically tractable. In reality, in many systems, we see damping are hysteretic in nature. Hysteresis is a strongly nonlinear phenomenon. Analytical study of tuned mass systems with hysteretic damping is therefore challenging. Here, we use a rate-independent hysteresis model developed by Biswas et al. in Int J Mech Sci 108:61–71, 2016, as the damper. We use numerical and semi-analytical approaches to study the dynamic responses of systems. The amplitude versus frequency curves shows two resonance peaks. The numerical and analytical results show good match.

Keywords Tuned mass damper · Hysteresis · Multi-frequency approximation

1 Introduction

A tuned mass damper (TMD) is a device that is commonly used in physical systems that undergo significant vibrations. Typically, a secondary mass, a spring, and a damper are attached to the primary structure in order to reduce the amplitude of the vibration. Figure 1 schematically shows a simple single degree of freedom TMD system. Here, m_1 is the primary mass that is supported by a spring of stiffness K_1; and m_2 is the secondary mass that is attached to the primary mass by means of a spring of stiffness K_2 and a viscous damper C.

The concept of TMD was first developed, by Frahm [1] in 1909, in order to reduce the rolling amplitude of ships. In 1928, a theory for the TMD was presented by Ormondroyd and Den Hartog [2]. Subsequently, an analytical study for the optimal design of TMD was given by Den Hartog [3] in 1940. TMD devices are extensively

K. K. Rathore (✉) · S. Biswas
Department of Mechanical Engineering, Indian Institute of Technology Jammu, Jammu 181221, India
e-mail: 2019rme0030@iitjammu.ac.in

A. Maurya et al. (eds.), *Recent Trends in Mechanical Engineering*, Lecture Notes in Mechanical Engineering, https://doi.org/10.1007/978-981-19-7709-1_10

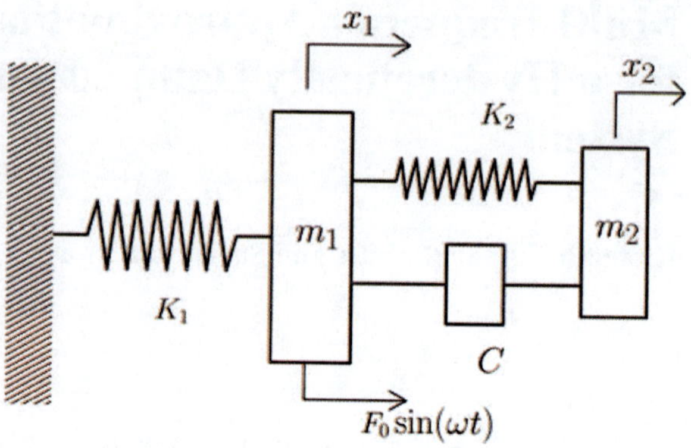

Fig. 1 A single degree of freedom TMD system

used in the area of vibration control. The applications of TMD are seen in tall buildings, high water tanks, large span bridges, etc. A significant number of papers have been published over the last several decades on the passive and active control of structures using TMD, see e.g., [4–6] and the references therein.

We note that the damper of the TMD device plays a very important role in reducing the amplitude of the primary structure near the resonant frequency. In the literature, researchers used both linear and nonlinear dampers in the TMD systems, see e.g., [7, 8]. Simple TMD systems with linear viscous damping are analytically tractable. By contrast, TMD systems with nonlinear damping are mathematically more challenging.

In recent times, several researchers have used nonlinear hysteretic dampers as the damping model in the TMD systems [9, 10]. In [11], the authors used hysteretic tuned mass damper for structural vibration reduction. In [12], the authors studied a hysteretic TMD described by the Bouc-Wen hysteresis model [13, 14]. The motivation of using hysteretic damper is that many mechanical systems show dissipation in the form of rate-independent hysteresis. We note that the inclusion of a rate-independent hysteresis model in the system makes it strongly nonlinear with *signum* nonlinearity. This makes analytical treatments very difficult, and numerical simulations are most feasible. However, numerical simulations often cannot provide clear insights into the solutions. A semi-analytical approach can give clearer insight.

With the above motivation, we study here a single degree of freedom TMD system, where the damping is defined by a recently developed rate-independent scalar hysteresis model in [15]. The hysteresis model is motivated by a study of an elastic plate with several frictional microcracks. The model is given as below:

$$\dot{\theta}(t) = \frac{\kappa}{|x(t)| + \in}\{\theta_a + \beta\,\mathrm{sgn}(x(t)\,\dot{x}(t)) - \theta(t)\}|\dot{x}(t)|, \qquad (1)$$

where x is the input displacement to the system and θ is an internal variable. Here, θ_a, k, β are model parameters, and ε is a small regularizing parameter. The hysteretic force is given by

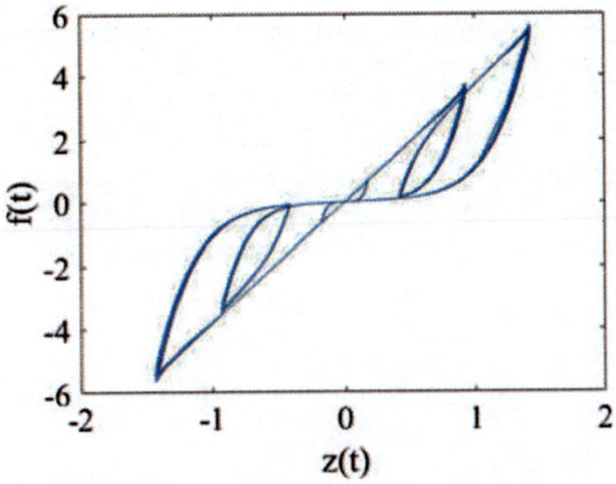

Fig. 2 Hysteretic model result for parameters $\kappa = 4$, $\theta_a = 2$, $\beta = 1.8$, $f(t) = \theta(t)x(t)$ and $\in = 10^{-6}$ with externally specified $x(t) = \text{Sin } 2\pi t + 0.5 \text{ Sin } 8\pi t$

$$f = \theta \cdot x. \tag{2}$$

Figure 2 shows a typical response of the hysteresis model given by Eqs. (1) and (2).

As discussed, we use the hysteresis model given by Eqs. (1) and (2) as the damper in a SDOF-TMD system. We calculate the response of the system via direct numerical calculation. Then, we use a semi-analytical multi-frequency approximation [16] to study the amplitude versus frequency of the system.

1.1 Tuned Mass System with Hysteretic Damping

We use the hysteresis model of [15] as the damper in the SDOF-TMD system as shown in Fig. 3. Here, m_1 is the mass of the primary system, K_1 is the stiffness of the spring on which m_1 is mounted, m_2 is the mass of the secondary structure which is attached to the primary mass with a spring of stiffness K_2 and a hysteretic damper indicated by f. A harmonic force F_0 Sin (ωt) is applied to the primary mass.

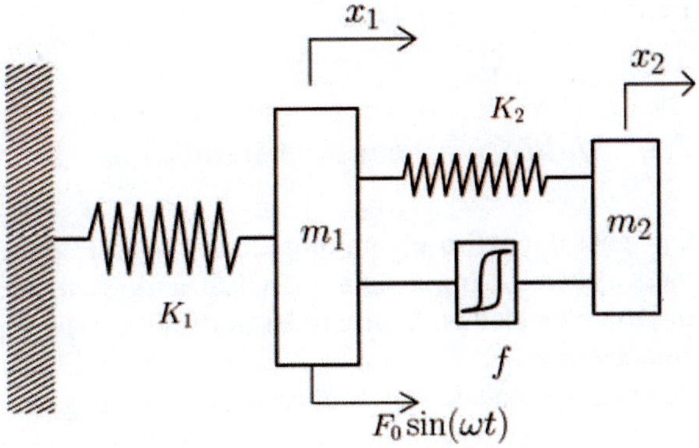

Fig. 3 Hysteretically damped tuned mass system

Equations of motion of the system are as follows:

$$m_1\ddot{x}_1 + (K_1 + K_2)x_1 - K_2x_2 - f = F_0 \sin(\omega t), \tag{3}$$

$$m_2\ddot{x}_2 - K_2x_1 + K_2x_2 + f = 0. \tag{4}$$

Here,

$$f = \gamma \cdot \theta \cdot (x_2(t) - x_1(t)), \tag{5}$$

where γ is a scalar multiplier that controls the level of damping. Note that the input displacement to the damper is $x_2(t) - x_2(t)$.

The rate of change of hysteretic internal state θ is given by

$$\dot{\theta}(t) = \frac{\kappa}{|x_2(t) - x_1(t)| + \in}\{\theta_a + \beta\,\mathrm{sgn}[(x_2(t) - x_1(t))(\dot{x}_2(t) - \dot{x}_1(t))] - \theta(t)\} \times |\dot{x}_2(t) - \dot{x}_1(t)| \tag{6}$$

where $x_2(t) - x_2(t)$ is the input displacement to the damper; and θ_a, β, and κ are model parameters, and ε is a small regularizing parameter.

2 Direct Numerical Simulation

We first calculate the response of the system given by Eqs. (3) through (6) using direct numerical simulations. We use MATLAB function *ode45* for numerical integrations.

Figure 4 (left) shows a response of the primary mass versus time. The response shows steady-state behavior after initial transients. Figure 4 (right) shows frequency sweep response of the primary mass. We use $\omega = 10^{-5}t$, for the frequency sweep calculation. Resonance peaks are seen in the frequency sweep response. The frequency sweep response indicates the amplitude versus frequency of the primary mass.

2.1 *Multi-frequency Approximation*

Our goal is now to obtain amplitude versus frequency plots for hysteretic tuned mass systems using a semi-analytical multi-frequency approximation as reported in [16]. The method is discussed in detail in [16] and briefly described below for completeness.

First, we consider a new distorted time scale given by

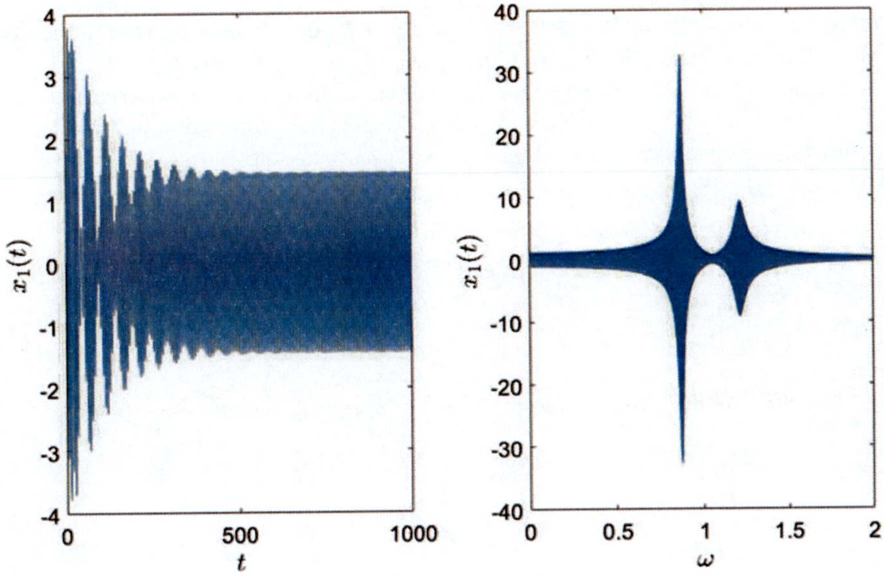

Fig. 4 Left: response of the primary mass versus time. Right: frequency sweep response of the primary mass. Resonance peaks are seen. Here, $m_1 = 1$, $m_2 = 1/10$, $K_1 = 1$, $K_2 = 1/10$, $F_0 = 10$ in suitable units. Hysteresis parameters: $\gamma = 0.01$, $\kappa = 4$, $\beta = 1.8$, $\epsilon = 10^{-6}$

$$\tau = \omega t + \sum_{k=1}^{n} [A_k \sin k\omega t + B_k(\cos k\omega t - 1)], \tag{7}$$

where ω is the fundamental frequency. Assuming solutions

$$x_1 = A \sin(\tau + \phi_1) \tag{8}$$

$$x_1 - x_2 = B \sin(\tau) \tag{9}$$

and the applied external force is redefined for convenience as

$$F = F_0 \sin(\tau + \phi_2). \tag{10}$$

Here, ϕ_1 and ϕ_2 are the phases. The hysteretic response of Eq. (1) is approximated as,

$$\theta(t)x(t) = f = \alpha_0 \sin(\tau) + \alpha_1 \cos(\tau) + \alpha_2 \sin(3\tau) + \alpha_3 \cos(3\tau). \tag{11}$$

The values of the coefficients α_0 through α_3 are fitted from numerical solutions obtained which are $\alpha_0 = 1.1253$, $\alpha_1 = 0.7632$, $\alpha_2 = -0.5624$, $\alpha_3 = -0.3816$.

Substituting Eq. (7) in Eqs. (8) through (10) and subsequently in Eqs. (3) and (4), we obtain long analytical expressions of the residuals R_1 and R_2. We consider the effective residual $R^2 = R_1{}^2 + R_2{}^2$. $R^2 = R_1^2 + R_2^2$ details omitted for the brevity. We call the residual as $R(A, B, A_1, A_2, .., A_n, B_1, B_2, .., B_n, \phi_1, \phi_2, \omega, t)$.

We have found $n = 2$ to be enough for our calculation. We also observe that only 'sine' terms in Eq. (7) are sufficient. Using a least square approximation, we get the following six equations.

$$\frac{\partial}{\partial A}\int_0^{2\pi/\omega} R^2 dt = \int_0^{2\pi/\omega} R\frac{\partial R}{\partial B}dt = 0, \int_0^{2\pi/\omega} R\frac{\partial R}{\partial \phi_1}dt = 0, \int_0^{2\pi/\omega} R\frac{\partial R}{\partial \phi_2}dt = 0 \tag{12}$$

Similarly, we get

$$\int_0^{2\pi/\omega} R\frac{\partial R}{\partial B}dt = 0, \int_0^{2\pi/\omega} R\frac{\partial R}{\partial A_1}dt = 0, \int_0^{2\pi/\omega} R\frac{\partial R}{\partial A_2}dt = 0, \tag{13}$$

$$\int_0^{2\pi/\omega} R\frac{\partial R}{\partial \phi_1}dt = 0, \int_0^{2\pi/\omega} R\frac{\partial R}{\partial \phi_2}dt = 0 \tag{14}$$

In Eqs. (12) through (14), ω is kept as a free parameter. All the integrals are evaluated numerically using Simpson's rule and further solved iteratively using Newton-Raphson method, with numerically estimated Jacobians and with the tolerance of 10^{-5}.

Figure 5 shows that the envelope of the sweep frequency response (in Fig. 4) is well captured using the multi-frequency approximation. The multi-frequency approximation is able to capture both resonance peaks accurately.

We will now use the multi-frequency approximation to plot amplitude versus frequency for a few cases. The results are shown in the next section.

3 Results

We show the amplitude versus frequency plots for several different cases as in Fig. 6. The figure contains four subplots with different parameters. (a) $m_1 = 1$, $m_2 = 1/10$, $K_1 = 1$, $K_2 = 1/10$, $F_0 = 10$, (b) $m_1 = 1$, $m_2 = 1/5$, $K_1 = 1$, $K_2 = 1/10$, $F_0 = 0.9$, (c), $m_1 = 1$, $m_2 = 1/12$, $K_1 = 1$, $K_2 = 1/15$, $F_0 = 2$, (d), $m_1 = 1$, $m_2 = 1/10$, $K_1 = 1$, $K_2 = 1/5$, $F_0 = 1.5$. The above parameters are given in suitable units. In each subplot, there are four different cases considred with $\gamma = 0.01, 0.02, 0.05$ and 0.1. Note that all amplitude versus frequency curves show two resonance peaks. For higher damping, the resonant amplitudes decrease.

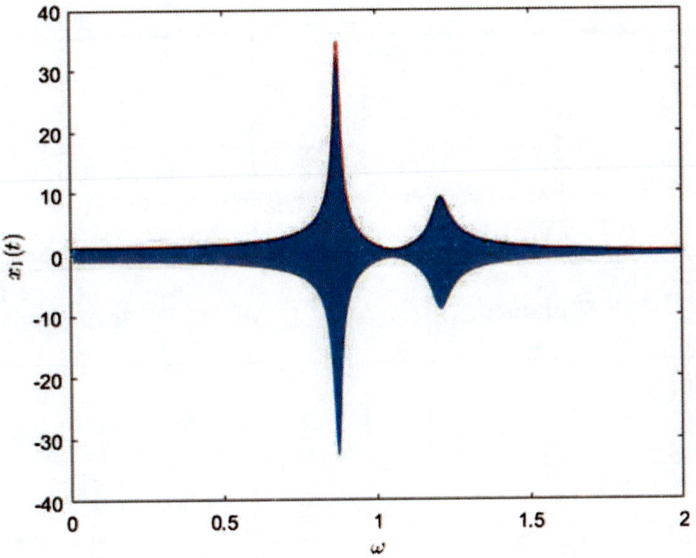

Fig. 5 Envelope of the frequency sweep response, obtained in Fig. 4, is captured using the multi-frequency approximation

4 Conclusions

In this paper, we have studied tuned mass systems with a hysteretic damping model developed in [15]. The goal of the paper was to study dynamic responses of the hysteretically damped TMD using direct numerical simulation. Since the system is strongly nonlinear with *signum* nonlinearity in the hysteresis equation, conventional harmonic balance method seems difficult to obtain the amplitude versus frequency of the system.

We have used a multi-frequency approximation [16] to describe amplitude versus frequency of the systems. We are not aware of any such prior study in the literature. We have found that the multi-frequency approximation describes the amplitude versus frequency curves very accurately, with expected two resonance peaks. We have demonstarted that the approximation can capture frequency sweep response.

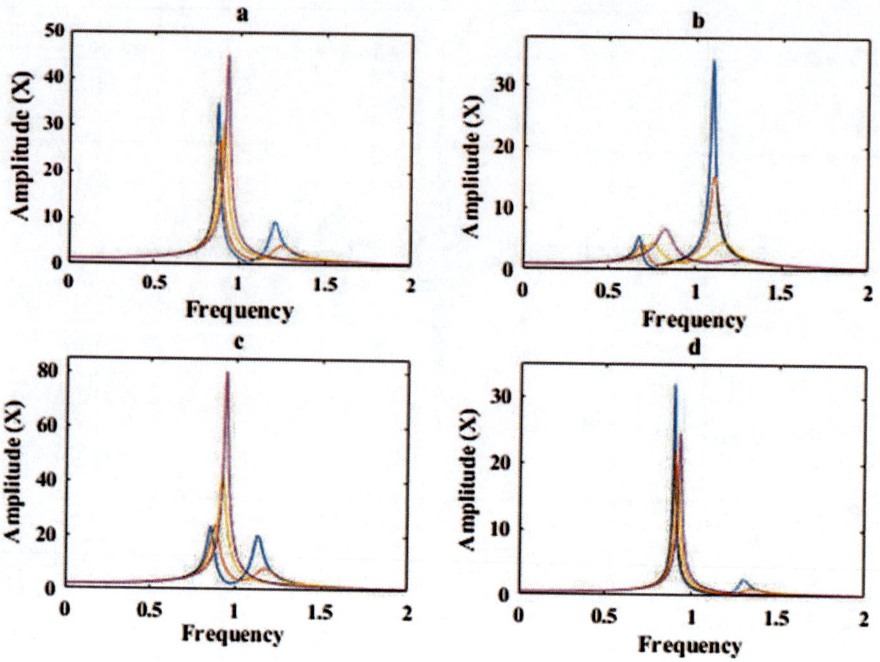

Fig. 6 Amplitude versus frequency of the primary mass. (**a**) $m_1 = 1$, $m_2 = 1/10$, $K_1 = 1$, $K_2 = 1/10$, $F_0 = 10$, (**b**) $m_1 = 1$, $m_2 = 1/5$, $K_1 = 1$, $K_2 = 1/10$, $F_0 = 0.9$, (**c**) $m_1 = 1$, $m_2 = 1/12$, $K_1 = 1$, $K_2 = 1/15$, $F_0 = 2$, (**d**) $m_1 = 1$, $m_2 = 1/10$, $K_1 = 1$, $K_2 = 1/5$, $F_0 = 1.5$. All subplots contain four different cases with $\gamma = 0.01$, 0.02, 0.05, and 0.1

References

1. Frahm H (1909) Device for damping vibrations of bodies. US Patent No 989, 958
2. Ormondroyd J, Den Hartog JP (1928) The theory of the dynamic vibration absorber. ASME J Appl Mech 50:9–22
3. Den Hartog JP (1947) Mechanical vibration. McGraw–Hill, New York
4. Jangid RS, Datta TK (1995) Seismic behavior of the base-isolated buildings: a state-of-the art review. Struct Build 110(2):186–203
5. Soto MG, Adeli H (2013) Tuned mass dampers. Arch Comput Meth Eng 20:419–431
6. Elias S, Matsagar V (2017) Research developments in vibration control of structures using passive tuned mass dampers. Annu Rev Control 44:129–156
7. Randall SE, Halsted DM, Taylor DL (1981) Optimum vibration absorbers for linear damped systems. J Mech Des (ASME) 103:908–913
8. Starovetsky Y, Gendelman OV (2009) Vibration absorption in systems with a nonlinear energy sink: nonlinear damping. J Sound Vib 324(3–5):916–939
9. Gerges RR, Vickery BJ (2005) Design of tuned mass dampers incorporating wire rope springs: part I: dynamic representation of wire rope springs. Eng Struct 27:653–661
10. Gerges RR, Vickery BJ (2005) Design of tuned mass dampers incorporating wire rope springs: part II: simple design method. Eng Struct 27:662–674
11. Carpineto N, Lacarbonara W, Vestroni F (2014) Hysteretic tuned mass dampers for structural vibration mitigation. J Sound Vib 333:1302–1318

12. Lacarbonara W, Vestroni F (2002) Feasibility of a vibration absorber based on hysteresis. In: Proceedings of the Third World Congress on Structural Control, Como, Italy
13. Bouc R (1967) Forced vibration of mechanical systems with hysteresis. In: Proceedings of the 4th Conference on Nonlinear Oscillation, p 315, Prague
14. Wen Y (1976) Method for random vibration of hysteretic systems. J Eng Mech Div 102(2):249–263
15. Biswas S, Jana P, Chatterjee A (2016) Hysteretic damping in an elastic body with frictional microcracks. Int J Mech Sci 108:61–71
16. Balija S, Biswas S, Chatterjee A (2018) Stability aspects of the Hayes delay differential equation with scalable hysteresis. Nonlinear Dyn 93(3):1377–1393

CFD Analysis of Particle Shape and Size on Impact Velocity and Effect of Stand-off Distance in the Cold Spray Process

Mohsin Khan, Mohammad Zunaid, and Qasim Murtaza

Abstract The computational fluid dynamics (CFD) analysis is the latest technology, and also the accuracy is very close to the experimental analysis. In the present work, the effect of shape and size of feedstock powder particles is analyzed upon impact velocity of cold spray (CS) coating via CFD analysis. The geometry for the work has been drawn by SolidWorks, and the analysis has been carried out through fluent. The analysis has been carried out with the best input parameters for CS coating. The pressure-based; axisymmetric model has been used to solve the CS nozzle. The most realistic two-equation realizable k-ε model has been taken for the analysis. In this analysis, there is a range of particle diameter or varying particle sizes, and with varying the standoff distance from the nozzle, exit/outlet has been taken. The analysis is carried out using copper as the spray powder particles and steel as the substrate material. It has been found that the spherical shape of powder particles is more reliable when sprayed with a standoff distance of 35 mm.

Keywords CFD simulation · Cold spray · Impact velocity · Standoff distance · Particle shape

1 Introduction

In the cold spray (CS) coating process, the feedstock powder is sprayed with the help of a nozzle to get a high impact velocity or jet towards the substrate upon which coating is to be performed. With this coating process technology number of metals, composite materials, polymers, and ceramics can be coated based upon the applications and requirements. Nowadays, this coating technology is growing in the manufacturing industry by creating a dedicated branch in additive manufacturing called cold spray additive manufacturing, solid-state powder-based deposition process [1–3], and cold gas dynamic manufacturing [4].

M. Khan (✉) · M. Zunaid · Q. Murtaza
Department of Mechanical Engineering, Delhi Technological University, Bawana Road, Delhi 110042, India
e-mail: mohsindtu@gmail.com

A. Maurya et al. (eds.), *Recent Trends in Mechanical Engineering*, Lecture Notes in Mechanical Engineering, https://doi.org/10.1007/978-981-19-7709-1_11

Decrease of particle size to the nano-level, domination of surface roughness observed parallelly [5], residual stresses increases upon grinding in nanostructured coatings [6]. Powder material heat treatment increases the efficiency of deposition about two times and decreases the porosity about four times [7]. Cladded powders can be developed by covering the alumina particles with a nickel layer using the hydrogen reduction method [8]. Using CS technology compact nanocrystalline grains produced due to impact, moiré pattern was observed due to preserved tamping effect produced by microstructure, and there was no size change found in nano-sized grains [9]. The feasible and simple way for determining particle distribution of cold sprayed coated composites found to be the Weibull distribution method [10].

Using crystal orientation of particles on the coating surface with the help of molecular dynamics simulation, the shape of particles at the bottom edge is found of square-shape, rectangle-shape, and hexagon-shape [11]. The aerosol deposition (AD) method is also a type of CS coating process technique, in which ceramic coating formation and bonding were analyzed practically as well as using CFD simulations [12]. The molecular dynamic simulation gives the results with asymmetric particle deformation at the interface that occurred during cold gas spray [13]. The particle size of 40 μm founds gives high impact velocity as compared to the 20, 60, 80, and 100 μm particle sizes [14].

CS coated surface has compressive residual stresses because of deformation and particle impact at high velocity [15]. The newly developed model based on Johnson-Cook (JC) model can effectively predict the stresses and deformed shape of particles [16]. In CS process technology, the temperature of the powder particles also depends upon the injector length, the higher the injector length, the lower will be the temperature of particles and vice-versa [17]. The CS coating process gives high thermal conductivity and sensible adhesion strength [18]. The surface roughness of sintered alumina in comparison with alumina found to be lower [19] which affects the porosity, hardness, and density in outer and middle regions due to grain size reduction. The wear rate increases with an increase in the load [20]. Laser remelting decreases the porosity of thermal ceramic coatings, a significant effect as compare to as-sprayed coating exhibits upon various mechanical properties like microhardness, elastic modulus, etc. [21]. Crater formation found negligible upon hard metal substrates [22]. CS coating using composite powder gives low porosity and high hardness and fracture-toughness [23].

CS coatings with low temperature (500 °C) give the best corrosion resistance. Post deposition heat treatment improves bonding and porosity [24]. In comparison with uncoated material with aluminium and chromium coated, aluminium coated found to be better oxidation resistance at high working temperature environment [25]. Pulse plasma nitriding is not sufficient to improve corrosion resistance [26]. The pre-heat temperature in CS coating technology improves the first layer deposition which improves the bonding strength [27].

Critical velocity and coating quality improve with improving the impact velocity well above the critical velocity. In the presence of oxide, the particles jetting and flattening decrease [28]. Velocity or impact velocity affects the CS coating to an extent

[27]. Iron-based amorphous/nanocrystalline composite coatings exhibit better corrosion resistance as compared to galvanized steel with improved corrosion resistance and microhardness than steel substrate [28, 29]. Copper and copper-titanium dioxide coating using low-pressure CS coating technique give metallic and amorphous structure, respectively [30].

This research aims to improve the impact velocity and temperature of powder particles by improving the geometrical parameters like standoff distance and powder shape and size range in CS coating process technology. This analysis was carried out with constant parameters which include propelling gas mixture, particle size range, pressure conditions, temperature conditions at inlet and outlet sections, etc. The best-suited shape of powder particles and standoff distance based upon the simulation results was discussed.

2 Model, Material, and Methodology

2.1 Geometry

The geometry required for the simulation of the CS process technology should have the same arrangement as that of the experimental model. The geometry used in this simulation process had been drawn using SolidWorks software, which consists of an injector-nozzle arrangement, the substrate has been placed in front of the nozzle outlet on the same axis as the nozzle, and the arrangement is made known in Fig. 1. The nozzle used in this work is of the circular cross-sectional area with diameters of 19 mm, 3 mm, and 7 mm at inlet, throat, and outlet, respectively. The length of the convergent-divergent nozzle has been taken as 230 mm with 55 mm convergent part and 175 mm divergent part. The barrel has been placed at the inlet of the convergent part of diameter 19 mm with 15 mm length. The injector has been positioned at the inlet of the nozzle of 2.5 mm diameter and 30 mm in length. A circular plate shape substrate, with 6 mm thickness and 60 mm diameter, has been placed in front of the nozzle outlet. All the parts of the geometry have been designed to simulate the work in 2D and using symmetry to reduce the computational time.

2.2 Computational Field and Boundary Condition

The computational field with boundary conditions with all the phases and domains has been demonstrated in Fig. 2. The geometry has been converted into several elements and nodes of quadrilaterals. The CFD analysis has been carried out using standoff distances of 15 mm, 25 mm, 35 mm, and 45 mm having 90,579, 92,499, 93,174, and 90,578 nodes, respectively. The geometry with 25 mm of standoff distance has been employed for the analysis of different particle shapes and size range. The CFD

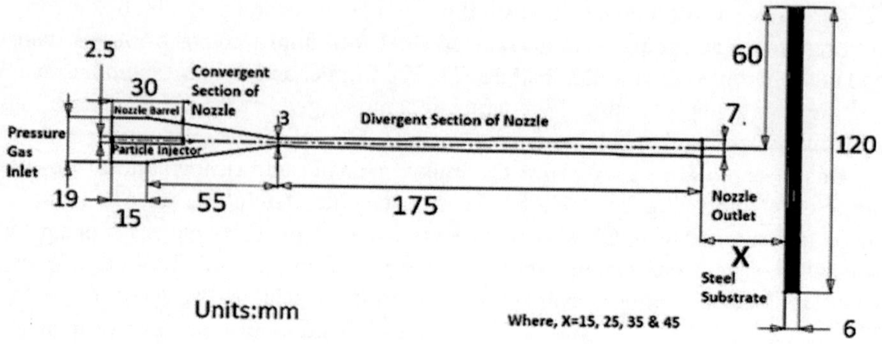

Fig. 1 Geometry with standoff distance ($X = 15$ mm, 25 mm, 35 mm, and 45 mm)

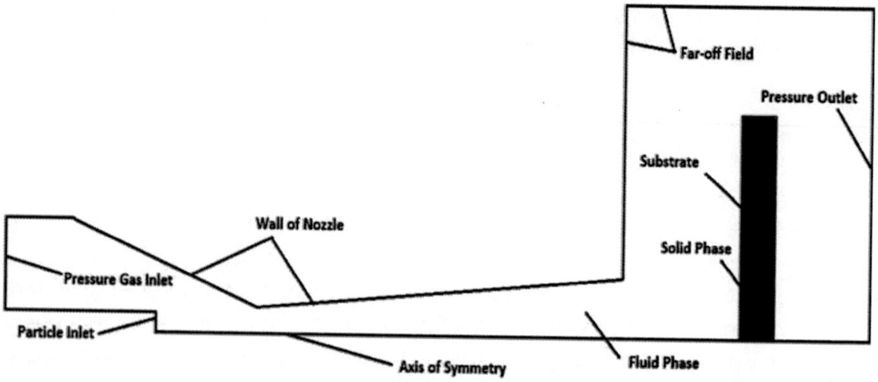

Fig. 2 Boundary conditions for CS coating CFD simulation

simulation meshing has been shown in Fig. 3, the chief regions which include the symmetry, the outlet of the nozzle including the substrate, throat area and inlet of the nozzle had been displayed separately via Fig. 3a–d, respectively.

2.3 Governing Equations and Material Properties

The gas phase is governed by ideal gas law by considering the compressibility effect. The gas flow embraces the continuity equations, energy equations, and momentum equations predominant to the flow of gases. The gases used in this simulation process have a mixture of nitrogen and helium, and the mixture of nitrogen and helium has been taken in the ratio of 4:6 (N_2: He). The powder particles have been kept as copper metal with different shapes and sizes, having a density of 8978 kg/m^3 and specific heat (Cp) of 381 J/Kg K. The shape of powder particles has been taken as spherical and non-spherical (SF = 0.8) (crushed sandstone shape) [31] with a size range of

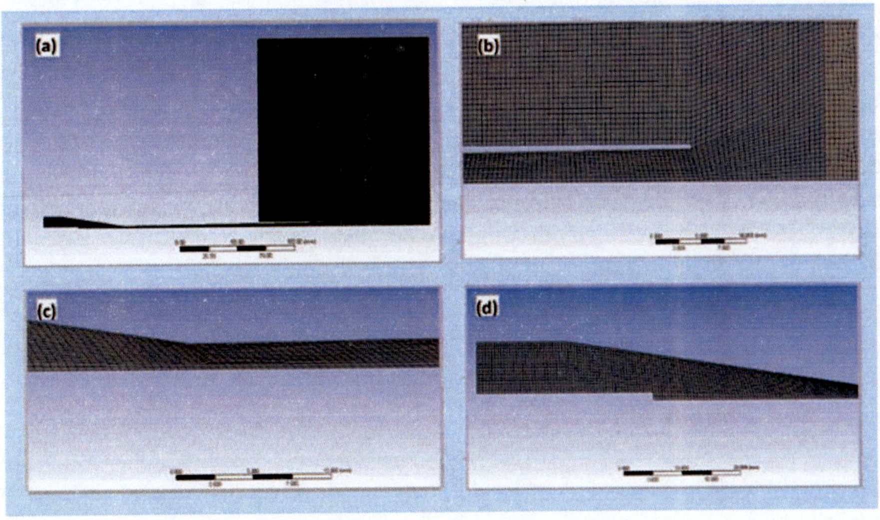

Fig. 3 CFD simulation mesh: **a** meshing of geometry, **b** meshing at nozzle outlet/near the substrate, **c** meshing at the nozzle throat, and **d** meshing at the particle and gas inlet

15–60 μm using the rosin-rammler method with the mean particle size of 30 μm and injected with the temperature of 300 K. The substrate material has been taken as steel at the temperature of 300 K, with density, specific heat (Cp), and the thermal conductivity of 8030 kg/m^3, 502.48 J/kg K, and 16.27 W/m K, respectively.

2.4 *Numerical Simulation*

The CFD analysis, employing the ANSYS software, has been carried out to escalate the role of particle shape and size and the effect of standoff distance before impact. The temperature and pressure of the propelling gas at the gas inlet are 675 K and 15 bar, whilst at particle, inlet has been taken as 325 K and 1 bar.

The simulation of CS process technology has been carried out using the pressure-based, axisymmetric model. The numerical simulation model has been taken as the two-equation realizable k-ε model with standard wall function, as this model is the most realistic as compared to other available models.

3 Outcomes and Discussion

Interpreting to the prior research work [14, 27], the pressure taken was 3–7 MPa to obtain the optimal impact velocity of particles having a diameter of 10 μm. Whereas, in the present study, the pressure is taken as 15 bar or 1.5 MPa with varying particle size and shape of powder particles. Consequently, this study implies the idea of shape effect on impact velocity accompanied by particle size. Secondly, the effect of particle shape on temperature and impact velocity is investigated.

3.1 Effect of Shape and Size on Impact Velocity and Temperature

The impact velocity and temperature simulation analysis for two different shapes of particles and size range upon impact velocity and temperature have been shown by Figs. 4 and 5. The simulation results are plotted in Figs. 6 and 9, and the velocity and temperature of the spherical-shaped particle found higher as compare to non-spherical, which shows that spherical shape of powder particles will certainly affect the coating quality in terms of porosity, hardness, and bond strength as compare to non-spherical shape powder particles, because of the velocity and temperature difference and shape difference. The impact velocity and temp. of the particles before impact have been indicated by the data label in Figs. 6 and 7 in which the velocities of spherical and non-spherical particles just before impact upon the substrate have been

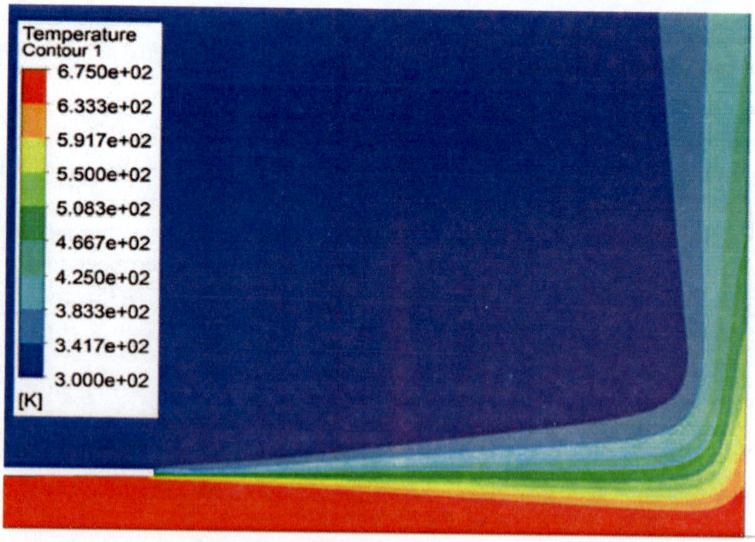

Fig. 4 Temperature contour near the substrate

Fig. 5 Velocity contour near the substrate

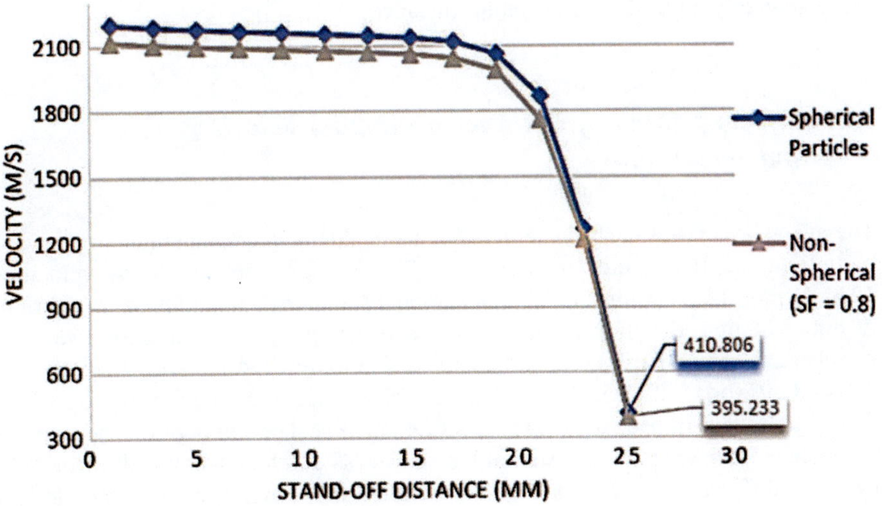

Fig. 6 Effect of particle shape on velocity

recorded as 1259.03 and 1211.87 m/s. The final impact velocity of spherical and non-spherical particles was 410.806 and 395.233 m/s. The temperature of spherical and non-spherical particles is 671.06 K and 621.81 K at the substrate. This indicates that

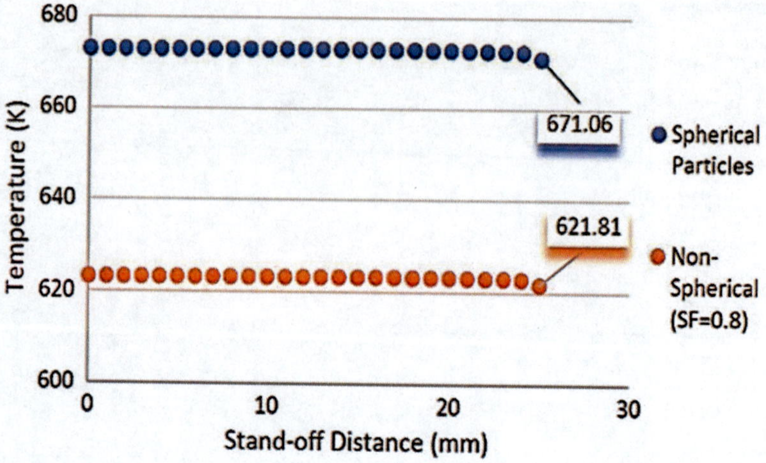

Fig. 7 Effect of particle shape on temperature

the spherical particles strike with high velocity and temperature as compare to non-spherical ones. All the simulations of different particle shapes have been performed using geometry with a 25 mm standoff distance.

3.2 *Effect of Standoff Distance on Impact Velocity and Temperature*

The analysis of standoff distance upon impact velocity and temperature of powder particles gives the results as shown in Figs. 8 and 9, in which the velocities 1234.84 m/s, 1259.03 m/s, 1325.06 m/s, and 1110.85 m/s were found for 15 mm, 25 mm, 35 mm, and 45 mm standoff distance, respectively. The final impact velocity of spherical and non-spherical particles was 403.806 m/s, 4010.806, 443.216, and 381.448 m/s for 15, 25, 35, and 45 mm standoff distances, respectively, upon which the particle gets impacted on the substrate surface. The temperature behaves in ascending order with that of standoff distance. The velocity before impact indicates the difference and found to be best with 35 mm standoff distance and lowest with 45 mm standoff distance. Considering velocity and temperature 35 mm standoff distance found efficient for CS coating.

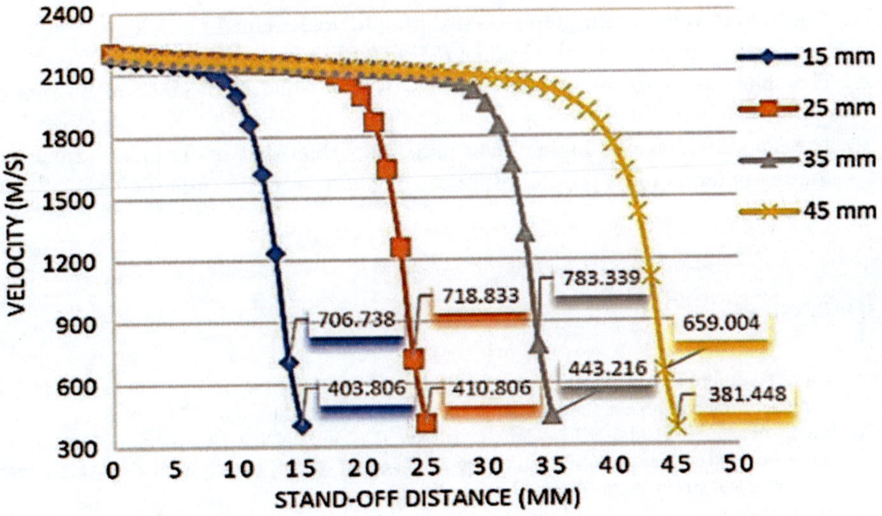

Fig. 8 Effect of standoff distance on velocity

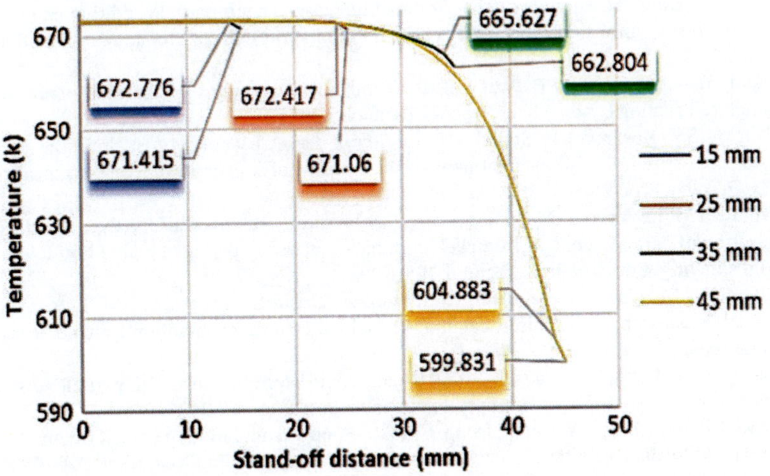

Fig. 9 Effect of standoff distance on temperature

4 Conclusion

The research findings to analyze the effect of particle shape and size on impact velocity and temperature and also the role of standoff distance in CS process technology indicates that:

1. The impact velocity and temperature of spherical-shaped particles are high as compare to the non-spherical shape (SF = 0.8) of particles.
2. The spherical shape of powder particles with a standoff distance of 35 mm is optimum to use in CS coating.
3. The standoff distance higher than and lower than that of findings, affect the coating in terms of its properties, porosity, bond strength, and efficiency.

References

1. Yin S, Cavaliere P, Aldwell B, Jenkins R, Liao H, Li W, Lupoi R (2018) Cold spray additive manufacturing and repair: fundamentals and applications. Addit Manuf 21:628–650
2. Liu Z, Wang H, Haché MJR, Chu X, Irissou E, Zou Y (2020) Prediction of heterogeneous microstructural evolution in cold sprayed copper coatings using local Zener-Hollomon parameter and strain. Acta Mater 193:191–201
3. Yeom H, Sridharan K (2021) Cold spray technology in nuclear energy applications: a review of recent advances. Ann Nucl Energy 150:107835
4. Pattison J, Celotto S, Morgan R, Bray M, O'Neill W (2007) Cold gas dynamic manufacturing: A non-thermal approach to freeform fabrication. Int J Mach Tools Manuf 47:627–634
5. Kumar M, Singh H, Singh N (2020) Effect of increase in nano-particle addition on mechanical and microstructural behaviour of HVOF and cold-spray Ni-20Cr coatings on boiler steels. Mater Today Proc 21:2035–2042
6. Das P, Bandyopadhyay PP, Paul S (2019) Finish form grinding of thermally sprayed nanostructured coatings. Adv Mater Process Technol 5:39–52
7. Klinkov SV, Kosarev VF, Shikalov VS, Vidyuk TM, Chesnokov AE, Smirnov AV (2019) Influence of preliminary heat treatment and ball milling of copper powder on cold spray process. Mater Today Proc 25:360–362
8. Winnicki M, Kozerski S, Małachowska A, Pawłowski L, Rutkowska-Gorczyca M (2021) Optimization of ceramic content in nickel–alumina composite coatings obtained by low pressure cold spraying. Surf Coatings Technol 405
9. Vilardell AM, Cinca N, Cano IG, Concustell A, Dosta S, Guilemany JM, Estradé S, Peiró F (2016) Dense nanostructured calcium phosphate coating on titanium by cold spray. J Eur Ceram Soc
10. Cao K, Yu M, Liang CM, Chen H (2020) Quantitative determination of SiC particles distribution of cold sprayed Al5056/SiC composite coatings. Surf Eng 36:1040–1048
11. Zhao P, Zhang Q, Guo Y, Liu H, Deng Z (2020) Atomic simulation of crystal orientation effect on coating surface generation mechanisms in cold spray. Comput Mater Sci 184:109859
12. Jami H, Jabbarzadeh A (2020) Molecular simulation of high-velocity deposition of Titanium dioxide nanoparticles on titanium. Appl Surf Sci 542:148567
13. Oyinbo ST, Jen TC, Zhu Y, Ajiboye JS, Ismail SO (2020) Atomistic simulations of interfacial deformation and bonding mechanism of Pd-Cu composite metal membrane using cold gas dynamic spray process. Vacuum 182:109779
14. Khan M, Zunaid M, Murtaza Q (2020) Examination of titanium powder with different particle sizes for velocity. Mater Today Proc 29:1–5
15. Wu H, Huang C, Xie X, Liu S, Wu T, Niendorf T, Xie Y, Deng C, Liu M, Liao H, Deng S (2020) Influence of spray trajectories on characteristics of cold-sprayed copper deposits. Surf Coat Technol 405:126703
16. Chakrabarty R, Song J (2020) A modified Johnson-Cook material model with strain gradient plasticity consideration for numerical simulation of cold spray process. Surf Coat Technol 397:125981

17. Kumar S, Zunaid M, Murtaza Q, Ansari N, Arora A (2015) Simulation of Injector in Cold Spray Process by Fluent-6. In: International conference on advanced research in innovation, pp 464–472
18. Singh S, Singh P, Singh H, Buddu RK (2019) Characterization and comparison of copper coatings developed by low pressure cold spraying and laser cladding techniques. Mater Today Proc 18:830–840
19. Kumar AS, Kar S, Bandyopadhyay PP, Paul S (2018) Grinding of ceramics-sintered ceramics versus ceramic coatings. Adv Mater Process Technol 4:538–547
20. Reddy BVR, Kummitha OR (2017) Characterization of spray formed and cold rolled Al–Pb alloy. Mater Today Proc 4:267–276
21. Das B, Bandyopadhyay PP, Nath AK (2018) An investigation on corrosion resistance and mechanical properties of laser remelted flame sprayed coating. Adv Mater Process Technol 4:660–668
22. Chakrabarty R, Song J (2020) Numerical simulations of ceramic deposition and retention in metal-ceramic composite cold spray. Surf Coat Technol 385:125324
23. Seraj RA, Abdollah-Zadeh A, Assadi H, Hajipour H, Kadkhodaee M (2020) Effect of substrate on the properties of cold sprayed coating of WC-10Ni. Adv Mater Process Technol 00:1–14
24. Tripathy S, Behera A, Pati S, Roy S (2020) Corrosion resistant nickel coating on mild steel by cold gas dynamic spraying. Mater Today Proc
25. Sabanayagam S, Chockalingam S (2020) Analysis of high temperature oxidation behaviour of SS316 by Al2O3 and Cr2O3 coating. Mater Today Proc 3–7
26. Mindivan F, Mindivan H (2016) Surface properties and tribocorrosion behaviour of a thermal sprayed martensitic stainless steel coating after pulsed plasma nitriding process. Adv Mater Process Technol 2:514–526
27. Khan M, Zunaid M, Murtaza Q (2021) Simulation of cold spray coating for powder pre-heat and impact velocity. Mater Today Proc. https://doi.org/10.1016/j.matpr.2021.01.780
28. Hemeda AA, Zhang C, Hu XY, Fukuda D, Cote D, Nault IM, Nardi A, Champagne VK, Ma Y, Palko JW (2020) Particle-based simulation of cold spray: influence of oxide layer on impact process. AdditManuf 101517
29. Kumar A, Nayak SK, Bijalwan P, Dutta M, Banerjee A, Laha T (2019) Mechanical and corrosion properties of plasma-sprayed Fe-based amorphous/nanocrystalline composite coating. Adv Mater Process Technol 5:371–377
30. Rutkowska-Gorczyca M (2020) X-ray diffraction and microstructural analysis of Cu–TiO2 layers deposited by cold spray. Mater Sci Technol (United Kingdom) 0836
31. Wang D, Fan LS (2013) Particle characterization and behavior relevant to fluidized bed combustion and gasification systems. In: Fluidized bed technologies for near-zero emission combustion and gasification, pp 42–76

Experimental Investigation of Evaporator Coil Performance with Ethylene Glycol

Manav Kumar, K. Kiran Kumar, and Sharifuddin Mondal

Abstract Heat exchangers are widely being used in various fields throughout the world. All cooling systems find various applications in industry, domestic, etc. Energy consumption reduction in cooling system is always a challenging task. An experimental setup is fabricated to study the heat transfer rate of an evaporator coil. Ethylene glycol is a secondary refrigerant and considered as a working fluid. The influence of the cooling liquid inlet temperature on air cooling is studied. The average temperature drop in inlet air is around 9 °C for maximum cooling liquid inlet temperature as −8 °C. The experimental results demonstrate that the ethylene glycol can be used in cooling system for small operating temperature range for different applications.

Keywords Heat exchanger · Experimental · Secondary refrigerant · Evaporator · Cooling system

1 Introduction

Heat exchanger is an equipment used to transfer the heat from one medium to another medium due to existing temperature difference in between. An evaporator is a type of heat exchanger used to absorb the heat from another surrounding medium. Heat exchangers are widely used for cooling and heating purposes in many industrial processes and engineering applications like power plants, boiler, automobile, food processing, refrigeration and air-conditioning systems, etc. [1]. The common objective in heat exchanger design is to transfer large amount of heat energy from low temperature difference. Due to shortage of fossil fuels and concern of environmental pollution, it is a demand of nature to develop better heat exchanger with improved efficiency and effectiveness. The energy saving in the current globalization of world has been a great challenge [2]. In the industrial applications of heat exchangers like

M. Kumar (✉) · S. Mondal
Department of Mechanical Engineering, National Institute of Technology, Patna 800005, India
e-mail: manavk.phd17.me@nitp.ac.in

K. K. Kumar
Department of Mechanical Engineering, National Institute of Technology, Warangal 506004, India

A. Maurya et al. (eds.), *Recent Trends in Mechanical Engineering*, Lecture Notes in Mechanical Engineering, https://doi.org/10.1007/978-981-19-7709-1_12

boilers and radiators, the energy may be saved by heating or cooling of such equipment and increased its working life [3]. It is important to make industrial processes as cost effective as possible. To reduce the cost of goods, energy-efficient methods must be used as much as possible in energy-consuming production facilities. The two most critical factors for energy savings in various processes are heat transfer enhancement and flow drag reduction strategies. The heat transfer enhancement can conserve energy by reducing energy consumption by increasing the heat transfer coefficient, while flow drag reduction can minimize pumping electricity consumption [4]. The demand for efficient, reliable and cost-effective heat exchanger equipment is growing rapidly around the world, particularly in large-scale power and process industries, as well as refrigeration and air-conditioning systems [5]. Ethylene glycol has been used in several heat exchanger applications as a standard working fluid due to its low and high temperature tolerance. Since ethylene glycol is compatible with water, hence with the composition of water and ethylene glycol is being used for the low and high temperature applications of heat exchangers [6]. Many research works have been presented with combination of ethylene glycol and with different types of nanoparticles for the enhancement of heat transfer performance. In [7], CuO/water nanofluid has been used as working fluid for investigating the thermal performance of double U-tube heat exchanger. An experimental investigation has been performed on double pipe heat exchanger with ethylene glycol water based TiO_2 nanofluid as the working fluid to find the heat transfer coefficient and friction factor with and without helical coil inserted in the inner pipe [8]. An experimental study has been presented for nanofluid heat transfer enhancement in a shell tube heat exchanger by using Fe_2O_3/water and Fe_2O_3/ethylene glycol as a working fluid [9]. In [10], ZnO and TiO_2 nanofluids in ethylene glycol/water have been used for studying the heat transfer performance in compact heat exchanger. The thermo-hydraulic performance has been investigated for a counter-current corrugated plate heat exchanger with MgO/ethylene glycol nanofluid as working fluid [11].

Different arrangement and configuration of heat exchangers have been investigated with several composition of water, ethylene glycol and nanoparticles in the literature but with this specific configuration of evaporator coil and ethylene glycol as working fluid has not been investigated so far with the best of our knowledge. The main effort made in this experimental work is to conduct the performance analysis of evaporator coil under various flow conditions. The mass flow rates for air and cooling liquid are kept constant. Air cooling capacity of evaporator coil is examined by varying the inlet temperature of liquid for different inlet temperature of air. The outlet temperature of air and liquid are measured for evaluating the performance of evaporator coil.

The rest of paper is organized as follows: Section 2 presents the test rig fabrication with different components of experimental setup. Section 3 represents experimental procedure and working fluid. Section 4 describes the experimental results. Finally, concluding remarks and scope for future work are incorporated in Sect. 5.

2 Test Rig Fabrication

For the performance analysis of evaporator coil, a setup is fabricated with the desired configuration. A rectangular duct of plywood material is made with one end open and other is closed. On the closed end of the rectangular duct, a fan is fitted at the center. A dimmer stat or regulator with fan is used for getting the different flow rate of air. The top face of the rectangular duct is covered with thermocol sheet. The evaporator coil is placed ahead of fan position by some distance so that uniform flow of air passes through each tube pass. The material of tube and fin of the coil is aluminum. The fins are rectangular in cross section and placed at some distinct pattern over the one pass of the tube length. The schematic diagram of the experiment setup is shown in Fig. 1.

A chiller system, which is itself a refrigeration system, is used to maintain the temperature of secondary refrigerant (ethylene glycol). The pump available in the chiller system supplies constant flow rate, but at different operating temperature, its flow rate is affected by density variation. Thermocouples are fitted at appropriate locations for temperature measurement and its data are collected from data acquisition system. Dry and wet bulb thermometer arrangements are fitted after the fan position to measure relative humidity of air at inlet section. Manometers and pressure gages are used to measure the pressure drop for refrigerant side and for air side, respectively. Digital anemometer is used to measure the face air velocity.

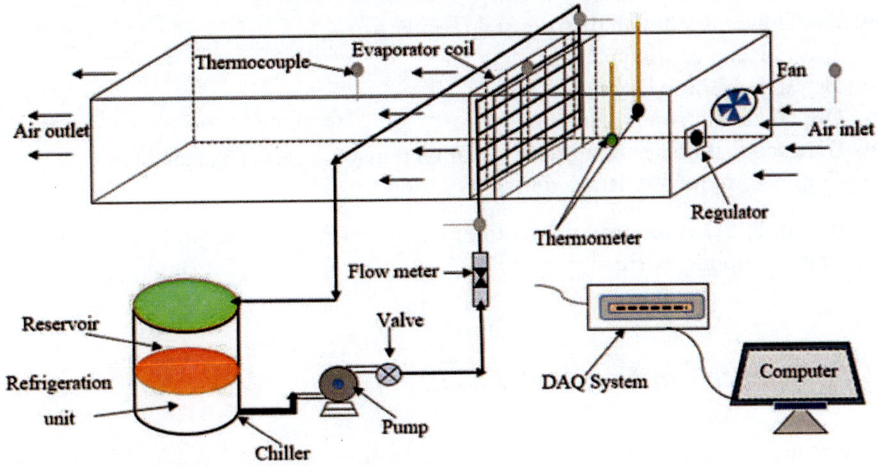

Fig. 1 Schematic diagram of experimental setup

Fig. 2 Laboratory view of complete experimental setup

2.1 Components of Experimental Setup

The experimental setup is fabricated by using different equipment and components as listed below:

- Evaporator coil;
- Chiller system fitted with pumping system: Make: Aditya, Cooling capacity: − 10 °C;
- Fan/blower fitted with dimmer stat (regulator);
- Thermocouples: K-type thermocouple;
- Dry and wet bulb thermometers;
- Digital anemometer, manometers;
- Data acquisition system: Make: - National instruments (16 channels);
- Computer to record data acquired by DAQ.

In Fig. 2, the laboratory view of complete experimental setup is shown and its different components are shown in Fig. 3.

2.2 Geometry of Evaporator Coil

The evaporator coil selected for the experimental work is commonly used in house hold refrigerator. The dimensions and other parameters of coil are measured with the help of proper measuring devices. Table 1 presents the specification of evaporator coil.

Fig. 3 Different components of experimental test rig (ordered from top left corner in clockwise direction, **a** dry and wet bulb thermometers, **b** fan with regulator, **c** data acquisition system and **d** front view of evaporator coil)

Table 1 Specification of evaporator coil

Parameter	Values
Duct size (W × H × L)	45 cm × 30 cm × 75 cm
Tube length for one pass	37 cm
Tube arrangement	Staggered
Fin type	Plain
Transverse pitch of tubes	2.8 cm
Longitudinal pitch of tubes	3 cm
Fin thickness	0.019 cm
Fin density (fins/meter)	70
Average fin spacing	1 cm
Parameter	Values
Duct size (W × H × L)	45 cm × 30 cm × 75 cm
Tube length for one pass	37 cm
Tube arrangement	Staggered
Fin type	Plain

3 Experimental Procedure

Ethylene glycol is an important organic compound and chemical intermediate used in many industrial processes [12]. Ethylene glycol is the single-phase fluid used as a cooling agent and its melting/freezing point is −12.5 °C. It is the most common antifreeze fluid for cooling or heating applications. To measure and record the temperatures, a data acquisition device that has a resolution of 0.1 °C is used. Chromel/constantan (K-type) thermocouples with a diameter of 0.5 mm and an uncertainty of ±1.1 °C are used to calculate all temperatures. Thermocouples are used to determine the temperature at the following locations: the temperature of the air-dry bulb before the coil of the evaporator, the temperature of the air at the coil outlet portion, the refrigerant outlet point and two thermocouples on the tube surface at the inlet and outlet section to measure the temperature of the wall. Inlet refrigerant temperature is taken from the reading taken from temperature indicator of the chiller system. The face air velocity at air inlet to the evaporator is measured using a portable digital anemometer with uncertainty of ±0.01 m/s.

The liquid is cooled at certain temperature with the chiller system, and then its flow rate is measured by collecting it in flask and measured time from stopwatch. The density of the liquid at a particular temperature is calculated by the correlation developed by polynomial curve fitting from data available for ethylene glycol. With dry and wet bulb temperatures recorded at atmospheric pressure, thermometers are used to calculate the relative humidity of inlet air. The face velocity of air is varied by the regulator fitted with fan and measured by the digital anemometer. A 16 channel data acquisition system (DAQ) of national instruments is used to record the data of measured temperature by thermocouples. All the thermocouples are connected to DAQ, which is connected to the computer that records the temperature readings. LabVIEW is used as the interface software. Each temperature measurement is done by thermocouples connected to data acquisition system.

4 Experimental Results

The reliability and accuracy of experimental setup is tested with water as cooling liquid before conducting the experiment with ethylene glycol. The experimental results demonstrate the effectiveness of the evaporator coil by cooling the ambient air. The inlet temperature of cooling liquid (fluid) is varied, and mass flow rates of air and liquid are kept constant to observe the outlet temperature of air and liquid. The experiment is conducted for 30 min, and required temperatures are recorded by DAQ system. The obtained result is validated by making the energy balance analytically for air and liquid side and found to be reasonably in good order.

Figure 4 shows the temperature at different locations namely outlet liquid temperature (T_{lo}), inlet air temperature (T_{ai}) and outlet air temperature (T_{ao}) at inlet liquid temperature −8 °C. The average inlet and outlet temperatures of air are, respectively,

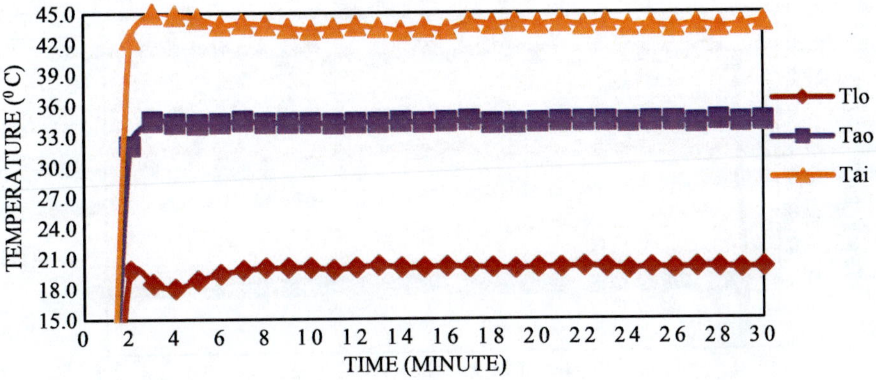

Fig. 4 Variation of air inlet temperature, air outlet temperature and liquid outlet temperature with respect to time at liquid inlet temperature −8 °C

43.8 and 34.2 °C. The average temperature of ambient air is dropped by 9.6 °C in the entire run of experiment.

Figure 5 shows the different temperature distribution such as T_{lo}, T_{ai} and T_{ao} along the time at inlet liquid temperature −6 °C. The average inlet and outlet temperatures of air are, respectively, 40.3 and 32.3 °C. The average temperature of ambient air is dropped by 8 °C in the entire run of experiment.

Figure 6 shows the different temperature distribution such as T_{lo}, T_{ai} and T_{ao} along the time at inlet liquid temperature −4 °C. The average inlet and outlet temperatures of air are, respectively, 42 and 33.1 °C. The average temperature of ambient air is dropped by ~9 °C in the entire run of experiment.

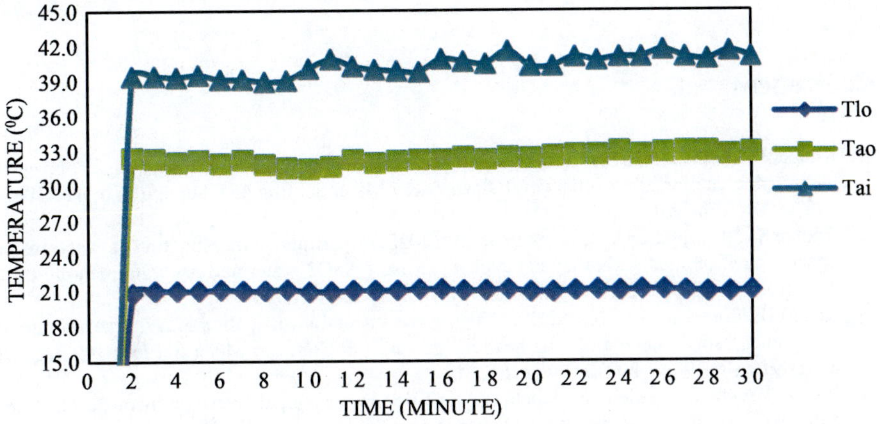

Fig. 5 Variation of air inlet temperature, air outlet temperature and liquid outlet temperature with respect to time at liquid inlet temperature −6 °C

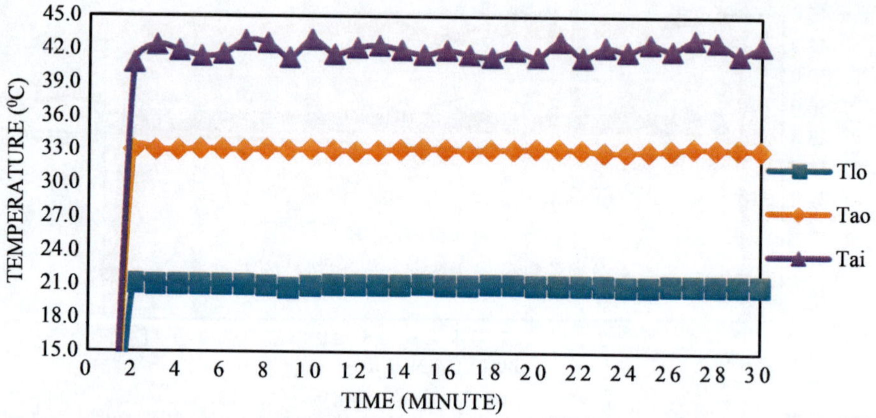

Fig. 6 Variation of air inlet temperature, air outlet temperature and liquid outlet temperature with respect to time at liquid inlet temperature −4 °C

5 Conclusions

The experimental investigation is carried out for evaluating the evaporator coil performance in cooling the ambient air. The flow of air is transverse to the flow direction of cooling liquid. The average temperature drop of ambient air is around 9.6 °C for inlet temperature of cooling liquid at −8 °C. For this significant temperature drop, it is concluded that ethylene glycol may be a good candidate as a cooling agent in the heat exchanger for air cooling applications. In the future, this work may be extended to improve the cooling performance of evaporator coil by adding nanoparticles to the ethylene glycol in order to enhance the heat transfer rate.

References

1. Khalaf BS, Falih AH (2018) Numerical and experimental investigation of heat transfer enhancement in double pipe heat exchanger. IOP Conf Ser Mater Sci Eng 454. https://doi.org/10.1088/1757-899X/454/1/012066
2. Sridhar SV, Karuppasamy R, Sivakumar GD (2020) Experimental investigation of heat transfer enhancement of shell and tube heat exchanger using SnO2-water and Ag-water nanofluids. J Therm Sci Eng Appl 12:1–6. https://doi.org/10.1115/1.4045699
3. Rao MSE, Sreeramulu D, Naidu DA (2016) Experimental investigation of heat transfer rate of nano fluids using a shell and tube heat exchanger. IOP Conf Ser Mater Sci Eng 149. https://doi.org/10.1088/1757-899X/149/1/012204
4. Raei B, Peyghambarzadeh SM, Salehi Asl R (2018) Experimental investigation on heat transfer and flow resistance of drag-reducing alumina nanofluid in a fin-and-tube heat exchanger. Appl Therm Eng 144:926–936. https://doi.org/10.1016/j.applthermaleng.2018.09.006
5. Meikandan M, Malarmohan K, Hemachandran E (2019) Experimental investigation on thermal performance of nano coated surfaces for air-conditioning applications. Therm Sci 23:175–175. https://doi.org/10.2298/tsci160825175m

6. Keklikcioglu O, Dagdevir T, Ozceyhan V (2020) Second law analysis of a mixture of ethylene glycol/water flow in modified heat exchanger tube by passive heat transfer enhancement technique. J Therm Anal Calorim 140:1307–1320. https://doi.org/10.1007/s10973-020-09445-w
7. Du R, Jiang D, Wang Y (2020) Experimental investigation on thermal performance of double-U-tube heat exchanger using CuO/water nanofluid as heat transfer fluid. E3S Web Conf 165:0–3. https://doi.org/10.1051/e3sconf/202016501022
8. Chandra Sekhara Reddy M, Vasudeva Rao V (2014) Experimental investigation of heat transfer coefficient and friction factor of ethylene glycol water based TiO2 nanofluid in double pipe heat exchanger with and without helical coil inserts. Int Commun Heat Mass Transf. https://doi.org/10.1016/j.icheatmasstransfer.2013.11.002
9. Kumar N, Sonawane SS (2016) Experimental study of Fe2O3/water and Fe2O3/ethylene glycol nanofluid heat transfer enhancement in a shell and tube heat exchanger. Int Commun Heat Mass Transf. https://doi.org/10.1016/j.icheatmasstransfer.2016.09.009
10. Manikandan SP, Baskar R (2018) Heat transfer studies in compact heat exchanger using zno and TiO2 nanofluids in ethylene glycol/water. Chem Ind Chem Eng Q. https://doi.org/10.2298/CICEQ170720003M
11. Arya H, Sarafraz MM, Arjomandi M (2018) Heat transfer and fluid flow of MgO/ethylene glycol in a corrugated heat exchanger. J Mech Sci Technol. https://doi.org/10.1007/s12206-018-0748-x
12. Yue H, Zhao Y, Ma X, Gong J (2012) Ethylene glycol: Properties, synthesis, and applications. Chem Soc Rev. https://doi.org/10.1039/c2cs15359a

Energy and Exergy Analysis of Combined Ejector Refrigeration Cycle Using Eco-Friendly Refrigerants

Aftab Anjum, Radhe Shyam Mishra, and Samsher

Abstract Exergy and energy analysis of combined ejector refrigeration cycle coupled with exhaust of solar energy, gas turbine, geothermal energy and industrial waste heat has been presented, using working fluids R141b, R152a, R600a and R717. The results of the study demonstrate that as evaporator temperature increases, COP increases and exergy efficiency decreases, though both continuously increases and decreases with ejector inlet pressure, condensing pressure and ejector inlet temperature, respectively. Refrigerant R717 gives the maximum COP of 0.836 and exergetic efficiency of 12.74% among others, at condensing pressure of 30 kPa and evaporation temperature of 258 K, respectively, and R600a gives minimum value of COP and exergetic efficiency. The results of the study signify that evaporator temperature, ejector inlet pressure, condenser temperature and ejector inlet temperature strongly influence the refrigeration effect, exergetic efficiency and entrainment ratio of such system.

Keywords Coefficient of performance · Exergy efficiency · Exergy destruction rate · Efficiency defect · Exergy destruction ratio · Entrainment ratio

1 Introduction

In the view of environment temperature increases, the need of refrigeration system is increasing in the few decades. With the latest techniques, that may solve the actual environmental problems, by enhanced operational techniques and lesser alleviation of the ozone depletion and global warming, Memet and Preda [1]. In spite of all efforts in study of energy production as well as its utilization in the world, the extreme usage of fossil fuels is responsible for global warming and climate imbalance. This may affect further melting glaciers and rainfall pattern changes, Elum and Momodu [2].

The amount of power consumption in such a process takes almost 15% of the entire power consumption universally, IIR. COP-15 [3].

A. Anjum (✉) · R. S. Mishra · Samsher
Department of Mechanical Engineering, Delhi Technological University, Shahabad Daulatpur, main Bawana Road, Delhi 110042, India
e-mail: aftabanjum915@gmail.com

A. Maurya et al. (eds.), *Recent Trends in Mechanical Engineering*, Lecture Notes in Mechanical Engineering, https://doi.org/10.1007/978-981-19-7709-1_13

Comparing to vapor absorption refrigeration cycle, ejector refrigeration cycle requires lesser installation and operational cost [4–6].

Different types of combined power and refrigeration process were projected and using first law analysis was examined [7–12]. Dai et al. [13] investigated the combined cycle incorporating the organic Rankine cycle with the ejector cooling cycle using thermodynamic analysis, in which the ejector is positioned between the condenser and steam turbine.

The flexibility of the ejector refrigeration system allows using various working fluids. Such as water (R718) provides numerous benefits [14–19]. Halocarbon compounds can have a refrigerating effect below 0 °C and also take advantages from low-quality energy sources at 60 °C and produces COP around (0.4–0.6), [20–21]. Besides the latest halocarbon compounds, its nominal impact on the environment makes them plausible alternatives [22–25]. Jiapeng et al. [26] proposed compressor-ejector-based cycle and compared the effectiveness of the given cycle with experimental results.

Yingjie et al. [27] projected a new enhanced compression-ejector cooling cycle using economizer (CERCE) of simple ejector-compressor cycle (CERC). Xingyang et al. [28] worked on a new combined power and ejector cooling cycle using zeotropic blend. Khaliq [29] obtained the first and second laws analysis to study solar-based integrated arrangement that gives two separate outcomes simultaneously using R141b as operating fluid. From the above-mentioned literature survey, it seems that there are numerous studies performed using energy analysis of the ejector-based cooling cycle and a slight attention is given to exergy analysis using various eco-friendly refrigerants. The combined cycle utilizes low temperature waste heat as a source of heat with R141b, R152a, R600a and R717 which shows excellent performance characteristics as to the high performance coefficient and low cost power consumption that makes it most relevant working substance as operating fluid. Therefore, the main purpose of the ongoing study to be suggested and investigated waste heat source-based combined ejector refrigeration cycle which works on heat source taken from flue gasses. A parametric investigation was conducted to examine the effect of system parameters including condensation pressure, evaporating temperature, ejector inlet pressure and ejector entrainment ratio. Numerical results are shown on graph and comments are given as below.

2 System Description

The waste heat-based combined cycle consists of heat recovery vapor generator (HRVG), evaporator, condenser, pump, expansion valve and the ejector as depicted in Fig. 1. A solar energy, turbine flue gas, geothermal energy or some other industrial waste heat may run the cycle. In the present analysis, flue gas is used as the heat source for the combined cycle simulation consisting of 96.16% of N_2, 3.59% of O_2, 0.23% of H_2O and 0.02% of $NO + NO_2$ per volume. R141b, R600a, R152a and R717 are taken as the working material since this refrigerant is known as eco-friendly.

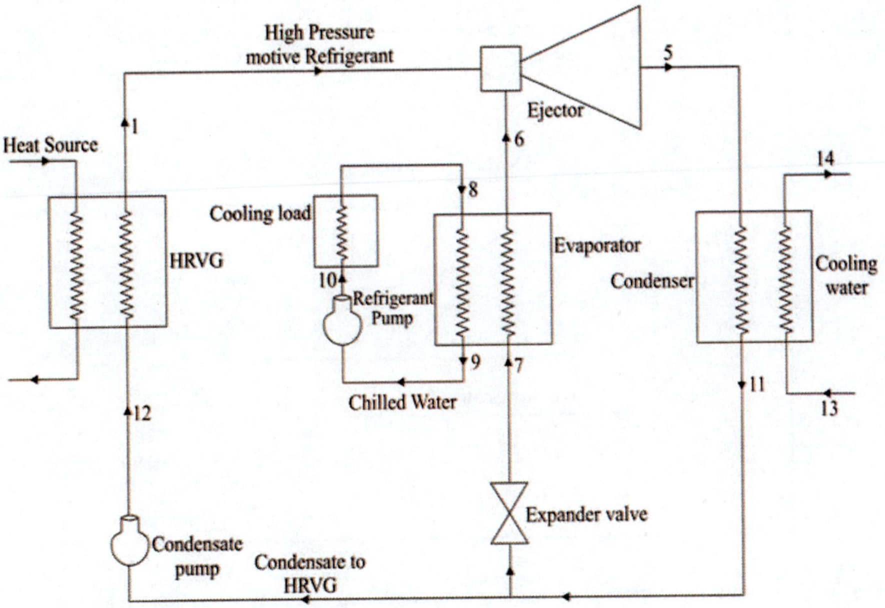

Fig. 1 Schematic diagram of combined ejector refrigeration cycle

The vapor velocity at the nozzle outlet is very high, creating vacuum at the inlet of the mixing chamber and enabling the secondary vapor from the evaporator inside the mixing chamber. In the mixing section, the primary and secondary vapors were mixed afterward it behaves like a transient supersonic flow. After it passes through the constant cross-sectional area, there seems to be occurrence of normal shocks, and there after rise in pressure.

The mixed stream finally exits from the ejector and enters the condenser where it becomes condensed and converted from vapor to liquid by transferring heat to the surrounding area. It bifurcates into two parts at the condenser outlet, the first part of the stream passes through the evaporator after moving via the throttling valve, and the second part allows to flow through the boiler (HRVG) via refrigerated pump, which increases its condenser pressure to the boiler. At the exit of expansion valve, the refrigerant has low temperature and pressure passes through the ejector evaporator, where heat is absorbed from its surrounding and as a result desired effect is produced. The temperature-entropy diagram for the detailed working of the given combined cycle is shown in Fig. 2. The assumptions of working parameters and important thermodynamic modeling requirements of the proposed cycle are described in Table 1 as given.

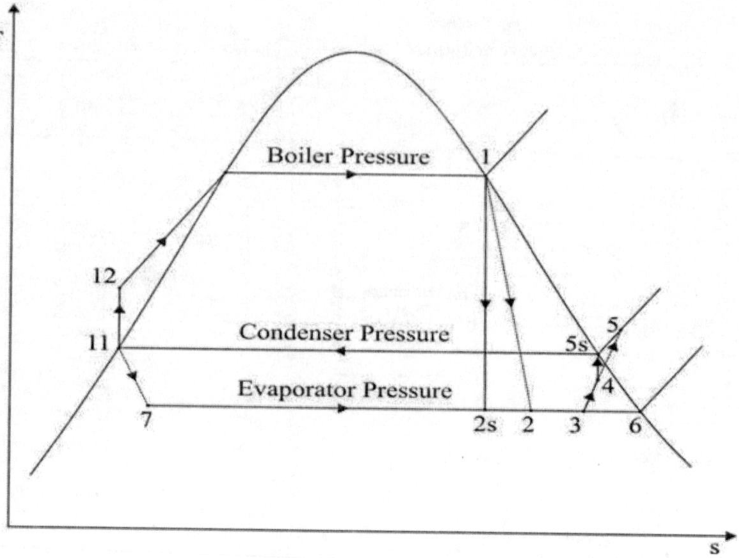

Fig. 2 Temperature-entropy representation of the combined ejector refrigeration cycle

Table 1 Assumptive operating condition considered for the cycle

Parameter	Value
Environment temperature (°C)	25
Environment pressure (Mpa)	0.101325
Ejector inlet pressure (Mpa)	0.3
Ejector inlet temperature (°C)	120
Heat source inlet temperature (°C)	130
Heat source outlet temperature (°C)	75
Ejector evaporator temperature range (°C)	−15 to −5
Heat source mass flow rate (kg/s)	10
Pump isentropic efficiency (%)	100
HRVG efficiency or effectiveness	100
Pinch point temperature difference (°C)	10
Nozzle efficiency (%)	90
Entrainment efficiency (%)	65
Diffuser efficiency (%)	80

3 Thermodynamic Modeling and Analysis

The model and simulation are performed using the Engineering Equation Solver (EES) software. Ejector is the main parts of the proposed arrangement, and its performance assessment is calculated by applying one dimensional flow model at constant pressure and the fundamental concept were elaborated by Keenan et al. [30]. The entrainment ratio is the ejector's key parameter and is described in terms of the magnitude of secondary refrigerant mass flow rate to the primary refrigerant mass flow rate that passes through the HRVG. The loss of energy across the ejector and the pressure at the mixing section is lesser than that of secondary flow which is taken with some logical concern, and the entrainment ratio can be calculated using the following mass, momentum and energy equation as follows [13, 21].

The conservation of the adiabatic nozzle energy equation and the primary steady flow is given as:

$$h_1 + \frac{v_1^2}{2} = h_2 + \frac{v_2^2}{2} \tag{1}$$

The efficiency of the nozzle is calculated as:

$$\eta_n = \frac{h_1 - h_2}{h_1 - h_{2s}} \tag{2}$$

Using the above Eqs. (1) and (2), the velocity at the outlet of primary flow is written as:

$$v_2 = \sqrt{2\eta_n(h_1 - h_{2s})} \tag{3}$$

The ejector's entrainment ratio is expressed as:

$$\mu = \frac{\dot{m}_{sf}}{\dot{m}_{pf}} = \frac{\dot{m}_6}{\dot{m}_1} \tag{4}$$

It is well known that the entrainment ratio (μ) of ejector is the key parameters to evaluate its performance. From the above equation, it can be derived as:

$$\frac{1}{\mu} = \frac{1}{\eta_e \eta_n \eta_d \frac{(h_1 - h_{2s})}{(h_{5s} - h_4)} - 1} \tag{5}$$

3.1 First Law Analysis

In the case of considered cycle, the performance of such a cycle may be calculated as the ratio of the desired effect to the rate of waste heat obtained from the thermal energy. The fundamental equations are used by applying laws of conservation of energy at the different section are expressed as follows.

The performance of the cycle is calculated by the COP and exergetic efficiency:

$$\text{COP} = \frac{R.E}{\text{Heat input}} = \frac{\dot{Q}_E}{\dot{Q}_{\text{in}}} \tag{6}$$

where $\dot{Q}_E$ the desired effect or refrigeration effect and $\dot{Q}_{\text{in}}$ is the cumulative heat input.

3.2 Second Law Analysis

Exergy is described as the most useful work possible which could be achieved from the system as it enters the environmental state. The equation of exergy balance under steady state over the control volume for every parts of the proposed cycle is expressed as following:

$$\sum_R \left(1 - \frac{T_0}{T_R}\right)\dot{Q}_R - \dot{W}_{C.V} - \sum_i \dot{m}_i e_i - \sum_e \dot{m}_e e_e - \dot{E}D = 0 \tag{7}$$

where subscript '0' and 'R' represents the physical property of a particular state with respect to ambient and D is the exergy destruction rate, 'e' is the exergy rate per unit mass flow and the subscription 'e' and 'i' represents the outlet and inlet position, respectively.

The specific exergy may also be described by using the classical concept, ignoring kinetic and potential energy, Vijayaraghavan and Goswami [8].

$$e = (h - h_0) - T_0(s - s_0) \tag{8}$$

Exergy efficiency is calculated as the ratio of exergy output to the exergy input entering the cycle.

$$\eta_{\text{exergy}} = \frac{\dot{W}_{\text{net}} + \Delta\dot{E}_E}{\dot{E}_{\text{in}}} \tag{9}$$

where $\dot{E}_{\text{in}}$ is the exergy of the heat input, that can be expressed as

$$\dot{E}_{\text{in}} = \dot{m}_h[(h_{hi} - h_{h0}) - T_0(s_{hi} - s_{h0})] \tag{10}$$

$\Delta\dot{E}_E$ is the exergy involved in the cooling effect that is measured as the difference between the exergy of the working fluid over the evaporator is described as below.

$$\Delta\dot{E}_E = \dot{m}_E[(h_6 - h_7) - T_0(s_6 - s_7)] \tag{11}$$

4 Results and Discussion

An energy and exergy analysis is being developed of the given combined ejector refrigeration cycle. The cycle performance shall be calculated on the basis of the working conditions listed in Table 1. A simulation method was used to determine the influence of certain design parameters affecting, including the ejector inlet pressure, ejector inlet temperature, condenser pressure and evaporator temperature on combined cycle output.

Figure 3 illustrates the outcome of varying evaporator temperatures on both COP and exergy efficiency. It is revealed that COP increases as the evaporator temperature increases, whereas the exergy efficiency decreases. Increasing the temperature of the evaporator leads to a substantial increment inside the secondary flow rate, whereas the primary flow rate is at constant pressure. It is further seen that as the evaporator temperature rises from (−15 to −5 °C) raises its coefficient of performance from 0.65 to 0.7 and exergy efficiency decreases from 10.3% to 7.51%, respectively, in case of fluid R141b. It can also be seen from above graph, R717 gives the highest COP followed by R152a, R141b and R600a, respectively. Similarly, R717 gives highest exergy efficiency followed by R152a, R141b and R600a, respectively.

Figure 4 presents the effect of varying ejector inlet pressure on the COP and the exergy efficiency of the combined cycle. It is observed that COP increases noticeably as the ejector inlet pressure rises. This increment is caused by the fact that the enthalpy change increases with increasing in the inlet pressure through the ejector. Alternatively, the exergy efficiency is increasing steadily as the ejector inlet pressure rises. It is because of the increased pressure of the ejector inlet results in reduction

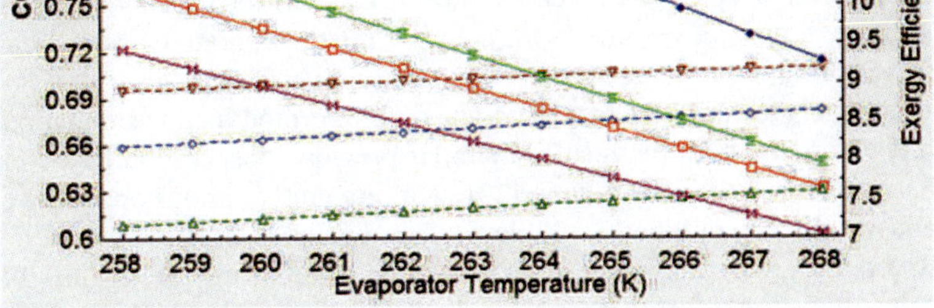

Fig. 3 Effect of evaporator temperature on coefficient of performance and exergy efficiency of the cycle

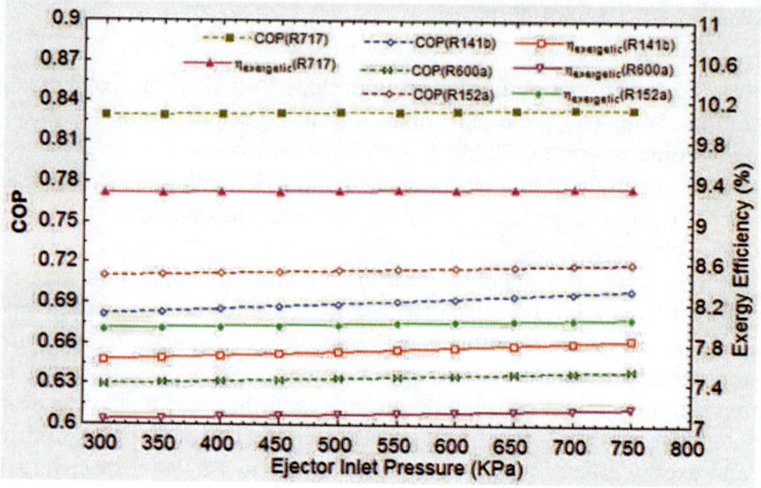

Fig. 4 Effect of ejector inlet pressure on coefficient of performance and exergy efficiency of the cycle

in temperature of primary flow as well as the flow rate at the exit of the nozzle, consequently, the secondary vapor entrainment decreases.

Figure 5 demonstrates the effect of change in condenser pressure on the coefficient of performance and exergetic efficiency. It reveals that the COP decreases appreciably as the condenser pressure rises. This is due to the rise in condenser pressure, which raises the temperature, means its compression ratio raises. Such increment in compression ratio reduces the mass flow rate of entrained vapor at the constant speed as a result, it decreases the refrigeration effect. Also, with increasing condenser pressure, the combined cycle exergy efficiency decreases. Further, it can also reveal that R717 gives better COP and exergy efficiency accompanied by R152a, R141b and R600a, respectively.

Figure 6 demonstrates the effect of ejector inlet temperature on the coefficient of performance and exergetic efficiency. It is found that the coefficient of performance decreases as ejector inlet temperature rises. This decrease is due to the rise of the primary flow inlet velocity as the ejector's inlet temperature also increases the pressure resulting in higher secondary vapor entrainment.

Consecutively to verify the results obtained, the thermodynamic simulation of the system is compared with the result of similar previous observation of a combined solar-based R141b system by Khaliq [29], working under various operating parameters and results are outlined in Table 5. The outcomes of a combined cycle as discussed during the simulation are validated with the results of Khaliq [29]; and the results trends show that there is a great similarity present among these results to those previously reported with the same working condition.

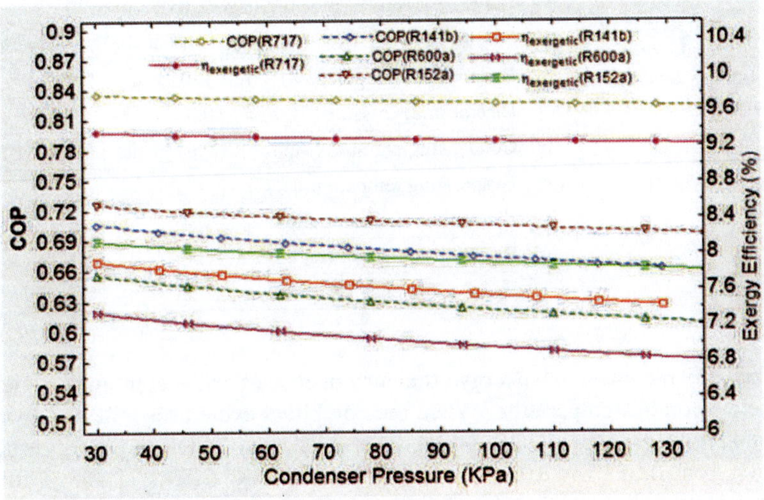

Fig. 5 Effect of condenser pressure on COP and exergy efficiency of the cycle

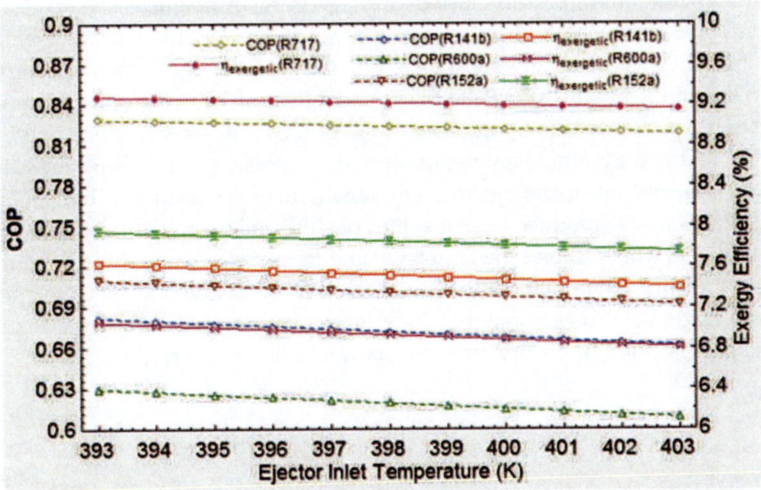

Fig. 6 Effect of ejector inlet temperature on coefficient of performance and exergy efficiency of the cycle

5 Conclusion

Thermodynamics first and second laws were used to establish the influences of various working conditions on cycle performance and exergy efficiency. From the ongoing study, following conclusions can be made:

Table 5 Comparison of computed results of proposed cycle with the data reported by Khaliq [29]

Parameter	Present study	Khaliq [29]
Ejector inlet temperature (°C)	120	120.7
Ejector inlet pressure (Mpa)	0.3	0.3
Condensing pressure (Mpa)	0.0775	0.07874
Evaporating temperature (°C)	−5	−5
Ejector entrainment ratio	0.367	0.269
Refrigeration effect (Kj/kg)	206.5	206.3
Exergy efficiency (%)	7.63	7.39

1. The COP increases and exergy efficiency decreases as the combined cycle raises the evaporator temperature. When the combined cycle uses R717 as a working fluid produces highest COP and exergy efficiency, i.e., COP and exergy efficiency vary from 0.652 and 10.3% at 258 K to 0.685 and 7.53% at 268 K followed by R152a, R141b and R600a take up the least value of exergy efficiency and coefficient of performance.
2. Increasing the ejector inlet pressure increases the combined ejector refrigeration cycle coefficient of performance as well as exergy efficiency. The results shows that cycle produces highest exergy efficiency and coefficient of performance using R717 as a working substance, i.e., COP and exergy efficiency vary from 0.6718 and 7.53% at 300 kPa to 0.712 and 7.89% at 750 kPa followed by R152a, R141b and R600a take up the least value of COP and exergy efficiency.
3. COP and exergy efficiency decrease with increasing condensing pressure and inlet temperature of the ejector. The results may revealed that both R717 and R152a relatively produce largest values of COP and exergy efficiency along with R141b and R600a shows the least COP and exergy efficiency value in both cases.
4. Based on the outcomes of the study, it may be concluded that the proposed combined cycle shows potential for efficient use of waste heat from the energy and exergy point of observation incorporating in ejector refrigeration cycle.

References

1. Memet F, Preda A (2015) An analysis of the performance of ejector refrigeration cycle working with R134a. IOP Conf Ser: Mater Sci Eng 95:Paper No. 012035
2. Elum ZA, Mmodu AS (2017) A discourages approach renew climate change mitigation and renewable energy for sustainable development in Nigeria. Sustain Energy Rev 76:72–80
3. International Institute of Refrigeration statement at COP-15 Copenhagen, vol 15, Denmark, pp 7–18
4. Megdouli K, TashtoushB M, Ezzaalouni Y, Nahdi E, Mhimid A, Kairoani L (2017) Performance analysis of a new ejector expansion refrigeration cycle (NEERC) for power and cold: exergy and energy points of view. Appl Therm Eng 122:39–48

5. Elbel S, Lawrence N (2017) Review of recent developments in advanced ejector technology. Int J Refrig 62:1–18
6. Chen X, Omer S, Worall M, Riffat S (2013) Recent development in ejector refrigeration technologies. Renew Sustain Energy Rev 19:629–651 (2013)
7. Besagni G, Meeru R, Inzloli F (2016) Ejector refrigeration: a comprehensive review. Renew Sustain Energy Rev 53:373–407
8. Vijayaraghavan S, Goswami DY (2006) A combined power and cooling cycle modified to improve resource utilization efficiency using a distillation stage. Energy 311:177–1196
9. Zheng D, Chen B, Qi Y, Jin H (2006) Thermodynamic analysis of a novel absorption power/cooling combined-cycle. Appl Energy 83:311–323
10. Zhang N, Lior N (2007) Methodology for thermal design of novel combined refrigeration/power binary fluid systems. Int J Refrig 30:1072–1085
11. Yu Z, Han J, Liu H, Zhao H (2014) Theoretical study on a novel ammonia-water cogeneration system with adjustable cooling to power ratios. Appl Energy 122:53–61
12. Grosu L, Marin A, Dobrovicescu A, Queiros-Conde D (2015) Exergy analysis of a solar combined cycle: organic Rankine cycle and absorption cooling system. Int J Energy Environ 95:155–168
13. Dai Y, Wang J, Gao L (2009) Exergy analysis, parametric analysis and optimization for a novel combined power and ejector refrigeration cycle. Appl Therm Eng 29:1983–1990
14. Elbel S, Hrnjak P (2008) Ejector refrigeration: an overview of historical and present developments with an emphasis on air-conditioning applications. In: Proceedings of international refrigeration and air conditioning conference at purdue (Publisher West Lafayette), vol 7, pp 14–17
15. Wang J, Dai Y, Zhang T, Ma S (2009) Parametric analysis for a new combined power and ejector-absorption refrigeration cycle. Energy 34:1587–1593
16. Ruangtrakoon N, Aphornratana SS (2014) Development and performance of steam ejector refrigeration system operated in real application in Thailand. Int J Refrig 48:142–152
17. Šarevski MN, Šarevski VN (2012) Preliminary study of a novel R718 refrigeration cycle with single stage centrifugal compressor and two-phase ejector. Int J Refrig 40:435–49
18. Angelino G, Invernizzi C (2008) Thermodynamic optimization of ejector actuated refrigerating cycles. Int J Refrig 31:453–463
19. Chen YM, Sun CY (1997) Experimental study of the performance characteristics of a steam-ejector refrigeration system. Exp Therm Fluid Sci 15:384–94
20. Tyagi KP, Murty KN (1995) Ejector-compression systems for cooling: utilizing low grade waste heat. J Heat Recov Syst 5:545–550
21. Shestopalov KO, Huang BJ, Petrenko VO, Volovyk OS (2015) Investigation of an experimental ejector refrigeration machine operates with refrigerant R245fa at design and off-design working conditions. Part 1. Theoretical analysis. Int J Refrig 55:01–11
22. Al-Khalidy N (1998) An experimental study of an ejector cycle refrigeration machine operating on R113: etude expérimentaled'une machine frigorifique à éjecteur au R113. Int J Refrig 21:617–625
23. Kasperski YJ, Gil B (2014) Performance estimation of ejector cycles using heavier hydrocarbon refrigerants. Appl Therm Eng 71:197–203
24. Zhang T, Mohamed S (2014) Conceptual design and analysis of hydrocarbon-based solar thermal power and ejector cooling systems in hot climates. J Sol Energy Eng 137:155–165
25. Pereira PR, Varga S, Soares J, Oliveira AC, Lopes AM, de Almeida FG (2014) Experimental results with a variable geometry ejector using R600a as working fluid. Int J Refrig 46:77–85
26. Jiapeng L, Wang L, Jia L, Zhen L, Hongxia Z (2017) A control oriental model for combined compression-ejector refrigeration system. Energy Convers Manag 138:538–546
27. Yingjie X, Ning JX, Xiaohong H, Wenqiao H, Ning M, Guangming C (2018) Performance evaluation and energy-saving potential comparison of heat-powered novel compression-enhanced ejector refrigeration cycle with an economizer. Appl Therm Eng 130:1568–1579

28. Xingyang Y, Zheng N, Zhao L, Deng S, Hailong L, Zhixin Y (2016) Analysis of a novel combined power and ejector-refrigeration cycle. Energy Convers Manag 108:266–274
29. Khaliq A (2017) Energetic and exergetic performance investigation of a solar based integrated system for cogeneration of power and cooling. Appl Therm Eng 112:1305–1316
30. Keenan H, Neumann EP, Lustwerk F (1950) An investigation of ejector design by analysis and experiment. J Appl Mech-Trans ASME 72:299–309

Emission Characteristics of Split Injection System in Low-Temperature Combustion Diesel Engine: A Review

Sadhu Pranava Sreedhar, Akash Venkateshwaran, and Bhisham Kumar Dhurandher

Abstract The necessity for reducing pollution due to increased population and its adverse effects on health and the environment has called for worldwide government policies. Researchers and engineers from automobile communities have been working on the advanced design and development of a diesel engine to meet these mandatory emission regulations. The split injection among these new developments has proven to be the key to reduce soot and NOx emissions in diesel engines. The split injection system is augmented with various low-temperature combustion (LTC) modes like homogeneous charge compression ignition engine (HCCI), premixed charge compression ignition engine (PCCI), and reactivity controlled compression ignition engine (RCCI) to counter the drawback of the traditional approach. This review paper provides a compilation of these recent innovations. A comprehensive overview of abundant research carried out for different split injection schemes in these LTC modes has been presented in the current research work. All these modern strategies and advancements in the split injection system are investigated and compared to expose and evaluate their strengths and weaknesses to arrive at a prominent solution.

Keywords CI engine · Split injection · Performance · Emission control · Low-temperature combustion

1 Introduction

Due to the NOx–PM trade, it is very difficult to achieve a simultaneous reduction of these two contaminants, the emission of NOx and PM by diesel engines is a significant problem and a challenge to solve. During these last two decades, various techniques including the homogeneous charge compression ignition (HCCI), premixed charge compression ignition (PCCI), and reactivity controlled compression ignition

S. P. Sreedhar (✉) · A. Venkateshwaran · B. K. Dhurandher
School of Mechanical Engineering, VIT Chennai, Chennai, India
e-mail: pranavbatman@gmail.com

B. K. Dhurandher
e-mail: Bhishamkumar.d@vit.ac.in

A. Maurya et al. (eds.), *Recent Trends in Mechanical Engineering*, Lecture Notes in Mechanical Engineering, https://doi.org/10.1007/978-981-19-7709-1_14

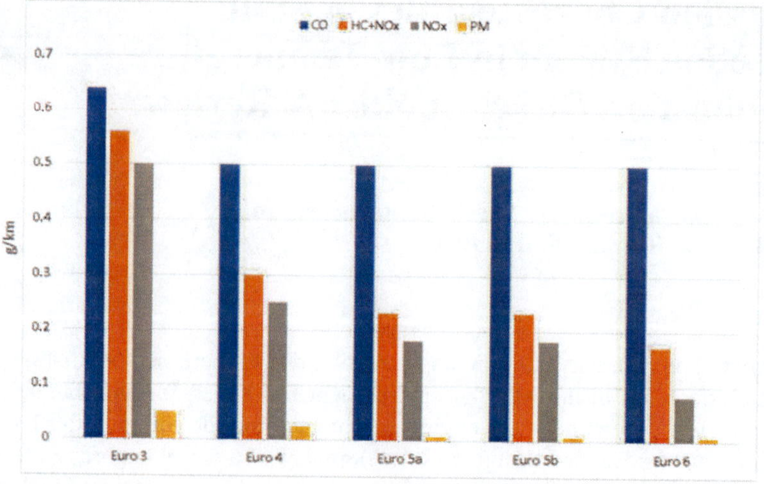

Fig. 1 Emission standards in diesel passenger car [6]

(RCCI) [1]. There are some flaws in the way these techniques are implemented. HCCI technology focused on a particular mixture of gasoline and diesel motors. HCCI is unpredictable and, in cases of variation in engine temperature, there is a dangerous risk of engine structural destruction. Also, this approach is efficient only in a small range. Knocking analysis showed that the combustion of the PCCI mode caused comparatively higher knock-and-burn noise in comparison with other modes of combustion. Another LTC technology "RCCI" has been developed because of the limits of HCCI and PCCI. In contrast to the baseline CI and the PCCI modes of combustion due to trapped methanol in crevices, RCCI mode combustion was more robust to minimize NOx but higher HC [2]. With EGR with RCCI, the rates between soot, NOx, and BSFC have a trade-off. The application of high EGR (lower cooling) also has a detrimental effect on the A/F ratio. It also has harmful effects on both oxidation and soot formation [3]. Moreover, multiple injection techniques are capable of the decline of NOx and soot emissions without significant motor output compromise [4]. The results from the experiments indicate the importance to find a desired pilot quantity and interval to apply the pilot injection to simultaneously reduce emissions and combustion noise [5]. European emission standards can be met using these techniques. An overview of emission limits is shown in Fig. 1.

1.1 LTC Strategies

The heterogeneity in combustion that produces high-temperature hotspots (responsible for greater emission of NOx) and fuel-rich areas locally (responsible for the

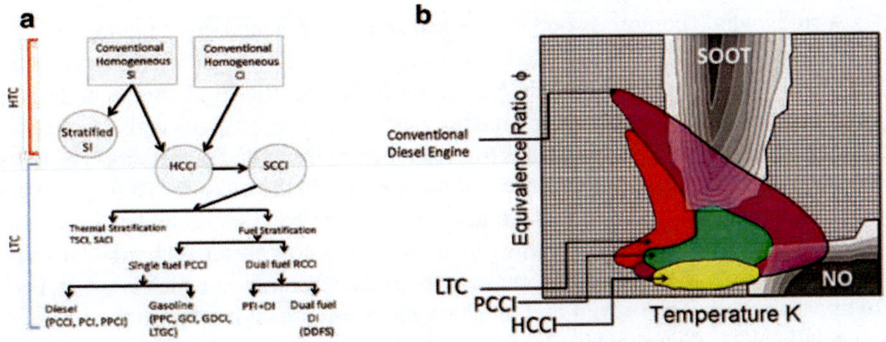

Fig. 2 **a** Evolution of LTC and **b** $T - \varphi$ correlation of various LTC types [7]

formation of PM) in the conventional diesel combustion. Although the temperature of flame and equivalence ratio heavily affects the emission of soot and NOx, flame temperature regulation and local equivalence ratios will concurrently reduce the flame and soot emissions.

LTC strategies have gained considerable interest and attention. Enhanced LTC fuel-air mixing prevents the local regions of the fuel-rich mixture and decreases the combustion temperatures under 2100 K, minimizing both NOx and PM emissions. It has several types which are mainly HCCI, PCCI, and RCCI, etc., as shown in Fig. 2.(a). In concern to fuel-air equivalence ratios (Φ), flame temperature, as well as the operating domains of various LTC modes, Fig. 2b indicates the individual regions of NOx and soot emissions [7]. Figure 2b shows also that LTC can be obtained at temperatures considerably below the region of NOx formation and local LTC equivalence ratios are far underneath the domain of soot formation.

1.2 Homogeneous Charge Compression Ignition (HCCI)

HCCI combustion is one of the low-temperature combustion modes in which significant research and studies have been conducted by the research community because of its intrinsic benefits. HCCI can satisfy good performance in terms of high efficiency at the same time without tolerating emissions for soot and NOx. However, there are a lot of technical obstacles while using HCCI which have to be resolved such as difficulty in controlling the timing of the ignition of the fuel as it has to automatically ignite and a very constrained and restricted range of operating the engine.

A zero-dimensional PDF-based, stochastic reactor modeling (SRM) technique is being used for two-stage injection into an HCCI engine. The role of increased load stratification induced by DI in prolonging the duration and increase the consumption rate of isooctane is demonstrated in systematic parametric studies relating to spray penetration, characteristic evaporation lag, and stratification level [8].

A study which comprises both numerical and experimental investigation on two-stage direct injection was executed. The numerical study was simulated in ANSYS FLUENT linked with a CHEMKIN. As the injection timing of the second stage was retarded, there was not much time for vaporization because the cooling effect of charge increases which overall leads to a reduction of the cylinder pressure. The early timing of the second injection causes inhomogeneity in the mixture. Thus late injection timing was implemented which also reduced the NOx emission by decreasing the temperature however due to this low-temperature non-premixed partial combustion took place that leads to an increase in the HC concentration which is a drawback to this technique. The second fuel injection has a significant impact on the emission especially NOx. Thus there is a trade-off which choosing an optimum injection point [9].

The work uncovers the combustion characteristics in two cases. One with single-stage ignition and the other with a double. High-octane fuel is used in the simulation, to get a better understanding of the ignition control mechanism in "two-zone HCCI" combustion using 3D CFD. The model provided temperature field, velocity distribution in the cylinder, and the streamline field. The injection timing and the combustion rate can be controlled by the dual-injection technique, and the HCCI operation range can be extended at a higher load [10].

A dual-stage injector is used to apply the HCCI combustion mode. The engine employed in this study was a single-cylinder four-stroke direct-injection engine. The paper initially explains the consequences of fuel injection timing variations and changes on the emission as well as the combustion characteristics. It was recognized that the injection timing of the 1st injection has an impact on the decline of smoke and NOx emission, whereas the 2nd injection solely affects the reduction of the smoke. For the multi-injection case, the injection timing increased the performance, and the smoke was extremely low [11].

This test was performed to analyze the double injection configuration with a narrow spray angle effect. The key combustion incident was relegated to the latter part of the delayed second injection timing, so that the emission of NOx was lower because of the low charging temperature, at the cost of the reduction in engine performance due to delayed combustion. For the operating conditions of the early first injection timing and later second injection timing, the emissions of NOx have been minimized. The timings of each injection have had a substantial effect on the HC emissions. For the first timing of injection, the emissions of HC have risen with a delay of the second injection time. Raises the in-cylinder temperature by diffusive combustion with advanced second injection, thereby significantly minimizing CO emissions [12].

The second injection time effect was investigated with ethanol and methanol blended fuel on the combustion and emission characteristics of the HCCI direct-injection petrol engine. Two key parameters play important roles in order to regulate auto-ignition and combustion in HCCI combustion. Firstly, the history of time and temperature, secondly, the character of fuel auto-ignition. Two phase direct-injection (TSDI) technique can monitor these two parameters. The second time of fuel injection in contrast with NOx emissions was less affected the UHC and CO emissions. The second fuel injection timing was delayed which lead to higher HC, CO, and lower NOx emissions. With the delay of the second injection of fuel due to a lower cylinder gas pressure, the NOx emissions decreased rapidly [13].

1.3 *Premixed Charge Compression Ignition (PCCI)*

PCCI has a great potential to incorporate low emission characteristics as well as high efficiency of HCCI. A low fuel/air combination heterogeneity and a decreased flame temperature represent PCCI. A split injection technique lowers the NOx and OM levels, and it gives PCCI a more homogeneous and leaner mixture with a lower combustion temperature. A wider load range is needed for PCCI-based combustion in order to transition between combustion modes without degradation of efficiency and exhaust emissions. The low temperature in PCCI unfortunately produces CO and HC.

A strategy for the regulation of injection and EGR is being drawn up to increase the load spectrum with low noise combustion and low emissions and good thermal performance [14]. Two sets of tests, which are single-stage and double-stage injections, using a single four-stroke cycle were conducted. The first-stage injection technique could expand the combustion range of the injections by delaying the time and reducing NOx emission level and high thermal performance and the second injection stage to lower this pressure raise rate. The careful selection of both injection and two-stage injection EGR conditions offers a dramatic reduction of pollution with a minor pressure increase equal to the pilot diesel process (Fig. 3).

Analysis is done for the emission and combustion characteristics of the PCCI strategy were analyzed by a 4-cylinder diesel engine with a two-stage injection. Furthermore to analyze PM forming and facilitate engine test research, the technique of flame visualization was used. To investigate the effect of various injection timings, two tests were conducted. A high fuel quantity proportion in the first injection leads

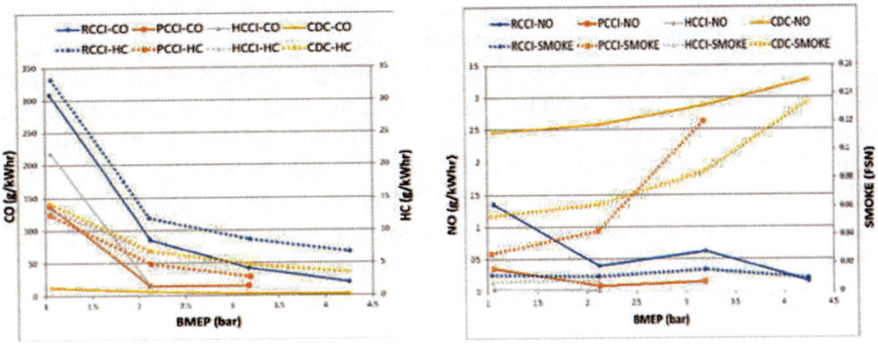

Fig. 3 Comparison of CO, HC, NO, and smoke emission in LTC and conventional engine [15]

to a quick burn before TDC, leading to a reduction in burning power and efficiency. The first injection mass part of 30% tends to be the most desirable ratio [16].

Analysis is done for the key difference from other experiments, which is the lower quantity of fuel used in the second injection. Non-luminous combustion can be seen. There is almost zero soot that characterizes this combustion. The second injection can serve as the PCCI combustion ignition promoter and as the controller for the combustion process. A new study of partial PCCI combustion with multiple injections has recently been conducted to tackle the disadvantages of PCCIs. This research is intended to explain the influence of two-stage of injection by eliminating the influence of cylinder volume changes and flow [18]. In a simulated partial PCCI combustion, a constant-volume vessel is used to analyze results of fuel quantity ratio used in the first and second injections as well as injecting period on soot formation, heat release, and nitrogen oxides (NOx) formations. For a small first quantity of fuel in the first injection, the peak of the initial heat release rate would be decreased with longer dwells of injection, whereas the second injection eliminates the first-injector inflammation of the mixture and raises the fuel mixture before ignition, leading to a higher initial thermal release in case of short dwell. The total NOx concentration is reduced by dual-stage injection with the released heat, particularly when the injection dwell period is long regardless of the fraction of the atmosphere oxygen mole [17].

A direct-injection diesel engine with four valves and a multi-injection technique for PCCI was examined using KIVA-3V code combined with comprehensive chemistry [19]. The results of the proportion of fuel break, inject timing, spray angles, and injection speed have been investigated. The k-β RNG turbulence model was used. The second injection timing and velocity fuel injection affect the combustibility and pollutants substantially. Low soot, CO, and UHC emissions can be accomplished with large injection angles of inclusion but have some NOx regulation drawbacks. The effect of the injection pressure on the combustion as well as mixing processes were simulated. Higher NOx emissions were observed at higher injection speeds. The injection velocity for cases of late second injection plays an important role in the improvement of soot emissions.

The LTC mode with primary reference fuel and ethanol were investigated using CFD. In a split injection technique similar to partial fuel stratification, more of the fuel was injected into a mixed load early at the beginning of the intake stroke, while a part of the load was injected directly near to ignition to trigger induced thermal stratification and equivalence ratios. This provides better control over the LTC combustion process by constraining the auto-ignition process through mixture stratification. The peak heat release rate and elongated range of the combustion process were able to achieve with split injection strategy due to the enhanced stratification. Raising fuel splitting by late injection reduces the peak thermal release rate and increases the time of burn by increasing the stratification of the mixture [20].

1.4 Reactivity Controlled Compression Ignition (RCCI)

A dual-fuel combustion engine that gives better control over the combustion process and has the ability to reduce fuel consumption and emissions drastically defines the RCCI engine. It uses an in-cylinder fuel alongside nothing less than two fuels of different reactivity and multiple injections to be able to control the reactivity and optimize the combustion phasing, magnitude, and duration. By choosing the reactivity of each fuel in the right manner, their fuel charges, relative amounts, timing, and combustion can be molded to reach the optimal power output at controlled temperatures like NOx along with controlled equivalence ratios. An inline 4-cylinder VW TDI engine with EGR and a 1.9 L cylinder were used for this study. An important conclusion that could be taken from this study was the relationship between the increase in GIE and the delay in combustion phasing which deserves attention. An increase in the PR to 85% and above at later stages, however, resulted in a drop in GIE and a rise in the CA50, which is significant. Also, we can conclude that when the NG in charge is more, lower temperatures in the reactive zones can be observed, which as a result produces more CO emissions. Also, when the SOI2 is being retarded, HC and CO are increased, and the NOx emission is decreased [21]. When the SOI1 is retarded toward the TDC in the split injection strategy, the combustion was seen to increase because of the local fuel reactivity mixture. Also, due to this, the NOx increases, and HC and CO decrease.

A numerical tool is used to simulate the RCCI combustion. Combustion characteristics, the performance of the engine, and emissions are determined in this study, fueled with gasoline/biodiesel. SOI timing is an important factor to shape the combustion mode as it determines the combustion core site in the cylinder. If the fuel injected through the 1st pulse is high, it might result in some high fuel deposition. However, if the fuel injected is less, it will evaporate and involve itself in the combustion process. CONVERGE CFD tool was used in this study to run the simulations of the RCCI engine. CO emission was seen to be increasing with increasing natural gas mass fraction. The CO emission is fundamentally related to two factors; they are formation and oxidation. The HC emission is seen to be increasing with an increase in the natural gas mass fraction. There are three factors, namely oxygen concentration, residence time, and gas temperature affect the NOx emission in the RCCI combustion. Moreover, the shorter duration of reaction time also impacts HC and CO oxidation. The combustion temperatures and the local fuel reactivity are increased when the local air equivalence ratio is increased [22].

A Kirloskar AV 1 single cylinder which is commercially available was modified to achieve a dual-fuel RCCI at a rated speed and different loads. Reduction in the in-cylinder temperature is caused by multiple injections, latent heat, and EGR which is lower than that of the NOx emission [23]. The emissions of CO and HC are seen to be lower when the injections are multiple when compared to a single injection. A noteworthy reduction of 91.5% was observed in the NOx emissions with EGR and multiple injections and when the fraction of n-pentanol increases from 20 to 40 percent. A single-cylinder, high-speed, direct-injection AVL diesel engine was modified accordingly to run a dual-fuel operation with CNG [24]. 3D CFD software CONVERGE was the one that was used in this study. Experiments were carried out with a Pin of exactly 0.8 bar, and the substitution rate was at 70%. The comparisons between the experiment and simulation were made based on in-cylinder pressure and heat release rate. It was observed that the simulation results clearly matched the experimental data. Also, the NOx concentrations distribution was observed to be consistent with the equivalence ratio and the temperature distributions. In the lean mixture, because of the low load condition the region in the cylinder with a slightly high fuel/air equivalence ratio is rich in oxygen, hence high in NOx formation. RCCI combustion was observed to obtain much fewer NOx emissions as expected as compared to the CDF combination. However, RCCI tends to yield higher CO emissions. As observed that RCCI generates way fewer NOx emissions [25].

2 Conclusion

Several LTC innovations and technologies have developed in recent decades and attracted worldwide considerable recognition from engineering researchers. The combustion control over a limited range of load, increased noise from combustion, and comparatively higher emissions of CO and HC are the various significant challenges for LTC. These issues were addressed by augmenting the split injection strategy with LTC. This paper helps in selecting the optimal dwell period and proportionality in each type of LTC for achieving the emission reductions as well as enhancing the performance. All the studies described in this paper have shown a vital reduction in emissions in LTC mode with a split injection strategy, and the engine output is also very similar to traditional combustion engines thus without any compromise. These strategies can also be more adapted to support LTC for industrial applications including automobiles.

References

1. Johnson TV (2009) Diesel emission control in review. SAE Int J Fuels Lubr 1(1):68–81. https://doi.org/10.4271/2008-01-0069
2. Singh AP, Kumar V, Agarwal AK (2020) Evaluation of comparative engine combustion, performance and emission characteristics of low temperature combustion (PCCI and RCCI) modes. Appl Energy 278(May):115644. https://doi.org/10.1016/j.apenergy.2020.115644
3. Hountalas DT, Mavropoulos GC, Binder KB (2008) Effect of exhaust gas recirculation (EGR) temperature for various EGR rates on heavy duty DI diesel engine performance and emissions. Elsevier 33:272–283. https://doi.org/10.1016/j.energy.2007.07.002
4. Okude K, Mori K, Shiino S, Yamada K, Matsumoto Y (2018) Effects of multiple injections on diesel emission and combustion characteristics 2007(724)
5. Sae technical a study of pilot injection in a DI diesel engine, no 724 (2018)
6. Singh AP, Agarwal AK (2018) Low-temperature combustion: an advanced technology for internal combustion engines

7. Akihama K, Takatori Y, Inagaki K, Sasaki S, Dean AM (2001) Mechanism of the smokeless rich diesel combustion by reducing temperature. SAE Tech Pap 724:2001. https://doi.org/10.4271/2001-01-0655
8. Mosbach S et al (2007) Dual injection homogeneous charge compression ignition engine simulation using a stochastic reactor model. Int J Engine Res 8(1):41–50. https://doi.org/10.1243/14680874JER01806
9. Coskun G, Soyhan HS, Demir U, Turkcan A, Ozsezen AN, Canakci M (2014) Influences of second injection variations on combustion and emissions of an HCCI-DI engine: experiments and CFD modelling. Fuel 136:287–294. https://doi.org/10.1016/j.fuel.2014.07.042
10. Wang Z, Shuai SJ, Wang JX, Tian GH (2006) A computational study of direct injection gasoline HCCI engine with secondary injection. Fuel 85(12–13):1831–1841. https://doi.org/10.1016/j.fuel.2006.02.013
11. Lee C, Chung J, Lee K (2017) Emission characteristics for a homogeneous charged compression ignition diesel engine with exhaust gas recirculation using split injection methodology. Energies 10(12). https://doi.org/10.3390/en10122146
12. Kim MY, Lee CS (2007) Effect of a narrow fuel spray angle and a dual injection configuration on the improvement of exhaust emissions in a HCCI diesel engine. Fuel 86(17–18):2871–2880. https://doi.org/10.1016/j.fuel.2007.03.016
13. Turkcan A, Ozsezen AN, Canakci M (2013) Effects of second injection timing on combustion characteristics of a two stage direct injection gasoline-alcohol HCCI engine. Fuel 111:30–39. https://doi.org/10.1016/j.fuel.2013.04.029
14. Horibe N, Harada S, Ishiyama T, Shioji M (2009) Improvement of premixed charge compression ignition-based combustion by two-stage injection. Int J Engine Res 10(2):71–80. https://doi.org/10.1243/14680874JER02709
15. Pandian MM, Anand K (2018) Comparison of different low temperature combustion strategies in a light duty air cooled diesel engine. Appl Therm Eng 142(July):380–390. https://doi.org/10.1016/j.applthermaleng.2018.07.047
16. Kim TY, Lee SH, Combustion and emission characteristics of wood pyrolysis oil-butanol blended fuels in a di diesel engine. Int J 13(2):293–300
17. Kook S, Bae C (2004) Combustion control using two-stage diesel fuel injection in a single-cylinder PCCI engine. SAE Tech Pap 724:2004. https://doi.org/10.4271/2004-01-0938
18. Horibe N, Annen T, Miyazaki Y, Ishiyama T (2010) Heat release rate and NOx formation process in two-stage injection diesel PCCI combustion in a constant-volume vessel. SAE Tech Pap. https://doi.org/10.4271/2010-01-0608
19. Peng Z, Liu B, Wang W, Lu L (2011) CFD investigation into diesel PCCI combustion with optimized fuel injection. Energies 4(3):517–531. https://doi.org/10.3390/en4030517
20. Boldaji MR, Sofianopoulos A, Mamalis S, Lawler B (2018) A CFD investigation of the effects of fuel split fraction on advanced low temperature combustion: comparing a primary reference fuel blend and ethanol. Front Mech Eng 4(July):1–12. https://doi.org/10.3389/fmech.2018.00006
21. Poorghasemi K, Saray RK, Ansari E, Irdmousa BK, Shahbakhti M, Naber JD (2017) Effect of diesel injection strategies on natural gas/diesel RCCI combustion characteristics in a light duty diesel engine. Appl Energy 199:430–446. https://doi.org/10.1016/j.apenergy.2017.05.011
22. Li J, Ling X, Liu D, Yang W, Zhou D (2018) Numerical study on double injection techniques in a gasoline and biodiesel fueled RCCI (reactivity controlled compression ignition) engine. Appl Energy 211(November 2017):382–392. https://doi.org/10.1016/j.apenergy.2017.11.062
23. Paykani A, Kakaee AH, Rahnama P, Reitz RD (2015) Effects of diesel injection strategy on natural gas/diesel reactivity controlled compression ignition combustion. Energy 90(x):814–826. https://doi.org/10.1016/j.energy.2015.07.112
24. Sabu VR, Thomas JJ, Nagarajan G (2020) Experimental investigation on the effects of multiple injections and EGR on: N-pentanol-biodiesel fuelled RCCI engine. RSC Adv 10(49):29498–29509. https://doi.org/10.1039/d0ra03723k
25. Meng X, Meng S, Cui J, Zhou Y, Long W, Bi M (2020) Throttling effect study in the CDF/RCCI combustion with CNG ignited by diesel and diesel/biofuel blends. Fuel 279(June):118454. https://doi.org/10.1016/j.fuel.2020.118454

Selection of Suitable Material for the Solar Panel Using SWARA and COPRAS Method

Chandra Prakash Pandit, Mukesh Chandra, Aashutosh Choubey, and Sonu Rajak

Abstract Renewable energy resources are becoming a major demanding area in recent years due to alarming global warming and climate change. This adverse situation demands minimal use of fossil fuel and maximum use of renewable energy including sunlight, wind, tides, biomass, hydropower, etc. The use of renewable energy resources also helps in overall sustainable development. This research work uses a hybrid approach of multiple-criteria decision making (MCDM) consisting of stepwise weight assessment ratio analysis (SWARA) and complex proportional assessment (COPRAS) for selection suitable material for the solar panel. The study has taken into consideration of twelve important criteria that affect the solar panel material and three alternatives of solar panel material available. The result shows that silicon metal is the most suitable material for solar panels compared to germanium and gallium phosphate. Consequently, this research could help solar panel manufacturers for proper selection of solar panel material and hence, effective utilization of renewable energy.

Keywords Material selection using MCDM · SWARA · COPRAS · Solar panel · Renewable energy

Abbreviations

$C_1, C_2, C_3, \ldots,$	Different criteria
P_j	Relative average
S_j	Comparative significance value
C_j	Coefficient value
S'_j	Corrected weight

C. P. Pandit · M. Chandra (✉) · A. Choubey
Department of Production and Industrial Engineering, BIT Sindri, Dhanbad 828123, India
e-mail: mchandra018@gmail.com

S. Rajak
Department of Mechanical Engineering, National Institute of Technology, Patna 800005, India

A. Maurya et al. (eds.), *Recent Trends in Mechanical Engineering*, Lecture Notes in Mechanical Engineering, https://doi.org/10.1007/978-981-19-7709-1_15

W_j	Final criteria weight
$\tilde{X}_{ij}$	Normalized value
D_{ij}	Weighted normalized value matrix;
Q_i,	Elative importance weight
S_{i+},	Sum of X_j, for useful criteria
S_{i-},	Sum of X_j, for useless criteria
P_i	Performance index

1 Introduction

As we have completed two decades of the twenty-first century, we must start using our resources efficiently and sustainably. For every human activity, energies are the prerequisite either in household applications or in transportation or all industrial activities. The non-renewable form of energy, i.e., fossil fuels, have served most of the energy demanded by humans for many decades. As a result, the environment got harshly affected. Climatic change-related phenomena like depletion in the ozone layer, melting of the iceberg, global warming, pollution, etc., are increasing rapidly. Decreasing the emission of greenhouse gases is an urgent necessity, and we will only achieve this by using renewable forms of energy. All these factors are responsible for shifting to renewable energy from non-renewable energy. The main benefits of using renewable energy resources, i.e., wind energy, geothermal energy, solar energy, hydropower, etc., are that they are found in abundance and consumption of these resources do not cause any adverse effect on the environment. Most importantly they help in sustainable development [1].

The multiple-criteria decision-making technique is a discipline of operation research that explicitly evaluates multiple conflicting criteria in decision making [2]. There are different methods of MCDM with different characteristics and approaches. They include deterministic, stochastic, or fuzzy multi-objective decision making (or MODM) and multi-attribute decision making (or MADM) methods [3].

Many researchers have used different MCDM approach for different industrial application and selection of suitable renewable energy resource. Arunodaya et al. [4] carried out novel multi-criteria intuitionistic fuzzy SWARA–COPRAS method for the selection of alternatives for the bioenergy production process. Yücenur et al. [5] used an integrated approach of SWARA and COPRAS methods for the selection of suitable cities in Turkey for biogas production. Maghsoodi et al. [6] presented an application of a hybrid MADM approach based on the SWARA approach with a hierarchical arrangement combined with the multi-objective optimization based on ratio analysis plus the full multiplicative form (MULTIMOORA) for selection of renewable energy technology. Later on, Maghsoodi et al. [7] used a hybrid approach of SWARA method and combinative distance-based assessment (CODAS) technique for the selection of dam construction material. Huang et al. [8] used TOPSIS method for the materials selection challenge for environmentally conscious design (ECD).

Vinodh et al. [9] used hybrid fuzzy-based MCDM approach for selection of Lean Six Sigma (LSS) project for the automotive component manufacturing industry and PROMETHEE for selection of the best sustainable concept. Mousavi et al. [10] used COPRAS and TOPSIS for the selection of tool material and suggested that these two methods are efficient MCDM methods for solving general material selection problems. Anojkumar et al. [11] carried out a comparative analysis of MCDM for the selection of the material of pipe for the sugar industry. Halil et al. [12] investigated PROMETHEE II, TOPSIS and VIKOR for the selection of material for the tool holder for milling.

In this research paper, solar energy which is the most important renewable energy source today [13], is considered for the study. Solar panel manufacturers face problems in the selection of proper material for the solar panel because of its unique properties. A hybrid approach of MCDM techniques including SWARA and COPRAS is used to find the most suitable material for the solar panel. The criteria that affect sustainable development and reduce environmental negative impact have been included in the study. The remaining paper is organized in the following manner. Section 2 explains the methodology, application of the integrated SWARA and COPRAS method calculations involved in the investigation. Section 3 includes the summary of the result and discussion, and Sect. 4 summarizes and concludes the work.

2 Methodology

a. SWARA Method

The stepwise weight assessment ratio analysis (SWARA) method is a multiple attribute decision-making (MADM) approach that provides a significant ratio of the attributes in the process of rational decision determination [14].

Steps involved in SWARA method

Step 1: Decision-makers (DMs) arrange the criteria, according to their priority by giving points (max. point 1.00 and steeply decreasing by 0.05). The points assigned to the criteria is written as

$$p_j^k, j = 1, k = 1, 1; 0 \leq p_j^{(k)} \leq 1$$

Step 2: Calculation of relative average (P_j) for all criteria using the Eq. (1).

$$P_j = \frac{\sum_{k=1}^{l} p_j^k}{l}, j = 1, 2, 3, 4 \ldots N \quad (1)$$

Step 3: Arranging all the P_j in decreasing order. In addition, a separate column is prepared which indicates comparative significance value (S_j), calculated using Eq. (2).

$$P_j = S_{j-1} - S_j \tag{2}$$

Step 4: Calculate coefficient value (C_j) using Eq. (3). The C_j value with the greatest S_j value is, $C_j = 1$

$$C_{j=}S_j + 1 \tag{3}$$

Step 5: Calculate corrected weight (S'_j) using Eq. (4). Value of S'_j in the first rank is one.

$$S'_J = \frac{S'_{j-1}}{C_j} \tag{4}$$

Step 6: Calculate final criteria weight (W_j) using Eq. (5).

$$W_j = \frac{S'_j}{\sum_{j=1}^{n} S'_j},\ j = 1, 2, 3 \ldots n \tag{5}$$

b. **COPRAS Method**

Complex proportional assessment (COPRAS) method was introduced by Zavadskasin [15]. It is an effective method in selecting alternatives based on their significance and degree of utility.

Steps involved in COPRA methods

Step 1: Evolution of decision matrix (D) having each alternative ($A_1\ A_2\ A_3$), and criteria ($C_1\ C_2 \ldots\ldots\ldots\ldots C_{12}$) are arranged in the matrix, shown in Eq. (6).

$$\text{D}= \begin{matrix} A_1 \\ A_2 \\ \\ A_3 \end{matrix} \begin{bmatrix} X_{11} & X_{12} & \ldots X_{1n} \\ X_{21} & X_{22} & \ldots X_{2n} \\ \vdots & \vdots & \vdots \\ X_{m1} & X_{m2} & \ldots X_{mn} \end{bmatrix} \tag{6}$$

Step 2: Normalized value of each criterion is obtained by Eq. (7) and are arranged in the matrix D_{ij} as in Eq. (8) normalized value ($\tilde{X}_{ij}$) of alternative (i) according to criterion (j), ($i = 1, 2, \ldots, m$ and $j = 1, 2, \ldots, n$).

$$\tilde{X}_{ij} = \frac{X_{ij}}{\sum_{i=1}^{m} \mathrm{X_{ij}}} \tag{7}$$

$$D_{ij}D = \begin{array}{c} \\ A_1 \\ A_2 \\ \vdots \\ A_m \end{array} \begin{array}{c} \begin{array}{cccc} C_1 & C_2 & \ldots & C_n \end{array} \\ \begin{bmatrix} \tilde{X}_{11} & \tilde{X}_{12} & \ldots & \tilde{X}_{1n} \\ \tilde{X}_{21} & \tilde{X}_{21} & \ldots & \tilde{X}_{2n} \\ \vdots & \vdots & & \vdots \\ \tilde{X}_{m1} & \tilde{X}_{m2} & \ldots & \tilde{X}_{mn} \end{bmatrix} \end{array} \quad (8)$$

Step 3: Multiply weight of criteria (W_j) obtained from SWARA method of Eq. (5) by normalized decision matrix. The D_{ij} of alternatives i according to criteria j ($i = 1, 2\ldots m$ and $j = 1, 2,\ldots n$).

$$D = \begin{array}{c} \\ A_1 \\ A_2 \\ \vdots \\ A_m \end{array} \begin{array}{c} \begin{array}{cccc} C_1 & C_2 & \ldots & C_n \end{array} \\ \begin{bmatrix} D_{11} & D_{12} & \ldots & D_{1n} \\ D_{21} & D_{22} & \ldots & D_{2n} \\ \vdots & \vdots & & \vdots \\ D_{m1} & D_{m2} & \ldots & D_{mn} \end{bmatrix} \end{array} \quad (9)$$

Step 4: Categorize the useful criteria as the maximum value and the useless criteria as the minimum value. Subsequently, obtain a matrix D^* and then re-write the column comprised of useful to useless according to their value in Eq. (10).

$$D_{ij}^* = \begin{array}{c} \\ A_1 \\ A_2 \\ \vdots \\ A_m \end{array} \begin{array}{c} \begin{array}{ccccccccc} C_{max1} & C_{max2} & \ldots & C_{maxk} & \ldots & C_{\min(k+1)} & \ldots & C_{\min n} \end{array} \\ \begin{bmatrix} D_{11} & D_{12} & \ldots & D_{1k} & \ldots & D_{1(k+1)} & \ldots & D_{1n} \\ D_{21} & D_{22} & \ldots & D_{2k} & \ldots & D_{2(k+1)} & \ldots & D_{2n} \\ \vdots & \vdots & & \vdots & & \vdots & & \vdots \\ D_{m1} & D_{m2} & & D_{mk} & & D_{m(k+1)} & & D_{mn} \end{bmatrix} \end{array} \quad (10)$$

Step 5: Calculate the sum of X_{ij} for useful criteria (S_{i+}) and the sum of X_{ij} for useless criteria (S_{i-}) using Eqs. (11) and (12), respectively.

$$S_{i+} = \sum_{j=1}^{k} D_{ij}, \ j = 1, 2, \ldots, k \quad (11)$$

$$S_{i-} = \sum_{j=(k+1)}^{n} D_{ij} \ j = k+1, \ k+2, \ \ldots, n \quad (12)$$

Step 6: Calculate relative importance weight (Q_i) using Eq. (13).

$$Q_i = S_{i+} + \frac{\sum_{i=1}^{m} S_{i-}}{S_{i-} \times \sum_{i=1}^{m} \frac{1}{S_{i-}}} \quad (13)$$

Step 7: The best value is chosen among all the alternatives whichever is having the highest value ($Q_{\max}$).

$$Q_{\max} = \max_i\{Q_i\}, i = 1, 2, 3 \ldots m \tag{14}$$

Step 8: Calculate performance index (P_i) for each alternative using Eq. (15).

$$P_i = \frac{Q_i}{Q_{\max}} \times 100\% \tag{15}$$

c. **Application of the integrated SWARA and COPRAS method**

The objective of this research is set to achieve, with the help of a hybrid approach of SWARA and COPRAS methods. The ranking and weight of twelve criteria taken into consideration for the study were evaluated using SWARA. In SWARA method, decision-makers play an important role in deciding about the rank of the criteria [16] and subsequently weights for different criteria are calculated. For this, we have selected four experts as DMs working in the area of renewable energy from the industry as well as the academia. A set of questionnaires was generated to take input from the experts. The benefit of the integrated approach of SWARA and COPRAS method is that the calculation merges in one process. Hence, the two MCDM techniques combine to get the results. In this calculation, steps 1–6 involves SWARA technique. Steps 7 to 11 involves COPRAS technique.

Twelve criteria and their designation are as follows; C1: availability; C2: cost; C3: power generation capacity; C4: malleability, C5: thermal conductivity; C6: forbidden energy gap; C7: e-waste; C8: corrosion; C9: recycling; C10: efficiency; C11; working environment and C1: effect on the environment. Similar three alternatives along with their designation are as follows A_{Si}: silicon; A_{Ge}: germanium and $A_{\mathrm{Ga}(p)}$: gallium phosphate. The detailed calculation stepwise of the integrated MCDM approach is as follows:

Step 1: Ranking of criteria based on their importance by DMs, shown in Table 1.

Step 2: DMs assign the score of different criteria and P_j importance is calculated.

Step 3: Arrangement of ranking of criteria. Determination S_j, represented in Table 2.

Step 4: Determination of C_j value for all criteria using Eq. (3), represented in Table 2.

Step 5: Determination of $S_j{}'$ by using Eq. (4), represented in Table 2.

Step 6: Determination of W_j by using Eq. (5), represented in Table 2.

Step 7: Create a matrix D having a maximum and minimum value of criteria, shown in Table 3.

Step 8: Normalized decision matrix is obtained using Eq. (8), shown in Table 4.

Step 9: Criteria were classified as useful and useless and their sum of the values in the weighted normalization decision matrix was obtained by Eqs. (11) and (12), shown in Table 5.

Table 1 Criteria rankings by decision-makers

Criteria	DM 1	DM 2	DM 3	DM 4
C1	4	4	5	4
C2	2	1	1	5
C3	1	3	3	3
C4	6	9	2	10
C5	3	5	9	6
C6	5	6	10	1
C7	8	11	8	7
C8	9	8	7	9
C9	7	10	4	8
C10	10	7	11	2
C11	12	12	6	11
C12	11	2	12	12

Table 2 Values of C_j, S'_j and W_j of the criteria

Criteria	Average	S_j	C_j	S'_j	W_j
C2	0.94	–	1	1	0.101
C3	0.93	0.01	1.01	0.99	0.101
C1	0.84	0.09	1.09	0.91	0.092
C6	0.78	0.06	1.06	0.0.86	0.087
C5	0.76	0.02	1.02	0.84	0.085
C4	0.71	0.05	1.05	0.80	0.081
C9	0.69	0.02	1.02	0.79	0.080
C10	0.68	0.01	1.01	0.78	0.079
C8	0.64	0.04	1.04	0.75	0.076
C7	0.63	0.01	1.01	0.74	0.075
C12	0.59	0.04	1.04	0.71	0.072
C11	0.54	0.05	1.05	0.68	0.069

Table 3 Values for each criterion for the selection of the most suitable alternative

	C1 max	C2 min	C3 max	C4 max	C5 max	C6 min	C7 min	C8 min	C9 max	C10 max	C11 max	C12 min
Wj	0.092	0.101	0.101	0.081	0.085	0.087	0.075	0.076	0.080	0.079	0.069	0.072
A_{si}	90	25	85	80	85	30	25	15	80	80	80	35
A_{Ge}	75	60	90	85	70	25	30	20	75	70	85	40
A_{ga}	65	40	70	65	65	40	35	25	70	65	75	25

Table 4 Normalized decision matrix

	C1	C2	C3	C4	C5	C6	C7	C8	C9	C10	C11	C12
A_{Si}	0.391	0.200	0.347	0.348	0.386	0.316	0.278	0.250	0.356	0.372	0.333	0.350
A_{Ge}	0.326	0.480	0.367	0.370	0.318	0.263	0.333	0.333	0.333	0.326	0.354	0.400
$A_{Ga(p)}$	0.283	0.320	0.286	0.283	0.295	0.421	0.389	0.417	0.311	0.302	0.313	0.250

Table 5 Separation of useful and useless criteria of to the left and right, respectively, of the decision matrix

	C1	C3	C4	C5	C9	C10	C11	C2	C6	C7	C8	C12
A_{Si}	0.036	0.035	0.028	0.033	0.028	0.029	0.023	0.020	0.027	0.021	0.019	0.025
A_{Ge}	0.030	0.037	0.030	0.027	0.027	0.026	0.024	0.048	0.023	0.025	0.025	0.029
$A_{Ga(p)}$	0.026	0.029	0.023	0.025	0.025	0.024	0.022	0.032	0.037	0.030	0.032	0.018

Table 6 Relative importance weight and ranking of the alternatives

Relative importance weight				Pi and the ranking			
	A_{Si}	A_{Ge}	$A_{Ga(p)}$		A_{Si}	A_{Ge}	$A_{Ga(p)}$
Q_i	0.377	0.324	0.298	P_i	100.00	85.94	79.05
				Ranking	1	2	3

Step 10: The Q_i was calculated using Eq. (13), shown in Table 6.

Step 11: Relative importance weight and ranking of the alternatives were obtained using Eqs. (14) and (15), respectively, shown in Table 6.

3 Result and Discussion

Proper selection of a material for a specific application is becoming a challenging task these days because of the discovery of new materials. The newer materials have better capabilities, performance, efficiency, more flexibility, etc., and hence result in overall customer satisfaction. This study was carried to select the suitable material for the solar panel for its better performance, efficiency and durability. The three alternatives of materials were silicon, germanium and gallium phosphate. An integrated MCDM approach of SWARA and COPRAS were used for raking the criteria and calculating their final weights, respectively.

Table 2 shows the ranking of the criteria given by the decision-makers. The panel cost and generation of electricity are the major concern for solar paper these days, and hence, decision-makers in the majority have given these two criteria top rank. Table 9 shows the relative importance of weight obtained with COPRAS. The relative importance weight in descending order is silicon, germanium and gallium phosphate having values 0.377, 0.324 and 0.298, respectively. The performance index obtained for gallium phosphate is 79.05, germanium is 85.94 and Silicon is 100. Many authors carried a similar study of the selection of material for different applications and fields including green material selection for sustainable development [17], selection of pipe material for the sugar industry [11], material selection for engineering components [18], etc. The common in all these studies is that they provide a very good theoretical background of material selection in a specific application.

4 Conclusion

Energy is the prior requirement and is increasing its demand regularly. In this research, we focused on the use of renewable energy particularly solar energy. The increasing demand for solar energy could be met only when an effective solar panel material is used in the solar power device. Using a hybrid MCDM approach of SWARA and COPRAS, the best suitable material for the solar panel is investigated. Among the three selected alternatives, silicon is found to be the most suitable material for the solar panel. This could help the solar panel manufacturers to easily decide the best material available for effective uses of solar energy. Hence, the surplus demand for renewable energy in the future can be fulfilled. The limitation of this investigation is that the study has taken into consideration of only three materials and twelve criteria. Therefore, future research may be carried out with different available materials that may also be sustainable for a solar panel.

References

1. Østergaard PA, Duic N, Noorollahi Y, Mikulcic H, Kalogirou S (2020) Sustainable development using renewable energy technology. Renew Energy 146:2430–2437
2. Zavadskas EK, Turskis Z (2011) Multiple criteria decision making (MCDM) methods in economics: an overview. Technol Econ Dev Econ 17(2):397–427
3. Triantaphyllou E, Shu B, Sanchez SN, Ray T (1998) Multi-criteria decision making: an operations research approach. Encycl Electr Electron Eng 15:175–186
4. Ghoushchi SJ, Rahman MNA, Raeisi D, Osgooei E, Ghoushji MJ (2020) Integrated decision-making approach based on SWARA and GRA methods for the prioritization of failures in solar panel systems under Z-information. Symmetry (Basel) 12(2)
5. Yücenur GN, Çaylak Ş, Gönül G, Postalcıoğlu M (2020) An integrated solution with SWARA&COPRAS methods in renewable energy production: city selection for biogas facility. Renew Energy 145:2587–2597
6. Ijadi Maghsoodi A, Ijadi Maghsoodi A, Mosavi A, Rabczuk T, Zavadskas E (2018) Renewable energy technology selection problem using integrated H-SWARA-MULTIMOORA approach. Sustainability 10(12):4481
7. Maghsoodi AI, Maghsoodi AI, Poursoltan P, Antucheviciene J, Turskis Z (2019) Dam construction material selection by implementing the integrated SWARA-CODAS approach with target-based attributes. Arch Civ Mech Eng 19(4):1194–1210

8. Huang H, Zhang L, Liu Z, Sutherland JW (2011) Multi-criteria decision making and uncertainty analysis for materials selection in environmentally conscious design. Int J Adv Manuf Technol 52(5–8):421–432
9. Groeneveld TAS (2010) Lean Six Sigma project selection using hybrid approach based on fuzzy DEMATEL–ANP–TOPSIS. Eletron Libr 34(1):1–5
10. Mousavi-Nasab SH, Sotoudeh-Anvari A (2017) A comprehensive MCDM-based approach using TOPSIS, COPRAS and DEA as an auxiliary tool for material selection problems, vol 121. Elsevier Ltd.
11. Anojkumar L, Ilangkumaran M, Sasirekha V (2014) Comparative analysis of MCDM methods for pipe material selection in sugar industry. Expert Syst Appl 41(6):2964–2980
12. Çalişkan H, Kurşuncu B, Kurbanoĝlu C, Güven ŞY (2013) Material selection for the tool holder working under hard milling conditions using different multi criteria decision making methods. Mater Des 45:473–479
13. Letcher TM (2018) Why solar energy? In: *A* Comprehensive guide to solar energy systems. Elsevier, pp. 3–16
14. Keršuliene V, Zavadskas EK, Turskis Z (2010) Selection of rational dispute resolution method by applying new step-wise weight assessment ratio analysis (SWARA). J Bus Econ Manag 11(2):243–258
15. Zavadskas EK, Kaklauskas A, Vilutiene T (2009) Multicriteria evaluation of apartment blocks maintenance contractors: Lithuanian case study. Int J Strateg Prop Manag 13(4):319–338
16. Zolfani SH, Saparauskas J (2013) New application of SWARA method in prioritizing sustainability assessment indicators of energy system. Eng Econ 24(5):408–414
17. Tian G, Zhang H, Feng Y, Wang D, Peng Y, Jia H (2018) Green decoration materials selection under interior environment characteristics: a grey-correlation based hybrid MCDM method. Renew Sustain Energy Rev 81:682–692
18. Ilangkumaran M, Avenash A, Balakrishnan V, Kumar SB, Raja MB (2013) Material selection using hybrid MCDM approach for automobile bumper. Int J Ind Syst Eng 14(1):20–39

Production of Sunflower Biodiesel as an Alternative Fuel for Compression Ignition Engine: A Review

Chamala Vaishnavi, Naveen Raj Srinivasan, and Bhisham Kumar Dhurandher

Abstract The fuels derived from conventional sources are becoming increasingly expensive on account of their growing demand and non-renewable nature. The fossil fuels find necessity in a range of industrial sectors pivotal to the economy, including but not limited to power generation, manufacturing, and transportation, which exacerbates their deleterious effects. This being the case, the search for alternative sources of fuel and their widespread integration in practice is imperative. To that end, biodiesel obtained from sunflower oil can be considered, owing to its many utilitarian qualities, like high oil content of about 40%, which leads to highly efficient yields of about 600 pounds of oil per acre. The review focuses on the wide range of production methods proposed for sunflower biodiesel, based on extant research data, and how the fuel output ensues from each of those methods, in terms of their properties, which have a bearing ultimately on their performance in a Compression Ignition (CI) Engine. Attention has been given to comparing a wide variety of catalysts employed in the transesterification process, which is predominantly used to produce the fuel. Data has been gathered on the properties exhibited by the different variants of fuel, in order to weigh up the relative merits of each option and select the one most suited to fulfill any requirement. A comprehensive research has been done to study all technical aspects involved in the production of sunflower biodiesel, in order to aid in the possible supplantation of fossil fuels by sunflower biodiesel.

Keywords Alternative fuel · Sunflower biodiesel · Production · Catalyst · Transesterification

C. Vaishnavi (✉) · N. R. Srinivasan · B. K. Dhurandher
School of Mechanical Engineering, VIT Chennai, Chennai, India
e-mail: chamalavaishnavi.j2018@vitstudent.ac.in

N. R. Srinivasan
e-mail: naveenraj.srinivasan2018@vitstudent.ac.in

B. K. Dhurandher
e-mail: bhishamkumar.d@vit.ac.in

A. Maurya et al. (eds.), *Recent Trends in Mechanical Engineering*, Lecture Notes in Mechanical Engineering, https://doi.org/10.1007/978-981-19-7709-1_16

1 Introduction

The energy resources of a country dictate and determine its economic wealth and status. Fossil fuels are a quintessential resource, used in almost all sectors of industrial activity, like agriculture, transport, power plant, and manufacturing. These petroleum-derived fuel reserves are on the wane with every passing day and almost on the verge of exhaustion. Because of their multitudinous and uninterrupted uses, coupled with the increasing rate of population and urbanization, it is predicted that the supply might no longer cater to the demand, shortly. Besides, these fuels also strain the environment by emitting monoxide (CO), carbon dioxide (CO_2), hydrocarbons (HC), and oxides of nitrogen (NOx). Among them, diesel is especially utilized for rapid industrialization, as a result of its many merits like high thermal efficiency, high compression ratio, low pumping requirement, and low fuel consumption. On the global front, India produces about 1% of the overall crude oil in use but consumes about 3.1% [1]. The current demand in India for High-Speed Diesel (HSD) is 102 billion liters, out of which 80% is covered through imports from the few, specific, highly concentrated oil reserves of the world. This demand is expected to grow by 95% by 2030 [2, 3]. This jeopardizes the country owing to the growing international voices of concern about its greenhouse gas emissions, in addition to exposing it to the risk of a foreign exchange crisis.

For all these reasons, research about alternative fuels has received increasing thrust and urgency, especially biodiesel (or Fatty Acid Methyl Ester)—a bio-decomposable, a non-toxic, and renewable variant of traditional diesel, which can be produced from indigenous feedstocks and used entirely, or in tandem with diesel, at regulated blend ratios [3]. The National Policy on Biofuels (2018) has made plans to supplant HSD using biodiesel, with a benchmark of 5% to be reached by 2030, which entails a production of 500 crore liters of biodiesel annually at that point in time [3].

In recent times, it is gaining favor by its superior properties like higher cetane number, higher flash point, and self-lubrication. Since it has high oxygen content and does not contain minerals like sulfur, it shows a low rate of emissions [5]. Another salient feature worth noting is its very high energy balance, as opposed to regular diesel, which has a negative value for the same. The American Department of Agriculture and the University of Idaho have published research which shows that for every unit of fossil energy, 5.54 units of energy in the form of biodiesel can be extracted [6]. Having said all these, it also requires mention that the compromise made between food and fuel restricts oil seeds for biodiesel consumption to not be 100% sustainable just yet. Yet, this problem is not insurmountable, thanks to the methods of utilization of food wastes for biodiesel production [7].

With the data presented in Tables 1, 2 and 3, an attempt has been made to provide a glimpse into the existing overall picture of sunflower production in India. India ranks among the top 15 producers of sunflower, at the world level, and thus, would be able to provide raw material for possible large-scale sunflower biodiesel production.

Table 1 Production of sunflower oil by top countries [8]

Rank	Country	Production (Metric Tonnes)
1	Ukraine	4,400,324
2	Russia	4,063,080
3	Argentina	931,700
4	Turkey	721,882
5	France	632,900
15	India	180,000

Table 2 India's nationwide production of sunflower oil from 2011 to 15 [8]

Season		2011–12	2012–13	2013–14	2014–15	2015–16
Karif	Area	2.602	2.740	2.481	2.052	1.578
	Production	1.474	1.463	1.541	1.109	0.663
	Yield	566	534	621	541	420
Rabi	Area	4.717	5.565	4.234	3.846	3294
	Production	3.693	3.979	3.499	3.233	2.300
	Yield	783	715	826	841	698
Total	Area	7.320	8.305	6.715	5.898	4.868
	Production	5.166	5.441	5.039	4.353	2.963
	Yield	706	655	750	738	609

Table 3 State wise distribution of sunflower production in India [8]

State	2014–15			2015–16		
	Area	Production	Yield	Area	Production	Yield
Karnataka	3.560	2.060	579	3300	1.515	459
Andhra Pradesh	0.500	0.400	800	0.270	0.230	852
Maharashtra	0.820	0.330	402	0.470	0.100	213
Odisha	0.217	0.259	1195	0.160	0.191	1189
Bihar	0.117	0.166	1428	0.114	0.162	1421
West Bengal	0.150	0.220	1467	0.120	0.170	1417

Also elucidated, is the season-wise distribution of sunflower production, the yield quantities, and their progression over the years. The major states contributing to this activity are also listed below, based on which possible facilities could be set up, if so necessitated.

2 Production Methodology

Biodiesels are a product of a chemical reaction termed "transesterification," which is essentially the reaction of vegetable oil (ester) with an alcohol, where their respective alkyl groups are interchanged. It has been identified that if the alcohol used was to be higher alcohol instead of a lower one, the biodiesel obtained would possess a higher cetane number, higher power, higher calorific value, lower emissions, lower boiling point, and more similarity overall, in levels of performance, to conventional diesel [9, 10] The oils used have been derived from a multitude of sources, like sunflower, jatropha, rice, mutton fat, microbial oils, mustard, barley, castor, soybean, cottonseed, groundnut, copra, algae, and waste oil. Yield optimization has been targeted by studying the impact of each reaction variable on the overall process. At present, a yield of 97.1% was found to be achieved with the most optimum operating conditions, viz. a reaction temperature of 60 °C, a methanol/oil ratio of 6:1, and a concentration of 1% (w/w) of catalyst, with an agitation provided for 120 min of the reaction time at Transesterification [4] process, as depicted in Fig. 1, is the most widely prevalent method of choice for biodiesel production, which is basically a chemical process that consumes an alcohol and a vegetable oil, to produce a fatty acid alkyl ester (FAAE), and glycerol, with an enhanced reaction rate by virtue of a catalyst. The FAAE part is the biodiesel, which can be derived from different kinds of oils such as the ones used for cooking, fats found in skin of animals, vegetable oils, as well as non-edible oils [12].

2.1 *Catalysts*

It is a real problem to dispose of the large quantities of used oils generated as a result of using vegetable oils for frying. An alternative to this could be their reuse in the process of biodiesel production, owing to their cost economy. Choosing the starting material as vegetable oils that originate directly from plants, for biodiesel production makes the reaction have a high conversion rate of triglyceride to methyl ester leading to reactions that are relatively faster [13].

Both heterogeneous and homogeneous catalysts have been reportedly possible to be used in the transesterification reactions [14, 15]. There are some considerations to be made while selecting either of these two kinds of catalysts for use in the transesterification process. The advantages of homogeneous catalysts lie in how they comprise plenty of active sites and spread easily, bringing about quicker completion of the reaction. They also have some undesirable characteristics such as the hassles involved in the separation of products, and the corrosive effects they show on the reactor. That being the case, heterogeneous catalysts are recommended owing to their appreciable thermal stability, relatively low preparation costs, low corrosive tendency, and the ability to be reused [16, 17].

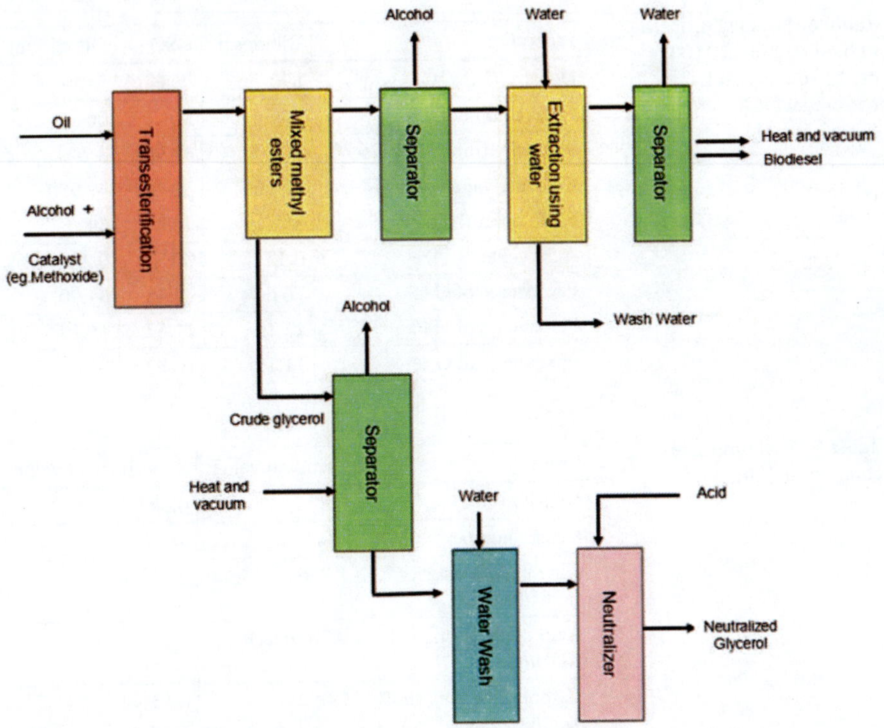

Fig. 1 Flowchart of production process [11]

2.1.1 Inedible Oil with Peroxidation Process

In this study, a comparison was made between the physical properties measured from diesel fuel and crude sunflower oil, with the source of the former being a commercial gas station, and an oil processing factory yielding the latter. The biodiesel fuel obtained from transesterification was subjected to a peroxidation process and used for measurements. As observed in Table 4, the physical properties of B100 fuel don't differ much from those of diesel. While the B100 fuel saw its viscosity and density decrease due to esterification, the heat capacity experienced an increment. Density, cetane number, viscosity, and flash point of B100 fuel were superior to those in the case of diesel, with only the calorific value being found to be less [18].

2.1.2 NaOH Catalyst in Edible Oil

Used sunflower cooking oil is converted to biodiesel by conventional transesterification process using monohydric alcohol with a catalyst of sodium hydroxide in various concentrations, to give mono-alkyl esters [19]. In this case, the best biodiesel

Table 4 Physical properties of biodiesel fuel (B100), crude sunflower oil, and diesel fuel [18]

Properties	Biodiesel	CSO	Diesel fuel
Density @ 26 °C (kg/m^3)	890	918	840
Cetane number	74	36	56
Viscosity (mm^2/s) @ 26 °C	4.5	33.98	3.2
Calorific value (kJ/kg)	40,565	39,342	42,980
Flash point (°C)	85	220	59
Acid value	0.13	0.15	0.22
Percentage of C (%)	76.66	77.46	84.90
Percentage of H (%)	12.19	11.67	15.10
Percentage of O (%)	11.15	10.87	–

Table 5 Fuel properties comparison [20]

Property	Standard value	Experimental value
Density (g/cc)	0.87	0.68
Cetane number	>47	48
Kinematic Viscosity (mm^2/s)	1.9–6.5	6
Acid number (mg KOH/g)	0.50 (max)	0.36
Carbon Residue (100% sample)	0.050	0.15

yield percentage of 80% was achieved at a temperature of 60 ± 1 °C for 1–3 h, a 6:1 by moles methanol/oil ratio, and NaOH catalyst (1%). Fuel properties analysis was carried out conferring to ASTM biodiesel standards and is displayed in Table 5. Therefore, the properties of biodiesel prepared from used vegetable oil were invariably close to those of commercial diesel. Thus, it can be concluded that sunflower vegetable oil is a good choice for large-scale production of biodiesel [20].

2.1.3 Hydrotalcites—Heterogeneous Catalysts

These catalysts are benign to the environment and are basically double-layered hydroxides. They are found to have a layered brucite-like structure [21]. Magnesium–aluminum hydroxy carbonates are the most widely used hydrotalcite as they exhibit excellent basic nature, which renders them ideal for transesterification process. Hydrotalcites also have a "memory effect" wherein they revert to their initial form once the mixed oxides interact with water [22].

A major role in this may be due to the method of preparation chosen for incorporating calcium, as well as the heat treatment designed for the support/catalyst. The number of basic sites determines the adsorption quantity of CO_2, and the strength of each basic site is determined by the temperature. A stronger basic site is indicated by a higher CO_2 desorption temperature [23]. FAME yield is visibly affected by the catalyst loading. An incomplete reaction ensues when there is insufficiency in the catalyst amount [24]. As the catalyst percentage increases, the yield increases. When the catalyst weight is increased, the number of sites of O^{2-} anion also increases, which results in increased adsorption H^+ from methanol, ultimately leading to the creation of active centers [25]. The reactants will therefore be in close contact with such active centers, thus improving the yield.

The ester conversion in response to transesterification continues to be completed if sufficient time is given [26]. The reaction begins slowly when the alcohol is dispersed into the oil, but gets accelerated over time. As the time of reaction goes on, the conversion rate increases considerably. It was experimentally found that the Ca600/Mg_4Al_2HT catalyst exhibited the best functioning as a catalyst and proved that the input of calcium is vital for the formation of active sites [27]. A number of experiments over Ca600/Mg_4Al_2HT helped to infer that the ideal conditions for production of biodiesel were as follows: A reaction time of 6 h, a catalyst/oil ratio of 2.5% by weight, and a methanol/oil ratio of 15:1 by moles, the product of which was a 95% FAME yield.

2.1.4 CaO Catalyzed with Crude Biodiesel as Cosolvent

CaO-based catalysts are being used increasingly far and wide for production of biodiesel, among the heterogeneous catalysts for they are inexpensive, extremely alkaline, capable of being prepared from natural or recyclable waste sources [28, 29]. Some of the disadvantages of CaO-based catalysts are leaching of calcium during the reaction that ruins the purity of all products, making their reusability questionable. The transesterification of sunflower oil catalyzed by CaO with cosolvent as crude biodiesel at moderate reaction conditions and atmospheric pressure was studied. The addition of biodiesel in its crude form poses certain desirable prospects such as faster separation of phases once the reaction ends [30]. Another point to note in presence of crude biodiesel was a continuous increase in the Fatty Acid Methyl Ester (FAME) content from the initial stage of the process. This can be attributed to the reactants becoming more miscible, thereby easing their way to the CaO active sites and accelerating the reaction [31–33]. Due to the positive effect that the reaction temperature has on the FAME formation and mass transfer, FAME content increased as the reaction temperature increased. The optimum conditions guaranteeing the most extreme FAME content were discovered as the accompanying: The concentration of catalyst being 0.74 mol/L, the molar proportion of methanol/oil being 7.1:1, and the reaction temperature being 52 °C [30].

2.1.5 Mg-Al-LDH Geopolymer

The transesterification of waste sunflower cooking oil, governed by a heterogeneous catalyst of an unconventional geopolymer—that was produced from synthetic

Mg/Al LDH, MCM-41, and natural metakaolin, also loaded with potassium—was performed and the results showed high Lewis's basicity ($11 < (H0) < 15$) with a value of 52 mmol of HCl/g. The expected optimal conditions via a statistical design, shown in Fig. 2, are reaction temperature of 117.5 °C, reaction time of 5 h, methanol/oil ratio of 16.4:1, and catalyst loading of 5.4% by weight, attaining a biodiesel output of 96.12%. The experimentally observed, suitable reaction conditions for this process, at the end of the study were, a reaction time of 4 h, reaction temperature of 120 °C, catalyst loading of 4% by weight, methanol/oil ratio of 15:1, all of which contributed to arrive at a biodiesel yield of 94.6%. The biodiesel produced exhibited properties adhering to the international standards of biodiesel [34].

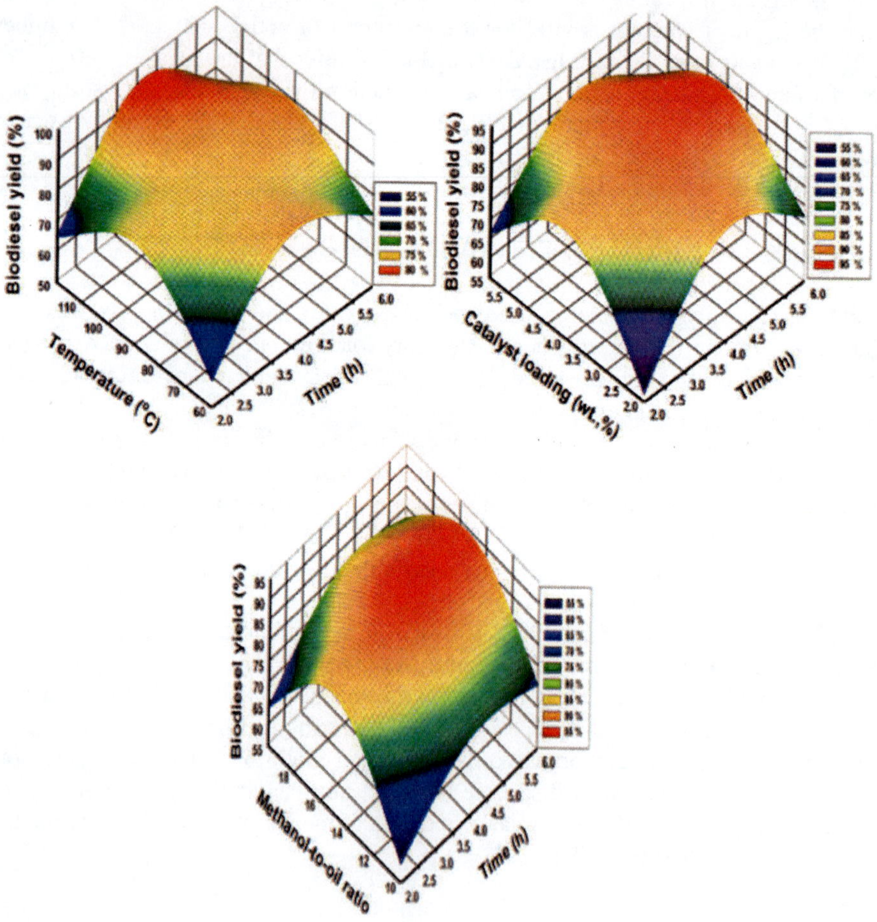

Fig. 2 Statistical design predicting biodiesel yield as a function of various parameters and finding optimal value [34]

Table 6 Reaction rate constants at different temperatures [35]

Temperature (K)	R2 value	Reaction rate constant k (min^{-1})
338	0.98709	0.01558
328	0.96467	0.00917
318	0.96943	0.0034
308	0.97152	0.00196

2.1.6 K_2CO_3 Catalyst

A biodiesel from sunflower oil was prepared by transesterification in methanol, employing a novel heterogeneous catalyst that was synthesized through the impregnation of K_2CO_3 onto Talc support in wet conditions. 40% by weight of K_2CO_3 species fabricated upon the Talc support showcased the best catalytic activity due to its higher basicity. Using a low catalyst loading of about 4% by weight, into the reactor and methanol/oil molar ratio of 6:1 maximum biodiesel yield of 98.4% could be achieved, being operated at 338 K. A kinetic study, whose results have been given in Table 6, proved that this transesterification reaction could be given a pseudo-first order reaction rate, with a rate constant of 0.01558 min^{-1} at 338 K [35].

Examination of the effects of calcination temperature upon the activity and properties of the 40% by weight K_2CO_3 loading catalyst was done. It was calcined at various temperatures, and the resultant catalysts and their catalytic activities were identified and tabulated in Table 7. It was observed from the table that the catalyst calcined at 823 K demonstrated the best catalytic activity and gave high yield of biodiesel while the ones calcinated at 1023 K and 723 K granted negative effects on the catalyst activity. It can be concluded from basic properties of the samples that the calcination temperature affects the basicity of the resultant catalyst, and so the sample calcined at 823 K exhibited the highest basicity. This proved to be a further evidence of its superiority in terms of catalytic activity.

Table 7 Study of calcination temperature effects upon the K_2CO_3(40%)/Talc catalyst [35]

S. no	Calcination temperature (K)	Biodiesel yield (%)	Basicity (mmol/g) [Standard error ± 0.05]	Basic strength (H_)
1	723	85.9 ± 1.7	2.63	9.8 < H < 15
2	823	98.4 ± 0.5	3.38	9.8 < H < 15
3	923	92.1 ± 1.3	3.21	9.8 < H < 15
4	1023	80.2 ± 0.9	2.24	9.8 < H < 15

3 Conclusion

In this review, studies based on a number of different methods of preparation of sunflower biodiesel have been compiled and presented. Each method has been discussed with an overview of the process requirements, the nature of process, and the expected biodiesel output. And for each method, data has been provided based on which the processes could be optimized to result in efficient performance, as well as optimum and good quality biodiesel output, wherever applicable. Comparisons of properties belonging to both sunflower biodiesel and the diesel produced from fossil fuels have been presented to infer that the biodiesel does not really differ much from the latter. This would mean that the biodiesel would not leave anything more to desire in terms of performance, in relation to conventional diesel, thereby meeting the primary requirement of any alternative fuel that aims to enter the mainstream market. An added incentive is that these alternatives do away with some of the less impressive qualities that the original features, like the toxic emissions for instance. The overall picture is that sunflower biodiesel would provide almost similar if not same performance output, with reduced gross greenhouse gas addition into the environment. It is hoped that this study proves to be efficacious in selecting a method of production for sunflower biodiesel, and in carrying it out with maximum efficiency. This work could be extended further in the direction toward seeking further improvements in the realm of alternative fuels for internal combustion engines.

References

1. Ministry of Petroleum and Natural Gas—Economics and Statistics Division, Government of India (2016). Indian Petroleum and Natural Gas Statistics
2. Hughes L (2013) Meeting India's Energy Requirements in 2030, Future Directions International—Strategic Analysis Paper
3. Ministry of Petroleum and Natural Gas, Government of India (2019). World Biofuel Day to be observed on 10th August 2019 [Press Release]
4. Adekunle AS et al (2020) Biodiesel potential of used vegetable oils transesterified with biological catalysts. Energy Rep 6:2861–2871
5. Melikoglu M (2014) Demand forecast for road transportation fuels including gasoline, diesel, LPG, bioethanol and biodiesel for Turkey between 2013 and 2023. Renew Energy 64:164–171
6. National Biodiesel Board (2020) Chapter 39—Biodiesel sustainability fact sheet, biomass to biofuels and waste to energy 2:749–752
7. Elkady MF et al (2015) Production of biodiesel from waste vegetable oil via KM Micromixer. J Chem 2015:9
8. Directorate of Oilseeds Development, Ministry of Agriculture and Farmers Welfare, Government of India (2017). Area, Production and Yield of Oilseed Crops from 2004–05 to 2015–16
9. Kumar BR, Saravanan S (2016) Use of higher alcohol biofuels in diesel engines: a review. Renew Sustain Energy Rev 60:84–115
10. Lahane S, Subramanian KA (2015) Effect of different percentages of biodiesel-diesel blends on injection, spray, combustion, performance, and emission characteristics of a diesel engine. Fuel 139:537–545

11. Caroline BC (2021) The reaction of biodiesel: transesterification, alternative fuels from biomass sources, Penn State's college of earth and mineral sciences, Creative Commons License. Accessed 2021
12. Taher H et al (2011) A review of enzymatic transesterification of microalgal oil-based biodiesel using supercritical technology. Enzyme Res 2011:25
13. Allen M (2002) Straighter than straight vegetable oils, pp 345–356
14. Vahid BR, Haghighi M (2017) Biodiesel production from sunflower oil over MgO/MgAl2O4 nanocatalyst: effect of fuel type on catalyst nanostructure and performance. Energy Convers Manag 134:290–300
15. Lani NS, Ngadi N, Yahya NY, Rahman RA (2017) Synthesis, characterization and performance of silica impregnated calcium oxide as heterogeneous catalyst in biodiesel production. J Clean Prod 146:116–124
16. Mardhiah HH et al (2017) A review on latest developments and future prospects of heterogeneous catalyst in biodiesel production from nonedible oils. Renew Sustain Energy Rev 67:1225–1236
17. Al Zoubi W et al (2020) Recent advances in hybrid organic-inorganic materials with spatial architecture for state-of-the-art applications. Prog Mater Sci 100663
18. Ilkilic C et al (2008) Biodiesel fuel obtained from sunflower oil as an alternative fuel for diesel engines, the Online Journal of Science And Technology - July 2017. Analysis of waste cooking oil a raw material for biofuel production, pp 82–83
19. Hossain ABMS, Boyce AN, Salleh A, Chandran (2010) Bio diesel production from waste soyabean oil biomass as renewable energy and environmental recycled process, pp 4323–4240
20. Thirumarimurugan M et al (2012) Preparation of biodiesel from sunflower oil by transesterification. Int J Biosci, Biochem Bioinform 2(6)
21. Ma Y, Wang Q, Zheng L, Gao Z, Wang Q, Ma Y (2016) Mixed methanol/ethanol on transesterification of waste cooking oil using Mg/Al hydrotalcite catalyst. Energy 107:523–531
22. Abello S, Medina F et al (2007) Aldol condensation of campholenic aldehyde and MEK over activated hydrotalcites. Appl Catal B Environ 70:577–584
23. Yu X, Wang N et al (2012) Carbon dioxide reforming of methane for syngas production over La-promoted NiMgAl catalysts derived from hydrotalcites. Chem Eng J 209:623–632
24. Xie W, Peng H, Chen L (2006) Calcined Mg-Al hydrotalcites as solid base catalysts for methanolysis of soybean oil. J Mol Catal A Chem 246:24–32
25. Marinkovic DM et al (2016) Calcium oxide as a promising heterogeneous catalyst for biodiesel production: current state and perspectives. Renew Sustain Energy Rev 56:1387–1408
26. Issariyakul T, Dalai AK (2014) Biodiesel from vegetable oils. Renew Sustain Energy Rev 31:446–471
27. Dahdah E et al (2020) Biodiesel production from refined sunflower oil over Ca-Mg-Al catalysts: effect of the composition and the thermal treatment. Renew Energy 146:1242–1248
28. Miladinović MR et al. Further study on kinetic modeling of sunflower oil methanolysis catalyzed by calcium-based catalysts. Chem Ind Chem Eng Q
29. Kesić Ž, Lukić I et al (2016) Calcium oxide based catalysts for biodiesel production: a review. Chem Ind Chem Eng Q 22:391–408
30. Todorovića ZB et al (2019) Optimization of CaO-catalyzed sunflower oil methanolysis with crude biodiesel as a cosolvent. Fuel 237:903–910
31. López Granados M et al (2009) Transesterification of triglycerides by CaO: increase of the reaction rate by biodiesel addition. Energy Fuel 23:2259–2263
32. Zhou H et al (2006) Solubility of multicomponent systems in the biodiesel production by transesterification of Jatropha curcas L. oil with methanol. J Chem Eng Data 51:1130–1135
33. Chueluecha N et al (2017) Enhancement of biodiesel synthesis using cosolvent in a packed-microchannel. J Ind Eng Chem 51:162–171

34. Sayed MR et al (2020) Synthesis of advanced MgAl-LDH based geopolymer as a potential catalyst in the conversion of waste sunflower oil into biodiesel: response surface studies. Fuel 282:118865
35. Zehtab Salmasi M et al (2020) Transesterification of sunflower oil to biodiesel fuel utilizing a novel K2CO3/Talc catalyst: process optimizations and kinetics investigations. Ind Crops Products 156:1128476

A Comprehensive Review of Cold Spray Coating Technique

Shailesh Kumar Singh, Somnath Chattopadhyaya, Qasim Murtaza, Shailesh Mani Pandey, R. S. Walia, Mohit Tyagi, and Satyajeet Kumar

Abstract The cold spraying, a solid-state deposition phenomenon relates to one of the thermal spray coating. It is an innovative coating technology mainly based on metals and ceramic particle's high-speed impact on different substrates in which solid powders are propagated towards a substrate. It is one of the promising technologies offering several technological advantages and utilizes kinetic. Several undesired effects can be avoided, like oxidation. For every spray material, there is a specific critical velocity below which the spray particle will not be able to form a proper bonding. The sprayed particle will adhere to the substrate due to impact and high kinetic energy. This paper presents an insight of different aspects of cold spraying process parameters related to mechanical properties and the history of the emergence of this process, and types are also reviewed.

Keywords Cold spray · Nano-composite · Metallic powder · Tribomechanical properties

S. K. Singh · S. Chattopadhyaya
Department of Mechanical Engineering, Indian Institute of Technology (ISM), Dhanbad, Jharkhand, India

Q. Murtaza
Department of Mechanical Engineering, Delhi Technological University, Delhi 110042, India

S. M. Pandey (✉) · S. Kumar
Department of Mechanical Engineering, National Institute of Technology, Patna, India
e-mail: smp.me@nitp.ac.in

R. S. Walia
Department of Production and Industrial Engineering, Punjab Engineering College Chandigarh, Chandigarh, India

M. Tyagi
Department of Mechanical Engineering, Dr. B. R. Ambedkar National Institute of Technology Jalandhar, Jalandhar, India

A. Maurya et al. (eds.), *Recent Trends in Mechanical Engineering*, Lecture Notes in Mechanical Engineering, https://doi.org/10.1007/978-981-19-7709-1_17

1 Introduction

Materials are the earth's rare resources, and it is imperative to shield for proper use. Surface treating processes like thermal spraying and cold spraying are many widespread phenomena industrialized at the Institute of Theoretical and Applied Mechanics in the mid of 1980s [1]. A wide range of materials, its alloy and composites, can be successfully deposited to the different substrate for variety of application [2]. Cold spray process also known as Cold gas dynamic spray (CGDS) in which the coating is formed by exposing the substrate at very high velocity of particle size ranges between (1–40 μm). These powder particles are accelerated by the compressed gas through supersonic jet. It is done at the lowest possible temperature by the appropriate combination of particle size, velocity, and temperature. Compressed gases with the spray particle are accelerated at a velocity of 400–1200 m/s. There is also a concept of preheating the gas at temperature of 600–1000 °C for increasing the flow velocity through the nozzle. Supersonic velocity is achieved through a converging/diverging nozzle by using compressed gas [3]. The contact time of the sprayed particle with the compressed gases is very less and gases cools rapidly. During the expansion through the nozzle, it is to be assumed that the particle temperature remains below initial gas preheat temperature for producing large deformation and high temperature and pressure, which produces a solid-state bonding. Kinetic energy of the sprayed particle is slightly below than the energy required melting the particle. Suggesting that the deposition mechanism is primarily, or perhaps entirely, a solid-state process. The most crucial factor of the cold spraying is the velocity of the in-flight particle, for each combination of substrate and coated material, there exists a critical particle velocity. Spraying can be done successfully only when the velocity of in-flight particles exceeds a certain critical velocity [4–7]. The system possesses a powder feeder including preheater with the electric heater carrying gas preheater as well as main gas preheater, gas pressure regulators, powder feeder, and spray gun. A personal computer is being used for monitoring and controlling the system. The nozzle will be used according to the application like required critical velocity and the supplied materials. The gas temperature and pressure in the pre-chamber were measured and monitored by the thermocouple and pressure gage mounted on the spray gun. There are two main types of cold spray system: one is high-pressure cold spray (HPCS) in which nano-metallic powders are injected just before to the spray nozzle throat under a high-pressure gas supply, whereas in low-pressure cold spray (LPCS) powders are injected in the diverging section of the spray nozzle under a low-pressure gas supply. Some specifications of both the system (refer Table 1) are given as [8, 9].

Table 1 Operating parameters of Low and High pressure cold spray system

Cold spray	Low-pressure cold spray (LPCS)	Gases usually O_2 or N_2
		Pressure 5–10 bar
		Preheated up to 550 °C
		Gas velocity 300–600 m/s
		Deposition efficiency max. 50%
	High-pressure cold spray (HPCS)	Gases He or N_2
		Pressure 25–30 bar
		Preheated up to 1000 °C
		Gas velocity 600–1200 m/s
		Deposition efficiency 50–90%

2 Bonding Mechanism and Impact Phenomena

In cold spraying bonding mechanism critical velocity plays a major role, at or above this velocity particle–substrate and particle–particle interface generated local adiabatic shear instabilities. Due to a particle impaction, a force is introduced which causes the shifting of material laterally due to generated shear load. Such type of shear load causes the localized shear straining under that condition leads to adiabatic shear instability. True bonding mechanism in cold spraying is not still very clear where the predominant bonding in this process is attributed to adiabatic shear. In such processing the in-flight particle undergoes plastic deformation due to impact consequently oxide film breaks at the surface and in turn, intimate conformal contact is achieved and combined with high contact pressure, promotes bonding with the target surface interaction of the particle, and substrate in this process plays major role in the bonding and effect on coating characteristics also. Cold spraying success depends upon mainly at the exact selection of the process velocity which will lie between the erosion and critical velocity [10–14].

3 Preparation of Powders

This section reviews different type of powders with combination of pure metal and intermetallic compounds for the cold spraying. In intermetallic compounds, different elements are ordered into different sites with different environment while in alloys, they substitute randomly in the crystal structure. Intermetallic compounds are different from alloys although they are both metallic phases having more than one element. Some powders which are being cold spray coated on substrate from the literature are given below: three powders were mixed so as to achieve a composition (wt %) of 72%X–8%Y–20%Z. The nano-crystalline powder was synthesized by blending three types of powders in a planetary ball mill. One of the powders was a commercially available Ni powder having 99.9% purity and 74 nm average particle size [15] Commercially available artificial diamond powder and Ni60alloy powder having average size 20–25 μm and 10–50 μm, respectively, were used for producing the diamond/Ni60 composite coating specimens. Ni60 as the binding phase is a Ni-based alloy [16]. Silicon and aluminium substrate are coated by the composite coating of Al-Al_2O_3 with the different particle size. Ceramic and metal and agglomerations are used in the ratio of 10:1 wt% and 1:1 wt%. Powders are sprayed by a specific type of nozzle, and air is used as a carrying gas instead of helium. Commercially, pure spherical titanium powders (particle size 29 mm average diameters) have been analysed by microstructure and morphology, used for cold spray coating. Hyung-Jun Kim et al. [4] revealed that powder particle of size1-50 μm are accelerated in supersonic jet at a speed of 500–000 m/s in cold spraying. Nano-composite WC–Co powders were deposited by this technique using helium and nitrogen gases. It is also concluded that nano-sized WC is better over micro-sized WC in cold spraying because higher particle velocity can be obtained with the same gas velocity [8].

4 Overview of Some Nano-Metallic Powders Used for Cold Spray Coating

Table 2 Characteristics features of cold spray coating materials with suitable area of applications

Nano-metallic powder	Specification of powder	Substrate	Application	Refs.
WC–Co	5–45 μm with preheat temperature of 200–500 according to gas	Stainless steel (SUS 304) of 5 mm thickness	Well known for their wear resistance applications	[8]
Cu powder	5 to 80 μm with an average diameter of 28 μm	Al5052, Al6063 and stainless steel 316 L	Coatings for adhesive strength test	[17]
Al2O3 particle-reinforced	Average particle size of 15 μm and 22 μm	Cast AZ91E alloy	Aerospace and aeronautic	[14]
Ni-20Cr	10–45 μm with combination of gases	T22 and SA 516 steels	Dimensional restoration and repair	[18]
NiO/Al2O3	5–50 μm in size	Stainless steel and aluminium	Dimensional restoration and repair	[18]

(continued)

Table 2 (continued)

Nano-metallic powder	Specification of powder	Substrate	Application	Refs.
Titanium powders	29 μm average diameter	Ti	Dimensional restoration and repair, medical application	[16]
Al 6061 Al 5083	Average size of 20 μm	Al 6061	Meso-scale machining	[19]
Al 2618 + Sc	5–40 μm in size	Al	Aerospace, aeronautic and automotive applications	[20]
Fe/Al	10–40 μm in size	Stainless steel	Several industrial applications for medium to high temperature	[21]
Sintered WC–12 wt.% Co powder and a Ni powder	40 μm	Mild steel substrates	Wear resistant applications particularly heavy machinery sector	[8]
Ni powder + Al_2O_3	10–50 μm	Inconel 600 substrate	Intermetallic compounds for possible structural applications	[21]
Al + Ni	45 μm and 65 μm	Substrates were Al-based alloy	Applications that need refractory materials, i.e lighting, tools, lubricants	[22]

5 Comparison of Cold Spray to Competitive Technologies and Advantages

Cold spraying offers many advantages over thermal spraying like HVOF (high velocity oxyfuel) and arc spraying, plasma spraying because it does not involve the use of high temperature source. Some important of them are given below. Microstructural change in the base material is nominal or very low because there will be small heat transfer to the coated part. Some air sensitive material like titanium and copper are cold sprayed without any material degradation because there will not be chemical reaction and in-flight material oxidation. Some materials intermetallic, nano-phase and amorphous materials, which are not amenable to conventional thermal spraying processes, can be cold sprayed easily [1, 23–26] (Table 3).

Table 3 Comparison of cold spray process to competitive technologies by listing out important process features [1, 22–26]

Process features	Arc spray	Plasma spray	HVOF	Cold spray
Bonding mechanism	Metallurgical	Metallurgical	Mechanical	Mechanical /Chemical
Max. thickness	0.1 mm	<0.5 mm	<1.5 mm	0.05–10 mm
Surface finish	2.0 μm Ra	13.0 μm Ra	1.3–2.0 μm Ra	<1 μm Ra
Deposition efficiency	55–65%	30–60%	50–70%	>95%
Wear resistance	6 mm^3	10 mm^3	27 mm^3	50mm^3
Bond strength	20–30 MPa	30–50 MPa	30–70 MPa	30–40 MPa
Spray velocity	2300 ft/sec	1500–2000 ft/sec	2200 ft/sec	2000 ft/sec
Powder feed rate	125–150 kg/hr	15 kg/hr	25 kg/hr	25–75 kg/hr
Power consumption	5–10 kW	30–100 kW	1–2 kW	5–15 kW

6 Conclusions

Cold spray gas technology (CSGT) is a new technique, but it is not the replacement of any of thermal spray coating. Cold spraying is needed to expend the application range for thermal spray coating as a greener alternative and health safety regulation and better environment. In preparing our survey, we are not focusing on the modelling phenomena and optimization parameters (like nozzle design, velocity, temperature and pressure, etc.) of cold spraying techniques. Literature is showing the gap towards the design parameters of the nozzle and mathematical modelling to optimize a greater number of parameters. Various materials have been deposited on different substrate by cold spraying from biomedical, decorative, power plants, automotive and space industry. There is also need to give the clear understanding on the spraying of hard and brittle ceramic materials. Currently boiler industry is facing many tube failure problems that causes major breakdown; we can work on the life of boiler tubes by preventing them from high temperature corrosion in sulphates-based and aggressive chlorine environment by cold spraying.

References

1. Papyrin A, Kosarev V, Klinkov S, Alkhimov A, Fomin VM (2006) Cold spray technology. Elsevier
2. Goyal T, Walia RS, Sidhu TS (2012) Surface roughness optimization of cold-sprayed coatings using Taguchi method. Int J Adv Manuf Technol 60(5–8):611–623
3. Ghelichi R, Guagliano M (2009) Coating by the cold spray process: a state of the art. Frattura ed Integrità Strutturale 3(8):30–44

4. Kim HJ, Lee CH, Hwang SY (2005) Fabrication of WC–Co coatings by cold spray deposition. Surf Coat Technol 191(2–3):335–340
5. Dykhuizen RC, Smith MF (1998) Gas dynamic principles of cold spray. J Therm Spray Technol 7(2):205–212
6. Li CJ, Wang HT, Zhang Q, Yang GJ, Li WY, Liao HL (2010) Influence of spray materials and their surface oxidation on the critical velocity in cold spraying. J Therm Spray Technol 19(1–2):95–101
7. Huang R, Ma W, Fukanuma H (2014) Development of ultra-strong adhesive strength coatings using cold spray. Surf Coat Technol 258:832–841
8. Villafuerte J (2010) Current and future applications of cold spray technology. Met Finish 108(1):37–39
9. Tinashe SE (2010) Conceptual design of a low pressure cold gas dynamic spray (LPCGDS) system, Doctoral dissertation, University of the Witwatersrand
10. Lima RS, Kucuk A, Berndt CC, Karthikeyan J, Kay CM, Lindemann J (2002) Deposition efficiency, mechanical properties and coating roughness in cold-sprayed titanium. J Mater Sci Lett 21(21):1687–1689
11. Grujicic M, Saylor JR, Beasley DE, DeRosset WS, Helfritch D (2003) Computational analysis of the interfacial bonding between feed-powder particles and the substrate in the cold-gas dynamic-spray process. Appl Surf Sci 219(3–4):211–227
12. Assadi H, Gärtner F, Stoltenhoff T, Kreye H (2003) Bonding mechanism in cold gas spraying. Acta Mater 51(15):4379–4394
13. Grujicic M, Zhao CL, DeRosset WS, Helfritch D (2004) Adiabatic shear instability based mechanism for particles/substrate bonding in the cold-gas dynamic-spray process. Mater Des 25(8):681–688
14. Stoltenhoff T, Kreye H, Richter HJ (2002) An analysis of the cold spray process and its coatings. J Therm Spray Technol 11(4):542–550
15. Yao J, Yang L, Li B, Li Z (2015) Beneficial effects of laser irradiation on the deposition process of diamond/Ni60 composite coating with cold spray. Appl Surf Sci 330:300–308
16. Ajaja J, Goldbaum D, Chromik RR (2011) Characterization of Ti cold spray coatings by indentation methods. Acta Astronaut 69(11–12):923–928
17. Koivuluoto H, Lagerbom J, Vuoristo P (2007) Microstructural studies of cold sprayed copper, nickel, and nickel-30% copper coatings. J Therm Spray Technol 16(4):488–497
18. Benenati G, Lupoi R (2016) Development of a deposition strategy in cold spray for additive manufacturing to minimize residual stresses. Procedia Cirp 55:101–108
19. Villafuerte J (ed) (2015) Modern cold spray: materials, process, and applications. Springer
20. Woo DJ, Sneed B, Peerally F, Heer FC, Brewer LN, Hooper JP, Osswald S (2013) Synthesis of nanodiamond-reinforced aluminum metal composite powders and coatings using high-energy ball milling and cold spray. Carbon 63:404–415
21. Spencer K, Fabijanic DM, Zhang MX (2009) The use of Al–Al2O3 cold spray coatings to improve the surface properties of magnesium alloys. Surf Coat Technol 204(3):336–344
22. Hall AC, Brewer LN, Roemer TJ (2008) Preparation of aluminum coatings containing homogenous nanocrystalline microstructures using the cold spray process. J Therm Spray Technol 17(3):352–359
23. Goyal T, Walia RS, Sidhu TS (2012) Study of cold spray process for coating thickness using Taguchi method. Int J Mater Manuf Process 27(2):185–192
24. Goyal T, Walia RS, Sidhu TS (2011) Low-pressure cold sprayed coatings process parameter optimization using Taguchi multi-response. Int J Surf Eng Mater Technol 1:5–11
25. Goyal T, Prince S, Walia RS, Sidhu TS (2011) Effect of nozzle geometry on exit velocity, temperature and pressure for cold spray process. Int J Mater Sci Eng 2:65–72
26. Goyal T, Sidhu TS, Walia RS. Cold sprayed copper coatings on brass: corrosion studies in simulated marine and industrial environment

A Comprehensive Review on Dynamics of Droplet Impact on Airfoil Surface and Its Adverse Impact on Airfoil Performance Characteristics

H. R. Praneeth, Amit Kumar Thakur, and P. S. Ranjit

Abstract Water droplets descend through the flow field and strike the airfoil surface during hostile conditions of torrential rainfall and icing. Rain affects aircraft performance like any other hazardous weather conditions such as gusts, precipitation, and shear winds. The increment in C_D and decrement in C_L value denotes the adverse impact of rain on aerodynamic characteristics. The review article describes the intricate mechanism of droplet dynamics which involves splashing of droplets and water film formation on airfoil surface. The changes in flow field characteristics in presence of rain due to fluid and solid interaction are also focused on. Effects of surface wettability and premature boundary layer transition due to boundary layer tripping on aerodynamic characteristics of the airfoil are visualized.

Keywords Surface wettability · Droplet–solid interaction · Splashed droplets · Water film · Airfoil

Abbreviations

C_L	Lift coefficient
DDB	Droplet deformation and breakup model
C_D	Drag coefficient
TAB	Taylor analogy breakup model
AOA	Angle of attack

H. R. Praneeth
Dayananda Sagar College of Engineering, Bangalore, India

A. K. Thakur (✉)
School of Mechanical Engineering, Lovely Professional University, Phagwara, Punjab 144411, India
e-mail: amit.25010@lpu.co.in

P. S. Ranjit
Department of Mechanical Engineering, Aditya Engineering College, Surampalem, Andhra Pradesh, India

A. Maurya et al. (eds.), *Recent Trends in Mechanical Engineering*, Lecture Notes in Mechanical Engineering, https://doi.org/10.1007/978-981-19-7709-1_18

SLD Supercooled large droplets

1 Introduction

Rain is one of the familiar meteorological conditions which causes havoc on airfoil aerodynamics. The studies conducted initially on the above subject consider that the effect of rainfall could be inculcated in the flow field by altering the air density while some of them presumed that droplets stay in the same form after they impinge onto the solid surface. In contrast to the above ideas, the experiments carried out in later stages provide evidence regarding the transformation of droplets into transient water layers as they impinge onto the solid surface [1]. A study was conducted by Thomson and Marrochello [2] to detect the spot where rivulets originate due to the flow of water over the NACA 4412 surface for a rainfall rate of 50–160 mm/hr. Wu's study contemplated the adverse effect on aircraft aerodynamics due to water droplet impact on airfoil surface in case of rain and icing [19, 20].

Laminar flow airfoils are most widely used in the case of sailplanes. The comparative study is conducted at lower Reynolds No of 3.1×10^5 on a dry and wet day (1000 mm/hr) for laminar Wortmann FX67-K710 airfoil, a transport category NACA 64-210 airfoil, and a symmetric NACA 0012 airfoil. The outcomes of the aforementioned comparative study exhibit that the laminar Wortmann airfoil suffered the major degradation of aerodynamic characteristics in comparison with other airfoils at higher AOA. The performance degradation is also pronounced at lower AOA, as the boundary layer transition occurs at the leading edge initially. At later stages, the onset of runback layers of water leads to roughened airfoil geometry, but this could be not an exact representation as the results obtained for the subscale models could be exaggerated [3]. Adelaida proposed a new shear breakup regime for the droplet in the vicinity of the leading edge of an airfoil [21]. Sor, Suthyvann et al. proposed a new DRD model similar to Clarke's and DDB model. All three models included only pressure, viscosity, and tension forces. The data obtained from all three models were in good agreement with experimental data for water droplets in the vicinity of leading edge along a stagnation line [22].

The paper focuses on droplet dynamics as it strikes onto the surface of the airfoil, flow field characteristics in presence of rainfall, lift and drag forces acting on the airfoil surface during rainfall, and droplet interaction mechanism with the solid airfoil surface. The understanding of the above concepts may help researchers to design and optimizing the shape of airfoils having enhanced aerodynamic characteristics during hazardous conditions of rainfall.

2 Dynamics of Droplet Strike on the Surface of the Airfoil

Generally, water droplets nearing an airfoil surface get deformed to a greater extent but in case the impact velocity is high enough it may also result in the breakup of drop in the vicinity of the airfoil [4]. The above occurrence is inspected during flight testing in adverse weather conditions of rainfall and SLD icing [5].

Figure 1a depicts a time-based deformation of a single large droplet in the vicinity of an airfoil. As seen in the figure, it could be noticed that initially that the droplet flattens at the front and bulges at the rear side. Further, as the droplet gets warped it becomes slimmer along the horizontal direction and stretched in an upright manner. The protrusion of the water droplet appears at the midsection on the side in the vicinity of the airfoil which marks the inception of the bag-type breakup process [6]. All the droplets vary in the manner how they deform based on their sizes as depicted in the figures below. The droplets with dia of 100 μm only portray a spherical deformation and the larger droplets could deviate from predefined modes of deformation as suggested by the TAB model [7], DDB models [8], and Clarke's analogy [9]. The slip velocity escalates in its value as droplets approach nearer to the airfoil surface.

In the culminating stages of droplet deformation, the rate of deformation exceeds the initial value wherein the droplet breaks into secondary droplets due to structure destabilization as shown in Fig. 1b. During the aforementioned breakup, the cross-section area is increased to twice the initial droplet diameter [10]. The droplets finally may or may not impinge onto the airfoil surface. The droplets which interact with the airfoil surface may either impact in a normal or in an oblique manner. In an oblique impact case, the droplets approach an airfoil at a certain incidence as shown in Fig. 1c. The secondary droplets formed in this process are smeared into the windward side of the impact boundary. The spreading of droplets onto the airfoil surface ceases in case of no corresponding velocity of water drops exists concerning air. The water droplets in a form of a sphere strike the airfoil surface normally resulting in axis-symmetric splash configuration. Figure 1d shows the initiation of the normal impact of droplet wherein droplet impinges onto water surface with a greater velocity which leads to droplet deformation with no splashing of droplets. It forms a crater wall which upon reaching a critical size gives rise to a mist of secondary droplets.

The development of water film on the top layer of airfoil surface is as depicted by Fig. 2a–d, wherein airfoil surface near leading edge gets roughened due to the presence of an uneven film of water. The water film turns wavy in shape downstream and the front turns into a sawtooth-shaped as it is driven downstream by the incoming airflow. At certain locations where the air pressure exceeds surface tension of water film present over airfoil surface in such cases, the frontal part of water film breaks up into several rivulets that run downstream.

The suction surface of airfoil witnesses accumulation of water droplets in the proximity of leading edge for higher wind velocities, but this distribution of the water film is non-uniform. This results in the formation of numerous rivulets flow downstream. The tangential shear of the air counteracts the tensioning surface of water film present over airfoil surface, thus securing the stability of rivulets flows

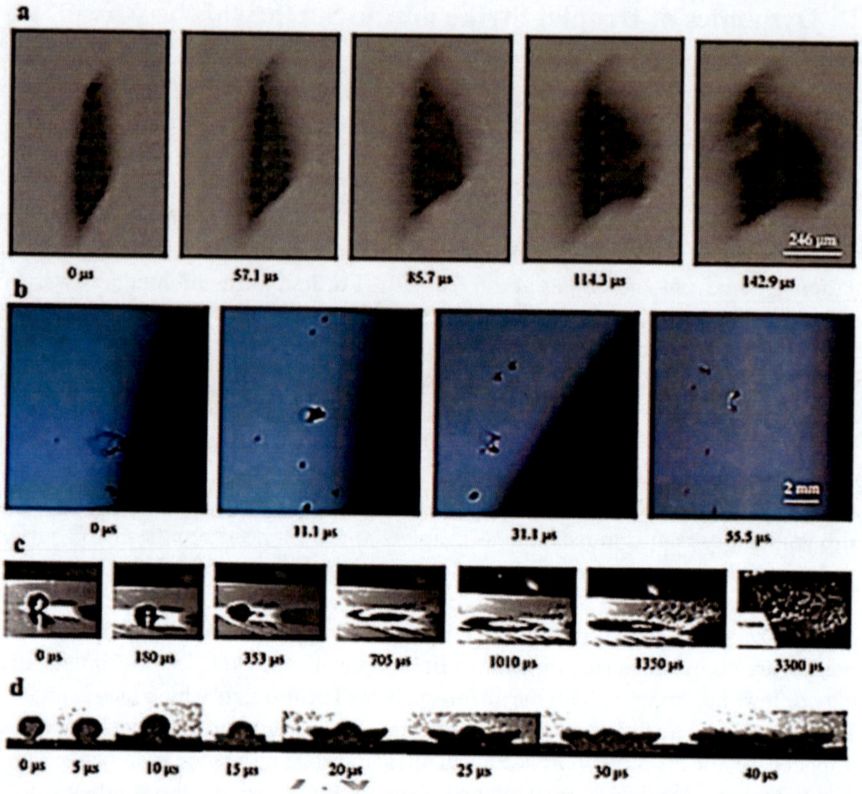

Fig. 1 Kinematic characteristics of the droplet as it approaches airfoil. **a** Chronological progression of deformation of water drop of initial diameter 490 μm before its breakup. **b** Chronological progression of rupture of water drop having an initial diameter of 250 μm in the vicinity of an airfoil. **c** Chronological progression of repercussions of water droplet striking airfoil surface having an initial diameter of 4 μm. **d** Chronological shadowgraphs of a spherical water drop impacting airfoil surface normally with an initial diameter of 4 μm [15]

and the water film. Formation of water film front and formulation of rivulets take place simultaneously as shown in Fig. 2e–h.

AOA is another predominant factor that defines water film attributes over airfoil surfaces. As seen from Fig. 3a at lower AOA, water coheres with wing surface to form surface water patterns such as impacting droplets, droplets smeared off from airfoil surface, formation of water film, and rivulets. Puddles of water could be noticed on the suction surface of the airfoil at higher AOA due to flow separation as shown in Fig. 3b, which is contradicting to nature of water droplets present on the wing at lower AOA. The flow separation on the suction surface of the airfoil does not affect the thin layer of water film present on the pressure surface of an airfoil in any manner.

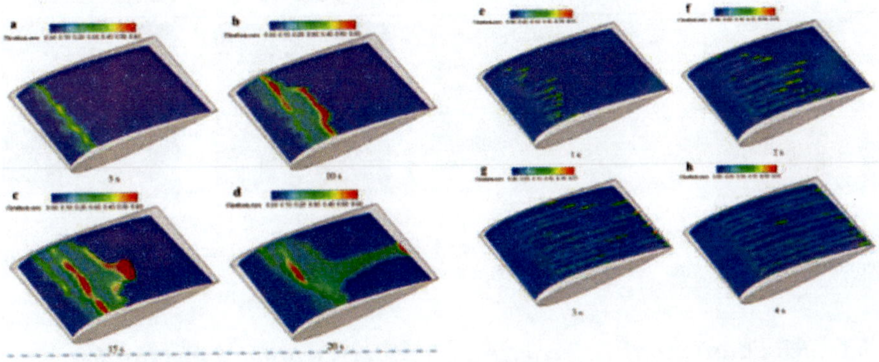

Fig. 2 **a–d** Chronological progression of thickness measurement of water film and the rivulet flow on the surface of NACA 0012 wing section at 10 m/s [16]. **e–h** Chronological progression of thickness measurement of water film and the rivulet flow on the surface of NACA 0012 wing section at 25 m/s [16]

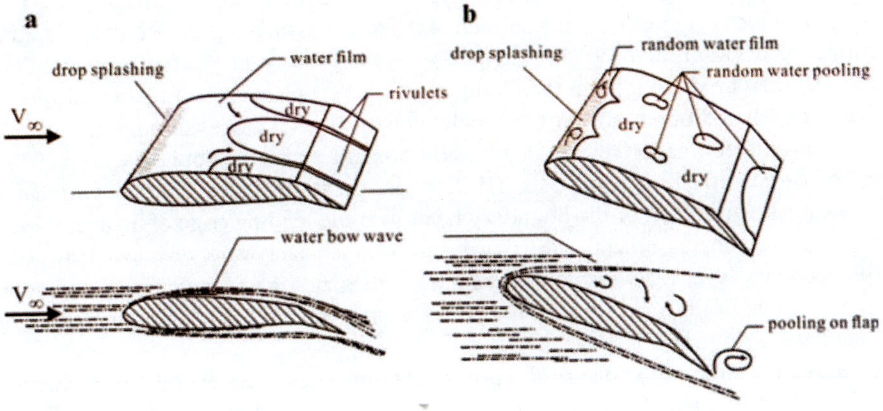

Fig. 3 Representation of traits of water droplets present of the surface of wing at **a** lower AOA and **b** higher AOA during rainfall [17]

3 Flow Field Characteristics in Presence of Rainfall

Water droplets indulge in noteworthy interactions with gas in terms of momentum and thermal transfer which results in predominant changes of velocity and pressure over airfoil surface. The droplets bring a substantial decrease in atmospheric pressure in comparison with gas alone [11]; they also reduce the pressure gradient which exists between the upper and lower surface of the airfoil resulting in reduced lift. The water droplets also cause early onset of boundary layer separation at the trailing edge as shown in Fig. 4a–d.

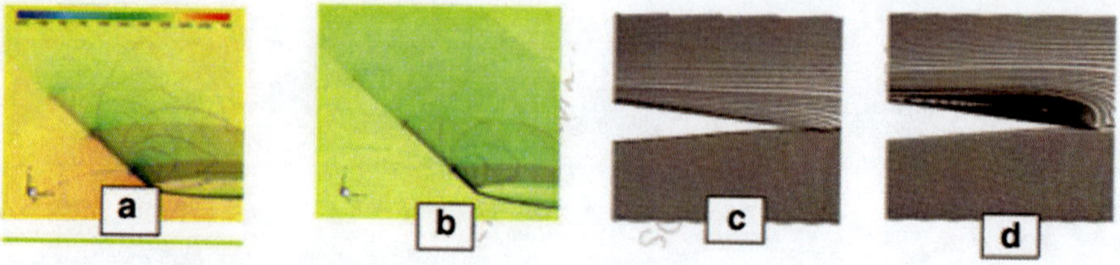

Fig. 4 **a, b** Pressure distribution over a NACA 64-210 wing on a dry and wet day [11], **c, d** Velocity contours over a 3D NACA 64-210 wing on a dry and wet day [11]

3.1 Mechanism of Interaction of Water Droplet with Solid Wall

The deceleration of the boundary layer is caused due to droplets that splash back after impinging onto airfoil, [12] also the airfoil surface becomes rugged owing to the existence of arbitrarily shaped layers of water due to the runback [13]. The droplets which impinge onto the surface of the airfoil with a high velocity lead to the formation of craters at the leading edge. These craters throw back a few large droplets into the flow field along with the cloud of smaller droplets up to a certain distance. These expelled droplets are further accelerated by air as it passes through the airfoil. The above phenomenon results in a loss of momentum by the boundary layer air as it gets de-energized in the process. The flow deceleration takes place predominantly in the upstream region of the boundary layer near the leading edge of an airfoil on a rainy day in comparison with a dry day due to splashed back water droplets. However, the boundary layer recovers downstream. De-energization of boundary layer results in decrement of lift, increment of drag, premature separation, stall, and transition of boundary layer [11].

Surface wettability is one of the prominent attributes of an airfoil surface which affects the drop impact dynamics and aerodynamics characteristics of an airfoil. For a reasonably wettable airfoil, surface water droplet spreads out in an even manner as a thin layer. The roughness of the airfoil, in this case, is due to the waviness of the water layer formed which results in increased drag and a reduced lift. For a non-wettable surface, however, water gets accumulated discretely as separate beads which increases roughness over airfoil surface leading to a substantial decrement in the C_L for all AOA. The wind tunnel results indicate that laminar Wortmann FX-67-K-170 airfoils coated with epoxy gel are least wettable, while a wax-coated surface has an air–water contact angle of 90° and lastly a soap-coated airfoils surface has intermediate wettability characteristics [14]. It could be noticed that with reduced wettability the slope of the lift curve decreases and also transits downwards in comparison with a dry condition as shown in Fig. 5a, b. The momentum transfer from water droplets to airfoil surface as a consequence of reduced wettability results in an increment of drag as seen from Fig. 5.

The tripping of the boundary layer at various sections along the chord length has been tested in wind tunnel experiments which prove that phenomenon of premature

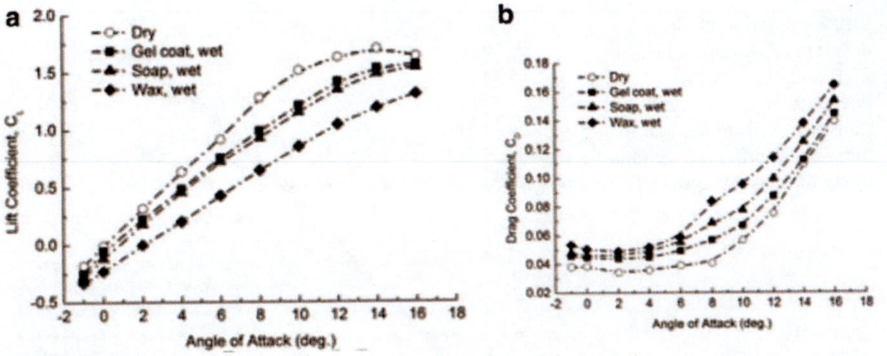

Fig. 5 Comparison of variation of (**a**) C_L and (**b**) C_D at different AOA for varied surface coatings on Wortmann FX-67-K-170 airfoil in presence of rainfall and dry condition [14]

boundary layer transition takes place owing to the existence of water film which converts an airfoil surface into a rugged one. The experimental test has been carried out in wind tunnel for Wortmann FX-67-K-170 airfoil by transforming the boundary layer from laminar condition to turbulent at various sections along the chordwise direction such as quarter chord, mid chord, and 1/16th chord locations, respectively [14]. The transformation of the boundary layer from laminar to turbulent near the leading edge causes a reduction in lift slope which supports the fact that in wet conditions boundary layer transition occurs near to the leading edge at about 1/4th of a chord for the airfoil chosen. Contrary to it on a dry day, the transition occurs behind 1/2 the chord length. Figure 6 depicts the increment in drag for gel and wax-coated airfoils as the transition point traverses in the forward direction toward the leading edge.

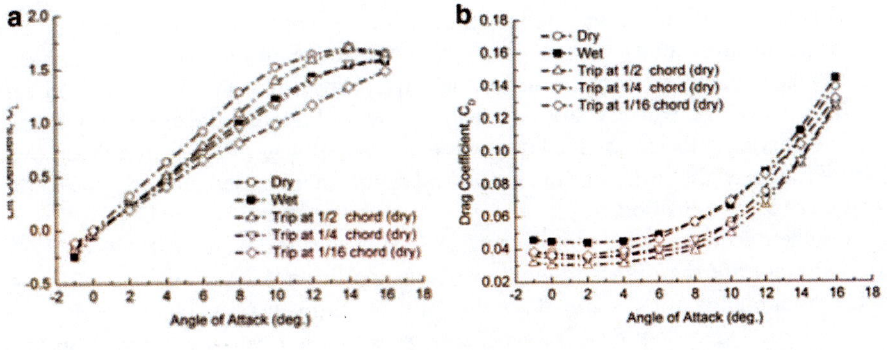

Fig. 6 Comparison of variation of (**a**) C_L and (**b**) C_D for boundary layer tripping at different sections along the chord and AOA for a Wortmann FX-67-K-170 airfoil in presence of rainfall and dry condition [14]

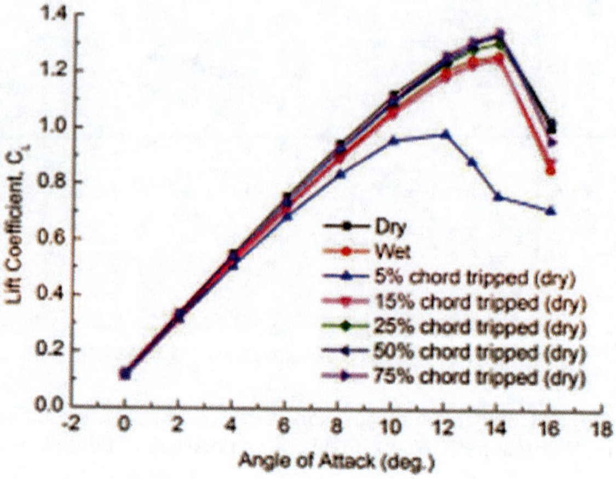

Fig. 7 Variation of C_L at different AOA for the NACA 23012 airfoil with the boundary layer artificially transformed from laminar to turbulent at different sections along the chord length [18]

The behavior of turbulent boundary layer formed due to tripping at 15% of the chord on the suction surface of airfoil on a dry day matched exactly with a NACA 23012 subjected to a wet condition on a rainy day. The computational approach validates the deterioration of the aerodynamic potential of the airfoil due to the early onset of boundary layer separation near the leading edge (Fig. 7).

4 Conclusion

Weber No decides the extent of deformation and breakup, and the droplet has undergone before it does impact onto airfoil surface as it travels through the flow field.

The two predominant occurrences during droplet and solid body interaction are the formation of secondary droplets which are smeared back into the flow field due to splashed back droplets from the airfoil surface and layer of water film over the airfoil surface. The splashed back droplets result in loss of boundary layer air momentum and de-energizes the boundary layer, while a layer of water film accumulated over airfoil roughens the surface of the airfoil and results in increased drag and premature boundary layer transition.

Droplets accumulated on airfoil surface form water film which exhibits different behavior at different AOA and wind velocities. At lower AOA, rivulets are formed, whereas at higher AOA, regional pooling of water droplets takes place.

The drop-laden flow field influences the pressure and velocity distribution over the airfoil surface which could result in premature boundary layer transition.

References

1. Zhang K, Johnson B, Rothmayer AP, Hu H (2014) An experimental investigation on wind-driven rivulet/film flows over a NACA0012 airfoil by using digital image projection technique. In: 52nd aerospace sciences meeting, p 0741
2. Thompson BE, Marrochello MR (1999) Rivulet formation in surface-water flow on an airfoil in rain. AIAA J 37(1):45–49
3. Hansman RJ Jr, Craig AP (1987) Low Reynolds number tests of NACA 64–210, NACA 0012, and Wortmann FX67-K170 airfoils in rain. J Aircr 24(8):559–566
4. Wu Z, Cao Y (2017) Dynamics of initial drop splashing on a dry smooth surface. PLoS ONE 12(5):e0177390
5. Tan J, Papadakis M, Sampath MK (2005) Computational study of large droplet breakup in the vicinity of an airfoil. US Department of Transportation, Federal Aviation Administration, Office of Aviation Research
6. Wang C, Chang S, Wu H, Xu J (2014) Modeling of drop breakup in the bag breakup regime. Appl Phys Lett 104(15):154107
7. O'Rourke PJ, Amsden AA (1987) The TAB method for numerical calculation of spray droplet breakup (No. 872089). SAE Technical Paper
8. Clark MM (1988) Drop breakup in a turbulent flow—I. Conceptual and modeling considerations. Chem Eng Sci 43(3): 671–679
9. Ibrahim EA, Yang HQ, Przekwas AJ (1993) Modeling of spray droplets deformation and breakup. J Propul Power 9(4):651–654
10. Vargas M, Feo A (2011) Deformation and breakup of water droplets near an airfoil leading edge. J Aircr 48(5):1749–1765
11. Wu Z, Cao Y (2014) Aerodynamic study of aerofoil and wing in simulated rain environment via a two-way coupled Eulerian-Lagrangian approach. Aeronaut J 118(1204):643–668
12. Wu Z, Cao Y (2018) Investigation of vertical axis wind turbine airfoil performance in rain. Proc Inst Mech Eng, Part A: J Power Energy 232(2):181–194
13. Wu Z, Cao Y (2015) Numerical simulation of flow over an airfoil in heavy rain via a two-way coupled Eulerian-Lagrangian approach. Int J Multiph Flow 69:81–92
14. Hansman RJ Jr, Barsotti MF (1985) Surface wetting effects on a laminar flow airfoil in simulated heavyrain. J Aircr 22(12):1049–1053
15. Feo A, Vargas M, Sor S (2011) Rotating rig development for droplet deformation/breakup and impact induced by aerodynamic surfaces (No. 2011-38-0087). SAE Technical Paper
16. Zhang K, Johnson B, Rothmayer AP, Hu H (2014) An experimental investigation on wind-driven rivulet/film flows over a NACA0012 airfoil by using digital image projection technique. In 52nd aerospace sciences meeting, p 0741
17. Campbell BA, Bezos GM (1989) Steady-state and transitional aerodynamic characteristics of a wing in simulated heavy rain. NASA Technical Paper 2932
18. Wu Z, Cao Y (2017) Airfoil performance degradation by coupling effects of supercooled raindrop icing and heavy rainfall. Proc Inst Mech Eng, Part G: J Aerosp Eng 231:2384–2395
19. Wu Z (2018) Drop "impact" on an airfoil surface. Adv Coll Interf Sci 256:23–47
20. Cao Y, Tan W, Wu Z (2018) Aircraft icing: an ongoing threat to aviation safety. Aerosp Sci Technol 75:353–385
21. Garcia-Magariño A, Sor S, Velazquez A (2017) Breakup criterion for droplets in the vicinity of a leading edge of an airfoil. In: 9th AIAA atmospheric and space environments conference
22. Sor S, García-Magariño A (2015) Modeling of droplet deformation near the leading edge of an airfoil. J Aircr 52(6):1838–1846

Thermodynamic Analysis of an Integrated OTEC-Based Multi-Generation Plant for Hydrogen and Freshwater Production

Aravind Ramachandran, U. B. Arun Shal, and Siddharth Ramachandran

Abstract This paper presents the energy analysis of a theoretical multi-generation plant which utilizes integrated solar thermal energy and ocean thermal energy for hydrogen and desalinated water production. The proposed novel plant consists of Ocean Thermal Energy Conversion (OTEC) unit coupled with a solar boosted Multi-Effect Distillation (MED-TVC) unit. The temperature difference of the seawater pumped from both surface and depth of the oceans is utilized to produce electricity at the OTEC unit and then fed as feedwater to the MED-TVC plant for desalination. For the first time in literature, a parabolic trough collector (PTC) field is incorporated to act as the heat source for the MED-TVC unit. The waste heat recovered from MED-TVC unit is utilized to superheat the OTEC plant to improve its thermal efficiency and electricity generation. The electricity generated from the combined system is entirely utilized to power a Polymer Electrolyte Membrane (PEM) electrolyser-based hydrogen production unit. Basic thermodynamic equilibrium equations for mass and energy were balanced for individual components of the system. A comprehensive parametric investigation is carried out for the proposed system and found that integrated plant can produce 2.471 kg/s of hydrogen and 89.21 kg/s of desalinated water under steady state operating conditions. Further, the energy efficiency of the multi-generation plant was found out to be 29.43%.

Keywords Renewable energy · Hydrogen production · Desalination · OTEC · Solar thermal

A. Ramachandran (✉) · U. B. Arun Shal
Government Engineering College, Kozhikode, Kerala 673005, India
e-mail: aravind12mea007@gmail.com

S. Ramachandran
Indian Institute of Information Technology, Design and Manufacturing, Kancheepuram, Tamil Nadu 600127, India

A. Maurya et al. (eds.), *Recent Trends in Mechanical Engineering*, Lecture Notes in Mechanical Engineering, https://doi.org/10.1007/978-981-19-7709-1_19

1 Introduction

The prosperity achieved by the present civilization is all based on the notion that resources are in fact non-depletable. With this scenario comes the need for renewable sources for energy and water to satisfy the basic human needs in the future. Sustainability is the only way forward for achieving this. In this study, both energy and potable water needs are addressed by putting forward an idea of a multi-generation plant with both hydrogen production and desalinated water for human consumption, purely from renewable sources [1]. A novel multi-generation plant is designed while keeping in mind, grid independence and fossil fuel free operation. For this an Ocean Thermal Energy Conversion (OTEC) plant is integrated with a Multi-Effect Distillation-Thermo Vapor Compressor plant (MED-TVC). An array of solar collector to act as heat source for the MED-TVC plant is incorporated for the first time in literature.

Ocean Thermal Energy Conversion Plants (OTEC) works on the principle of organic Rankine cycle having the heat source and heat sink of the system to be warm surface seawater and cold deep seawater, respecyively [2]. The warm surface seawater is pumped to the evaporator of the OTEC system, where the heat gets transferred to the working fluid (R32). The working fluid gets evaporated at very low temperature and is converted into steam. This steam is then expanded in a low-pressure turbine generator system to produce electricity. A condenser with cold deep seawater acts the heat sink [3]. Post-electricity production, the seawater is usually dumped back into the oceans. In this study, these seawaters are fed as feedwater to the MED-TVC system [4]. The distillation of this water occurs at various towers known as effects, where the seawater is sprayed onto tubes containing steam [5]. The seawater gets evaporated and is utilized to vaporize the seawater in the subsequent effects. The distilled water gets condensed at the tubes once the heat transfer takes place gets collected in a large storage from which it can be used for human consumption. A PTC-1000-type solar collector heat exchanger is incorporated into the multi-generation system to act as the heat source for the MED-TVC system [3]. Additionally, a part of the hot water that gets condensed in the first effect is utilized to superheat heat the working fluid of the OTEC system. This is done to reduce the heat loss and to maximize the power output of the MED-TVC system and OTEC system, respectively. Figure 1 depicts the multi-generation plant with parabolic trough collectors and warm seawater acting as heat source. The electricity produced is supplied to a PEM-based plant for continuous supply of hydrogen production.

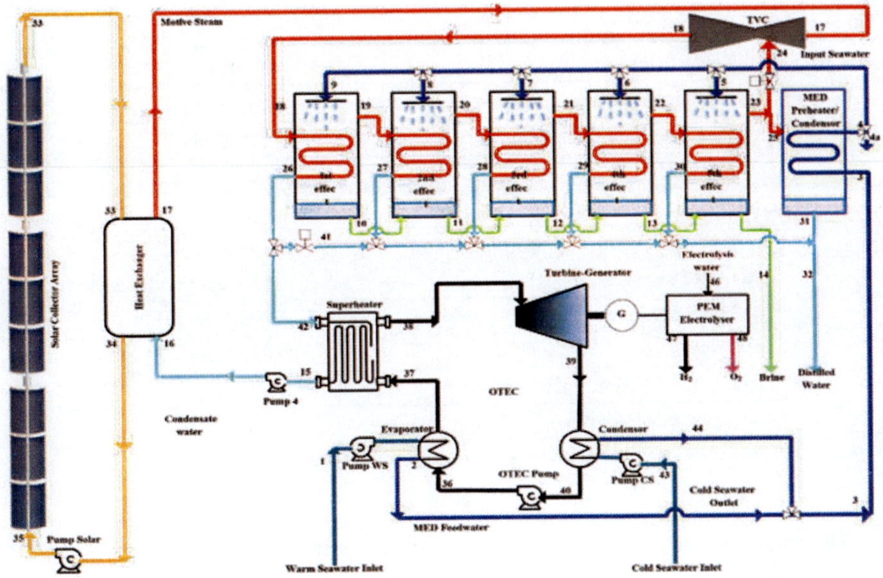

Fig. 1 Proposed multi-generation power plant

2 Energy Analysis

In order to conduct the energy analysis, first law of thermodynamics is considered. Table 1 denotes the initial input parameters taken into consideration for conducting the energy analysis of the proposed multi-generation system. In this study, the mass, energy, and entropy balance equations are described accordingly. The analyses are performed by using Engineering Equation Solver (EES) software package.

2.1 PTC-1000 Solar Collector

The collector efficiency of a can be taken as a function of the mean temperature across the solar collector (T_m), the ambient temperature (T_a), incident solar irradiation (G_B) from the characteristic curve of PTC type solar collector and is given as [7, 8],

$$\eta_{ptc} = \eta_{opt} - C_1(T_m - T_a) - C_2\frac{(T_m - T_a)}{G_B} - C_3\frac{(T_m - T_a)^2}{G_B} \tag{1}$$

where $C_1 = 0.0000045$ W/m^2, $C_{2,} = 0.039$ W/m^2, $C_{3,} = 0.0003$ W/m^2, and $T_m = T_{33} + T_{35}/2$.

Table 1 Initial input parameters

Sl. No	Input parameters	Values
1	Warm seawater inlet temperature	299 K [1]
2	Cold seawater inlet temperature	278 K [1]
2	Inlet seawater pressure	101.32 kPa [1]
3	Specific heat of seawater	4.025 kJ/kg-K [1]
4	Initial mass flowrate of warm seawater	194 kg/s [2]
5	Salinity of seawater	35 parts per thousand (ppt)
6	TVC inlet pressure	2000 kPa [5]
7	Solar irradiation	850 W/m^2 [5]
8	Pinch point difference across boiler	50 K [5]
9	TVC compression ratio	2.1 [5]
10	Electricity consumed for 1 kg hydrogen production	32.7 kWh [6]

The total area of the collector can be computed by its energy balance equations and is given by the Eq. (2) as

$$A_{ptc} = Q_u \big/ \eta_{ptc} G_b = Q_{Solar} \big/ G_b \tag{2}$$

2.2 *Ocean Thermal Energy Conversion Unit (OTEC)*

The warm seawater at the surface of the ocean is pumped to the evaporator of the OTEC. Here, it acts the heat source for its working fluid and gets evaporated to its saturated vapor condition. The cold seawater is pumped to the condenser of the OTEC where it acts as the heat sink The energy balance of the process is given by [3, 9] as

$$Q_{E(C),otec} = \dot{m}_{wf}(h_{37(40)} - h_{36(39)}) = \dot{m}_{ws(cs)} c_{ps}(T_{2(44)} - T_{3(43)}) \tag{3}$$

where $\dot{m}_{wf}$ is the mass flow rate of the organic working fluid, h stands for the enthalpy values, T is the temperatures of the seawater, and c_{ps} is the specific heat of the seawater.

The working fluid leaving the evaporator is superheated with the aid of the heat recovered from the working fluid of the MED-TVC desalination unit, and the energy balance of the process is given by [3, 9] as

$$Q_{Sup,otec} = \dot{m}_{wf}(h_{38} - h_{37}) = \dot{m}_{42} c_{pw}(T_{42} - T_{15}) \tag{4}$$

where $\dot{m}_{42}$ is the mass flow rate of the saturated liquid water that reaches superheater and c_{pw} is the specific heat of water.

The superheated working fluid from the superheater of the OTEC is fed to inlet of the turbine-generator, where it expands isentropically, and the power output of the turbine-generator set is given as [3]

$$W_{TG,otec} = \dot{m}_{wf}\eta_G\eta_T(h_{38} - h_{39}) \tag{5}$$

where η_G and η_T represent the generator and isentropic turbine efficiency, respectively.

The various pumps employed in OTEC system are the working fluid pump, the warm and cold seawater pumps, and its power consumed is given by [3]

$$W_{Pwf,otec} = \dot{m}_{wf}(h_{36} - h_{40}) = \frac{\dot{m}_{wf}v_{wf}(P_{36} - P_{40})}{\eta_{P,wf}} \tag{6}$$

where v_{wf} is specific volume of the working fluid and $\eta_{P,wf}$ is the isentropic efficiency of the OTEC working fluid pump. the power consumption equations of warm and cold seawater pumps are taken as [10]

$$W_{Pws(Pcs),otec} = \frac{\dot{m}_{ws(cs)}(g\Delta H)}{\eta_{ws(cs),otec}} \tag{7}$$

Here, g represents the acceleration due to gravity, ΔH represents head differences which is taken from previous literatures, and $\eta_{ws(cs),otec}$ are the efficiencies of the warm and cold seawater pumps.

2.3 *Multi-Effect Distillation Plant (MED)*

The energy balance equations of various components of the MED-TVC unit are given from equation. The energy balance across the TVC sub-system is given by [4, 11, 12] as

$$\dot{m}_{ds}h_{ds} = \dot{m}_{ms}h_{ms} + \dot{m}_{en}h_{en} \tag{8}$$

where $\dot{m}_{ds}$, $\dot{m}_{ms}$, and $\dot{m}_{en}$ represent the mass flow rates of the discharged steam, motive steam, and entrained vapor from the nth effect of the MED Distillation unit. The discharged steam from the TVC acts as the heat source to first effect of the MED subsystem, and its energy balance equations is given by

$$\dot{m}_{ef1}(h_{ef1v} - h_{ef1c}) = \dot{m}_{ef2}(h_{ef2v} - h_{feed}) + \dot{m}_{efb}c_p(T_{ef1b} - T_{feed}) \tag{9}$$

where $\dot{m}_{ef}$ represents the various mass flow rates inside the effects.

The vapor and brine from the preceding effects are utilized as heat sources for the subsequent effects; hence, the energy balance up to the 5th effect is given by the Eq. (10) as

$$\begin{aligned} &\dot{m}_{ef,n}(h_{efv,n} - h_{efc,n}) + \dot{m}_{efb,(n-1)}c_p(T_{efb,(n-1)} - T_{efb,n}) \\ &= \dot{m}_{ef,n+1}(h_{ef,n+!} - h_{feed}) + \dot{m}_{efb,n}c_p(T_{efb,n} - T_{feed}) \end{aligned} \tag{10}$$

Some of the vapor from the 5th effect is entrained by the TVC, and rest is condensed at the condenser of the MED system. The Eq. (11) denotes the energy balance at the condenser as

$$\dot{m}_{25}(h_{25} - h_{31}) = \dot{m}_{3a}c_p(T_{3a} - T_4) \tag{11}$$

The net distillate production ($\dot{m}_d$) of the MED-TVC unit is given by

$$\dot{m}_d = \dot{m}_{41} + \dot{m}_{27} + \dot{m}_{28} + \dot{m}_{29} + \dot{m}_{30} + \dot{m}_{31} \tag{12}$$

$$GOR = \dot{m}_d/\dot{m}_{17} \tag{13}$$

where GOR is the gain output ratio of the MED-TVC system. As we have mentioned earlier, the TVC entrains some vapor from the nth effect of the MED system. The amount of entrained vapor is found out using the following relation [13]

$$ER = 0.296\frac{(P_s)^{1.19}}{(P_{ev})^{1.04}}\left[\frac{P_m}{P_{ev}}\right]^{0.015}\left[\frac{PCF}{TCF}\right] \tag{14}$$

where ER is entrainment ratio defined as the ratio of the mass of motive steam per unit mass of entrained vapor. Here, P_s, P_m,P_{ev} are discharge steam pressure, motive steam pressure, and pressure of the entrained vapor, respectively. PCF and TCF are the pressure correction factor and temperature correction factor of TVC.

The compression ratio of the TVC is given as

$$CR = P_s/P_{ev} \tag{15}$$

2.4 Boiler

A boiler-type heat exchanger is employed to transfer the heat from the working fluid of the solar collector to produce motive steam. The energy balance of boiler evaporator, economizer, and superheater is based on the pinch method that are given by the Eqs. (16, 17 and 18), respectively, [14].

$$\dot{m}_s c_{pt}(T_{33} - T_a) = \dot{m}_{steam}(h_{17} - h_d) \tag{16}$$

$$\dot{m}_s c_{pt}(T_a - T_b) = \dot{m}_{steam}(h_d - h_c) \tag{17}$$

$$\dot{m}_s c_{pt}(T_b - T_{34}) = \dot{m}_{steam}(h_c - h_{16}) \tag{18}$$

The work consumed at the boiler heat exchanger water pump is given by

$$w_{P,BHX} = m_{16}(h_{16} - h_{15}) \tag{19}$$

2.5 *Overall Energy Efficiency*

The overall energy efficiency of the multi-generation plant is given by equation [6]

$$\eta_{Multi} = \frac{Output}{Input} = \frac{m_{distill}h_{distill} + W_{net}}{Q_{solar} + m_{ws}h_{ws}} \tag{20}$$

3 Results and Discussion

In order to find the performance variation of the multi-generation plant, design parameters such as warm seawater temperature and warm seawater mass flow rate are varied.

Figure 2 shows the effects of warm seawater flowrate on various parameters of the multi-generation system. As the mass flow rate of the inlet warm seawater increases, the heat input to evaporator of the OTEC system increases. This results in the increase of net power output of the system and hydrogen produced. Since the same seawater is utilized as feedwater for the MED system, the amount of feed entering each effect increases, thereby resulting in increased evaporation and condensation of the distilled water. The increased distilled water production also results in the increase of gained output ratio (GOR). Even though there is an increase in the vapor produced in the nth effect, the amount of entrained vapor remains constant due to constant entrainment ratio set in the TVC subsystem. So as the distilled water and hydrogen production increased, the increased inlet flowrate of warm seawater has led to decrease in the overall energy efficiency of the system.

Figure 3 shows the effects of warm seawater temperature on various parameters of the multi-generation system. As the temperature increases the heat input to evaporator of OTEC subsystem increases. This results in the increase in mass flowrate of the

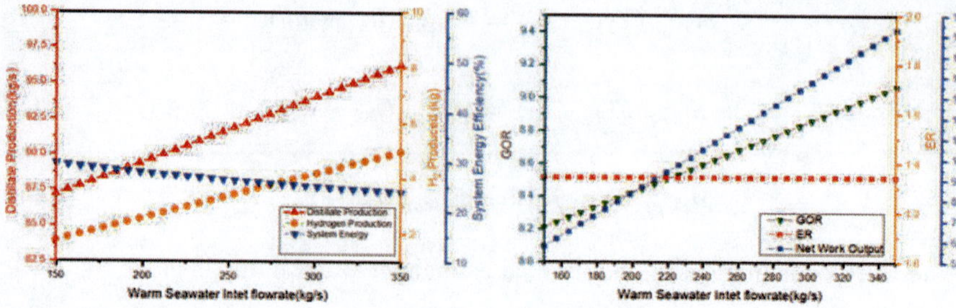

Fig. 2 Variation of performance parameters of the multi-generation plant when warm seawater flowrate of the OTEC is varied

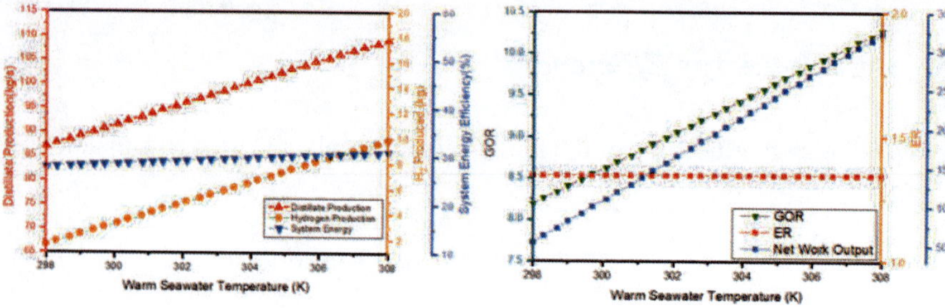

Fig. 3 Variation of performance parameters of the multi-generation plant when warm seawater inlet temperature at the OTEC is varied

working fluid inside the subsystem. This change along with superheating of the working fluid results in increased net power output of the system.

This change along with superheating of the working fluid results in increased net power output of the system. Also, while increasing the temperature, the heat input to the condenser of the MED subsystem increases. This results in the increased condensation at the condenser which leads to increase in GOR and distill out of the system. The increase in net power output and distill output of the system results in the increase of overall energy efficiency of the system.

4 Conclusions

A multi-generation plant based on OTEC and MED-TVC was studied for hydrogen and potable water production with waste heat recovery and for the first time in literature, to improve system performance, heat source has been designed using PTC collectors. Performance of the multi-generation plant has been investigated via a detailed parametric study, and found that

1. Increased warm seawater inlet temperature improves the performance of the integrated system, as it improves distilled water output and hydrogen production results in improved overall energy efficiency.
2. Increasing the warm seawater flow rate increases the distilled water and hydrogen production but decreases the overall energy efficiency.

Further, this investigation can be used as a basis for a detailed exergo-economic analysis and multi-objective multi-variable optimization studies to find optimal operating conditions.

References

1. Li Z, Siddiqi A, Anadon LD, Narayanamurti V (2016) Towards sustainability in water-energy nexus: ocean energy for seawater desalination. Renew Sustain Energy Rev 82:3833–3847. https://doi.org/10.1016/j.rser.2017.10.087
2. Muralidharan S (2012) Assessment of ocean thermal energy conversion, p 113. http://dspace.mit.edu/handle/1721.1/76927#files-area
3. Aydin H, Lee HS, Kim HJ, Shin SK, Park K (2014) Off-design performance analysis of a closed-cycle ocean thermal energy conversion system with solar thermal preheating and superheating. Renew Energy 72:154–163. https://doi.org/10.1016/j.renene.2014.07.001
4. You H, Han J, Liu Y (2019) Performance assessment of a CCHP and multi-effect desalination system based on GT/ORC with inlet air precooling. Energy 185:286–298. https://doi.org/10.1016/j.energy.2019.06.177
5. Gholinejad M, Bakhtiari A, Bidi M (2016) Effects of tracking modes on the performance of a solar MED plant. Desalination 380:29–42. https://doi.org/10.1016/j.desal.2015.11.015
6. Atiz A, Karakilcik H, Erden M, Karakilcik M (2019) Assessment of electricity and hydrogen production performance of evacuated tube solar collectors. Int J Hydrogen Energy 14137–14144. https://doi.org/10.1016/j.ijhydene.2018.09.100
7. Yuksel YE, Ozturk M, Dincer I (2016) Thermodynamic performance assessment of a novel environmentally-benign solar energy based integrated system. Energy Convers Manag 119:109–120. https://doi.org/10.1016/j.enconman.2016.04.040
8. Calise F, Dentice d'Accadia M, Piacentino A (2015) Exergetic and exergoeconomic analysis of a renewable polygeneration system and viability study for small isolated communities. Energy 92:290–307. https://doi.org/10.1016/j.energy.2015.03.056
9. Wang T, Ding L, Gu C, Yang B (2008) Performance analysis and improvement for CC-OTEC system. J Mech Sci Technol 22(10):1977–1983. https://doi.org/10.1007/s12206-008-0742-9
10. Nafey AS, Sharaf MA (2010) Combined solar organic Rankine cycle with reverse osmosis desalination process: energy, exergy, and cost evaluations. Renew Energy 35(11):2571–2580. https://doi.org/10.1016/j.renene.2010.03.034
11. Buabbas SK, Al-Obaidi MA, Mujtaba IM (2020) A parametric simulation on the effect of the rejected brine temperature on the performance of multieffect distillation with thermal vapour compression desalination process and its environmental impacts. Asia-Pacific J Chem Eng 1–14. https://doi.org/10.1002/apj.2526
12. Ameri M, Jorjani M (2016) Performance assessment and multi-objective optimization of an integrated organic Rankine cycle and multi-effect desalination system. Desalination 392:34–45. https://doi.org/10.1016/j.desal.2016.04.009
13. Ruíz AAB, El-Dessouky HT, Ettouney HM (2015) Fundamentals of salt water desalination. Elsevier 2002 3(2)
14. Sharaf MA, Nafey AS, García-Rodríguez L (2011) Thermo-economic analysis of solar thermal power cycles assisted MED-VC (multi effect distillation-vapor compression) desalination processes. Energy 36(5):2753–2764. https://doi.org/10.1016/j.energy.2011

Performance Evaluation and Comparison of Machine Learning Algorithms for Prediction of Electrodeposited Copper Ions

Vimal Kumar Deshmukh, Mridul Singh Rajput, and H. K. Narang

Abstract Machine learning predictions algorithms have been used in different scenarios by earlier researchers. This article found the application of these algorithms for predictions of the material deposition rate of copper ions through high-speed selective jet electrodeposition (HSSJED). The presented article emphasizes the selection of the best suitable machine learning prediction algorithms among Gaussian process regression (GPR), support vector machine (SVM), and linear regression (LR) for prediction and compares them with each other for selection of the best algorithm which agrees with experimental data. The comparison of performances of these machine learning algorithms has been done taking root-mean-square error (RMSE), coefficient of determination (R2), mean squared error (MSE), and mean absolute error (MAE) as a basic measure for the same, and also various graphs have been plotted and discussed for better understanding. The dataset has been collected by performing the HSSJED experiment on various parameters like electrolyte composition, electrode gap, and applied dc potential. It has been found that GPR predictions are better than the other two which also agree with the experimental output as the quantitative comparison has been done, and results are discussed.

Keywords Machine learning · HSSJED · GPR · SVM · LR

1 Introduction

In this scenario, human life is surrounded by electronic gadgets and devices such as smartphones, smartwatches, laptops, television, and many more. Day by day these electronic devices are getting smaller and smaller (using micro and nanoscale components), and performance is enhancing continuously, so complexity and difficulties to precisely channel electric charges are increasing. The electrodeposition technique can be useful to construct microelectronic and macro-electronic devices at a low cost [1]. It is feasible to deposit a very thin solid layer (i.e., less than a micron) with

V. K. Deshmukh (✉) · M. S. Rajput · H. K. Narang
Department of Mechanical Engineering, National Institute of Technology, Raipur 492010, India
e-mail: deshmukh.vimal1920@gmail.com

A. Maurya et al. (eds.), *Recent Trends in Mechanical Engineering*, Lecture Notes in Mechanical Engineering, https://doi.org/10.1007/978-981-19-7709-1_20

the use of high-speed jet electrodeposition [2, 3]. In this paper, machine learning deploys three different supervised learning algorithms to generate the best model to predict the electrodeposition rate in a high-speed jet electrodeposition machine. The independent input variables for each algorithm are voltage ranging from 10 to 30 V, the concentration of $CuSO_4$-$5H_2O$ minimum 9 g/l to 22 g/l maximum, and the gap between nozzle and substrate varying 2 mm to 4 mm. Experimentally obtained electrodeposition rate of copper ions is used as a dependent target for supervised learning. The algorithms used in this work are GPR, SVM, and linear regression, as these algorithms are used by many researchers for good prediction with small datasets [4–10]. The best model has been identified by evaluating the observed model statistics and illustrating the different plots generated by each algorithm. The model statistic consists of a parameter such as RMSE, R^2, MSE, and MAE. Response plot predicted vs actual response plot and residual plot are generated by these algorithms.

2 Data Collection

The essential data, for the prediction of electrodeposition rate for this paper, is gathered from previous works [2], where a detailed experimental setup is explained. Table 1 shows the experimental outcome of the deposition rate of copper ions with variable input parameters.

Table 1 HSJED experiment input parameter and deposition rate

S. No	Voltage (V)	Concentration of $CuSO_4 \cdot 5H_2O$ (gm/liter)	Inter Electrode gap (mm)	Deposition rate (mg/min.)		
1	10	15	3	0.33	0.21	0.21
2	15	15	3	0.25	0.35	0.44
3	20	15	3	0.35	0.63	0.34
4	25	15	3	0.75	0.58	0.47
5	30	15	3	1.01	1.32	1.27
6	20	9	3	0.045	0.012	0.018
7	20	12	3	0.25	0.17	0.2
8	20	15	3	0.35	0.49	0.48
9	20	18	3	0.64	0.5	0.45
10	20	21	3	0.46	0.69	0.59
11	20	15	2	0.55	0.47	0.48
12	20	15	2.5	0.48	0.5	0.47
13	20	15	3	0.48	0.44	0.44
14	20	15	3.5	0.33	0.39	0.4
15	20	15	4	0.34	0.27	0.23

3 Prediction Model and Analysis

In this study, the regression learner app in MATLAB version R-2018a is used for training and validation of the recorded data. As both input and output data are available, supervised machine learning algorithms are used to predict new inputs. There are several supervised machine learning algorithms such as Naive Bayes, GLM, SVR, and neural network. In this present work, GPR, SVM, and LR algorithms are used for prediction.

3.1 Gaussian Process Regression (GPR)

The GPR model does not count on the assumption to any particular distribution, and this model can accurately predict for small datasets [5, 6].

The prediction function for the linear GPR model is as follows:

$$y_* = \beta_0 x_* + \beta_1 + \varepsilon_t$$

where $\varepsilon_t \sim N\left(0, \sigma_n^2\right)$;

The relocating probabilities for the GPR model use the Bayesian approach [7] on the parameter (z) by defining the prior distribution p(z) using Bayes' rule:

$$p(z|y, x) = \frac{p(y|x, z)p(z)}{p(y|x)}$$

where $p(y|x) \sim N(0, K_N + \sigma_n^2 I)$

The GPR predictive distribution:

$$p\left(y^*|x^*, y, x\right) = \int_z p\left(y^*|x^*, z\right)p(z|y, x)dz$$

$$p(y_*|x_*, x, y) \sim N\left(\mu_*, \sigma_*^2\right)$$

where $\mu_* = K_{*N}\left(K_N + \sigma_n^2 I\right)^{-1} y$

$$\sigma_*^2 = K_{**} - K_{*N}(K_N + \sigma_n^2 I)^{-1} K_{N*}$$

Nomenclature

- x_* = test input matrix
- y_* = test output matrix

- ε_t = noise term
- K_N = covariance matrix for the training set
- I = identity matrix
- σ_n^2 = noise term variance
- x = training input

3.2 Support Vector Machine (SVM)

SVM is a supervised learning type machine learning algorithm mainly used for classification, but it can also do regression for linear as well as nonlinear challenges by employing kernel technique [8, 9]. The kernel transforms low dimensional input data to high dimensional data space [10]. In this algorithm, classification is performed by segregating the data points plotted in n-dimensional space, using a suitable hyperplane with maximum margin (distance between data points of different classes). The present SVM model is trained for the prediction of the deposition rate of the copper ion, and the training data consist of input parameters and experimentally obtain deposition rate, which is given in Table 1.

3.3 Linear Regression

This supervised machine learning algorithm is used for predicting dependent variables using independent variables. In this study, the electrodeposition rate is predicted based on three independent variables (voltage, the concentration of $CuSO_4.5H_2O$, and electrode gap). This model draws a best-fitted regression line by establishing the relationship between various input variables and given output variables.

Hypothesis function:

$$y = \theta_1 + \theta_2.x \tag{1}$$

θ_1 and θ_2 can be calculated by the following steps. Taking sum (y) the equation becomes

$$\sum y = n * \theta_1 + \theta_2 \sum x \tag{2}$$

where n is the number of input variable x. Multiply x on both sides will result

$$\sum xy = \sum x * \theta_1 + \theta_2 \sum x^2 \tag{3}$$

Solving both equations will give θ_1 and θ_2, and the solution will be given by

$$\theta_1 = \frac{(\sum y)(\sum x^2) - (\sum x)(\sum xy)}{n * (\sum x^2) - (\sum x)^2}$$

$$\theta_2 = \frac{n(\sum xy) - (\sum x)(\sum y)}{n*(\sum x^2) - (\sum x)^2}$$

Putting the value of θ_1 and θ_2 in Eq. (1) will give a solution for y for every input of x.

Nomenclature:

- x = input data
- y = output data
- θ_1 = intercept
- θ_2 = coefficient of input data.

4 Result and Discussion

After training the model, the best model is chosen by the best score of root mean square error (RMSE) among linear regression, SVM, and GPR. On the validation set, the best RMSE is 0.057658 for the GPR model, and RMSE for SVM and linear regression is 0.12295 and 0.16019, respectively. Table 2 shows the statistic of RMSE, coefficient of determination (R^2), mean squared error (MSE), and mean absolute error (MAE) for GPR, SVM, and linear regression models.

Figure 1 shows the response plot which displays the result of regression models after training the models. This figure shows the predicted response concerning the record number. The vertical line shows the error between the predicted response and the record number. One end of the vertical line is showing the experimental value (dark blue dot) and another end is showing the predicted value (light orange dot). To check the performance of different models the predicted vs actual response graph is plotted and shown in Fig. 2. In Fig. 2, the line represents the perfect prediction, and points represent the actual observations, the vertical distance between the perfect prediction line and observation point shows the error. For a good prediction model, the error should be small, and the points must be scattered near the perfect prediction line.

Table 2 Model statistics for GPR, SVM, and LR

Statistic	Tips	GPR	SVM	LR
RMSE	Smaller value is better	0.057668	0.12295	0.16019
R^2	Value close to 1 is better	0.96	0.78	0.58
MSE	Smaller value is better	0.0032432	0.016704	0.31695
MAE	Smaller value is better	0.045733	0.078151	0.12418

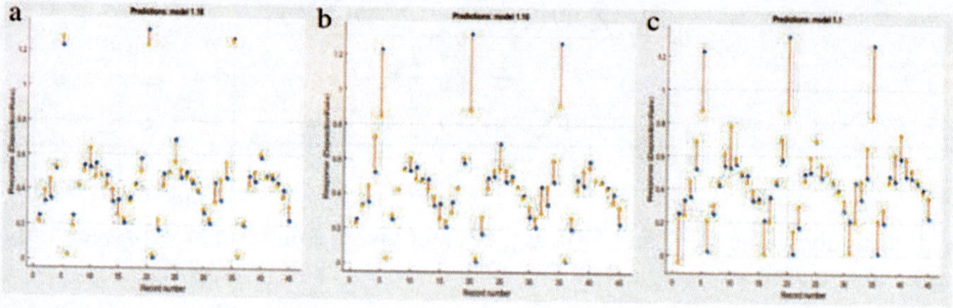

Fig. 1 Response plot for **a** GPR, **b** SVM, and **c** linear regression

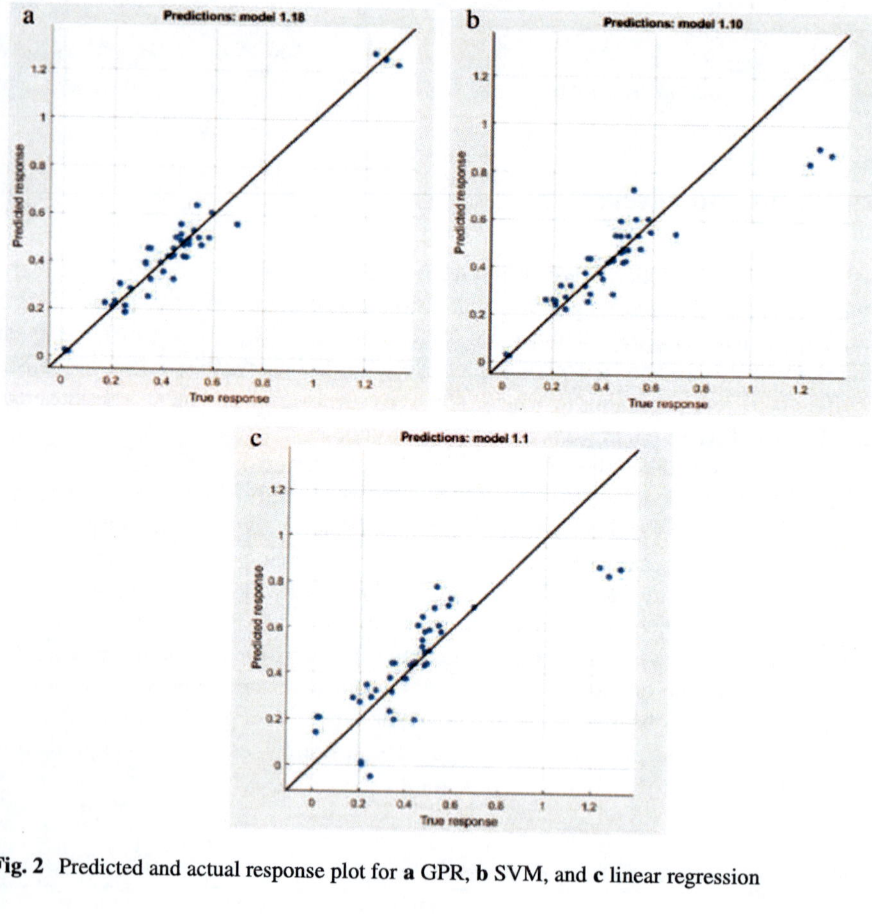

Fig. 2 Predicted and actual response plot for **a** GPR, **b** SVM, and **c** linear regression

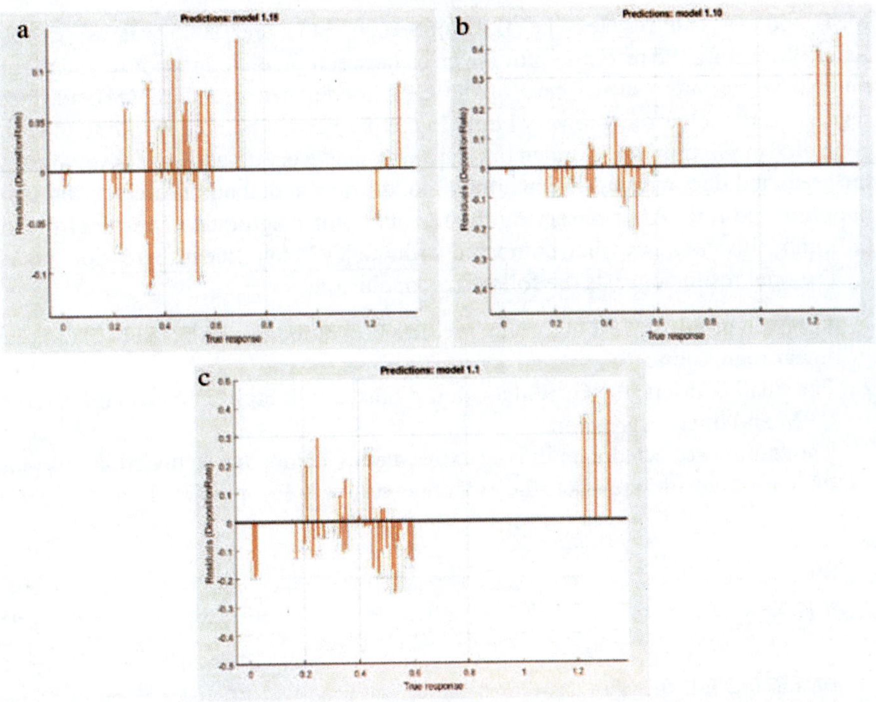

Fig. 3 Residual plot for **a** GPR, **b** SVM, and **c** linear regression from left to right

After seeing Fig. 2 for the GPR model (Fig. 2a) all the points are scattered near to the perfect prediction line that indicates.

For the SVM model (Fig. 2b) few points are a little far from the line, and in the case of linear regression (Fig. 2c), many points are away from the line.

Residual value is a measure of how much a regression line vertically misses the data point. Figure 3 shows the residual plot concerning true response for GPR, SVM, and linear regression. In the GPR model (Fig. 3a), residual is not symmetrically distributed, and the magnitude of residual is small (below 0.15). In SVM initially, magnitude is smaller, than for few data it is larger (above 3). In the linear regression model, most of the residuals lie on one side and the variation in magnitude is very frequent, also the magnitude of many points is much large (above 3.5).

5 Conclusion

In this paper, the electrodeposition rate of copper ions has been predicted using GPR, SVM, and linear regression algorithms to identify the best-fitted model. The model statistic table showed the GPR model is best suited, compared to SVM and linear regression for the present problem with R^2 value closed to 1, i.e., 0.96, also RMSE,

MSE, and MAE are very less, i.e., 0.057668, 0.0032432, and 0.045733, respectively. After illustrating the response plot the error between predicted response and record number is minimum in the case of the GPR model, whereas in SVM and linear regression model, errors are very large. In the predicted vs actual response plot, the generated prediction line is drawn in such a way that every actual point is very near to the predicted line, whereas in remaining models, few actual points are very far from the predicted line. After observing the residual plot magnitude of residual for the GPR model is very less when compared with the SVM and linear regression model.

The final result suggests the following conclusion:

1. Forgiven problem GPR algorithm predicts best outcomes followed by SVM then linear regression
2. For small datasets, the GPR algorithm predicts with high accuracy compared to SVM and linear regression.
3. Though the collected data having experimental errors due to uncertain reasons, the GPR algorithm predicts the outcome with very few predicted errors.

References

1. Dharmadasa IM, Haigh J (2006) Strengths and advantages of electrodeposition as a semiconductor growth technique for applications in macroelectronic devices. J Electrochem Soc 153(1):G47–G52
2. Rajput MS, Pandey PM, Jha S (2015) Micromanufacturing by selective jet electrodeposition process. Int J Adv Manuf Technol 76(1–4):61–67
3. Rajput MS, Pandey PM, Jha S (2015) Modelling of high-speed selective jet electrodeposition process. J Manuf Process 17:98–107
4. Lee I, Shin YJ (2019) Machine learning for enterprises: applications, algorithm selection, and challenges. Business Horizons
5. Cai H, Jia X, Feng J, Li W, Hsu YM, Lee J (2020) Gaussian process regression for numerical wind speed prediction enhancement. Renew Energy 146:2112–2123
6. Zeng A, Ho H, Yu Y (2019) Prediction of building electricity usage using Gaussian Process Regression. J Build Eng 101054
7. van der Heide EMM, Veerkamp RF, van Pelt ML, Kamphuis C, Athanasiadis I, Ducro BJ (2019) Comparing regression, naive Bayes, and random forest methods in the prediction of individual survival to the second lactation in Holstein cattle. J Dairy Sci 102(10):9409–9421
8. Leong WC, Kelani RO, Ahmad Z (2019) Prediction of Air Pollution Index (API) using support vector machine (SVM). J Environ Chem Eng 103208
9. Yan M, Wang X, Wang B, Chang M, Muhammad I (2019) Bearing remaining useful life prediction using support vector machine and hybrid degradation tracking model. ISA Trans
10. Shine P, Scully T, Upton J, Murphy MD (2019) Annual electricity consumption prediction and future expansion analysis on dairy farms using a support vector machine. Appl Energy 250:1110–1119

Analysis of Physical Properties and Tribological Wear Behavior of Al-Based Composite Alloy Using Unidirectional Tribo Tester

Satyajeet Kumar, Shailesh Mani Pandey, and Jay Shankar Kumar

Abstract The life span of many engineering components depends upon their surface properties. The improved surface properties of the materials are essential for enhancing the mechanical and tribological performance of the material. The tribological behaviors of sintered composites under dry sliding condition were analyzed by using pin on disk tribometer (Model no: DUCOM-TR-20-M100). Particle size of elemental powder of $Al_{70}Cu_{22}Ti_8$ was measured by Malvern laser particle size analyzer. PSM shows the variation in median size with milling time. The effect of density and hardness on wear behavior of composite has been analyzed. DSC examines the physical behavior of Al-based intermetallic as a function of time and temperature. DSC study shows the presence of small endothermic peak at 535 °C for formation of Al_2Cu and two exothermic peaks at 701 °C and 850 °C for formation of other intermetallic like $AlTi_3$, Cu_9Al_4, and AlCu. A maximum of 81% theoretical density was obtained for the composite powder mixture sintered at 1100 °C for 1 h. Wear study shows that TiO_2 and Y_2O_3 dispersed base alloy have higher wear resistance than base alloy.

Keywords Tribological behavior · Wear analysis · Tribometer · DSC · Hardness and density

S. Kumar (✉) · S. M. Pandey
Department of Mechanical Engineering, National Institute of Technology Patna, Patna, Bihar 800005, India
e-mail: Satyajeetkumar59@gmail.com

S. M. Pandey
e-mail: smp.me@nitp.ac.in

J. S. Kumar
Department of Mechanical Engineering, Technocrats Institute of Technology, Madhya Pradesh, Bhopal 462021, India

A. Maurya et al. (eds.), *Recent Trends in Mechanical Engineering*, Lecture Notes in Mechanical Engineering, https://doi.org/10.1007/978-981-19-7709-1_21

1 Introduction

Bauxite is the primary ore of aluminum which is abundantly found in the earth crust. The method of processing of aluminum alloys is quite simple as compared to most of the structural materials. These alloys have high specific strength, light in weight (2.77 g/m^3) and widely used for production and structural development after well-known steel [1]. Likewise in today's scenario, all the structural and practical usage cannot be fulfilled by any single material whether it is polymer or ceramic and metal alloys. So, it is required to have better properties of materials in combined state. This amalgamation of material properties can be incorporated by the development of various intermetallic compounds [2]. Among the category of advanced materials, intermetallic is a new set of material. The structural distribution of intermetallic compounds is different as that of parent material. Intermetallic compounds form when the bonding strength of like atoms (Al-Al or Cu-Cu) is lower that of dissimilar atoms (Al-Ti). Since the distribution of atoms in these compounds is in ordered manner, hence its properties lie between metal and ceramics [3]. The important properties of AI based intermetallic compounds include good oxidation resistance, decreased density with and high T_m makes these compounds superior for structural applications [4–7]. Intermetallic can be processed at very high temperature and can even substitute stainless steel and super alloys [8]. The elastic moduli of Al_3Ti intermetallic is 216 GPA that is greater than other oxides of titanium. Most of the super alloys lie in this range of modulus of elasticity; hence, it may substitute super alloy. High operating temperature and reduction in weight make them well suited for structural application in jet engines. Apart from FeAl, other intermetallic compound like Al_3Ti possesses less density [3.31 g/cm^3], toughness, and good oxidation resistance [9]. Intermetallic possesses very few dislocations movement and low diffusivities. Al_3Ti intermetallic possess ordered body center tetragonal structure. Hence, it is less ductile. Al_3Ti intermetallic is brittle at low temperature. But when this compound comes in contact to air, there is a formation of layered aluminum oxide (Al_2O_3). Hence, the oxidation resistance increases. [9, 10]. It is required to increase the material property like ductility for various structural and automotive industry. Ductility of Al_3Ti intermetallic can be enhanced by micro-alloying of ordered BCT structure and addition of period four elements like chromium (Cr), iron (Fe), nickel (Ni), copper (Cu), zinc (Zn), etc. Because of addition of these elements with micro-alloyed powder, the crystal structure changes from BCT to SC [11–15]. Centrifugal force is developed in a planetary ball mill due to rotation. The direction of both supporting disk and vials is opposite to each other resulting in producing centrifugal effect. Since the grinding balls are present inside the vials, thus the particles get trapped after every impact. The striking force splits the particle and generates further uncontaminated surface. Around 1000–1500 particles are produced during each collision [16].

The present research work aims at synthesis of Al-based (Al, Ti, and Cu) intermetallic compounds by mechanical alloying for various industrial applications like aerospace, turbine, automobile components, etc. The idea of the current research work is to inspect the mechanical and physical properties such as density, hardness,

Table 1 Composition (wt. %) of the powder mix for mechanical milling

S. No	Material and its percentage composition (wt%)				Name of alloy
1	Al	Cu	Ti		Base alloy
	70	8	2		
2	Al	Cu	Ti	Tio2	Base alloy + TiO_2
	69	8	2	1	
3	Al	Cu	Ti	Y_2O_3	Base alloy + Y_2O_3
	69	22	8	1	

and particle size of Al-based intermetallic compound. Tribological wear resistance and thermal conductivity of the intermetallic have also been studied.

2 Material Used

The material used for the preparation of Al-based intermetallic compounds can be can be summarized in a tabular form (Table 1).

3 Methodology

3.1 Synthesis of (Al, Cu)3Ti Powder Mix

Al-based intermetallic compound $(Al, Cu)_3Ti$ is synthesized in two steps: (a) Firstly, the elemental powder (base alloy) of different composition (wt %) was milled for 0–50 h at an interval of 10 h in a planetary ball mill, (b) and secondly, these compounds were subjected to conventional pressure less sintering. During milling, toluene ($C_6H_5CH_3$) was used to avoid oxidation of powders. Various milling parameters used are mentioned in Table 2. The fine and homogeneous base powders obtained after milling for 50 h were mixed independently with Y_2O_3 and TiO_2 powders (1 wt. % each) for 1 h. Cylindrical punch and die of 15 mm diameter were used to make pellets through hydraulic press with a load of maximum up to 450 MPa. The cold compacted cylindrical specimen was then sintered at different temperature but well below its melting point under argon atmosphere.

Table 2 Milling process parameters (Al, Cu)$_3$ Ti

Milled parameter	Value
Rotational speed (rpm)	275–300
Ball to power ratio (in weight)	5:1
Elemental powder (Al: Cu: Ti) (wt%)	70:22:8
Material of ball	Steel
Time of milling (h)	0–50 h (interval of 10 h)
Grinding media	Toluene
Container capacity	250
Diameter of steel ball (mm)	15 and 10

3.2 *Investigation of Physical, Mechanical Properties, and Wear Study*

In the present research, green, sintered, and theoretical density were calculated. Malvern laser particle size analyzer was used to investigate the particle size of elemental powder of $Al_{70}Cu_{22}Ti_8$ (Wt %). The hardness of the compacts was measured with the help of Vickers hardness tester using diamond indenter. The maximum load given to most of the compacts prepared through powder metallurgy route is 100 gf force. In order to have consistent result, seven readings were recorded at a load of 50 gf. DSC examines the physical behavior of Al-based intermetallic as a function of time and temperature. This has been done when the powder is heated in an inert atmosphere from ambient temperature to 1000 °C with a heating rate of 10 °C. Pin-on-disk tribometer (Model no: DUCOM-TR-20-M100) was used to investigate the wear characteristics of Al-based intermetallic. Different tribo test conditions are like load, rotation sliding time, and lubrication.

4 Results and Discussion

It can be seen from the plot that both theoretical and green density of all three alloy almost remains same. Sintering is a temperature dependent phenomenon. As the melting temperature of the alloy increases, the rate of diffusion and its sintered density also increases. Both mass transfer and neck formation occur during sintering mechanism. Since the dispersed particles are bonded and fused together, ultimately porosity decreases and leads to the increase in density with temperature. It can be seen from the graph that with an increase in sintering temperature densification parameter also increases. The higher rate of diffusion and densification at a given sintering temperature results in shrinkage of alloy particles (Fig. 1). With an addition of Y_2O_3 and TiO_2 to the base alloy, the value of DP increases. Variation in densification parameter, green, theoretical, and sintered density of Al-based intermetallic compound at

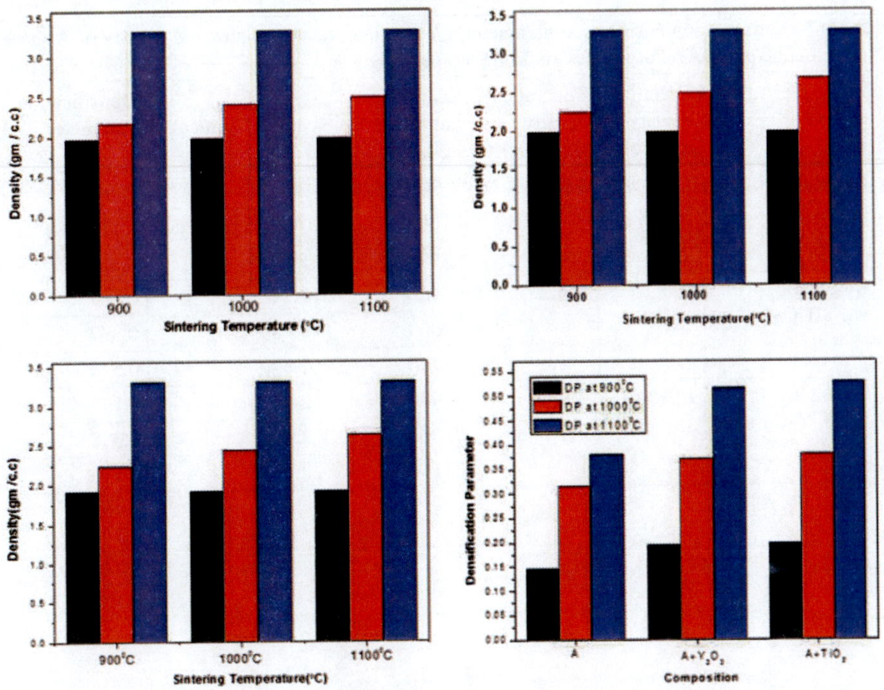

Fig. 1 **a–c** Shows the green, sintered and theoretical density of base alloy along with 1 wt% of yttrium and titanium oxide sintered at 900, 1000, and 1100 °C, respectively. **d** Variation of densification parameter of Al-based alloy

various sintering temperature is given in Table 3. Figure 2 shows the plot of microhardness value versus sintering temperature of Al-based intermetallic compound. The hardness value of the compound increases with increase in sintering temperature. However, the rate of increase of hardness of base alloy is less as compared to Y_2O_3 and TiO_2 dispersed base alloy. As both yttrium and titanium oxide are hard and brittle in nature, hence addition of these elements to the base alloy leads to the increase in hardness value. It is evident from the graph that the hardness is lowest for base alloy and highest for titanium oxide dispersed alloy. Figure 3 shows the comparative DSC study of base ($Al_{70}Cu_{22}Ti_8$) powder and 50 h mechanically alloyed powder.

At a heating rate of 10 °C/min the powders were heated to a maximum temperature of 1000 °C under argon atmosphere. Presence of endothermic peak indicates the fusion of pure aluminum at 660 °C in case of base alloy. This pattern is not similar in case of mechanically alloyed powder. 50 h milled powder is characterized by two exothermic peaks at 701 and 850 °C and an endothermic peak at 535 °C. Exothermic peaks are due to the presence of $AlTi_3$, Cu_9Al_4 and AlCu intermetallic compounds. Also, Al_2Cu is present in case of endothermic transformation. Figure 4 shows the variation in wear depth with sliding time of $Al_{70}Cu_{22}Ti_8$ (base alloy), Y_2O_3, and

Table 3 Variation in densification parameter, green, theoretical, and sintered density of Al-based intermetallic compound at various sintering temperature

Name of sample	Green density (gm/cc)	Theoretical density (gm/cc)	Sintering temperature (°C)	Sintered density(gm/cc)	Densification parameter
Base alloy – A ($Al_{70}Cu_{22}Ti_8$)	1.977	3.316	900	2.171	0.147
			1000	2.398	0.314
			1100	2.487	0.381
Base alloy + 1 wt.% TiO_2	1.989	3.302	900	2.246	0.197
			1000	2.489	0.377
			1100	2.679	0.525
Base alloy + 1 wt. % Y_2O_3	1.927	3.306	900	2.235	0.195
			1000	2.425	0.370
			1100	2.635	0.516

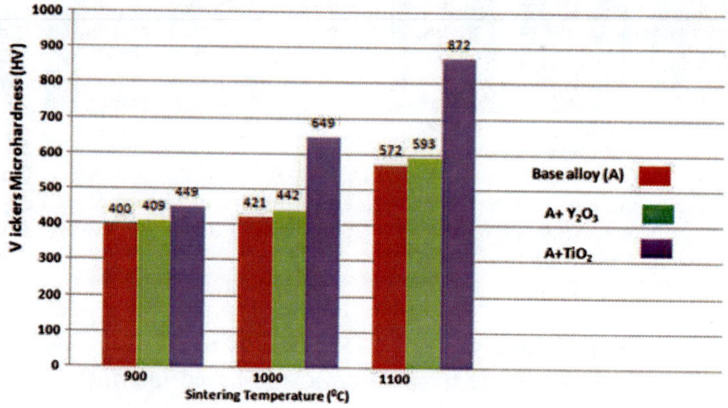

Fig. 2 Plot of micro-hardness value versus sintering temperature

TiO_2 dispersed base alloy sintered at 1000 and 1100 °C, respectively. The material removal rate and wear depth decrease with the addition of Y_2O_3 and TiO_2 to the base alloy. The oxides of yttrium and titanium are distributed uniformly over the Al base powder, hence reduces the chances of pore formation. The wear resistance of compound increases due to the formation of strong interfacial bond over the layers. Since addition of TiO_2 increases the hardness property of alloy. Therefore, compacts sintered at 1100 °C possess good wear resistance properties than another specimen. The maximum wear depth recorded is 200 and 340 μm for base alloy sintered at 1100 °C and 1000 °C, respectively. A graph below (Fig. 5) shows the weight loss of sintered compacts during wear analysis. Material removal rate of dispersed TiO_2 alloy is low while on the other hand base alloy shows high MRR. This can also be co-related from the hardness value of TiO_2 with respect to sintering temperature.

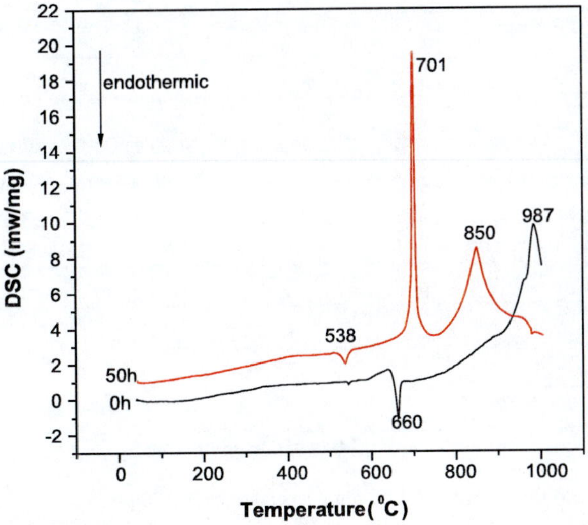

Fig. 3 DSC plot of milled $Al_{70}Cu_{22}Ti_8$ powder

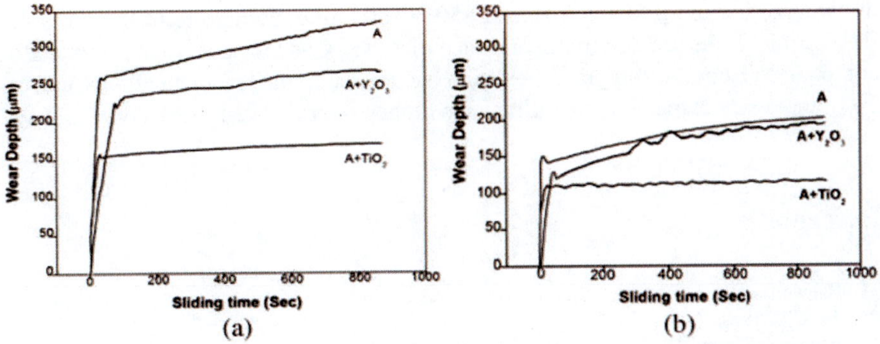

Fig. 4 **a** and **b** wear depth versus sliding time of sintered compacts at 1000 and 1100 °C

5 Conclusion

The present study focuses on analysis of physical properties and tribological wear behavior of Al-based composite alloy. Both theoretical and green density of all three alloys lies in the range of 1.9–303 g/cc. As the melting temperature of the alloy increases, the rate of diffusion and its sintered density also increases. The higher rate of diffusion and densification at a given sintering temperature results in shrinkage of alloy particles. With an addition of Y_2O_3 and TiO_2 to the base alloy, the value of DP increases. A maximum micro-hardness value of 449, 649, 872HV were obtained for base alloy, Y_2O_3 and TiO_2 dispersed alloy sintered at 1100 °C, respectively. DSC study examines the physical behavior of Al-based intermetallic as a function

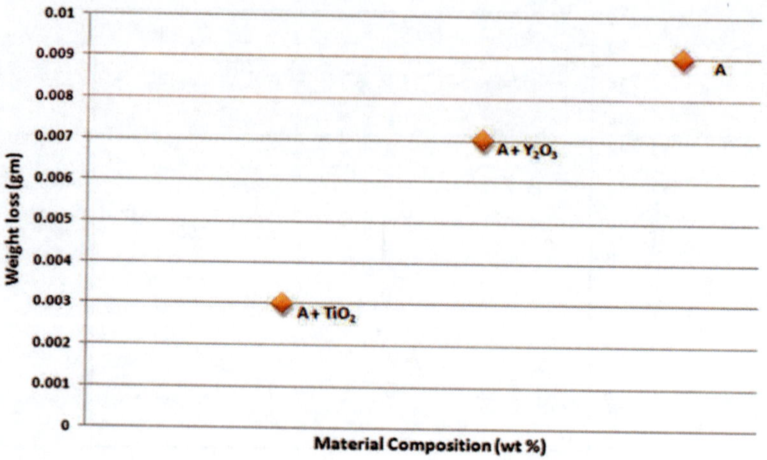

Fig. 5 Material loss during wear test of compacts sintered at 1000 °C

of time and temperature. Presence of endothermic peak indicates the fusion of pure aluminum at 660 °C in case of base alloy. 50 h milled powders are characterized by two exothermic peaks at 701 and 850 °C and an endothermic peak at 535 °C. Exothermic peaks are due to the presence of $AlTi_3$, Cu_9Al_4, and AlCu intermetallic compounds. Endothermic transformation occurs due to the presence of Al_2Cu. Both TiO_2 and Y_2O_3 dispersed base alloy have higher wear resistance than base alloy.

References

1. Schweitzer PA (2003) Metallic materials: physical, mechanical, and corrosion properties, vol 19. CRC press
2. Kumar S, Singh SK, Kumar J, Murtaza Q (2018) Synthesis and characterization of Al-alloy by mechanical alloying. Mater Today: Proc 5(2):3237–3242
3. Pithawalla YB, El Shall MS, Deevi S (2000) Vapor phase synthesis of intermetallic nanoparticles. In: Cluster and nanostructure interfaces, pp 175–183
4. Lee JK, Park JY, Oh MH, Wee DM (2000) Microstructure control in two-phase Al–21Ti–23Cr alloy. Intermetallics 8(4):407–416
5. Darolia R, Lewandowski JJ, Liu CT, Martin PL, Miracle DB, Nathal MV (1993) Structural intermetallics
6. Varin RA, Zbroniec L, Wang ZG (2001) Fracture toughness and yield strength of boron-doped, high (Ti+ Mn) L12 titanium trialuminides. Intermetallics 9(3):195–207
7. Shull RD (1993) Nanocrystalline and nanophase materials. Nanostruct Mater 2(3):213–216
8. Nayak SS, Murty BS (2004) Synthesis and stability of L12–Al3Ti by mechanical alloying. Mater Sci Eng, A 367(1–2):218–224
9. Yamaguchi M, Umakoshi Y, Yamane T (1987) Plastic deformation of the intermetallic compound Al3Ti. Philos Mag A 55(3):301–315
10. Kumar S, Kumar JS, Tiwari S, Prakash O, Singh SK (2021) Mechanical characterization & phase evaluation of ODS aluminum alloy. In: IOP conference series: materials science and engineering, March, vol 1104, No 1. IOP Publishing, p 012023

11. Baker I, Munroe PR (1988) Improving intermetallic ductility and toughness. JOM 40(2):28–31
12. Kogachi M, Kameyama A (1995) Composition dependence of site occupancies in the ternary Ll2 compound Al3Ti Cr. Intermetallics 3(4):327–334
13. Dimiduk DM, Miracle DB, Ward CH (1992) Development of intermetallic materials for aerospace systems. Mater Sci Technol 8(4):367–375
14. Mabuchi H, Hirukawa KI, Nakayama Y (1989) Formation of structural L12 compounds in TiAl3-base alloys containing Mn. Scr Metall 23(10):1761–1765
15. Suryanarayana C, Ivanov E, Boldyrev VV (2001) The science and technology of mechanical alloying. Mater Sci Eng, A 304:151–158
16. Rousselot S, Bichat MP, Guay D, Roué L (2009) Structure and electrochemical hydrogen storage properties of Mg-Ti based materials prepared by mechanical alloying. ECS Trans 16(42):91

Recognition of Prior Learning in INDIA: A Case Study

Roshan Lal Tamrakar, Vimal Kumar Deshmukh, and Suraj Kumar Mukti

Abstract The objective of the present work is to study the recognition of prior learning (RPL) in the Indian scenario. Through an examination and assessment of a current RPL, it tries to toss light on the capability of RPL to help support skills development, and the difficulties around its inauguration in India, and the approach and practice that can bolster future advancement in this area. India's vast informal segment incorporates a great many individuals with unrecognized skills for whom RPL could be a truly necessary advancement once more into education and training, from which they may have been barred till date. Through a progression of contextual investigation, it was planned to assess the GAPs regarding creating and managing RPL and the perspectives of the initiatives from Certifying Agencies on the achievement and success of the programme. Given the nascent stage of development, this initiative is at, the assessment does not conclude the post-assessment impact on job roles, access to training or other forms of progression. An innovative approach has been built up to assessment design in the absence of funds/targets that form the basis of RPL and have effectively created assessments that can be actualized on a gigantic scale. The research converges to outlining and building up another model for RPL that may start to lay the basis for a more extensive presentation of RPL and it as effective as possible in contributing to skilling India.

Keywords RPL · Skills development · Skills training

1 Introduction

In recent years, the idea of recognition of prior learning has picked up enthusiasm amongst the policymakers in the zone of vocational education framework and skill development. The RPL has created specific interest for India setting because of two highlights: to draw in individuals in learning who have beforehand been barred from formal instruction for some reason, it especially considers, and in the Indian

R. L. Tamrakar (✉) · V. K. Deshmukh · S. K. Mukti
Department of Mechanical Engineering, NIT, Raipur 492010, India
e-mail: nitrpr.rt@gmail.com

A. Maurya et al. (eds.), *Recent Trends in Mechanical Engineering*, Lecture Notes in Mechanical Engineering, https://doi.org/10.1007/978-981-19-7709-1_22

economy where the greater part of the workforce is informally employed, it offers the likelihood of certifying skills present in them [1]. India intends to skill 500 million individuals by 2022, a huge extent of individuals who are as of now working in different sectors. Of these, numerous individuals belong to economically, socially, or other distraught gatherings for whom entering any type of education or training might be an overwhelming undertaking. Under these conditions, interest in RPL as a device for connecting with these segments of individuals is obvious [2–4].

India is having 40% of people aged 13–35 years. Since the development of skills and utilization has a great scope with the RPL. The Indian government has practised aligning the skilled workforce by initiating various schemes, e.g. Modular Employable Scheme - Skills Development (MES-SDIS). After clearing the respective examinations, these schemes are not only offering direct entry to various modules, short courses but also providing opportunities for direct assessment and obtain a National Trade Certificate. Whilst the reason for RPL is undisputed and endeavours to place it by and by having been ongoing all around for few years; yet practically speaking, the majority of these endeavours have confronted various difficulties and shifting levels of accomplishment. We might endeavour to look at a portion of these as contextual investigations to evoke important discovering that could be used in the Indian context.

India, along with other emerging economies, however, faces momentous challenges in introducing RPL—principally the flow of funds to certify informally skilled large workforce employed. India faces an immediate challenge of up-skilling millions of people already working in the informal sector, in the process of developing policy/frameworks. The RPL is not a simple system to get right; nor is it a "handy solution" arrangement. Notwithstanding, through our endeavours, we may build up an arrangement of RPL in India that can encourage enhanced access to additional training and job for millions and improve utilization of the abilities effectively present in the nation's workforce.

2 Initiatives in India Leading Towards RPL

Ministry of Education formerly known as MoHRD began action on RPL in September 2012 as part of the National Vocational Education and Qualification Framework (NVEQF). The NVEQF encourages people to obtain certifications in a variety of ways. The National Institute of Open Schooling's (NIOS) RPL Manual provides a framework for evaluating the competences and skills gained through information [5–7]. NVEQF also defines the mapping between the levels acquired via RPL and through formal education. This additionally provides the student with an option to switch between the two forms of education. But till date, the framework has not been adopted by any industry or assessment body. The NVEQF also establishes a relationship between the levels attained through RPL and those attained through formal education. Additionally, the learner has the option of switching between the two types of schooling. However, no industry or assessment agency has accepted the approach to far.

2.1 *All India Council for Technical Education (AICTE)—Skill Knowledge Provider (SKP) Scheme*

The AICTE has launched a Skill Knowledge Provider programme (SKP). To build a credit-based approach to vocational higher education, the programme relies on a system of co-creation of skills and general academics. The first two levels are equivalent to school grades IX and X. Every year, approximately, 1000 h of study and training are required for each level. [8]. So, if one has the abilities and has passed your 10th grade exam, he/she can go to a skill knowledge provider and can have talents assessed for a certificate. The system will be implemented in 12 different industries, including information technology, telecommunications, banking, tourism, autos, finance, marketing, entertainment, and construction. The sector has contributed to the creation of curriculum and assessments. The skill credit can be transferred to a vocational training score at the university where the candidate is registered for pursuing formal education. The central university or the technical board committee can also adopt and facilitate the same for a certain certificate course. Certificate courses also allow the candidate to pursue a professional career and acquire additional credit simultaneously with distance/part-time programmes for degree courses [7, 9, 10].

2.2 *National Skill Development Corporation (NSDC)—Pradhan Mantri Kaushal Vikas Yojana (PMKVY)*

The core interest of the skill certification is to benefit the skilled youth population at large. Providing a platform to get industry required skills and help them to make a better livelihood. The skilled individuals who have prior learning experience from informal training can be assessed and get certified by the RPL scheme. The three core objectives of RPL are (i) to line up the unorganised unregulated workforce populations with the NSQF standardisations, (ii) to increase the job opportunities and employability in the relevant industry at an individual level and open the ways to pursue higher education through proper channels, (iii) to mitigate the inequality on basis of giving privilege to one over the other for not having prior experience from the proper channel [7].

3 RPL Process: "The RPL Process Comprises of Five Steps, Specified in Fig. 1"

3.1 *Mobilization*

The engagement of NGO/training cum testing agencies (training partners)/association by project implementation agency (PIA) can be done in mobilizing the rpl candidate to the training centres.

3.2 *Counselling and Pre-screening*

TOT passed qualified trainers approved by the SSC will facilitate the counselling and pre-screening of the RPL candidate for all types of projects discussed above. The pre-screening stage comprises of two parts named below:

Part 1: Collection of supporting documentation and evidence from the candidate.

Part 2: Candidate self-assessment.

3.3 *Orientation*

For all project types, the following orientation activity would be adopted.

3.4 *Final Assessment*

The common guidelines for the assessment of all types of projects can be adopted as follows.

PIA will initiate the final assessment of the candidate at the accredited assessment agencies (AA). The AA will be declared by the National Board for skill certification (NBSC) body or the SSC. Overlapping of duties of RPL facilitator and AA will

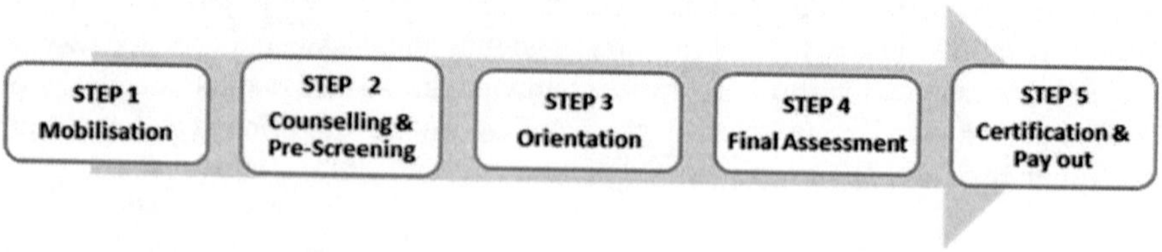

Fig. 1 5-step RPL process

be avoided. The assessment process will likely be done in the same manner as the short-term training component of PMKVY and in compliance with NSQF. SSC shall earmark the scoring system as CORE for a qualified person (QP)NOSs and non-core. The core NOSs part will have 70% weightage, and the remaining belong to non-core as per NSQF level 4.

3.5 *Certification and Payout*

Once the assessment agency uploads the computed result, the skill service centre will validate and approve it. The registered candidates can avail the mark sheet and the certificate at the facilitated centre after the final declaration of result by the PIAs. PIAs can also notify the candidates by sending SMS and emails regarding examination and result schedules. The skill certificate will have the grades A/B/C as per the marking system. The candidate will be facilitated with a skill certificate and mark sheet and payout (INR 500) following the eligibility of the candidate.

4 PPP-Based RPL Model

Certification Agency and PPP Partner will work through PPP mode as shown in Fig. 3. Step wise procedures are given in Table 1. Through PPP model, network of training cum testing centres will be established across with special focus on remote and inaccessible areas. The hierarchy is given in Fig. 2.

5 Conclusion

India is a developing country with moderate growth in the industrialization and education sector. A major of the population who belongs to the informal sector are lacking basic education but they have acquired certain labour skills for their livelihood. Inaccessibility of proper channels is a barrier to obtain skills. Here, RPL plays an important role. RPL encircle all these issues and provide a framework through which aids skilled labours to get the benefits available to the skilled person. RPL fills the gap between vocation and formal learning. The RPL facilitates the professional gain to the employee who obtained skills by informal channels, e.g. master craftsman, automobile mechanics, etc., and also increases the socio-economic mobility of the skilled labour.

Table 1 Proposed RPL model

Steps	Activities	Responsible agency	to	Engage entity
Step 1	Target	Certification Agency (CA)	to	Implementing Agency (IA)
Step 2	Revenue generation	Candidate Employers RPL Centres Company Social Responsibility (CSR) Fund	to	Implementing Agency (IA)
Step 3	Engagement of partners	Implementing Agency (IA)	to	Testing Cum Training Provider (TTP) Assessment Agency (AA)
Step 4	Mobilization	Testing Cum Training Provider (TTP)	to	Candidate
Step 5	Counselling screening of candidate			
Step 6	Orientation			
Step 7	Assessment	Assessment Agency (AA)	to	Candidate
Step 8	Skill evaluation report of candidate	Assessment Agency (AA)	to	Implementing Agency (IA)
Step 9	Verification of evaluation report	Implementing Agency (IA)	to	Certification Agency (CA)
Step 10	Certification	Certification Agency (CA)	to	Candidate
Step 11	Distribution of fund	Implementing Agency (IA)	to	Certification Agency (CA) Testing Cum Training Provider (TTP) Assessment Agency (AA)

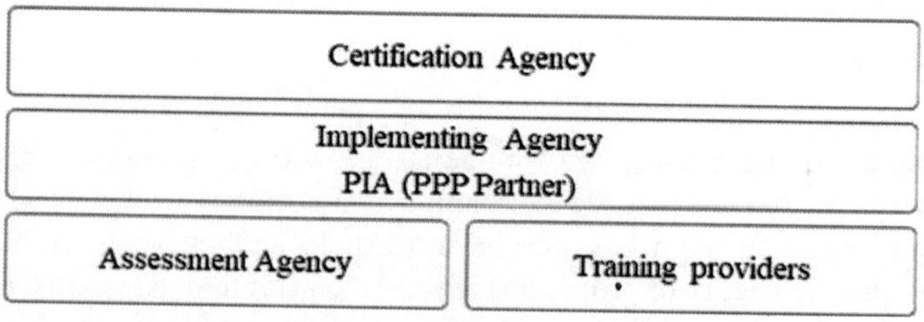

Fig. 2 Structure of PPP model

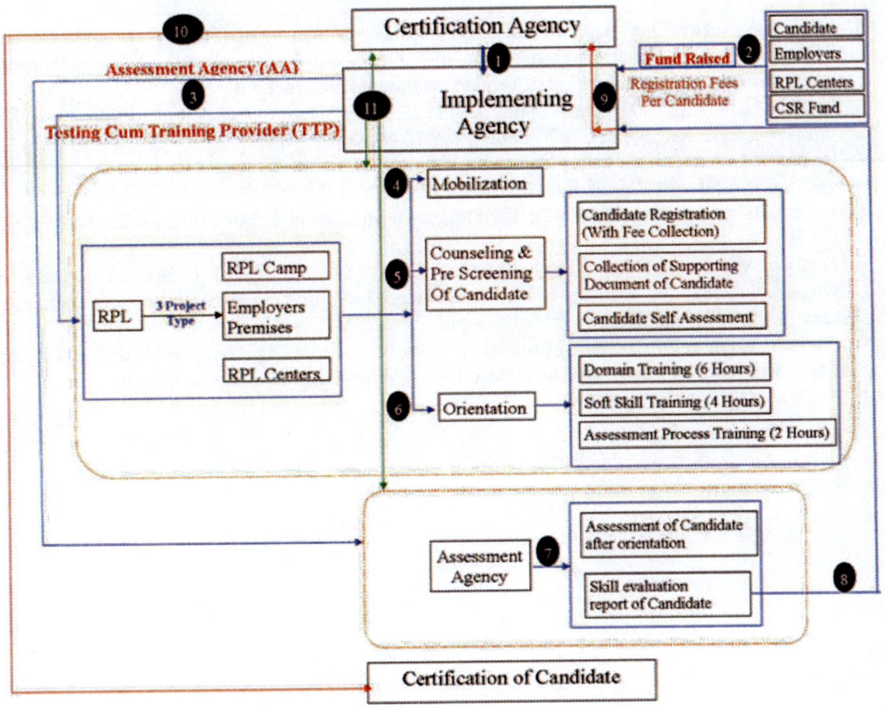

Fig. 3 Route map of proposed RPL model

Acknowledgements Authors are grateful to the Federation of Skill Development Organization (FSDO) and Chhattisgarh Swami Vivekanand Technical University, Bhilai (CSVTU Bhilai), a state government university of Chhattisgarh state in India for signing MoU to implement the proposed PPP-based RPL model named as Swami Vivekanand Yuva Kaushal Setu (SVYKS).

References

1. Ministry of Skill Development and Entrepreneurship, Government of India, NSQF Notification (2018). http://www.skilldevelopment.gov.in/assets/images/gazetteeonsqf.pdf
2. Annual Report (2016–2017) English, 1 Ministry of Skill Development and Entrepreneurship, Government of India. http://www.skilldevelopment.gov.in/assets/images/annual%20report/annual%20report%202016-2017%20-%20english.pdf
3. Annual Report (2015–2016) English, 1 Ministry of Skill Development and Entrepreneurship, Government Of India. http://www.skilldevelopment.gov.in/assets/images/annual%20report/annual%20report%202015-16%20eng.pdf
4. Common Norms 2nd Amendments (2017) 1 Ministry of Skill Development and Entrepreneurship, Government of India. http://www.skilldevelopment.gov.in/assets/images/notification/28.02.2017%20common%20norms%20amendment%20notification.pdf

5. Common Norms 1st Amendments, 20/05/2016, Ministry of Skill Development and Entrepreneurship, Government of India. http://www.skilldevelopment.gov.in/assets/images/notification/1st%20modification%20to%20common%20norms.pdf
6. Pradhan Mantri Kaushal Vikas Yojana _Guidelines_Booklet. http://msde.gov.in/assets/images/pmkvy/state%20engagement%20guidelines%20-%20pmkvy.pdf
7. Pradhan Mantri Kaushal Vikas Yojana Guidelines (2016–2020). http://msde.gov.in/assets/images/pmkvy/pmkvy%20guidelines%20%282016-2020%29.pdf
8. Union Budget of India (2017–2018). https://www.indiabudget.gov.in/budget2017-2018/budget.asp
9. National Vocational Education Qualifications Framework (NVEQF), F. No. – 1-4/2011-VE Government of India, Ministry of Human Resource Development. https://www.aicteindia.org/downloads/nveqf_notification_mhrd_goi.pdf
10. All India Council For Technical Education, www.Aicte-India.Org. National Skill Qualification Framework (NSQF). https://www.aicte-india.org/downloads/annexure%20ii.pdf

The Skill Development Ecosystem: A Brief Study on Policies and Projects in Context of India

Roshan Lal Tamrakar, Manish Pandey, and Suraj Kumar Mukti

Abstract India is the home of second largest population in the world. A huge section of India's unorganized workforce is unskilled and semi-skilled. Maximum people picked up skills, trainings, and knowledge from an informal setup by observing skilled workers and by working under their supervision or through complete self-learning. The skill development ecosystem directly affects the Human Development Index (HDI) factor which is an universal indicator of overall development of any country or society. It emphasizes the availability of livelihood opportunities to household of any nation. The need of skill development in India and its impact is emerging with time since the first introduced Apprenticeship Act 1961 till Skill Development Policy 2015 and now the New Education Policy 2020. This review document will cover the significant, essential and relevant research, reviews, policies, policies reviews, and govt. notifications in order to understand the current stand of skill development scenario in INDIA.

1 Introduction

India is the seventh largest country in terms of area; however, we are second largest country in terms of population.

India is home of 121 crore peoples, and huge population brought several other challenges which are associated with the basic needs of common people. Government is trying to provide equal and appropriate means of livelihood for improving rank of India in Human Development Index (HDI) which is 129th among 189 countries in the world [16]. But, on the other hand, we can consider this situation as a boon in terms of available manpower, if our population is skilled enough to match the international requirement and standards [4, 5].

R. L. Tamrakar (✉) · S. K. Mukti
Department of Mechanical Engineering, NIT Raipur, Raipur, Chhattisgarh, India
e-mail: nitrpr.rt@gmail.com

M. Pandey
Department of CSE, CSVTU, Bhilai, Chhattisgarh, India

A. Maurya et al. (eds.), *Recent Trends in Mechanical Engineering*, Lecture Notes in Mechanical Engineering, https://doi.org/10.1007/978-981-19-7709-1_23

With many government schemes like "Make in India" to "Atmanirbhar Bharat" and policy reforms such as "NEP 2020" are examples that the country is trying hard to become an skill capital of world [6].

Many schemes and policies were announced and implemented since the beginning of this decade. To implement National Skill Development Policy, the government setup "National Skill Development Corporation (NSDC)" and introduced a framework called "National Skill Qualification Framework (NSQF)" for certifying and mapping of informal skill education with the formal school and higher education. The NSQF was divided into 10 levels, where level 1 is equivalent to primary school education, and level 10 denotes to Ph.D. level of education. NSQF promotes the parallel form of education instead of traditional top bottom approach where a person is forced to attend the formal school and higher education within a stream. NSQF suggests multi-entry and exit points with multidisciplinary education and skill learning options [6–8, 23].

With the gazette notification in the year 2013 (The Gazette of India, Extraordinary, Part-I, Section-2, December 2013), government suggested to all the central and state boards/universities for mapping formal school and higher education with its NSQF counterpart within 5 years. This is the point where the biggest challenge begun when the responsibility of mapping had been imposed to all individuals (boards/universities) across country instead of suggesting standardized mapping pattern [10]. As an ancient custom and a popular saying, it is believed that "everyone's responsibility is no one's responsibility", true to that no one came forward to take the challenge. Even with the "New Education Policy (NEP) 2020", the situation is same except one thing that the certification patterns of college education are now mapped with NSQF standards; still, most of the mapping challenges remain unaddressed like mapping of NSQF with school and vocational education, mechanism of entering or exiting in/from any course, update in recruitment norms, etc. [7, 8].

1.1 *Skill Development Ecosystem in India*

The Indian skill development ecosystem can be considered as large, diverse, and complex in terms of providing various levels of skills across the country. Skill development in India can be broadly classified into 2 broad category as **Education** and **Vocational Training. The illustrated arrangement** exhibits here is the comprehensive framework of Skill Development in India [25, 26] (Figs. 1 and 2).

1.2 *Skill Development Hierarchy in India*

The 3 major ministries in India constitute the skill development structure and provide framework for the same.

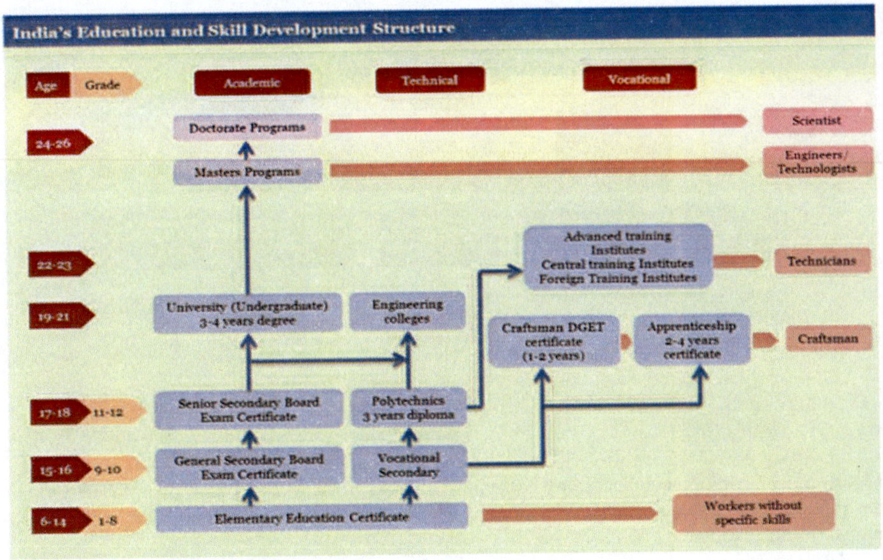

Fig. 1 India's education and skill development structure. Obtained from "Skill Development in India" Report by FICCI [19]

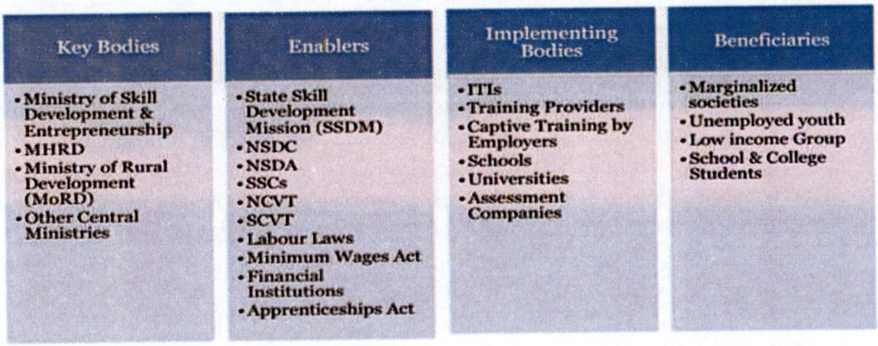

Fig. 2 Skill development ecosystem in India. Obtained from "Skill Development in India" Report by FICCI [19]

i. **MHRD**

The Ministry of Human Resource Development is regulating ministry for NCERT, UGC, and AICTE which are engaged in school education and higher education in both tech and non-tech education. In India, skills can be attained in both formal and informal ways. Formal vocational training is imparted in both public and private sector [1]

ii. **MoRD**

Ministry of Rural Development is regulating ministry for DDU-GKY (Deen Dayal Upadhyay Gram Kaushalya Yojna) which is nationwide skill training

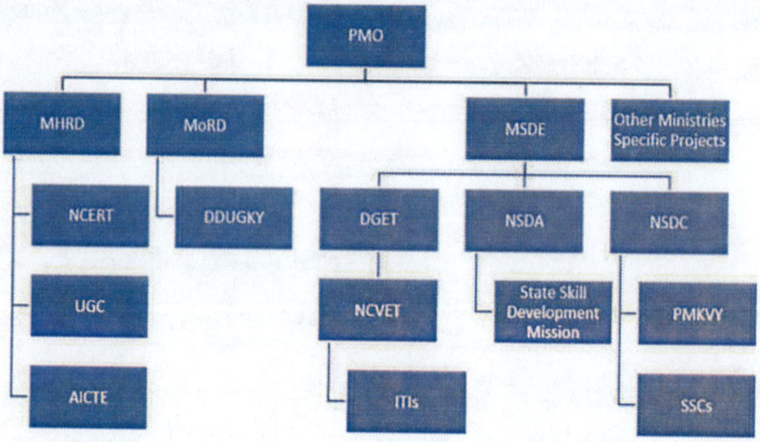

Fig. 3 Skill Development Hierarchy in India. Diagram of hierarchy based on information and descriptions available on official Websites of MHRD, MSDE, NSDC along with other various ministries and agencies

and development program focused to skill the rural population of India with aim to improve overall HDI aspects in the country [13].

iii. **MSDE**

Ministry of Skill Development and Entrepreneurship acts as the largest implementing agency for skill development in India. Also, that it is responsible for regulation 3 different and core entities as DGET, NSDA, and NSDC. Director General of Employment and Training (DGET) regulates National Center for Vocational Education and Training (NCVET) which is affiliating body for ITIs. On the other side, National Skill Development Agency (NSDA) and National Skill Development Corporation (NSDC) are entities which design, execute, and regulate various skill development schemes across India and eventually credited for training and certification of more than 80% (approx.) skilled workforce out of the total people skilled and certified in past 7 years [1–6, 8, 12, 13] (Fig. 3).

However, other 21 Central Ministries under the central government are working for improving skill development status of India. Setting up their own specific training centers and formulating Public Private Partnership (PPP) models are their core objectives.

2 Overview of Key Policies and Reports

2.1 National Skill Development Policy 2015

It discusses the population to skilled workforce ratio and compares the current skilled workforce in India with international average and standards to which suggests that less than 5% skilled workforce in the country is not sufficient to achieve the production driven economy. This policy is an expansion of National Skill Policy, 2009 having vision to create an ecosystem of empowerment by skilling on a large scale at speed with high standards and justifies the demand to supply chain of skilled workforce among 24 key sectors [9, 20–24].

2.2 *National Skill Qualification Framework (NSQF)*

This notification included all key definitions related to NSQF, like—Competence, Credit, Knowledge, Learner, Learning Outcome, NSQC, RPL, Sector, Skills, Trainer, and Training Provider. This notification briefly introduced the need for framework, objective of NSQF, level descriptor, process, qualification register, functions of various stockholder, implementation mechanism, etc. It also indicates that implementation schedule within a span of 5 years and that time limit has already passed during the year 2018 for adopting NSQF at all levels of education, training, and placement rules [7].

2.3 *Annual Report by NSDC, 2015–16*

This annual report is mirror of budget allocation, sector, and state wise current skill certification status with future targets and expected incremental job opportunities from 2016–22. Further, it includes the information about establishment of Sector Skill Councils (SSCs), National Skill Development Funds, Major Skill Development Schemes like Pradhanmantri Kaushal Vikas Yojna (PMKVY), Udaan, etc. [11].

2.4 *Annual Report of 2016–17 by MSDE*

The changes and improvement from its previous year counterpart are very nominal or minimal, however, number of sector skill councils were increased from 28 to 34. In terms of International Engagements, Qatar and Japan have also included. The whole report is based on promising future of skill ecosystem in India but somehow its lacking the pace in ground development which observed very low [11].

2.5 *New Education Policy (NEP) 2020*

The new policy also emphasizes the need to develop new and excellent skills and ensure recognition of acquired skills at various levels and through it to increase employment prospects. According to the current estimate, less than 5% of Indians in the age group of 19–24 received formal education, which is much lower than the world average, which clearly shows the need for rapid development in this area. For this, a target has been set to give 50% of the students experience of vocational education by 2025 [27].

3 Overview of Key Skill Development Projects

3.1 Deen Dayal Upadhyaya Grameen Kaushalya Yojana (DDU-GKY)

Project is focused on rural youth of economically weak families of young age group. As an important part of Skill India campaign, it plays an instrumental role in supporting programs like Make in India, Digital India, Smart Cities and Start-Up India, etc. [13].

3.2 Pradhan Mantri Kaushal Vikas Yojna (PMKVY)

PMKVY is focuses on turning skill development as a key element for future development of INDIA by the year 2022. This program focus on short-term training program, RPL, special project, kaushal and rozgar mela, placement, and monitoring activities [12].

3.3 Udaan

Udaan is the Special Industry Initiative (SII) for Jammu and Kashmir (J&K) which is funded by Ministry of Home Affairs and implemented by National Skill Development Corporation (NSDC). The target was to reach out to 40,000 youth in J&K over a period of 6 years [14].

3.4 Seekho Aur Kamao (Learn and Earn)

The scheme specifically introduced for minorities section of India. The scheme ensures placements of minimum 75% trainees, out of which at least 50% placement is in organized sector [15].

3.5 *Other State Level Projects*

Various states in India also offer skill development programs to their youth and unemployed population to cater the local needs as per geography, demography, and means of livelihood opportunities [8, 14].

4 Reviews by International Bodies of Indian Skill Development Ecosystem and Policies

4.1 *The Australian Education International of Australian Government*

It was an initial policy review by any foreign entity. Objectives of this review were associated with understanding the Indian skill policies and ecosystem for further using it as critical understanding in developing strategic collaborations and partnership opportunities [17].

4.2 *European Center for the Development of Vocational Training*

This 2011 report is a good reference for understanding skill development structure of developed countries of EU. This European research review of the advantages of vocational training and training (VET) presents the results of research administered from 2005 to 2009 in 21 European countries [18].

5 Conclusion

Key Finding: The complex and diverse skill ecosystem in India is mainly divided in formal and informal education system. Potential workforce in India is around 46 crore people but only 4.69% among them have gone through the formal education or skill development and certification system which is creating a huge skill gap in terms of demand to supply ratio.

Element of Advantage and Current Limitations: The availability of huge workforce would be a boon for India in all round development of country but establishment of various authorities at central and state level is causing coordination gap at some extent. The annual reports of central ministries and few international entities are also indicating the same. However, the central authorities and state bodies have taken many corrective measure to fill this gap, because unification of various projects and well defined coordination between various stakeholders is need of hour.

Future Perspective of Skill Ecosystem in India: The proposed "New Education Policy 2020" has shown the hope for removing the existing deficiencies and to set up a much better education and skill development mechanism across the country which will move forward side by side. It is expected that a new credit-based mechanism will smoothen the transitions between formal and informal education system in order to enhance means of skill upgradation and better livelihood.

References

1. Ministry of Human Resource Development. https://www.education.gov.in/en
2. All India Council for Technical Education. https://www.aicte-india.org/about-us/history
3. University Grant Commission. https://www.ugc.ac.in/page/genesis.aspx
4. Ministry of Skill Development and Entrepreneurship. https://www.msde.gov.in/about-msde
5. Directorate General of Training. https://dgt.gov.in/About_DGT
6. National Skill Development Agency. https://www.nsda.gov.in/nsda-about-us.html
7. National Skill Qualification Framework Notification, 2013, published in The Gazette Of India: Extraordinary [Part I—Sec. 2], Dated 27th Dec 2013, Point 14 (iii & iv), Page 27,28 by Ministry of Finance (Department of Economic Affairs). https://www.nsda.gov.in/nsqf.html#:~:text=The%20National%20Skills%20Qualifications%20Framework,non%2Dformal%20or%20informal%20learning
8. National Skill Development Corporation. https://nsdcindia.org/sector-skill-councils
9. National Skill Policy (2015) Ministry of Skill Development and Entrepreneurship. https://msde.gov.in/reports-documents/policies/national-policy-skill-development-and-entrepreneurship-2015
10. Ministry of Micro, Medium, Small Enterprises.http://msme.gov.in/
11. Annual Skill Development Reports, National Skill Development Corporation, 2015–16, 2016–17. https://nsdcindia.org/annual-reports
12. Pradhanmantri Kaushal Vikas Yojna. http://pmkvyofficial.org/
13. Deen Dayal Upadhyay Gram Kaushalya Yojna. http://ddugky.gov.in/content/about-us-0
14. Udaan, National Skill Development Corporation. https://nsdcindia.org/udaan#:~:text=Udaan%2C%20the%20Special%20Industry%20Initiative,Skill%20Development%20Corporation%20(NSDC).&text=Udaan%20program%20is%20focused%20on,or%20three%20year%20diploma%20engineers
15. Seekho Aur Kamao, Ministry of Minority Affairs. http://seekhoaurkamao-moma.gov.in/About.aspx?from=H
16. United Nations Development Report (2019). http://hdr.undp.org/en/content/2019-human-development-index-ranking
17. Policy overview of vocational education and training in India (2016) The Australian Education International of Australian Govt
18. The Benefits of Vocational Education and Training, European Center for the development of vocational training (2011)
19. Skill Development in India Report, FICCI (2015)
20. Anand YK (2009) National qualifications framework in TVET for India—Issues and Challenges presented in International Conference on "Harnessing Qualifications Framework Towards Quality Assurance in TVET" December 2009, Manila, Philippines
21. Human Resource and Skill Requirement in the Education and Skill Development Services sector (2011) – A Report by NSDC, India

22. Human Resource and Skill Requirements in the Unorganized Sector (2011) – A Report by NSDC, India
23. Mitra A (2008) India Non-formal education, Country profile prepared for the Education for National Policy for Skill Development Entrepreneurship (2015), MSDE
24. Pradhan Mantri Kaushal Vikas Yojna (PMKVY) Guideline (2016) Psacharapoulos G (1997) Vocational education and training today: Challenges and responses. Journal of Vocational Education and Training, 49(3), 285–393. Planning Commission, Government of India. (2008). Eleventh Five Year Plan (2007–2012). Sage Publications, New Delhi. Planning Commission, Government of India (2013). Twelfth Five Year Plan (2012–2017). Sage Publications, New Delhi
25. Sharma YP (2010) Skill development programs in India. Study visit on skills and TVET: India and China, November 2010
26. Tilak JBG (2002) Determinants of household expenditure on education in rural India. National Council of Applied Economic Research, Working Paper Series No. 88. New Delhi: NCAER
27. New Education Policy 2020. https://www.education.gov.in/sites/upload_files/mhrd/files/NEP_Final_English_0.pdf

Prediction of CI Engine Exhaust Emissions and Performance Using ANN

S. Charan Kumar and Amit Kumar Thakur

Abstract Automotive engines fueled by fossil fuel have contributed to a surge in ambient air pollution and the exhaustion of these fuels. To overthrow flaws posed by fossil fuels, a substitute is required. Fuels, namely CNG, LPG, alcohols, hybrid fuels, and bio-diesel, can be a substitute for fossil fuel for CI engines. To study alternate fuel viability in CI engine, investigators performed experiments which are deficient and prolonged time absorbing; therefore, modeling approaches have been adopted for better analysis. One of such modeling techniques is ANN. ANN can resolve complex tasks where numerical or traditional methods fail or are ineffective. ANN overview, application, and forecast of ANN on CI engine emission, performance is assessed. The study revealed that the majority of the investigations yield desirable engine characteristics prediction outcomes using ANN. The contrast between predicted and experimental results revealed a strong correlation coefficient, showing that the ANN model could effectively predict CI engine emissions and performance characteristics.

Keywords Alternate fuels · Artificial neural network · Bio-diesel · CI engine

Abbreviations

CNG	Compressed Natural Gas
NOx	Nitrogen Oxides
LPG	Liquefied Petroleum Gas
BTE	Break Thermal Efficiency
LNG	Liquefied Natural Gas
CO_2	Carbon dioxide
ANN	Artificial Neural Network
UHC	Unburnt Hydro Carbons
MLP	Multilayer Perceptron

S. Charan Kumar · A. K. Thakur (✉)
Mechanical Engineering, LPU, Phagwara, Punjab, India
e-mail: amit.25010@lpu.co.in

A. Maurya et al. (eds.), *Recent Trends in Mechanical Engineering*, Lecture Notes in Mechanical Engineering, https://doi.org/10.1007/978-981-19-7709-1_24

H-L	Hidden Layer
BNN	Biological Neural Network
CO	Carbon Monoxide
SFC	Specific Fuel Consumption
NON	Number of Neurons
H/O	Hidden/Output Layer
A.F.	Activation Function
MIMO	Multiple Input Multiple Output
L.F.	Loss Function
Alg	Algorithm

1 Introduction

Energy demand began to rise significantly during the 18th and 19th centuries due to the industrial revolution, and this demand is kept on increasing till now because of increment in motorization. Completely relying on crude (fossil fuels), their reserves will be exhausted in the upcoming time, and an increment in atmospheric pollution is taking place, using petro-diesel automotive vehicles. It is therefore important to go for alternate fuels like CNG, LPG, LNG, bio-diesel, hybrid fuel. Modifications in engine are necessary if CNG, LPG, and LNG are used as a substitute for diesel. Without altering the engine, it is possible to use bio-diesel. Bio-diesel is a nontoxic, and biodegradable fuel made from resources that are renewable and domestic. Bio-diesel fuel is free from sulfur content and rich in oxygen when compared with diesel which helps decrement in gases such as carbon monoxide, hydrocarbons, and acid rain. These are further categorized in various generations depending on feed biomass. The first generation utilizes edible biomass, which includes mustard, coconut, peanut, sunflower, rapeseed, and palm. The second generation utilizes non-edible crop biomass, which includes castor oil, karanja, neem, mahua, and jatropha. The third generation utilizes biomass of microorganisms, which includes algae. The fourth generation utilizes engineered algae which are genetically engineered (modified) as feed-stock. Emission, performance determination of fuels in CI engine, is time-consuming, expensive, and intricate (complex). To terminate the above flaws, modeling techniques are used. These techniques reduce reliance on laboratory results (experimental) [1, 2]. ANN is one of those techniques. ANNs are increasingly becoming crucial and frequently utilized strategies in handling a wide range of industrial challenges. For investigating combined and single aspects of experimental factors which contribute to output responses, ANN is found to be cost-effective and efficient.

1.1 ANN Feasibility

- It is learning capability in solving relations that are nonlinear in nature between variables of input–output.
- Predicts new findings from the previous trends.
- Due to its fault tolerance, it predicts output variables (findings) with deficient data.
- Helps in solving tasks for non-existence, expensive algorithmic methods.
- It is ability in executing information with incredible speed.

2 ANN Introduction

Warren Mc Culloch and Walter Pitt's modeled uncomplicated neural networks with the help of electrical circuits in 1943 to report the working of biological neurons in the brain. The above paradigm showed the way for the separation of study which includes studying biological process and neural network approach to AI. Rosenblatt created a learning system known as perceptron for the neuron model of Warren Mc Culloch, Walter Pitts in 1958. Models, namely MADALINE and ADALINE, were developed in 1960 by Marcian Hoff and Bernard Widrow. Minsky introduced MLP in 1969. For multilayer network training, in 1974, Werbos found backpropagation. It is a model of machine learning used to predict and validate the output variables based on input variables, which originated from a principle of human brain simulation, i.e., its architecture is based on neurons that are present in the brain (human). When these neurons are interconnected, they will perform tasks with incredible speed. The time taken for the human brain to complete the task is few milliseconds. ANN mimics the network of human brain neurons [3, 4]. Biological neurons have mainly four parts, namely axon, soma, synapse, and dendrites. Dendrites receive data from neurons from which it is connected (other neurons), that is, it acts as an input. Soma is a neuron's cell body whose function is to receive information from dendrites and process that information. Axon acts like a transmitter through which neurons information is sent. The connection between axon and dendrite is known as a synapse. Figure 1 shows the relationship between BNN and ANN.

2.1 ANN Structure

ANN is composed of three layers such as 1. input, 2. hidden, and 3. Ooutput layers [4–6] in Fig. 2. Input is fed into the input layers which consist of one or more than one variable (input variables). The output is obtained through an output layer which consists of single or multivariables (output variables). Hidden layer is placed in between the input–output layer. A given task may consist of single or multihidden layers. ANN consists of three modes, namely training, testing, and validating modes. Suppose we have X data results, the first training of model is performed on 70–80%

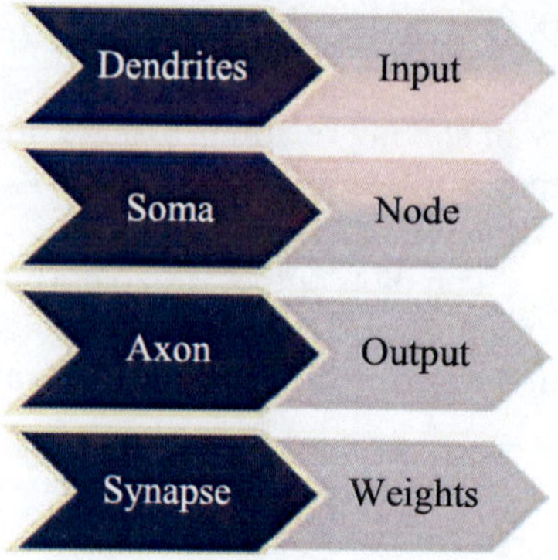

Fig. 1 Relationship between BNN and ANN [11]

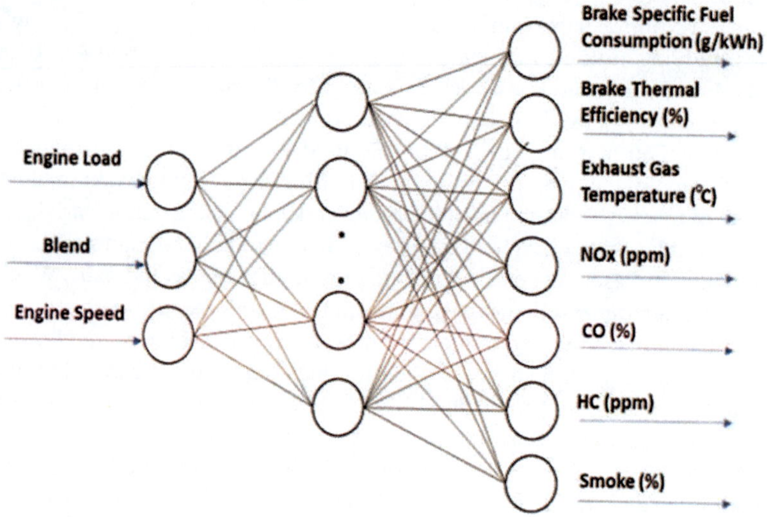

Fig. 2 ANN structure [7]

of the data results, and the remaining 30–20% data is tested and validated to predict model performance. The task (problem) using ANN is executed in four steps, shown in Fig. 3.

2.2 *Activation Function*

It defines output of a node that has to go as input to neurons in the next layer (Fig. 4), which is calculated based on mathematical formula [3] given below.

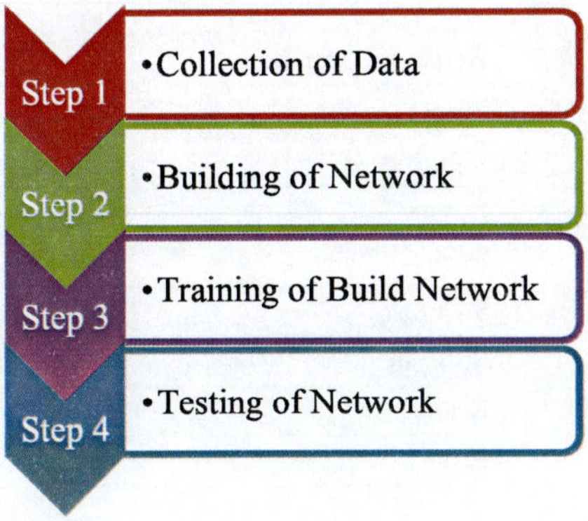

Fig. 3 Steps for using ANN for a given task [4]

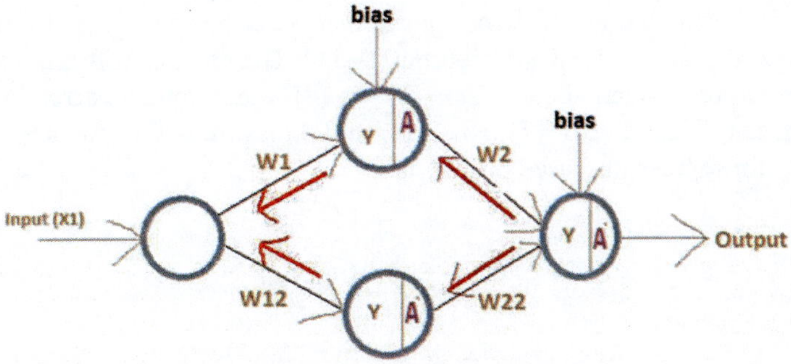

Fig. 4 Network with activation function and backpropagation [10]

$$Y = (\text{weight} \times \text{input}) + \text{bias} \tag{1}$$

The Y values range between $-\infty$ and $+\infty$. Activation function decides at what values, whether or not neuron should trigger [8]. Activate function is needed to solve the nonlinearly task (problems). It is classified into three main categories, 1. linear, 2. non-linear, and 3. binary-step function (threshold). Non-linear functions are further classified into 1. sigmoid or logistic, 2. rectified linear units (ReLUs), 3.tanh or hyperbolic, 4. parametric rectified linear units (P-ReLU), 5. exponential linear units (ELUs), 6. soft plus, and 7. softmax. Ranges of the various activation functions are given in Table 1. Sigmoid, tanh, and ReLU activation functions are used more often.

Table 1 Various activation function ranges

Activation function	Range
Linear	$-\infty$ to $+\infty$
Sigmoid	0 to 1
Binary	$-\infty$ to $+\infty$
Tanh	-1 to 1
Softmax	0 to 1
Soft plus	0 to $+\infty$
ReLU	0 to $+\infty$
ELU	$-\infty$ to $+\infty$

2.3 Loss Function and Learning Algorithm

Loss function evaluates how well the algorithm models the given data, used to predict error between obtained and desired output [9–11]. The one that offers the least error (loss) is preferred. Mean square error (MSE) [4], mean relative error (MRE) [7], and root mean squared error [7] (RMSE) are the loss functions that are most used commonly for regression-based problems.

$$MSE = \frac{1}{N}\sum_{i=1}^{N}(a_i - p_i)^2 \tag{2}$$

$$MRE = \frac{1}{N}\sum_{i=1}^{N}\left|100 \times \frac{(a_i - p_i)}{a_i}\right| \tag{3}$$

$$RMSE = \sqrt{\frac{1}{N}\sum_{i=1}^{N}(a_i - p_i)^2} \tag{4}$$

Selection of loss function depends on the output variable that is predicting output as a numerical value, binary, etc.…The learning and training of algorithms are a most important part which is conducted by updating bias (constant) and weights, to predict the acceptable output within errors. These are widely categorized into three groups such as 1. reinforcement, 2. unsupervisied, and 3. supervised learning. Backpropagation [12–16] (Fig. 4) is the most commonly used algorithm (learning, training) which includes Quasi-Newton, Levenberg–Marquardt (trainlm), scaled conjugate gradient, etc., which comes under supervised learning. The accuracy obtained from training and testing is compared by correlation coefficient (R). R ranges from -1 to $+1$. If it is near to $+1$, there is a strong relation between two factors (variables).

2.4 ANN Application in Mechanical Engineering

- Fault diagnosis of equipment.
- Structural analysis.
- Validation of data.
- Prediction of thermal properties.
- Geometrical modeling, optimization, and design.

3 Performance and Emissions Using ANN in CI Engine

3.1 Bio-Diesel

Vairamuthu et al. [1] examined calophyllum inophyllum derived bio-diesel of proportion (B100) by varying opening pressure of a nozzle (250, 240, 230, 220 bar) in CI four stroke engine. They stated SFC increment and CO, HC decrement for B100. They have done validation of the above through ANN. Authors chosen nozzle opening pres: (NOP), blend, load as input, output variables were smoke, SFC, NO_x, BTE, CO_2, UHC, EGT, CO. 15% and 15% and 75% of data were distributed for testing and validating and training purpose. Transfer function selected by the authors was logsigmoid/linear for the hidden/output layer. Training was executed using different algorithms, namely trainrp, traingdm, trainscg, trainlm. Result was correlated with R

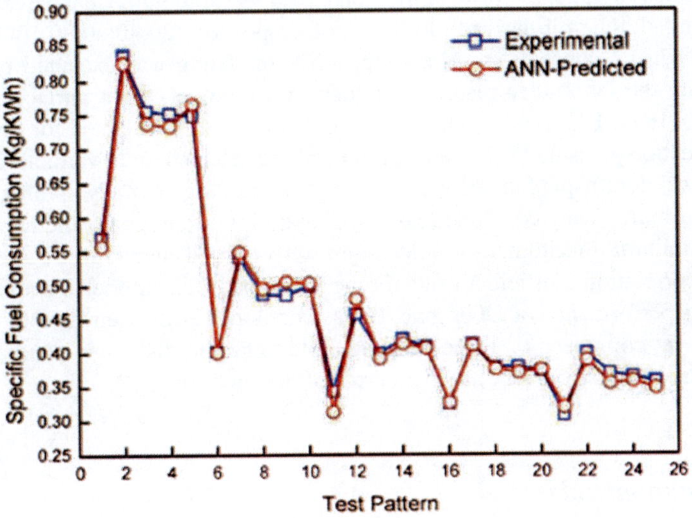

Fig. 5 SFC with ANN [1]

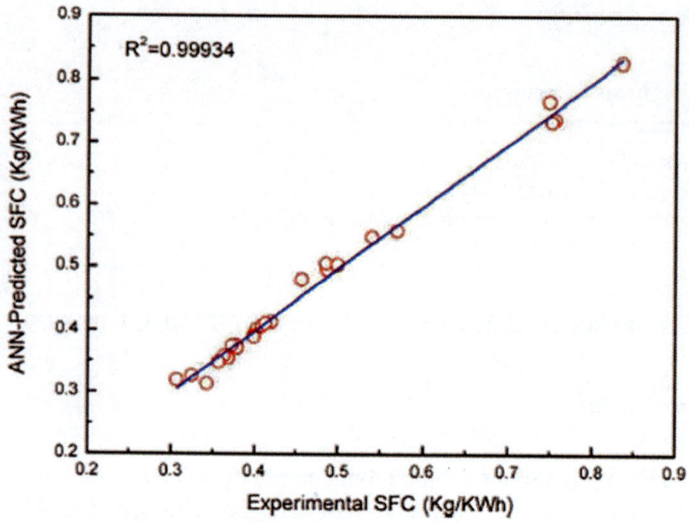

Fig. 6 SFC and its correlation coefficient value [1]

(in Figs. 5 and 6). They conclude least error and high R for hidden layers composed of 23 neurons with trainlm algorithm.

An investigation was conducted by Giwa et al. [2] to predict properties of fuel using ANN model. Their results demonstrated high accuracy for log sigmoid/linear as hidden/output layer transfer function with trainlm algorithm. Their ANN model has predicted high degree of R in the range of 0.991 to 0.994. Channapattana et al. [3] examined hone oil derived bio-diesel of proportion (20,40,60,80,100) to know feasibility in CI constant speed engine. ANN modeling was executed by authors to validate the laboratory results. For their ANN model, eight variables, namely BTE, CO, EGT, HC, BSFC, NOx, CO_2, and smoke, were selected for output. CR, static injection pressure (SIT), load, and blend are selected input variables. Network training was done using four different algorithms (training), namely trainlm, traingdx, trainscg, trainrp. They conclude least MSE and high accuracy for the H–L with 28 neurons, trainlm algorithm. In another major study carried out by Javed et al. [4] yields a strong correlation coefficient value for the developed ANN model to map emissions as well as performance of CI engine. It was concluded from their investigation that hidden layer composed of 16 neurons are able to predict the engine characteristics (emissions and performance) with an error of not more than 5%.

3.2 Nanoparticles

Saree et al. [5] tested nanoparticle cerium oxide of proportions 10, 20, 40 ppm to know its viability in CI engine when blended in diesel. Investigators observed BSFC, HC,

NOx decrement with increase in proportions of nanoparticle. Investigators validated laboratory results using ANN. Input variables SFC, nanoaddition, engine speed, and output variables, namely CO, HC, power, and NOx, were chosen by authors. Investigators conclude least MSE and high R, respectively, 0.000172 and 1 for 3-12-4 neural architecture with trainlm implemented training rule and activation function logsigmoid. An ANN model was developed to predict the engine characteristics of a CI engine powered with blends of nanoalumina in diesel [6]. The forecast was determined by means of the R. Strong correlation between expected and obtained values was observed for NOx, CO, HC, fuel consumption, power, with R as 0.9998, 0.99977, 0.98506, 0.99912, 0.99999, respectively.

3.3 *Alcohols and CNG*

Uslu et al. [7] examined diethyl ether fuel of proportions by volume (10, 7.5, 5, 2.5%) to determine emission and performance in CI air cooled engine. They observed 56% NOx decrement, 45% CO decrement, and 10% BSFC increment for DEE10 against diesel. ANN was employed in order to validate the experimental data. Authors distributed data in proportions 75% for training and 25% for testing. By means of RMSE, MRE loss parameters were determined which was found to be in the range 0.51% to 4.8%. They conclude good correlation between predicted and laboratory results using ANN. Yusuf et al. [8] evaluated the model of ANN to predict output variables, namely exhaust temp: O_2, CO_2, CO, NOx, BTE, torque, BSFC, BP based on input variables diesel-CNG ratio and engine speed. Authors disturbed their data into 70%, 30% for training, testing purpose. MATLAB was selected by investigators for simulating ANN. Training was executed using different algorithms, namely traingdx, trainlm. Various transfer functions were selected by the authors logsigmoid/linear, sigmoid/axon, softmax/linear, tanx/linear for hidden/output layer. Investigators observed least error and strong R for hidden layer composed of 19 neurons with logsigmoid/linear as activation function for hidden/output layer and trainlm algorithm. The R values were 0.9934, 0.9705, 0.9964, 0.9359, 0.9570, 0.9289, 0.9884, 0.9838, 0.9808, respectively, for exhaust temperature, O_2, CO_2, CO, NOx, BTE, Torque, BSFC, and BP. ANN model of various authors is given in Table 2. Correlation coefficient findings of various authors are shown in Table 3.

4 Conclusion

Present paper evaluates experimental work executed by various investigators employing ANN on engine emissions and performance of alternate fuels. ANN can

Table 2 Various authors findings-ANN model

Refs.	A.F. -H/O	Training Alg:	L.F	H-L (NON)
Javed et al. [4]	Tan Sig/Tan Sig	Resilient	MSE	15
	Linear/Tan Sig	Backpropagation	MSE	16
	Log Sigmoid/Tan Sig		MSE	8
	Tan Sig/Tan Sig		MSE	20
Saraee et al. [5]	Log Sigmoid/Linear	trainlm	MSE	12
Tosun et al. [10]	Log Sigmoid/Linear	trainlm	MAPE	7
Tosun et al. [10]	Log Sigmoid/Linear	trainlm	MAPE	9
Tosun et al. [10]	Log Sigmoid/Linear	trainlm	MAPE	13
Uslu et al. [11]	Log Sigmoid	trainlm	MSE	10
Özgür et al. [12]	Log Sigmoid/Linear	trainlm	MAPE	3, 4

Table 3 Correlation coefficient findings of various authors

Refs.	CO	BTE	NOx	BSFC	HC
Saree et al. [5]	0.99899	–	0.99989	–	0.99899
Uslu et al. [7]	0.9825	0.9829	0.9831	0.985	0.964
Yusaf et al. [8]	0.9359	–	0.95707	0.9838	–
Roy et al. [9]	–	0.9966	0.99947	0.9999	0.9997
Kullolli et al. [13]	1	0.996	0.99	–	0.99
Rao et al. [14]	0.98	0.999	0.999	0.98	0.985
Syed et al. [15]	0.9975	0.9996	0.9996	0.9968	0.9898
Joshi et al. [16]	0.9682	0.96858	0.99044	0.95232	0.9646
Charudatta et al. [17]	0.9998	0.9978	0.9986	0.9904	0.9923
Raghuvaran et al. [18]	0.9989	0.9994	0.99899	0.999	0.9883
Hosamani et al. [19]	0.9800	0.9989	0.9332	0.9971	0.9700

forecast an accurate relationship between input and output variables. According to the literature analysis, the following conclusions are drawn.

- MIMO topology was used.
- Hidden layers found to be double or single, and algorithm trainlm was found to be foremost suited for training the network.
- Strong correlation can be seen for predicted and obtained values using ANN.

References

1. Vairamuthu G, Thangagiri B, Sundarapandian S (2018) Experimental and artificial neural network based prediction of performance and emission characteristics of DI diesel engine using Calophyllum inophyllum methyl ester at different nozzle opening pressure. Heat Mass Transf 54:99–113
2. Giwa SO, Adekomaya SO, Adama KO, Mukaila MO (2015) Prediction of selected biodiesel fuel properties using artificial neural network. Front Energy 9:433–445
3. Channapattana SV, Pawar AA, Kamble PG (2017) Optimisation of operating parameters of DI-CI engine fueled with second generation bio-fuel and development of ANN based prediction model. Appl Energy 187:84–95
4. Javed S, Satyanarayana Murthy YVV, Baig RU, Prasada Rao D (2015) Development of ANN model for prediction of performance and emission characteristics of hydrogen dual fueled diesel engine with Jatropha Methyl Ester biodiesel blends. J Nat Gas Sci Eng 26:549–557
5. Soukht Saraee H, Taghavifar H, Jafarmadar S (2017) Experimental and numerical consideration of the effect of CeO2 nanoparticles on diesel engine performance and exhaust emission with the aid of artificial neural network. Appl Therm Eng 113:663–672
6. Saraee HS, Jafarmadar S, Alizadeh-Haghighi E, Jamal Ashrafi S (2015) Experimental investigation of pollution and fuel consumption on a CI engine operated on alumina nanoparticles—Diesel fuel with the aid of artificial neural network. Environ Progr Sustain Energy 35(2):540–546
7. Uslu S, Celik MB (2018) Prediction of engine emissions and performance with artificial neural networks in a single cylinder diesel engine using diethyl ether. Eng Sci Technol Int J 21:1194–1201
8. Yusaf TF, Buttsworth DR, Saleh KH, Yousif BF (2010) CNG-diesel engine performance and exhaust emission analysis with the aid of artificial neural network. Appl Energy 87:1661–1669
9. Roy S, Banerjee R, Das AK, Bose PK (2014) Development of an ANN based system identification tool to estimate the performance-emission characteristics of a CRDI assisted CNG dual fuel diesel engine. J Nat Gas Sci Eng 21:147–158
10. Tosun E, Aydin K, Bilgili M (2016) Comparison of linear regression and artificial neural network model of a diesel engine fueled with biodiesel-alcohol mixtures. Alex Eng J 55:3081–3089
11. Uslu S (2020) Optimization of diesel engine operating parameters fueled with palm oil-diesel blend: Comparative evaluation between response surface methodology (RSM) and artificial neural network (ANN). Fuel 276
12. Özgür C, Tosun E (2017) Prediction of density and kinematic viscosity of biodiesel by artificial neural networks. Energy Sour Part A: Recov, Utiliz, Environ Eff 39:985–991
13. Kullolli S, Sakthivel G, Ilangkumaran M (2016) A neural network model for the prediction of compression ignition engine performance at different injection timings. Int J Ambient Energy 37:227–236
14. Prasada Rao K, Victor Babu T, Anuradha G, Appa Rao BV (2017) IDI diesel engine performance and exhaust emission analysis using biodiesel with an artificial neural network (ANN). Egypt J Pet 26:593–600
15. Syed J, Baig RU, Algarni S, Murthy YVVS, Masood M, Inamurrahman M (2017) Artificial Neural Network modeling of a hydrogen dual fueled diesel engine characteristics: an experiment approach. Int J Hydrogen Energy 42:14750–14774
16. Joshi MP, Thipse SS (2019) Combustion analysis of a compression-ignition engine fuelled with an algae biofuel blend and diethyl ether as an additive by using an artificial neural network. Biofuels 7269
17. Kshirsagar CM, Ramanathan A (2017) Artificial neural network applied forecast on a parametric study of Calophyllum inophyllum methyl ester-diesel engine out responses. Appl Energy 189:555–567

18. Raghuvaran S, Ashok B, Veluchamy B, Ganesh N (2021) Evaluation of performance and exhaust emission of C.I diesel engine fuel with palm oil biodiesel using an artificial neural network. Mater Today: Proc 37(2):1107–1111. https://doi.org/10.1016/j.matpr.2020.06.344
19. Hosamani BR, Syed Abbas Ali, Vadiraj K (2021) Assessment of performance and exhaust emission quality of different compression ratio engine using two biodiesel mixture: artificial neural network approach. Alex Eng J 60(1):837–844. https://doi.org/10.1016/j.aej.2020.10.012

Energy and Exergy Analysis of a Solar Dish Stirling Heat Engine with Bottoming Organic Rankine Cycle

M. S. Ashwin, Siddharth Ramachandran, and Naveen Kumar

Abstract Effective utilization of solar thermal energy is one of the thrust full areas of research for a sustainable future. Stirling engines are a promising technology which is widely used to produce kW level electric power from solar energy using parabolic dish concentrators. However, due to design imperfections, typical Stirling engines approximately have 51% heat loss. This energy can be utilized for additional power generation via coupling an organic Rankine cycle (ORC) with a waste heat recovery unit. The previous investigations shown a 41.5% overall efficiency of combined Stirling-ORC systems and adding ORC to Stirling cycle can improve power output and efficiency by 4% and 8%, respectively. However, these systems have not yet been coupled with solar thermal energy, and its real-time feasibility has to be evaluated energetically and exergetically. In the present investigation, a typical solar parabolic dish system is considered as a heat source to the Stirling heat engine. The effect of real-time solar irradiation and absorber temperature on performance of solar Stirling-ORC system has been quantified. It was found that 48% of the total energy lost by the Stirling cycle can be recovered by combining it with ORC. With the intention of design emphasis, a component wise exergy destruction rate is evaluated and illustrated. Further, a selection criterion for working fluids of ORC is also evolved.

Keywords Renewable energy · Waste heat · Stirling-ORC · Design · Solar thermal

1 Introduction

Solar energy is one of the least polluting sources of energy. In addition to it is renewable in nature, recent advancements in these technologies have made solar energy reasonably affordable to the people [1]. Sequentially, this has resulted in solar energy being used as one of the mainstream electricity sources. Solar thermal energy conversion has a wider range of application than photovoltaics, because solar radiation can

M. S. Ashwin (✉) · S. Ramachandran · N. Kumar
Indian Institute of Information Technology, Design and Manufacturing, Kancheepuram, Tamil Nadu 600127, India
e-mail: mfd16i006@iiitdm.ac.in

A. Maurya et al. (eds.), *Recent Trends in Mechanical Engineering*, Lecture Notes in Mechanical Engineering, https://doi.org/10.1007/978-981-19-7709-1_25

be concentrated, collected, and transmitted effectively. Major applications of solar thermal energy conversion are for water heating, cooking, home heating and cooling, or generating electricity [2–4]. Stirling engine is a kind of external combustion engine that is used to convert the solar thermal energy into kinetic (or) mechanical energy, and this is done by simultaneous heating and cooling of the working gas sealed inside the cylinders [4]. Detailed energetical and exergetical analysis with design optimization of these systems provide an in-depth knowledge about the feasibility of system [5].

In theory, the thermal efficiency of Stirling engine can reach as high as a Carnot cycle which has the highest possible thermal efficiency theoretically [6]. Therefore, literature in this genre focuses mainly on two areas for improvement, those are reduction of losses, increase power output [7, 8]. However, there is a considerable amount of heat lost from the cold side heat exchanger of the Stirling engine, i.e., for an engine operating at a temperature of 900 K and 1 kW, almost 700 W of heat is rejected from the cold side [9].

This waste heat can be effectively utilized for utility purposes or additional power generation to improve the overall efficiency of the solar dish Stirling engine system [9, 10]. Recently, Bahrami et al. [11] have shown that the total power output may be improved with the proper implementation of a combined power cycle with a bottoming organic Rankine cycle (ORC). There is enough amount of heat available for running a bottom cycle which can improve the effective efficiency of the combined cycle (see Fig. 1) [9–11]. However, so far in the literature, these systems are not yet explored energetically and exergetically. In this study, a detailed thermodynamic model comprises of energy and exergy balance which simulate the waste heat recovery from the Stirling engine via a bottoming organic Rankine cycle (ORC) is evolved. The developed thermodynamic model is used for detailed parametric study. The effect of parameters such as absorber temperature and concentration ratio on performance of solar Stirling-ORC is discussed. Further, performance of different ORC working fluids is also incorporated as a selection criterion.

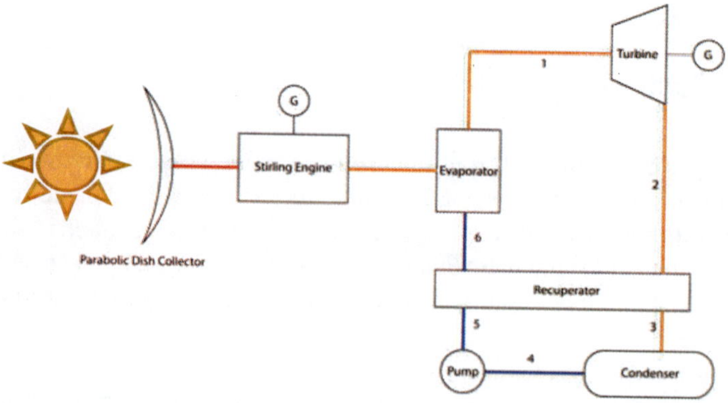

Fig. 1 Schematic of combined solar Stirling-ORC system

2 Solar Stirling-ORC Combined Cycle Thermodynamic Model

The whole of the system was modeled in assuming steady-state conditions. The thermal efficiency of solar dish collector was calculated using Eq. 1 [3],

$$\eta_{solar-collector} = \eta_0 - \frac{1}{IC}[h(T_H - T_0) + \varepsilon\partial(T_H^4 - T_0^4)] \tag{1}$$

where η_0 is the optical efficiency of the collector, I is the direct flux solar intensity, h is the conduction/convection coefficient, ε is emissivity factor of the collector, T_0 is the ambient temperature, and ∂ is the Stefan Boltzmann constant.

Considering isentropic efficiencies for all of the components, also the heat exchanger effectiveness was considered to be ideal. All of the calculations were done for the cold side temperature of Stirling engine of 140 °C. The Stirling engine was modeled with a fraction of Carnot (f_{carnot}) 0.5, and the thermal efficiency of Stirling engine was calculated using Eq. 1.

$$\eta_{Stirling} = f_{carnot}\eta_{carnot} \tag{2}$$

$$\eta_{carnot} = 1 - \frac{T_C}{T_H} \tag{3}$$

where η is the thermal efficiency, T_C is the cold side temperature and T_H is the hot side temperature (K).

The Stirling cycle power output, W_{st}, is calculated by multiplying the Stirling engine efficiency and the heat input to the Stirling engine, $Q_{in,St}$. The remaining heat, $Q_{out,St}$, is rejected out from the Stirling engine. This rejected heat is used as input heat for the bottoming organic Rankine cycle (ORC).

$$W_{St} = \eta_{St}\, Q_{in,St} \tag{4}$$

$$Q_{out,St} = W_{St} - Q_{in,St} \tag{5}$$

The cycle is modeled by evaluating values at different state points of the various components. The highest temperature of ORC is assumed to be 5 °C pinch point difference from cooler side Stirling cycle temperature, and isobaric conditions are used [9].

$$\dot{m}_{ORC}(h_1 - h_6) = Q_{out,St} \tag{6}$$

$$h_2 = h_1 - \eta_{s,\exp}(h_1 - h_{2s}) \tag{7}$$

where $\dot{m}$ is the mass flow rate(g/s), h is the specific enthalpy at any state point and $\eta_{s,\exp}$ is the Rankine isentropic efficiency which is taken as 0.9.

The recuperator is considered if the pump outlet temperature (point 5) is less than or equal to the outlet turbine temperature (point 2). For the present study, recuperator effectiveness (ε_{rec}) is assumed to be 0.9 [9, 11].

$$\varepsilon_{rec} = \frac{h_2 - h_3}{h_2 - h(P_2, T_5)} \tag{8}$$

$$h_5 = h_4 + \frac{1}{\eta_{s,pump}}(h_{5s} - h_4) \tag{9}$$

where P_2 is the pressure at state point 2 and T_5 is the temperature at state point 5. The pump was modeled with a constant isentropic efficiency of 50%, which accounts for motor efficiency and the actual isentropic efficiency of the pump.

The net work done by the system (W_{net}) is calculated using (10) [10],

$$W_{net} = W_{St} + W_{ORC} - W_{pump} \tag{10}$$

where W is the work done, St is Stirling and ORC organic Rankine cycle. The total efficiency and the 2nd law efficiency are calculated using Eqs. (11) and (12), respectively.

$$\eta_{total} = \frac{W_{net}}{Q_{in}} \tag{11}$$

$$\eta_{2law} = \frac{\eta_{total}}{\eta_{carnot}} \tag{12}$$

The exergic efficiency of the system is given by [3, 6],

$$\eta_{ex} = \frac{W_{net}}{W_{net} + T_0 S_{gen}} \tag{13}$$

where S_{gen} is the entropy generation rate given as

$$S_{gen} = \frac{Q_{cond}}{T_{cond}} - \frac{Q_{is,St}}{T_H} \tag{14}$$

Equations (1)–(13) are solved simultaneously in Engineering Equation Solver (EES), and results are generated.

3 Results and Discussion

The enthalpies, entropies, and other properties like pressure and temperature at different state points were calculated using the EES software. Results showed that there is an increase in efficiency when we incorporate an organic Rankine bottoming cycle to the Stirling engine output.

Parameter	Parameter description	Value	Units
η_{St}	Stirling engine efficiency	28.77%	–
$\eta_{s,\exp}$	Expander isentropic efficiency	90%	–
$\eta_{s,pump}$	Pump isentropic efficiency	50%	–
η_{total}	Total efficiency	41.32%	–
$Q_{in,St}$	Stirling engine heat input	1000	kJ/s
Q_{cond}	Condenser heat	551.6	kJ/s
$Q_{out,St}$	Stirling engine heat rejected	712.3	kJ/s
W_{St}	Work done Stirling engine	287.7	kJ/s
W_t	Work done turbine	133.9	kJ/s
W_{pump}	Work done pump	8.437	kJ/s
W_{ORC}	Work done organic Rankine cycle	133.0.9	kJ/s
W_{net}	Net work done	413.2	kJ/s

The efficiency increased from 28.77% without bottoming cycle to 41.32% with ORC bottoming cycle (See Fig. 2).

The simulated numerical model was run for different values of absorber temperature. Figures 2 and 3 give us an idea about the effect of absorber temperature on the Stirling-ORC efficiency, and it is found out that maximum efficiency is at an absorber temperature of around 1130 K with maximum efficiency of 36.9% and power output

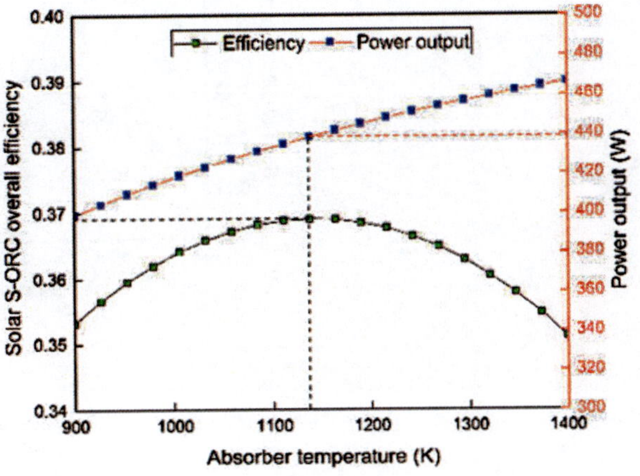

Fig. 2 Effect of absorber temperature on collector efficiency and power output of Stirling-ORC

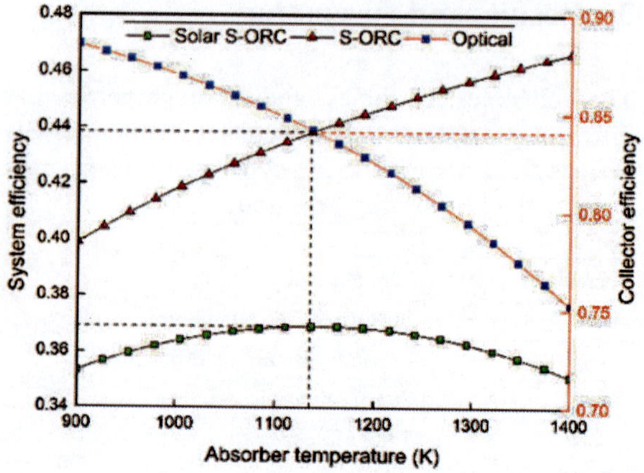

Fig. 3 Effect of absorber temperature on collector efficiency, S-ORC efficiency, solar S-ORC efficiency, and Stirling-ORC

of around 440 W for heat input of 1 kW. The selection criteria for working fluid for optimum efficiency are selected by studying the effect of various working fluids like FC72, FC87, HFE7000, and HFE7100. As shown in Fig. 3, it was found that HFE7100 has the highest efficiency and power output compared to the other fluids at any given temperature (Fig. 4).

Figure 4 shows the variation of the solar S-ORC efficiency with respect to the absorber temperature for different values of concentration ratio (C). Change in concentration ratio affects the optical efficiency of the solar collector which, in turn, affects the overall solar S-ORC efficiency significantly (Fig. 5).

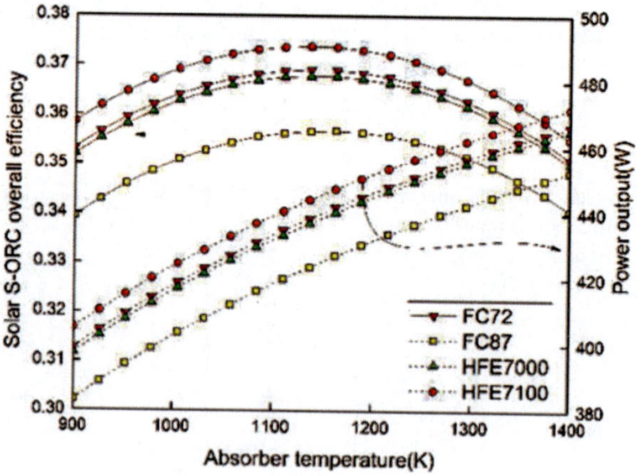

Fig. 4 Selection of optimum working fluid

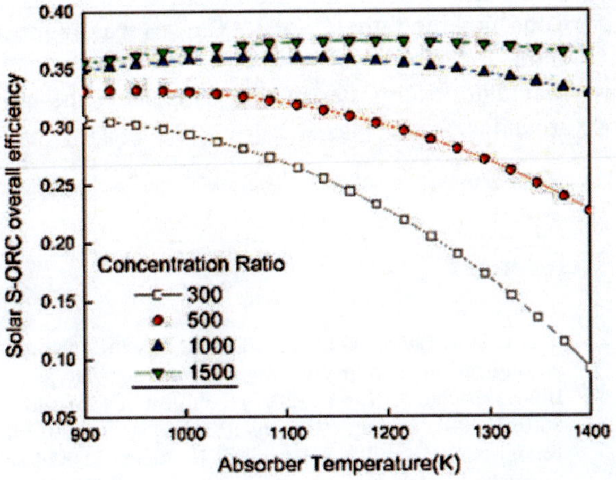

Fig. 5 Effect of concentration ratio (C) on solar S-ORC efficiency for different absorber temperature

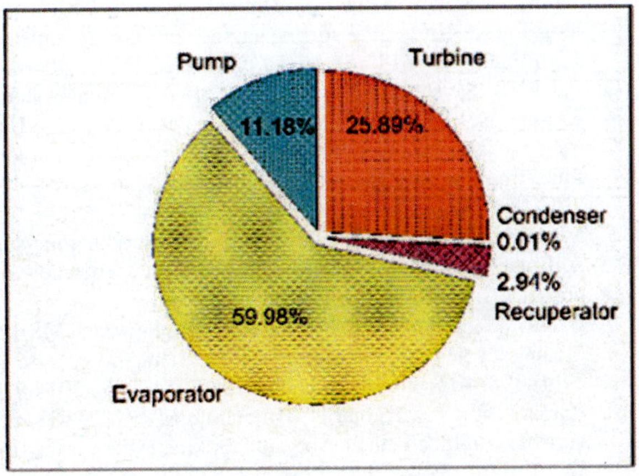

Fig. 6 Component wise percentage exergy destruction in waste heat recovery ORC

4 Conclusion

Waste heat recovery from the Stirling engine to run the ORC has increased the overall efficiency of solar dish Stirling engine from 28.77 to 41.32%. Exergy and energy analysis carried out in the present investigation give an insight about identifying the optimum working fluid for ORC and performance prediction. This investigation can be used as a basis for a detailed design and optimization of the solar Stirling-ORC system. From the detailed parametric study, it was found that

- Solar Stirling engine absorber temperature of 1130 K yields maximum efficiency for the solar S-ORC system.
- HFE7100 is the optimum working fluid among the four with respect to efficiency (0.375) and power output of 450 W at optimum temperature of 1100 K.

- Concentration ratio (C) of 1500 gives maximum efficiency at any given temperature.
- Maximum exergy destruction happens at the evaporator in the ORC which is around 60% (see Fig. 6).

References

1. International Energy Outlook (2019) Energy Information Administration. Outlook, vol 0484, no July, pp 70–99. https://www.eia.gov/outlooks/ieo/pdf/ieo2019.pdf
2. Boutammachte N, Knorr J (2012) Field-test of a solar low delta-T Stirling engine. Sol Energy 86(6):1849–1856. https://doi.org/10.1016/j.solener.2012.03.001
3. Kalogirou SA, Karellas S, Braimakis K, Stanciu C, Badescu V (2016) Exergy analysis of solar thermal collectors and processes. Prog Energy Combust Sci 56:106–137. https://doi.org/10.1016/j.pecs.2016.05.002
4. Ramachandran S, Kumar N, Timmaraju MV (2020) Thermodynamic analysis of solar low-temperature differential stirling engine considering imperfect regeneration and thermal losses. J Sol Energy Eng 142(5). https://doi.org/10.1115/1.4046629
5. Patel VK, Savsani VJ, Tawhid MA (2019) Thermal system optimization
6. Klein S, Nellis G (2011) Thermodynamics. Cambridge University Press, Cambridge
7. Yaqi L, Yaling H, Weiwei W (2011) Optimization of solar-powered Stirling heat engine with finite-time thermodynamics. Renew Energy 36(1):421–427. https://doi.org/10.1016/j.renene.2010.06.037
8. Martaj N, Grosu L, Rochelle P (2007) Thermodynamic study of a low temperature difference stirling engine at steady state operation. Int J Thermodyn 10(4):165–176. https://doi.org/10.5541/IJOT.1034000200
9. Bahari SS, Sameti M, Ahmadi MH, Haghgooyan MS (2016) Optimisation of a combined Stirling cycle–organic Rankine cycle using a genetic algorithm. Int J Ambient Energy 37(4):398–402. https://doi.org/10.1080/01430750.2014.977497
10. Ramachandran S, Kumar N, Timmaraju MV (2021) Thermodynamic investigation of an irreversible combined stirling-organic rankine cycle for maximum power output condition. J Eng Gas Turbines Power. https://doi.org/10.1115/1.4049775
11. Bahrami M, Hamidi AA, Porkhial S (2013) Investigation of the effect of organic working fluids on thermodynamic performance of combined cycle stirling-ORC. Int J Energy Environ Eng 4(1):1–9. https://doi.org/10.1186/2251-6832-4-12

A Review on Tribological Properties of Mild Steel Improved by Laser Cladding Process Using Different Coating Powder

Sujeet Kumar and Anil Kumar Das

Abstract Mild steel possesses a unique combination of mechanical properties, hence frequently used as a structural material. While poor tribological properties of mild steel restricted to use of a wide range of applications; therefore, laser cladding process gained much attention to improve tribological properties of mild steel by providing a composite layer near the surface. The properties of the composite layer influenced by the use of matrix as well as reinforcement materials. In the present work, authors have discussed in detail the effect of coating materials as well as process parameters of cladding techniques such as current, voltage, scan speed, stand-off distance (SOD), beam focal position, and feeding ways of coating materials (in case of laser cladding) on microstructure, wear resistance, coefficient of friction, and microhardness of a composite layer of mild steel. The methods to providing the coating powder during cladding on the mild steel are summarized. Problems create during cladding on mild steel; some solution and preparation tendencies are also reviewed.

Keywords Cladding · Microstructure · Wear resistance · Microhardness

1 Introduction

Mild steel is a very essential material for many industries such as infrastructure, manufacturing industries, and shipbuilding to make the components due to their high strength and low cost. However, low corrosion resistance and low hardness lead to wear out and damage of the components; therefore, these drawbacks reducing their service life. There are various surface modification processes such as laser cladding, tungsten inert gas (TIG) cladding, and electron beam (EBM) cladding, and these

S. Kumar (✉) · A. K. Das
Department of Mechanical Engineering, National Institute of Technology Patna, Patna 800005, India
e-mail: sujeetk65@gmail.com

A. K. Das
e-mail: akdas@nitp.ac.in

A. Maurya et al. (eds.), *Recent Trends in Mechanical Engineering*, Lecture Notes in Mechanical Engineering, https://doi.org/10.1007/978-981-19-7709-1_26

cladding processes are responsible to enhance the surface properties and tribological properties of the materials such as hardness, corrosion resistance, wear resistance, and friction coefficient without changing and altering its bulk material properties [1, 2]. Nowadays, researchers are attracted to the laser cladding field because of the following reason, it provides dense and refined microstructure with the less heat-affected zone (HAZ) as well as lower dilution rate and small thermal deformation [3]. Laser cladding is a unique cladding process that provides a thick coating layer with a good metallurgical bond between the coating material and substrate [4, 5].

Wear is the most frequent failure mode of mechanical components. It attacks the upper surface of the components which is under moving condition, and it indicates that wear is a surface phenomenon. It is expected that for excellent wear resistance the substrate, materials should have high hardness and high toughness [6]. These properties of materials are achieved by producing reinforced metal matrix composite using surface modification process due to synergetic effect of the combination of ductile matrix and hard reinforcement. The coefficient of friction or friction force of metal matrix composite (MMC) of any substrate is high because of the presence of a metallic binder [7]. In machinery industries, reduction of friction drag and wear rate is challenging tasks. Using liquid lubricants, it is capable to reduce friction and wear damage but liquid lubricants inefficient to retain their properties at high temperatures during operating conditions. Therefore, a solid lubricant is used to overcome these difficulties [8]. Lu et al. [9] produced the clad layer with the mixture of nickel and hexagonal boron nitride (h-BN) which act as self-lubricating in the substrate layer [9].

In this paper, the authors give details about the development status of the cladding layer on the mild steel by the laser cladding process. This paper also focused on the influence of process parameters of laser on the tribological properties, hardness, and microstructure of the coating.

2 Laser Cladding Procedure

2.1 Preparation of Substrate and Coating Powder

Several researchers are interested to enhance the surface properties of mild steel. There are following mild steel grades such as AISI 1010, 1020, 1030, 1040, 1050, and 16Mn steel that are selected as substrate by the researchers. For the present review, mild steel was also selected as substrate materials. The authors used emery paper to remove the oxide layer from the substrate for proper bonding and then bathe the substrate in acetone.

2.2 Methods to Use Coating Powder

There are two methods to use the coating powder for the formation of metal matrix composite on the substrate layer.

(i) Ex situ method: In this method, the ceramic powder such as TiN, TiC, and TiB_2 adding directly for the formation of a coating layer on the substrate. The composite coating produced by the ex situ method leads to poor adhesion and a poor metallurgical bond between the coating powder and substrate. Zhang et al. [10] synthesized the Ni-based WC and TiC composite coating by ex situ method means WC and TiC were used directly as coating materials [10].

(ii) In situ method: In this method, ceramic phases such as carbide, boride, and nitride are synthesized by reaction during the cladding process. In situ cladding produced good bonding between coating powder and substrate. When compared with the ex situ process, the in situ process is the more economical process [11]. Du et al. [4] were also developed in situ TiB_2 and Fe composite coating using ferrotitanium and ferroboron as coating powder. They successfully produced a crack-free and smooth cladded layer using laser cladding [4]. Meng and Ji [12] produced the TiN-TiB2/Ni composite Coating on 16Mn steel plate by the in situ process using argon arc cladding [12]. They used titanium and BN powder and synthesized the TiN-TiB2/Ni composite coating that was verified in the result of X-ray diffraction result.

2.3 Feeding Methods of Cladding Materials

There are four methods to feeding the cladding materials that are shown in Fig. 1.

(i) Paste form feeding: Fig. 1a showing the laser cladding arrangement in which the coating materials provide in paste form. In this method, it make the molten pool of the selected coating powder and then provide it just before the laser beam at the same time.

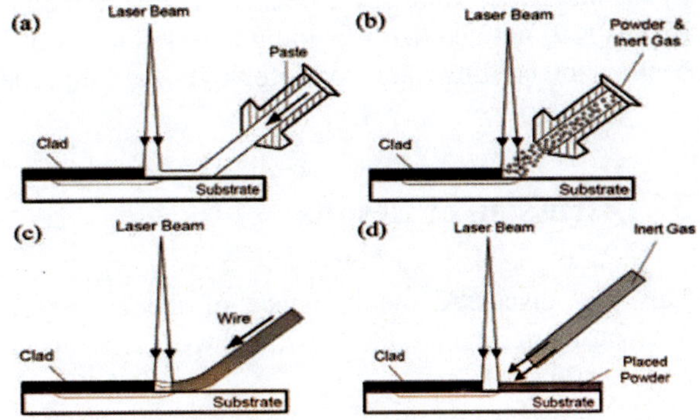

Fig. 1 Feeding methods of coating materials [14]

(ii) Direct powder feed method: The second method is shown in Fig. 1b; in this method, the coating powder is supplied with inert gas attached with the laser cladding machine. Zhang et al. [13] feed the powder directly with the powder feed rate 0.618 g/s attached with a laser cladding machine and investigated the microhardness of the cladding layer and weight loss of the materials from the cladding layer [13].

(iii) Wire feeding: The coating materials may also provide in the form of wire, shown in Fig. 1c. But there is a limitation in this method because there are limited materials available in the form of wire.

(iv) Preplacement Method: Forth method is preplaced method which is shown in Fig. 1d. Preplaced method is two steps process, in which the first step is to provide the semi-solid solution of coating powder on the prepared substrate and then leave it for the cure. The second step is to provide the laser on the preplaced layer to melt the coating powder with the substrate to make the laser cladding layer [14]. Iravani et al. [15] successfully deposited the metal matrix diamond composite using the preplaced method by laser cladding [15]. Manna et al. [16] also used the preplaced method to deposit the composite coating of Fe–B–C, Fe–B–Si and Fe–BC–Si–Al–C on plain carbon steel using laser cladding. They investigated the microhardness as well phase presented in composite coating [16].

2.4 *Parameters of the Laser Cladding Process*

After preparation of the coating powder, the laser beam uses as an external heating source with optimized parameters such as type of powder, diameter of laser beam, scanning scan, powder delivery velocity, and gas flow rate. The specific energy can be calculated by the following formula [17].

$$\text{Specific Energy} = (\text{Laser Power})/(\text{Scan Speed} \times \text{Beam Diameter}) \quad (1)$$

During the laser cladding, the shape and thickness of the clad layer may be controlled by the interaction time of the beam and powder feed rate [18]. Dutta Majumdar and Manna [19] noticed that providing high laser power during cladding achieved high heating and cooling rate; therefore, high surface properties were achieved [19].

3 Discussion of Results

This part discussed the influence of the process parameters on the results of microstructure images, hardness, and wear resistance of the coating layer deposited on the mild steel.

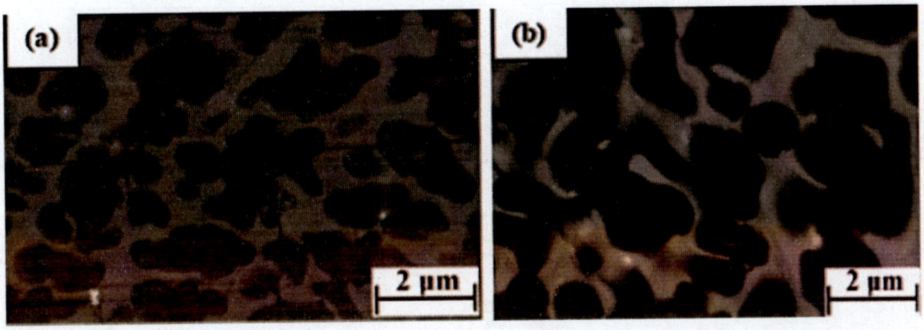

Fig. 2 FESEM micrograph of the cross section of the coating developed at **a** AISI 1020 and **b** AISI 304 steel, with the same parameters [21]

3.1 Microstructure and Phase Analysis of the Coated Layer

A scanning electron microscope (SEM) is conducted to show the crack and pore of the coating, distribution of coating powder. The coated sample consists of three regions such as clad layer, HAZ, and substrate [4]. Wang et al. [20] in their study concluded that the microstructure features or microstructure size depend on the solidification velocity, cooling rate, and temperature gradient at the interface of solidification. Temperature gradients of the materials depend on their properties such as thermal diffusivity, thermal conductivity, and absorbing power of heat [20]. Masanta et al. [21] also compared the microstructure images of composite coating on the mild steel AISI 1020 Fig. 2a and the stainless steel AISI 304 Fig. 2b. They concluded that the thermal conductivity of the AISI 304 has lower than that of the AISI 1020 steel; therefore, lower temperature gradient and cooling rate developed at the melt and interface. As a result, a lower degree of refinement was achieved in the case of AISI 304 steel (Fig. 2) [21].

X-ray diffraction plays an important role in the coating field. XRD identifies the intermetallic compounds present in the coating layer. Figure 3 shows that before cladding Fe_2Ti, FeB and Fe phases present. These phases are the constituents of the coating powder of ferrotitanium and ferroboron. While after laser cladding, only TiB_2 and α-Fe phases were detected which formed in situ analysis.

3.2 Microhardness of the Composite Coating

The microhardness of the composite layer is found maximum at the top surface of the coating and decreases gradually along with the depth of the clad layer. The microhardness found minimum at bottom of the composite layer due to attributed of a lower area fraction of coating materials as compared to the top surface of the composite [22]. Du et al. [4] concluded that the TiB_2 was formed by nucleation growth mechanism and separated from the melt during cooling. Due to the presence of a high

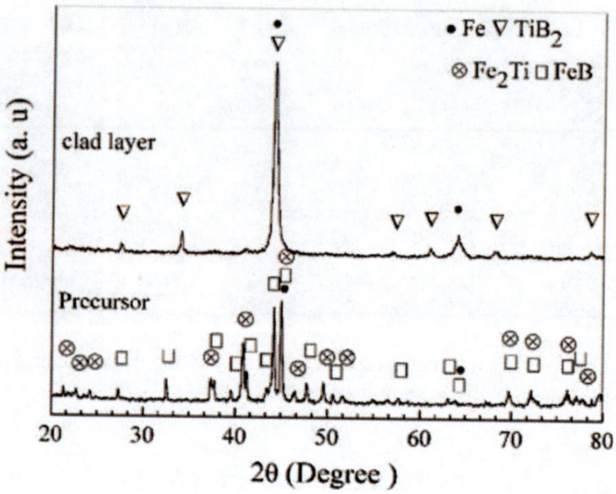

Fig. 3 XRD pattern of the top surface of the coating layer and precursor [4]

volume fraction of TiB_2, microhardness approaches about 1000 HV that is very high. Ding et al. [23] deposited the clad layer using Ti, VN, and Co-based powder. They reported that the addition of the titanium powder promoted the complete formation of intermetallic compounds.

Figure 4 reveals that the hardness value increases with increasing the amount of the titanium contents because more Ti contents lead to the formation of titanium nitride (TN) and VC. It can also be noticed that the microhardness decreases from top to bottom [23].

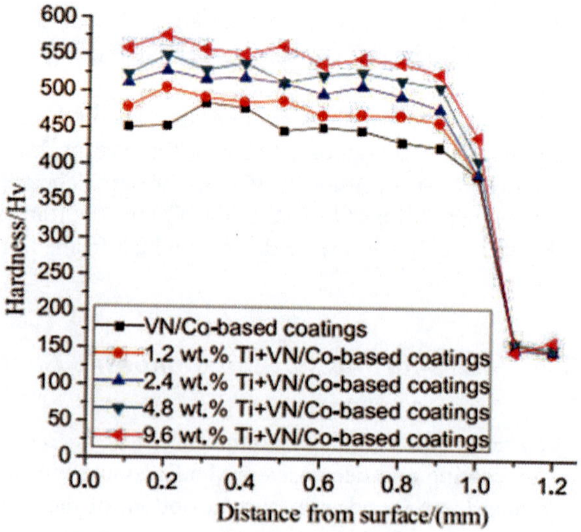

Fig. 4 Hardness distribution of composite coating with different titanium content [23]

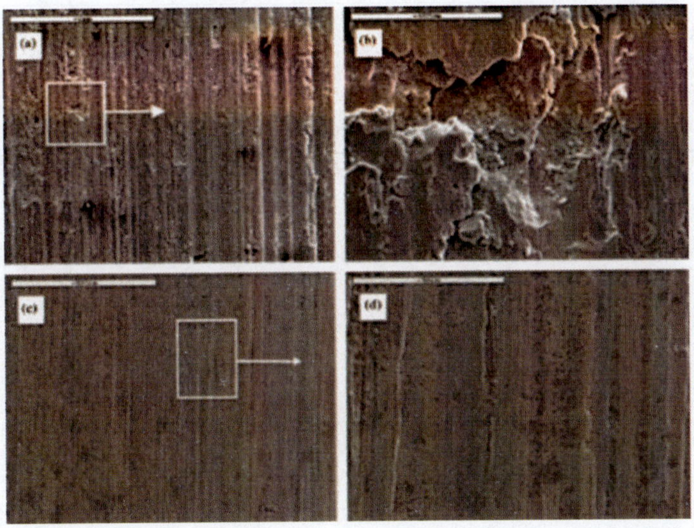

Fig. 5 SEM images of the worn-out substrate **a** at low, **b** at high magnification, and cladded layer **c** at low, **d** at high level [4]

3.3 About the Wear Resistance of the Coating

Du et al. [4] deposited the TiB_2 and Fe composited coating and conducted the wear test on the coated as well as substrate samples. Figure 5 illustrates that the craters and deep grooves were produced during the dry sliding wear test, whereas at the composite surfaces, only smooth scratches present after the dry sliding test. The authors suggested that TiB_2 presents in the coating oppose the wear attacks against the metallic counterparts. They also noticed that TiB_2 formed in situ synthesis which leads to a strong interfacial bond between the TiB_2 and α-ferrite matrix, as a result, the pullout of TiB_2 particles absent during the sliding wear test [4].

4 Conclusion

In the present study, reviewed many papers and the following conclusions have been made that help to understand the mechanism of the laser cladding process to increase the tribological properties as well as mechanical properties of the mild steel. In situ synthesis is a more economical process than ex situ synthesis. It provides good bonding between the coating powder and substrate. Preheating of the substrate may occur with high laser power, which decreases the temperature gradients as a result crack, and pore-free composite coating can be found. The microhardness depends upon the properties of the formation of the new intermetallic phases during laser cladding and the presence of reinforced particles of the coating powder. Hardness also depends upon the laser scan speed, and higher scan speed provides a higher cooling rate which leads the more refinement of the structure. More refinements structure provides higher hardness value. The wear resistance of the coating depends on the properties of coating powder in which the matrix enhances the binding forces during cladding. The wear resistance of the composite coating increases with an increase in the scan speed of the laser and the amount of the hard ceramic materials.

Acknowledgements The authors would like to thank the National Institute of Technology Patna to allow me doing this research. I also thankful to the Elsevier publisher for giving the permission to republish the figures from published journals.

References

1. Ding L, Hu S, Quan X, Shen J (2017) Effect of aging treatment on microstructure and properties of VN alloy reinforced Co-based composite coatings by laser cladding. Mater Charact 129:80–87
2. Farahmand P, Kovacevic R (2015) Corrosion and wear behavior of laser cladded Ni–WC coatings. Surf Coat Technol 276:121–135
3. Candel JJ, Jimenez JA, Franconetti P, Amigó V (2014) Effect of laser irradiation on failure mechanism of TiCp reinforced titanium composite coating produced by laser cladding. J Mater Process Technol 214:2325–2332
4. Du B, Zou Z, Wang X, Qu S (2008) Laser cladding of in situ TiB2/Fe composite coating on steel. Appl Surf Sci 254(20):6489–6494
5. Yang S, Liu W, Zhong M, Wang Z (2004) TiC reinforced composite coating produced by powder feeding laser cladding. Mater Lett 58(24):2958–2962
6. Agarwal A, Dahotre NB (2000) Comparative wear in titanium diboride coatings on steel using high energy density processes. Wear 240(1–2):144–151
7. Anal A, Bandyopadhyay TK, Das K (2006) Synthesis and characterization of TiB2-reinforced iron-based composites. J Mater Process Technol 172(1):70–76
8. Alazemi AA, Dysart AD, Shaffer SJ, Pol VG, Stacke LE, Sadeghi F (2017) Novel tertiary dry solid lubricant on steel surfaces reduces significant friction and wear under high load conditions. Carbon 123:7–17
9. Lu XL, Liu XB, Yu PC, Zhai YJ, Qiao SJ, Wang MD, ... Chen Y (2015) Effects of heat treatment on microstructure and mechanical properties of Ni60/h-BN self-lubricating anti-wear composite coatings on 304 stainless steel by laser cladding. Appl Surf Sci 355:350–358
10. Zhang M, Li M, Chi J, Wang S, Ren L, Fang M, Zhou C (2019) Microstructure evolution, recrystallization and tribological behavior of TiC/WC composite ceramics coating. Vacuum 166:64–71
11. Singh A, Dahotre NB (2004) Laser in-situ synthesis of mixed carbide coating on steel. J Mater Sci 39(14):4553–4560
12. Meng J, Ji Z (2013) Microstructure and technology research of in-situ synthesis TiN-TiB2/Ni composite coating by argon arc cladding. Phys Procedia 50:253–260
13. Zhang Z, Yu T, Kovacevic R (2017) Erosion and corrosion resistance of laser cladded AISI 420 stainless steel reinforced with VC. Appl Surf Sci 410:225–240
14. Liu J, Yu H, Chen C, Weng F, Dai J (2017) Research and development status of laser cladding on magnesium alloys: a review. Opt Lasers Eng 93:195–210
15. Iravani M, Khajepour A, Corbin S, Esmaeili S (2012) Pre-placed laser cladding of metal matrix diamond composite on mild steel. Surf Coat Technol 206(8–9):2089–2097

16. Manna I, Majumdar JD, Chandra BR, Nayak S, Dahotre NB (2006) Laser surface cladding of Fe–B–C, Fe–B–Si and Fe–BC–Si–Al–C on plain carbon steel. Surf Coat Technol 201(1–2):434–440
17. Rashid RR, Abaspour S, Palanisamy S, Matthews N, Dargusch MS (2017) Metallurgical and geometrical characterisation of the 316L stainless steel clad deposited on a mild steel substrate. Surf Coat Technol 327:174–184
18. Zhenda C, Chew LL, Ming Q (1996) Laser cladding of WC-Ni composite. J Mater Process Technol 62(4):321–323
19. Majumdar JD, Manna I (2003) Laser processing of materials. Sadhana 28(3–4):495–562
20. Wang W, Wang M, Jie Z, Sun F, Huang D (2008) Research on the microstructure and wear resistance of titanium alloy structural members repaired by laser cladding. Opt Lasers Eng 46(11):810–816
21. Masanta M, Shariff SM, Choudhury AR (2011) A comparative study of the tribological performances of laser clad TiB2–TiC–Al2O3 composite coatings on AISI 1020 and AISI 304 substrates. Wear 271(7–8):1124–1133
22. Majumdar JD, Chandra BR, Nath AK, Manna I (2008) Studies on compositionally graded silicon carbide dispersed composite surface on mild steel developed by laser surface cladding. J Mater Process Technol 203(1–3):505–512
23. Ding L, Hu S, Quan X, Shen J (2018) Effect of Ti on the microstructure evolution and wear behavior of VN alloy/Co-based composite coatings by laser cladding. J Mater Process Technol 252:711–719

A Numerical Study for Estimation of the Solar Irradiance on Dome Shaped Solar Collectors/Stills

T. R. Adithyan, Siddharth Ramachandran, and Naveen Kumar

Abstract Solar stills are considered as a cheaper solution for potable water production in developing and underdeveloped nations. Numerous experimental studies have been done in this area to show that use of dome shaped or hemispherical solar still top cover gives better performance and distillate output. However, most of the investigations rely on experimentally developed empirical relations for different types of conventional flat plate solar still. These models come into use whilst accurately predicting the performance of a flat plate type solar still but not for dome shaped or hemispherical shaped collectors. In the present study, emphasis has been given to developing an accurate model for estimation of the solar irradiance on dome shaped collectors/stills. Since the dome shape is not a conventional shape, the amount of radiation incident on the surface is different from flat surface. A comparative investigation is carried out between four types of dome shapes with different cap angle and flat plate, for a period of twelve months geographical location (Chennai - 13°N, 80°16′E). It was found that a dome shaped collector receives more sunlight compared to flat plate throughout the day. Typically, an increased irradiance of 50–100 W/m^2 was got during morning and evening time and a slight decrease of 10–40 W/m^2 during noon. The computation model uses finite element method and also takes into consideration of the dome surface area under shadow which is not facing sun directly. The developed model will be useful for predicting the performance of dome shaped solar collectors/stills.

Keywords Renewable energy · Dome shape · Modelling · Design · Solar still

T. R. Adithyan (✉) · S. Ramachandran
Department of Mechanical Engineering, Indian Institute of Information Technology, Design and Manufacturing, Kancheepuram, Tamil Nadu 600127, India
e-mail: mfd16i002@iiitdm.ac.in

N. Kumar
Department of Physics, Indian Institute of Information Technology, Design and Manufacturing, Kancheepuram, Tamil Nadu 600127, India

A. Maurya et al. (eds.), *Recent Trends in Mechanical Engineering*, Lecture Notes in Mechanical Engineering, https://doi.org/10.1007/978-981-19-7709-1_27

1 Introduction

The solar still which replicates the principle by which the nature makes rain is an environment friendly and cost effective way to produce distilled water [1]. The solar still is usually used in geographical locations where drinking water scarcity is prevailed. Only 1% of the total water on earth is usable for drinking, and 97% of the water is either saline water or mixed with impurities [2]. Due to the ever-increasing population of the human race, it has become much difficult to meet demands for pure water for domestic uses. Desalination using the solar still is a viable method to overcome this difficulty. The solar still work by converting solar radiation to heat energy. Water filled in the basin of the still is evaporated due to solar heat energy, which condenses on the top cover of the still, and is collected. The conversion of solar radiation to useful heat depends on many factors like top cover material, basin material, energy absorptivity, total solar insolation on the still, etc., [1]. Experimental studies done by Arunkumar [3], Ismail [4], and Dhiman [5] have shown the effectiveness of a dome shaped top cover over conventional flat or single sloped top cover. The solar radiation estimation model for this type of solar stills is based on that for flat plate solar still. In this study, an attempt is made to relate the incident solar radiation to the shape of the top cover, quantitative comparison between different types of the dome shaped top cover of the solar still and validate the findings, showing that the complex shape of the top cover has its benefits. A solar radiation estimation model is established for comparing different dome shaped top covers for a particular geographical location.

2 Estimation of Hourly Solar Radiation on Curved Surface

The solar radiation is generally estimated for a horizontal surface, but to get better advantage, any collector surface is arranged with an inclination [6]. The solar radiation is usually estimated as an average value for a given period of time. Mainly, there is two periods for which the radiation is estimated, daily average solar radiation, and hourly average solar radiation [7]. In the following sections, the model to estimate hourly average solar radiation falling on a curved surface is explained.

2.1 Mathematical Modelling of Radiation Influx

For this study, two types of curved surfaces have been considered first hemispherical and second one dome shaped which are the main two types used as a top cover of solar still. The dome shapes are classified based on the cap angle. As studies have proven the inclination and orientation of the surface have a major role in determine the amount of solar radiation that gets incident on the basin [8]. When we consider a curved surface like a dome, the inclination is changing according to the height of the dome surface from base. The calculation is done assuming the dome which is split into small flat surfaces with different degree of inclination, using the method of finite element and integrating over the total area that is lit by the sunlight [6]. The area is approximated using the Riemann summation for integration.

2.2 Radiation on Hemispherical Surface

A hemispherical surface is a special case of dome surface, where the cap angle is 90°. As we know the points on the sphere can be expressed in terms of the polar co-ordinate, the area of a sector can be calculated using the polar co-ordinate given the θ and ϕ of the sector. The surface is split into infinitesimal portion of $\Delta\theta$ and $\Delta\phi$. Now considering this sector approximately equal to flat plate, the solar angular orientation of the surface can be written in terms of the polar coordinates. The angular orientations that are used to determine the incident angle are [6]

$$\begin{aligned} i &= angle\, of\, incidence \\ \beta &= tilt\, angle \\ \gamma &= surface\, azimuth\, angle \end{aligned} \tag{1}$$

where the tilt β and surface azimuth γ can be written in terms of the ϕ and θ shown in Fig. 1:

$$\begin{aligned} \beta &= 90^\circ - \phi \\ \gamma &= \theta \end{aligned} \tag{2}$$

The angle of incidence can be calculated using the Eq. (3) [8]

$$cos i = sin\delta\, sin L\, cos\beta - sin\delta\, cos L\, sin\beta\, cos\gamma$$

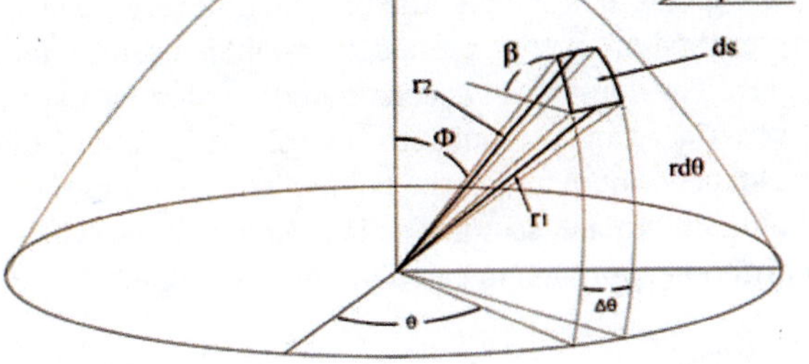

Fig. 1 Sector on dome surface [9]

$$+ cos\delta\, cosL\, cos\beta\, cos\omega + cos\delta\, sinL\, sin\beta\, cos\gamma\, cos\omega + cos\delta\, sin\beta\, sin\gamma\, sin\omega \tag{3}$$

where δ, L, ω are angle of declination, latitude angle, and hour angle, respectively. Now, the Eq. (3) for a section can be written by substituting Eq. 2 in Eq. (3) [7, 9]:

$$\begin{aligned} cosi &= sin\delta\, sinL\, sin\phi - sin\delta\, cosL\, cos\phi\, cos\theta \\ &+ cos\delta\, cosL\, sin\phi\, cos\omega + cos\delta\, sinL\, cos\phi\, cos\theta\, cos\omega \\ &+ cos\delta\, cos\phi\, sin\theta\, sin\omega \end{aligned} \tag{4}$$

The beam radiation incident on tilted surface can be related to beam radiation on a horizontal surface using Eq. (5) [8]:

$$\mathrm{I_b} = \frac{I_b}{cos\theta_z} cosi \tag{5}$$

The total radiation on the hemisphere can be calculated using the surface integral given by Eq. (6) [9]:

$$I_t = \iint (I_d + I_b)dA = \int_0^{2\pi} \int_0^{\frac{\pi}{2}} (I_d + I_b)r^2 cos\phi d\phi d\theta \tag{6}$$

In Eq. (6), I_b and I_d are beam radiation and diffuse radiation, respectively. The double integral expressed by Eq. (6) is computed for all values of θ and ϕ up to the point where cos$i$ becomes either zero or negative. When cosi ≤ 0, it corresponds to portion of surface behind the sun or under shadow [9, 10]. The double integral is computed using finite element method and Riemann sum using MATLAB programming language.

2.3 *Radiation on Dome Surface*

As explained previously, the study is done on dome shapes that vary in cap angle. The cap angle is the parameter which differentiate the dome from each other. The representation of the geometric variables used to define a dome is explained in the figure. The dome that we considered is spherical caps of a sphere. When we consider a spherical cap, we run into one obstructed, and the radius of each sector is not constant as compared to the hemispherical dome. The complexity creates a different tilt angle for each sector [9]. The radius varies with ϕ; hence, we should formulate a different approach to calculate the tilt angle.

$$\beta = \theta' - \phi \tag{7}$$

The Eq. (2) has to made change accordingly to accommodate this change. The 90° is changed to θ', where the θ' is the angle between surface element and normal axis. The infinitesimal element $|ds|$ is calculated from the two radii r_1 and r_2 which are the starting and ending radius of the surface element. The expression is got from the triangle made by r_1, r_2, and $|ds|$ as shown in the Fig. 1.

$$|d_s| = \sqrt{r_1^2 + r_2^2 - 2r_1r_2cos\Delta\phi} \tag{8}$$

The relations between the tilt and ϕ are derived from the figure as:

$$\frac{sin\theta'}{r_2} = \frac{sin\Delta\phi}{ds} \tag{9}$$

$$\theta' = sin^{-1}\left(\frac{r_2}{ds}sin\Delta\phi\right) \tag{10}$$

The area of the element is calculated by approximating the dome element surface to an equivalent hemispherical element with radius $\bar{r}$ [9].

$$dA = r^2 sin\phi\, d\phi\, d\theta \tag{11}$$

The double integral for I_t is computed for the dome surface by substituting the Eqs. (7) and (11), following the same algorithms as the previous case.

3 Results and Discussion

The study was done to come up with an accurate model to estimate the solar radiation on a domed surface. The solar radiation in Chennai was calculated using the models. A total of six cases were taken into account, which are flat plate, flat plate with tilt 30° and four types of domed surfaces. The estimation done by taking the latitude of Chennai as 13° N and the climate condition is tropical climate, and beam diffuse is isotropic in nature. The four types of dome surfaces that are selected are based on the cap angle (90°, 60°, 45° and 30°). Solar radiation data were calculated for the twelve months of the year but here in this study, focus on solar radiation of three months. The months that are selected are considering the three different times of year ranging from summer, monsoon, and winter. These data calculated are not to be taken to predict the daily climatic conditions of that location, since it does not account to the unpredictable nature of environment (for example, the day can be cloudy and partially cloudy or due to erratic climate change and could altogether be different). The data simulated can be accurate for a clear sky day. The above said data on the three months are plotted as shown in Fig. 2.

As per the estimated solar radiation in Dhahran, an attempt was made to calculate the efficiency of the hemispherical solar still using the distillate data got from the

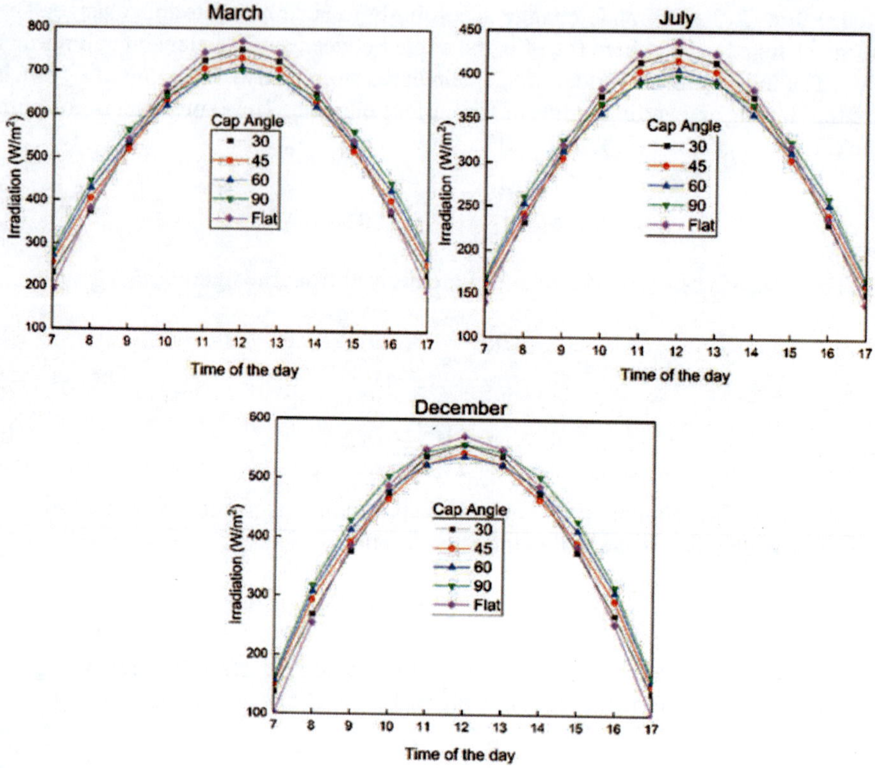

Fig. 2 Solar radiation plot for month March, July, and December (Chennai)

papers of Ismail [4]. The distillate output for still was got from the evaporative heat flux (Q_{ew}). The distillate was calculated from the equation of convective heat transfer for a flat plate solar still [2, 4, 11]. The efficiencies were calculated and tabulated as shown below. The experiment done by the author was for a period of 6 days, and we calculated and compared the efficiency for all the even number days. The calculated efficiencies are plotted as shown in Fig. 3.

The comparison study shows that the estimated values follow the trend as expected. Even though the irradiation influx received during the noon time on the dome shapes is less when compared to the flat plate, the overall influx is more for dome shaped receiver (see Fig. 2). The Q-Q plot of proposed model and experimental value was plotted as shown in Fig. 4 and found that the error in prediction is less (R-squared value of 0.97244).

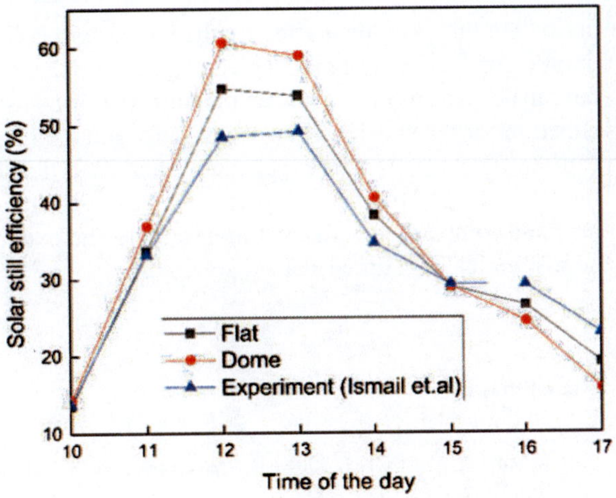

Fig. 3 Efficiency—hour of the day plot [4]

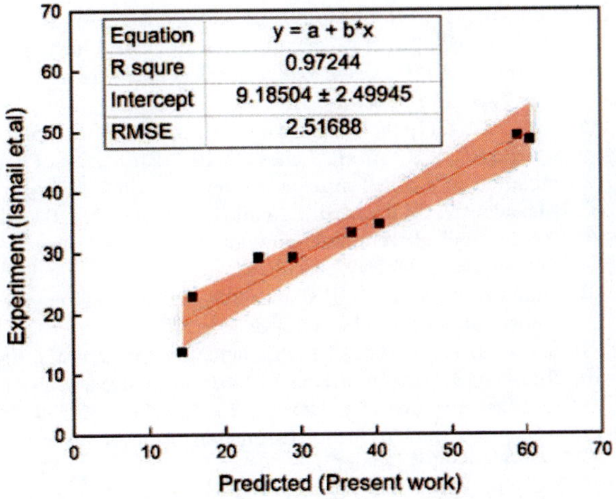

Fig. 4 Q-Q plot of experiment and predicted

4 Conclusion

A model for estimation of the solar radiation and technique for accurately calculate the radiation on a dome surface was developed and investigated. The model was constructed based on the location of sun at particular time of day and the incident angel of sun's rays on a flat surface at various orientation. The numerical analysis related to the model was done using MATLAB and can be easily scaled up and changed for other geographical locations. The numerical analysis results were plotted, and it was found that the dome shaped has the ability to collect more solar irradiance during forenoon and afternoon when compared to a flat and sloped surface. The dome surface shows a small decrease in collected irradiance during the noon period but this is compensated by the increase in collection in morning and evening duration of the day. The model gives a good estimate of the solar radiation on surfaces but can be made more accurate in the future study by using better empirical correlation equation for the specified geographic location [7, 12], and the model and technique can be validated using the experimental and analytical methods. However, the present model developed predicts the performance of distillation units with reasonable accuracy and comparable with existing literature and experimental study done on various hemispherical and dome shaped solar still.

References

1. Kumar S, Tiwari GN (2009) Life cycle cost analysis of single slope hybrid active solar still. Appl Energy 86:1995–2004
2. Tiwari GN (2017) Advanced solar-distillation systems basic principles, thermal modeling and its application. Springer Singapore
3. Arunkumar T, Jayaprakash R, Denkenberger D, Ahsan A, Okundamiya MS, Kumar S, Tanaka H, Aybar HS (2012) An experimental study on a hemispherical solar still. Desalination
4. Ismail BI (2009) Design and performance of a transportable hemispherical solar still. Renew Energy 34:145–150
5. Dhiman NK (1988) Transient analysis of a spherical solar still. Desalination 69:47–55
6. Maleki SAM, Hizam H, Gomes C (2017) Estimation of hourly, daily and monthly global solar radiation on inclined surfaces: models re-visited. Energies
7. Makade RG, Chakrabarti S, Jamil B, Sakhale CN (2020) Estimation of global solar radiation for the tropical wet climatic region of India: a theory of experimentation approach. Renew Energy 146:2044–2059
8. Duffie JA (Deceased), Beckman WA (2016) Solar engineering of thermal processes, photovoltaics and wind. John Wiley & Sons, 2013
9. Sabzevari AAG (2009) Solar radiation on domed roofs. Energy Build 41:1238–1245
10. Runsheng T, Meir IA, Etzion Y (2003) An analysis of absorbed radiation by domed and vaulted roofs as compared with flat roofs. Energy Build 35:539–548
11. Kumar S, Tiwari GN (1996) Estimation of convective mass transfer in solar distillation systems. Sol Energy 57:459–464
12. Srivastava RC, Pandey H (2013) Estimating Angstrom-Prescott coefficients for India and developing a correlation between sunshine hours and global solar radiation for India. ISRN Renew Energy

Design and Vibrational Analysis of Ceramic-Based Nose Cone Using ANSYS

Pranav Rajesh, R. Sudarshan, M. Sreedharan, and Lokavarapu Bhaskara Rao

Abstract When nose cone travels in hypersonic flow, it experiences high vibration due to which the nose cones are prone to damage. The secondary issue is that the material used for manufacturing of nose cone, i.e., HRSI is less abundant. The main aim of this work was to design a nose cone suitable for hypersonic flow and to propose an appropriate material that can be used in the nose cones. On further study of various researches, it was found that among nose cones, parabolic nose cones with fineness ratio (L/D) greater than 1.2 are more efficient for hypersonic flow because it produces less heat flux as well as lesser drag force. At present, high-temperature reusable surface insulation (HRSI) ceramics are one among the popular choices for materials used in the nose cones. But HRSI ceramics have lesser availability compared to other ceramics. The main objective here is to provide a viable and cost-effective solution. Therefore, alternative ceramics with good mechanical as well as good refractory properties and a metal alloy were selected. Two ceramic materials, namely zirconia, mullite, and a metal alloy, namely alpha–beta titanium aluminum alloy, were chosen. Finally, vibrational analyzes of these three nose cones were made using ANSYS software. The deformation results of these nose cones were compared, and the suitable material was selected. It was found that the nose cone made up of zirconia ceramic was found to have the least deformation. Therefore, zirconia was proposed as a better alternative ceramic material that can be used in hypersonic nose cones.

Keywords Nose cone · Hypersonic · Deformation · Vibration · Zirconia

P. Rajesh · R. Sudarshan · M. Sreedharan · L. Bhaskara Rao (✉)
School of Mechanical Engineering, Vellore Institute of Technology, Chennai Campus, Vandalur-Kelambakkam Road, Chennai, Tamil Nadu 600127, India
e-mail: bhaskarbabu_20@yahoo.com

P. Rajesh
e-mail: pranavrajesh2000@gmail.com

A. Maurya et al. (eds.), *Recent Trends in Mechanical Engineering*, Lecture Notes in Mechanical Engineering, https://doi.org/10.1007/978-981-19-7709-1_28

1 Introduction

Nose cones are incorporated in almost every high-speed vehicles including rockets, missiles, fighter jets, formula one cars, etc. It is generally designed in a way to reduce the force exerted by the surrounding atmosphere. The outer surface of the nose cone is built to withstand higher heat flux generated at the outer surface of the earth. In aircrafts, nose cone acts as a radome, shielding the weather radar from the forces exerted by the atmosphere. In hypersonic flow, nose cones are made up of refractory materials because of the extreme temperature conditions involved. Some of the preferred material for nose cone are HRSI, pyrolytic carbon, and reinforced carbon–carbon composite [1]. The process of monitoring the level and pattern of vibrational signals to detect the abnormal vibrations reflecting from the object with the help of Fourier transform on time waveform is called vibrational analysis. The vibrational analysis can be found out by using various methods such as crest factor, skewness, standard deviation, peak amplitude, kurtosis, and root-mean-square (RMS). When an aircraft travels under aerodynamic conditions, it will definitely undergo a stress with varying load factors which thereby affects the natural vibration system. The ideal design and material where almost all regions inside nosecone exhibit less vibration for hypersonic travel are required. Vibration limits are defined using long-term operation, maintenance history, or through referring to established standards. If the overall condition of the machine is degrading and defects start to develop, it could mean that the limit is crossed. Irregular vibration change can be controlled based on the vibrational analysis depending on the force applied [2]. Nose cones are made up of refractory materials because of their ability to withstand extreme temperature conditions. At present, HRSI ceramics are one among the popular choices for materials used in the nose cones. But HRSI ceramics have lesser availability compared to other ceramics. According to United States Geographical Survey published in the year 2014, there are 3400 ceramic reserves in India, and 67,000 ceramic reserves across the world. A nose cone of titanium—(Ti-6Al-4V) material was designed, and its structural analysis was done [3, 4]. A nose cone design was done, and deformation analysis of the nose cone made up of two different metal alloys was done under different conditions like pressure, velocity, and altitude [5]. Geometrical 3D model of the missile was designed, and modal analysis was performed on the model. Experimental modal analysis was performed under free boundary conditions on the geometric model which was fabricated. The natural frequencies were compared with the computational results, and therefore, the mode shapes were found [6]. New design of nose cones based on honey bee abdomen was designed, and structural analysis was done [7].

All these papers focused on the design and structural analysis of the nose cone. Aircrafts and rockets especially while traveling at hypersonic or supersonic speeds tend to vibrate. It is important to study whether the nose cone undergoes failure under vibration. This was one of the major literature gaps in these papers. Counterattacking to this problem with a practical as well as a reliable solution would reduce the overall cost and increase the production. Therefore, alternative ceramics with good mechanical as well as good refractory properties were chosen and compared with a metal alloy (titanium alloy) based on the results obtained from vibrational analysis. Two ceramic materials, namely zirconia, mullite, and a metal alloy, namely titanium alloy, were selected. Post analysis, the material with lowest deformation, was chosen as alternative for HRSI.

2 Mathematical Formulae

The speed of hypersonic rockets is 5 times that of the speed of sound, i.e., Mach number is greater than 5. This results in greater drag force and high vibrations on the surface of the nose cone. To combat this issue, a better geometrical design suitable for hypersonic flow is required. The full body of revolution of the nose cone is formed by rotating the profile around C/2 [8]. For all available nose cone geometry, L is the overall length, R is the radius of base of the nose cone, and y is the radius at any point x, as x varies from x to 0(to the tip and L). Generally, the standard nose cone diameter values are 20 mm, 25 mm, 30 mm, and 35 mm. Usually for all kinds of nose cones, fineness ratio will be twice the diameter. For parabolic nose cones, k is equal to 1.

$$\text{For } 0 \leq K' \leq 1 : y = R\left(\frac{\frac{x}{L} - 2(x/L)^2}{2 - K'}\right) \tag{1}$$

Taking d = 35 mm, radius = 17.5 mm, length = 70 mm (based on fineness ratio). Substituting these values in the above equation,

$$y = -0.00186 \times x^2 + 0.4 \times x \tag{2}$$

On solving (2), we get

$$x_1 = 0 x_2 = 70$$

Substituting these values in the Spline Equation Driven line, we get a curve as shown Fig. 1.

3 Modeling and Analysis

According to CFD, the nose cones used in the missiles during the hypersonic flow should be parabolic in nature. The Mach number which is the ratio of speed of missile to speed of sound is always around 5. Aerodynamically, parabolic nose cones are preferred over sharp and pointed nose cones due to ease of manufacturing, safety,

Fig. 1 Sketching the nose cone with desired dimensions

resistance to handling and flight damage. When the missile enters the atmosphere, pressure waves or shock waves begin to form at the bow end of the nose cone. These shock waves tend to travel at the same speed of the nose cone. As the missile starts to flow in hypersonic flow, these shock waves start to settle on top of the bow end of the nose cone. These shock waves are called bow shock waves. When the shock waves are compressed in a bow, a lot of heat is generated. The heat generated is generally more destructive to a sharp and pointed nose cone than to a parabolic nose cone because the bow shock wave tends to be closer to the bow end of the sharp and pointed nose cones than parabolic nose cones. Due to these, parabolic nose cones are preferred over sharp and pointed nose cones in hypersonic flow travel. It is also found that for hypersonic flow, parabolic noses with fineness ratio greater than 1.2 are more efficient because it reduces the heat flux generated as well as reduces the drag force. Fineness ratio (L/D) > 1.2 for parabolic nose cones is the most efficient. Taking all this into consideration, a parabolic nose cone of fineness ratio equal to 2 is designed and modeled in Solidworks software [9]. The model of the nose cone and the design measurements is given below in Figs. 2 and 3.

The vibrational analysis was done using ANSYS software after designing the CAD model [10]. It was done considering the varying force constraints and simulating the frequency results. Random vibration analysis was done where displacement was taken as a factor. The main objective of this study was to predict the displacement of the nose cone when the required frequency was given. Based on which the results

Fig. 2 Nose cone design

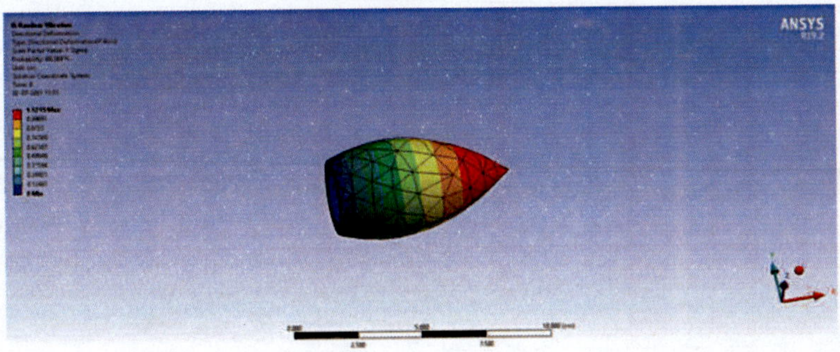

Fig. 3 Directional deformation for zirconia based nose cone

were computed in following table (Frequency vs. m^2). Before proceeding with the analysis, validation process was done with flow analysis of hypersonic nose cone with titanium aluminum alloy material, and comparison was made.

Structural analysis was performed on the proposed design by taking reference from the literature Devendra Aditya et al. [11]. Deformation, equivalent stress, and equivalent elastic strain values have been obtained as shown in Table 1. These values are compared with the results obtained by the literature Devendra Aditya et al. The obtained results were in the range of values obtained by Devendra Aditya et al. Therefore, the proposed design is valid.

The ANSYS workbench was opened, and the nosecone made up of zirconia model was imported. The required force constrains were applied, followed by meshing of the model. Meshing was done in order to gather the different results for different regions. Triangular mesh was used, and it was differentiated. The modal analysis was combined with stress analysis and with the option of inducing the stress parameter onto the model. The results were obtained. The number of supports are determined by the force constrains. Generally, random vibration analysis is compatible with fix support. Here, introducing other supports will change the properties of the material as well as force constrains applied to the nose cone. In random analysis, PSD displacement was chosen primarily because vibration may lead to displacement of the object.

Table 1 Comparison of drag force

Parameters	Present		Devendra Aditya et al.	
	Minimum	Maximum	Minimum	Maximum
Deformation (m)	0	0.079371	0	0.07725
Equivalent stress (Pa)	3.200e+9	2.807e+10	4.9078e+9	2.023e+10
Equivalent elastic strain	1.01e−5	0.00084	0.049	0.00169

Table 2 Properties of materials used in manufacturing of nose cone

Properties	Zirconia	Mullite	Alpha–beta titanium aluminum alloy (Ti-6Al-4 V)	HRSI
Density	6090 kg/m^3	2800 kg/m^3	4405 kg/m^3	9200 kg/m^3
Bulk modulus	9.436e+10 Pa	1.5278e+11 Pa	1.0075e+11 Pa	Not specified
Poisson's ratio	0.259	0.26	0.323	Not specified
Shear modulus	5.4805e+10 Pa	8.7302e+10 Pa	4.4038e+10 Pa	
Young's modulus	1.38e+11 Pa	2.2e+11 Pa	1.07e+11 Pa	
Thermal conductivity	2.7 W/m*K	6 W/m*K	7.1 W/m*K	

4 Materials

Materials used on the nose cone for hypersonic flow were decided based on the following mechanical properties:

- Good thermal shock and stress resistance
- Low thermal conductivity
- Good strength
- Wear and tear resistant
- Usable to high temperatures
- Low electrical conductivity
- Refractory properties.

Considering the above properties, two ceramics such as zirconia and mullite were chosen and were applied on the nose cone during the vibrational analysis. The mechanical properties of zirconia, alpha–beta titanium aluminum alloy, and mullite are given in Table 2 [12–19].

5 Result and Discussions

The 6 different nodes were created to undergo vibrational analysis in a detailed way. The maximum deformed value for these 6 different nodes is shown in Table 3. The deformation values were squared (as cm^2) as per PSD displacement condition and noted down as shown in Table 3. The directional deformation was applied, and the region with maximum and minimum deformation was obtained for zirconia-based nose cone as shown in Fig. 3. Figure 4 represents the normal stress values for zirconia-based nose cones.

The steps followed for vibrational analysis of mullite are very similar to the vibrational analysis done to zirconia. The values were noted as shown in Table 4. Figures 5 and 6 show the deformation and stress values of different regions for mullite-based nose cones.

Table 3 Deformation versus frequency for zirconia-based nose cone

Mode	Frequency (Hz)	Deformation (cm)	cm^2	cm^2/Hz
1	5412.6	0.17787	0.031638	5.874e−06
2	5415.2	0.17786	0.031634	5.841e−06
3	14,327	0.12696	0.016119	1.125e−06
4	17,312	0.15817	0.025018	1.445e−06
5	17,435	0.15935	0.025392	1.456e−06
6	18,523	0.13321	0.017745	9.579e−07

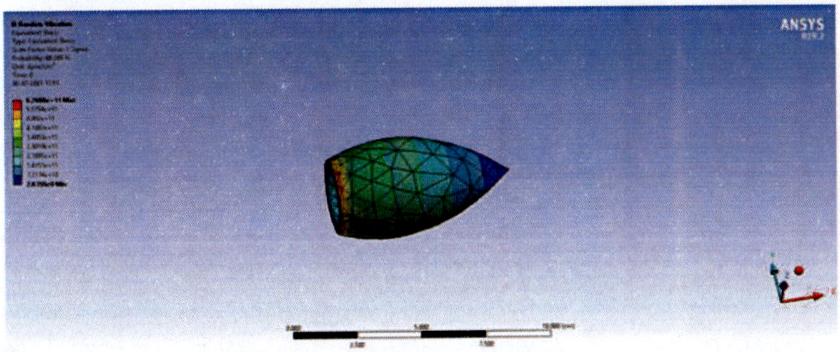

Fig. 4 Equivalent stress analysis for zirconia-based nose cones

Table 4 Deformation versus frequency values of mullite-based nose cone

Mode	Frequency (Hz)	Deformation (cm)	cm^2	cm^2/Hz
1	10,078	0.26231	0.068807	6.827e−06
2	10,082	0.26229	0.068796	6.823e−06
3	26,667	0.18724	0.035059	1.314e−06
4	32,236	0.23326	0.05441	1.687e−06
5	32,464	0.23499	0.05522	1.701e−06
6	34,487	0.19644	0.038589	1.118e−06

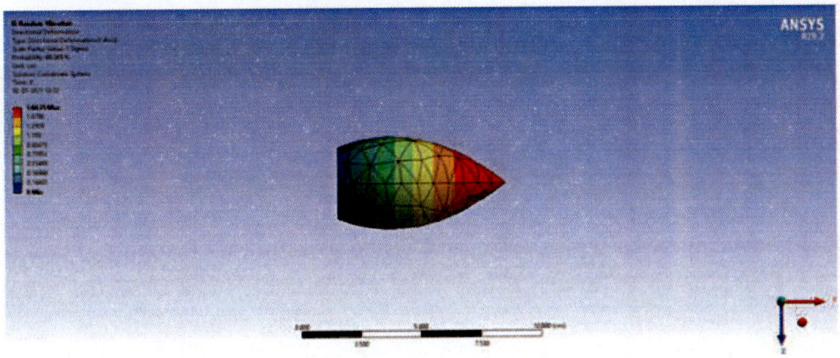

Fig. 5 Directional deformation of mullite-based nose cone

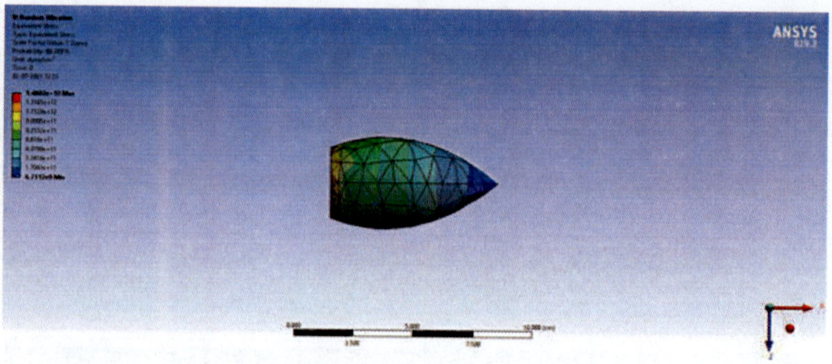

Fig. 6 Normal stress of mullite-based nose cone

Table 5 Deformation versus frequency of alpha–Beta titanium aluminum alloy (Ti-6Al-4 V)

Mode	Frequency (Hz)	Deformation (cm)	cm^2	cm^2/Hz
1	5563.4	0.20864	0.043531	7.824e−06
2	5566.7	0.20864	0.043531	7.819e−06
3	14,465	0.14928	0.022285	1.540e−06
4	17,915	0.18568	0.034477	1.924e−06
5	18,042	0.18649	0.034779	1.927e−06
6	19,025	0.15621	0.024402	1.282e−06

Similar steps were followed for the alpha–beta titanium aluminum alloy (Ti-6Al-4 V), and the values were noted down which contained the frequency, deformation, and deformation square as shown Table 5. Figures 7 and 8 show the deformation and stress values of different regions for alpha–Beta titanium aluminum alloy-based nose cones.

Table 6 shows that convergence of frequency versus deformation for zirconia-based nose cone. When frequency is varied, the resulting deformation of nose cone lies within the range of 3.2–3.5 cm. This is done to show that the process performed is valid.

Based on the results, the maximum deformation of zirconia-based nose cone is 1.1215 cm, whereas the mullite and titanium aluminum alloy have deformation value to be almost equal to 1.6635 cm and 1.3159 cm, respectively. To conclude, zirconia-based nose cone is ideal for hypersonic travel. Also, to note that since the jets and missile revolve around high velocity environment considering that the stress would huge. In the case of zirconia, the maximum stress value would be 6.2688e+10 Pa which is comparatively less with mullite and titanium alloy which would be 1.4802e+11 Pa and 5.13e+10 Pa, respectively. The order of preference would be zirconia, alpha–beta titanium aluminum alloy, and mullite.

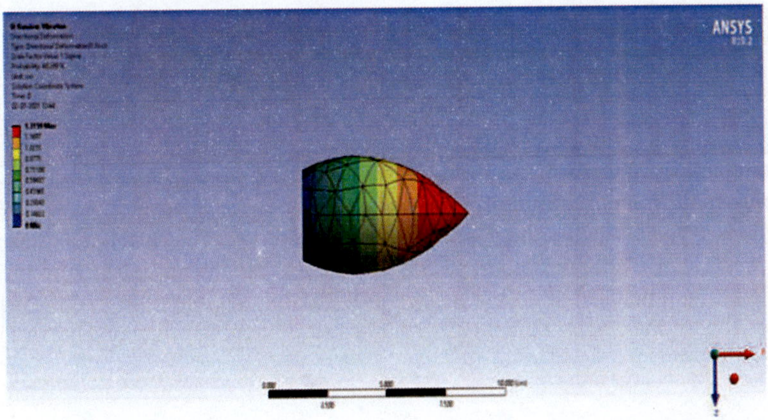

Fig. 7 Directional deformation of titanium aluminum alloy-based nose cone

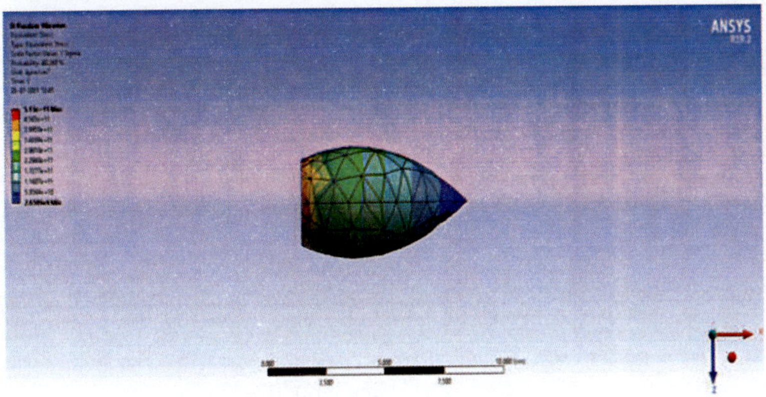

Fig. 8 Stress analysis of titanium aluminum alloy-based nose cone

Table 6 Convergence of frequency versus deformation

Mode	Frequency	Deformation
1	0.701	3.2589
2	0.885	3.4257
3	0.889	3.4543
4	0.991	3.4587
5	0.997	3.5656
6	1.000	3.5660

6 Conclusion

A hypersonic nose cone was designed, and the computational fluid dynamics theory is employed to decide on the type of nose cone, i.e., a parabolic nose cone with dimensions L = 70 mm, D = 35 mm, and fineness ratio equal to 2 was modeled. Zirconia, mullite, and aluminum alloy were applied to the nose cone. Vibrational analysis was performed on the nose cones, and the deformations in each of the nose cones were compared. Zirconia-based nose cone with the lowest deformation is more reliable and can be used as an alternative to HRSI-based nose cone that is currently employed with maximum deformation value approximately equal to 1.1215 cm.

References

1. Pike J (2000) Ballistic missile basics. Special Weapons Primer. Federation of American Scientists
2. Norton MP, Nelson FC (1990) Fundamentals of noise and vibration analysis for engineers. J Acoust Soc Am 88(4):2044–2044
3. Bedinger GM (2016) Zirconium and Hafnium U.S. Geological Survey, Mineral Commodity Summaries, pp 194–195
4. Sreenivasula Reddy M, Keerthi N (2017) Design and structural analysis of missile nose cone. Austr J Basic Appl Sci 11(11):30–40
5. Venkata Suresha P, Ayyappaa K, Anjalee Kumaria V, Koteswararaoa M, Venumurali J (2015) Design of an aircraft nose cone and analysis of deformation under the specified conditions with different materials using ANSYS. Int J Eng Res Technol 4(2):268–272
6. Sharma N, Maheshwar Reddy U, Yashwanth KV (2016) Modal analysis of typical missile configuration. Int J Eng Res Technol 5(8):119–123
7. Zhao J, Yan S, Deng L, Huang H, Liu Y (2017) Design and analysis of biomimetic nose cone for morphing of aerospace vehicle. J Bionic Eng 14(2):317–326
8. Narayan A, Narayanan S, Kumar R (2017) Hypersonic flow past nose cones of different geometries: a comparative study. Simulation: Trans Soc Model Simul Int 1–16
9. Solid works 2019 Tutorial – SDC Publications
10. Ansys User Manual – Workbench User's Guide 2013
11. Devendra Aditya PV, Rajesh Kumar Y (2018) Design and structural analysis of missile nose cone using different materials. Int J Adv Technol Innov Rcs 10(12):1147–115
12. Engineering materials 1: an introduction to their properties and applications by David R. H. Jones and Michael F.
13. Mechanical properties of materials, Textbook by Joshua Pelleg
14. Díaz M, Smirnov A, Gutiérrez-González CF, Estrada D, Bartolomé JF (2020) Microstructure and mechanical properties of Zirconia (3Y-TZP)/Zr composites prepared by wet processing and subsequent spark plasma sintering Ceramics, MDPI, 53–646
15. Liu Y, Lian W, Su W, Luo J, Wang L (2019) Synthesis and mechanical properties of mullite ceramics with coal gangue and wastes refractory as raw materials. Int J Appl Ceram Technol 1–6
16. Anggono J (2005) Mullite ceramics: its properties, structures and synthesis. J TEKNIK MESIN 7(1):1–10
17. Della Bona A, Pecho OE, Alessandretti R (2015) Zirconia as a dental biomaterial. Materials (Basel) MDPI 8(8):4978–4991
18. Harikrishnan R, Rao Lokavarapu B (2021) Design and analysis of rocket nozzle. Elsevier 38(5):3365–3371

19. Narayan A, Subramanian N, Kumar R, Kumar R (2018) Hypersonic flow past parabolic and elliptic nose cone configurations: a comparative study. In: 7th international and 45th national conference on fluid mechanics and fluid power (FMFP)
20. Naveen Kumar G, Velliangiri M, Kamesh Adithya SB, Kirubakaran A, Kishore M (2020) Design and CFD analysis of hypersonic nose cone. Int J Res Appl Sci Eng Technol (IJRASET) 8(7):1526–1538

Additive Manufacturing in Industry 4.0: A Review

Pratyush Srivastava and Pankaj Sahlot

Abstract Additive manufacturing is one of the imperative components in the Industry 4.0 framework. It is widely used in concurrent engineering for effective product development and many other vital domains. Extensive research is being carried out in a continuum of Industry 4.0 fields, namely Internet of things, cloud computing, cybersecurity, autonomous robotics, big data, simulation, and augmented reality. AM has a crucial impact in these fields. This review paper provides a comprehensive study in various spheres of additive manufacturing and Industry 4.0. This research also provides a profound inter-association of these systems in the domain of product development and optimization of process parameters, which can be later employed in biomedical, aerospace, and construction applications. The exploration would foster a crucial review of the subject and reveal the matter to understand the future latitude in the domain. The study embraces the amalgamation of various ecosystems in Industry 4.0 and frameworks to substantially affect the research and development in numerous industries. This review study will support industries to keep themselves alongside the advancements in additive manufacturing and Industry 4.0.

Keywords Additive manufacturing · Industry 4.0 · Internet of things · Product development · Optimization · Machine learning

1 Introduction

Studying numerous papers on additive manufacturing (AM) demonstrates that it is certainly a "disruptive technology" and promises to positively impact the product development (PD) industry. It allows for flexibility in design and is an important exponent in concurrent engineering and it reduced inventory cost and reduced waste

---

P. Srivastava · P. Sahlot (✉)
Mechanical Engineering, School of Technology, Pandit Deendayal Energy University, Gandhinagar, GJ 382426, India
e-mail: pankaj.sahlot@sot.pdpu.ac.in

P. Srivastava
e-mail: pratyush.smc17@sot.pdpu.ac.in

A. Maurya et al. (eds.), *Recent Trends in Mechanical Engineering*, Lecture Notes in Mechanical Engineering, https://doi.org/10.1007/978-981-19-7709-1_29

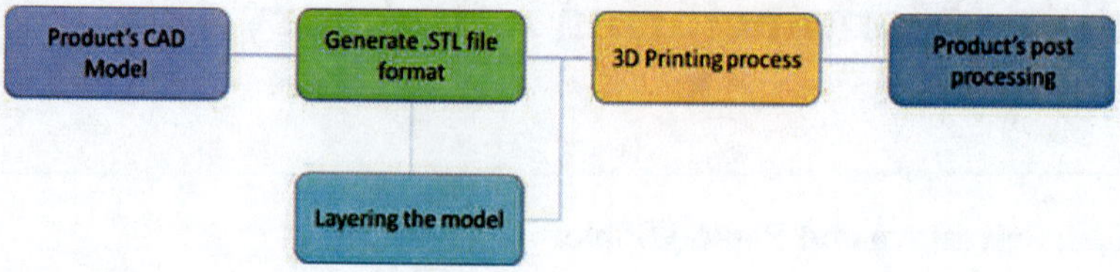

Fig. 1. 3D printing manufacturing process [6]

substantially making it more sustainable. It is a lucrative method that does not consume much energy or can be optimized according to the application. The "adaptability" of the system towards cloud data processing allows the AM to be integrated with Industry 4.0-based technologies such as artificial intelligence, cloud computing, big data, cybersecurity, AR, and virtual reality, and these disruptive techniques can, in turn, be used for further applications, and hence, AM has a plethora of opportunities in the innovation in numerous domains [1]. Figure 1 shows the rudimentary 3D printing manufacturing process, and various facets of the process can be leveraged by Industry 4.0 technologies. Few prevalent domains of application are biomedical, cybersecurity, material science, aerospace, etc. The adaptability of the AM system also allows several algorithms to enhance the system using cyberspace techniques, namely machine learning algorithms like neural network (NN). Qi et al. [2] studied system supervising, quality examination, etc. NNs solve complicated formulations and preventing them in most case. AM per se has a significant impact on a variety of research areas such as space exploration. Martínez et al. [3] worked habitats on the Moon using powder-based fusion AM technique as this can be controlled efficiently when linked with cyber system and can facilitate on-demand fabrication of structures with the required degree of customization.

There are many circumstances when remote PD is required. These situations are usually epidemics, e.g. COVID 19 pandemic, where the use of Ventilators is exceedingly crucial. Paszkiewicz et al. [4] have suggested a new model incorporating Industry 4.0, and the digital twin of the model can be made which would accelerate the prototype fabrication time. Kim et al. [5] worked on an innovative design of nuclear power plant design and operation. AM can expedite the fabrication of crucial components, namely turbomachinery and valves, and make the components more economical as compared to the case when they are manufactured in a conventional setup.

2 AM in the Industry 4.0 Framework

Industry 4.0 allows physical and virtual systems to interact and cooperate as a single entity. It uses various physical/wired networks and wireless connections to collect data from physical systems. AM provides a suitable approach for product manufacturing, as this process is controlled by computer systems, and hence, this data can be used in the cyber/virtual systems [7]. Moreover, this method is suitable for the production of components in the aerospace, biomedical, and other domains and can produce “customized” products so this proves to be a suitable technique to be used in the Industry 4.0 framework. This approach of controlling can facilitate customization of products, namely form, hue, materials, etc. [8]. For cyber systems, data delivery, customer data, and various others are related to logistics and inventory [7].

“Big data” is an important component of Industry 4.0 and will have a substantial impact on AM systems as it will enrich the system by enhancing equipment work, reducing energy requirements and hence cost, and enhancing the quality of production. Big data is the compendium of data gathered by the machines or physical systems, logistics data, etc., which may or may not be expedient to the industry [8]. Wang et al. [9] have worked on the impact of big data in cyber-physical systems, digital manufacturing, and Industry 4.0. They have accentuated the fact that big data will have a substantial impact on the digital manufacturing and smart manufacturing framework.

“Cloud computing”, which is an important component of Industry 4.0, allows the use of the product life cycle (PLC) management systems as per the user’s need. Guo et al. [10] focussed on the application of cloud computing in the manufacturing domain. They developed a framework that has four layers, namely interaction layer, service layer, virtual resource tool, manufacturing resource, and their access. Mai et al. [11] devised another framework by dividing the framework into five sections, namely device layer, adapter layer, servitization layer, management layer, and application layer. It allows manufacturers to contribute to innovation and take advantage of various PD processes.

Machine learning (ML) can facilitate fabricating parts with multifaceted characteristics, viz. nozzle path development, safety, printing viability, etc. In the biomedical domain, a tailor-made model for patients using the CT scan approach can be developed. We can accomplish multi-material printing and by training the ML algorithm to study from the large dataset for mechanical properties for different organ systems. Tissue Engineering is yet another important area where ML and AM can play an important role [12]. It is also used in the building and construction domain as well. However, the right materials must be used to bear the compressive and tensile loads [13].

“Augmented reality (AR)” is also a vital component in the Industry 4.0 framework and is also one of the important concepts linked with the virtual PLC management systems. Nee et al. [14] worked on the AR-based design and manufacturing framework and provided us with an outline on the basis and prospects of AR in this domain. It can give the user an instinctive means to manipulate related facets, namely design, analysis, and maintenance, and offer the user an amicable interface to develop products and with reduced production costs and increased production rates [15].

Ceruti et al. [16] focussed on the application of AM and AR in maintenance in the aeronautical domain. The framework provides the user with a user-friendly guidebook to perform modifications in the system or can instruct the user to perform maintenance jobs by using reverse engineering methods to make the CAD model

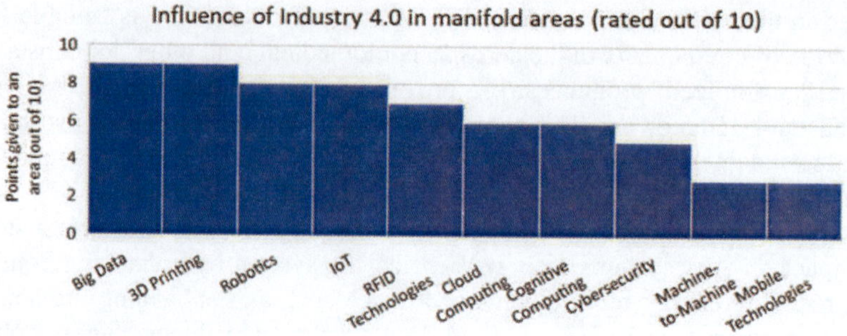

Fig. 2 Influence of Industry 4.0 in manifold areas [18]

of the part, 3D-print it, and then install it by using the guidebook. The use of AM and AR is also impacting rather general domains like the education sector and other social realms [17].

Figure 2 indicates the influence of Industry 4.0 on various technological areas.

3 Exhuming the Potential of AM in a Multitude of Explorations

AM and Industry 4.0 can create an impact on various aspects of a system, namely materials, usability, disruptive techniques, design, so on, and so forth.

3.1 Product Development (PD)

Industry 4.0 has a pivotal role in PD because it works and interacts with the physical system and can control the processes online. PD in the defence domain is one of the applications of AM. Anand et al. [19] focussed on the feasibility aspect of smart manufacturing in the defence domain in India. The survey found out that it would influence the productivity and quality in the defence manufacturing companies, the framework can be incorporated in the currently used system and will be economical and productive. However, challenges related to the security of data in cyberspace, training of the plant labour, enhancement of network infrastructure to incorporate the smart manufacturing system.

Goguelin et al. [20] worked on the use of the manufacturing system in cyberspace to enhance the PD process and abet the designer to improve the part design so that it can be printed in the AM systems and reduce the waste materials during fabrication and decrease the cost associated with the process when the part is not fabricated as per the requirement. The product life cycle (PLC) in the PD process has numerous stages which are important in the decision-making process in the smart manufacturing

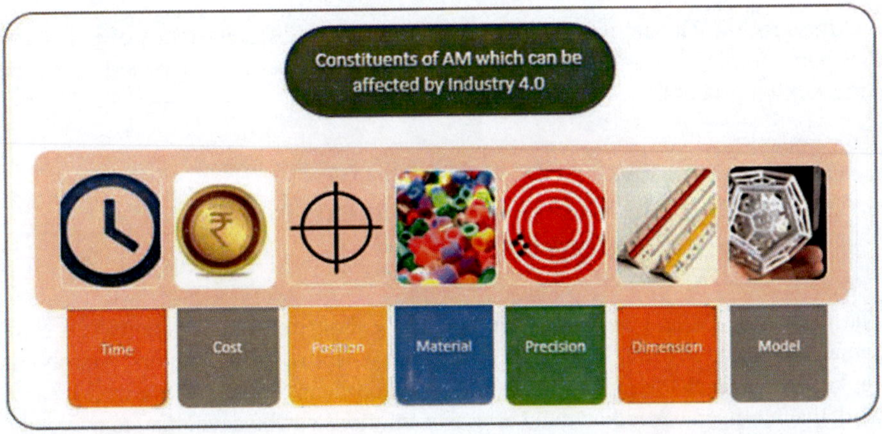

Fig. 3 Ways in which an AM system can be enhanced by Industry 4.0 [23]

framework to meet customer's needs. Ahmed et al. [21] worked on a framework that consists of three components that recommend the optimized solution based on the given constrictions, and this can be stored in the system to solve queries if they arise in the further stages.

Design is an important exponent in the PD process, and it is rather important when it comes to employing concurrent engineering for effective PD. One of the applications which deal with this domain is the "interface" [15] which will enhance productivity and is important as much as design is. Conde et al. [22] worked on a customer-centric AM framework where they tried to bring the CAD system closer to the customer and making them a part of the system will reduce the net process time. It offers a "multi-touch" UX and UI which asserts to have enhanced usability of the system and making the process more efficient (Fig. 3).

3.2 *Process Optimization*

Several techniques use different materials for various applications but require different parameters and designs to be considered before they can be used to produce tangible products. They can be obtained using the cyber system in Industry 4.0 framework. To select appropriate process parameters for 3D printing, artificial NN (ANN), which is an ML technique, is used. This can be made possible by using suitable algorithms, namely Garson's algorithm [24], to find out which parameter has a prevalent effect on the final expected product form. ML can optimize the performance of the final product, and various techniques can be used to optimize the parameters required to obtain the best results. In AM of materials, optimization is required to find out the melting state which would give anticipated properties of materials [25–27]. The main feature of the optimization is to reduce the porosity that is formed during the melting process. Hence, it is advantageous to analyze different types of pores and that pores formed under thermal conditions provide a better understanding of the optimization process.

4 Summary of the Current Developments in AM Technologies and Industry 4.0

The following Table 1 is a synopsis that talks about the developments that have happened in recent years. It consists of the core objectives of the papers, namely the application, and challenges. This would give a collective insight into the research work that has been carried out, and effective solutions can be developed to overcome these challenges.

5 Conclusion, Challenges, and Future Scopes

From the study of different articles and review papers, it is evident that there is an enormous scope of AM in various facets of Industry 4.0 to enhance the PD process. The systems are capable of incorporating cloud-based product life cycle management tools which are crucial for the usage of the information in the cyberspace. The imminent disruptive technologies expedite design optimization, "customized" products fabrication, controlling, and monitoring of the entire process. The systems are designed to consider the human–machine interaction (HMI) aspect of the system and enhance its usability. Certain challenges are present, viz. the network infrastructure must be robust enough to work efficiently, and the system must facilitate data sharing and ensure the utmost security of the data. Different materials must be studied to incorporate into this system, the process is slow compared to the standard processes, and parts with only limited build size can be manufactured. There are social challenges, viz. the inclusion of this framework might replace humans from the workplace in a majority of domains if not all and the current working labour force must be trained to work in an unaccustomed environment. The use of the Industry 4.0 framework in AM can indubitably enhance the workplace in various functionalities if employed with ingenuity.

Table 1 Outline of the current developments in AM Technologies and Industry 4.0

Authors	Research area	Application	Common challenges
Conde et al. [22]	AM, CAD	Smart factory, handheld devices	Improve upon the computational cost and robustness of the system
Anand et al. [19]	DPD in smart manufacturing	Defence	Cybersecurity-related risks, digital, and physical infrastructure cost are high, educating the workforce
Goguelin et al. [20]	DPD of manufacturability assistant for digital PD	Smart factories	Algorithms supporting CAD model files in the AM system without converting in the ".STL" by the user
Zhang et al. [15]	AR and design	DPD domain	Enhancing the monitoring and repairing system to abet AM process
Ceruti et al. [16]	AR and design	Aerospace maintenance	The cost of AM and AR systems is huge, and a robust UX and UI system has to be made to support the operations
Pan et al. [6]	AM	Biomedical	Integration of IoT and AM has to be studied in the medical domain
Wang et al. [9]	Big data and cybersecurity	AM systems for PD	Lack of accuracy is there, slow process, a robust system is needed
Baturynska et al. [28]	Machine learning	AM systems for PD	Software like Python is required by the user to make requests
Snell et al. [25]	ML algorithms	Material science	Algorithms have to be studied to get the optimized result
Godina et al. [29]	Sustainability	Sustainability in business models	Challenge in reducing the cost per part. Lack of accuracy of the part

References

1. Vinodh S, Wankhede VA (2020) Application of fuzzy DEMATEL and fuzzy CODAS for analysis of workforce attributes pertaining to Industry 4.0: a case study. Int J Qual Reliab Manag. https://doi.org/10.1108/IJQRM-09-2020-0322
2. Qi X, Chen G, Li Y, Cheng X, Li C (2019) Applying neural-network-based machine learning to

additive manufacturing : current applications, challenges, and future perspectives. Engineering 5(4):721–729. https://doi.org/10.1016/j.eng.2019.04.012
3. Labeaga-Martínez MN, Sanjurjo-Rivo JD-Á, Martínez-Frías J (2017) Additive manufacturing for a Moon village Additive manufacturing for a Moon village. Procedia Manuf 13:794–801. https://doi.org/10.1016/j.promfg.2017.09.186
4. Paszkiewicz A, Bolanowski M, Budzik G, Sowa P (2020) Applied sciences remote design and manufacture through the example of a ventilator
5. Kim KM, Bang IC (2020) Design and operation of the transparent integral effect test facility, URI-LO for nuclear innovation platform. Nucl Eng Technol. https://doi.org/10.1016/j.net.2020.08.006
6. Pan H, Yang X (2019) Application of Internet of Things technology in 3D medical image model. IEEE Access 7:5508–5518. https://doi.org/10.1109/ACCESS.2018.2886223
7. Dilberoglu UM, Gharehpapagh B, Yaman U, Dolen M (2017) The role of additive manufacturing in the era of Industry 4. 0. Procedia Manuf 11(June):545–554. https://doi.org/10.1016/j.promfg.2017.07.148
8. Mehrpouya M, Dehghanghadikolaei A, Fotovvati B, Vosooghnia A, Emamian SS, Gisario A (2019) The potential of additive manufacturing in the smart. Appl Sci:34
9. Lidong W, Guanghui W (2016) Big data in cyber-physical systems, digital manufacturing and industry 4.0. Int J Eng Manuf 6(4):1–8. https://doi.org/10.5815/ijem.2016.04.01
10. Guo L, Qiu J (2018) Combination of cloud manufacturing and 3D printing : research progress and prospect
11. Mai J, Zhang L, Tao F, Ren L (2015) Customized production based on distributed 3D printing services in cloud manufacturing. Int J Adv Manuf Technol. https://doi.org/10.1007/s00170-015-7871-y
12. Khan Z, Kahin K, Rauf S, Ramirez-Calderon G, Papagiannis N, Abdulmajid M, Hauser C (2019) Optimization of a 3D bioprinting process using ultrashort peptide bioinks. Int J Bioprint 5(1):3–6. https://doi.org/10.18063/ijb.v5i1.173
13. Tan K (2018) The framework of combining artificial intelligence and construction 3D printing in civil engineering. MATEC Web Conf 206:1–5. https://doi.org/10.1051/matecconf/201820601008
14. Nee AYC, Ong SK, Chryssolouris G, Mourtzis D (2012) CIRP annals—Manufacturing technology augmented reality applications in design and manufacturing. CIRP Ann Manuf Technol 61(2):657–679. https://doi.org/10.1016/j.cirp.2012.05.010
15. Zhang Y, Kwok TH (2018) Design and interaction interface using augmented reality for smart manufacturing. Procedia Manuf 26:1278–1286. https://doi.org/10.1016/j.promfg.2018.07.140
16. Ceruti A, Marzocca P, Liverani A, Bil C (2019) Maintenance in aeronautics in an Industry 4.0 context: the role of augmented reality and additive manufacturing. J Comput Des Eng 6(4):516–526. https://doi.org/10.1016/j.jcde.2019.02.001
17. Pence HE (2020) Education sciences how should chemistry educators respond to the next generation of technology change?
18. Liao Y, Rocha E, Deschamps F, Brezinski G (2018) The impact of the fourth industrial revolution: a cross-country/region comparison. Production 5411:18. https://doi.org/10.1590/0103-6513.20180061
19. Anand P, Nagendra A (2019) Industry 4.0: India's defence industry needs smart manufacturing. Int J Innov Technol Explor Eng 8(11) Special Issue:476–485. https://doi.org/10.35940/ijitee.K1081.09811S19
20. Goguelin S, Colaco J, Dhokia V, Schaefer D (2017) Smart manufacturability analysis for digital product development. Procedia CIRP 60:56–61. https://doi.org/10.1016/j.procir.2017.02.026
21. Bilal M, Sanin C, Szczerbicki E, Ahmed MB, Sanin C, Szczerbicki E (2019) ScienceDirect Smart Virtual Product Development (SVPD) to Enhance Product Smart Virtual Product Development (SVPD) to enhance product manufacturing in industry 4. 0 manufacturing in industry 4. 0. Procedia Comput Sci 159(2018):2232–2239. https://doi.org/10.1016/j.procs.2019.09.398
22. Rodriguez-Conde I, Campos C (2020) Towards customer-centric additive manufacturing: making human-centered 3d design tools through a handheld-based multi-touch user interface. Sensors (Switzerland) 20(15):1–28. https://doi.org/10.3390/s20154255

23. Goh GD, Sing SL, Yeong WY (2020) Potential, and challenges, 0123456789. Springer Netherlands
24. Zhang Y et al (2020) A parametric study of 3D printed polymer gears, pp 4481–4492
25. Snell R et al (2020) Methods for rapid pore classification in metal additive manufacturing. JOM 72(1):101–109. https://doi.org/10.1007/s11837-019-03761-9
26. Wankhede V, Jagetiya D, Joshi, Chaudhari R (2019) Experimental investigation of FDM process parameters using Taguchi analysis. Mater Today Proc 27:2117–2120. https://doi.org/10.1016/j.matpr.2019.09.078
27. Soni H, Gor M, Rajput G, Sahlot P (2021) Thermal modeling of laser powder-based additive manufacturing process. In: Mathematical modeling, computational intelligence techniques and renewable energy. Advances in Intelligent Systems and Computing, pp 1–8
28. Baturynska I, Martinsen K (2021) Prediction of geometry deviations in additive manufactured parts: comparison of linear regression with machine learning algorithms. J Intell Manuf 32(1):179–200. https://doi.org/10.1007/s10845-020-01567-0
29. Godina R, Ribeiro I, Matos F, Ferreira BT, Carvalho H, Peças P (2020) Impact assessment of additive manufacturing on sustainable business models in industry 4.0 context. Sustain 12(17):1–21. https://doi.org/10.3390/su12177066

A Comparison of Mechanical Properties of 3D Printed Specimens Based on Filament Quality

Omkar Bankar, Nikhil Mule, Saurabh Prabhune, and Vipin B. Gawande

Abstract Fused filament printing is the most popular process used in most of the 3D printing machines as compared to other techniques. Polylactic acid (PLA) filament is primarily used for product development in desktop 3D printers due to its lower printing temperature. A measurable amount of material is wasted during 3D printing during the handling of various parameters. This waste filament material along with raw material is used in a 3D printer extruder machine to create recycled PLA. 3D printed parts are printed using the imported filament and filament produce from the extruder machine. A set of tensile tests on three specimens are carried out to check the mechanical properties of these two specimens. A comparative analysis of mechanical properties (tensile) is represented in the paper. Results show that using recycled filament, there is a slight decrease in the yield strength, ultimate tensile strength, and elongation in the recycled filament part.

Keywords PLA · Mechanical properties · 3D printing · Extruder machine · Recycling

1 Introduction

Additive manufacturing (AM) is the process of joining materials to make objects from computer-aided design (CAD) model data, usually layer upon layer, as opposed to subtractive manufacturing methods. This tool-less manufacturing method can produce fully dense metallic and plastic parts in short time, with high precision. A 3D model is created in the modeling software, and .STL file is imported in the 3D printer software. The .STL file contains all the information regarding the 3D model to be printed using 3D printer. Printing parameters are set in the software. One of the major parts of additive manufacturing in 3D printing is filament. This filament is called as virgin filament. The virgin filament used for this study is polylactic acid (PLA). PLA is produced from plant materials like corn starch and sugar cane.

O. Bankar (✉) · N. Mule · S. Prabhune · V. B. Gawande
Department of Mechanical Engineering, VPKBIET, Baramati 413133, India
e-mail: omkar.bankar.mech2017@vpkbiet.org

A. Maurya et al. (eds.), *Recent Trends in Mechanical Engineering*, Lecture Notes in Mechanical Engineering, https://doi.org/10.1007/978-981-19-7709-1_30

The advantage of using PLA is that it is biodegradable. A wide use of 3D printing generates a significant amount of plastic waste. The main reason for this being the failed parts and discarded support structures. A huge amount of filament material is utilized for developing prototypes, as 3D printing offers quick part production in short time. The scrap PLA is managed through several methods like combustion, recycling, composting, or it is dumped in landfills. Among these method, recycling of PLA is a best option as PLA is biodegradable.

The scrap PLA is used as a raw material in the extruder machine developed in our laboratory to produce recycled PLA filament [1]. The raw PLA is first crushed in a mechanical crusher and then send to heater assembly which has a rotating auger drill bit. The filament of desired diameter comes out through nozzle through extrusion process. The filament is then cooled and is checked for desired diameter. The final product is suitable for 3D printer as an input material.

Literature review shows that [1–5], mechanical properties of filament play a vital role in the part development. Although these properties are available for imported filament (virgin), very little data are available for the mechanical properties of recycled PLA material. This paper presents a study to evaluate the mechanical properties of recycled PLA and its comparison with the virgin PLA filament.

2 Methodology

The tensile test specimen was manufactured according to American Society of Testing Materials (ASTM) standard ASTM D368-14 Type-IV using a 3D printer, with the use of recycled PLA filament under melting temperature of 210 °C. The parameters like layer height, shell thickness, fill density, print speed, printing temperature, and bed temperature are set in Cura software [6, 7]. The STL file thus created is send to the printer for printing the specimen as shown in Fig. 1. The test specimen printing took almost 2 h.

The printed part then was tested under UTM, as shown in Fig. 2, for the calculation of mechanical properties of the element prepared by the recycled PLA filament. The testing was done until the part had a failure. The mechanical properties were calculated based upon testing parameters.

3 Results and Discussion

Three tensile specimens, created from virgin filament and recycled filament, were tested for yield strength, ultimate tensile strength, percentage elongation, tensile failure stress, and percentage strain. The summary of these results is listed in Tables 1, 2 and 3.

The tensile load versus percentage elongation for three virgin and recycled specimens is shown in Fig. 3. The values evaluated from the study show that there is

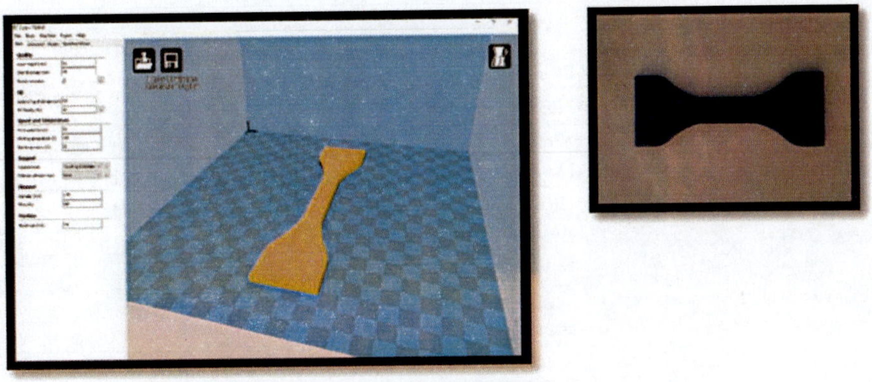

Fig. 1 Parameter setting for test specimen and final printed specimen

Fig. 2 Ultimate testing machine (FUN400)

Table 1 Comparison of tensile properties for virgin and recycled test specimen 1

Mechanical properties	Virgin specimen	Recycled specimen
No. of specimens	01	01
Yield strength (Ys)	24.03 MPa	22.01 Mpa
Ultimate tensile strength	32.928 Mpa	28.86 Mpa
% Elongation	0.72%	0.80%
Tensile failure stress	0.72 Mpa	0.58 Mpa
% Strain	0.12%	0.09%

Table 2 Comparison of tensile properties for virgin and recycled test specimen 2

Mechanical properties	Virgin specimen	Recycled specimen
No. of specimens	02	02
Yield strength (Ys)	24.03 MPa	23.86 Mpa
Ultimate tensile strength	32.928 Mpa	30.66 Mpa
% Elongation	0.72%	0.78%
Tensile failure stress	0.72 Mpa	0.62 Mpa
% Strain	0.12%	0.10%

Table 3 Comparison of tensile properties for virgin and recycled test specimen 3

Mechanical properties	Virgin specimen	Recycled specimen
No. of specimens	03	03
Yield strength (Ys)	24.03 MPa	23.57 Mpa
Ultimate tensile strength	32.928 Mpa	31.42 Mpa
% Elongation	0.72%	0.70%
Tensile failure stress	0.72 Mpa	0.65 Mpa
% Strain	0.12%	0.09%

very slight difference in the mechanical properties of 3D printed test specimens using virgin material and recycled filament. Average mechanical properties of test specimens made from recycled filament differ by 8.77–13.16% when compared with specimens made from virgin material. This led to the conclusion that the parts produced from recycled filament have consistent mechanical properties and show close resemblance in terms of mechanical properties. The degradation in the properties of recycled filament shows a failure as tested under UTM as shown in Fig. 4.

The decrease in mechanical properties in recycled material may include the process used for recycling. The process includes several parameters like melting temperature and impurities that may appeared in the recycled parts that are used for recycling. The other possibilities might be the process parameters set for developing a part using a 3D printer. Though a close tolerance is maintained during extrusion process and in 3D printing process, a detailed comparison is suggested by author for this study using other forms of materials like ABS and others.

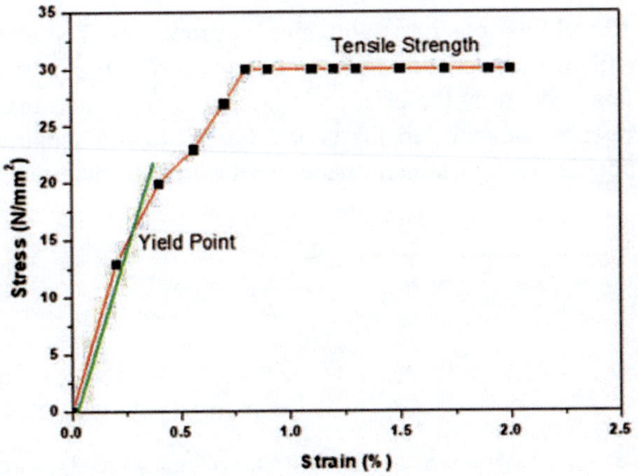

Fig. 3 Tensile load versus percentage elongation for specimen 1

Fig. 4 Failure of specimen from recycled filament

4 Conclusions

In this study, 3D printed tensile specimens created from virgin filament and recycled filament. Three specimens were tested for comparison of their mechanical properties. Study shows that, recycled specimens revealed similar mechanical properties as that of virgin filament. However, the recycled filament causes clogging during printing and emits ultrafine particles. Instead of this, the study shows that the recycled filament could be used to yield various parts with similar properties. The recycling of scrap PLA material will also save cost of printing, energy, and CO_2 emission. In the future study, researchers may study the effect of filament strength and printability on the mechanical properties of recycled filament. There is also a scope to compare the mechanical properties using filament made up of different materials.

References

1. Aris MAIZ, Mat S, Sam MS, Ramli FR, Alkahari MR, Abdul Kudus SI, Universiti Teknikal Malaysia Melaka Pembuatan (2020) Design and development of 3D printer filament extruder. Proc Mech Eng Res Day 2020: 293–294
2. Lanzotti A, Martorelli M, Maietta S, Gerbino S, Penta F, Gloria A (2019) A comparison between mechanical properties of specimens 3D printed with virgin and recycled PLA. Procedia Cirp 79:143–146
3. Farah S, Anderson DG, Langer R (2016) Physical and mechanical properties of PLA, and their functions in widespread applications—A comprehensive review. Adv Drug Deliv Rev 107:367–392
4. Math, Kadadevara RS, Goutham R, Srinivas Prasad KR (2018) Study of effects on mechanical properties of PLA filament which is blended with recycled PLA materials. In: IOP conference series: materials science and engineering, vol 310, no 1. IOP Publishing, p 012103
5. Anderson I (2017) Mechanical properties of specimens 3D printed with virgin and recycled polylactic acid. 3D Print Addit Manuf 4(2):110–115
6. Galeta T, Raos P, Stojšić J, Pakši I (2016) Influence of structure on mechanical properties of 3D printed objects. Procedia Eng 149:100–104
7. Fernandez-Vicente M, Calle W, Ferrandiz S, Conejero A (2016) Effect of infill parameters on tensile mechanical behavior in desktop 3D printing. 3D Print Addit Manufg 3(3):183–192

Analysis of Wind Energy for Power Generation in India

Nitin Kumar and Om Prakash

Abstract India is the fastest-growing country in the world, and the demand for energy increases day by day. Renewable energy sources play an important role to fulfill the energy demand. International Energy Agency identified wind energy as the main source of renewable energy. Wind energy is clean and pollution-free energy. Wind power production capacity in India increased in recent years. At present, India has the fourth-highest wind energy installed capacity in the world. Wind energy virtually is a form of solar energy when the surface of the earth differential heating then creates pressure difference and flow the wind from high pressure to low pressure. The wind has kinetic energy, and these converted into electric energy with the help of a wind turbine. This article discussed the potential of wind energy in the different states of India and compared wind energy production with the world.

Keywords Wind energy · Renewable energy · Wind turbine · Wind velocity

1 Introduction

Renewable energy is the fastest-growing sector in the energy production sector of the world because of the pollution increases day by day in our environment. So, all over the world every country searching for an alternative source of electrical energy production, renewable energy is the alternative source of electrical energy production for the reason that it is pollution-free energy [1]. Among various renewable energy sources, wind energy offers the greatest potential for energy production. It has a significant alternative source of power generation [2]. Figure 1 shows the different renewable energy sources to produce electricity such as wind energy, solar power ground-mounted, solar power rooftop, small hydropower, biomass (bagasse) cogeneration, biomass (non-bagasse) cogeneration/captive power, and waste power. The

N. Kumar (✉) · O. Prakash
Department of Mechanical Engineering, National Institute of Technology Patna, Patna, India
e-mail: nitink.pg18.me@nitp.ac.in

O. Prakash
e-mail: om.prakash@nitp.ac.in

A. Maurya et al. (eds.), *Recent Trends in Mechanical Engineering*, Lecture Notes in Mechanical Engineering, https://doi.org/10.1007/978-981-19-7709-1_31

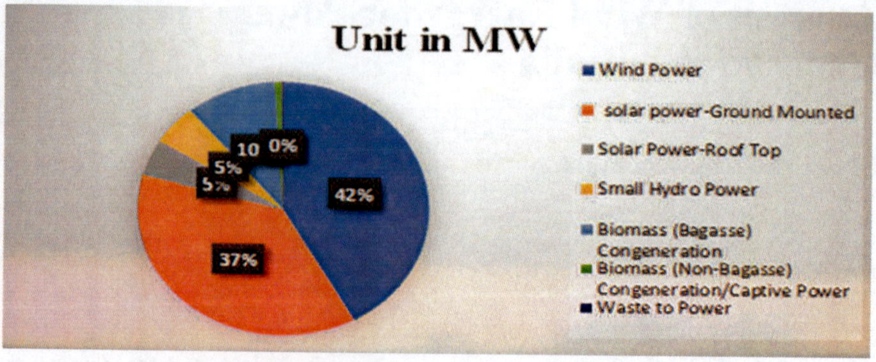

Fig. 1 Total installed capacity of grid-interactive renewable power [3]

total install capacity of grid-interactive renewable energy is 92550.74 MW, and the contribution of wind energy is 38683.65 MW or 42% [3]. The International Energy Agency can experience a 6.9% compound annual growth rate of renewable energy consumption in the new policies situation between 2014 and 2040. In the world, some countries have been increased the consumption of renewable energy between 2002 to 2017; the compound annual growth rate was China 39%, India 23%, Ireland 21%, and Great Britain 20% [4]. In the world, China is the highest production of wind energy whereas, in this list, India is the fourth position.

2 Global Scenario of Wind Energy

At present, wind energy is the fastest-growing sector of non-conventional energy sources in the world, and it is the most widely used alternative source of energy [5]. Wind energy is the fastest-growing sector in the last decades, and it continues at a faster rate [6].

As per the Global Wind Energy Report 2019, the total installed onshore wind energy capacity is 621 GW. China is the largest producer of wind energy is 229.77 GW, and it is produced 36% of the total wind energy in the world (Figs. 2 and 3, Table 1).

India had a record year, and globally, it was the fourth position in the term of an annual addition and the cumulative capacity in 2018. 2191 MW new wind power was added and reached 35,129 MW at the end of 2018. In the year 2019, India has the fourth position globally in cumulative capacity 37,506 MW and third position on the new annual addition 2377 MW of wind energy at the end of the year 2019. India wind power installed capacity is 6% share in the global market [8].

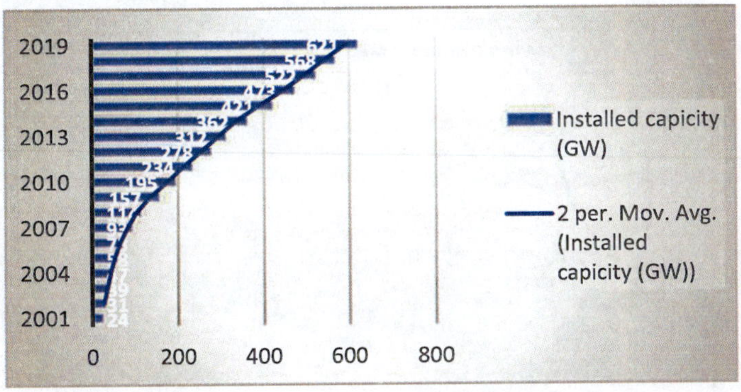

Fig. 2 Evolution of wind energy in the world [7]

Fig. 3 Wind energy distribution in the world [7]

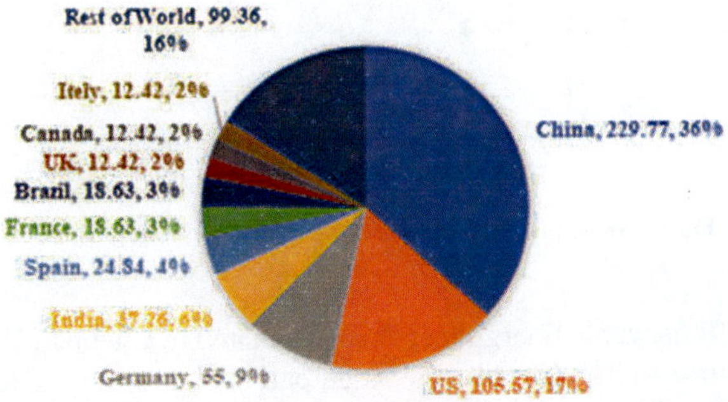

Table 1 Development of total installation of wind energy [7]

Country	New installation 2017 (MW)	Total installation 2017 (MW)	New installation 2018 (MW)	Total installation 2018 (MW)	New installation 2019 (MW)	Total installation 2019 (MW)
China	18,499	185,604	20,200	205,804	23,760	229,564
US	7017	89,047	7588	96,488	9143	105,436
Germany	5334	50,779	2402	52,932	1078	53,913
India	4148	32,938	2191	35,129	2377	37,506

3 Wind Map of India

In the early 1960s, National Aeronautical Laboratory (NAL) systematically studies the data from the Indian Metrological Department (IMD) from the energy point of view. Based on the data available, the National Aeronautical Laboratory prepared the wind map of India [9]. The effective production of the wind depends on many factors such as the wind speed (m/s), wind power density (w/m^2), and site availability. The National Institute of Wind Energy (NIWE) under the Ministry of New and

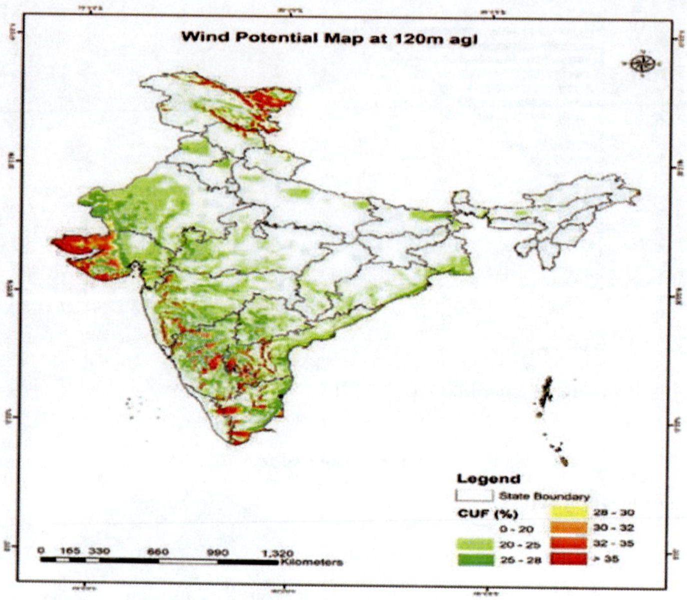

Fig. 4 Wind potential map of India at 120 m AGL [10]

Renewable Energy (MNRE) is situated in Chennai, Tamil Nadu. It is self-sustaining research and development center that conducts wind assessment programs across the country and successfully draws Indian wind energy maps of different heights [10] (Fig. 4).

On the map, you can see clearly the Western Ghats of India is a high wind potential, and also, the southern part of India is a good range in the wind power potential. Indian states with high wind potential are Rajasthan, Gujarat, Madhya Pradesh, Maharashtra, Karnataka, Kerala, Tamil Nadu, Andhra Pradesh, and some parts of Telangana (Table 2).

4 Wind Power of Turbine

The wind is generated from the sun. Since the sun does not heat the same amount of land, the sea, and the atmosphere, there is a difference in the temperature and pressure. When the pressure difference occurs in the environment, the air is flowing from high pressure to low pressure. The power from the wind turbine is generated when the blades rotate due to blowing the air; then, mechanical energy is used to generate electrical energy [12]. The power of the wind turbine is the kinetic energy of the wind passing through the swept area of the blade perpendicular to the wind speed in a given time. The power of the wind turbine can be generated

Table 2 Wind monitoring station with MAWS & MAWPD in Tamil Nadu [11]

SL. no	Station Tamil Nadu	District	Commenced on	Mast height (m)	Elevation (masl)	MAWS at 18/20/25/30/50/78/80/100/120 m (m/s)	MAWPD at 18/20/25/30/50/78/80/100/120 m (W/m^2)
1	Kayathar	Tuticorin	29/08/13	120	88	6.29	319.2
2	Kalimandayam*	Dindigul	13/12/13 reconfigured on 20/08/19	10	302	5.92	279.36
3	Akkanayakanpatti*	Tuticorin	19/12/13 reconfigured on 01/02/17	10	72	7.42	468.29
4	TNPL Perungudi#	Tirunelveli	30/07/16	50	45		
5	Dalmia (Muppandal)	Kanyakumari	01/11/17	50	121		
6	Kavalkinaru (ABAN 1)	Tirunelveli	20/01/18	100	72		
7	Veeranam	Tirunelveli	24/01/18	100	122		
8	Radhapuram (ABAN 2)	Tirunelveli	22/01/18	100	49		

$$P_{WT} = \frac{1}{2} C_P(\lambda, \theta) \rho A u^3$$

where ρ is the density of air (kg/m^3), CP is the power coefficient of wind turbine which is the function of two-parameter, one is the tip speed ratio (λ), and the second is the blade angle (θ). A is the swept area of the blade (m^2), and u is the velocity of the wind (m/s).

(a) Efficiency
The overall efficiency of the wind turbine is the function of the power coefficient (Cp) and mechanical efficiency (η) of the wind turbine [13]. It is calculating as

$$\eta_{\text{overall}} = \frac{P_{\text{out}}}{\frac{1}{2}\rho A u^3} = \eta C_P$$

5 Potential of Wind Energy

Wind as you know was nothing but air in motion this happens around the globe all 24 h and all through the day and night. It can be peaceful, violent, and powerful. We have harness wind for thousands of years for transportation, agriculture to grained grace for water pumping and now for generating electricity. The wind is the power full source of energy on our planet. India has lots of potentials to produce electrical energy through renewable sources of energy because India has a large coastal length, i.e., 7516.6 km; it is a vast coastal length; this coastal area is a source of good fresh wind. Near the coastal area, the wind speed is high. According to the Global Wind Energy Council report 2019, India has a 37.26 GW total installed capacity. India has nine major states (i.e., Rajasthan, Gujrat, Madhya Pradesh, Maharashtra, Andhra Pradesh, Telangana, Karnataka, Kerala, and Tamil Nadu) the high potential of production of electricity through the wind source [14] (Fig. 5).

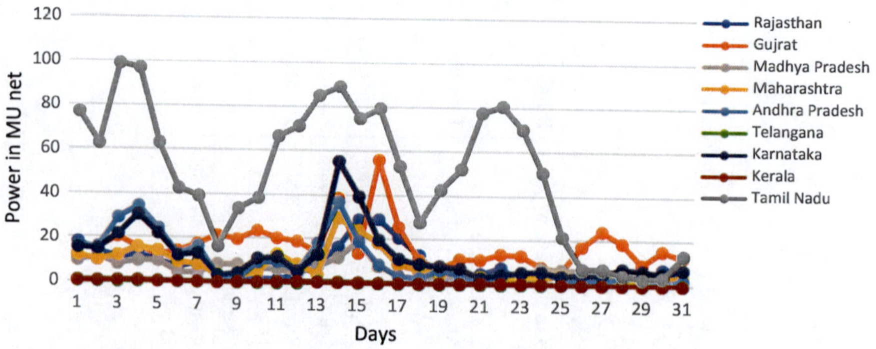

Fig. 5 Wind power production in different states in October 2020

6 Conclusion

This article shows the power generation through wind energy in India. Compare the wind energy installed capacity of India to the world and the total installed wind energy capacity of the world is 621 GW, and India has 37.26 GW or 6% shear of the total wind energy. India has total installed wind energy capacity is 35.6 GW as of 31st March 2019 and produced around 52.66 Billion units during 2017–18. Study the wind map of India and the potential of wind energy in the different states of India. Measure the wind speed with the help of wind monitoring stations installed all over the country by the National Institute of Wind Energy. In India, nine major states produce wind energy, and Tamil Nadu is the higher producer of wind energy.

Acknowledgements We would like to thank The National Institute of Technology, Patna for allowing this study to be undertaken.

References

1. Jain A, Balasubramanian R, Tripathy S (2011) Reliability analysis of wind embedded power generation system for Indian Scenario. Int J Eng Sci Technol 3(5):93–99. https://doi.org/10.4314/ijest.v3i5.68570
2. Carolin Mabel M, Fernandez E (2008) Analysis of wind power generation and prediction using ANN: a case study. Renew Energy 33(5):986–992. https://doi.org/10.1016/j.renene.2007.06.013
3. MNRE/Physical progress, MNRE I Physical Progress. Grid interactive power. https://mnre.gov.in/the-ministry/physical-progress. Accessed 22 Feb 2021
4. Sadorsky P (2020) Wind energy for sustainable development: driving factors and future outlook. https://doi.org/10.1016/j.jclepro.2020.125779
5. Dincer F (2011) The analysis on wind energy electricity generation status, potential and policies in the world. Renew Sustain Energy Rev 15(9):5135–5142. https://doi.org/10.1016/j.rser.2011.07.042
6. Kumar Y et al (2016) Wind energy: trends and enabling technologies. Renew Sustain Energy Rev 53:209–224. https://doi.org/10.1016/j.rser.2015.07.200
7. Wind G, Council E (2019) Gwec global wind report 2019. http://www.gwec.net/wp-content/uploads/vip/GWEC_IWEO_2019_LR.pdf
8. Global Wind Energy Council (2016) Indian Wind Energy, p 20. http://www.gwec.net/wp-content/uploads/vip/GWEC_IWEO_2016_LR.pdf
9. Rakesh MP (2014) Wind resource assessment in India : a program with a difference. Int J Environ 37–41. https://doi.org/10.1080/00207230701747499
10. Sharma S, Sinha S (2019) Indian wind energy & its development-policies-barriers: an overview. Environ Sustain Indic 1–2(May):100003. https://doi.org/10.1016/j.indic.2019.100003
11. Maiti and Bidinger (2021) Wind monitoring station. J Chem Inf Model 53(9):1689–1699
12. Köktürk G, Tokuç A (2017) Vision for wind energy with a smart grid in Izmir. Renew Sustain Energy Rev 73(January):332–345. https://doi.org/10.1016/j.rser.2017.01.147
13. Kumar A, Prasad S (2010) Examining wind quality and wind power prospects on Fiji Islands. Renew Energy 35(2):536–540. https://doi.org/10.1016/j.renene.2009.07.021
14. Daily Renewable Generation Report—Central Electricity Authority. https://cea.nic.in/daily-renewable-generation-report/?lang=en. Accessed 02 Mar 2021

Modeling of Portable Graphene Water Filter

Shantnu Chawla, Puneesh, Piyush Verma, and Ravinderjit Singh Walia

Abstract Graphene is a two-dimensional, nuclear scale, the hexagonal grid in which one particle frames every vertex. It can likewise be considered as an inconclusively enormous fragrant particle, a definitive instance of the group of level polycyclic sweet-smelling hydrocarbons. Due to the unique properties, graphene membranes have numerous applications in fields like Biomedical, Composites, Coatings Electronics, Energy, Membranes, Sensors, and Water Filtration. The present disclosure relates to the process of developing a portable water filter using CAD modeling and ANSYS simulation. The idea is to use vertically aligned graphene membranes to purify water. Graphene is produced using different methods and is extracted from the substrate by etching it in a specific chemical. The graphene membranes are placed on one and each other to form a thick foam-like structure. The graphene foam-like structure can be embedded into any shape using 3D designing and modeling. The pore size of the graphene membranes is small enough that it only allows water molecules to pass through it. The dirt and unwanted particles get blocked by the membranes. This technique can be used for the desalination of seawater. The uniqueness of this product is that the membranes are replaceable, and the cost is very low as compared to the other water filter. There is no requirement for water storage and no use of electricity for the filtration as a result the energy is also conserved. The wastage of water during the filtration is negligible.

Keywords Portable · Water filtration · CAD modeling · ANSYS simulation · Vertically aligned graphene membranes

1 Introduction

Graphene is one of the hottest materials in the scientific community due to its unique properties such as structure, electrical conductivity, mechanical, and catalytic properties. Generally, graphene can be synthesized from raw material like graphite,

S. Chawla (✉) · Puneesh · P. Verma · R. S. Walia
Department of Production and Industrial Engineering, PEC, Chandigarh, India
e-mail: Shane01041999@gmail.com

A. Maurya et al. (eds.), *Recent Trends in Mechanical Engineering*, Lecture Notes in Mechanical Engineering, https://doi.org/10.1007/978-981-19-7709-1_32

using chemical vapor deposition (CVD) [1]. But, one of the major challenges in synthesizing graphene is how to produce high-quality material in a reproducible manner, on a large scale, and at a low cost. The present research progresses on the synthesis of vertically aligned graphene sheets using edible and non-edible oils without involving any complex machine-like CVD, thus helps in reducing the cost of production. Graphene-based nanomaterials have revealed unique features, and there are highlighted new routes for the easy and proficient preparation of graphene-based nanoparticles and vertically aligned sheets with applications in various fields. Due to these unique properties, graphene has numerous applications in fields like Biomedical, Composites, Coatings Electronics, Energy, Membranes, Sensors, and Water Filtration. The present research focuses on the application of water filtration. The prototype designing was done on the Autodesk Inventor software, and simulation is done on ANSYS.

2 Literature Review

Presently, the graphene sheets are primarily fabricated using chemical vapor deposition (CVD) [2, 3] and plasma-enhanced CVD (PECVD) [4]. In these methods, hydrocarbons (acetylene, methane, or any other carbon source) are dissociated at very high temperatures (>1000 °C) and vacuum inside a chemical reactor in an inert atmosphere using a catalytic layer on the substrate [5, 6]. The catalytic layer is reduced using some reducing gas (H_2 or NH_3) before dissociating hydrocarbon to grow graphene sheets [7, 8]. The challenges associated with CVD and PECVD are the use of precursor gases like argon, hydrogen, and acetylene/methane makes the growth process very sensitive and costly. The purity of precursor gases is also a critical parameter. The growth process parameters such as vacuum, temperature, and gas flow rates are difficult to control precisely and thus adversely affect the quality of grown graphene sheets. These are very time-consuming processes and require a highly skilled operator to carry out the growth process [9, 10].

Due to all these limitations, CVD and PECVD processes are not economical for mass production of vertically aligned graphene sheets [11]. This has led to the development of alternate processes. In a closer relation to the proposed method, graphene sheets are grown in the air using soybean oil on Ni foils [12, 13]. However, this method is limited on Ni foil alone, making its use rather limited. The limitations of CVD and PECVD have paved the way to develop a new method to grow vertically aligned graphene sheets using several edible oils on various types of substrates such as Ni foil and Cu foil without the use of CVD. The potential of graphene to serve as a key material for advanced membranes comes from two major possible advantages of this atomically thin two-dimensional material: permeability and selectivity. Graphene-based membranes are also hypothetically attractive based on concentration polarization and fouling, and graphene's chemical and physical stability. Using these properties of graphene, it can be transformed into a mesh-like structure that can be modified to filter water.

3 Materials and Methodology

Vertically aligned graphene sheets are very promising as their electrical and thermal properties and the surface-area-to-volume ratio are better than conventional graphene sheets. Vertically aligned graphene sheets are synthesized using PECVD alone to date. The low yield and longer process time with PECVD [10, 11] make the graphene quite expensive. Consequently, there is a need to develop alternative ways suited for the mass production of vertically aligned graphene sheets have recently grown multilayer graphene using edible oil at 800 °C temperature in air.

Figure 1 shows the different oils, and Fig. 2 shows the substrates used in the experimental process. The reasons for choosing these oils are shown in Table 1. However, the method only works with Ni foil with very little or no growth on Cu or silicon substrates. In current innovation, vertically aligned graphene sheets are grown on silicon wafers, Cu, and Ni foil using different types of edible oils using a very simple process in a laboratory furnace at 800 °C. The developed method is promising for mass production of vertically aligned graphene sheets with varying number layers. The number of layers in graphene sheets is controlled by the quantity of oil used and adjusting the annealing time during the growth process.

Table 1 shows the comparison between the different edible oils based on different fatty acids present in them. Linoleic acid ($\omega - 6$) is present more than 50% in the corn and soya bean oils, and the value for other fatty acids is close to each other. Hence, it is assumed, both corn oil and soya bean oil will exhibit a similar type of grown graphene structure. The experimental results prove it to be true.

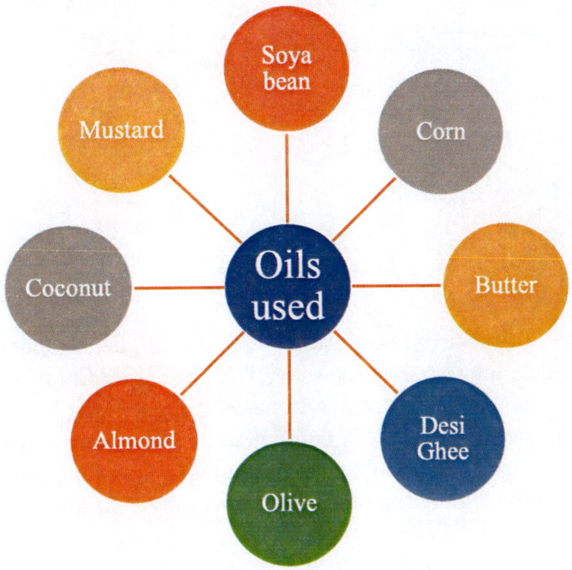

Fig. 1 Oils used in the experiment

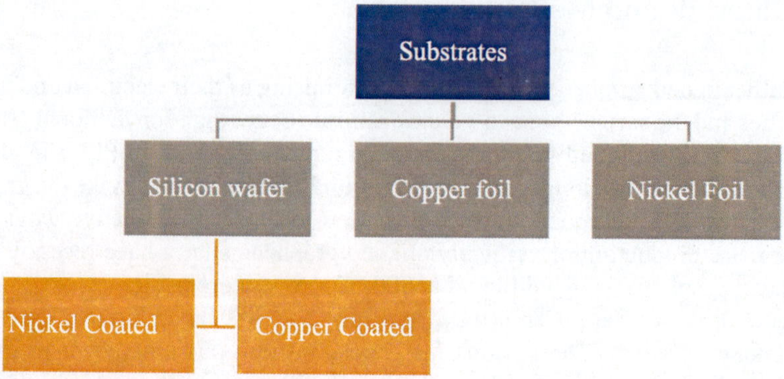

Fig. 2 Substrates used for the experiment

Table 1 Nutrition properties present in different type of edible used in the present invention [14]

Vegetable (edible) oils

Type	Saturated fatty acids	Monounsaturated fatty acids		Polyunsaturated fatty acids			Smoke point
		Total mono	Oleic acid (ω – 9)	Total poly	Linolenic acid (ω – 3)	Linoleic acid (ω – 6)	
Soybean	15.6	22.8	22.6	57.7	7	51	238 °C
Almond	8.2	69.9	N.A	17.4	N.A	N.A	221 °C
Mustard	11.5	59.1	N.A	21.2	N.A	N.A	249 °C
Coconut	82.5	6.3	6	1.7	N.A	N.A	175 °C
Corn	12.9	27.6	27.3	54.7	1	58	232 °C
Olive	13.8	73.0	71.3	10.5	0.7	9.8	193 °C
Clarified butter (Desi Ghee)	61	28.7	N.A	3.7	N.A	N.A	250 °C
Butter	51	21	N.A	3	N.A	N.A	150 °C

To make the best use of vertically aligned graphene sheets, a water filter can be made out to purify seawater. The pore size of graphene is small enough that it only allows water molecules to pass through it and blocks the salt particle. Hence, the desired water filter is being made keeping many instructions in mind. The filter is designed and simulated assuming it is used for commercial purposes.

The simulation of the design was done on the Ansys software for stress analysis. Ansys is a finite element analysis software that is used for product design, testing, and operation. Ansys is used so that we can determine how the product will function or deliver results with various specifications, loads, temperature, and other operating

Table 2 Meshing senses and the count of the elements

Nodes	25,432
Elements	13,184
Element size	0.002 m
Smoothing	Medium

conditions without actually building prototypes and conducting crash tests on the prototypes, for example, Ansys software can be used to simulate how much load can a chair hold when the person with different mass sit on it with different positions, how to design a swing for children which will use less material but will also provide necessary safety standards. Finite element analysis or FEA is the simulation of a physical phenomenon using a numerical mathematic technique referred to as the finite element method, or FEM. It also is one of the key principles used in the development of simulation software. It helps to reduce the number of physical prototypes by running virtual experiments which help in optimizing the design of the product by saving time as well as reducing cost (Table 2).

Figure 3 shows the mesh generation in prototype for a water filter was designed, and FEA was carried out on the prototype. The mesh was created on the prototype, and later on, the constraints were applied.

Figure 4 shows the constraints that applied on the prototype design, and a pressure 2×10^6 Pa was applies on the inner surface normal to it (Table 3).

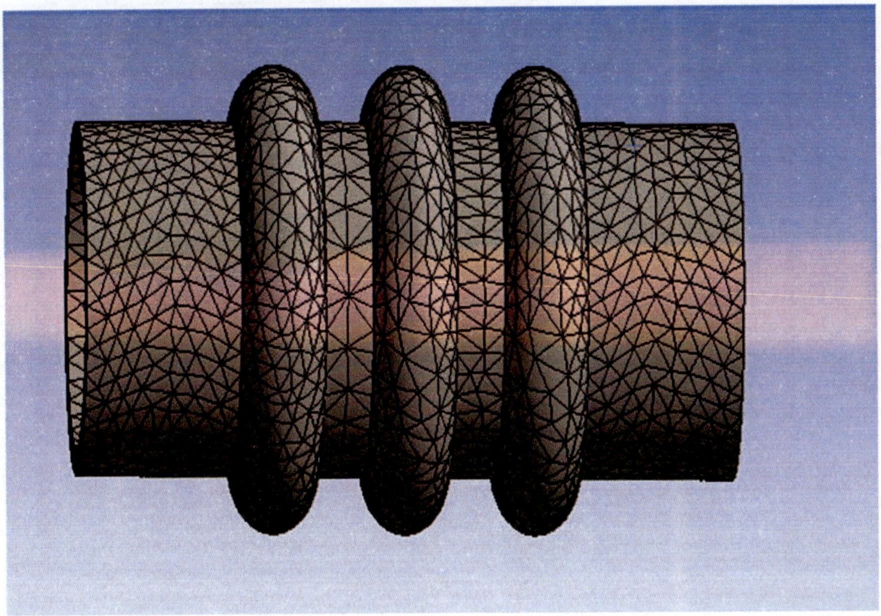

Fig. 3 Mesh generation on the prototype

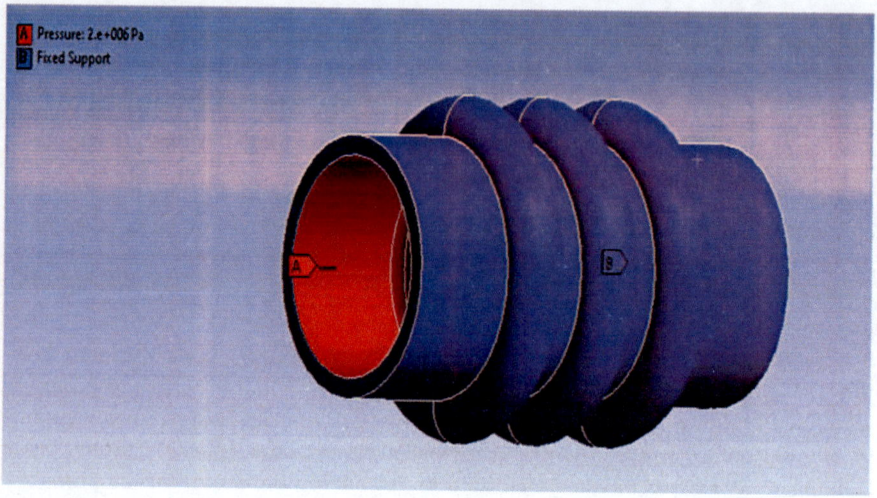

Fig. 4 Inner and outer surface of the prototype

Table 3 Magnitude and direction of water pressure on the prototype

Fixed surface	The outer surface (Blue)
Pressure applied on	Inner surface (Red)
Magnitude of pressure	2×10^6 Pa
Direction of pressure	Normal to inner surface (Red)
Material	Stainless steel

Governing Equations [15]

The governing equation of fluid flow is

(1) **Conservation of mass or Continuity equation**
$A_1V_1 = = A_2V_2$ where A and V are area of cross-section and velocity, respectively.
(2) **Conservation of momentum or Newton's Second Law**
$F = m.a$ where F, m, and a are the force, mass, and acceleration, respectively.
(3) **Conservation of Energy or the Bernoulli's Equation**
$P_1 + 0.5\rho V_1{}^2 + \rho g h_1 = P_2 + 0.5\rho V_2{}^2 + \rho g h_2$
where P_1 and P_2 are the pressures, V_1 and V_2 are the velocities, h_1 and h_2 are the heights, ρ is the density of the fluid, and g is the acceleration due to gravity at points 1 and 2.

4 Results and Discussion

In the developed process, graphene sheets are grown on silicon wafers and Cu and Ni foils using edible oils inside a simple laboratory furnace. Thus, the current innovation can address the following issues related to currently used processes; current process uses edible oils and thus avoids the use of dangerous gases such as acetylene, hydrogen, or NH_3 to grow graphene sheet at 800 °C and thus avoids vacuum or other parameters which are difficult to control.

The SEM analysis of grown graphene sheets is shown in Fig. 5. The applicability of developed process is demonstrated by growing graphene sheets on Ni foil using corn oil with a fixed annealing time of 3 min. In current technology, due to the simpler method of production, there is no requirement of a highly skilled operator which is needed in CVD and PECVD operations. In the new method of production of vertically aligned graphene sheets, there is no requirement of any extra catalyst layer. The total process time of current technology (15–30 min) is significantly lower than conventional CVD and PECVD techniques (3–4 h). All these improvements in current inventions make it a cheaper growth process in comparison to CVD and PECVD process. Thus, it is suited for the mass production of vertically aligned graphene sheets. A prototype is being made based on the applications of graphene to filter saline water. Simulation was done on the prototype, and the results of the simulation are then looked upon for various other factors like material selection (Table 4).

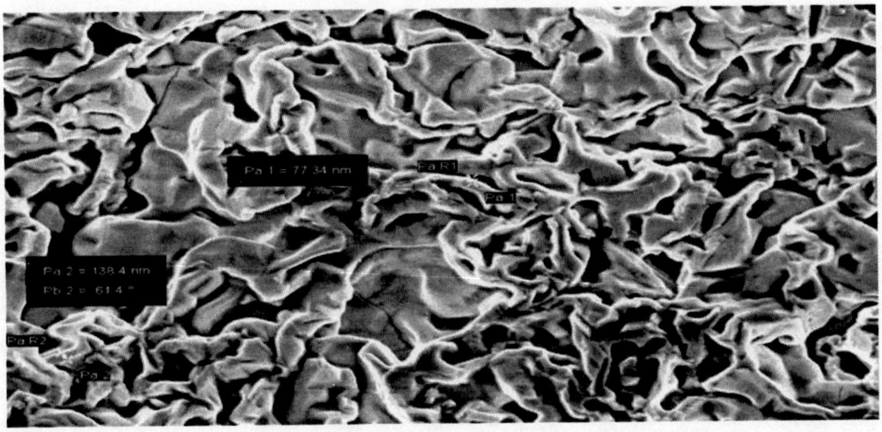

Fig. 5 SEM image of vertically aligned graphene sheets at 3 min of annealing time using corn oil on nickel foil

Table 4 Magnitude of maximum water pressure on the prototype

Pressure applied on	Inner surface
Magnitude of maximum stress	1.26×10^6 Pa

Figure 6 shows the results of the simulation on the final design in the form of stress analysis. The stress analysis shows that the maximum principal stress will have a magnitude of around 1.26 MPa and will be experienced between the budged region (for graphene foam) (Table 5).

Figure 7 shows the results of the simulation on the final design in the form of deformation analysis. The deformation analysis shows that the maximum deformation will be in the region which will experience the maximum stress and will have the magnitude of 26.42 nm, and the outer surface will experience almost no deformation.

Fig. 6 Stress analysis of outer body of filter

Table 5 Magnitude of maximum stress magnitude on the prototype

Pressure applied on	Inner surface
Maximum deformation	2.642×10^{-8} m

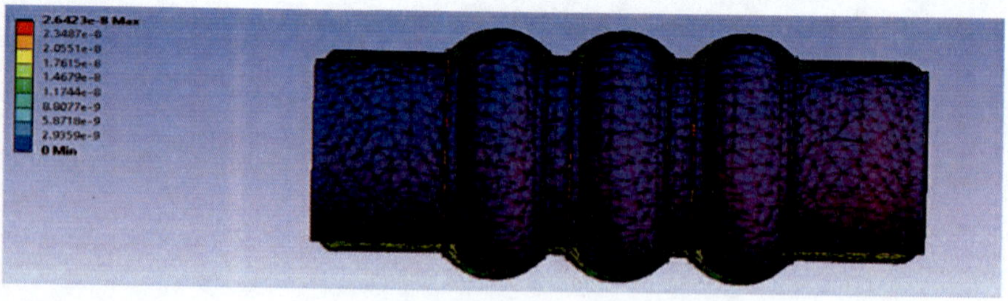

Fig. 7 Deformation analysis of outer body of filter

5 Conclusion

The graphene water filter can be an advantage over the existing filters. The graphene water filter does not require any electricity for use, unlike the existing ones. Moreover, in a graphene filter, there is no need for water storage. The water is instantly purified as soon as the water is passed through the membranes. It also helps to save the water that is being disposed of in the existing water filters during the filtration process. In existing water filters, there is an outlet from where the water is injected out during filtration and thus becomes waste. It is economical and sustainable. The simulation of the product design was done on Ansys software which shows that the magnitude of maximum stress will be 1.26×10^6 Pa, and the maximum deformation will be 2.642×10^{-8} m. The assumption taken here was that the difference between the height of filter and the water reservoir (Water tank) is 200 m which resulted in 2×10^6 Pa of pressure exerted by water on the filter neglecting the frictional losses. When 2×10^6 Pa of pressure is applied on inner surface of filter yields maximum stress of 1.26×10^6 Pa, which is much less than the yield stress of most common materials like stainless steel. So, the experiment shows the product is sustainable to hold large pressure, and hence, it can be used to purify water using vertically aligned graphene sheets.

References

1. An H, Lee W-J, Jung J (2011) Graphene synthesis on Fe foil using thermal CVD. Curr Appl Phys 11:S81–S85
2. Dhingra S, Hsu J-F, Vlassiouk I, D'Urso B (2014) Chemical vapor deposition of graphene on large-domain ultra-flat copper. Carbon 69:188–193
3. Naghdi S, Rhee KY, Park SJ (2018) A catalytic, catalyst-free, and roll-to-roll production of graphene via chemical vapor deposition: Low temperature growth. Carbon 127:1–12
4. Han C-P, Veeramani V, Hsu C-C, Jena A, Chang H, Yeh N-C, Hu S-F, Liu R-S (2018) Vertically-aligned graphene nanowalls grown via plasma-enhanced chemical vapor deposition as a binder-free cathode in Li–O2 batteries. Nanotechnology 29:505401
5. Fu M, Quan B, He J, Yao Z, Gu C, Li J, Zhang Y (2016) Ultrafast terahertz response in photoexcited, vertically grown few-layer graphene. Appl Phys Lett 108:121904
6. Zhu L, Huang Y, Yao Z, Quan B, Zhang L, Li J, Gu C, Xu X, Ren Z (2017) Enhanced polarization-sensitive terahertz emission from vertically grown graphene by a dynamical photon drag effect. Nanoscale 9:10301–10311
7. Ghosh M, Anand V, Gowravaram MR (2018) Wetting characteristics of vertically aligned graphene nanosheets. Nanotechnology 29:385703
8. Ma Y, Jang H, Kim SJ, Pang C, Chae H (2015) Copper-assisted direct growth of vertical graphene nanosheets on glass substrates by low-temperature plasma-enhanced chemical Vapour deposition process. Nanosc Res Lett 10
9. Amade R, Muyshegyan-Avetisyan A, Martí González J, Martí Pino AX, György E, Pascual E, Andújar JL, Serra EB (2019) Super-capacitive performance of manganese dioxide/graphene nano-walls electrodes deposited on stainless steel current collectors. Materials 12:483
10. Seo DH, Pineda S, Fang J, Gozukara Y, Yick S, Bendavid A, Lam SKH, Murdock AT, Murphy AB, Han ZJ, Ostrikov K (2017) Single-step ambient-air synthesis of graphene from renewable precursors as electrochemical genosensor. Nat Commun 8

11. Ghosh S, Polaki SR, Kumar N, Amirthapandian S, Kamruddin M, Ostrikov K (2017) Process-specific mechanisms of vertically oriented graphene growth in plasmas. Beilstein J Nanotechnol 8:1658–1670
12. Ghosh S, Ganesan K, Polaki SR, Ilango S, Amirthapandian S, Dhara S, Kamruddin M, Tyagi AK (2015) Flipping growth orientation of nanographitic structures by plasma enhanced chemical vapor deposition. RSC Adv 5:91922–91931
13. Nikolskaya E, Hiltunen Y (2018) Determination of carbon chain lengths of fatty acid mixtures by time domain NMR. Appl Magn Reson 49:185–193
14. Vegetable Oil. Wikipedia, Wikimedia Foundation. https://en.wikipedia.org/wiki/Vegetable_oil
15. Fluid Dynamics. Wikipedia, Wikimedia Foundation. https://en.wikipedia.org/wiki/Fluid_dynamics

Design and Analysis of Polymer Heat Sink for Li-Ion Battery Thermal Management System

Anirban Sur, Swapnil Narkhede, Ajit Netke, and Hritik Palheriya

Abstract The thermal analysis of polymer heat sink for thermal management of prismatic Li-ion battery used in electric vehicles is presented. A battery pack consisting 22 prismatic cells has been considered for this analysis. On the sides of battery, coolant channels made up of polymer are placed referred as heat sink here. Because of the contact between the battery surface and the channel, heat transfer takes place between the battery and the coolant flowing through the polymer channels. The system proposed here considers polymer as heat sink material to reduce the weight of the system. Transient simulations based on the multi-scale multi-domain model implemented in ANSYS-Fluent have been carried out at constant discharge current for different C-rates. For the different charging/discharging rates (0.5–2 C), the maximum battery temperature rise of a single prismatic cell is observed to be 8 °C; however, the maximum temperature difference across a hole battery pack reached 40 °C. Results of the study depict that the heat sink or heat exchanger made up of polymer by additive manufacturing process shows significant capability to be implemented in battery thermal management system in electric vehicles.

Keywords Battery thermal management system · Prismatic cell · Li-ion battery · Polymer heat sink · Polymer heat exchanger

---

A. Sur (✉) · S. Narkhede · A. Netke · H. Palheriya
Symbiosis Institute of Technology, Symbiosis International University, Pune, India
e-mail: anirbansur26@gmail.com

S. Narkhede
e-mail: swapnil.narkhede.phd2018@sitpune.edu.in

A. Netke
e-mail: 18070125512@sitpune.edu.in

H. Palheriya
e-mail: 18070125514@sitpune.edu.in

A. Maurya et al. (eds.), *Recent Trends in Mechanical Engineering*, Lecture Notes in Mechanical Engineering, https://doi.org/10.1007/978-981-19-7709-1_33

1 Introduction

Emission from gasoline vehicles generates a large amount of thermal pollution, which has serious consequences for global warming. To avoid exacerbating the abovementioned problems, the governments of the European countries have recently announced the timetable for stopping the production of gasoline vehicles, most of which are from 2025 to 2040 [1, 2]. In the foreseeable future, electric vehicles will largely replace gasoline vehicles. Rechargeable batteries are the main components of electric vehicles and require high performance. As lithium-ion batteries high power energy, long service life, high energy compare to other available rechargeable batteries (nickel-cadmium batteries, nickel-metal hydride batteries, and lead-acid batteries) [3, 4], they are very popular for EV applications. However, electric vehicles (EVs) applications require a huge amount of battery power; therefore, battery packs (large nos of prismatic or cylinder batteries connect with series and parallel combinations) are used [5, 6]. The heart of the EV is the battery pack. The life, cost, and performance of EVs depend on the battery pack. During charging and discharging of the battery, due to the transformation of chemical energy to electrical energy, heat generates inside the individual battery. This accommodating heat increases the temperature inside of the battery pack [7–10]. Therefore, the battery should operate within the temperature range with the best performance and longevity [11, 12].

Because of lower specific heat and thermal conductivity of air, air cooled BTMS has some limitations and may result in thermal runaway of the battery pack. In contrast to that, liquids have better thermal performance, but because of requirement of additional components in liquid cooled thermal management systems such as a pump and a metallic heat exchanger, the system becomes heavy and consumes much larger space. Recently, the PCM-based BTMS has been the subject of research among the researcher community. The major problem with such a system is the lower thermal conductivity of the PCM material which makes it difficult to spread the heat inside the PCM to melt it. To address this issue, plate fins as thermal conductivity enhancers have been implemented. But, this again increases the weight of the system. In order to address these issues, here, we propose a polymer-based liquid cooled heat sink for effective thermal management of batteries.

2 Design of a Polymer-Based BTMS System

After reviewing the above problems in BTMS. The heat sink material is a polymer matrix composite. Initially, to understand the nature of heat generation inside a lithium-ion battery, transient simulations have been carried out at constant discharge currents using multi-scale multi-domain model implemented in ANSYS-Fluent. The inputs provided to this model are the capacity, lower cut off voltage, upper cut off voltage of the battery, and the discharge rate. Figure 1 shows the temperature contour

Fig. 1 Single battery temperature analysis for 1 C (12 V Li-ion)

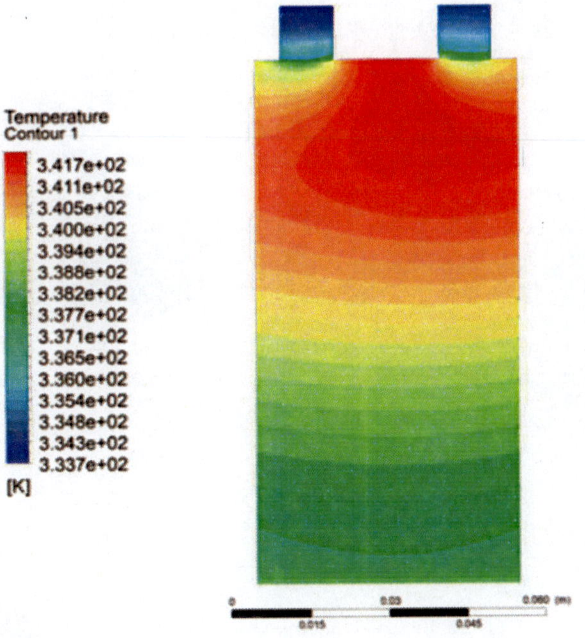

on the surface of single cell at the end of discharge cycle at 1C rate. The similar analysis has been carried out at different discharge rates (0.5–2 C), and the maximum temperature rise on the surface of the battery is observed to be between 4° and 8 °C. But, in case of battery pack consisting 22 Li-ion batteries (1 s 22P), temperature rise is observed between 40 and 50 °C for different C-rates. To reduce the rise in the temperature of the battery pack by means of effective heat dissipation from the battery, a polymer-based heat sink is designed and analyzed as shown in Fig. 2. The blue and red tubes shown in Fig. 2 are the inlet and outlet manifolds to distribute the fluid to and from the polymer channels. The total number of channels are 48 and each having square cross-section of dimensions (3 × 3) mm. These channels are touching the surface of the battery, and the heat transfer takes place between the battery and the coolant water. After absorbing the heat from battery, the hot water from the channels is collected in the outlet manifold (Red color Fig. 2) and is cooled again and sent back to the inlet and the process repeats.

3 Domain Discretization, Boundary Conditions and Assumptions

The computational domain consists of solid volume of polymer and the fluid volume. Both the domains are discretized with unstructured mesh having 12,17,125 elements. The meshing has been to achieve minimum orthogonal quality of 0.2 and maximum

Fig. 2 Battery thermal management system with batteries and hot and cold fluid flow channels

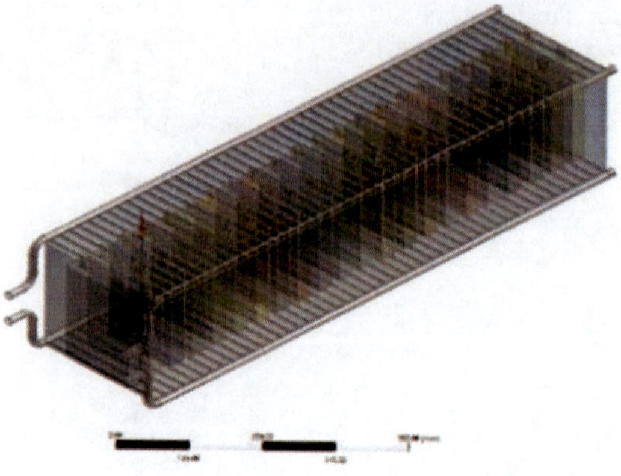

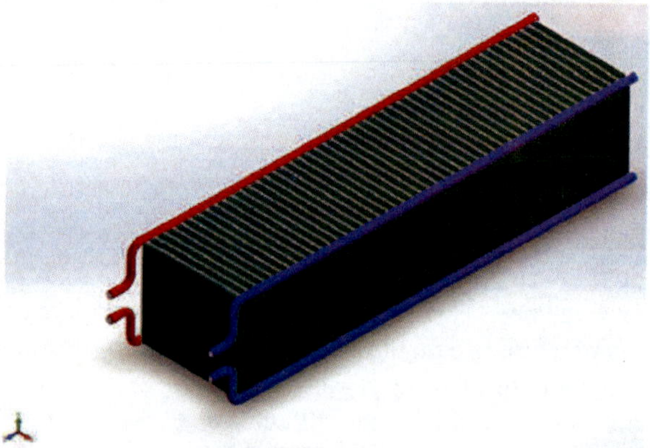

skewness less than 0.94. The mass flow rate (1 LPM) and the temperature of water (15 °C) at inlet and the pressure outlet at the outlet are specified as the boundary conditions for the present analysis. The polymer material considered for the construction of polymer heat sink is thermoset polystyrene. The thermal conductivity of polystyrene is 245 W/m k and density of 1.9 g/cm^3 and specific heat 871 J/ kg°C. During the analysis, following assumptions are imposed: (i) The coolant fluid is considered incompressible; (ii) The flow is considered steady and laminar (Fig. 3).

4 Results and Discussion

The temperature distribution along the heat sink is shown in Fig. 4. Near inlet to the channels, the temperature rise of fluid is observed to be more because of low temperature gradient. Along the length of channels, the heat transfer gradually decreases

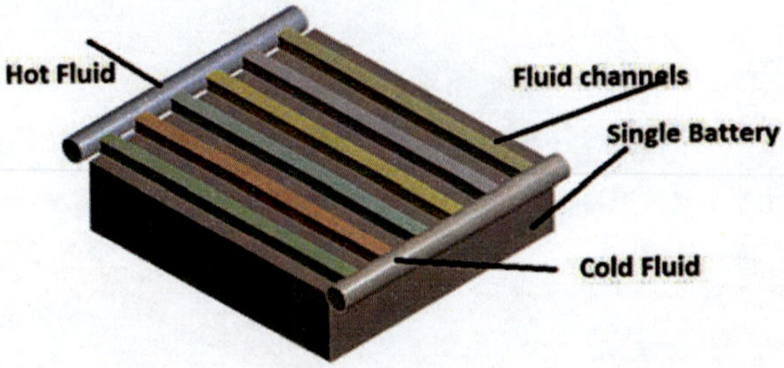

Fig. 3 Geometry of single battery and heat exchanger

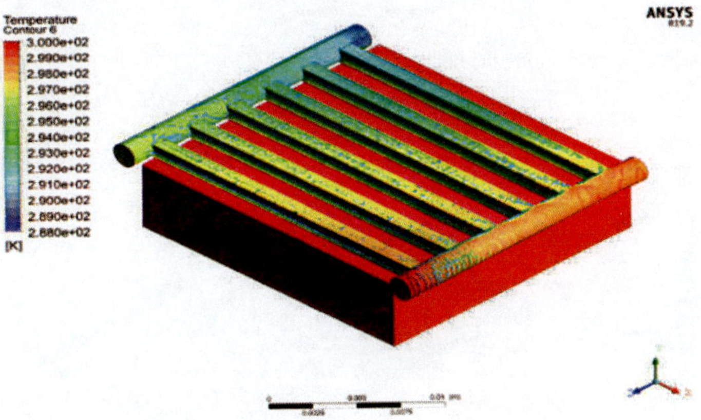

Fig. 4 Temperature distribution along the heat exchanger

because of lower temperature gradient between the battery surface temperature and the fluid temperature.

Figure 5 shows the pressure distribution of the coolant along the length of the heat sink. The pressure drop mainly occurs because of reduction in cross-section from inlet manifold to the channels. Also, Fig. 6 shows the velocity distribution in the fluid domain. The velocity seems to increase when it enters the channels from the manifold because of smaller cross-section. Also, the velocity in all the channels is not uniform depicting the flow maldistribution. This flow maldistribution affects the performance of the heat sink. To reduce the flow maldistribution inside the channels, manifold designs need to be modified.

5 Conclusion

The thermal performance of a polymer heat sink for BTMS is presented here. Three-dimensional numerical simulations have been carried out using ANSYS-Fluent. This approach uses coupled solver which is composed of multi-scale multi-domain model

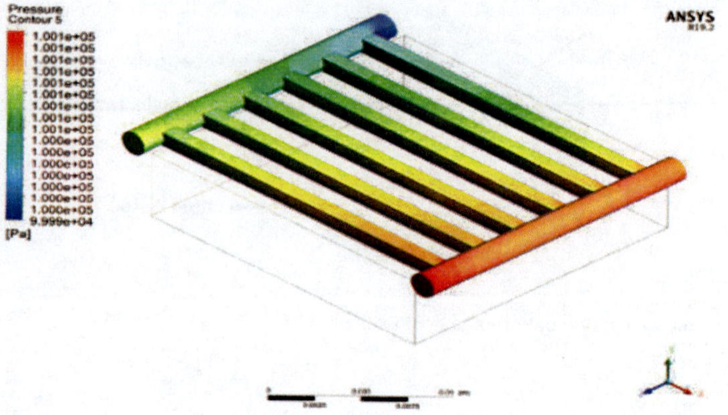

Fig. 5 Pressure variation along the heat exchanger

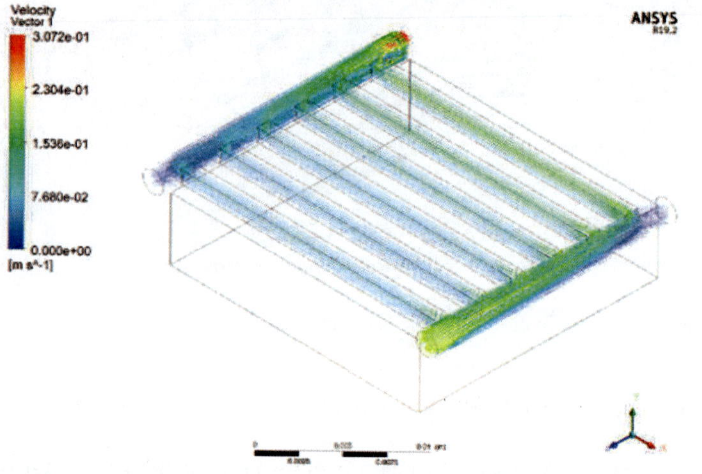

Fig. 6 Velocity of coolant variation along the heat exchanger

and CFD to solve heat generation in Li-ion battery and temperature and pressure distribution inside the coolant domain. The analysis presented here demonstrates the potential of polymer-based system for heat sink or heat exchanger application. The polymer heat sink utilized for thermal management of battery has a benefit to reduce the overall weight of the system. Also, the polymer heat sink or heat exchanger can be manufactured through additive manufacturing process which makes use of intricate geometries possible thereby increasing the heat transfer performance.

References

1. Huang X, Li G, Zhang J, Zhang F, He Y (2018) Li, Experimental investigation of the thermal performance of heat pipe assisted phase change material for battery thermal management system. Appl Therm Eng. https://doi.org/10.1016/j.applthermaleng.2018.06.04
2. Chen K, Song M, Wei W, Wang S (2018) Structure optimization of parallel air-cooled battery thermal management system with U-type flow for cooling efficiency improvement. Energy 145:603–613
3. Chen K, Wang S, Song M, Chen L (2017) Structure optimization of parallel air-cooled battery thermal management system. Int J Heat Mass Transf 111:943–952
4. De las Heras A, Vivas FJ, Segura F, Redondo MJ, Andújar JM (2018) Air-cooled fuel cells: keys to design and build the oxidant/cooling system. Renew Energy 125:1–20
5. Han JED, Qiu A, Zhu H, Deng Y, Chen J, Zhao X, Zuo W, Wang H, Chen J, Peng Q (2018) Orthogonal experimental design of liquid-cooling structure on the cooling effect of a liquid-cooled battery thermal management system. Appl Therm Eng 132:508–520
6. Rao Z, Qian Z, Kuang Y, Li Y (2017) Thermal performance of liquid cooling based thermal management system for cylindrical lithium-ion battery module with variable contact surface. Appl Therm Eng 123:1514–1522
7. Xie Y, Tang J, Shi S, Xing Y, Wu H, Hu Z, Wen D (2017) Experimental and numerical investigation on integrated thermal management for lithium-ion battery pack with composite phase change materials. Energy Convers Manage 154:562–575
8. Zou D, Ma X, Liu X, Zheng P, Hu Y (2018) Thermal performance enhancement of composite phase change materials (PCM) using graphene and carbon nanotubes as additives for the potential application in lithium-ion power battery. Int J Heat Mass Transf 120:33–41
9. Sun Z, Fan R, Yan F, Zhou T, Zheng N (2019) Thermal management of the lithium-ion battery by the composite PCM-Fin structures. Int J Heat Mass Transf 145:118739
10. Raijmakers LHJ, Danilov DL, Eichel RA, Notten PHL (2019) A review on various temperature-indication methods for Li-ion batteries. Appl Energy 240:918–945
11. Feng X, Zheng S, Ren D, He X, Wang L, Cui H, Liu X, Jin C, Zhang F, Xu C, Hsu H, Gao S, Chen T, Li Y, Wang T, Wang H, Li M, Ouyang M (2019) Investigating the thermal runaway mechanisms of lithium-ion batteries based on thermal analysis database. Appl Energy 246:53–64
12. Kim J, Oh J, Lee H (2019) Review on battery thermal management system for electric vehicles. Appl Therm Eng 149:192–212

Implications of Volumetric Porosity-Based Interpretation of Mechanical Properties Associated to Structures with Constant Engineered Porosity

Atul Chauhan and Amba D. Bhatt

Abstract It is known that bulk porosity of structures is negatively correlated to their mechanical properties (e.g., Young's moduli). But, in reference to bone tissue engineering (BTE), favorable bulk porosity levels are also known. Thus, in reference to design porous structures as scaffolds, bulk porosity is not much of much use. Therefore, to the design the porous lattice structures as scaffolds to assist in BTE, it is required to quantify the effect of secondary architectural features (like shape of pores) on mechanical properties through a general robust mathematical framework. With the help of micromechanics and available scaling laws, this work reports the following findings. In the form of general mixture rule (GMR), we have a computationally validated unified general framework to map the change in Young's moduli (stiffness) of porous lattice structures with change in shape of pores at given bulk porosity levels. GMR may be useful to guide the design of porous lattice structures as scaffolds with desired stiffness to minimize the phenomena of stress shielding and thus may provide assistance in bone tissue engineering.

Keywords Micromechanics · Stiffness · General mixture rule · Design · Scaffolds · Bone tissue engineering

1 Introduction

In reference to rational mixture theory, porous lattice structures can be considered as biphasic mixture of solid and pores. These can be designed and additively manufactured as scaffolds to assist in bone tissue engineering (BTE) [1]. To minimize the phenomena of stress shielding, Young's moduli of scaffolds should be almost equal to Young's modulus of host bone tissue [2]. In this regard, porous lattice structures

A. Chauhan (✉) · A. D. Bhatt
Department of Mechanical Engineering, Motilal Nehru National Institute of Technology Allahabad, Allahabad, India
e-mail: rme1705@mnnit.ac.in

A. D. Bhatt
e-mail: abhatt@mnnit.ac.in

A. Maurya et al. (eds.), *Recent Trends in Mechanical Engineering*, Lecture Notes in Mechanical Engineering, https://doi.org/10.1007/978-981-19-7709-1_34

which comprise inner architecture (unit cell) at length scales intermediate between micro and macro which is now a 'new degree of freedom' in material design (M.F. Ashby) for multi-objective properties [1, 3]. Porosity–property relationships (scaling laws) correlate bulk porosity of lattice structures to their Young's moduli [4–6]. Consequently, it is difficult to quantitatively explain the variation of Young's moduli of porous lattice structures at their constant bulk porosity levels. Thus, absence of any general framework to map the variation in Young's moduli with inner architectural details poses bottlenecks in the design of porous materials. Therefore, this work should be considered as an attempt in this regard.

It may be useful to note that mechanical behavior can be quantified in terms of mechanical properties like Young's modulus, shear modulus and Poisson's ratio, among others [4]. Moreover, set of inner architectural characteristics comprises bulk porosity, size, shape, orientation and distribution (continuity and connectivity) of pores [4, 5]. One particular set of inner architectural features governs the magnitude and anisotropy of Young's moduli of porous lattice structure [4–7, 13–15]. In this work, Young's moduli in three mutually orthogonal directions have been computed through micromechanics [3, 9]. Several porosity–property models (like Maxwell, Coble–Kingery, power law, Pabst–Gregorová and general mixture rule among others) help to correlate bulk porosity and Young's modulus through an empirical constant, in general [4–6]. The aim of this work is to test the efficacy of these models to provide quantified reasons behind the observed variation in Young's moduli of porous lattice structure with the change in shape of pore at constant bulk porosity levels.

Rest of the article has been divided into the following sections. Section 2: problem description and solution method. Section 3: results and discussion with future perspectives of this work. Section 4: conclusions.

2 Problem Description and Solution Method

2.1 Background

Mechanical behavior sensu stricto is the coefficient in linear constitutive relations (e.g., Hooke's law) [4]. It is known that one particular set of inner architectural features governs the magnitude and anisotropy of Young's moduli of porous lattice structure [4–7, 13–15]. Therefore, it is reasonable to consider mechanical property-inner architecture dependence. In this regard, both journal articles and monographs have been found to be profound with relations that map mechanical properties to bulk porosity of porous materials [4–6]. Pabst and Gregorová have emphasized (in contrast to common belief evoked by many publications) that set of predictive admissible relations reduces to essentially following five types, namely linear relation, Maxwell model, Coble–Kingery relationship, power law, Pabst–Gregorová [4] and Ji et al. in the form of one general mixture rule (GMR) [5]. In mathematical form, these models can be expressed as following.

$$\boldsymbol{Linear\ type}: \left(\boldsymbol{P^H/P_s}\right) = (\mathbf{1} - \boldsymbol{\varphi}) \quad (1)$$

$$\boldsymbol{Maxwell\ type}: \left(\boldsymbol{P^H/P_s}\right) = (\mathbf{1} - \boldsymbol{\varphi})/(\mathbf{1} + (\boldsymbol{\chi} - \mathbf{1})\boldsymbol{\varphi}) \quad (2)$$

$$\boldsymbol{Coble - Kingery\ type}: \left(\boldsymbol{P^H/P_s}\right) = \mathbf{1} - \boldsymbol{\chi\varphi} + (\boldsymbol{\chi} - \mathbf{1})\boldsymbol{\varphi}^2 \quad (3)$$

$$\boldsymbol{Power\ law}: \left(\boldsymbol{P^H/P_s}\right) = (\mathbf{1} - \boldsymbol{\varphi})^{\boldsymbol{\chi}} \quad (4)$$

$$\boldsymbol{Pabst - Gregorová}: \left(\boldsymbol{P^H/P_s}\right) = \boldsymbol{exp}(-\boldsymbol{\chi\varphi}/(\mathbf{1} - \boldsymbol{\varphi})) \quad (5)$$

$$\boldsymbol{General\ Mixture\ Rule}\ (\boldsymbol{GMR}): \left(\boldsymbol{P^H/P_s}\right) = (1 - \boldsymbol{\varphi})^{1/\boldsymbol{J}} \quad (6)$$

where P^H denotes the homogenized mechanical property, P_s denotes corresponding solid phase material property, χ and J are the empirical constants and φ denotes bulk porosity of porous materials.

Literature suggests that empirical constants (χ and J) represent the effect of secondary architectural features on mechanical properties, where secondary architectural features include size, shape, orientation and distribution (continuity and connectivity) of pores [4, 5].

2.2 Problem Statement

A general mathematical framework that correlates bulk porosity to Young's moduli of structure in wide range of bulk porosity favorable for bone tissue engineering do not yet exist [6]. It is also known that bulk porosity of structures is negatively correlated to mechanical properties (Young's moduli) [4, 5]. But, in reference to BTE, favorable bulk porosity levels are known; therefore, bulk porosity is not much useful in this regard [2]. Therefore, to the design the porous lattice structures as scaffolds to assist in BTE, it is required to quantify the effect of secondary architectural features (like shape) on mechanical properties through a general robust mathematical framework.

2.3 Objective

The objective of this study is to test the efficacy of available porosity–property models for explaining the change in Young's moduli of porous lattice structures with change in shape of pore at given bulk porosity level.

2.4 Solution Method

Young's moduli of porous lattice structures have been calculated through strain energy-based homogenization technique (micromechanics) [3, 9] with the help of structure mechanics module available in COMSOL Multiphysics [11]. To discuss the micromechanics in gross details is beyond the scope of this article, but for the sake of completeness, a brief overview of micromechanics has been given as follows. According to linear elasticity theory [3], it has been well known that the static equilibrium of unit cell (UC) without body forces can be represented by Eq. (7), where comma denotes partial derivative with respect to coordinate system attached to UC.

$$\sigma_{ij,j} = 0 \tag{7}$$

Homogenization of heterogeneous medium implies that we need to obtain constitutive relation ($\overline{\sigma}_{ij} = E^H_{ijkl}\overline{\epsilon}_{kl}$) which can be inverted and expressed as $\overline{\epsilon}_{ij} = S^H_{ijkl}\overline{\sigma}_{kl}$. Note that, $\overline{\sigma}_{ij}$ and $\overline{\epsilon}_{ij}$ are stress and strain fields of the homogeneous material, respectively. Here, E and S are known as stiffness and compliance tensor, respectively, and the relation $S = E^{-1}$ has been known to exist [3]. For gross details regarding the fundamentals of micromechanics, authors of this work have referred few recent articles [3, 9, 12] and monograph authored by Kachanov and Sevostianov [10]. In finite element framework, Eq. (7) has to be solved over unit cell of porous lattice structures subjected to macroscopic stress ($\overline{\sigma}_{ij}$) or macroscopic strain ($\overline{\epsilon}_{ij}$) or their combinations in the form of periodic boundary conditions (PBCs) subjected to Hill-Mandel macro homogeneity condition, so that homogeneous material is energetically equivalent to actual heterogeneous material [3, 8, 9]. Note that, each displacement under PBC yields one column of 6×6 square matrix of homogenized stiffness coefficients ($\overline{\sigma}_{ij} = E^H_{ijkl}\overline{\epsilon}_{kl}$) given by Eq. (8).

$$\begin{Bmatrix} \overline{\sigma}_{11} \\ \overline{\sigma}_{22} \\ \overline{\sigma}_{33} \\ \overline{\sigma}_{23} \\ \overline{\sigma}_{13} \\ \overline{\sigma}_{12} \end{Bmatrix} = \begin{bmatrix} E^H_{1111} & E^H_{1122} & E^H_{1133} & E^H_{1123} & E^H_{1113} & E^H_{1112} \\ E^H_{2211} & E^H_{2222} & E^H_{2233} & E^H_{2223} & E^H_{2213} & E^H_{2212} \\ E^H_{3311} & E^H_{3322} & E^H_{3333} & E^H_{3323} & E^H_{3313} & E^H_{3312} \\ E^H_{2311} & E^H_{2322} & E^H_{2333} & E^H_{2323} & E^H_{2313} & E^H_{2312} \\ E^H_{1311} & E^H_{1322} & E^H_{1333} & E^H_{1323} & E^H_{1313} & E^H_{1312} \\ E^H_{1211} & E^H_{1222} & E^H_{1233} & E^H_{1223} & E^H_{1213} & E^H_{1212} \end{bmatrix} \begin{Bmatrix} \overline{\epsilon}_{11} \\ \overline{\epsilon}_{22} \\ \overline{\epsilon}_{33} \\ \overline{\epsilon}_{23} \\ \overline{\epsilon}_{13} \\ \overline{\epsilon}_{12} \end{Bmatrix} \tag{8}$$

2.5 Contribution

Accomplishment of objective of this work will provide validated unified general framework to map the change in Young's moduli of porous lattice structures.

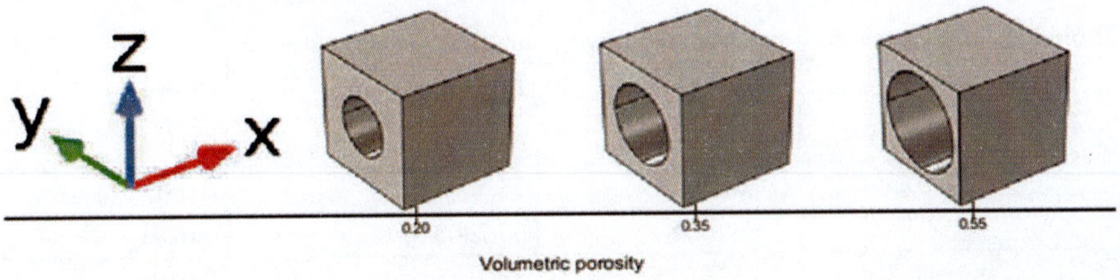

Fig. 1 Unit cells for validation of applied methodology

3 Results and Discussion

3.1 Numerical Results for Validation of Applied Methodology

Unit cells (generated through Boolean difference) similar to the recent published work [12] in terms of geometry (Fig. 1) and material ($E = 70$ GPa, $v = 0.3$) have been adopted in order to validate the strain energy-based homogenization method for calculating the mechanical properties of lattice structures (Sect. 2.4).

Directions X, Y and Z represent the orthogonal coordinate directions used to define the computational domain (unit cells) shown in Fig. 1. Mechanical properties have been calculated with respect to X, Y and Z directions throughout this article. Table 1 compares the mechanical properties calculated through applied method (Sect. 2.4) and averaging homogenization (AH) from an article [12]. Thus, it can be said that from the validation point of view, error column of Table 1 proves the fidelity of applied method, and hence, it can be applied frequently as per the research objective of this work.

3.2 Mapping of Young's Moduli of Porous Lattice Structure with Empirical Parameters Related to Pores Shape at Constant Bulk Porosity Level

Unit cells have been generated through Boolean operations. Extruded cross-section is an ellipse with varying aspect ratio (a) leading to change in pore shape (Fig. 2). Required material properties have been kept same as they are in Sect. 3.1. Bulk porosity of porous materials (or equivalently their unit cells) has been kept constant at 0.20. Other obvious data required for unit cells generation have not been elaborated for the sake of brevity. Table 2 shows the mapping of Young's moduli in three mutually orthogonal coordinate directions (X, Y and Z) with change in empirical pore shape coefficient available in Eqs. (2–6). It is obvious that linear models cannot be used for such mapping (Eq. (1)).

Table 1 Variation of mechanical properties with bulk (volumetric) porosity (φ)

E^H_{ijkl}	φ	AH [7]	Applied method	Error (%)
E^H_{1111}	0.20	68.9164	68.8865	0.0434
	0.35	53.8373	53.8044	0.0611
	0.55	35.7980	35.7908	0.0201
E^H_{3333}	0.20	53.3923	53.3845	0.0166
	0.35	36.5371	36.5341	0.0082
	0.55	20.4995	20.5053	−0.0182
E^H_{1133}	0.20	21.5274	21.5138	0.0663
	0.35	13.8954	13.8855	0.0712
	0.55	7.1633	7.1633	0
E^H_{2233}	0.20	18.3657	18.3488	0.0920
	0.35	9.7809	9.7746	0.0644
	0.55	3.3784	3.3787	−0.0089
E^H_{1212}	0.20	17.5138	17.9438	−2.4552
	0.35	12.4030	12.9138	−4.1184
	0.55	7.0355	7.4595	−6.0567
E^H_{2323}	0.20	13.0961	13.4367	−2.6008
	0.35	6.3101	6.6251	−4.9920
	0.55	1.6902	1.8161	−7.4488

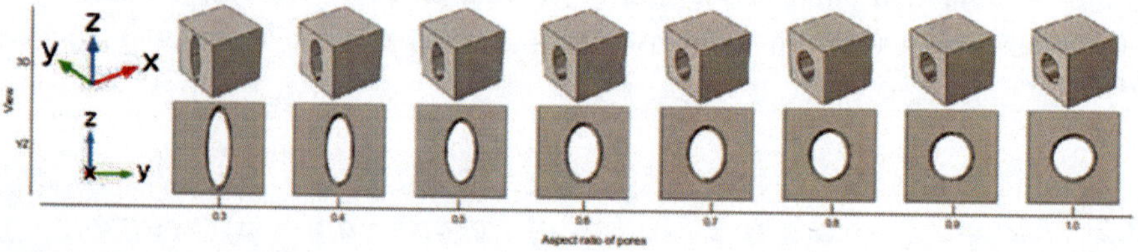

Fig. 2 Unit cells with pores of varying aspect ratio

In reference to data shown in Table 2, authors have made an attempt to put forward following remarks in perspective of framed objective of this work as follows. Pore shape coefficients (χ and **J**) in Eqs. (2–6) have the potential to provide a quantified reasoning behind the observed variation in Young's moduli (irrespective of direction) of porous lattice structures due to change in shape of pores at constant bulk porosity levels (0.20). For the purpose of mapping, Eqs. (2–6) are equivalent. Moreover, one single equation of GMR (Eq. (6)) can represent the all other Eqs. (2–5) in this regard. Thus, previous attempt to scale down the number of admissible expressions that can predict the mechanical properties to just five has been further improved. Moreover, GMR is preferable because of simple mathematical symmetry, and additionally, it postulates no assumptions on either.

Table 2 Mapping of structure's stiffness with pore shape coefficients (χ and J), aspect ratio (a) and its bulk porosity (φ), where XX, YY and ZZ denote the orthogonal coordinate directions

a			0.3	0.4	0.5	0.6	0.7	0.8	0.9	1.0
φ			0.20	0.20	0.20	0.20	0.20	0.20	0.20	0.20
χ	Equation (2)	XX	1.0006	1.0006	1.0025	1.0025	1.0013	1.0019	1.0006	1.0013
		YY	11.4500	6.3413	4.6806	3.8094	3.2635	2.8859	2.6094	2.3837
		ZZ	1.1807	1.3533	1.5203	1.6883	1.8582	2.0332	2.2121	2.3837
χ	Equation (3)	XX	1.0006	1.0006	1.0025	1.0025	1.0013	1.0019	1.0006	1.0013
		YY	4.3819	3.5825	3.1200	2.7987	2.5581	2.3694	2.2175	2.0837
		ZZ	1.1744	1.3300	1.4713	1.6050	1.7325	1.8563	1.9756	2.0837
χ	Equation (4)	XX	1.0006	1.0006	1.0022	1.0022	1.0011	1.0017	1.0006	1.0011
		YY	6.0558	4.2567	3.4722	2.9982	2.6735	2.4342	2.2506	2.0948
		ZZ	1.1591	1.3060	1.0871	1.5780	1.7099	1.8418	1.9728	2.0948
χ	Equation (5)	XX	0.8931	0.8931	0.8946	0.8946	0.8936	0.8941	0.8931	0.8936
		YY	5.4053	3.7994	3.0992	2.6762	2.3863	2.1727	2.0088	1.8698
		ZZ	1.0346	1.1657	1.2885	1.4085	1.5262	1.6439	1.7608	1.8698
J	Equation (6)	XX	0.9994	0.9994	0.9977	0.9977	0.9989	0.9983	0.9994	0.9989
		YY	0.1651	0.2349	0.2880	0.3335	0.3741	0.4108	0.4443	0.4774
		ZZ	0.8628	0.7657	0.6927	0.6337	0.5848	0.5430	0.5069	0.4774
E^H/E_s		XX	0.7999	0.7999	0.7996	0.7996	0.7998	0.7997	0.7999	0.7998
		YY	0.2589	0.3868	0.4608	0.5122	0.5507	0.5809	0.6052	0.6266
		ZZ	0.7721	0.7472	0.7246	0.7032	0.6828	0.6630	0.6439	0.6266

Physical properties or process (e.g., iso-stress or iso-strain) or microstructure (e.g., over simplified unit cells) [5]. Therefore, GMR becomes the potential expression to explain observation in lieu of all other expressions that also satisfy material limits (i.e., $\boldsymbol{E^H/E_s \rightarrow 1}$ **as** $\boldsymbol{\varphi \rightarrow}$ 0 and $\boldsymbol{E^H/E_s \rightarrow}$ 0 ***as*** $\boldsymbol{\varphi \rightarrow}$ **1**). Empirical constant J in GMR (Eq. 6) captures the phenomena of transition in deformation mechanism (stretching to bending) by changing shape of pores at constant bulk porosity. It is known that stretching dominated structures offer more stiffness than the bending counterparts. Thus, this work can guide to design porous lattice structures with desired stiffness (Young's modulus) at given bulk porosity because stiffness has been a parameter of critical importance to avoid stress shielding in bone implants [2]. Rigorous theoretical analysis through mathematical mechanics is further required to provide physical meaning to GMR [5], and hence, it should be regarded as one of future aspects of this work, among others, like cross-property bounds and explanation of symmetry in elastic engineering constants.

4 Conclusions

In the form of general mixture rule (GMR), we have a computationally validated unified general framework to map the change in Young's moduli (stiffness) of porous lattice structures with change in shape of pores at given bulk porosity levels. GMR may be useful to guide the design of porous lattice structures as scaffolds with desired stiffness to minimize the phenomena of stress shielding and thus may provide assistance in bone tissue engineering.

Acknowledgements The authors gratefully acknowledge the financial support from the Ministry of Education (formerly the Ministry of Human Resource Development), Government of India.

References

1. Pan C, Han Y, Lu J (2020) Design and optimization of lattice structures: a review. Appl Sci 20:6374, 1–36. https://doi.org/10.3590/app10186374
2. Abbasi N, Hamlet S, Love RM, Nguyen N (2020) Porous scaffolds for bone tissue regeneration. J Sci: Adv Mater Dev 5:1–9. https://doi.org/10.1016/j.jsamd.2020.01.007
3. Dirrenberger J, Forest S, Jeulin D (2019) Chapter 4: Computational homogenization of architectured materials. In: Estrin et al (eds) Architectured materials in nature and engineering, springer series in material science, vol 282. Springer Nature, Switzerland, pp 89–139. https://doi.org/10.1007/978-3-030-11942-3_4
4. Pabst W, Gregorová E (2016) Critical assessment 18: elastic and thermal properties of porous materials-rigorous bounds and cross-property relations. Mater Sci Technol 31(15):1801–1808. https://doi.org/10.1080/02670836.2015.1114697
5. Ji S, Gu Q, Xia B (2006) Porosity dependence of mechanical properties of solid materials. J Mater Sci 41, 1757–1768. https://doi.org/10.1007/s10853-006-2871-9
6. Choren JA, Heinrich SM, Silver-Thorn MB (2013) Young's modulus and volume porosity relationships for additive manufacturing applications. J Mater Sci 48:5103–5112. https://doi.org/10.1007/s10853-013-7237-5
7. Kang J, Dong E, Li D, Dong S, Zhang C, Wang L (2020) Anisotropy characteristics of microstructures for bone substitutes and porous implants with application of additive manufacturing in orthopaedics. Mater Des 191:108608, 1–10. https://doi.org/10.1016/j.matdes.2020.108608
8. Hollister SJ, Kikuchi N (1992) A comparison of homogenization and standard mechanic analyses for periodic porous composites. Comput Mech 10:73–95. https://doi.org/10.1007/BF00369853
9. Refai K, Montemurro M, Brugger C, Saintier N (2019) Determination of the effective elastic properties of titanium lattice structures. Mech Adv Mater Struct 1–18. https://doi.org/10.1080/15376494.2018.1536816
10. Kachanov M, Sevostianov I (2018) Micromechanics of materials, with applications. © Springer International Publishing AG, part of Springer Nature 2018. https://doi.org/10.1007/978-3-319-76204-3
11. COMSOL Multiphysics 5.4. https://www.comsol.co.in/documentation
12. Dutra TA, Ferreira RTL, Resende HB, Guimarães A, Guedes JM (2020) A complete implementation methodology for asymptotic homogenization with finite element commercial software: pre-processing and post-processing. Compos Struct 245(112305):1–19. https://doi.org/10.1016/j.compstruct.2020.112305

13. Feng J, Liu B, Lin Z, Fu J (2021) Isotropic octet-truss lattice structure design and anisotropy control strategies for implant application. Mater Des 203(109595):1–11. https://doi.org/10.1016/j.matdes.2021.109595
14. Torres-Sanchez C, McLaughlin Bonallo R (2018) Effect of pore size, morphology and orientation on the bulk stiffness of a porous Ti35Nb4Sn alloy. JMEPEG 27:2899–2909. https://doi.org/10.1007/s11665-018-3380-0
15. Maskery I, Aremu AO, Parry L, Wildman RD, Tuck CJ, Ashcroft IA (2018) Effective design and simulation of surface based lattice structures featuring volume fraction and cell type grading. Mater Des 155:220–232. https://doi.org/10.1016/j.matdes.2018.05.058

Intake Boosting Techniques in Internal Combustion Engines to Increase Engine Performance

S. Ashish, M. Rishie Aravind, R. Abhinav, and Bhisham Kumar Dhurandher

Abstract The consistently expanding demand for transport is supported by both spark and compression ignition engines. Due to globalization and rapidly expanding economies, the necessity for transport energy is enormous and constantly expanding across the world. IC engines till date prove to be the best source of powering an automobile. Even though they have been in use for many decades, there is always scope for improvement in technologies related to it. IC engines waste a huge amount of fuel energy via exhausts. Technologies like turbo-compounding prove to be useful to reduce fuel consumption and CO_2 emissions. Other modifications made to the design of intake manifold also result in changes to torque and horsepower. There are not many fully developed elective choices that may replace the internal combustion engines as they are in developing stages. As of 2020, 99.8% of the transportation worldwide is powered by IC engines and 95% of the energy comes from petroleum-based fuels. The researchers and engineers have to develop technologies to enhance the power output of the IC engines in order to fulfill the needs. This paper reviews enhancement of the IC engine by various intake boosting techniques in terms of better performance, progressive combustion and improved emissions while discussing future trends.

Keywords Brake specific fuel consumption · Turbochargers · Intake air temperature · Engine downsizing · Transient performance · Turbo-lag compensation

1 Introduction

In recent years, the use of automobiles has been increasing exponentially, and IC engines are being used vastly for powering the same. IC engine vehicles boast longer driving range before refueling, better performance and lower cost of production and maintenance. Also, potential competitors to IC engines which include battery

S. Ashish (✉) · M. R. Aravind · R. Abhinav · B. K. Dhurandher
School of Mechanical Engineering, VIT Chennai, Chennai, India
e-mail: ash25022001@gmail.com

A. Maurya et al. (eds.), *Recent Trends in Mechanical Engineering*, Lecture Notes in Mechanical Engineering, https://doi.org/10.1007/978-981-19-7709-1_35

powered, fuel cell electric and hybrid vehicles are not yet being used commercially [1]. As of 2018, the worldwide production of LDVs was approximately 70 million and that of commercial vehicles was approximately 25 million. In developing countries, the number of vehicles is increasing continuously, and by 2040, it is estimated that the production will be around 1.7–2 billion [2]. Currently, the brake thermal efficiency of spark ignition systems is approximately about 30–36% and that of compression ignition systems is about 40–47%. Roadmap of road vehicles surpasses the average fuel consumption of road vehicles by 30%–50% by 2030 [3]. Currently, all modes of transport are powered by combustion engines, flights run on jet engines and land and marine transport run on internal combustion engines. These engines also serve important roles in power generation and other industries.

Many alternatives for IC engines are being developed right now due to the concerns that arise with the emissions of CO_2, nitrogen oxides, carbon monoxide, etc. In fact, in many countries, there has been a heavy criticism of ICEs by the media, and some of the politicians believe that the elimination of ICE is about to happen. The desire for economic growth, energy independence and energy scrutiny has influenced the transport policy in many countries. There are many alternatives for conventional fuels as well like biofuels, natural gas, hydrogen, etc., but battery or a fuel cell is considered to be one of the main alternatives for internal combustion engines [4, 5]. Biodiesel has been used extensively only in diesel engines. It is seen that a mixture of 20% biodiesel and 80% diesel fuel gives a comparable performance and emission to that of a conventional diesel engine [6]. Biodiesel has been identified to be one of the most successful alternative fuels for diesel engine. *Euglena sanguinea*, fresh water microalgae when blended with conventional diesel fuel showed that for partial load applications, it would be advantageous. Also, smaller proportions of this microalgae when mixed yield better brake-specific CO and HC emissions [7]. However, all the other options to IC engines start from a very low base and face a lot of difficulties in rapid growth and development. If these alternatives are enforced prematurely, there will be serious economic and environmental consequences [8]. For example, the impact created by greenhouse gases of battery electric vehicles (BEVs) can be much worse than conventional vehicles if electricity generation and energy consumed in battery manufacture arc not properly decarbonized. Hence, IC engines continue to serve as the best option to sustain transportation [9]. Therefore, it is indispensable to improve the already existing technologies.

There are many ways to enhance the performance of IC engines. Recent developments and design modifications have proved to be instrumental for better efficiency and performance. One of the more effective ways is intake manifold boosting. The main aim of boosting is to increase the density of air flowing into the cylinder. Boosting enables a smaller engine to perform at the same level of a larger naturally aspirated engine, which in turn makes the engine more efficient [10]. Proposing advancement ideas regarding IC engines can be approached in two different ways. One is sacrificing fuel consumption to increase the engine performance, and the other is to increase fuel efficiency by limiting the engine performance. In an ideal world, the mix between these two approaches is highly necessary to make the vehicle suitable at different terrains and conditions [11]. This paper reviews the latest technological advancements in different intake boosting techniques to improve performance, efficiency and control emissions. Various types and setups of turbochargers have been analyzed. Turbochargers are extensively used in all types of commercial vehicles. Turbo-lag, one of the main problems associated with turbochargers is the lack of transient response at lower engine speeds. Potential solutions to turbo-lag have been discussed in detail. The influence of temperature of intake air on BSFC, emissions and engine performance are studied.

2 Turbochargers

A turbocharger is a device that uses exhaust gases to run a turbine coupled to a compressor which is used to push extra air into the intake manifold, thereby increasing the efficiency and power output of an IC engine. Using a turbocharger to forcefully inject air at greater pressure into the intake manifold is one of the best ways as it utilizes only the exhaust gases.

2.1 Electrically Turbocharged Engine

For the cutting edge ICE, the turbocharger is a compelling setup to improve economy and power density, by recuperating exhaust energy and compressing intake charge. But, a turbocharger cannot exert its full potential to retrieve exhaust energy due to the restricted intake pressure which results in the redundant exhaust gas which bypasses the turbine under wide open throttle conditions. Another issue of the turbocharger is the poor transient reaction during acceleration, called turbo-lag [12]. This delay in transient response time can be reduced to a great extent by using an electrically assisted turbocharger. Rodman et al. [13] researched the response of a diesel engine when an electric motor was connected to the shaft of a conventional turbocharger. This modification reduced the time needed for transient power increase from 3.9 to 1.7 s. Martinez-Botas and Nicola Terdich [14] tested a customized variable geometry electrically assisted turbocharger for non-road conditions. The results show a peak turbine efficiency of 69% for 0.65 velocity ratio and 60% for vane opening. A maximum shaft power of 5.4 KW in generating mode and 3.5 KW in motoring mode was obtained.

Petitjean et al. [15] compared turbocharged and non-turbocharged engines of sedan vehicles over a period of ten years. It was seen that by using electric motors, engines can be downsized by 30% for the same power and fuel economy can be increased by 8–10%. A more specific study was made by Burke et al. [16] on a 2.0 L SI engine. The main focus is to study the impact of electrically assisted turbochargers during lower engine speeds, which represents a more practical driving condition. For various engine speeds and powers, the transient response time was improved by 70–90%. Also, fuel consumption was decreased by 1.8% in the e-turbocharged engine.

From all these studies, we can say that electrically assisted turbochargers are one of the best options (in single-stage boosting) for engine downsizing, compensation of turbo-lag and improving fuel economy.

2.2 *Dual-Stage Turbocharger*

At low engine speeds, single-stage turbocharging makes it difficult to achieve both high boost pressure and heavy EGR rate due to limited overall turbocharging efficiency [17]. Boost pressure in a single-stage turbocharger is limited to 2.5 bar. Two-stage turbochargers have many advantages over single-stage turbochargers. It provides high intake manifold pressure and correspondingly high brake mean effective pressure and greater turbocharging efficiency at high overall compressor pressure ratio [18]. Since the high-pressure turbocharger is smaller in size, the transient behavior is upgraded so that at low speeds, both turbochargers can simultaneously operate [19]. Lee et al. [20] studied a thermodynamic zero-dimensional model in which a bypass valve was added to the high-pressure turbine and a waste gate valve was added to the low-pressure turbine. This system for a 4.5 L V6 engine yielded a maximum boost of 4 bars and an increment in power density by 29%. At lower torques, a bypass valve configuration proves advantageous over a waste gate configuration because of the smaller high-pressure turbine.

Liu et al. [21] put forth a matching method for a two-stage turbocharging system, which concluded that the contribution of compression ratios, turbine flow capacity, the cooler efficiency and bypass flow has an influence on exhaust energy utilization. As altitude increases, the temperature and pressure of air decrease. There is also a significant rise in fuel consumption as altitude increases. The most efficient method of power recovery at high altitudes is by using turbochargers. The influence of altitude on a two-stage turbocharger was studied by Yang et al. [22]. The inferences were: (i) at higher altitudes, the pressure ratio of low-pressure turbines increases directly with total pressure ratio, while the pressure ratio of high-pressure turbines decreases gradually with total pressure ratio (ii) The available flow energy increases by a small amount. However, the high increase in deficit means that available flow energy needs to be increased to facilitate power recovery as altitude increases. The two-stage turbocharger is hence a good option to attain boosting at lower torques than a conventional single-stage turbocharger. There is also a significant increase in power density of the intake manifold. This enables downsizing of the engine, i.e., a smaller turbocharged engine would provide the same power as a larger naturally aspirated one.

2.3 *Turbo-Compounding*

Turbo-compounding is a setup that uses the energy from exhaust gases to deliver power to the crankshaft directly. In mechanical turbo-compounding, the turbine run by the exhaust gases is directly connected to the crankshaft of the engine. In electrical turbo-compounding, the mechanical energy is converted to electrical energy with the help of a generator and afterward used to re-energize the battery and provide an additional boost. The former can be used in heavy-duty vehicles for fuel consumption and the latter in smaller vehicles because of its high flexibility and immediate availability of power [23–25]. The fuel savings obtained by turbo-compound method ranged approximately from 2 to 6%. Turbo-compounding method has a short paid pack period which is favorable for heavy-duty vehicle [26]. From parametric sweep study, it can be stated that the peak turbine rpm of a turbo-compounding system was 120,000 rpm at low speeds, 60,000 rpm at medium speeds and 95,000 rpm at high speeds [27]. One of the main downsides of turbo-compounding is the high exhaust back pressure in the exhaust manifold. This means that more work is to be done by the engine to expel exhaust gases [28]. To overcome the shortcomings of a turbo-compounding system, a new setup called electric booster and turbo generator (EBTG) system was designed. Since the coupling of the compressor and turbine leads to turbo-lag, they both were dissociated and were made to work independently. A motor was used to drive the compressor to avoid turbo-lag. The turbine was operated at high efficiency using speed optimization [12]. EBTG system works best under heavy-duty and high-speed conditions due to decrease in available exhausts and irreversibility of turbine. Under low-speed and light-duty conditions, EBTG is not ideal. About 2.6% of BSFC reduction was attained under 4,400 rpm.

3 Dual Intake Manifold

In this setup, the intake manifold plenum is divided into two sectors. Each plenum is connected to every other cylinder. Thus, each side of the manifold is exposed to pulses from alternative cylinders in correct firing order. The dual-plane manifold only sees an acceptable induction pulse every 180°, unlike the single-plane manifold which has covering pulses every 90° [29]. Generally, at idle and cruising rpm, dual-plane manifolds make better power. From the intake valve, they transmit the induction pulses to the carburetor in a better manner, which improves fuel atomization at low rpm. This dual-plane manifold can also be used as a restriction for higher rpm. This design is more suited for engines which are used for street driving. If the rpm is low, then the acceleration is better after a stop [30]. In a regular V8 engine design with a single manifold, the cylinders in the corners are further away from the plenum area. In engines with dual-plane manifolds, the two plenums have individual runners that are longer and closer. Greater fuel atomization and air-fuel charge are achieved because of the natural pressure-wave pulses. This boosts torque at idle and low load conditions [29]. Milling the partition between the two planes of the manifold helps in increasing the RPM. This mix of both single- and dual-plane characteristics gives greater torque at lower loads and also higher RPM. Additionally, a spacer at the top of this setup: (i) increases the plenum volume of the manifold, (ii) slightly cools down the charge, and (iii) improves the atomization of the fuel as it goes down into the manifold [30].

4 Air Intake Temperature

Recent studies show that the temperature of intake air dictates fuel consumption and emissions. Oxygen availability is key to determine the emission characteristics. Oxygen in excess leads to unstable combustion, misfiring, ignition delay, slow burning rate and knocking. Also, deficiency in oxygen leads to unburnt gaseous fuel and faster burning rate. Air intake temperature largely influences fuel consumption, combustion process and exhaust emissions.

4.1 Brake-Specific Fuel Consumption (BSFC)

Figure 1 shows the BSFC for different engine speeds and air intake temperatures at constant engine load. Engine speed ranged from 1,500 to 3,000 rpm and temperature of air was 20, 25 and 30 °C. The peak value of BSFC in this range was 380 g/kW h which were at 1,500 rpm and 30 °C. This was 4% higher to when the intake air temperature was 20 °C for the same engine speed. This trend remained the same for all engine speeds and showed that BSFC reduces with reduction in intake air temperature. Due to higher oxygen availability at low temperatures, the ignition delay is reduced. This allows combustion to occur at the end of compression stroke and the beginning of expansion stroke which reduces the wastage of fuel. For an intake air temperature of 30 °C, the BSFC from 380 g/kW h at 1,500 rpm reduces by 22% to 298 g/kW h at 3,000 rpm. At higher engine speeds, a larger amount of gaseous fuel involved in the oxidation process results in lower BSFC [31].

4.2 Carbon Monoxide (CO) Emissions

From Fig. 2, it is seen that air fuel ratio influences CO emissions to a great extent. Concentration of CO increases when there is insufficient oxygen for CO to form CO_2. The engine speed ranged from 1,500 to 3,000 rpm. At an engine speed of 3,000 rpm and an intake air temperature of 20 °C, CO emissions were recorded to be 341 ppm. Results prove that CO emissions increased with increase in engine speeds.

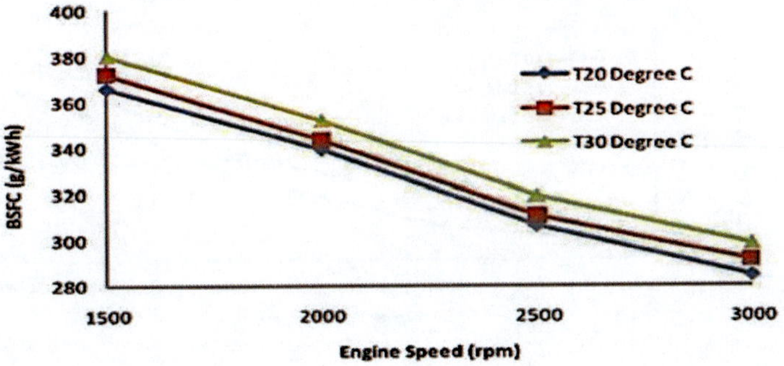

Fig. 1 BSFC versus engine speed at varying air intake temperatures [31]

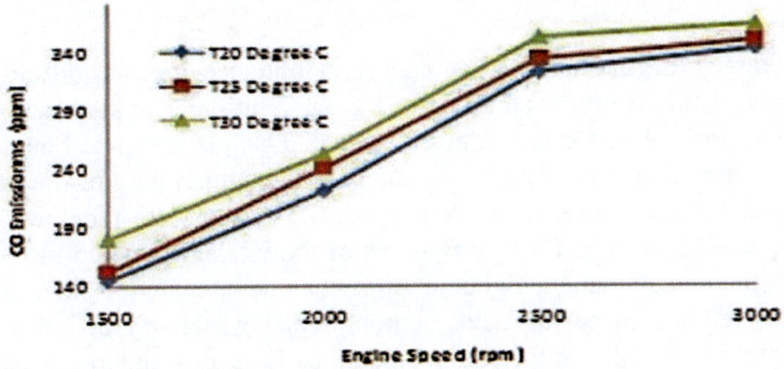

Fig. 2 CO emissions versus engine speed at varying air intake temperatures [31]

Also, intake air temperature was directly proportional to CO emissions. Hence, one can conclude that cold air intake can help reduce CO emissions [31].

4.3 Unburnt Hydrocarbons' (UHCs) Emissions

Figure 3 shows the unburned hydrocarbons' (UHCs) emission for varying engine speeds and intake air temperatures. Generally, fuel mixes with air before the combustion process. This mixing is dependent on fuel droplet size and availability of heat and oxygen [32]. Ten percent higher UHCs' emissions were observed at 30 °C than at 20 °C because of poor mixing and incomplete combustion. Pre-mixing and high rate of oxidation of UHCs at low temperatures result in lower UHCs emissions. The UHCs' emission at 3,000 rpm was found to be 85% higher than at 1,500 rpm. This trend is observed due to the high amount of injected fuel and reduced oxygen supply to the combustion chamber at higher engine speeds [31].

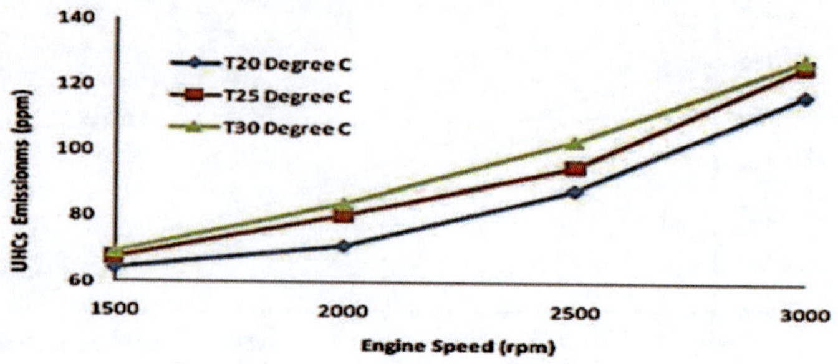

Fig. 3 UHC emissions versus engine speed at varying air intake temperatures [31]

4.4 Cold Air Intake

With the temperature of intake air having a great influence on emission and efficiency of an engine, cold air intake could be one great solution. The air filter of the cold air intake setup is placed away from the engine. This placement is beneficial since the temperature of the hot engine may increase the temperature of the intake air if the air filter is placed close to it. Cold air intake systems use tubes with minimal bends and large diameter. These properties of the intake tube help in reducing air resistance, thereby allowing a higher amount of air to enter [33]. Cold air intakes are built exclusively for the performance without focusing more on the noise reduction. Cold air intake filters, due to the cold temperature, lasts long and thus can be washed and reused unlike dry air intake filters which should be replaced every 15,000 miles. The engines running on cold air intake systems have an increased lifespan compared to the engines running on dry air intake engines [33]. More fuel burnt means more horsepower. The cold air intake systems have increased engine efficiency and performance [34]. The main risk associated with cold air intake systems is hydrolocking. This occurs when the engine absorbs moisture instead of air. The water trapped in the engine cylinders damages the piston and connecting rods [33].

5 Conclusion

A review of various intake boosting techniques was carried out with focus on improving engine performance, efficiency and controlling emissions. Various techniques and methods to improve performance at transient conditions were analyzed and compared. Turbochargers are used in all vehicle categories as they offer both fuel economy and reduced emissions. Although turbochargers are being used extensively, one main disadvantage associated with it is turbo-lag, which is the delay in response time between pressing the throttle and the turbo delivering the boost. An electrically assisted turbocharger proved to be the best solution for turbo-lag, with transient response time increasing by 70–90%. Also, 30% smaller engine size can be achieved for the same power output. One-third of the wasted energy is used as input fuel energy in modern diesel engines and light-duty SI engines. A waste heat recovery system like turbo-compounding is capable of recovering up to two-thirds of the wasted energy. However, due to losses and low efficiency of the power turbine, achieving maximum limits is not practical. Turbo-compounding enabled the engine to attain fuel savings of approximately 2–6%. High exhaust back pressure which was the main shortcoming of turbo-compounding can be overcome by using an electric booster and turbo generator system.

Dual-plane intake manifolds are a great way to increase average power and torque. At low rpm, the smaller plenums increase airflow which in turn results in better fuel atomization than a single-plane intake manifold. Customizations made to the plenum volume yielded better results for street performance. The temperature of intake air had great influence on BSFC, CO emissions and UHCs' emissions. At lower speeds and at lower intake air temperature BSFC, carbon monoxide emissions and unburnt hydrocarbon emissions seemed to be reduced when compared to higher intake air temperatures. This can be achieved by using cold air intake filters which increase engine performance and also have greater durability than normal air filters. Unlike cold air intake system, short ram intake draws air from inside the engine bay which is warmer air than what cold air intake system draws in. For complete combustion of the fuel, this warmer air is better suited than cold air. Hence, short ram intake gives better mileage than cold air intake system, but the latter can provide a significant boost in horsepower. Many developments and innovations have been made for the improvement of the performance of IC engines in the recent past, and several other new developments continue to take place. The key is to find a perfect balance between improving the performance and reducing the emissions. With IC engines being the main source of powering an automobile, it is essential to adopt these techniques and also explore advancements to enhance the overall output.

References

1. Wagner RM (2015) Engines of the future. Am Soc Mech Eng. https://www.asme.org/topics-resources/content/engines-of-the-future
2. OICA (2020) Motor vehicles production statistics
3. Liu H, Ma J, Tong L, Ma G, Zheng Z, Yao M (2018) Investigation on the potential of high efficiency for internal combustion engines. Energies 11(3)
4. Kalghatgi G (2018) Is it really the end of internal combustion engines and petroleum in transport? Appl Energy 225(May):965–974
5. Senecal PK, Leach F (2019) Diversity in transportation: why a mix of propulsion technologies is the way forward for the future fleet. Results Eng 4(November):100060
6. Sharma A, Dhakal B (2013) Performance and emission studies of a diesel engine using biodiesel tyre pyrolysis oil blends. SAE Technical Paper, vol 2, no (Ci)
7. Papu NH, Lingfa P, Dash SK (2020) Euglena Sanguinea algal biodiesel and its various diesel blends as diesel engine fuels: a study on the performance and emission characteristics. Energy Sour Part A Recov Util Environ Eff 00(00):1–13
8. Leach F, Kalghatgi G, Stone R, Miles P (2020) The scope for improving the efficiency and environmental impact of internal combustion engines. Transp Eng 1:100005

9. Khan TMY (2020) A review of performance-enhancing innovative modifications in biodiesel engines. Energies 13(17)
10. Baines N (2014) Intake boosting. Encycl Automot Eng. 2(1):1–15
11. Sawant P, Bari S (2018) Effects of variable intake valve timings and valve lift on the performance and fuel efficiency of an internal combustion engine. SAE Technical Paper, vol 2018, pp 1–11
12. Zi D, Zhang L, Chen B, Zhang Q (2019) Study of the electric-booster and turbo-generator system and its influence on a 1.5 L gasoline engine. Appl Therm Eng 162(August):114236
13. Katrašnik T, Rodman S, Trenc F, Hribernik A, Medica V (2003) Improvement of the dynamic characteristic of an automotive engine by a turbocharger assisted by an electric motor. J Eng Gas Turbines Power 125(2):590–595
14. Terdich N, Martinez-Botas R (2013) Experimental efficiency characterization of an electrically assisted turbocharger. SAE Technical Paper, vol 6
15. Petitjean D, Bernardini L, Middlemass C, Shahed SM (2004) Advanced gasoline engine turbocharging technology for fuel economy improvements, SAE Technical Paper, no 724
16. Dimitriou P, Burke R, Zhang Q, Copeland C, Stoffels H (2017) Electric turbocharging for energy regeneration and increased efficiency at real driving conditions. Appl Sci 7(4)
17. Hoyer KS, Sellnau M, Sinnamon J, Husted H (2013) Boost system development for gasoline direct-injection compression-ignition (GDCI). SAE Int J Engines 6(2):815–826
18. Ketata A, Driss Z (2017) Influence of rotor speed on mixed flow turbine performance. Am J Mech Eng 5(5):205–210
19. Winkler N, Ångström HE (2008) Simulations and measurements of a two-stage turbocharged heavy-duty diesel engine including EGR in transient operation, SAE Technical Paper, vol 2008, no 724. https://doi.org/10.4271/2008-01-0539
20. Filipi Z, Lee B, Jung D, Assanis D (2016) ICES2008-1692, pp 1–11
21. Zheng Z, Feng H, Mao B, Liu H, Yao M (2018) A theoretical and experimental study on the effects of parameters of two-stage turbocharging system on performance of a heavy-duty diesel engine. Appl Therm Eng 129:822–832
22. Yang M, Gu Y, Deng K, Yang Z, Liu S (2018) Influence of altitude on two-stage turbocharging system in a heavy-duty diesel engine based on analysis of available flow energy. Appl Therm Eng 129:12–21
23. Bin Mamat AMI, Martinez-Botas RF, Rajoo S, Romagnoli A, Petrovic S (2015) Waste heat recovery using a novel high performance low pressure turbine for electric turbocompounding in downsized gasoline engines: Experimental and computational analysis. Energy 90:218–234
24. Briggs I, McCullough G, Spence S, Douglas R (2014) Whole-vehicle modelling of exhaust energy recovery on a diesel-electric hybrid bus. Energy 65:172–181
25. Mamat AMI, Romagnoli A, Martinez-Botas RF (2014) Characterisation of a low pressure turbine for turbocompounding applications in a heavily downsized mild-hybrid gasoline engine. Energy 64:3–16
26. Zhao R, Zhuge W, Zhang Y, Yang M, Martinez-Botas R, Yin Y (2015) Study of two-stage turbine characteristic and its influence on turbo-compound engine performance. Energy Convers Manag 95:414–423
27. Andwari AM, Pesiridis A, Esfahanian V, Salavati-Zadeh A, Karvountzis-Kontakiotis A, Muralidharan V (2017) A comparative study of the effect of turbocompounding and ORC Waste Heat Recovery systems on the performance of a turbocharged heavy-duty diesel engine. Energies 10(8):1–17
28. Aghaali H, Ångström HE (2015) A review of turbocompounding as a waste heat recovery system for internal combustion engines. Renew Sustain Energy Rev 49:813–824
29. Kimbrough B (2012) Intake manifolds: Single plane or dual plane?. https://www.enginelabs.com/engine-tech/engine/intake-manifolds-single-plane-or-dual-plane/
30. Which style manifold is better?. https://help.summitracing.com/app/answers/detail/a_id/5235/~/which-style-manifold-is-better%3F
31. Abdullah NR, Ismail H, Michael Z, Rahim AA, Sharudin H (2015) Effects of air intake temperature on the fuel consumption and exhaust emissions of natural aspirated gasoline engine. J Teknol 76(9):25–29

32. Cinar C, Uyumaz A, Solmaz H, Sahin F, Polat S, Yilmaz E (2015) Effects of intake air temperature on combustion, performance and emission characteristics of a HCCI engine fueled with the blends of 20% n-heptane and 80% isooctane fuels. Fuel Process Technol 130(C):275–281
33. Knowling M (2007) Shielded Sucker: a cheap and easy air intake improvement—With no CAI pipe in sight. AutoSpeed
34. Birtok-Băneasă C, Rațiu S, Hepuț T (2017) Influence of intake air temperature on internal combustion engine operation. IOP Conf Ser: Mater Sci Eng 163:012039

Buckling Analysis of Square Composite Plate with Rectangular Cutout

Prathamesh Dehadray, Sainath Alampally, and Bhaskara Rao Lokavarapu

Abstract Composite materials are used in aeronautical, automobile, healthcare and marine industries due to their high stiffness, higher strength-to-weight ratio and long fatigue life. Thin plate plays vital role in manufacturing of engineering structures. Components made from composite materials often subjected damage while working. Buckling can cause loss of stability of component which subsequently leads to failure of the entire structure. This paper deals with the effect of rectangular cutout on the buckling behavior of composite square plate. Cutouts are generally used for ventilation and to reduce the weight of component. This study investigates the critical buckling load for square plate with rectangular cutout of various aspect ratios and different stacking sequences. Classical laminated plate theory (CLPT) is used for analytical calculations, and the obtained values are compared with the results of FEA carried out in ANSYS.

Keywords Buckling analysis · CLPT · Finite element analysis

1 Introduction

Thin plates are extensively used in many complex automobile, aviation, civil and marine structures. Generally, cutouts are made to reduce weight, to give access for fuel lines and for maintenance. If there are any imperfections, then buckling load highly depends upon cutout geometry and size.

Ghannadpour et al. [1] studied the effect of elliptical and circular cutout on the buckling tendency of composite rectangular plates. They concluded that as diameter and width of plate ratio increase, buckling load decreases and buckling load is highest in transverse direction for same cutout area of elliptical geometry. Lakshmi Narayana et al. [2] investigated the effects of fiber and resin types, cutout dimension, volume fraction and aspect ratio of plate on the symmetric laminated rectangular plate. Eigen value analysis is used to study critical thermal buckling temperature. They found out

P. Dehadray · S. Alampally · B. R. Lokavarapu (✉)
School of Mechanical Engineering, Vellore Institute of Technology, Tamil Nadu, Vandalur-Kelambakkam Road, Chennai 600127, India
e-mail: bhaskarlokavarapu@gmail.com

A. Maurya et al. (eds.), *Recent Trends in Mechanical Engineering*, Lecture Notes in Mechanical Engineering, https://doi.org/10.1007/978-981-19-7709-1_36

that graphite fiber–polyester resin plate can sustain the highest buckling temperature, whereas E-glass fiber–polyester resin plate sustains the lowest temperature. Buckling behavior of the quasi-isotropic rectangular plate made up of graphite/epoxy material is analyzed by Laxmi Narayana et al. [3] with the introduction of various shapes of cutouts. They found out that buckling load for CC boundary condition is twice than CS boundary condition for same geometry. Kumar and Singh [4] studied the effects of flexural boundaries on buckling behavior of isotropic laminate having different cutouts. Using FSDT and von Karman's assumptions, they formulated the FEM model. First ply failure loads are calculated for different geometries and compared with existing literature. They concluded that laminates with clamped boundary conditions have highest buckling strength and laminates with simply supported boundary conditions have lowest buckling strength irrespective of cutout shape and size.

Ahmet Erklig and Eyup Yeter [5] studied the effect of various cutouts such as circular, square, triangular, elliptical on the buckling load for square polymer matrix composite plates. The plate is made up of glass polyester material, and buckling analysis is carried out analytically and compared with experimentally. They concluded that as dimension of plate increases, buckling load decreases. Most critical buckling load is obtained at 45° fiber orientation angle, and plate with elliptical cutout gives maximum buckling load. Zhao Jing [6] analyzed the buckling behavior of rectangular orthotropic plate subjected to axial compression with simply supported boundary condition. Optimization of stacking sequence is found out with maximizing the buckling load with constant thickness as well as minimizing the plate thickness by keeping buckling load as constant parameter. Cappello and Tumino [7] investigated the buckling behavior of the composite plate for uniaxial compressive load. They studied the effect of position, stacking sequence and length of delamination on the critical buckling load in global, local and mixed modes. Sandeep Singh et al. [8] studied the elastic buckling tendency of simply supported and clamped thin rectangular isotropic plates with and without central circular cutout subjected to uniaxial partial edge compression by first-order deformation theory.

Ovesy et al. [9] investigated the compressive buckling behavior of composite laminates with embedded delamination for variation in cutout, for variation in thickness and for different stacking sequences. Prasun Jana (1) [10] analyzed the linear elastic buckling behavior of simply supported rectangular plate for uniform axial compression with circular cutout for various aspect ratios. He found out the optimal location of cutout for maximum buckling load by combining ANSYS with MATLAB. Singh et al. [11] studied that the buckling analysis of laminated composite plates is carried out by using an efficient C0 finite element model developed based on higher-order zigzag theory. Analysis is carried out for different objectives such as stacking sequence, aspect ratio, thickness ratio and boundary conditions. Djamel Ouinas and Belkacem Achour [12] investigated the buckling behavior of thin composite square plate made up of boron/epoxy without and with cutout subjected to uniaxial compression load. Orientation of cutout is varied from 0 to 90° with increment of 15° to find the critical buckling load. They found out that growth of buckling load is much faster when orientation of notch is more than 45°. Lopatin and Morozov [13] analyzed the buckling tendency of orthotropic composite rectangular plate under uniform compressive load. They applied clamped-clamped-free-free boundary conditions, and the results are evaluated. They have formulated the analytical solution for buckling load by combined Kantorovich and Galerkin methods. The obtained results are compared with finite element solution. Wankhade and Niyogi [14, 15] studied the refined plate theory to predict behavior of composite plates with different boundary conditions for buckling and performed the stability analysis on three-layered composite plate (0/90/0). They evaluated critical buckling loads for variation in ratio of elastic moduli considering simply supported and clamped boundary conditions. Abolfazl Shirkavand et al. [16] studied the effect of orientation of cutout on the buckling tendency of composite cylinder using experimental as well as numerical methods. They took glass/epoxy material under consideration. They concluded that buckling load is highly influenced by cutout geometry, stacking sequence and orientation of cutout. Galerkin–Kantorovich methodology is used by Michael Ebie Onyia et al. [17] to investigate elastic buckling behavior of thin plates with clamped and simply supported boundary conditions under uniaxial compressive loading. They concluded that the critical buckling load is obtained at first buckling mode. Nagendra Singh Gaira et al. [18] studied the factors affecting the buckling behavior of carbon/epoxy plate with clamped-free boundary conditions. They analyzed the behavior of plate with varying aspect ratio, cutout shape and effect of stress concentration.

Very few literatures are available that consider buckling analysis plates with different orientation of cutouts; hence, this study is focused on the buckling behavior of glass/epoxy composite plate with and without cutout with simply supported boundary conditions.

2 Analyzed Model

The geometry of plate is made in ANSYS 2020 R2, and eigen value buckling analysis is carried out. A square laminated composite plate with four plies at different ply orientations is considered. Material properties of glass/epoxy are mentioned in Table 1. The geometry of plate is shown in Fig. 1.

Dimensions of plate are 400 mm * 400 mm * 5 mm with ply thickness of 1.25 mm. A cutout is made centrally in the plate with length "*a*" mm and width "*b*" mm as shown in Fig. 1. All four edges of plate are simply supported, and uniaxial compressive load "Nx" is applied along *x* direction. From the literature survey, input parameters are

Table 1 Properties of glass/epoxy

Young's modulus (MPa)		Shear modulus (MPa)		Poisson's ratio	
E_1	45,000	G_1	5000	μ_{12}	0.3
E_2	10,000	G_2	3846	μ_{23}	0.067
E_3	10,000	G_3	5000	μ_{31}	0.3

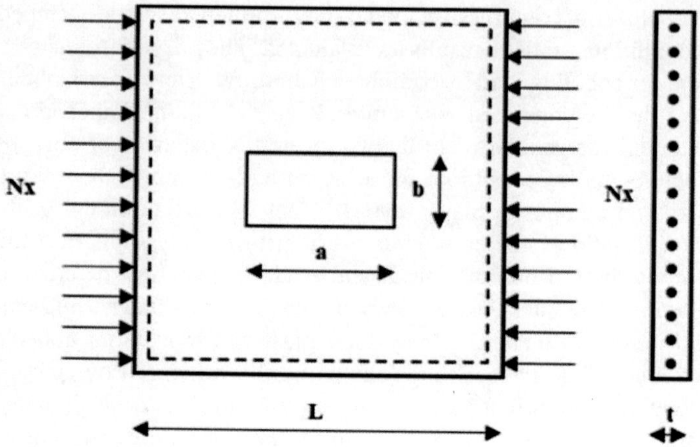

Fig. 1 Analyzed model

found out and impact of the change in those parameters on the buckling load is studied. Considering the literature gap, parameters to be varied for the analysis are fixed. Buckling behavior of plate is studied, and influence of stacking sequence, dimensions of plate, aspect ratio and orientation of cutout on the buckling capacity of the plate is found out.

Mathematical formulations are made in MATLAB for laminated composite plate without cutout using classical laminated plate theory [19]. Reduced stiffness matrix Q is given by,

$$Q = \begin{bmatrix} Q_{11} & Q_{12} & 0 \\ Q_{12} & Q_{22} & 0 \\ 0 & 0 & Q_{66} \end{bmatrix}$$

where

$$Q_{11} = E_1/(1 - \mu_{12} * \mu_{21}),$$

$$Q_{12} = E_2/(1 - \mu_{12} * \mu_{21}),$$

$$Q_{22} = E_2 * \mu_{12}/(1 - \mu_{12} * \mu_{21}),$$

$$Q_{66} = G_1$$

Let $\cos\theta = m$ and $\sin\theta = n$, where θ is angle of ply orientation. Then, transformed reduced matrix $\vec{Q}$ is given by,

$$\vec{Q} = \begin{bmatrix} \overrightarrow{Q_{11}} & \overrightarrow{Q_{12}} & \overrightarrow{Q_{16}} \\ \overrightarrow{Q_{21}} & \overrightarrow{Q_{22}} & \overrightarrow{Q_{26}} \\ \overrightarrow{Q_{61}} & \overrightarrow{Q_{62}} & \overrightarrow{Q_{66}} \end{bmatrix}$$

where

$$\overrightarrow{Q_{11}} = Q_{11} * m^4 + 2 * (Q_{12} + 2 * Q_{66}) * m^2n^2 + Q_{22} * n^4$$

$$\overrightarrow{Q_{12}} = (Q_{11} + Q_{22} - 4 * Q_{66}) * = (Q_{11} + Q_{22} - 4 * Q_{66}) * m^2n^2 + Q_{12} * \left(m^4 + n^4\right)$$

$$\overrightarrow{Q_{22}} = Q_{11} * n^4 + 2 * (Q_{12} + 2 * Q_{66}) * m^2n^2 + Q_{22} * m^4$$

$$\overrightarrow{Q_{16}} = (Q_{11} - Q_{12} - 2 * Q_{66}) * m^3n + (Q_{11} - Q_{12} - 2 * Q_{66}) * n^3m$$

$$\overrightarrow{Q_{26}} = (Q_{11} - Q_{12} - 2 * Q_{66}) * n^3m + (Q_{11} - Q_{12} - 2 * Q_{66}) * m^3n$$

$$\overrightarrow{Q_{66}} = (Q_{11} + Q_{22} - 2 * Q_{12} - 2 * Q_{66}) * m^2n^2 + Q_{66} * \left(m^4 + n^4\right)$$

The bending stiffness matrix is given by $D_{ij} = \sum_{1}^{n} Q_{ij} * \frac{1}{3} * \left(Z_{k-1}^3 - Z_k^3\right)$,

$$D = \begin{bmatrix} D_{11} & D_{12} & D_{16} \\ D_{12} & D_{22} & D_{26} \\ D_{16} & D_{26} & D_{66} \end{bmatrix}$$

The critical buckling load Ncr (N/mm) for square laminated composite plate without cutout for mode 1 is as follows:

$$\text{Ncr} = (\pi/L)^2 * (D_{11} + 2 * (D_{12} + 2 * D_{66}) + D_{22})$$

Table 2 Variation in stacking sequence

Stacking sequence	Critical buckling load (N/mm)	
	MATLAB	FEA
$[0/90]_2$	52.80	49.16
$[45/-45]_2$	72.14	65.03
$[30/60]_2$	67.32	53.36
$[0/45]_2$	62.50	53.97

3 Results and Discussion

3.1 Effect of Ply Orientation

Ply orientation has huge impact on the buckling behavior of composite laminates. The rotation angle of ply defines fiber orientation within each ply of laminate. Critical buckling load obtained for glass/epoxy material with four different stacking sequences is shown in Table 2.

Eigen value buckling analysis is carried out in ANSYS software [20] as well as calculations are made in MATLAB using classical laminated plate theory and found out that $[45/-45]_2$ stacking sequence gives maximum buckling capacity, whereas $[0/90]_2$ stacking sequence yields minimum value of buckling load.

3.2 Effect of Size of the Plate

As critical buckling load is obtained at $[45/-45]_2$ stacking sequence, the study is carried out by varying the length of plate from 400 to 800 mm with increment of 100 mm for the same stacking sequence. Comparison of results obtained in MATLAB and using finite element analysis is shown in Table 3. Variation of critical buckling load with respect to change in size of the plate is shown in Fig. 2. As size of the plate increases, buckling capacity of plate decreases.

Table 3 Variation in size of plate

Plate dimensions	Critical buckling load (N/mm)	
	MATLAB	FEA
400 * 400	72.14	65.03
500 * 500	46.17	41.83
600 * 600	32.06	29.14
700 * 700	23.55	21.46
800 * 800	18.03	16.46

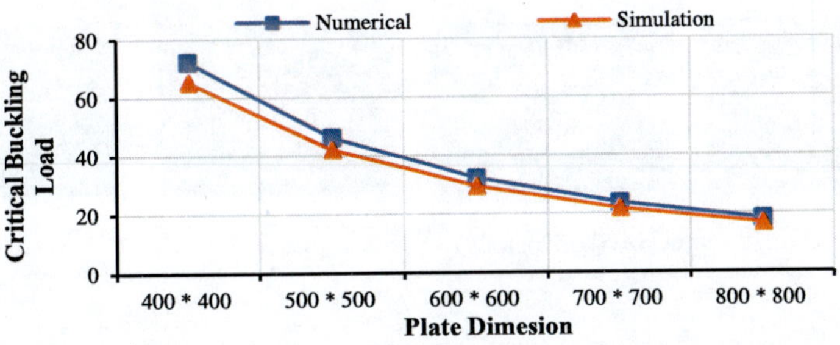

Fig. 2 Variation in size of the plate

3.3 *Effect of Cutout*

A rectangular cutout of 100 mm length and 50 mm width is made centrally, and buckling analysis is performed on the plate with and without cutout for different stacking sequences. With the introduction of cutouts, buckling loads are decreased by 13–17% (Fig. 3 and Table 4).

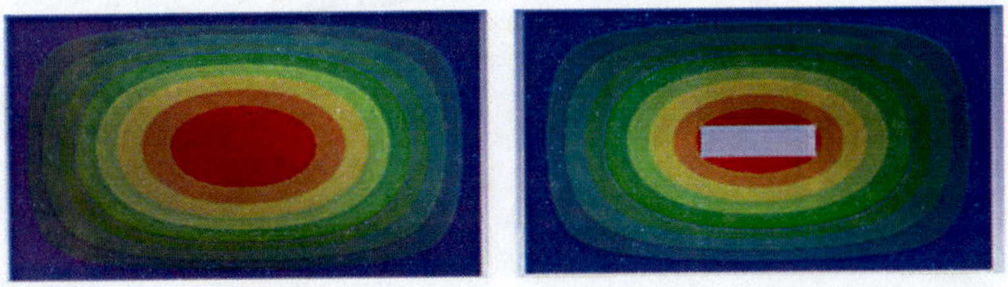

Fig. 3 Without and with cutout

Table 4 Without and with cutout

Stacking sequence	Critical buckling load (N/mm)	
	Without	With
$[0/90]_2$	49.16	40.48
$[45/-45]_2$	65.03	56.61
$[30/60]_2$	53.36	45.17
$[0/45]_2$	53.97	45.92

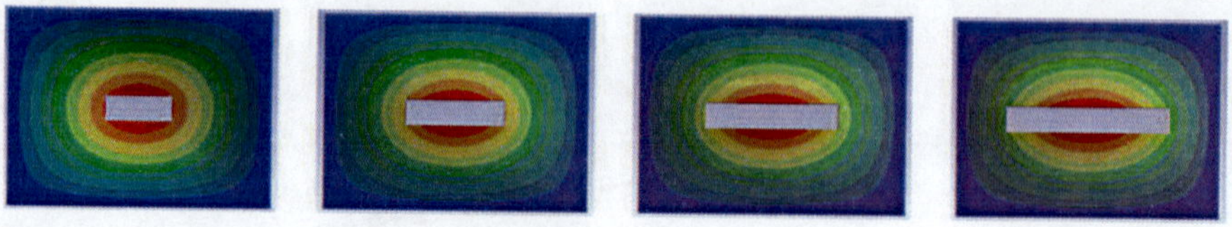

Fig. 4 Variation in cutout size (horizontally)

Table 5 Variation in aspect ratio of cutout (horizontally)

Cutout dimension	Cutout aspect ratio	Stacking sequence			
		$[45/-45]_2$	$[0/90]_2$	$[30/60]_2$	$[0/45]_2$
100 * 50	2	56.61	40.48	45.17	45.92
150 * 50	3	51.81	36.03	40.73	42.42
200 * 50	4	47.94	32.38	37.20	39.70
250 * 50	5	44.77	29.52	34.34	37.53

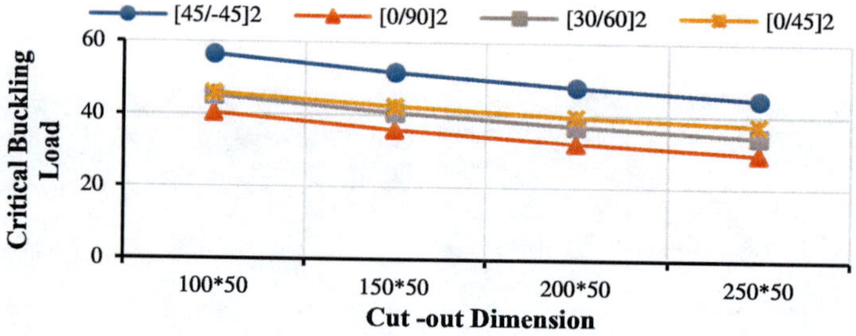

Fig. 5 Variation in aspect ratio of cutout (horizontally)

3.4 *Effect of Variation in Aspect Ratio of Cutout (Horizontally)*

Cutout aspect ratio (a/b) is varied horizontally from 2 to 5 as shown in Fig. 4, and results are obtained for various stacking sequences as listed in Table 5. Figure 5 shows that with increase in aspect ratio of the cutout horizontally, critical buckling load decreases linearly.

3.5 *Effect of Variation in Aspect Ratio of Cutout (Vertically)*

Cutout aspect ratio (a/b) is varied vertically from 2 to 5, and critical buckling load for different stacking sequences is calculated as shown in Table 6. Figure 7 shows that as cutout aspect ratio increases, critical buckling load increases (Fig. 6).

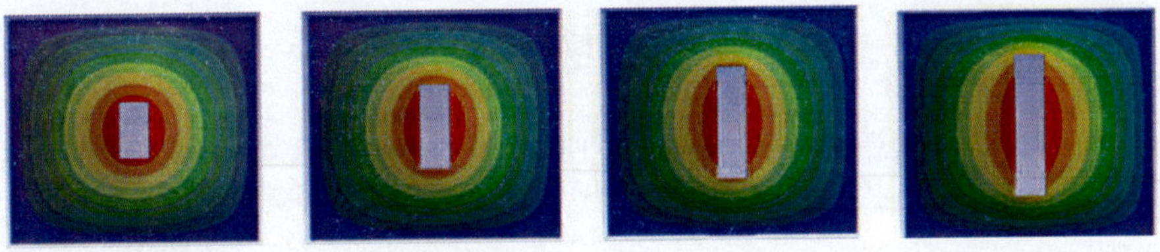

Fig. 6 Variation in cutout size (vertically)

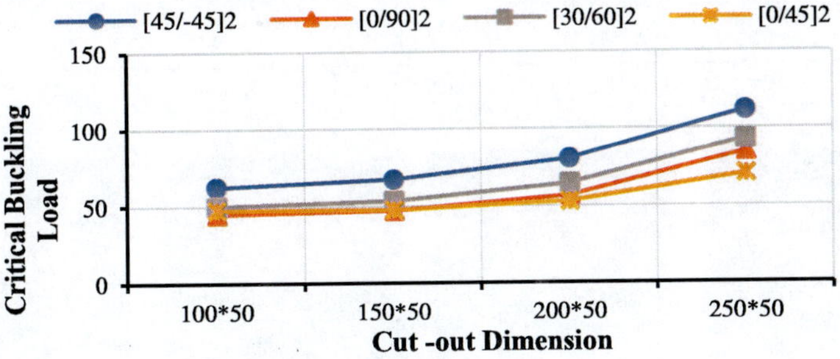

Fig. 7 Variation in aspect ratio of cutout (vertically)

Table 6 Variation in aspect ratio of cutout (vertically)

Cutout dimension	Cutout aspect ratio	Stacking sequence			
		$[45/-45]_2$	$[0/90]_2$	$[30/60]_2$	$[0/45]_2$
100 * 50	2	62.02	44.13	49.53	47.26
150 * 50	3	66.84	46.59	53.01	47.27
200 * 50	4	80.07	56.23	64.07	52.49
250 * 50	5	111.40	84.34	92.76	70.14

3.6 *Effect of Orientation of Cutout*

To study the effect of orientation of the cutout on the buckling behavior of laminated composite plate, cutout of aspect ratio 2 is taken, and results are calculated for orientation of 0, 15, 30, 45, 60, 75 and 90° for $[45/-45]_2$ ply orientation as shown in Table 7. Figure 9 shows that with increase in orientation of cutout, critical buckling load increases (Fig. 8).

Table 7 Variation in orientation of cutout for $[45/-45]_2$ stacking sequence

Cutout orientation	Cutout dimension			
	100 * 50	150 * 50	200 * 50	250 * 50
0	56.61	51.81	47.94	44.77
15	57.10	52.49	48.81	45.69
30	57.78	54.65	52.05	49.85
45	59.28	58.43	59.11	61.50
60	60.71	62.81	69.00	82.54
75	61.51	65.86	77.11	103.22
90	62.02	66.84	80.07	111.40

Fig. 8 Variation in cutout orientation from 0 to 90°

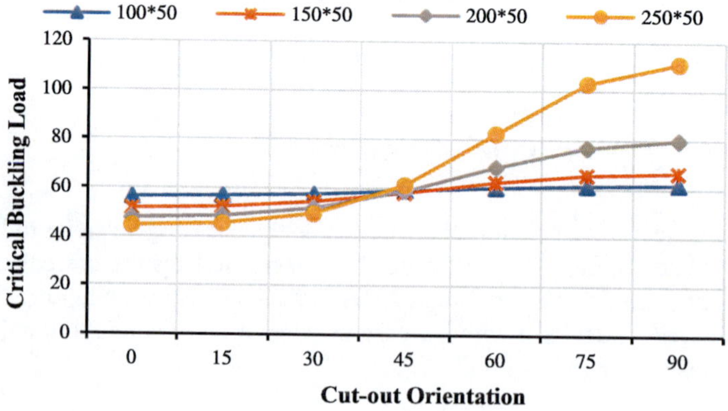

Fig. 9 Variation in orientation of cutout for $[45/-45]_2$ stacking sequence

4 Conclusion

Components with cutouts and holes are used in cabinet and fuselage of aircraft; hence, this study can be extended to curved plates or panels with holes, and also, importance of optimization is increased in industries; hence, the present study can be integrated with the artificial neural networks in future. Buckling behavior of laminated composite square plate with simply supported boundary condition under uniaxial loading is studied. Classical laminate plate theory is used, and results are obtained for different stacking sequences. Obtained results are validated with FEA results. The conclusions are as follows:

- Maximum buckling load is obtained for $[45/-45]_2$ stacking sequence for plate with and without cutout.
- Minimum buckling load is obtained for $[0/90]_2$ stacking sequence for plate with and without cutout.
- Size of the plate is inversely proportional to the buckling load. As size of plate increases, buckling capacity decreases.
- For horizontal cutout, critical buckling capacity decreases with increase in aspect ratio of cutout.
- For vertical cutout, critical buckling load increases gradually with increase in aspect ratio of cutout.
- As orientation of cutout increases from 0 to 90°, critical buckling load increases.

References

1. Ghannadpour SAM, Najafi A (2006) On the buckling behaviour of cross-ply laminated composite plates due to circular/elliptical cutouts. Compos Struct 75(3):6–16
2. Lakshmi Narayana R, Vijaya Kumar G, Rao K, Effect of Volume fraction on the thermal buckling analysis of laminated composite plate with square/rectangular cutout. Mater Today 5:5819–5829
3. Narayana AL, Rao K, Kumar RV (2014) Buckling analysis of rectangular composite plates with rectangular cutout subjected to linearly varying in-plane loading using FEM. Indian Acad Sci 39(3):583–596
4. Kumar D, Singh SB (2010) Effects of boundary conditions on buckling & post buckling responses of composite laminate with various shaped cutouts. Compos Struct 92:769–779
5. Erkliğ A, Yeter E (2012) The effects of cutouts on buckling behaviour of composite plates. 19:323–330
6. Zhao J (2020) Semi-analytical optimal solution for maximum buckling load of simply supported orthotropic plates. Int J Mech Sci 187
7. Cappello F, Tumino D (2006) Numerical analysis of composite plates with multiple delamination subjected to uniaxial buckling load. Compos Sci Technol 66:264–272
8. Singh S, Kulkarni K, Pandey R (2012) Buckling analysis of thin rectangular plates with cut-outs subjected to partial edge compression using FEM. J Eng Des Technol 10(1):128–142
9. Ovesy HR, Kharazi M, Taghizadeh M (2010) Semi-analytical buckling analysis of clamped composite plates containing embedded rectangular and circular delaminations. Mech Adv Mater Struct 17:343–352

10. Jana P (2016) Optimal design of uniaxially compressed perforated rectangular plate for maximum buckling load. Thin-Walled Struct 103:225–230
11. Singh SK, Chakrabarti A (2012) Buckling analysis of laminated composite plates using an efficient C0 FE model. 9
12. Ouinas D, Achour B (2013) Buckling analysis of laminated composite plates containing an elliptical notch. Compos: Part B 55:575–579
13. Lopatin AV, Morozov EV (2020) Buckling of a rectangular composite orthotropic plate with two parallel free edges and the other two edges clamped and subjected to uniaxial compressive distributed load. Euro J Mech/Solids 81(103960)
14. Wankhade RL, Niyogi SB (2020) Buckling analysis of symmetric laminated composite plates for various thickness ratios and modes. Innov Infrastruct Solut
15. Niyogi SB, Wankhade RL, Gajbhiye PD (2020) Buckling analysis of laminated composites considering the effect of orthotropic material. J Phys: Conf Ser 1706(012188)
16. Shirkavand A, Taheri-Behrooz F, Omidi M (2019) Orientation and size effect of a rectangle cut-out on the buckling of composite cylinders. Aerosp Sci Technol
17. Onyia ME, Rowland-Lato EO, Ike CC (2020) Galerkin—Kantorovich method for the elastic buckling analysis of thin rectangular SCSC plates. Int J Eng Res Technol 13(4):613–619
18. Gaira NS, Maurya NK, Yadav RK (2012) Linear buckling analysis of laminated composite plate 2(4):886–891
19. Jones RM, Mechanics of composite materials, 2nd edn
20. ANSYS version 19.2 user manual

Stochastic Fracture Analysis of FGM Plate with Edge Crack Under Mechanical Loadings using XFEM

Kundan Mishra, Achchhe Lal, and B. M. Sutaria

Abstract The second-order perturbation method is utilized for mixed (first and second) mode stress intensity factor (MMSIF) of functionally graded materials (FGMs) plate with edge crack under mechanical loadings. Extended finite element method (XFEM) is utilized for the fracture analysis of cracked FGM plate, and the stochastic based analysis is done by second-order perturbation technique (SOPT) for computation of mean and coefficient of variance (COV). The uncorrelated random parameters' material properties, crack length and crack angle are utilized in this present work. The purpose of present study is to find the critical random parameters, which are affecting more on MMSIF. The numerical results are evaluated for different gradient coefficients, crack angles, crack lengths with random system properties. The MATLAB [R2015a] environment is utilized for this study.

Keywords FGMs' plate · XFEM · MMSIF · COV · SOPT

1 Introduction

FGMs are the special type of composite materials, in which properties of materials can be changed along definite direction as per application. The purpose of using such material is to reduce the limitations of conventional composite. In this present paper, the effect of various fracture parameters on MMSIF of cracked FGM plate is observed. So, apart from deterministic approach, stochastic-based fracture analysis of materials is essential to understand the effect of individual parameter on MMSIF of cracked material plate. Kim and Paulino [1] studied the normalized MMSIF of FGM plate by conventional FEM, where they used J-integral, crack closer integral and displacement correlation method. Man et al. [2] investigated the MMSIF of cracked FGM plate by utilizing weight function method. Bayesteh and Mohammadi [3] presented fracture analysis of orthotropic materials by utilizing XFEM based, where it is confirmed that the orthotropic XFEM requires less degree of

K. Mishra · A. Lal (✉) · B. M. Sutaria
Mechanical Engineering Department, SVNIT Surat, Surat, India
e-mail: lalachchhe@yahoo.co.in

A. Maurya et al. (eds.), *Recent Trends in Mechanical Engineering*, Lecture Notes in Mechanical Engineering, https://doi.org/10.1007/978-981-19-7709-1_37

freedoms (DOFs). Belytschko et al. [4] have proposed XFEM for modeling the cracks without remeshing by presenting discontinuous enrichment functions. Reddy and Rao [5] investigated stochastic fracture analysis, which involves fractal finite element method, and the proposed method is compared and verified with the Monte Carlo simulation. Lal et al. [6] determined second-order critical stress intensity factor in composite plates with edge notch by utilizing random material properties under uniaxial tensile loadings. The basic formulation is done by utilizing HSDT for the evaluation of MMSIF. Lal et al. [7] presented fracture analysis with reliability of center crack composite plate under tensile loading. In this present work, SOPT and MCS are implemented for the evaluation of statistics of MMSIF, and the effect of random system parameters is also examined. Khatri and Lal [8, 9] investigated stochastic XFEM-based numerical analysis of isotropic plate with emerging crack from hole under uniaxial and biaxial loading. This analysis is carried out by SOPT. It is observed that even smaller crack emerging from hole is very critical. Wang and Zhou [10] investigated the fracture analysis of FGM plate by FEM with field enrichment. This method is further utilized to analyze crack propagation. The effectiveness of this method is also verified from some examples.

From the previously published research papers, it is noticed that researchers are giving more effort in fracture analysis of different materials. FGM is the newly developed material, and very less work is reported on stochastic-based fracture analysis of these materials. In this paper, SOPT is utilized for the analysis of cracked FGM plate under uniform tensile loading. Here, the COV of individual random properties is calculated to find the critical fracture parameter. After finding the critical random properties, the randomness on critical parameters is considered to find the MMSIF. The present numerical analysis is carried out by utilizing XFEM in MATLAB [R2015a] environment.

2 Problem Formulation

Modeling of crack growth by FEM is not as much suitable as by XFEM. In XFEM, remeshing is not required for every step of crack growth, and enrichment functions are used to model discontinuities. In this section, XFEM-based mathematical formulation is presented. The displacement equation is represented as [8]:

$$U^h(x) = \sum_{i=1}^{n} P_i(x)\overline{u_i} + \sum_{i=1}^{n_c} P_i(x)H(x)a_i + \sum_{i=1}^{n_{t1}} P_j(x)\sum_{\alpha=1}^{4}\Phi_\alpha^1(x)b_j^{\alpha 1} + \sum_{i=1}^{n_{t2}} P_j(x)\sum_{\alpha=1}^{4}\Phi_\alpha^2(x)b_j^{\alpha 2} \quad (1)$$

Here, $\overline{u_i}$ (conventional DOF), a_i (DOF of enrichment function), $b_j^{\alpha 1}$ (crack tip enrichment function) and $b_j^{\alpha 2}$ (crack face enrichment functions) are utilized in Eq. 1.

Fig. 1 Body with crack under traction

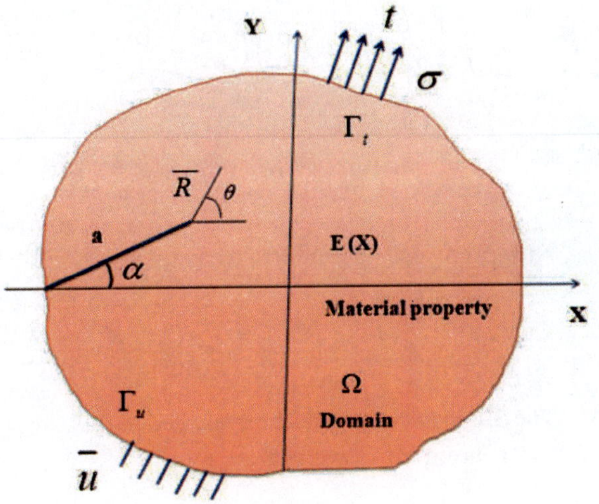

Figure 1 represents a body having area Ω with outer boundary Γ and a crack. The crack surface is assumed as traction free.

$$DOFs = size(\rho) + size(\rho_c) + size(\rho_t) \tag{2}$$

Heaviside function $H(x)$ is used for modeling crack face. φ_α^1 and φ_α^2 are the crack tip asymptotic functions. The asymptotic function at crack tip is:

$$\varphi_\alpha^{t1} = \sqrt{\overline{R}}\sin\left(\frac{\theta}{2}\right), \varphi_\alpha^{t2} = \sqrt{\overline{R}}\cos\left(\frac{\theta}{2}\right) \tag{3}$$

$$\varphi_\alpha^{t3} = \sqrt{\overline{R}}\sin\theta\cos\left(\frac{\theta}{2}\right), \varphi_\alpha^{t4} = \sqrt{\overline{R}}\sin\theta\cos\left(\frac{\theta}{2}\right) \tag{4}$$

J-integral can be represented in terms of SIF as [8]:

$$J = \frac{K_1^2 + K_2^2}{E_e} \tag{5}$$

Here, U is the strain energy, n_j is the outward normal to Γ and δ_{1j} is Kronecker delta. For the jth component, u_i is displacement field and σ_{ij} is stress tensor. A crack body is represented by two states, and Eq. (7) is formed by summation of these states.

$$J = \int_\Gamma \left(U\delta_{1j} - \sigma_{ij}\frac{\partial u_i}{\partial x_1}\right)n_j d\Gamma \tag{6}$$

$$J^{(1+2)} = \int_{\Gamma}\left[\frac{1}{2}\left(P_{ij}^{(1)} + P_{ij}^{(2)}\right)\left(\varepsilon_{ij}^{(1)} + \varepsilon_{ij}^{(2)}\right)\delta_{1j} - \left(P_{ij}^{(1)} + P_{ij}^{(2)}\right)\frac{\partial(u_i^{(1)} + u_i^{(2)})}{\partial x_i}\right]n_j d\Gamma \tag{7}$$

$$J^{(1+2)} = J^{(1)} + J^{(2)} + \frac{2}{E_e}(K_1^{(1)}K_1^{(2)} + K_2^{(1)}K_2^{(2)}) \tag{8}$$

Now, from Eqs. (3) and (6), we get,

$$I^{(1,2)} = \frac{2}{E_e}(K_1^{(1)}K_1^{(2)} + K_2^{(1)}K_2^{(2)}) \tag{9}$$

The SIF in two states can be evaluated as utilizing $K_1^{(2)} = 1$, $K_2^{(2)} = 0$ and $K_1^{(2)} = 0$, $K_2^{(2)} = 1$. From Eq. 7, we get,

$$K^{(1)} = \frac{M^{(1,Mode\ \mathrm{I})}E_e}{2} K_2^{(1)} = \frac{M^{(1,Mode\ \mathrm{II})}E_e}{2} \tag{10}$$

Here, $I^{(1,\mathrm{Mode}\ 1)}$ and $I^{(1,\mathrm{Mode}\ 2)}$ are the interaction integrals.

3 Stochastic Analysis Using Perturbation Method

Here, in this present section, mean variance of $K_I$and K_{II} by considering different random parameter is presented [8]. The MMSIF for two modes is:

$$K = f(K_n), \text{ where n } = \begin{cases} 1 \text{ for model } 1 \\ 2 \text{ for model } 2 \end{cases} \tag{11}$$

The value of K_n depends on random parameters (c_i) which can be correlated by means of mean γ_{b_i} and standard deviation σ_{b_i}. The mean values of K_I or K_2 can be obtained by solving Taylor series.

$$\begin{aligned} K =& K_1(\gamma_{c_1}, \gamma_{c_2}, \ldots, \gamma_{b_n}) + \sum_{i=1}^{n}(x_i - \gamma_{c_i})\frac{\partial K_I}{\partial a_i} \\ &+ \frac{1}{2}\sum_{i=1}^{n}\sum_{j=1}^{n}(x_i - \gamma_{c_i})(x_j - \gamma_{cj})\frac{\partial^2 K_1}{\partial c_i c_j} \end{aligned} \tag{12}$$

The mean of K_I and K_{II} is represented by $E\ (K')$

$$E(K_1) \approx K_1(\gamma_{c_1}, \gamma_{ac}, \ldots, \gamma_{c_i}) \text{ and } E(K_2') \approx K_2(\gamma_{c_1}, \gamma_{c_2}, \ldots, \gamma_{c_i}) \tag{13}$$

Similarly, variance of K_I and K_2can be represented as:

$$\mathrm{Var}(K_1) = \sum_{i=1}^{n}\sum_{j=1}^{n} \frac{\partial K_1}{\partial c_i}\frac{\partial K_1}{\partial c_j}\,\mathrm{cov}\big(c_i, c_j\big) \tag{14}$$

$$\text{and } \mathrm{Var}(K_2) = \sum_{i=1}^{n}\sum_{j=1}^{n} \frac{\partial K_2}{\partial c_i}\frac{\partial K_2}{\partial c_j}\,\mathrm{cov}\big(c_i, c_j\big) \tag{15}$$

4 Result and Discussion

The mean and COV of MMSIF of cracked FGMs' plate under tensile loading are evaluated by MATLAB environment. The XFEM approach and SOPT are utilized to evaluate the normalized MMSIF. The effect of different fracture parameters with random system parameters is examined.

The system variables (c_i) for edge crack FGM plate are taken as c_1 (for E_1), c_2 (for E_2), c_3 (for ν_{12}), c_4 (for a), c_5 (for α), where E_1, E_2, a, and α are Young's moduli at ends of left and right side, crack length, and crack angle, respectively.

The Young's modulus follows the exponential function written as Man et al. [2]:

$$E(x) = E_1 e^{\beta x}, \qquad 0 \le x \le W \tag{16}$$

$$\beta = \ln(E_2/E_1) \tag{17}$$

The following normalized mean MMSIF can be represented as:

$$K_{\mathrm{I}} = K_1/P\sqrt{\pi\, a},\; K_{\mathrm{II}} = K_2/P\sqrt{\pi a} \tag{18}$$

Figure 2 shows the comparative study for K_1 of edge-cracked FGM plate under mechanical (tensile) loading. The result obtained from present model is near to the result from reference. The dimension and the material properties of edge cracked FGM are $L = 8$, $W = 1$, $E_1 = 1.0$, $E_2 = E(W)$ for this study.

In Table 1, effect of individual random variables with respect to modulus ratio is shown. In this present work, it is clearly observed that the crack length and crack angle are more critical parameters.

The geometry and the material properties of edge-cracked FGM are $L = 8$, $W = 1$, $E_1 = 1.0$, $E_2 = E(W)$ and $E_2/E_1 = (0.1\text{–}10)$, $G_{12} = 0.5E/(1 + \nu)$, $\nu = 0.3$. Figure 3a, b shows the geometry of cracked FGM plate under mechanical loadings and the finite model with enrichment, respectively.

Figure 4 shows the variation of normalized mean and COV $\{c_i$, $(i = 4$ and $5)$ $= 0.10\}$ with respect to crack length at $\alpha = 45^{\circ}$ for different modulus ratios. It is

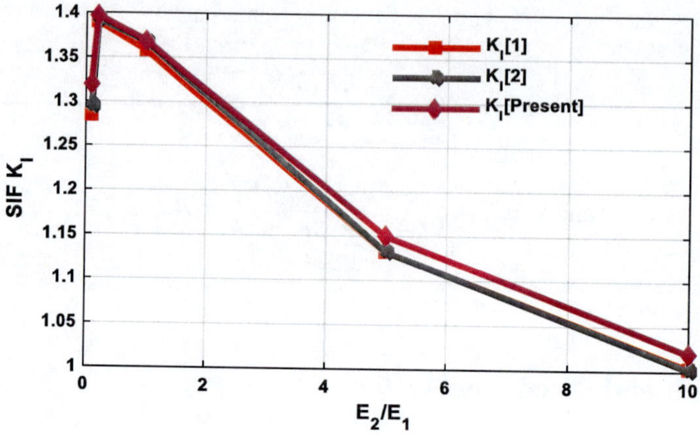

Fig. 2 Variation of K_I with respect to E_2/E_1

Table 1 Effect of random variables with modulus ratios on the normalized mean and

Random variables	SIF	$E_2/E_1 = 0.10$		$E_2/E_1 = 0.50$		$E_2/E_1 = 1$	
		Mean	COV	Mean	COV	Mean	COV
E_1	K_I	3.17	0.001	3.08	0.002	2.99	0.004
	K_{II}	0.39	0.002	0.37	0.004	0.36	0.005
E_2	K_I	3.17	0.002	3.08	0.005	2.99	0.008
	K_{II}	0.39	0.003	0.37	0.007	0.36	0.011
υ_{12}	K_I	3.17	0.001	3.08	0.003	2.99	0.003
	K_{II}	0.39	0.000	0.37	0.00	0.36	0.001
a	K_I	3.34	0.181	3.24	0.174	3.16	0.169
	K_{II}	0.41	0.183	0.39	0.169	0.38	0.158
α	K_I	3.17	0.004	3.08	0.003	2.99	0.003
	K_{II}	0.39	0.097	0.37	0.097	0.36	0.097

noticed that as crack length increases, mean and COV of MMSIF also increase for same modulus ratio. Crack length is the major critical parameters for the fracture analysis of the materials. Crack length with small size is necessary to increase the reliability of cracked structures. It is also observed that crack length is more sensitive for lower modulus ratio of mean and COV. Hence, higher modulus ratio is necessary for safety of FGM plate with crack.

Figure 5a, b shows the variation of normalized mean and COV ($c_i = \{i = 4$ and $5\} = 0.1$) of MMSIF of FGMs' plate for different modulus ratios, with $a = 0.45$. It is observed that as crack angle increases, the mean of K_I decreases, and for K_{II}, it increases for same modulus ratio. The randomness in first-mode SIF shows very less effect. Apart from crack length, crack angle is also the major factors in fracture analysis.

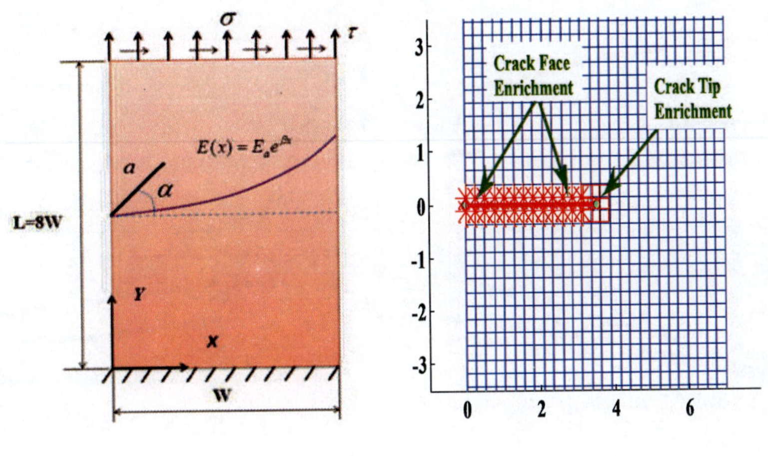

(a) (b)

Fig. 3 **a** Dimension and loading of cracked FGM plate, **b** crack enrichments

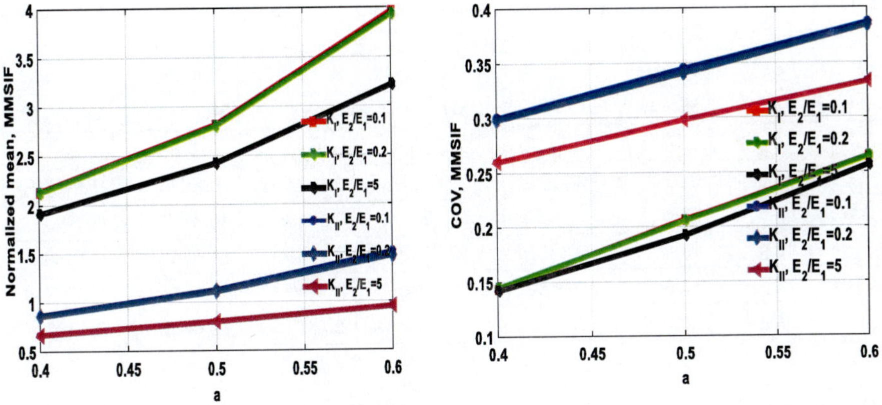

(a) (b)

Fig. 4 Variation of MMSIF with respect to crack length for different modulus ratios

Tables 2 and 3 represent the variation of MMSIF, mean and COV ($c_i = \{i = 4 \text{ and } 5\} = 0.1$) for different crack lengths and crack angles, respectively. In this study, randomness is considered in crack length and crack angle. In Table 2, crack angle = 45° and $E_2/E_1 = 0.4$, and in Table 3, crack length = 0.45 and $E_2/E_1 = 0.45$ are considered. It is observed that the maximum mean MMSIF is maximum for combined loading. The mean (K_{I} and K_{II}) and COV increase as a/W increases for tensile loading, whereas as crack angle changes from 0° to c, the mean of K_{I} and K_{II}

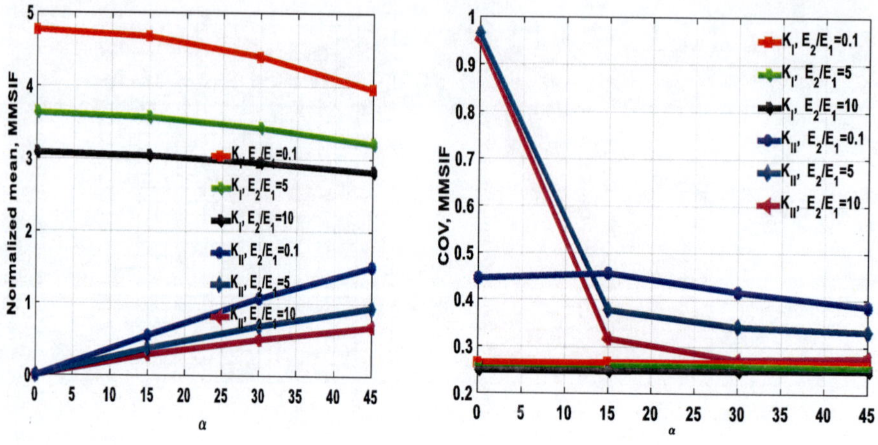

(a) (b)

Fig. 5 Variation of MMSIF with respect to crack angle (α) for different modulus ratios

Table 2 Effect of a/W on the normalized mean and COV ($c_i = \{i = 4 \text{ and } 5\} = 0.1$) of K_{I} and K_{II} angle (a) = 45. $E_2/E_1 = 0.4$

a/W	SIF	Tensile		Shear		Combined	
		Mean	COV	Mean	COV	Mean	COV
0.4	K_{I}	2.1075	0.1452	32.1395	0.0874	34.2465	0.0881
	K_{II}	0.8350	0.2950	9.0422	0.1936	9.8774	0.1911
0.5	K_{I}	2.7613	0.2036	37.5761	0.1012	40.3370	0.0993
	K_{II}	1.0691	0.3335	10.0152	0.1949	11.0844	0.1889
0.6	K_{I}	3.8567	0.2640	46.8090	0.0849	50.6653	0.0797
	K_{II}	1.4014	0.3802	11.5467	0.1817	12.9482	0.1701

decreases and increases, respectively. The COV decreases as crack angle increases for tensile loading. The COV shows random nature for shear and combined loading.

5 Conclusions

In this study, SOPT is utilized for MMSIF of the edge crack FGMs' plate subjected to mechanical loadings. Here, XFEM-based numerical analysis is carried out. Following are the observations of the present study.

- All random parameters as mentioned in Table 2 are analyzed for the fracture analysis of cracked FGM plate under uniform tensile loading, and it is observed

Table 3 Effect of crack angle on the normalized mean and COV ($c_i = \{i = 4 \text{ and } 5\} = 0.1$) of K_I and K_{II} angle (a) = 0.45. $E_2/E_1 = 0.45$

α	SIF	Tensile		Shear		Combined	
		Mean	COV	Mean	COV	Mean	COV
0	K_I	2.9340	0.2257	39.0510	0.1044	41.9850	0.1018
	K_{II}	0.0003	0.7474	1.7017	0.1625	1.7017	0.1626
15	K_I	2.8766	0.2216	38.9442	0.1042	41.8206	0.1017
	K_{II}	0.3565	0.3643	2.3189	0.3151	2.6757	0.3080
25	K_I	2.7741	0.2140	38.1777	0.1040	40.9514	0.1019
	K_{II}	0.5795	0.3478	4.9271	0.2471	5.5069	0.2429
45	K_I	2.4198	0.1817	34.8815	0.1003	37.3008	0.0995
	K_{II}	0.9477	0.3110	9.5604	0.1941	10.5081	0.1902

that crack length and crack angle are more critical parameters for the evaluation of MMSIF.

- As randomness in random parameter increases, the COV in MMSIF also increases.
- The effect of randomness in second-mode SIF is more severe. Hence, proper control of second-mode SIF is highly needed particularly for higher crack angles.

References

1. Kim JH, Paulino GH (2002) Finite element evaluation of mixed mode stress intensity factors in functionally graded materials. Int J Numer Methods Eng 53:1903–1935
2. Man S, Huaping W, Long L, Guozhong C (2014) Calculation of stress intensity factors for functionally graded materials by using the weight functions derived by the virtual crack extension technique. Int J Mech Mater Des 10:65–77
3. Bayesteh H, Mohammadi S (2013) XFEM Fracture analysis of orthotropic functionally graded materials. J Compos Part B 44(1):8–25
4. Belytschko T, Black T (1999) Elastic crack growth in finite elements with minimal remeshing. Int J Num Meth Eng 45(5):601–620
5. Reddy RM, Rao BN (2008) Stochastic fracture mechanics by fractal finite element method. Comput Methods Appl Mech Eng 198:459–474
6. Lal A, Mulani B, Kapania RK (2020) Stochastic critical stress intensity factor response of single edge notched laminated composite plate using displacement correlation method. Mech Adv Mat Struct 27(14):1223–1237
7. Lal A, Palekar SP, Mulani B, Kapania RK (2017) Stochasticextended finite element implementation for fracture analysis of laminated composite plate with a central crack. Aerosp Sci Technol 60:131–151
8. Khatri K, Lal A (2017) Stochastic XFEM fracture and crack propagation behaviour of an isotropic plate with hole emanating radial cracks subjected to various in-plane loadings. Mech Adv Mater Struct 25(9):732–755
9. Khatri K, Lal A (2018) Stochastic XFEM based fracture behaviour and crack growth analysis of a plate with a hole emanating cracks under biaxial loading. Theor Appl Fract Mech 96:1–22
10. Wang LF, Zhou XP (2021) Fracture analysis of functionally graded materials by the field-enriched finite element method. Eng Fract Mech 253:107875

Developments in Hybrid Abrasive Flow Machining: A Review on Models and Analyses

S. Mehta, P. Gauba, S. Kaushal, P. Ali, M. Dhanda, and R. S. Walia

Abstract Abrasive flow machining (AFM) has been imperative for the finishing of parts with varied materials and geometries. Hence, many researchers have tried to enhance aspects like surface finish and material removal rate to improve the efficiency and efficacy of the conventional process. This paper focuses on the reviews of all hybridizations conducted, their mathematical models, results of simulations, and varied experimental conditions to practically come up with a much enhanced and cost-effective machining process. It also discusses the comparison of these models with the actual experimental results and also talks about the authenticity of all these models. In the end, research gaps and areas for future scope have been found out from various research papers. It was concluded that processes like MAAFM, CFAAFM and UAAFM have immense practical machining advantages, over other processes, and modeling of TACAFM has been the most efficient till now.

Keywords Hybrid abrasive flow machining · Material removal rate · Geometry optimization · Surface finishing

1 Introduction

Surface finishing plays a major role in obtaining the necessary functional performance of the components and products. Considering the current market scenario in the manufacturing industries, there is a huge requirement for processes that can produce components with high levels of surface accuracy. One of the most suited unconventional processes to perform nano-finishing of holes and complex cavities is abrasive flow machining (AFM). This process includes passing a non-Newtonian media through the paths or cavities which are restrictive. The media is passed forcibly

S. Mehta · P. Gauba · S. Kaushal · M. Dhanda · R. S. Walia (✉)
Production and Industrial Engineering Department, Punjab Engineering College (Deemed to Be University), Sector 12, Chandigarh 160012, India
e-mail: waliaravinder@yahoo.com

P. Ali
Department of Mechanical Engineering, Delhi Technological University, Shahbad Daulatpur, Main Bawana Road, Delhi, NCR 110042, India

A. Maurya et al. (eds.), *Recent Trends in Mechanical Engineering*, Lecture Notes in Mechanical Engineering, https://doi.org/10.1007/978-981-19-7709-1_38

through the complex cavities by applying high pressure, thus making that the cutting edges of the abrasive media abrade the workpiece surface and provide an excellent surface finishing. The process involves a wide range of applications which includes finishing molds, dies, automobile parts, and medical equipment.

Although there is a lot of work happened and happening in the area of hybridization of AFM, still there are some gray areas and gaps which need to be focused upon. This paper reviews the work in this direction and mainly focuses upon recent developments in hybrid AFM. Section 2 briefly explains the recent developments and research work carried out in hybrid AFM processes. It mainly focuses on different developed mathematical models and hybridizations in CFAAFM, followed by the conclusion in Sect. 3.

2 Recent Developments in Hybrid AFM

Many researchers have worked in the direction to improve MRR and surface finish of the AFM process. Loveless et al. [1] studied the effects of AFM on machined surfaces and concluded that the process affects both surface finish and MR rate. Jain and Adsul [2] further conducted experiments with various process parameters of abrasive media on aluminum and brass and found that the concentration of abrasives, mesh size, number of cycles, and speed of media flow are the dominant factors for the AFM process. After investigating different process parameters, researchers began to combine the AFM process with other processes, in an attempt to improve efficiency. Jha and Jain [3] combined the AFM process with magnetorheological polishing (MRP) under the influence of the magnetic field, confirming the rheological behavior of MRP fluids, and showed an improvement in surface roughness, with an increase in the magnetic field. Dabrowski [4] used polymeric electrolytes and found that the surface roughness was affected by the type of abrasive paste being applied. They further created various polymer pastes (a schematic shown in Fig. 1) with varying ion conductivities and discovered that the surface roughness is mostly controlled by the type of abrasive paste used.

As shown in Fig. 1, electrochemical assistance is provided via the workpiece and conductivity of abrasive paste, and this significantly reduced the number of passes of machining.

Moving onto more niche hybridizations, Walia et al. [5] introduced centrifugal force to abrasive media by inserting a centrifugal rod in the media flow path. This led to a reduction of machining time by 70–80% as compared to the AFM process. Walia et al. [6] further studied process parameters in the CFAAFM process and concluded that the speed of the centrifugal rod had the highest contribution (39.70%). After this point, researchers began with the development of mathematical models and the performance of simulations, before getting onto the actual experimental conditions. Walia et al. [7] performed FEA of the non-Newtonian fluid used in CFAAFM and successfully predicted values of resultant pressure, velocity, and radial stress during the flow of media. Walia et al. [8] further applied the utility theory and Taguchi's

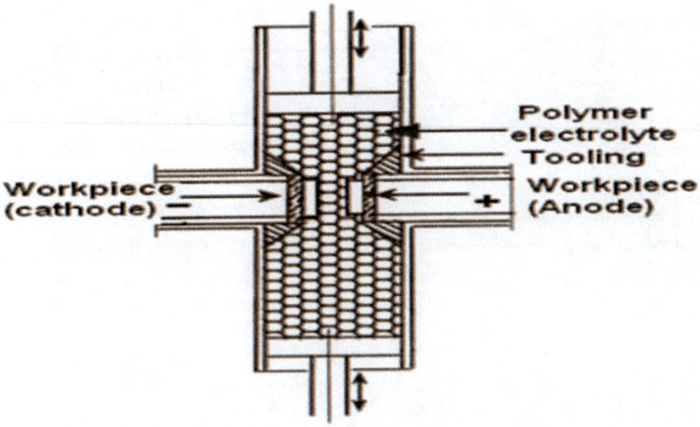

Fig. 1 Electrochemical abrasive flow machining (ECAFM) developed by Dabrowski et al. [21]

quality loss function and concluded that all input parameters significantly enhanced the utility function. Vaishya et al. [9] carried forward the hybridization and enhanced the MR rate by combining electrochemical machining with CFAAFM. Chahal et al. [10] further conducted experiments on electrochemical-assisted AFM (ECAAFM) and found that voltage contributes the most to material removal (MR). Singh and Shan [11] diversified by applying a magnetic field around the workpiece during AFM and developed a relationship between MRR and percentage improvement in surface roughness. Brar et al. [12] employed a standard helical drill bit which increased the MRR by a factor of 2.5. Venkatesh et al. [13] employed an ultrasonically assisted AFM to finish bevel gears and found a significant improvement in MR and surface roughness.

Ali et al. [14], in an attempt to improve the existent process, developed a mathematical model for the EDM-assisted CFAAFM process (TACAFM), which showed significant improvement in MR and surface finish. Ali et al. [15] developed a model for TACAFM, which showed double the MR as compared to conventional AFM while considering the Coriolis effect. Sankar et al. [16] improved the performance of abrasive flow finishing (AFF) by providing rotary motion to the workpiece and found that the rotational speed of the workpiece had a significant impact on the surface roughness.

Sankar et al. [17] hybridized the abrasive flow finishing (AFF) process by introducing a helical drill (schematically shown in Fig. 2). The figure shows the drill bit in the finishing zone to reshuffle the abrasive particles by randomizing their motion. Modeling was carried out for this process using nonlinear multivariable analysis and artificial neural networks. The experimental results were found to agree with the computational results.

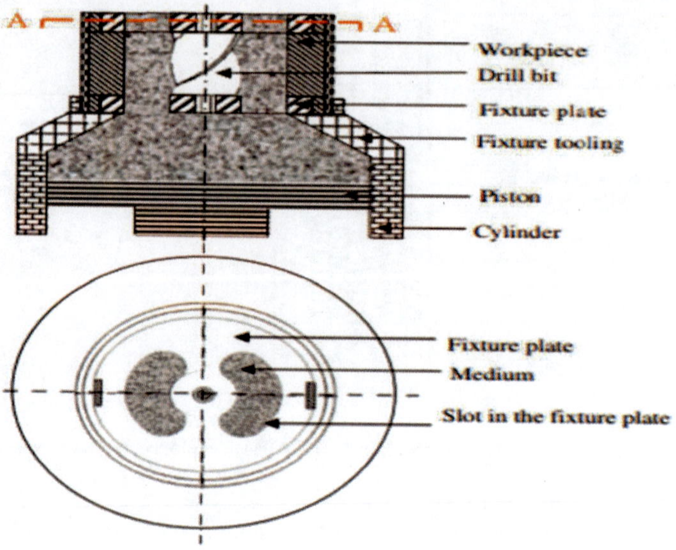

Fig. 2 DBG–AFF splitting media in the drill bit region with a double-slotted fixture plate [22]

2.1 Mathematical Models and Experimental Analysis of Hybrid AFM Processes

Figure 3 shows the flow chart of hybridization of the AFM process. Researchers have come up with various models to simulate hybridizations of AFM to predict MR (material removal) and identify different process parameters.

Different mathematical models of hybrid AFM processes with their results outcome and research gaps are presented in Table 1. Researchers who are exploring this field will help by their research outcome and gaps.

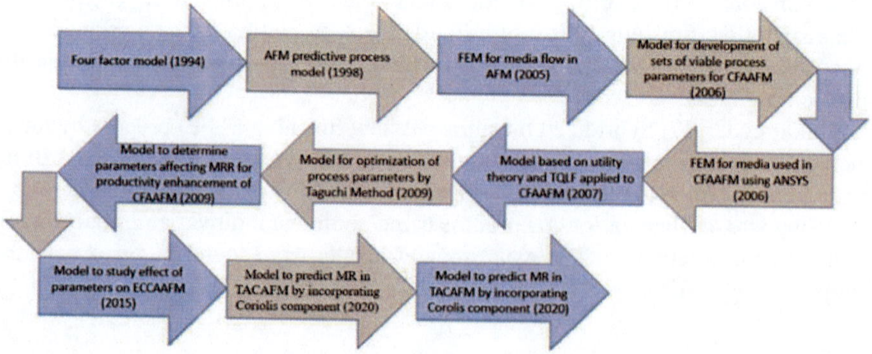

Fig. 3 Flow chart of hybridization of AFM process

Table 1 Different mathematical models and experimental analysis of hybrid AFM processes

Proposed model/topic	Authors and year	Results/outcomes	Research gaps
Four-factor model	Loveless et al. [1]	DDS modeling green's function plots indicated the absence of signs of machining	Judgment for AFM performance was limited to MR and percentage improvement in SR
AFM predictive process model	Kimberly et al. [20]	Predicted results were within an error of 5 Ra for polishing 0.001' for surface removal applications	The model fails to provide avenues for applications: prediction of products with varying shapes, technical objectives, etc.
Finite element model for the flow of media in AFM	Dabrowski et al. [4]	Actual MR is less than theoretical MR	Assumptions regarding the shape of the workpiece, viscous flow, isotropy, etc. restricted analysis
Model for the development of viable process parameters for CFAAFM	Walia et al. [5]	Average MR and SSR were well within the confidence interval of predicted optima	Variations in media flow rate, flow volume, viscosity, and temperature were not considered
Finite element model for media used in CFAAFM using ANSYS	Walia et al. [7]	Resultant pressure distribution and velocity distribution patterns were as expected	The 3D shape of abrasive particles and changes in machining conditions were not taken into account
Model for optimization of process parameters by Taguchi method	Walia et al. [6]	Average change in $\triangle$Ra was 28.39%, in adherence to CI of predicted $\triangle$Ra	Variations in polymer/gel ratio, abrasive concentration, workpiece material, etc., were not considered
Model-based on utility theory and TQLF applied to CFAAFM	Walia et al. [8]	Average of values falls within 95% CI of range of response characteristic	All quality characteristics were assumed to be equally important
Model to determine parameters affecting MRR for productivity enhancement of CFAAFM	Singh et al. [11]	S/N ratios for the number of cycles, shape, and rotational speed of CFG rod are within CI of optimal range data	The study was restricted to brass, and the L/D ratio was decided on basis of a recommendation from past literature

(continued)

Table 1 (continued)

Proposed model/topic	Authors and year	Results/outcomes	Research gaps
Developments in AFM: A review	Manjot et al. [12]	MAAFM works better on nonferrous materials, CFAAFM provides higher MRR in cylindrical components, UAAFM works better for complex geometries	Focus only on MRR and surface finish. No information on the action of abrasive particles on workpiece and tool
Model to study the effect of parameters on EC^2A^2FM	Vaishya et al. [9]	Average MR = 29.84 mg and results ranged from 28.63 to 31.69 mg	Variation of the number of cycles, media flow volume, etc., were not considered
Mathematical model for MR at a given thermal energy pulse in TACAFM	Ali et al. [14]	About 11.95% deviation in terms of accuracy of the model from experimental results	Assumptions regarding penetration depth, angle of attack, etc., the restrict analysis
Model to predict MR in TACAFM by Coriolis component	Ali et al. [15]	About 18.78% deviation in MR from experimental results	Neglection of plastic deformation of abrasive particles
Critical review of past research and advances in abrasive flow finishing process	Petare and Jain [23]	Design of fixtures and type of medium drive the efficiency of AFF process	Results of simulations were not taken into account
Research trends in abrasive flow machining: A systematic review	Dixit et al. [24]	Processed responses in AFM depend on interdependent parameters	Comparison of experimental analysis and simulations was not taken into account

2.2 Models on AFM and Its Hybrid Forms

Many researchers have worked on analyzing the results of AFM, and different hybrid processes have been established to enhance the efficiency of these AFM processes.

Loveless et al. [1] studied machining characteristics of AFM on surfaces produced by different machining processes. Analysis of variance was used to determine the effects of independent variables on metal removal. Dabrowski et al. [4] was the first to develop a FEM for the flow of abrasive media in AFM and replaced the conventional abrasive paste with a polymeric electrolyte. The results obtained from tests were compared with the results of basic research, and it was found that flow speed has a very low impact on material removal even after 100 cycles of media flow. Singh and Shan [11] diversified by applying a magnetic field around the workpiece, providing magnetic force to abrasive particles. Material removal was maximum at 0.4 T of the magnetic field, MR, post which the MR reduces sharply.

Brar et al. [12] applied Taguchi's quality engineering approach to the HLX–AFM process and concentrated on optimization of various process parameters and developed a robust machining process with significantly enhanced MR. The results show that the presence of a stationary drill bit increases the material removal by a factor of 2.5. S/N ratio analysis of the Taguchi method showed that the number of cycles and presence of stationary drill bit have a significant impact on the response parameters.

The UAAFM process is vividly different from all the processes discussed till now, and it consists of a hydraulic pressure setup that bombards abrasive media onto the workpiece via tooling. Venkatesh et al. [13] presented a finite element simulation of media during the finishing of bevel gears using UAAFM. A 3D model was constructed to simulate the flow of media through the outer wall of the tooth surface using the computational fluid dynamics (CFD) approach. The finishing of bevel gears was found to be more effective than the conventional AFM process in terms of finishing time and improvement in surface roughness. Rapid improvement in the quality of the surface was observed, though MRR remained unchanged.

Sharma [18] further diversified by proposing a method that combined the magnetorheological finishing with the rotational abrasive flow finishing process. A mathematical model was developed for the calculation of surface roughness and volumetric MR (VMR), and analysis was conducted using Taguchi method. Multiple regression analysis was done to find the correlation of other parameters with VMR and reduction in SR. The proposed model was found to be significant.

Walia et al. [5] developed a system to determine a set of viable process parameters for CFAAFM. Average MR and scatter of surface roughness were found to be 41.39 mg and 0.85 μm, which were well within the confidence interval of predicted optima of MR and SSR. Walia et al. [7] came up with a finite element model of media used in the CFAAFM process to evaluate resultant pressure, velocity, and radial stresses during a working cycle and performed all FEM analyses using ANSYS. The resultant pressure and velocity distribution patterns were as expected. Velocity was maximum near CFG rod, and it further decreased on moving normal to the axis of the workpiece.

Walia et al. [6] further performed optimization of finishing processes in CFAAFM by using the Taguchi orthogonal array. The average change in surface roughness (ΔRa) was in adherence to the confidence interval of predicted ΔR_a. The factors in decreasing order of significance were CFG rod, extrusion pressure, and grit size. Walia et al. [8] performed further optimization of process parameters of CFAAFM by applying utility theory and Taguchi quality loss function. Overall average of observed values fell within 95% CI of the optimal range of respective response characteristics.

Singh and Walia [19] lately presented the effects of 'Shape of CFG rod', 'Rotational Speed of the rod', and 'No. of Cycles' on MR using Taguchi L_9 OA technique. The results indicated that the CFG rod has the highest contribution toward response characteristics, followed by the shape of the CFG rod and number of cycles.

2.3 Hybridizations in CFAAFM

Many researchers have worked on hybridizing the niche process of CFAAFM with other unconventional machining processes. Vaishya et al. [9] developed EC^2A^2FM to enhance the efficiency and quality of the finished surface. The average observed MR was 29.84 mg, whereas predictions ranged from 28.63 to 31.69 mg. The optimum operating pressure was observed to be 6 N/mm^2, enhancing the MRR. Ali et al. in 2020 [14] further developed a model for the calculation of the amount of force required during machining and MR at a given thermal energy pulse in TACAFM. The predicted model showed a deviation of 11.95% from the results. The optimal value of residual stress was obtained at 4 Amp of current, 194 rpm of the electrode, 17 MPa of the extrusion pressure, and 0.3 fractions of abrasive concentration. Ali et al. recently in 2020 [15] developed a model to predict MR and indentation in TACAFM by incorporating the effect of the Coriolis component. The results show 44.34% and 39.74% improvement in MR and improvement in surface finish, respectively. The predictions show an 18.78% average deviation from the results. The optimum values of material removal and percentage improvement in the surface finish were found to be 36.57 mg and 42.39%, respectively.

Different mathematical models of hybrid AFM processes with their results outcome and research gaps are presented in Table 1.

This paper brings out the different directions/gaps to the interested researchers who are working in the areas of abrasive flow machining (AFM). Although different mathematical models are developed with their experimental validation, still some issues need to be resolved. In particular, enhancement of existing mathematical models is also required.

3 Conclusion

This review concludes that abrasive flow machining is an efficient and effective machining process to achieve nano-finishing of parts with complex geometries. Significant development has occurred in recent years with the introduction of the CFG rod, giving rise to more than 50% of the total hybridizations of AFM. MAAFM, CFAAFM, and UAAFM processes have significant machining advantages over others and hence are more suited for use in the industry. Different mathematical models have also been proposed by various researchers; yet, TACAFM has by far been the most efficient, having a relatively low deviation compared to other models. There is scope for certain other hybridizations in future, as well as for the enhancement of existing mathematical models and their experimental conditions.

References

1. Loveless TR, Williams RE, Rajurkar KP (1994) A study of the effects of abrasive-flow finishing on various machined surfaces. J Mater Process Technol 47(1–2):133–151
2. Jain VK, Adsul SG (2002) Experimental investigations into abrasive flow machining (AFM). Int J Mach Tools Manuf 40(7):1003–1021 (Development of magneto abrasive flow machining process)
3. Jha S, Jain VK (2004) Design and development of the magnetorheological abrasive flow finishing (MRAFF) process. Int J Mach Tools Manuf 44(10):1019–1029
4. Dabrowski L, Marciniak M, Szewczyk T (2006) Analysis of abrasive flow machining with an electrochemical process aid. Proc Inst Mech Eng Part B: J Eng Manuf 220(3):397–403
5. Walia RS, Shan HS, Kumar P (2006) Parametric optimization of centrifugal force-assisted abrasive flow machining (CFAAFM) by the Taguchi method. Mater Manuf Process 21(4):375–382
6. Walia RS, Shan HS, Kumar P (2009) Optimisation of finishing conditions in Centrifugal force assisted abrasive flow machining using Taguchi method. Int J Manuf Technol Manag 18(1):79–93
7. Walia RS, Shan HS, Kumar PK (2006) Finite element analysis of media used in the centrifugal force assisted abrasive flow machining process. Proc Inst Mech Eng Part B: J Eng Manuf 220(11):1775–1785
8. Walia RS, Shan HS, Kumar P (2006) Multi-response optimization of CFAAFM process through Taguchi method and utility concept. Mater Manuf Processes 21(8):907–914
9. Vaishya R, Walia RS, Kalra P (2015) Design and development of hybrid electrochemical and centrifugal force assisted abrasive flow machining. Mater Today: Proc 2(4–5):3327–3341
10. Chahal B, Gupta R, Vaishya RO (2015) Experimental investigation to optimize process parameters in electrochemical assisted abrasive flow finishing of Al-6061 alloy using Taguchi method. i-Manager's J Mater Sci 3(1):14
11. Singh S, Shan HS (2002) Development of magneto abrasive flow machining process. Int J Mach Tools Manuf 42(8):953–959
12. Brar BS, Walia RS, Singh VP, Sharma M. A robust helical abrasive flow machining (HLX-AFM) process. J Inst Eng (India): Ser C 94(1):21–29
13. Venkatesh G, Sharma AK, Kumar P (2015) On ultrasonic assisted abrasive flow finishing of bevel gears. Int J Mach Tools Manuf 1(89):29–38
14. Ali P, Walia RS, Murtaza Q, Ranganath MS (2020) Modeling and analysis of developed thermal additive centrifugal abrasive flow machining process. Surf Topogr Metrol Prop 8(3):035013
15. Ali P, Walia RS, Murtaza Q, Singari RM (2020) MR analysis of hybrid EDM-assisted centrifugal abrasive flow machining process for performance enhancement. J Braz Soc Mech Sci Eng 42:1–28
16. Sankar MR, Jain VK, Ramkumar J (2009) Experimental investigations into rotating workpiece abrasive flow finishing. Wear 267(1–4):43–51
17. Sankar MR, Mondal S, Ramkumar J, Jain VK (2009) Experimental investigations and modeling of drill bit-guided abrasive flow finishing (DBG-AFF) process. Int J Adv Manuf Technol 42(7–8):678–688
18. Sharma VK (2021) Modeling and analysis of a novel rotational magnetorheological abrasive flow finishing process. Int J Lightweight Mater Manuf
19. Singh R, Walia RS (2012) Selection of control parameters for hybrid magnetic force assistant abrasive flow machining process. In: International conference on advances in manufacturing technology, vol 15, p 17th
20. Petri KL, Billo RE, Bidanda B (1998) A neural network process model for abrasive flow machining operations. J Manuf Syst 17(1):52–64
21. Cheema MS, Venkatesh G, Dvivedi A, Sharma AK (2012) Developments in abrasive flow machining: a review on experimental investigations using abrasive flow machining variants and media. Proc Inst Mech Eng Part B: J Eng Manuf 226(12):1951–1962

22. Rajesha S, Some studies to enhance the capabilities of abrasive flow machining process (Doctoral dissertation, Ph.D. dissertation, IIT Roorkee)
23. Petare AC, Jain NK (2018) A critical review of past research and advances in abrasive flow finishing process. Int J Adv Manuf Technol 97(1):741–782
24. Dixit N, Sharma V, Kumar P (2021) Research trends in abrasive flow machining: a systematic review. J Manuf Process 1(64):1434–1461

Multichannel Sustainable Supply Chain Network Design: Review and Research Directions

T. Niranjan, Sonu Rajak, and P. Parthiban

Abstract A powerful, effective, and strong supply chain network system is a feasible upper hand for nations and firms and helps them to adapt to expanding natural turbulences and more exceptional aggressive weights. The supply chain network modeling is an important research area in the current scenario, and a lot of work is being done to optimize the network design to maximize profit and minimize the costs as well as dealing with uncertainties in the market. Recently, the supply chain network design is also being studied under the spectrum of environmental impacts and sustainability. Today, a mix of a physical channel and online channel serves customer needs more adequately than utilizing a single channel. This multichannel supply chain network provides choice, flexibility, and better responsiveness for firms. This review covered the available literature focusing on all these various aspects under one umbrella and addressed few research directions in the multichannel perspective to incorporate the changes in the supply chain models.

Keywords Multiple channels · Supply chain network · Sustainable · Optimize

1 Introduction

Supply chain network modeling and design are responsible for managing the distribution centers, suppliers, manufacturing plants, retail stores, and related entities to fulfill the market demand along with customer satisfaction as well as earning profit.

T. Niranjan (✉)
Mahatma Gandhi Institute of Technology, Kokapet, Gandipet, Hyderabad, Telangana, India
e-mail: tniranjan_mct@mgit.ac.in

S. Rajak
Department of Mechanical Engineering, National Institute of Technology Patna, Patna, Bihar, India
e-mail: sonu.me@nitp.ac.in

P. Parthiban
Department of Production Engineering, National Institute of Technology, Tiruchirappalli, Tamil Nadu, India
e-mail: parthiban@nitt.edu

A. Maurya et al. (eds.), *Recent Trends in Mechanical Engineering*, Lecture Notes in Mechanical Engineering, https://doi.org/10.1007/978-981-19-7709-1_39

Because of the large number of entities involved, it is a very critical task and usually focuses on one particular area which is the primary aim of the industry/organization, and other entities are optimized accordingly. The traditional methodology is to focus on minimizing the costs and increasing the profits.

Closed-loop supply chain comprises both the forward production network and turnaround inventory network with the significance of monetary advancement without influencing nature. A typical forward (dynamic) store network has a system of providers, makers, wholesalers and to convey an item or thing to clients. The turnaround store network includes the development of utilized/unsold items from the client to the upstream inventory network, from gathering utilized items, observing the procedure, reusing, reprocessing, revamping, and arranging off. It has been discovered that coordination of forward inventory network and turnaround store network can bring down the general expenses.

With the increasing competition due to industrial growths and globalization, the industries are changing their approach to supply chain network design from profit oriented to market oriented. This includes focusing on the uncertainties in the market, maximizing customer satisfaction, adapting to customer demands, increasing the market reach, minimizing the delivery times, etc. Also, now, the companies are taking up the social causes too with green supply chain design and supply chain focused on sustainability and environment to enhance their brand value as well as comply with the government norms.

The Internet channel gives more choices on the range of items to customers in the convenience of their living room reducing their search cost. Social media had been predominantly used for entertainment, but today, it is increasingly connecting with business than ever before. Therefore, multichannel supply is an essential approach to consumer firms today as it offers a single unified shopping experience.

This review goes through the various literatures discussing the different goals of various supply chain network designs and the novel optimization methods involved. It further looks into the network designs trying to tackle the uncertainties and the probabilistic activities. Further, we delve into the recent research on closed-loop green supply chain modeling and multichannel network designs considering sustainability.

2 Literature Review

Supply chain network modeling has been an important area of research considering its critical importance in the industrial world. A well-designed supply chain models can work same as the spine of the industry by satisfying the customers by its fast responsive nature, by simplifying and smoothing the flow of physical entities as well as information, and by minimizing the costs involved. We went through some recent research work done in the area of modeling and design of supply chain network in the past few years and tried to cover the different scenarios being focused by the researchers. We have categorized the research work under four major goals for the network design: (1) design to optimize and design against uncertainty, (2) green and sustainable design supply chain network, (3) closed-loop supply chain network, and (4) multichannel supply chain network design.

2.1 Design to Optimize and Against Uncertainty

Building up a diagnostic model emerging in the plan of supply chain network wherein numerous providers and distribution centers (DC) managing different items are attempting to satisfy the individual client requests has been discussed [1]. A two-level scientific model for provider determination in building up an inventory network has been exhibited in his paper. A mixed integer programming method has been proposed to manage multi-destinations and empower the leaders for assessing a more noteworthy number of option arrangements [2]. A contextual analysis of a plastic creating Turkish organization wishes to outline supply chain network (SCN) for their principle crude material PVC which incorporates choosing the providers by considering three goals such as minimization inbound and outbound dissemination costs, expansion of client administrations (scope), and boost of limit usage for DCs. The correlation demonstrated that genetic algorithm (GA) beat multi-objective simulated annealing (SA) as per normal number of pareto-ideal arrangements as well as nature of pareto-ideal arrangements. A novel model with all the while streamlines area, designation, limit, stock, and directing choices in a stochastic supply chain framework has been considered [3]. Every client's request is questionable and takes after a typical conveyance, and every circulation focus keeps up a specific measure of security stock. A heuristic technique construct is built up in light of combination of SA with Tabu search in their work. In another paper, an oligopolistic market harmony issues in the planning of firm's supply chain network have been demonstrated [4]. An imaginative encoding–decoding methodology installed inside a hereditary calculation (GA) to limit the aggregate strategic cost has been presented [5]. A novel arrangement of Lagrange multipliers methodology has been streamlined and displayed for the portrayal of monetary exercises related with supply chain networks, specifically, fabricating, circulation, and also stockpiling, which were used to plan both designing supply chain network, and overhaul issues have been proposed [6].

A critical change in the business condition (for example, client requests and transportation costs) has impelled an enthusiasm for outlining versatile and uncertain supply chains. An improvement shows for dealing with the inborn vulnerability of information; for example, client requests, in a shut circle supply chain network design issue, have been proposed and solved by CPLEX optimizer [7]. The unverifiable parameters mulled over are-client requests return amount, limits of various supply chain substances, settled and variable expenses in transportation and punishments has been considered in their linear integer model and analyzed by same CPLEX solver. It is expected that every DC can charge diverse costs to investigate the ability to purchase in various areas [8]. This is pertinent for some global organizations, since clients from various nations have distinctive purchasing forces, and deal costs are ordinarily not the same as nation to nation. Taking the instance of an organization that delivers a solitary item, this item is transported to various DCs after creation. A multi-objective stochastic programming approach has been produced by [9] for supply chain design under instability, and a pareto-optimal solution was obtained by goal attainment method. Primary vulnerability components are requests, supplies, preparing, transportation, deficiency allotment, and limit portion augmenting costs which were considered in their work. An independently planned work to request and build to stock configurations, regarding the determination of providers, parallel sourcing has fetched favorable circumstances under supply chain interruptions, and decomposition-based solution method for solving stochastic problems has been discussed [10] in their work.

2.2 Green and Sustainable Design Supply Chain Network

Carbon exchanging or outflow exchanging enables organizations to purchase emanations credits from the individuals who contaminate less with a specific end goal to remain underneath their discharges farthest point or top witch typically forced by a focal specialist (government or universal body). Data envelopment analysis (DEA) has been used to build a strategic system with assembling transportation, utilization, and end-of-life items which are identified as goals [11] in their paper. Building up a maintainable production network by Lagrange multipliers for organizing configuration display takes into consideration the assessment of ecological multi-criteria base leadership [12]. They mentioned transportation exercises are critical wellsprings of air contamination and ozone harming substance outflows, with the previous known to effect sly affect human well-being and in charge of a dangerous atmospheric deviation. A green supply network design chain has been presented for demonstrating in light of the traditional office area issue for the company's vital arranging [13]. The recognizing highlight of their model is its thought of ecological component which incorporates natural level of office and ecological impact in the taking care of transportation. It is a multi-target model and preferences are obtained by posteriori articulation, which comprises limiting aggregate cost and natural impact. The cost of carbon discharges is proposed where the outflows cost depends on experimental information that displays an inward relationship which has been considered [14]. Lagrangian approach is utilized to tackle the subsequent minimization objectives in their work.

By taking the number of attributes which might be associated with dealing a sustainable supply chain [15], has built up a structure that can be utilized as an integrative, multidimensional supportability instrument to investigate the connections and exchange offs among such qualities. A multi-level multi-objective model has been proposed by [16], showing that joins cost (speculation, operational), natural (discharge amount, squander age), social (business openings, enhanced work conditions, organizing group improvement), and the solution for the model has been attained by goal programming with epsilon constraint method. They likewise talked about in detail the uses of this model for worldwide producers. Triple bottom line optimization modeling has been proposed by [17], for the outlining and arranging of shut circle feasible supply chain under vulnerability. Monetary, ecological, and social components of maintainability are tended to be optimized using GAMS and CPLEX solvers. An examination on supportability in a wine inventory network in Australia was completed by [18] wherein another territory of production network was brought under scrutiny augmenting point of view for professionals and specialists in key area choices and both network models optimized by ε constraint method in CPLEX solver. Government direction and customers green (natural) concerns are the motivating force that fuses the firm's store network to break down it's tasks into system choices through the improvement of a coordinated production network [19]. This examination has been done by Lagrange multipliers method that solution helps in evaluating the monetary execution of an ecological maintainable inventory network in decreasing natural externalities. Usage of eco-productivity idea has been proposed by [20] to choose the lean and green practices that best add to decreasing organization's negative ecological effects and to enhancing financial execution, while not trading off the inventory network's general eco-proficiency. In another paper, goal programming has been used for solving a built up model that incorporates a secluded item plan for encouraging speedier assembling, remanufacturing, revamping, dismantling, and repairing utilizing new subassemblies [21]. It has been exhibited a complete answer for outlining dissemination systems, particularly in sustenance supply chains [22]. The model streamlines the system in three measurements of expenses (especially, in transportations), responsiveness, and ecological protection in their paper, and NSGA II has been proposed for solving the model.

The sustainable supply chain model designs a reasonable module development and a creation strategy for segments and items, and transportation and appropriation courses for acquiring an ideal business execution, and to thinks about the ecological worries for the unsafe emanations and the spent vitality.

2.3 *Closed-Loop Supply Chain*

Reverse logistics in a closed-loop supply chain has been strategically modeled and analyzed by genetic and particle swarm optimization algorithms [23]. This paper differs from the rest in the sense that it is designed for a built-to-order supply chain environment. A method to design a robust model in case of uncertainty in inputs has been proposed by [24]. Mixed integer linear programming model is implemented using CPLEX optimization software, whose robust counterpart is produced using developments in optimization theory. This facilitates in supporting both recovery and disposal activities in the closed-loop supply chain. The robust counterpart takes into consideration the uncertainties in returned products, demand for recovered ones, and logistics costs. A new aspect of closed-loop SC has considered by [25], where two different closed-loop supply chains (SCs) are compared. One is an existing SC supplying new products to the market, and the other SC is designed to supply both new and remanufactured products to the same market. An improved projection algorithm method involved finding an equilibrium point of flow of materials in both the supply chains. Kaya and Urek [26] came up with a procedure to determine the most optimal facility location, new product pricings, inventory quantities, and incentive values for the recovery of used products. The key objective of the procedure is to maximize the overall profit of the supply chain. The methodology is mixed integer nonlinear facility location–inventory–pricing model. A hybrid metaheuristics with variable neighborhood search of the model were also developed and checked with numerical methods for effectiveness. A closed-loop SC from a retailer's perspective has been analyzed by [27]. The paper mainly focuses on recollecting end-of-life machinery. It relies on mixed integer linear model and applied an improved hybrid genetic algorithm, and the solution is evaluated using LINGO optimization software. When compared to other studies in the field, this paper considers locations of dismantling centers too. It has been proposed a quality-based contingent division of profits, expansion of an auxiliary SC as a reusing outlet for the misuse of the essential SC [28]. Also, this paper gives point-by-point affectability investigation on closed-loop supply chain (CLSC) which was done by CPLEX optimizer in a material organization where three recuperation alternatives are considered: repairing, remanufacturing, and reusing.

In another paper, it has been proposed a system plan modeled by fuzzy goal programming and solved by CPLEX software for the improvement of a multi-time, multi-item, multi-echelon shut circle (closed-loop) inventory network in a dubious domain [29]. The vulnerability considered incorporates sick-known parameters which are taken care of with fluffy numbers in their paper. In another paper [30], they endeavored to manage ecological issues in a closed-loop supply chain arrange. They defined a bi-target nonlinear programming issue solved using CPLEX solver. The consequence of their model recommends that a venture can extend an eco-accommodating picture of their item which altogether lessens their utilization of transportation in the two bearings and results in an expansion in their request. A tire manufacturing supply chain system has been built up by [31] that incorporates various items, plants, request markets, recuperation innovations, and accumulation focuses. The model can decide number and areas of open offices and streams of items in the system. Aside from limiting the aggregate cost, a multi-target display considering minimization of imperfection rates and time of activities in accumulation was analyzed by GAMS solver.

2.4 *Multichannel Supply Chain*

A novel vital model has been proposed for planning inventory network systems with numerous dispersion channels. Coordination of three parts, i.e., the need based encoding plan, the pareto-optimality, and the artificial bee colony (ABC) algorithm, includes the tackling philosophy, in a spearheading approach [32]. It was a multi-target advancement for supportable production network arrange configuration considering different appropriation channels. A simulated invulnerable framework calculation in view of clonal selection algorithm (CLONALG) and Taguchi technique is connected to unravel, and hereditary and memetic calculations are utilized to approve. This model advantages clients by giving direct items and administrations from accessible offices rather than the customary stream of items and administrations. Demonstrating destinations are add up to cost, benefit separation, and condition impact. In their paper [33], distinguishes areas of preparations and shipment amount by abusing the exchange off among expenses, and emanations for a double channel production network organize. Orderly quantitative proof of eco-advancement has provided on two channels (immediate and aberrant), through which the age and dissemination of green innovations influence natural execution, and contextual investigation of European businesses was considered in this model [34].

Double-channel (the customary channel and E-trade channel) production network arrange plan has been examined under data vulnerability [35]. This fuzzy multi-objective model using exploit algorithm tries to take care of the issues in combination arrange outline and limit the production network task cost and expand the level of fulfillment between the coordination request and the supply chain (SC) hubs at the same time. It has been proposed a blended number straight programming model by [36] to handle the coordination and inventory network organize outline of a multi-item, multi-arrange, and multi-period dissemination and transportation framework issue solved by CPLEX software, while at the same time, the model limits the working, transportation, and the dealing with costs through every one of the levels of a store network arrangement utilizing hereditary calculation-based strategy. For the extension of this model, a multichannel multi-period closed-loop maintainable inventory network thinking about numerous items has been proposed by [37]. Supportable supply chain has been executed by having financial and natural targets in this model. They also modeled a multichannel supply chain network for a dairy firm in another paper and solved by modified TLBO algorithm [38]. The gathering channel and the generation choices in a CLSC with one retailer and one overwhelming maker have been contemplated from the perspective of both the firm benefit and framework robustness in their paper [39].

Researchers have identified several important variables which are decisive in multichannel decisions to manage multiple channels more efficiently. The problems related to integrated multichannel (omnichannel) supply chain design and coordination among various platforms of the chain were discussed by using Stackelberg game theory [40]. A major conclusion derived was that a display showroom-boosted sales form both online and offline channels. A holistic conceptual framework of a reverse omnichannel for a retail supply chain, the barriers involved, and the relation between the barriers was discussed using systematic literature review by [41]. The customers' perceived compatibility of omnichannel shopping was analyzed by structural equation modeling (SEM) and found that it is increased by enhancing connectivity. It is also concluded that customers' perceived risk factor should be reduced by enhancing personalization and consistency [42]. An approach that helps firms to categorize existing supply chain into any of the four levels and adopting methods to move to the next higher level toward pure omnichannel marketing was studied using systematic literature review by [43]. Barriers for such transitions and measures to overcome them are also identified. Firms were suggested to improve workforce by training and recruitment, and supply chain managers should develop multi-skill abilities and business and political ties to support Omnichannel implementations, studied using SEM [44]. Recommendations and theoretical and practical implications for adopting omnichannel implementation in China were made by [45]. Yan et al. [46] were studied an analytical model and found that omnichannel strategy reduced the cost of high-end products more than mass luxury products. Online profit of luxury goods firms increases and offline profit reduces, indicating that omnichannel strategy is profitable only when their online profit is high. Pishvaee et al. [7] studied on omnichannel marketing strategy using SEM approach and found that firms actively promoting products on social media draw customer trust. Most of the research demonstrates how the omnichannel framework addressing the different challenges when synchronizing multichannel to an omnichannel supply chain. There is very less research which was happened in omnichannel supply chain modeling.

Multichannel supply chain managers need to understand the critical important factors and their interrelationships with each other in different distribution situations to create channel strategy for their business to reduce distribution costs.

3 Research Directions

After going through the above literature review, a few gaps were identified in the research done in the field of supply chain network design. There can be two major categorizations in this regard: (1) research gaps in the specific domain, viz optimization, uncertainty, and green design, and (2) research gaps in between the domains.

In the research work done in problems pertaining to optimization, most of the work is done only to increase the profit and reduce the costs. The other significant work done is to determine the optimum number of distribution centers and facilities for quick responsive service. There is a scope to make a trade-off between the number of DCs/facilities and fiscally cost-effectiveness in the network design. In the uncertain spectrum, the work done has been to tackle the uncertainties considering the high competition in the global market.

The green supply chain network design and sustainable design are considerably a new area of research in this field and has mainly focused on the assessment of emissions through the network, thereby efforts are made to keep them under the "cap". This segment now should work with the focus to minimize (try to eliminate) the emissions by incorporating green technology like electric vehicles, non-conventional energy sources, etc.

Also, one of the major research gaps in the supply chain network design is the integration of various segments. The optimization problems should also consider the uncertainties in the market for the longer run as well as the concepts of green network design should be integrated together in the design to cater to various customer and regulatory needs.

Firms today are making a decent attempt to consolidate manageability in every one of the three fronts in their production network, to be specific, social, financial, and natural. In a true situation, an organization store network can include a wide range of items. Likewise, with the concoct of web-based showcasing, organizations are giving various channels to clients to enhance the responsiveness and give clients a few alternatives to look over. Thing reusing is given high significance in shapes like cars, batteries, and cookware expanding the significance of shut circle chains. Very few supply chain models in the literature tried to incorporate a multi-product, multichannel scenario in their work. It is concluded from the literature that firms has to deal with various challenges and uncertainties in moving from single-channel to multichannel supply chain network.

4 Summary

Supply chain plays a very crucial role in the functioning of industries. A well-designed supply chain network can elevate the business growth and profits, reduce service times as well as enhance customer satisfaction, and strengthen brand value of the company. Globalization and ever-increasing market competition further are pushing the industries to adopt better supply chain network design specially tailored to their needs. A lot of research work has also been done to tackle the vast unpredictability and uncertainty inherent in the demand and supply activities. We did a literature review of various new methodologies proposed for supply chain network design, optimizing it considering various parameters. We have also reviewed the most recent trends in the green supply chain network design, where utmost importance is given to the environmental impacts in terms of carbon footprints and where sustainability is an important goal.

This review has attempted to cover all the different spectrums relating the supply chain network design through the study of recently published work. The different methodologies developed to optimize the supply chain to maximize the profit, minimize the costs, tackle uncertainties, and also consider the environment impacts have been verified in the research work. These have also generated scope for further improvements and optimization to enhance the supply chain network design to a large extent.

The review highlighted the gaps in the research work done till now, out of which the major problem which was analyzed was the unavailability of such a model which can feature all the different aspects under a single umbrella. The need for a supply chain network is felt where all the different cost, risk, and environmental criteria are balanced against each other to develop a holistic model. It was found from the review that MILP models of multichannel closed-loop sustainable supply chain which considered multiple products are not thoroughly explored much in the existing literature.

With ever-increasing global competition, it is very important for any business to have a very efficient and robust supply chain network. But with so many dimensions to be considered and different models for each dimension, the major question that arises is where to start designing the multichannel supply chain network design model from? and how to integrate the different channels for making omnichannel supply chain network. At present, there is no such model which takes into consideration all the different aspects for the integrated multichannel supply chain (omnichannel) network design. There is scope to develop an omnichannel supply chain network design model which considers different aspects of optimization, uncertainty, sustainability, and environment effects together under a single umbrella.

References

1. Cakravastiaa A, Tohab IS, Nakamuraa N (2002) A two-stage model for the design of supply chain networks. Int J Prod Econ 80:231–248
2. Altiparmak F, Gen M, Lin L, Paksoy T (2006) A genetic algorithm approach for multi-objective optimization of supply chain networks. Comput Ind Eng 51:196–215
3. Javid AA, Azad N (2010) Incorporating location, routing and inventory decisions in supply chain network design. Transp Res Part E 46:582–597
4. Nagurney A (2010) Supply chain network design under profit maximization and oligopolistic competition. Transp Res Part E 46:281–294
5. Costa A, Celano G, Fichera S, Trovato E (2010) A new efficient encoding/decoding procedure for the design of a supply chain network with genetic algorithms. Comput Ind Eng 59:986–999
6. Nagurney A (2010) Optimal supply chain network design and redesign at minimal total cost and with demand satisfaction. Int J Prod Econ 128:200–208
7. Pishvaee MS, Rabbani M, Torabi SA (2011) A robust optimization approach to closed-loop supply chain network design under uncertainty. Appl Math Model 35(2):637–649
8. Zuo-Jun MS (2006) A profit-maximizing supply chain network design model with demand choice flexibility. Oper Res Lett 34:673–682
9. Azaron KN, Brown SA, Tarim MM (2008) A multi-objective stochastic programming approach for supply chain design considering risk. Int J Prod Econ 116:129–138
10. Lin C-C, Wang T-H (2011) Build-to-order supply chain network design under supply and demand uncertainties. Transp Res Part B 45:1162–1176
11. Neto JQ, Bloemhof-Ruwaard JM, van Nunen JA, van Heck E (2008) Designing and evaluating sustainable logistics networks. Int J Prod Econ 111:195–208
12. Nagurney A, Nagurney LS (2010) Sustainable supply chain network design: a multicriteria perspective. Int J Sustain Eng 3(3):189–197
13. Wang F, Lai X, Shi N (2011) A multi-objective optimization for green supply chain network design. Decis Support Syst 51:262–269
14. Elhedhli S, Merrick R (2012) Green supply chain network design to reduce carbon emissions. Transp Res Part D 17:370–379
15. Ahi P, Searcy C (2015) An analysis of metrics used to measure performance in green and sustainable supply chains. J Clean Prod 86:360–377
16. Arampantzi C, Minis I (2017) A new model for designing sustainable supply chain networks and its application to a global manufacturer. J Clean Prod 156(2017):276–292
17. Mota B, Gomes MI, Carvalho A, Barbosa-Povoa AP (2018) Sustainable supply chains: an integrated modeling approach under uncertainty. Omega 1–26
18. Varsei M, Polyakovskiy S (2017) Sustainable supply chain network design: a case of the wine industry in Australia. Omega 6:236–247

19. Ding H, Liu Q, Zheng L (2016) Assessing the economic performance of an environmental sustainable supply chain in reducing *environmental externalities*. Eur J Oper Res 255:463–480
20. Carvalho H, Govindan K, Azevedo SG, Cruz-Machado V (2017) Modelling green and lean supply chains: an eco-efficiency perspective. Resour Conserv Recycl 120:75–87
21. Das K, Posinasetti NR (2015) Addressing environmental concerns in closed loop supply chain design and planning. Int J Prod Econ 163:34–47
22. Musavi M, Bozorgi-Amiri A (2017) A multi-objective sustainable hub location-scheduling problem for perishable food supply chain. Comput Ind Eng
23. Kannan G, Noorul Haq A, Devika M (2009) Analysis of closed loop supply chain using genetic algorithm and particle swarm optimisation. Int J Prod Res 47(5):1175–1200
24. Rezapour S, Farahani RZ, Fahimnia B, Govindan K, Mansouri Y (2015)Competitive closed-loop supply chain network design with price-dependent demands. J Clean Prod 93:251–272
25. Kaya O, Urek B (2016) A mixed integer nonlinear programming model and heuristic solutions for location, inventory and pricing decisions in a closed loop supply chain. Comput Oper Res 65:93–103
26. Yi P, Huang M, Guo L, Shi T (2016) A retailer oriented closed-loop supply chain network design for end of life construction machinery remanufacturing. J Clean Prod 124:191–203
27. Masoudipour E, Amirian H, Sahraeian R (2017) A novel closed-loop supply chain based on the quality of returned products. J Clean Prod 151:344–355
28. Jindal A, Sangwan KS, Saxena S (2015) Network design and optimization for multi-product, multi-time, multi-echelon closed-loop supply chain under uncertainty. Procedia Cirp 29:656–661
29. Garg K, Kannan D, Diabat A, Jha PC (2015) A multi-criteria optimization approach to manage environmental issues in closed loop supply chain network design. J Clean Prod 100:297–314
30. Amin SH, Zhang G, Akhtar P (2017) Effects of uncertainty on a tire closed-loop supply chain network. Expert Syst Appl 73:82–91
31. Zhang S, Lee CKM, Kan W, Choy KL (2016) Multi-objective optimization for sustainable supply chain network design considering multiple distribution channels. Expert Syst Appl 65:87–99
32. Barzinpour F, Taki P (2016) A dual-channel network design model in a green supply chain considering pricing and transportation mode choice. J Intell Manuf. https://doi.org/10.1007/s10845-015-1190-x
33. Costantini V, Crespi F, Marin G, Paglialunga E (2017) Eco-innovation, sustainable supply chains and environmental performance in European industries. J Clean Prod 155:141–154
34. Yu J, Gan M, Ni S, Chen D (2018) Multi-objective models and real case study for dual-channel FAP supply chain network design with fuzzy information. J Intell Manuf 29(2):389–403
35. Biswas T, Samanta S (2016) A strategic decision support system for logistics and supply chain network design. Sādhanā 41(6):583–588
36. Niranjan T, Parthiban P, Sundaram K, Jeyaganesan PN (2019) Designing a omnichannel closed loop green supply chain network adapting preferences of rational customers.Sādhanā 44(3):1–10
37. Niranjan T, Parthiban P (2019) Modelling and analysing an integrated multi channel food supply chain distribution of an Indian dairy firm using modified TLBO algorithm
38. Han X, Wu H, Yang Q, Shang J (2017) Collection channel and production decisions in a closed-loop supply chain with remanufacturing cost disruption. Int J Prod Res 55(4):1147–1167
39. Liu L, Feng L, Xu B, Deng W (2020) Operation strategies for an omni-channel supply chain: who is better off taking on the online channel and offline service? Electron Commer Res Appl 39:100918
40. De Borba JLG, De Magalhães MR, Filgueiras RS, Bouzon M (2020) Barriers in omnichannel retailing returns: a conceptual framework. Int J Retail Distrib Manag
41. Shi S, Wang Y, Chen X, Zhang Q (2020) Conceptualization of omnichannel customer experience and its impact on shopping intention: a mixed-method approach. Int J Inf Manag 50:325–336

42. Berman B, Thelen S (2018) Planning and implementing an effective omnichannel marketing program. Int J Retail Distrib Manag
43. Song G, Song S, Sun L (2019) Supply chain integration in omni-channel retailing: a logistics perspective. Int J Logist Manag
44. Ye Y, Lau KH, Teo LKY (2018) Drivers and barriers of omni-channel retailing in China. Int J Retail Distrib Manag
45. Wei Y, Li F (2020) Omnichannel supply chain operations for luxury products with conspicuous consumers. Transp Res Part E: Logist Transp Rev 137:101918
46. Yan B, Chen YR, Zhou XT, Fang J (2019) Consumer behavior in the omni-channel supply chain under social networking services. Ind Manag Data Syst

Solar Bubble Dryer Preliminary Simulation Studies

Shankar R. Daboji, P. P. Revankar, K. N. Patil, and Sandeep I. Akki

Abstract Among the renewable energy, solar bubble dryer is cost-effective drying technologies for rural India. Solar bubble dryers' technical and scientific study for Indian conditions with respect to this simulation studies was performed. To dry agricultural produce in this technology, we are using solar PV panel to blow the air thus making complete renewable energy based and to reduce environmental concerns associated with post-harvesting technologies. Solar bubble dryer 3D environmental study results indicated 340–360 K (6 h, 9 AM–5 PM). There were two stagnation areas at the inlet section, and further, it has increased the temperature in these zones up to 380 K in the mid of the day 1 PM–3 PM. As the studies were preliminary, one only concentrated on steady-state analysis. Air sufficiently alleviated to dry agricultural produce in solar bubble dryer. Further, it can be extended to time-dependent study.

Keywords Solar bubble dryer · Computational fluid dynamics · Temperature and flow distribution

1 Introduction

Agricultural produce, i.e., crops must be stored at the recommended level of moisture, which varies for different types of grains. Excess moisture in the crops promotes the growth of microorganisms and can rot the crops which result in huge losses. In order to maintain the good quality of crop, drying of crop must be done after harvesting. Open sun drying is primitive method of drying the crops. Mechanical dryers use electrical energy; it is not viable from the point of view of economic aspects and availability of the electrical energy in rural fields [1]*. The Solar bubble dryer is a tunnel-type dryer which has a bed made up of a black PVC sheet on which crops are spread.

S. R. Daboji (✉) · K. N. Patil · S. I. Akki
SDM College of Engineering &Technology Dharwad, Visvesvaraya Technological University, Belagavi, Karnataka 580002, India
e-mail: shankardaboji@gmail.com

P. P. Revankar
KLE Technological University Hubli, Karnataka 580031, India

A. Maurya et al. (eds.), *Recent Trends in Mechanical Engineering*, Lecture Notes in Mechanical Engineering, https://doi.org/10.1007/978-981-19-7709-1_40

The crops are covered by a transparent cover. Basic principle behind working of the dryer is the selected crop is spread on the dryer bed; the transparent cover traps the solar radiation energy entering the dryer. The collector converts the radiation energy to heat energy, which increases temperature inside the dryer. The moisture content in the crops is evaporated due to increased dryer temperature. A blower is attached to bubble dryer which blows air into it sourced by PV panel. Hence, the moisture of crops is removed in controlled manner, and crops are dried to required level. There are several alternatives exist such as biomass, solar, or hybrid source-based dryers. The important aspect of these systems is their renewability. Solar energy available abundantly in rural India is a high potential for implementation of the large dryers. Agriculture waste as biomass resource is also one impressive viable alternative to the solar and electrical energy for rural India [2–4].

2 Literature Review

A detailed review of the solar dryers was done. The range of temperature of solar collectors for drying is 40–75 °C. Storage units are considered and studied for continuous operation during off-sunshine hours. Selective coating for absorber surface for improving heat absorption capacity and improving the thermal performance of solar collectors. Hybrid system utility was also studied for supplementing the heat for drying such as biomass, waste heat, and thermal energy from fossil fuels [2]. Green house solar dryer's operation active mode is better than passive mode. Higher moisture content products dry and having better quality, color, taste, and nutritious value better in forced convection mode than open sun drying. PV/T-integrated greenhouse dryer and use of thermal storage material increases inside temperature and reduces drying time [3].

Solar dryer components such as flat plate collectors are reviewed for performance, constructional features with different materials for absorption and storage. Exergy and energy-based analysis on different types of collectors also reviewed. Also, hybrid mode dryer such as biomass, solar, and industrial waste heat-based systems are studied. Lesser literature is available on exergo-economic, exergo-environmental studies; computational software simulations and numerical method analysis and short- and long-term performance are needed. Hybrid systems such as greenhouse dryers with solar collector for their performance and economic analysis are to be studied [4].

Solar tunnel dryer was used to dry turmeric mixed mode drying process. First and 2nd law efficiency was studied and reported to be around 50% and 22%, respectively. Also, 11 statistical models were studied for drying kinetics [5]. Indirect cabinet solar dryer's microscopic energy and exergy analysis and models were developed. Dynamic mathematical model for performance analysis also considered exergy destruction and efficiency were two important parameters [6]. Normally, thermodynamically inefficient due to much of the energy is vented in exhaust air. This analysis carried out for potato slices of 5 mm-thick gas-based band-type continuous dryer for velocity of air and energy utilization ratio [7]. From the literature, several solar dryers were used and studied. Literature available indicates CFD analysis aids in design development and testing of the dryers.

3 Development and Thermal Analysis of Prototype Bubble Drier

The design of dryer is based on the temperature fluid flow pattern on geometries and solar insolation on the place to be implemented. The bed of the solar bubble dryer absorbs as much as solar radiation as possible; in this prototype model, black PVC material is selected for bed material. The outer cover of the dryer covers grains and transmits maximum amount of radiation that falls on it, so it should be highly transparent. The sheets used for dryer must withstand temperature developed and pressure built inside dryer. Bubble dryer is a tunnel shape dryer in which crops are placed and air is blown. To obtain tunnel shape, the cross-section of dryer is taken as 1/3–1/2 of circle. The following are the geometrical specifications for the bubble dryer. Radius of arc (r) = 0.405 m, height of arc = h = 0.2025 m, length of arc = l = 1.45 m, width of arc = 2a = w = 0.7 m, and area of bed = $A = w * l$ = 1.189 m^2. With above dimensions, the bed of dryer is made with black PVC sheet, and transparent PVC sheet is attached with zip to get tunnel shape such that the opening and closing of dryer is possible to place food grains to dry. The air is blown from the one end of dryer by fan, which is made to go out through another end. The air blows the transparent cover of the dryer and gives it tunnel shape. Figures 1 and 2 of solar dryer model are attached below.

There are various software packages available for CFD studies. In this model, simulation carried out in ANSYS Fluent domain for flow and temperature distribution. Steady-state K-ε model with energy equations along with hourly solar radiations was used to simulate. Grid independence study was performed by using the mesh size of 18,654, 21,284, & 26,682. It has been observed that the computational results

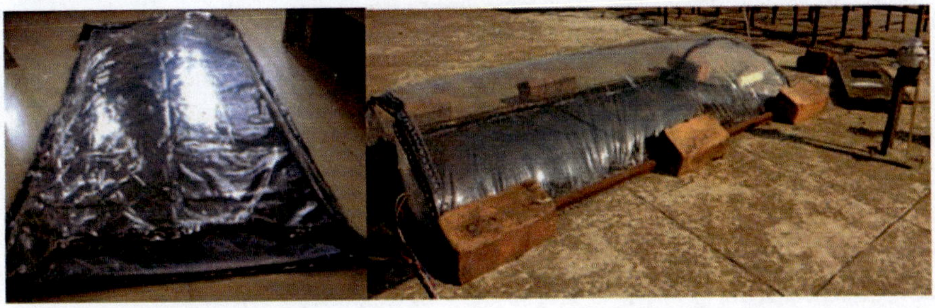

Fig. 1 Developed model of solar bubble dryer (SBD) without inflation and with inflated state

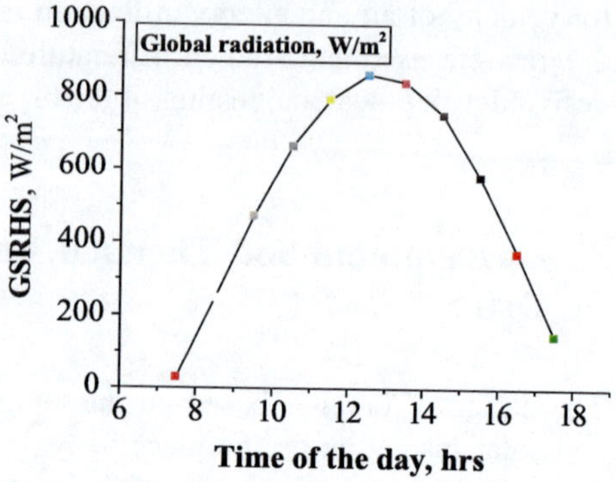

Fig. 2 Plot of solar insulation versus time of the day for a typical day in month

obtained by 21,284 & 26,682 mesh are almost same. Further to do the CFD simulation, 21,284 (minimum) mesh size is used. Figure 2 indicates the solar insolation trend on a day at Dharwad (15.5 deg N, 75 deg E) in the month of October and this is taken as one of the boundary conditions for simulations.

Figure 3 shows the geometry and meshed model of the geometry consisting of 21,284 elements.

The following images indicate the temperature of absorber plate for solar insolation on a day at time interval of everyone hour from 7.30 AM to 5.30 PM.

Figures 4, 5, 6, 7, 8, 9, 10 and 11 indicate the absorber plate temperature and air outlet temperature every one-hour interval of time for steady-state condition. The results are shown in Table 2.

4 Results and Discussion

Results of simulations are shown in the Table 2. By using these results, variation of ambient, absorber, and air outlet temperature has been plotted in Fig. 12. It can observed that the variation is much guided by the solar insolation which normal trend in the solar-based dryer. The maximum temperature rise on the simulation results was around 36 K with constant air flow rate of 4 m/s evident form Table 2.

The absorber temperature has risen to 373 K at mid of day, and it has increased during the start of the day till mid, and then, it has decreased. From Figs. 4, 5, 6, 7, 8, 9, 10, 11 and 12, indicate at the inlet side of the SBD, there is a high temperature zone of 380 K.

5 Conclusions

The present solar bubble dryer simulation studies indicate the following.

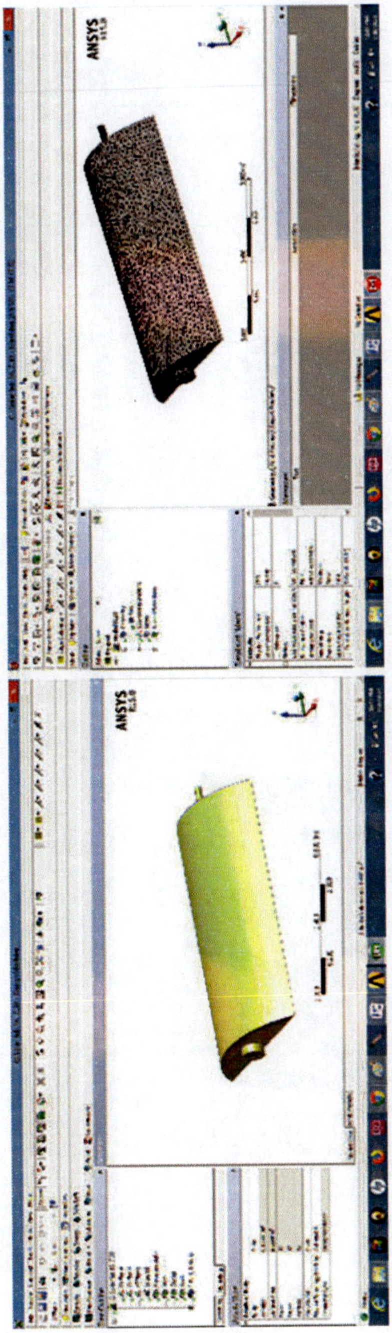

Fig. 3 Geometrical model and meshed model of fluid domain of SBD

Fig. 4 Absorber plate model of fluid domain of SBD at 7.30 AM

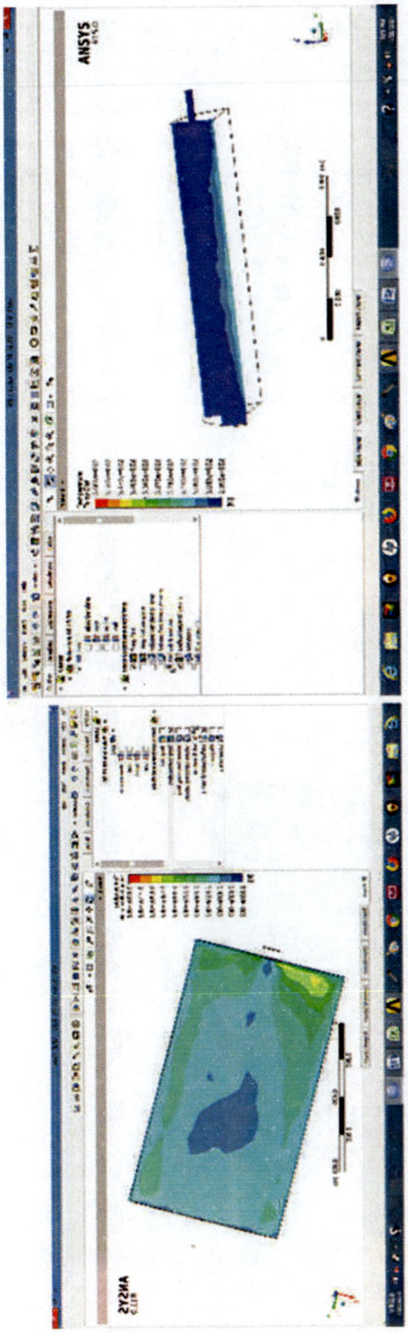

Fig. 5 Absorber plate model of fluid domain of SBD at 8.30 AM

Fig. 6 Absorber plate model of fluid domain of SBD at 9.30 AM

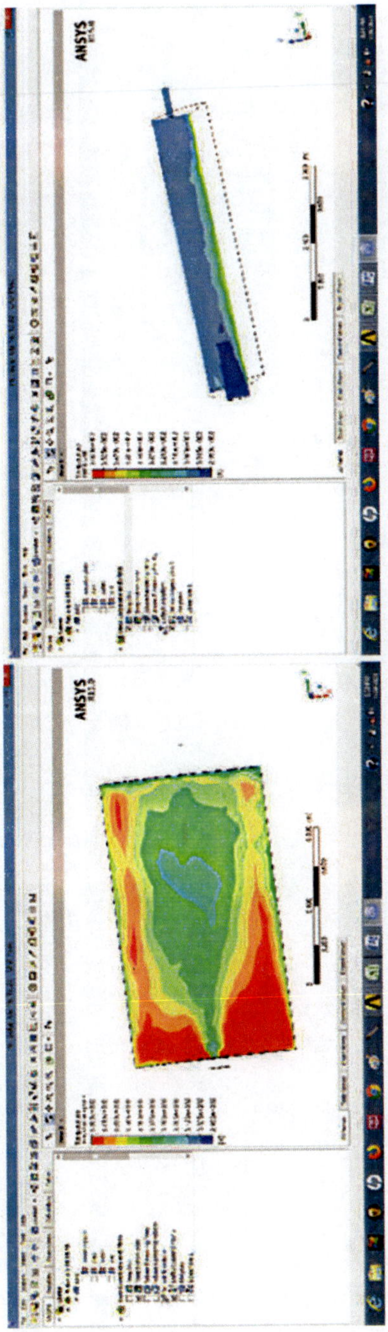

Fig. 7 Absorber plate model of fluid domain of SBD at 10.30 AM

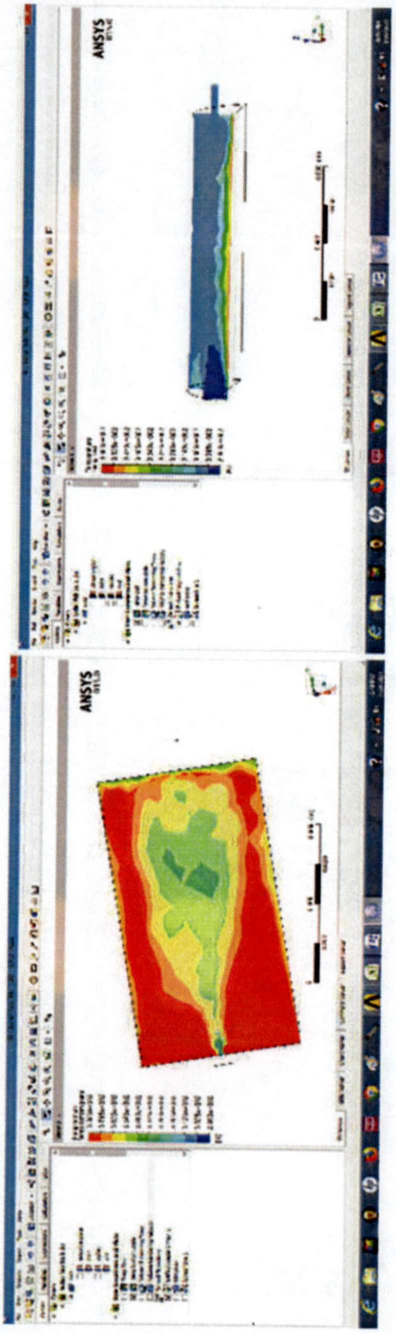

Fig. 8 Absorber plate model of fluid domain of SBD at 1.30 PM

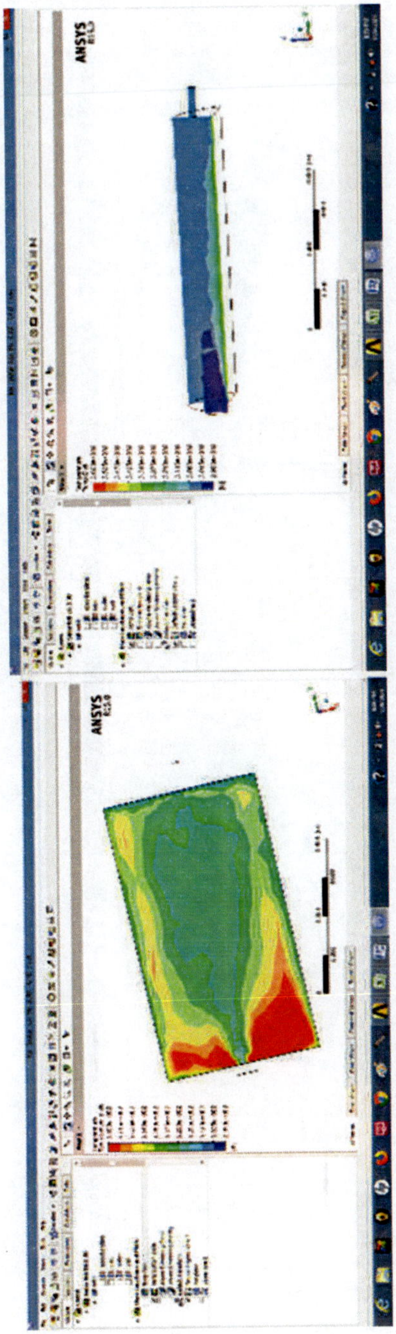

Fig. 9 Absorber plate model of fluid domain of SBD at 3.30 PM

Fig. 10 Absorber plate model of fluid domain of SBD at 4.30 PM

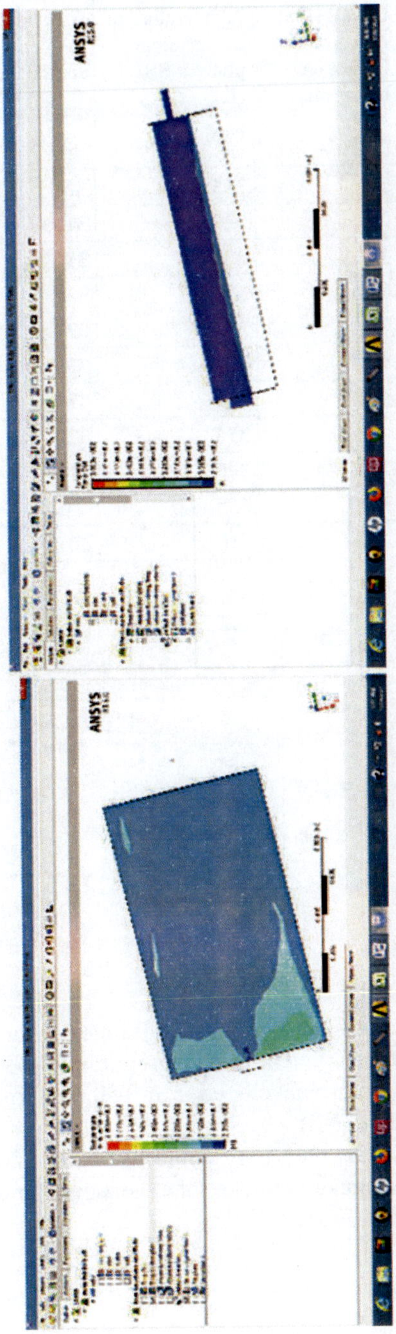

Fig. 11 Absorber plate model of fluid domain of SBD at 5.30 PM

Table 2 Simulation results of SBD absorber plate and air outlet temperature

Time of the day	Ambient temperature (K)	Global solar radiation falling on horizontal surface (W/m^2)	Temperature of the absorber plate of SBD (K)	Air outlet temperature for SBD (K)	Rise of air temperature in SBD (K)
7.5	295	31	315	301	6
8.5	295.5	249	320	310	14.5
9.5	296	472	338	315	19
10.5	296.5	663	350	321	24.5
11.5	297	790	363	327	30
12.5	297.5	859	373	334	36.5
13.5	302	837	371	334	32
14.5	301	749	354	322	21
15.5	300	580	345	315	15
16.5	299.5	372	335	305	5.5
17.5	298	147	310	301	3

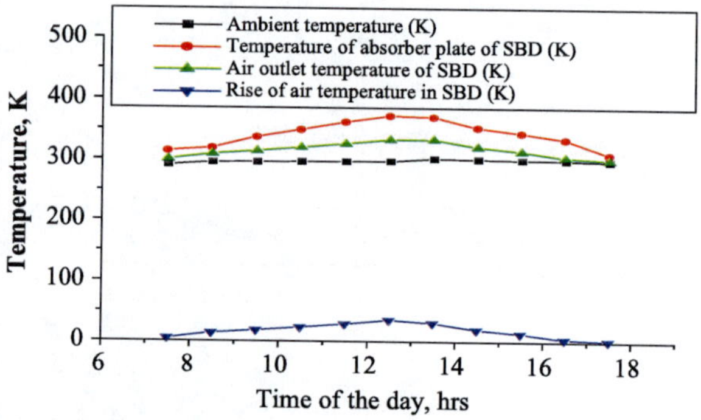

Fig. 12 Plot of variation of absorber and air outlet temperature from the SBD

- Temperature of 36–40 K above the ambient is possible during normal clear sunny day in solar bubble dryer is sufficient to dry agricultural produce.
- The absorber plate temperature developed is sufficiently higher in the range of 340–360 K for 9 AM–5 PM (6 h).
- Geometrical modification of bubble dryer at inlet and outlet of the prototype model simulated to remove the stagnation pockets and heat transfer and exergy analysis need to be done with unsteady state.

References

1. Solar energy. In: Sukathme SP, Nayak JP (eds), 3rd edn. TATA McGraw Hill publication
2. Salvatierra-Rojas A, Nagle M, Gummert M, de Bruin T, Müller J (2017) Development of an inflatable solar dryer for improved postharvest handling of paddy rice in humid climates. Int J Agric Biol Eng 10(3):269–282
3. Fudholi A, Sopian K (2019) A review of solar air flat plate collector for drying application. Renew Sustain Energy Rev 102:333–345
4. Chauhan YB, Rathod PP (2018) A comprehensive review of the solar dryer. Int J Ambient Energy. https://doi.org/10.1080/01430750.2018.1456960
5. Karthikeyan AK, Murugavelh S (2018) Thin layer drying kinetics and exergy analysis of turmeric (Curcuma longa) in a mixed mode forced convection solar tunnel dryer. Renew Energy 128:305–312
6. Sami S, Etesami N, Rahimi A (2011) Energy and exergy analysis of an indirect solar cabinet dryer based on mathematical modeling results. Energy 36:2847–2855
7. Aghbashlo M, Mobli H, Rafiee S, Madadlou A (2008) Energy and exergy analyses of thin-layer drying of potato slices in a semi-industrial continuous band dryer. Dry Technol 26(12):1501–1508. https://doi.org/10.1080/07373930802412231

CFD Analysis of Cabinet Dryer for Optimum Air and Temperature Distribution

Sandeep I. Akki, K. N. Patil, and Shankar R. Daboji

Abstract For the efficient working of a cabinet dryer, air distribution and temperature distribution inside the cabinet plays an significant role in obtaining better drying efficiency so that specific energy requirement for drying can be reduced. It becomes inevitable to study and optimize these distributions in the dryer prior to the development of the dryer and finalizing the pattern of tray arrangement. In view of this, the paper deals with study of different parametric optimization using computation fluid dynamics (CFD) coupled with heat transfer in cabinet dryer has been studied. CFD simulation included the study of number of air inlet, orientation of inlet, tray arrangements to study the velocity and temperature distribution in 3D environment in the dryer. The CFD simulation study showed that zig-zag orientation of tray with horizontal single air inlet and multiple air outlets has better velocity and temperature distribution rather than other cases of study. Single inlet vertical bottom intake for staggered orientation of tray is least preferred one from the point of temperature and velocity distribution.

Keywords Cabinet dryer · Computational fluid dynamics · Temperature and flow distribution

1 Introduction

Dryers are classified as batch dryers and continuous dryers, and later, based on the source of energy used to generate heat, they are classified as solar dryer, mechanical dryer, electrical dryer, poly-house dryer, microwave dryer [1].

One of the impacting parameters of dryers is temperature and flow distribution. This needs to be studied for selecting and developing dryers. Simulation studies help to analyze and develop dryers with optimum air circulation were studied. Rajkotia

S. I. Akki (✉) · K. N. Patil · S. R. Daboji
Mechanical Engineering Department, SDM College of Engineering and Technology, Dharwad 580002, India
e-mail: sandeepakkimech@gmail.com

Visvesvaraya Technological University, Belagavi, Karnataka, India

A. Maurya et al. (eds.), *Recent Trends in Mechanical Engineering*, Lecture Notes in Mechanical Engineering, https://doi.org/10.1007/978-981-19-7709-1_41

et al. [2] carried out a CFD analysis involving all the parameters including temperature, velocity, mass flow rate, etc. The CFD analysis gives the approximate but near solution which enables the researcher to analyze the optimum design and the overall performance of the natural convection solar dryer. AchintSanghi et al. [3] have worked on a computational fluid dynamics model to simulate the corn drying process for solar cabinet dryers and validated the experimental results. Jagadeesh et al. [4] have worked on experimental and CFD simulation for transient thermal performance of a single-pass solar air dryer and studied the effect of mass flow rate; the outlet temperature of the air is computationally analyzed in comparison with the experimental work.

The literature available is limited. From the above literature, it can be observed that rather than developing a dryer system directly and conducting experiments, it is better to visualize the system using some simulation tools to identify the better system in terms of temperature and flow distribution in the system. Temperature and flow distribution are among the top parameters to be studied in the dryers as these parameters impact the quality and quantity of dried products. The present work deals with study CFD simulation of cabinet dryer for temperature and velocity distribution within it.

2 Simulation Studies

In the present work, CFD study has been carried out for uniform flow velocity of 4 m/s, hot air temperature of 333°K, and also, inlet area was maintained constant for all combinations, with following combinations: number of the inlet for air entry (single inlet and double inlet), direction of inlet air entry (Vertical from the bottom and horizontal from the front), and orientation of trays (Uniform tray, zig-zag tray, and staggered type). The analysis was carried out in ANSYS software using a Fluent solver with an energy model, and for turbulence, $k - €$ model was used for analysis.

Figure 1 shows the geometry with a uniform tray arrangement with different inlet air entry conditions. Figure 2 shows the geometry with zig-zag tray arrangement with different inlet air entry conditions. Figure 3 shows the geometry with a staggered tray arrangement with different inlet air entry conditions. Figure 4 shows the typical mesh view of geometry; a similar type of mesh was done for all simulation studies.

3 Results and Discussion

Figure 5 shows the temperature and velocity distribution for uniform tray type, bottom double inlet. It can be seen that the air moves vertical upward direction hits the bottom baffle and reduces its velocity, but due to momentum in the air which makes it to move upward which helps the hot air to reach top tray, but this also creates a chance of hot air to exit without drying the products.

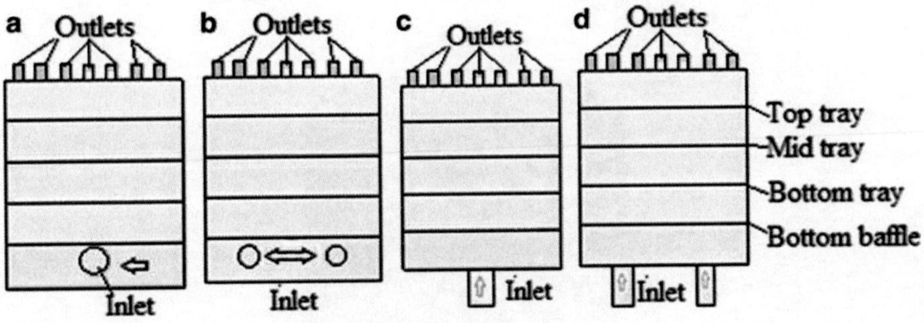

Fig. 1 Uniform tray type: **a** front single inlet, **b** front double inlet, **c** bottom single inlet, and **d** bottom double inlet

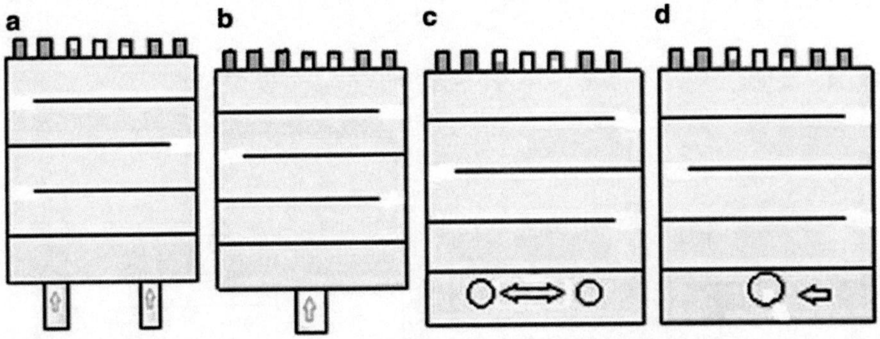

Fig. 2 Zig-zag tray type: **a** bottom double inlet, **b** bottom single inlet, **c** front double inlet, and **d** front single inlet

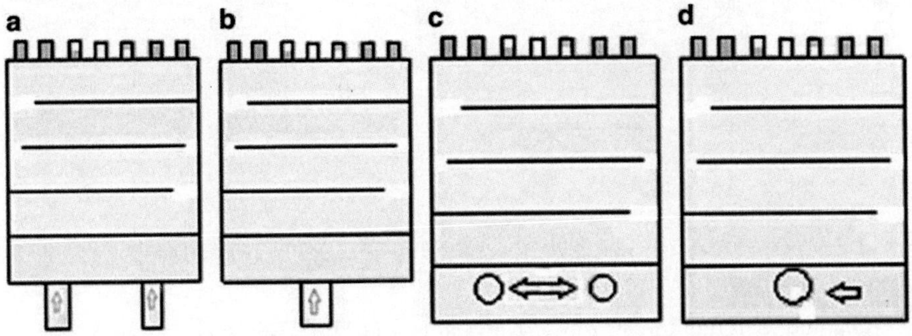

Fig. 3 Staggered tray type: **a** bottom double inlet, **b** bottom single inlet, **c** front double inlet, and **d** front single inlet

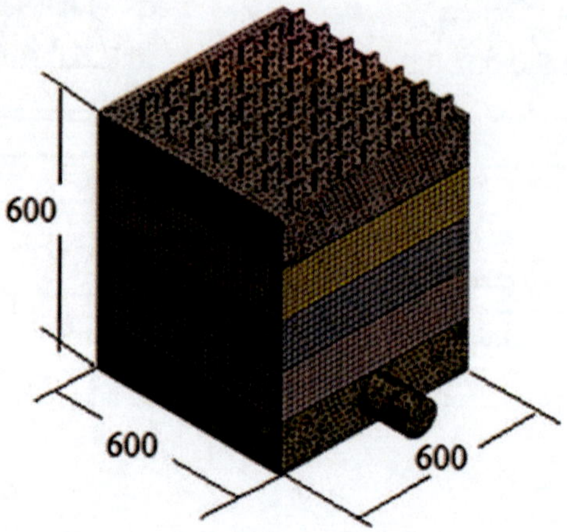

Fig. 4 Meshed geometry with dimensions in mm

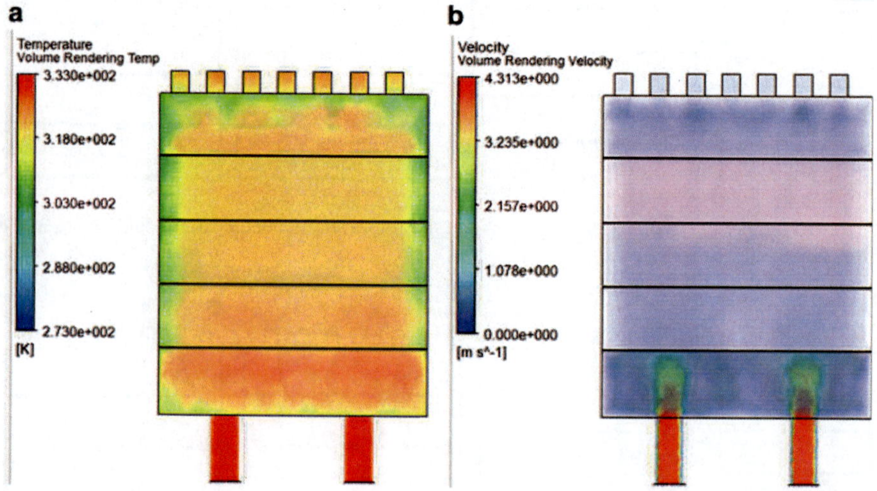

Fig. 5 Uniform tray type, bottom double inlet, **a** temperature distribution **b** velocity distribution

It can be seen that the air moves vertical upward direction hits the bottom baffle and reduces its velocity, but due to momentum in the air which makes it to move upward which helps the hot air to reach top tray, but this also creates a chance of hot air to exit without drying the products. Also, due to a single entry, 2 stagnation pockets above the baffle plate have been observed.

Figure 7 shows the temperature and velocity distribution for uniform tray type, front double inlet. It can be seen that the air moves in a horizontal direction hits the opposite wall and reduces its velocity, also, it has to change the direction to move upward; this further reduces the velocity of air, but the top tray gets slightly lesser hot air compared to mid and bottom tray; also, the chances of hot air exiting the cabinet without drying the products are less.

Figure 8 shows the temperature and velocity distribution for uniform tray type, front single inlet. It can be seen that the air moves in a horizontal direction hits the opposite wall and reduces its velocity; also, it has to change the direction to move upward; this further reduces the velocity of air, but the top tray gets slightly lesser hot air compared to mid and bottom tray; also, the chances of hot air exiting the cabinet without drying the products are less.

Figure 9 shows the temperature and velocity distribution for the zig-zag tray type, bottom double inlet. It can be seen that the air moves vertical upward direction hits the bottom baffle and reduces its velocity, but due to momentum in the air which makes it to move upward and as the presence of an alternative gap between the wall and tray the air flows in the least resistance path and follows a pattern of "S" between the trays.

Figure 10 shows the temperature and velocity distribution for zig-zag tray type, bottom single inlet. It can be seen that the air moves vertical upward direction hits the bottom baffle and reduces its velocity, but due to momentum in the air which makes it to move upward and as the presence of an alternative gap between the wall and tray the air flows in the least resistance path and follows a pattern of "S" between the trays.

Figure 11 shows the temperature and velocity distribution for the zig-zag tray type, front double inlet. It can be seen that the air moves in a horizontal direction hits the opposite wall and reduces its velocity; also, it has to change the direction to move upward; this further reduces the velocity of air, and as the presence of an alternative gap between the wall and tray, the air flows in the least resistance path and follows a pattern of "S" between the trays.

Figure 12 shows the temperature and velocity distribution for the zig-zag tray type, front single inlet. It can be seen that the air moves in a horizontal direction hits the opposite wall and reduces its velocity; also, it has to change the direction to move upward; this further reduces the velocity of air, and as the presence of an alternative gap between the wall and tray, the air flows in the least resistance path and follows a pattern of "S" between the trays.

Figure 13 shows the temperature and velocity distribution for staggered tray type, bottom double inlet. It can be seen that the air moves vertical upward direction hits the bottom baffle and reduces its velocity, but due to momentum in the air which makes it to move upward and as the presence of a gradual varying gap between wall and tray creating a converging section on one side and the diverging section on the other side, the air flows in the least resistance path and tries follows a pattern of "S" between the trays. Also, the temperature of the air inside the cabinet is less compared to earlier studies.

Figure 14 shows the temperature and velocity distribution for staggered tray type, bottom single inlet. It can be seen that the air moves vertical upward direction hits the bottom baffle and reduces its velocity, but due to momentum in the air which makes it to move upward and as the presence of a gradual varying gap between wall and tray creating a converging section on one side and the diverging section on the other side, the air flows in the least resistance path and tries follows a pattern of "S" between the trays. Also, the temperature of the air inside the cabinet is less compared to earlier studies.

Figure 15 shows the temperature and velocity distribution for staggered tray type, front double inlet. It can be seen that the air moves in a horizontal direction hits the opposite wall and reduces its velocity; also, it has to change the direction to move upward; this further reduces the velocity of air, and as the presence of a gradual varying gap between wall and tray creating a converging section on one side and the diverging section on the other side, the air flows in the least resistance path and tries follows a pattern of "S" between the trays. Also, the temperature of the air inside the cabinet is less compared to earlier studies.

Figure 16 shows the temperature and velocity distribution for staggered tray type, front single inlet. It can be seen that the air moves in a horizontal direction hits the opposite wall and reduces its velocity; also, it has to change the direction to move upward; this further reduces the velocity of air, and as the presence of a gradual varying gap between wall and tray creating a converging section on one side and the diverging section on the other side, the air flows in the least resistance path and tries follows a pattern of "S" between the trays. Also, the temperature of the air inside the cabinet is less compared to earlier studies. The results of the simulation study are shown in Table 1.

4 Experimentation

A cabinet dryer had been developed with a front single inlet, with a uniform tray arrangement to study the actual flow parameters. Figure 17 shows the cabinet dryer, with a temperature sensor and weighing machine. And to continuously record the data, Arduino (Mega) board controller 2560 was used with the following sensors: temperature sensor (LM 35), humidity sensor (DHT 21), and portable weighing machine. Results obtained were shown below in Table 2. Comparing the CFD studies for uniform tray, front single inlet (Fig. 8) with experimental results, it can be seen that the temperature variation is between 1 to 3°.

5 Conclusions

From the above simulation study, the following remarks can be drawn:

- Air velocity inside the cabinet is higher in the case of the bottom inlet (single and double), so hot air reaches the top tray, but this also creates the chance of hot air leaving the cabinet without drying the products.

Table 1 Summary of simulation studies

Reference figure no	Title of the figure	Minimum temperature in K	Maximum temperature in K	Remarks
Figure 5	Uniform tray type, bottom double inlet	318	325	Single stagnation pocket at the middle above the baffle plate
Figure 6	Uniform tray type, bottom single inlet	318	320	Two stagnation pockets, on both the sides above the baffle plate
Figure 7	Uniform tray type, front double inlet	317	320	Uniform air velocity in the drying tray region
Figure 8	Uniform tray type, front single inlet	320	324	Uniform air velocity in the drying tray region
Figure 9	Zig-zag tray type, bottom double inlet	322	327	Presence of alternative gap between wall and tray the air flows in least resistance path and follows a pattern of "S" between the trays
Figure 10	Zig-zag tray type, bottom single inlet	321	324	Presence of alternative gap between wall and tray the air flows in least resistance path and follows a pattern of "S" between the trays
Figure 11	Zig-zag tray type, front double inlet	322	328	Presence of alternative gap between wall and tray the air flows in least resistance path and follows a pattern of "S" between the trays

(continued)

Table 1 (continued)

Reference figure no	Title of the figure	Minimum temperature in K	Maximum temperature in K	Remarks
Figure 12	Zig-zag tray type, front single inlet	321	328	Presence of alternative gap between wall and tray the air flows in least resistance path and follows a pattern of "S" between the trays
Figure 13	Staggered tray type, bottom double inlet	316	320	Uniform velocity distribution, converging and diverging effect in flow can be seen, between tray and side wall
Figure 14	Staggered tray type, bottom single inlet	316	320	Uniform velocity distribution, converging and diverging effect in flow can be seen, between tray and side wall
Figure 15	Staggered tray type, front double inlet	316	324	Uniform velocity distribution, converging and diverging effect in flow can be seen, between tray and side wall
Figure 16	Staggered tray type, front single inlet	319	324	Uniform velocity distribution, converging and diverging effect in flow can be seen, between tray and side wall

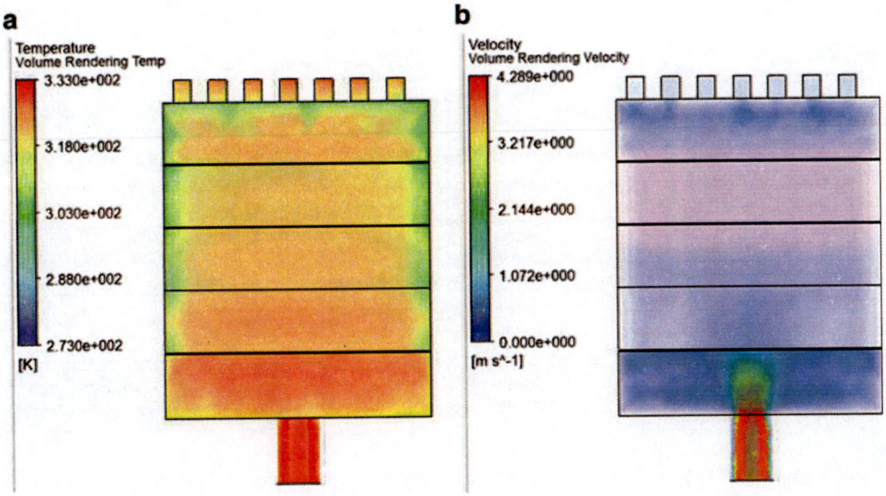

Fig. 6 Uniform tray type, bottom single inlet, **a** temperature distribution, **b** velocity distribution

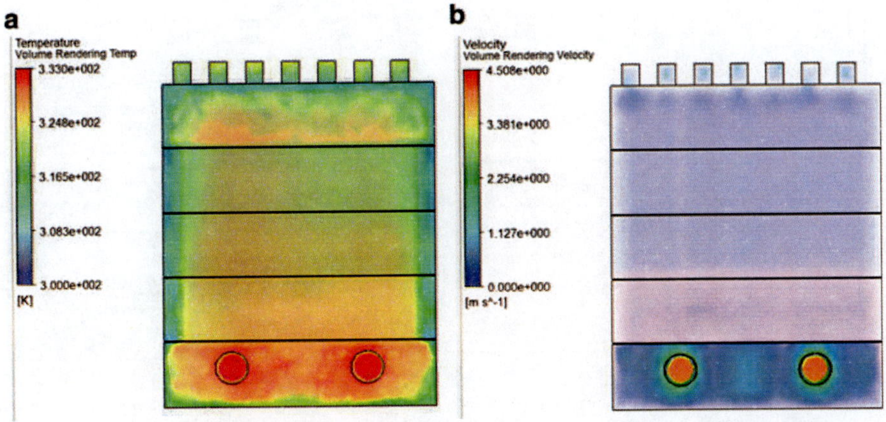

Fig. 7 Uniform tray type, front double inlet, **a** Temperature distribution, **b** velocity distribution

- In uniform tray arrangement, there is no gap between tray and walls; this forces air to move through trays; temperature distribution is uniform within the cabinet and better than staggered type tray arrangement.
- In zig-zag tray arrangement, the presence of an alternative gap between wall and tray the air flows in the least resistance path and follows a pattern of "S" between the trays. Also, the temperature distribution is uniform within the cabinet and better than the uniform type tray and staggered type tray arrangement.
- In staggered tray arrangement, the presence of a gradual varying gap between wall and tray creating a converging section on one side and the diverging section on the

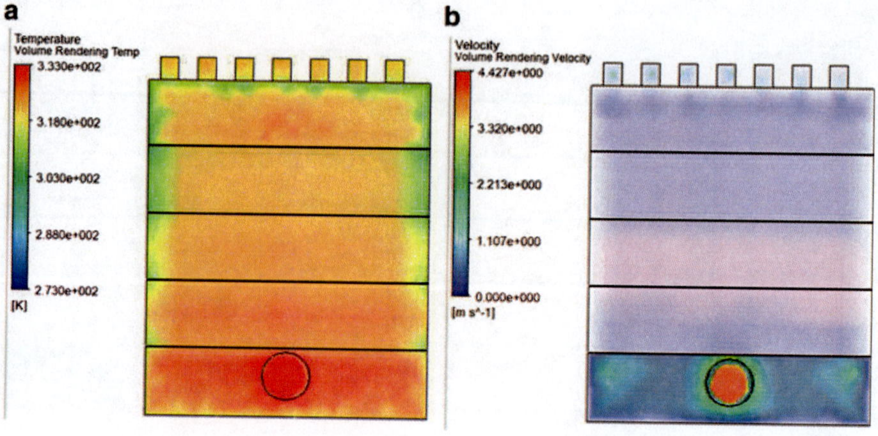

Fig. 8 Uniform tray type, front single inlet, **a** temperature distribution. **b** Velocity distribution

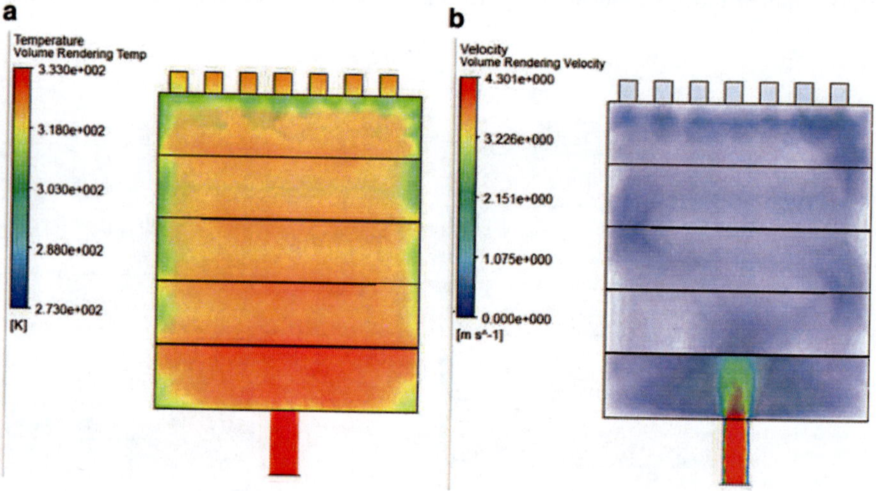

Fig. 9 Zig-zag tray type, bottom double inlet, **a** temperature distribution, **b** velocity distribution

other side, the air flows in the least resistance path and tries to follow a pattern of "S" between the trays. Also, the temperature distribution is non-uniform within the cabinet.

So, from the above study, we can conclude that zig-zag tray arrangement with front single and double inlet gives better higher temperature air distribution within the cabinet, compared to other cases of study.

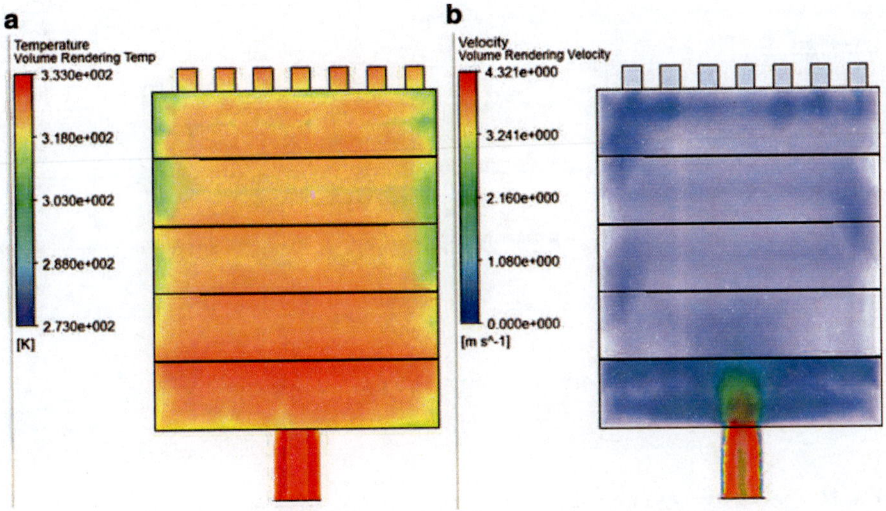

Fig. 10 Zig-zag tray type, bottom single inlet, **a** temperature distribution, **b** velocity distribution

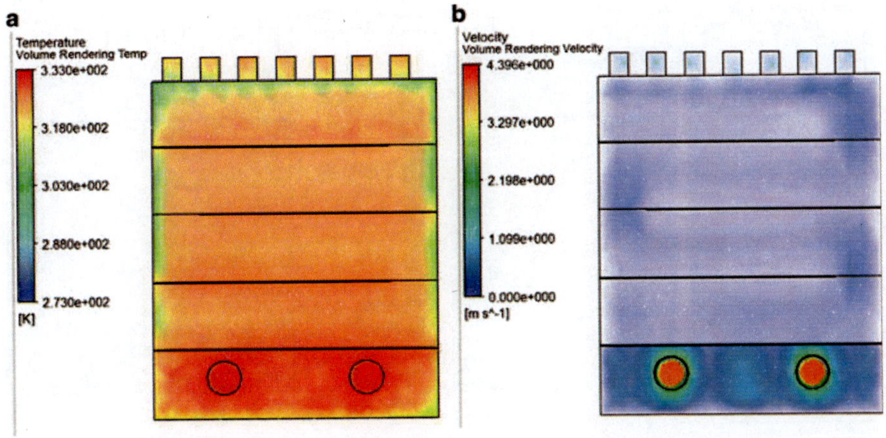

Fig. 11 Zig-zag tray type, front double inlet, **a** temperature distribution, **b** velocity distribution

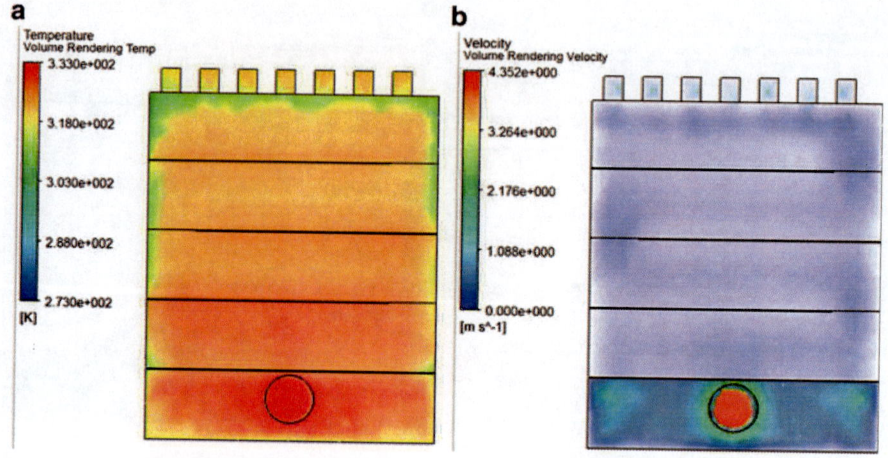

Fig. 12 Zig-zag tray type, front single inlet, **a** temperature distribution, **b** velocity distribution

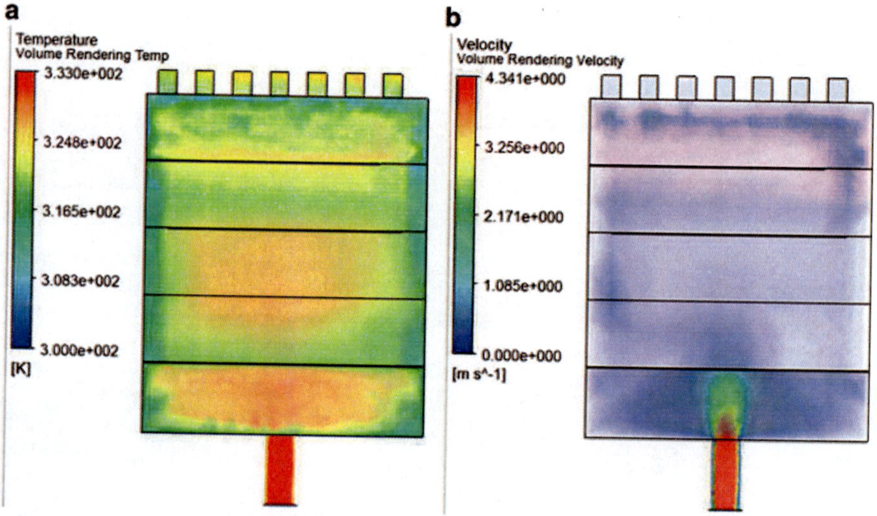

Fig. 13 Staggered tray type, bottom double inlet, **a** temperature distribution, **b** velocity distribution

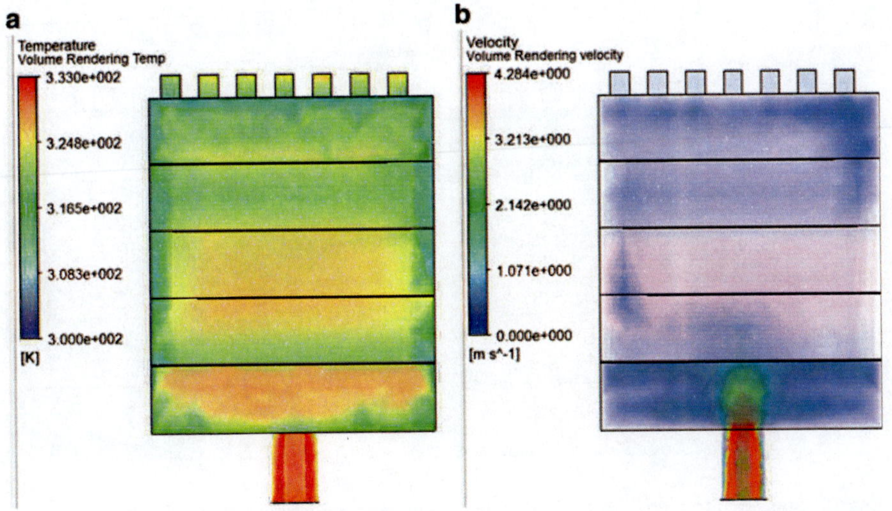

Fig. 14 Staggered tray type, bottom single inlet, **a** temperature distribution, **b** velocity distribution

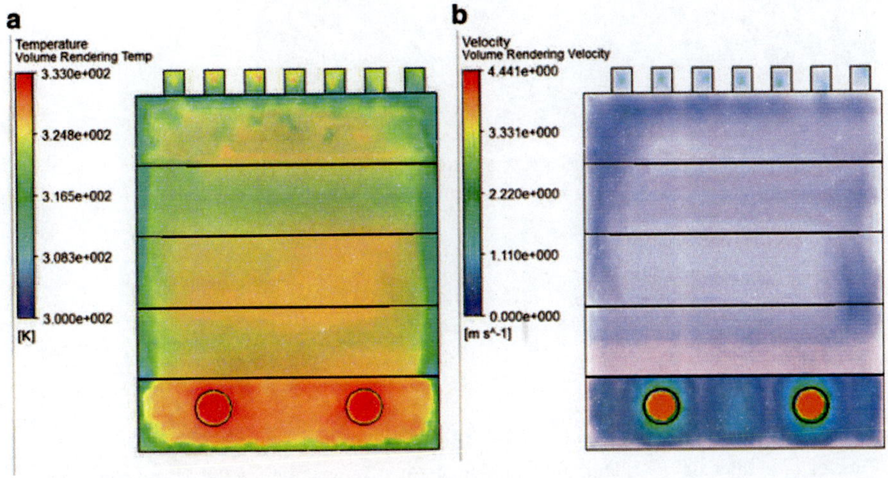

Fig. 15 Staggered tray type, front double inlet, **a** temperature distribution, **b** velocity distribution

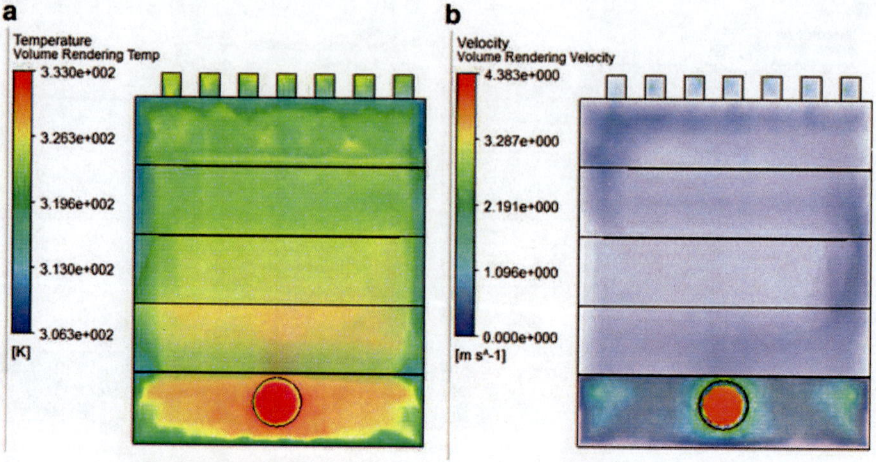

Fig. 16 Staggered tray type, front single inlet, **a** temperature distribution, **b** velocity distribution

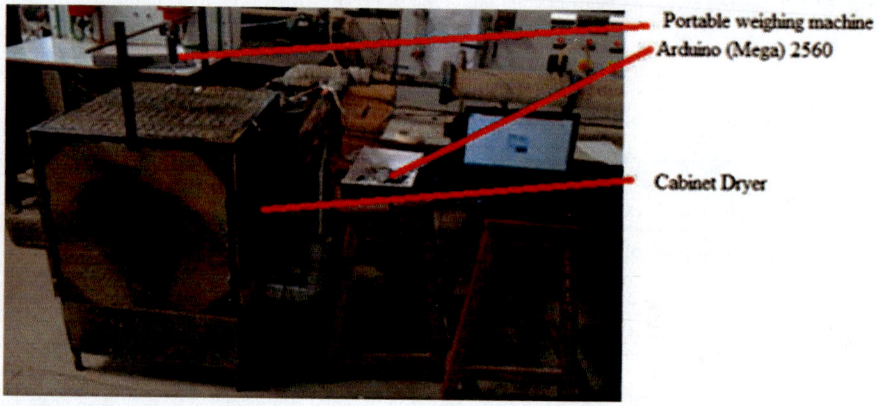

Fig. 17 Cabinet dryer

Table 2 Experimental results

Time of conduction	Inlet air temperature	Humidity of inlet air	Velocity of inlet air	Temperature in the drying region (°C)		
(AM)	(°C)	(%)	m/s	Above bottom tray	Above mid tray	Above top tray
10.30	59.1	38.8	4	26.5	25.5	25.0
10.45	59.6	44.4	4	44.5	41.5	40.5
11.00	60.0	47.0	4	48.5	46.5	45.5
11.15	60.0	47.1	4	49.5	47.5	46.0

References

1. Aravindh MA, Sreekumar A, Solar drying—A sustainable way of food processing
2. Rajkotia SD et al (2013) Performance improvement of solar dryer. Int J Sci Res 2(5):142–144
3. Sanghi A, Ambrose RK, Maier D (2018) CFD simulation of corn drying in a natural convection solar dryer. Dry Technol 36(7):859–870. https://doi.org/10.1080/07373937.2017.1359622
4. Jagadeesh D et al (2018) Transient computational fluid dynamics investigations on thermal performance of solar air heater with hollow vertical fins. Therm Sci: Year 22(6A):2389–2399

Experimental and Computational Investigation of Fluid Flow Through an Elbow

K. S. Srikanth and Jayaraj Yallappa Kudariyawar

Abstract In this research work, experimental investigations and computational studies on curved pipe were performed. CFD analysis of turbulent flow through a curved pipe has been performed using standard different turbulence models. Various flow parameters such as velocity, discharge, pressure difference and Reynolds number have been calculated by using experimental observations. The pressure difference value using abovementioned turbulence models has been calculated. Then, the comparison between experimental and data obtained from CFD analysis was carried out. Computational investigations with different turbulence models help to understand the flow pattern in an elbow. CFD simulations were performed using ANSYS FLUENT CFD software. Obtained computational results are matching well with experimental observations. Some important flow patterns such as secondary flow were explained with the help of CFD results.

Keywords Fluid flow · CFD · Turbulence model · Elbow · Secondary flow

1 Introduction

Transportation of fluids from one place to another place is usually through pipe networks. In these pipe networks, there will be pressure loss. This pressure loss may vary depending on the type of components in the pipe network, pipe material, the fluid that is being transported through the network and pipe fitting. Pipe network analysis is necessary to understand the flow pattern through pipe network. Many engineering problems are dealt with it.

Elbow and other fittings are used in a plumbing systems or pipe network to connect pipes or regulating and measuring flow fittings. Based on nature of flow, fluid flow can be classified as laminar and turbulent flow. If the flow is irregular fluctuations and mixing, then it is a turbulent flow. In such kind of flow, both magnitude and

K. S. Srikanth · J. Y. Kudariyawar (✉)
Department of Mechanical Engineering, SDM College of Engineering & Technology Dharwad, Dharwad 580 002, India
e-mail: jayarajyk@gmail.com

A. Maurya et al. (eds.), *Recent Trends in Mechanical Engineering*, Lecture Notes in Mechanical Engineering, https://doi.org/10.1007/978-981-19-7709-1_42

direction of flow are continuously changing. Turbulent flow occurs if the Reynolds number is greater than 4000 for flow through a pipe.

Secondary flow exhibits in curved pipes. Secondary flow is mainly due to the distortion of axial velocity profile and the peak velocity point is shifted to the outside. Distortion of velocity profile is mainly due to the centripetal force. This is acting at normal to the main direction of flow. Generation of secondary flow and its overall study helps to understand flow phenomenon in pipe bends, cooling and heating coils, blade passages of turbo machinery and aircraft intakes.

2 Literature Review

Patankar and Spalding [1] estimated turbulent flow in curved pipes. They found new difficulty, namely that of 'modeling' the turbulence phenomena. They used turbulence model of mixing-length type. Patankar et al. [2] performed theoretical analysis of turbulent flow through curved pipe by employing finite-difference procedure. They used two-equation turbulence model. Comparison of analytical values with the experimental data was performed and obtained good agreement between them. Such kind of studies has been performed by Rodi and Spalding [3], Hanjalic [4] and others. Later, comparison analytical results with experimental results for the developing flow in a 180° bend were performed by Rowe [5] and Hogg [6]. Mori and Nakayama [7] studied on fully developed flow. Fully developed flow region friction factors are compared with empirical correlation of Ito [8].

Athanasia Kalpakli [9] investigated experimentally turbulent flow with and without an additional motion, swirling or pulsating, superposed on the primary flow. They observed strong secondary flow. The three-dimensional flow field depicting varying centerline patterns has been captured under swirling and pulsating flow conditions.

Many researchers restricted their work on fully developed region and average friction factor. To understand the detailed measurement of turbulence quantities and Reynolds stress tensor in curved pipes, there is a need of detail study to use the turbulence models to simulate flow through an elbow.

CFD simulation with appropriate turbulence model may predict the accurate flow phenomenon in an elbow. In the present study, experimental investigations and CFD analysis using standard $k - \omega$ and also $k - \varepsilon$ [10] turbulence models have been performed, and results are discussed.

3 Experimental Study

Figure 1 shows the experimental facility to measure pressure drop across the bend and various fittings.

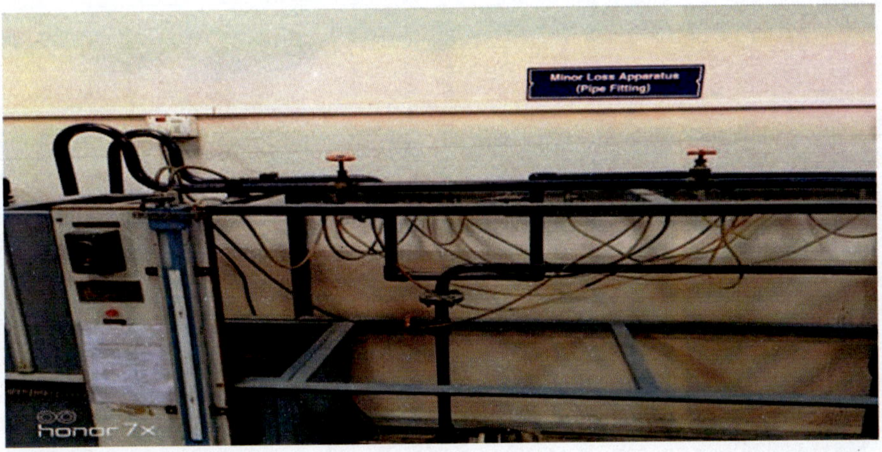

Fig. 1 Experimental facility

Bends in the pipe are provided to change the direction of flow through the pipe. An additional loss of head will occur due to bend. The observed loss of head due to bend was 123.56 pascals.

4 CFD Analysis

Fluid flow simulation has been performed by using CFD. In this, fluid flow governing equations are solved by numerical methods with appropriate initial and boundary conditions. Nowadays, due to the advancement of computational technology and development of mathematical models, complex problems like turbulent flow can be simulated. Various parameters of the fluid flow can be determined with the help of more accurate CFD tool. Turbulence model is a mathematical model to solve the system of mean flow equations of turbulence. Turbulent flows may be computed using several different approaches, either by solving the Reynolds-averaged Navier-Stokes equations (RANS) with suitable models for turbulent quantities or by computing them directly. Two-equation turbulence models are very widely used, as they offer a good compromise between numerical effort and computational accuracy.

Flow in straight pipe is two-dimensional, but flow in an elbow is three-dimensional. The fluid flow is governed by a set of conservation equations. RANS equations are adopted in turbulent flow models.

4.1 k − ε Turbulence Model

Standard $k-\varepsilon$ turbulence model relates Reynolds stresses to mean velocity gradient and the turbulent viscosity using gradient diffusion hypothesis. The most disturbing weakness of this model is lack of sensitivity to adverse pressure gradients; another shortcoming is numerical stiffness when equations are integrated through the viscous sub-layer, which are treated with damping functions that have stability issues. This model is valid for flows without separation and for fully turbulent flow.

4.2 k − ω Turbulence Model

This model allows for a more accurate near wall treatment with an automatic switch from a wall function to a low-Reynolds number formulation based on grid spacing. Demonstrates superior performance for wall-bounded and low-Reynolds number flows. Shows potential for predicting transition. Options account for transitional, free shear and compressible flows. The eddy viscosity ν_T, as needed in the RANS equations, is given by: $\nu_T = k/\omega$.

The $k-\omega$ model developed by Wilcox is the only two-equation model which can be integrated to the wall without using damping equations or the distance to the wall and hence should behave well numerically. Effects of free stream turbulence, mass injection and surface roughness are easily included in the model. Finally, transition can be simulated using low-Reynolds number of the model. It works on low-Reynolds number.

4.3 Modeling and Meshing of Computational Domain

In this study, a pipe of 25 mm diameter and 432.5 mm length is extruded, and a bend of having a radius 45 mm at an angle 90° is modeled. Further, it is extruded to a length of 300 mm. The diagram (Fig. 2) shown below is of the model used in this study.

The computational domain is meshed to solve the governing equations accurately. Figure 2 shows the meshed computational model. Analysis is done for every meshed area, and the summation of all the areas shows the total property gradient of the model. One can control the meshing by selecting different properties of mesh like size of mesh area, meshing style, mesh thickness, etc. Figure 2 illustrates a meshed model of the pipe used.

After creating the geometry and meshing as per requirement, then analysis has done with turbulent case of flow. Now, after completing the first three steps of modeling, meshing and setup, the main body of analysis starts from here. In this

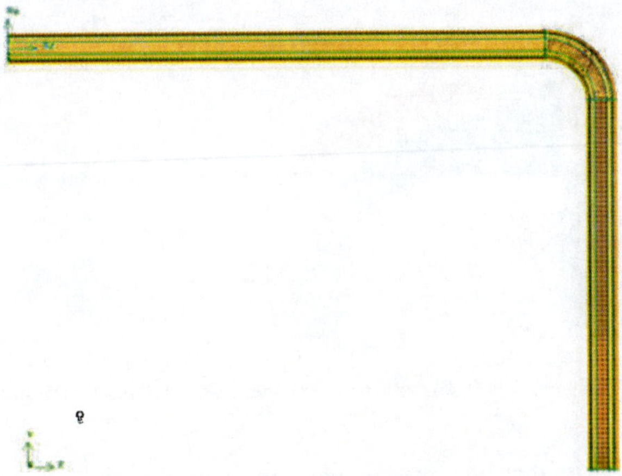

Fig. 2 Front view of a model with mesh

part, all the necessary steps are defined like types of flow, materials, models and boundary conditions, etc.

5 Results and Discussions

After creating the geometry and meshing as per requirement, then analysis has been performed with turbulent case of flow. Turbulence flow analysis has been performed with both viscous-$k - \varepsilon$ and viscous- $k - \omega$ turbulence models. Fluid materials of water are incorporated. Boundary condition such as velocity inlet: velocity—0.23 m/s, temperature 293.15 K, turbulent intensity 5%, hydraulic diameter 25 mm and pressure outlet: gauge pressure—0 Pa.

By computation, the pressure difference in $k - \varepsilon$ and $k - \omega$ models is found. They are as follows: $k - \omega = 131.66$ Pa, $k - \varepsilon = 151.38$ Pa.

Comparative analysis of experimental results (i.e., head loss due to bend is 123.56 pascals) and CFD results gives that $k - \omega$ model predicts more accurate results compared to $k - \varepsilon$ model for turbulent flow in curved pipe.

Further, CFD results are compared with predictions of flow through elbow available in the literature [11]. It has been observed that computational results predict the flow behavior, and there is less than 10% error between CFD, and literature data was found (Fig. 5).

Figure 3a, b shows the predicted pressure contours in the developing region of curved pipe using CFD analysis. The standard $k - \omega$ model gives more accurate result compared to $k - \varepsilon$ turbulence model. Through this comparative analysis, it is inferred that $k - \omega$ model is the best suitable model for predicting the turbulent flow in curved pipes among the two models used. The $k - \omega$ turbulence model is more nonlinear, and it is low-Reynolds number model. Generally, low-Reynolds number models are more accurate. Therefore, the $k - \omega$ model is the suitable model

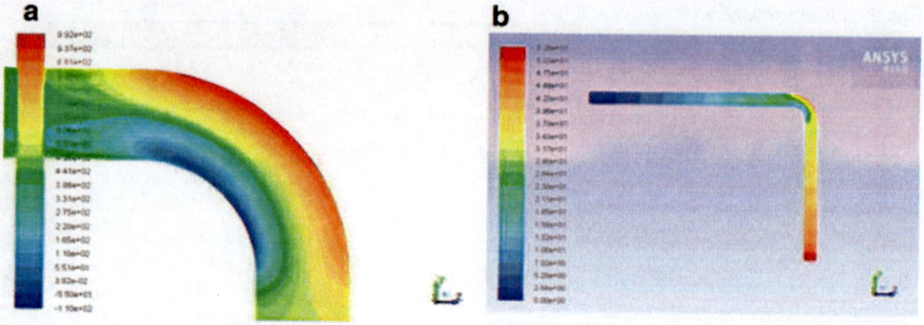

Fig. 3 Contour of static pressure obtained by using $k - \omega$ model

to predict flow phenomenon in internal flows, separated flows and jets and flow in strong curvature. Turbulence model $k - \omega$ is useful in cases, where turbulence model $k - \varepsilon$ is not accurate.

Figure 3a, b shows presence of secondary flows in an elbow. These 3D secondary effects change the dynamics of flow. The flow in curved pipes is different than flow in straight pipe principally by exhibiting a secondary flow in planes normal to the main flow. The transverse pressure gradient at the bend is predicted by CFD and is shown in Fig. 3b. Transverse pressure gradient provides centripetal force to the fluid elements to change the direction. Pressure gradient required for the fluid near the center of the pipe to follow the curved pipe is greater than that required for the slower moving fluid near the wall. This results in the fluid near the center of the pipe moving toward the wall of pipe (outward direction) and fluid near the wall moving inward, which leads to the secondary flow. Figure 4 shows the secondary flow pattern. Secondary flow will leads to the turbulence. CFD results (Fig. 3a, b) show that, $k - \omega$ turbulence model able to predict secondary flow phenomena in an elbow (Fig. 5).

Fig. 4 Secondary flow pattern

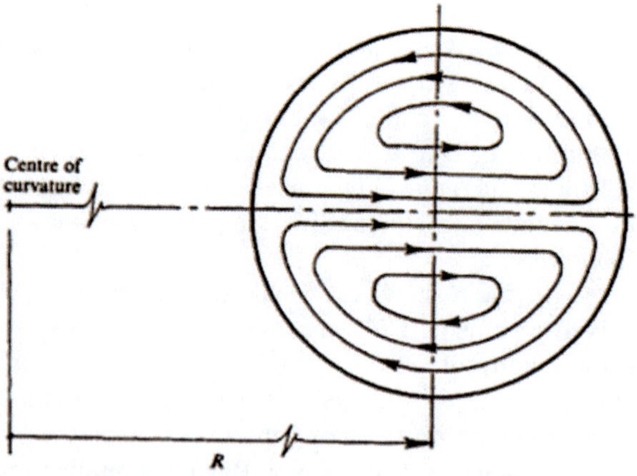

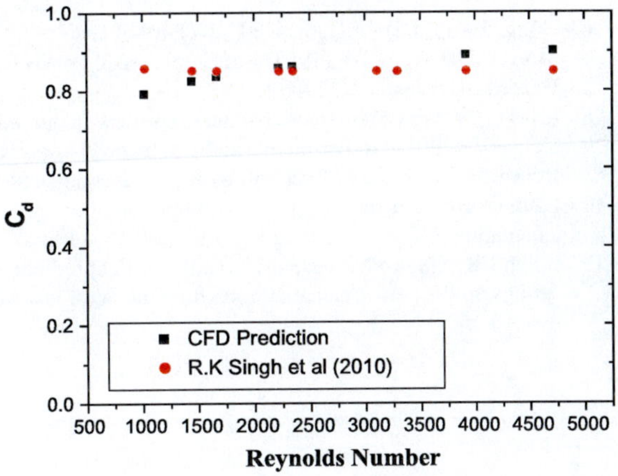

Fig. 5 Comparison of CFD data with experimental data available in the literature

6 Conclusions

An experimental and computational investigation on flow through an elbow was performed. It has been observed that CFD results are able to predict experimental observations. CFD results are also validated with the available experimental data in the literature. Following observations were made in this research work.

- CFD results compared with experimental data and show good agreement with experimental data.
- The standard $k - \omega$ model gives more accurate result compared to $k - \varepsilon$ turbulence model.
- Comparison of CFD results with available experimental data in the literature shows the accuracy of CFD model in predicting the flow characteristics.
- CFD results are helpful to understand the flow characteristics in an elbow.
- Secondary flow in an elbow has been predicted by CFD model.
- This study also deals with an understanding of turbulence models and their governing equations.

References

1. Patankar SV, Spalding DB (1972) A Calculation procedure for heat, mass and momentum transfer in tree-dimensional parabolic flows. Int J Heat Mass Transfer 15:1787–1806
2. Patankar SV, Pratap VS, Spalding DB (1975) Prediction of turbulent flow in curved pipes. J Fluid Mech 67(part 3):583–595
3. Rodi W, Spalding DB (1972) A two-parameter model of turbulence and its application free jets. Warme-and Stoffubertragung 35:20–30
4. Hanjalic K (1970) Ph. D. thesis. London University
5. Rowe M (1966) Some secondary flow problems in fluid dynamics. Ph.D. thesis. Cambridge University

6. Hogg GW (1968) Ph.D. thesis. University of Idaho
7. Mori Y, Nakayama W (1967) Study on forced convective heat transfer in curved pipes. Int J Heat Mass Transfer 10:37–59
8. Ito H (1982) Friction factors for turbulent flow in curved pipes. J Basic Eng 123–132
9. Kalpakli A (2012) Experimental study of turbulent flows through pipe bends. Technical Reports from Royal Institute of Technology KTH Mechanics SE-100 44 Stockholm, Sweden
10. Launder BE, Sharma BL (1974) Application of energy-dissipation model of turbulence to the calculation of flow near spinning disc. Lett Heat Mass Transfer 1:131–138
11. Singh RK, Singh SN, Seshadri V (2010) CFD prediction of the effects of the upstream elbow fittings on the performance of cone flow meters. Flow Meas Instrum 21:88–97

Bending Strength of 3d Printed Composite Gears

S. A. Megha Anand, Y. Arunkumar, M. S. Srinatha, and A. R. Rajesha

Abstract Gears are used to transmit the power from one shaft to another by means of successively engaging teeth. The materials used for gears range from metal to polymer composite materials. Composite materials are widely used in structures with weight as a critical factor, especially in aerospace industry and in many other industries. Recently, additive manufacturing technology has gained lot of importance in making composite materials. The amount of power transmitted by gears mainly depends on the bending strength of gear tooth. This paper deals with experimental determination of bending stress in teeth of 3D printed gear. The main purpose of the test is to determine the maximum tensile bending stress developed in gear using single tooth specimen.

Keywords 3D printed gear · Bending strength · Additive manufacturing

1 Introduction

Additive manufacturing (AM) involves manufacturing of a component by laying one material layer over the other. In automotive and aerospace industries, the traditional manufacturing of parts using composite material is a common practice by laying continuous reinforcing fiber within a matrix. Nevertheless, the use of composite materials in additive manufacturing seems to be relatively new. It significantly increases the structural applicability of parts fabricated by additive manufacturing. The introduction of composite materials to additive manufacturing takes the technology from prototyping stage further to the fabrication of strong functional parts [1]. In this paper,

S. A. M. Anand (✉)
Department of Mechanical Engineering, PES University Electronic City Campus, Bengaluru 5600100, India
e-mail: megha1914@gmail.com

Y. Arunkumar · M. S. Srinatha
Department of Industrial and Production, Malnad College of Engineering, Hassan 573202, India

A. R. Rajesha
Department of Mechanical Engineering, Govt. Polytechnic, Kushalnagar 571234, India

A. Maurya et al. (eds.), *Recent Trends in Mechanical Engineering*, Lecture Notes in Mechanical Engineering, https://doi.org/10.1007/978-981-19-7709-1_43

the terms '3D printer' refers specifically for 3D additive manufacturing machine and '3D printing' for the additive manufacturing process by 3D printer. Gears can be manufactured using additive manufacturing techniques like 3D printing which will reduce manufacturing time and reduces noise during meshing. Generally, gears are made up of steel or cast iron. Because of sufficient mechanical properties like tensile strength, thermal strength, etc., composite materials can be used in the process of manufacturing gears. These properties are similar to that of steel, CI, aluminum composites. Due to this reason, metallic gear can be replaced by composite materials gear [2]. Gears are kinematic elements which are used worldwide in many applications. The design features built into the gears, rule its operating characteristics. The power that gears can transmit depends on the maximum permissible tooth load during mesh. Modern industrial drives require gears that can transmit heavy power with still good performance [3]. Gear stress analysis and understanding the effect of misalignment and micro-geometry is important for gear designers and for those who work in gear maintenance. Bending stress and contact stress are the two important types of stresses that the spur gears undergo when subjected to several stress, from the design point of view [5].

1.1 Objective

- To characterize the mechanical properties of 3D printed polymer matrix composite material.
- Additive manufacturing of spur gear.
- Experimental study of load distribution on onyx reinforced fiber glass and onyx reinforced carbon fiber spur gear tooth.
- Determination of experimental bending stresses in 3D printed polymer matrix composite spur gear.
- Comparison of load distribution and bending stress in onyx reinforced carbon fiber gear tooth with onyx reinforced fiber glass.

2 3D Printing

Additive manufacturing (AM), also referred to as 3D printing, is a manufacturing process that produces three-dimensional (3D) parts from computer-aided design (CAD) software. From the CAD software, a file is generated in Standard Tessellation Language (STL), which is then imported into software called a Slicer, which slices or discretizes the model into layers and generates the instructions used by the 3D printer. The desired parts are fabricated by feeding instructions onto the 3D printer which can also be called as G code loading. Fused deposition modeling (FDM) is one of the categories of 3D printing [6]. Many materials can be used for FDM printing such as Kevlar, glass fiber, carbon fiber, nylon, onyx [7]. In this work, the focus will be on onyx and onyx reinforced carbon fiber.

2.1 Materials for 3D Printing

The carbon fibers, which are alternatively called as CF or graphite fiber, are about 5–10 μm in diameter, which is composed mostly of carbon atoms. These carbon fibers have several advantages such as high stiffness, high tensile strength, low weight, high chemical resistance, high temperature tolerance, and low thermal expansion. Thus, all these properties have made carbon fiber very popular in aerospace, civil engineering, military, and motorsports [4].

Glass fiber combines high strength, high stability, transparency, and resilience at a very reasonable cost-weight performance. The utilities of high-strength glass fiber composites are compared by physical, mechanical, electrical, thermal, acoustical, optical, and radiation properties. Fibber glasses continuous providing high strength at an accessible price. It is 2.5 times stronger and eight times stiffer than onyx. Fiberglass reinforcement results in strong, robust tools [4].

Onyx is thermoplastic which offers a chemical resistance, heat resistance, and high strength and surface finish. 3D printing material onyx used as a thermoplastic matrix. Onyx used in the 3D printing can be reinforced with the continuous fibers or printed to give strength similar to aluminum. Today, there are more than a million onyx 3D printed parts in the field transforming manufacturing [4].

2.2 Material Properties of Polymer Matrix Composite

The specimen analyzed in this study was modeled with SolidWorks computer-aided design (CAD) software (SolidWorks 2016). The geometry of the specimen was defined as per ASTM D3039. The test machine and specimen dimensions are shown in Fig. 1. The modeled specimen was exported from SolidWorks as a Standard Tessellation Language (STL) file and imported into the appropriate slicer. Markforged Mark Two printer was used to print the specimen. The 3D printed part was fabricated using Onyx, continuous carbon fiber. The part fabricated using reinforced carbon fiber yielded the largest increase in mechanical strength.

The tensile test speed for ASTM D3039 specimen can be determined by the material specification or time to failure (1–10 min). The speed for the standard test specimen 2 mm/min (0.078 in/min) be maintained throughout the test. To determine elongation and tensile modulus, an extensometer or a strain gauge is used. Table 1 and Fig. 2 show the tensile test results. The test results show the peak stress of onyx reinforced carbon fiber is more compared to onyx.

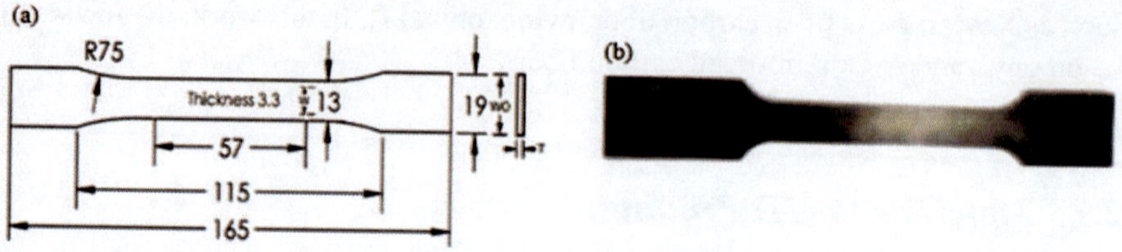

Fig. 1 **a** ASTM D3039 dimension. **b** 3D printed test specimen

Table 1 Tensile test results

Properties	Onyx + Fiber glass	Onyx + Carbon fiber
Peak stress (MPa)	45.72	54.16
Peak load (kN)	2.78	3.293
Yield strain (%)	1.61	1.44
Yield load (kN)	1.174	2.333
Modulus (GPa)	1.328	3.16
Upper yield point (MPa)	19.313	38.38
Lower yield point (MPa)	37.461	2.831
Limit of proportionality (MPa)	15.005	31.025
Elongation at break (Using strain) (%)	5.80	2.681
Reduction in area at break (User input) (%)	100	100
Strain % at max. load	5.755	2.658

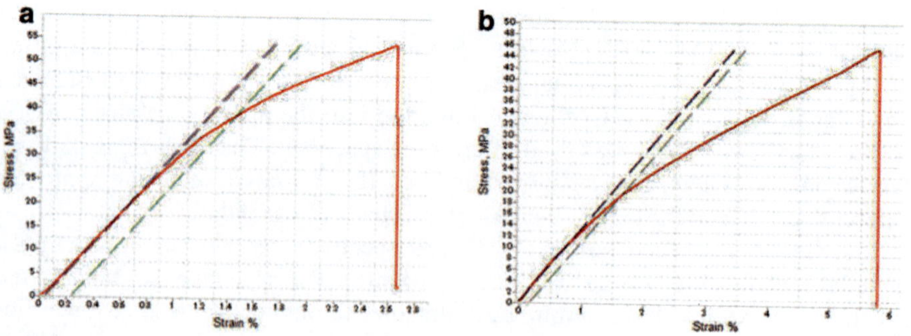

Fig. 2 **a** Stress strain curve for onyx reinforced carbon fiber. **b** Stress strain curve for onyx reinforced fiber glass

3 Spur Gear Design

Design specification of polymer matrix composite gear tooth is shown in Table 2.

The onyx reinforced carbon fiber and onyx reinforced fiber glass spur gears of above specification was modeled using SolidWorks and 3D printed using Markforged Mark Two 3D printer. Figure 3 shows the strain gauge mounted 3D printed onyx reinforced carbon fiber and onyx reinforced fiber glass spur gears.

3.1 *Experimental Stress Analysis as Applied to Gear Design Problem*

The amount of power transmitted by gears largely depends on the bending strength of the gear tooth. Experimental investigations in engineering practice build confidence among engineering and research fraternity. Testing of a gear tooth for its bending

Table 2 Design specification of gear tooth

Parameters	Design details
Teeth (Z)	50
Module (M)	4 mm
Pressure angle (α)	20°
Pressure angle at tip	29.53°
Tooth height	8.184 mm
Tooth thickness	6.996 mm
Normal tooth load	218.51 N

Fig. 3 3D printed onyx reinforced glass fiber and onyx reinforced carbon fiber spur gear tooth

strength is essential before it is put into service. As such, this chapter deals with experimental determination of bending stress in the teeth of standard dimensions. The main purpose of the test is to determine the maximum tensile bending stress developed in the fillet of the representative gear tooth using a single tooth specimen; each specimen employs strain gauge to perform stress analysis.

For this purpose, a gear tooth bending test (GTBT) fixture is developed in which a single tooth specimen of the gearing is held. A strain gauge is bonded around the fillet region of the tooth which is calibrated to measure the strain, and a strain indicator gives the digital display of the strain induced. Applying the load on the tip of the tooth, the strain induced in the fillet is obtained by the strain indicator.

3.2 *Estimation of Experimental Bending Stress*

The variables in this experiment are applied load and the induced strain. Experimental bending stress setup is shown in Fig. 4. The measurement of strain is the most important task of the experimental investigation. This is done using a strain gauge which is bonded on the specimen tooth and the strain indicator which indicates the strain, making up the complete instrumentation. Thus, on applying the load, the calibrated strain indicator directly displays the strain (micro-strain) that greatly simplifies the task. The load applied ranges from zero to 218.51 N in steps of 19.62 N, and the strain corresponding to the load is noted. This indicated strain is further used to estimate the bending stress in the gear tooth using Eq. (1)

$$\text{Experimental bending stress } \sigma_{\text{EXP}} = \text{e.E (MPa)} \tag{1}$$

where

e indicated micro-strain ($\times 10^{-6}$)
E elasticity of material.

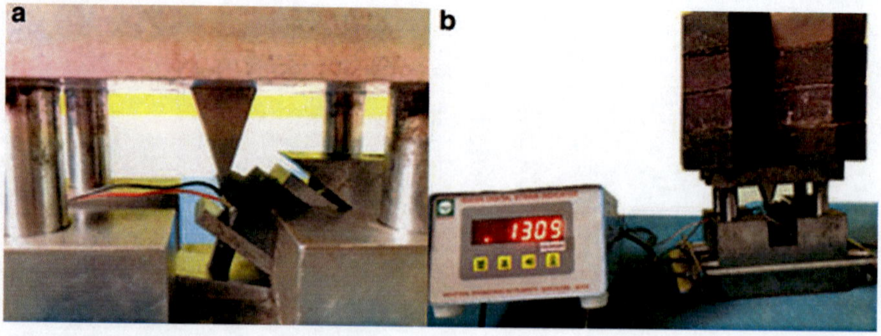

Fig. 4 **a** Gear tooth bending fixture (GTBF). **b** Experimental setup

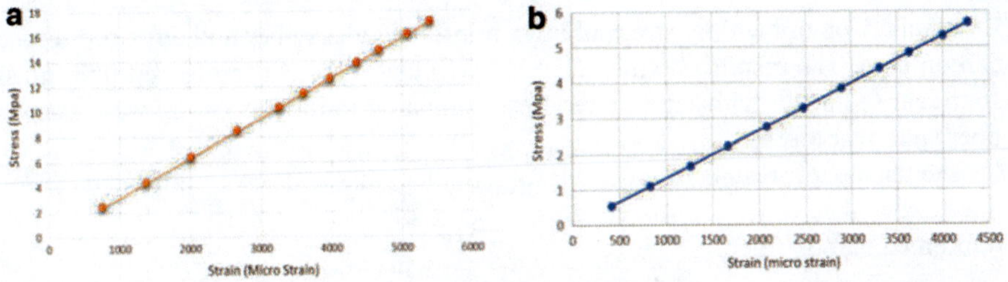

Fig. 5 Stress–strain curve for **a** onyx reinforced with carbon fiber, **b** onyx reinforced fiber glass

Table 3 Bending stress in onyx and onyx reinforced carbon fiber

Load (N)	Bending stress of onyx reinforced fiber glass (Mpa)	Bending stress of onyx + carbon fiber (Mpa)
22	0.555104	2.389
20	1.099584	4.326
39	1.665312	6.380
59	2.213776	8.431
78	2.764896	10.292
98	3.2868	11.414
118	3.841904	12.583
137	4.38904	13.797
157	4.824624	14.782
177	5.312	16.078
196	5.6772	17.086

4 Comparison of Experimental Results

In this section, the experimental results obtained are presented to draw a proper inference. The plots in Fig. 5 show the stress–strain curve for onyx reinforced fiber glass and onyx reinforced carbon fiber spur gear tooth. The results obtained by the experimental method show that the onyx reinforced fiber glass spur gear tooth can withstand more stress compared to the onyx reinforced carbon fiber spur gear tooth.

Table 3 shows the load distribution and variation of the bending stresses in onyx reinforced glass fiber and onyx reinforced carbon fiber spur gear tooth.

5 Conclusion

The composite spur gear tooth was manufactured using 3D printer. A strain gauge bonded in the fillet region is used for measuring the strains which are indicated by well calibrated digital indicator. The stresses computed using the strains are plotted for comparison of bending stress in onyx reinforced glass fiber and onyx reinforced carbon fiber. The results obtained clearly show that the onyx reinforced fiber glass spur gear tooth can withstand more stress compared to the onyx reinforced carbon fiber spur gear tooth.

The tensile test results of onyx reinforced fiber glass and onyx reinforced carbon fiber show that onyx reinforced glass fiber tooth has more strength than onyx reinforced carbon fiber.

References

1. Ghebretinsae F, Mikkelsen O, Akessa AD (2019) Strength analysis of 3D printed carbon fibre rein-forced thermoplastic using experimental and numerical methods. In: IOP conference series: materials science and engineering, vol 700
2. Ingale YA, Kadam AP, Bhosale DG (2016) Comparative study of steel, nylon 66 and delrin helical gears used in steering gearbox. Int J Res Eng Technol 05(06)
3. Rajesh AR, Gonsalvis DJ, Venugopal DK (2015) Investigation of tooth load distribution along the path of contact in altered tooth-sum gears. Int J Modern Eng Res (IJMER) 5(4)
4. https://markforged.com/3d-printers
5. Rajesh Kumar S, Sivasankar M, Muruganandam A, Sathishkumar S, Arun Kumar S (2016) Design and contact analysis of plastic spur gears using finite element analysis. Int J Adv Res Trends Eng Technol (IJARTET) 3(19)
6. Ning F, Cong W, Qiu J, Wei J, Wang S (2015) Additive manufacturing of carbon fiber reinforced thermoplastic composites using fused deposition modelling. Compos Part B Eng 80:369–378
7. Dickson AN, Barry JN, McDonnell KA, Dowling DP (2017) Fabrication of continuous carbon, glass and Kevlar fiber reinforced polymer composite using additive manufacturing. Addit Manuf 16:146–152

An Efficient Robot with Wireless Control

Suresh Kurumbanshi and Shashikant Patil

Abstract Recent advances in motion control technology placing various challenges in terms of accuracy & speed of operations. Development of accurate model & algorithm becomes the important aspect in robotics and control operation. It is seen that, 3–4 degrees of freedom robotic arms are the most common in industries and mostly used by medium and large scaled industries. In industries, there is a great need of constantly moving the goods from one place to another in logistics, warehouses, etc. This task is trivial and does not require human decision-making skills. But, employing robotic arms is expensive and requires skilled workers to operate them. Currently most of the Micro, Small & Medium Enterprises use human workers to accomplish this task. A robotic arm can perform this function with negligible amount of human interference. Proposed robotic arm will make sure that the workers can perform other skilful tasks also. With the help of 5 degrees of freedom, it also makes the task of goods manipulation easy and moves around in the facility with the help of its car. It performs the required tasks at different places which attempts to eliminate the need of incorporating static robotic arms. Proposed robotic is controlled with a smartphone and saves the motion of robotic arm as well as the motion of the car which in turn helps in autonomous operation. The developed algorithm has the ability to control the motion of the robot in the indoor designated area.

Keywords Motion control technology · Degrees of freedom · Path planning · Wireless control

S. Kurumbanshi (✉)
Department of Electronics & Telecommunication Engineering, MPSTME, NMIMS University, Shirpur Campus, Savalade, India
e-mail: suresh.kurumbanshi@nmims.edu

S. Patil
Computer Science and Engineering Department (AI&ML), ViMEET, Khalapur, Raigad, Maharashtra, India

A. Maurya et al. (eds.), *Recent Trends in Mechanical Engineering*, Lecture Notes in Mechanical Engineering, https://doi.org/10.1007/978-981-19-7709-1_44

1 Introduction

Robotics have been developed and drawn attention in industries widely for several applications. Tele-operated mobile robots are used recently in the robot research, and they are utilized in many fields and provided better efficiency. Nowadays Internet of things enabled robotics are involved for computing and networking purposes. Intelligent robotics are utilized to reduce human resources labour intensity and improves the efficiency. Robots have very important role in various applications such as nuclear sites, space exploration, military fields and telemedicine. Robots are being used in many dangerous environments for ensuring the workers' safety. So tele-operated mobile robots have catched the attention in the era of intelligent robotics. In recent years, developments have taken place in Wi-Fi network technology. Therefore, if Wi-Fi and robot technology are integrated, it has been developing as a new direction of robot research. Advances in sensor technology, perception and developing control algorithm are focussed in the next-generation industrial robotics. It will enhance the data processing capability and efficiency of robotics operation. There are many challenges in developing these robotics applications for precise movement and accuracy [1–6].

2 Related Work

Autonomous driving may raise serious controversies, especially in recent deadly accidents. Autonomous vehicles are gaining popularity and attracted may people. Motion planning techniques have been developed over the couple of years with a focus on highway planning. Motion planning is required for path generation and subsequent decision making. There are certain limitations during high speed and small curvature roads with limitations of driver rules [7]. Recent developments in multicontact planning and control, embedded simultaneous localization and mapping bipedal walking, multisensory task-space optimization control, contact detection and safety proposes that the humanoids is the best solution for automation in large-scale manufacturing sites. Challenge comes in integrating these scientific and technological advances into existing humanoid platforms. These are the torque-controlled robot and position-controlled Human Robotics [8]. The prostheses in robots have been instrumental in restoring their joint functions and enables them to perform diversified challenging tasks. Although it is observed that ankle prosthesis and robotic knee prosthesis have purpose of restoring these joint functions for lower-limb amputees, they have been identified as a distinct, standalone devices [9]. Robotics are also incorporated to mop the floor using the sponge, and use this method in designing a mopping module framework of a multifunction floor cleaning robot. Parameters are optimized by analysing the amount of the pumping out and recycling water [10]. Artificial Intelligence has introduced many required feature in Industry 4.0. Recent robotics uses theses feature for performing the tasks collaboratively. Currently industrial robotics have been deployed for automation in highly-controlled environments. Intelligent robots are developed to help people in daily life. An ARM vehicle robot is developed using STM32F103ZET6 microcontroller, along with a tracking obstacle avoidances facility, camera platform and a SG90 steering gear, and combines them to develop an ARM vehicle robot. Robot realizes the real-time video viewing and established robot control through APP interface of mobile phone [11]. Wireless sensor NRF24L01 is used to build a multi-robot cooperative system. Central server would be controlling robot by sending control commands show that the system can run successfully, the Central Server can control each robot by sending control commands. The host computer can control the robots. The host computer controls the robots and would be able to display the information of multi robots on real-time basis [12]. Mobile robotics with IOT is developed for farming applications. Master and slave configuration in robots has built wireless sensor network and those are connected via the NRF protocol for reliable sensor data sharing. The master robot effectively transmits the data to the IoT server established [13–15].

3 Methodology

Need of automatic system for manipulation of goods and control mechanism has incorporated 3–4 DOF robotic arm and mostly used in medium and large scaled industries. Additional of 1 DOF is incorporated for better goods manipulation. The mobility feature is also incorporated so that the same arm can be used in different sections of the warehouse. There are 4 pick and place facilities and the arm needs to perform pick and place operation in those facilities. The facilities 2 and 3 have complex pick and drop operation for this reason the 5th degree of freedom is added to the robotic arm. The mobility feature is incorporated through car which helps the robotic arm to move to respective facilities with ease and perform the operation. The arm is controlled by the application. The speciality of this application is that only one time input of user is needed that is the user needs to put the desired operation in application and the operation will perform endlessly without the involvement of the user (Fig. 1).

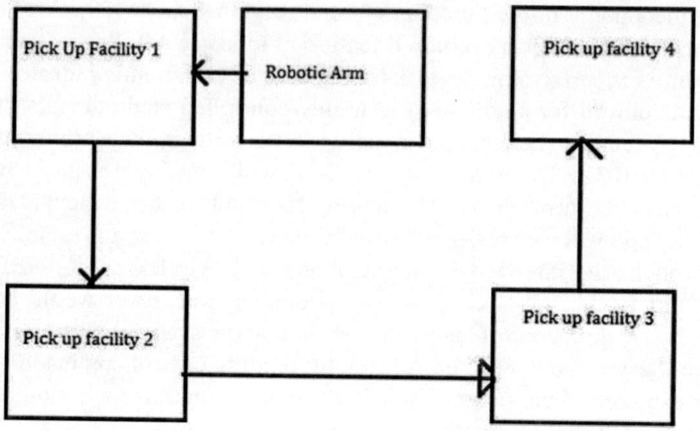

Fig. 1 Analysis diagram

3.1 Desigining of the Prototype

SketcH of robotic arm with 5 degrees of freedom is created and a CAD model is build using the rough sketch in Fusion 360 designing software. According to the sketch, torque calculation and centre of gravity calculation is done and selected MG996R servo motors and SG90 servo motors for the efficient operation of robotic arm. 100 rpm DC motors have been used for the car on which the arm is mounted. An android application is developed for the control of 6 servo motors and 4 DC motors. The links of robotic arm is printed using 3D printer which is an ABS material. The robotic arm is mounted on the car with proper CG which we calculated previously. Testing of the robotic arm and the car have been done separately with their respective electronics circuit. Once both the assemblies works fine individually, we could mount the arm on the car.

4 Results and Implementation

SketcH of robotic arm with 5 degrees of freedom with a CAD model was build using the rough sketch in Fusion 360 designing software. According to the sketch, torque calculation and centre of gravity calculation was done and MG996R servo motors and SG90 servo motors were selected for the robotic arm. 100 rpm DC motors are used for the car on which the arm is mounted. An android application is made for control of 6 servo motors and 4 DC motors. The links of robotic arm is printed using 3D

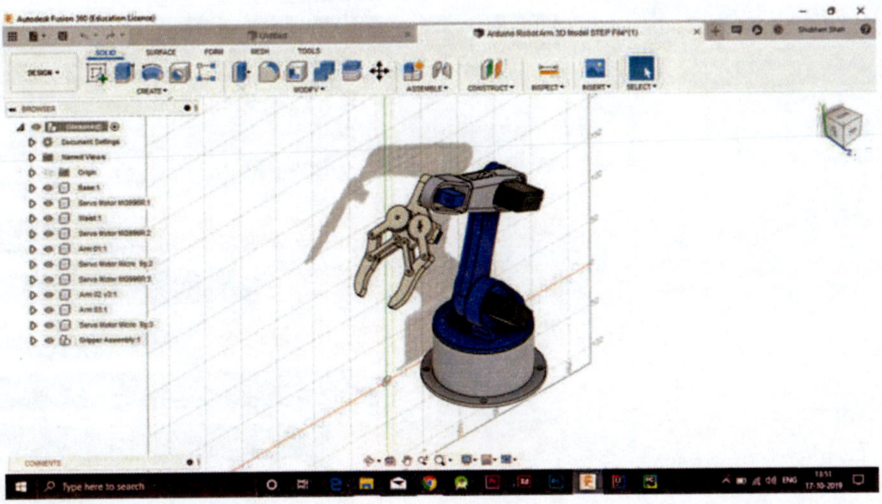

Fig. 2 Robotic arm CAD drawing

printer which is an ABS material. The robotic arm is mounted on the car with proper CG which we calculated previously. Testing of robotic arm and the car is required. Once both the assemblies are working fine individually, arm can be mounted on the car (Figs. 2, 3 and 4).

Built a circuit for servo motion by Bluetooth module and android app.

5 Results and Discussion

A versatile robotic arm is developed for 5 Degrees of freedom. The gripper is designed looking into the requirements of flawlessly goods manipulation.

Stable robotic arm is build considering the shape of links and its other physical attributes such as Mass, Centre of Mass, length, density, volume. For long operational hours a big battery or an efficient system is to make which is achieved by wisely choosing the servo motors used for the motion, the mode of communication, mobility system and the steering mechanism. Chassis can be changed and which can be used for different purposes, which include domestic and industrial purposes. By implementing other changes, it can also make goods manipulation can also more efficient.

Fig. 3 Mobile application

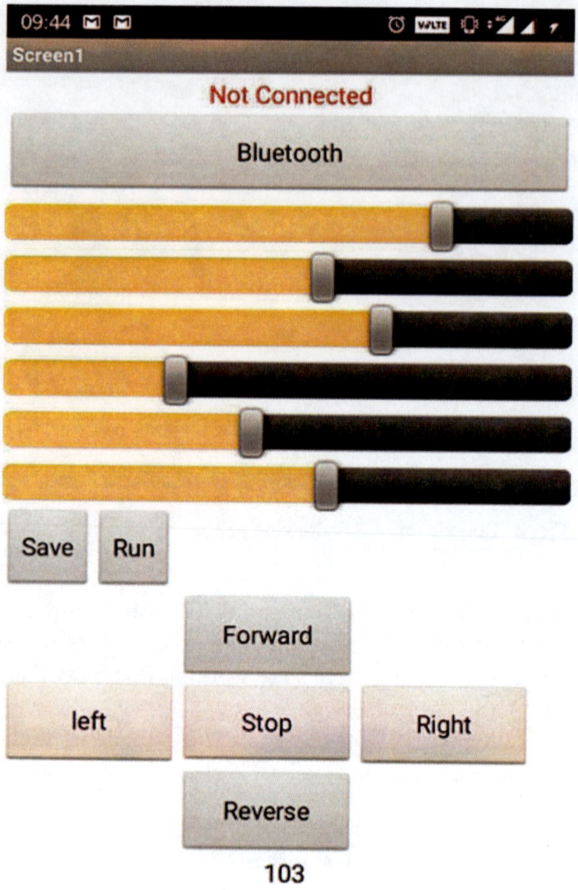

6 Future Scope

Machine learning and deep learning techniques can be incorporated by which it can identify objects by itself and do required action. Mecanum wheels can be incorporated to eliminate the need of mechanism by which it can be used at facilities with less space. With the help of its object identifying quality it can be used in mines to identify threats during exploration and can deal with it if possible. By adding the ability to swap gripper or with a universal gripper design it can hold any object. Mode of communication can be changed from Bluetooth to RF communication and can be used in military and other defence explorations.

Fig. 4 Test circuit

Acknowledgements The proposed work is carried out as a part of undergraduate project and developed under my mentorship by Mr. Vidyul Shah, Shubham Jayesh Shah and Ms. Anuja Sudhir Umrotkar.

References

1. Omijeh BO (2014) Design analysis of a remote controlled pick and place robotic vehicle. Int J Eng Res Dev 10(5):57–68
2. Nair SR (2012) Design of a robotic arm for picking and placing an object controlled using LabVIEW. Int J Sci Res Publ 2(5):1–4
3. Singh P (2013) Design of a robotic arm with gripper & end effector for spot welding. Univ J Mech Eng 1(3):92–97
4. Kadir WMHW, Samin RE, Ibrahim BSK (2012) Internet controlled a robotic arm. Procedia Eng
5. Al-Busaidi AM (2012) Development of an educational environment for online control of a biped robot using MATLAB and Arduino, MECHATRONICS), 9th France-Japan

6. Juang S, Lurrr KY (2013) Design and control of a two-wheel self-balancing robot using the Arduino microcontroller board. Control Autom ICCA
7. Claussmann L, Revilloud M, Gruyer D, Glaser S (2020)A review of motion planning for highway autonomous driving. IEEE Trans Intell Transp Syst 21(5):1–23
8. IEEE Robot Autom Mag (2019)
9. Haque MR, Shen X (2020) A unified knee and ankle design for robotic lower-limb prostheses. In: 2020 IEEE/ASME international conference on advanced intelligent mechatronics (AIM), Boston, USA (Virtual Conference)
10. Tantawi KH, Sokolov A, Tantawi O (2019) Advances in industrial robotics: from industry 3.0 automation to industry 4.0 collaboration. In: The 2019 technology innovation management and engineering science international conference (TIMES-iCON2019)
11. Zhang Q, Wang J, Yinghu LU, Wang W (2019)Design of control system for ARM vehicle robot. In: 2019 international conference on robots & intelligent system (ICRIS)
12. Zhang Y, Xia Q, Xu Y (2019)Multi-robot cooperation system based on wireless network. In: 2019 IEEE SmartWorld, ubiquitous intelligence & computing, advanced & trusted computing, scalable computing & communications, cloud & big data computing, pp 162–165
13. Khan A, Aziz S, Bashir M, Khan MU (2020)IoT and wireless sensor network based autonomous farming robot. In: 2020 international conference on emerging trends in smart technologies (ICETST)
14. Sharma D, Chauhan U (2020) War spying robot with wireless night vision camera. In: 2020 2nd international conference on advances in computing, communication control and networking (ICACCCN), pp 550–555
15. Liu Q et all (2020) Research on key technologies on intelligent fire fighting robot based on Zigbee network. In: 2020 international conference on artificial intelligence and computer engineering (ICAICE), pp 50–53

A Recent Development in Indirect Type Solar Dryer: A Comprehensive Review

Jyoti Singh Parihar, Harish Kumar Ghritlahre, and Manoj Verma

Abstract Drying of agriculture products is a process which demands energy. Due to environmental aspects, high cost and limited availability of fossil fuel an alternate solution must be found. Solar energy is most abundant, clean and sustainable energy source among all the form of green energy. In solar thermal system, solar dryer is a very important device for crop/vegetables drying. Moisture removal from the product is the basic operation of dryer. The process of drying reduces the bacterial growth on the products and as the growth reduces it extends the preservation time of the product. In solar dryers, indirect type solar dryer one of the important types of solar dryer in which solar radiation is absorbed by the absorber plate and transferred to the flowing air through the duct, finally the heated air is used for drying of products. The aim of present work is to review on the different types of investigations carried out in the field of development of indirect type forced convection solar dryer combined with solar air heater. In addition to this, research gap has been also identified for future work.

Keywords Solar energy · Energy · Solar dryer · Energy analysis · Exergy analysis

1 Introduction

India is a country of growing population that clearly indicates our dependency on energy and leads us to the energy crisis. Our reliance on energy makes us more responsible to look over the other alternate energy sources, i.e., solar energy, tidal energy, wind energy, nuclear energy, geothermal energy, etc. [1]. Solar energy is the most promising one among all the forms of renewable energy resources or green energy as it is free, clean and safe for the environment [2]. Utilization of solar energy

J. S. Parihar (✉) · H. K. Ghritlahre · M. Verma
Department of Mechanical Engineering, CSVTU, Bhilai, C.G. 491107, India
e-mail: pariharjyoti5@gmail.com

H. K. Ghritlahre · M. Verma
Energy and Environmental Engineering Department, University Teaching Department, CSVTU, Bhilai, C.G. 491107, India

A. Maurya et al. (eds.), *Recent Trends in Mechanical Engineering*, Lecture Notes in Mechanical Engineering, https://doi.org/10.1007/978-981-19-7709-1_45

increases by the application of different solar thermal devices either by collecting the thermal radiations or by directly converting it into electrical energy by the use of PV panels. Solar thermal systems comprise of solar dryer, solar cooker, solar air heater, solar pump, solar distillation, solar pond, solar air conditioning, etc. [3]. Food demand of India is increasing everyday as a result of human population growth. This demand can be accomplished either by increasing the crop cultivation/production or by minimizing the harvest loss. Therefore reducing the harvest losses hold immense significance. It depends upon the condition of storage and water content. Ultimately storage and moisture removal are the process of immeasurable importance for the food safety and drying is one the most adequate processes of moisture removal so that among all the forms of solar thermal technologies, researchers are mainly focusing toward the solar dryers [4, 5]. Solar dryer is a device that reduces the moisture content (amount of water) and thus minimizes the losses which take place during the storage and maintenance of the crop [6].

The aim of the present study is to review the recent developments of indirect type solar dryers (ITSD). This review paper makes it easy to understand the different design pattern and methods for the improvement of efficiency of the indirect type solar dryer. This review article may be very helpful to increase the performance of solar dryers and also the important research gaps have been reported in this paper.

1.1 Classification of Solar Dryers

Solar dryers can be broadly classified as open sun dryers and controlled dryers on the basis of use of solar energy as described by Prakash and Kumar [7].

Open Sun Dryer

In open sun drying we spread the material to be dried in the floor. It has been adopted in India since ancient times irrespective of the number of disadvantages: (a) large space requirement for drying crop and vegetables, (b) drying is uncontrolled, results in over drying or under drying, (c) unexpected weather condition, foreign particles and attack of insects reduces the quality of the product.

Controlled Dryers

Controlled dryers operate at variable parameters like temperature, pressure; velocity and humidity. It can control these parameters which is very difficult in open sun drying process. Classification of controlled dryers is as mentioned in Fig. 1.

Direct Type Solar Dryer

Direct type solar dryers are most widely used in rural areas. In direct type of solar dryer, the radiations directly fall on the grains, through a transparent mostly glass cover. Direct type solar dryers are basically either cabinet type or greenhouse type dryers. It improves the quality and also reduces the contaminations by foreign particles as compared to conventional open sun drying process.

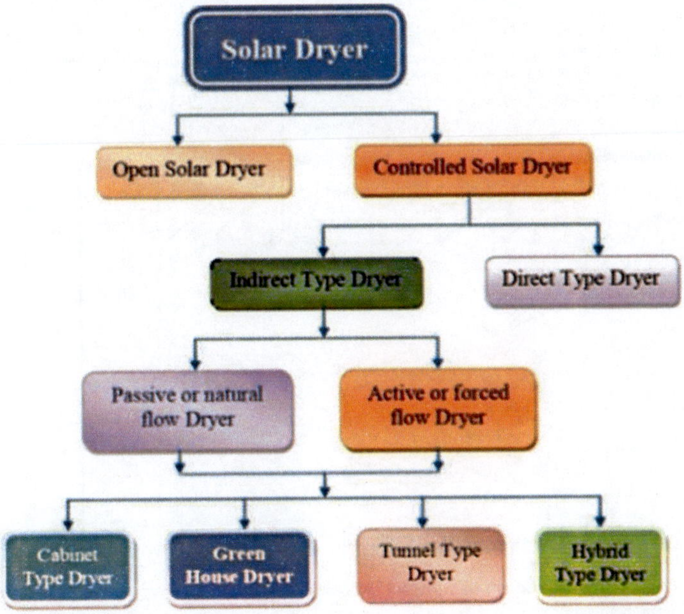

Fig. 1 Classification of solar dryer as mentioned by Prakash and Kumar [7]

Indirect Type Solar Dryer

Indirect type solar dryer separately collects the solar energy through the solar air heater and the hot air is directly fed into the drying chamber by means of various duct arrangements. Schematic diagram is represented in Fig. 2.

It can be further classified as passive and active indirect type solar dryers according to nature of flow of air. In passive solar dryer, also known as natural solar dryers, the flow of air through the passage of heating chamber is due to the natural circulation. In active or forced circulation solar dryer blowers are used use to increase the flow of air through the passage.

Performance study of solar dryer

Performance of a solar drying system can be evaluated by the following equations as given below [8];

Amount of moisture removed

$$\boldsymbol{M}_{\mathbf{evp}} = \frac{\boldsymbol{M}_i(\mathbf{mc}_i - \mathbf{mc}_f)}{(100 - \mathbf{mc}_f)} \quad (1)$$

where M_{evp} is amount of moisture removed, M_i is initial mass of the sample, mc_i is initial moisture content and mc_f is moisture content on dry basis.

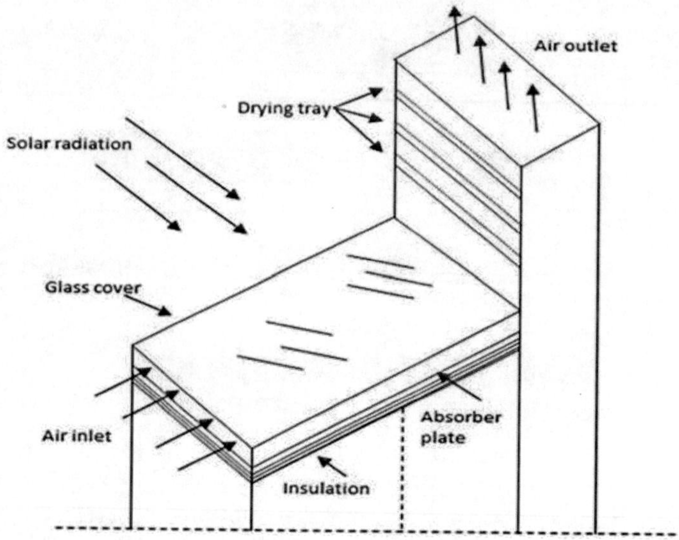

Fig. 2 Schematic representation of indirect type solar dryer

Drying Rate

It is the rate at which moisture from the crops or vegetables gets evaporated to the surrounding atmosphere.

$$\mathrm{DR} = \frac{\mathrm{mc}_{t+dt} - \mathrm{mc}_t}{dt} \quad (2)$$

where DR is drying rate, mc_{t+dt} is moisture content at time t + dt and mc_t is moisture content at time *t*.

Moisture ratio

It is the comparison of moisture content at any instant of time to the initial moisture content.

$$\mathrm{MR} = \frac{\mathrm{mc}_t - \mathrm{mc}_e}{\mathrm{mc}_i - \mathrm{mc}_e} \quad (3)$$

where MR is moisture ratio and mc_e is equilibrium moisture content.

Drying Efficiency

Drying efficiency is the ratio of heat or energy utilized for moisture removal to the overall energy provided. The drying efficiency can be evaluated for active as well as passive convection.

$$\eta_{DN} = \frac{M_{evp} \times h_{fg}}{A_{SC} \times I_R} \quad (4)$$

$$\eta_{DF} = \frac{M_{evp} \times h_{fg}}{A_{SC} I_R \times P_{ex}} \quad (5)$$

where, η_{DN}, η_{DF} natural and forced drying efficiency, h_{fg} is latent heat of vaporization, A_{SC} is area of solar collector, I_R is solar intensity, P_{ex} is power supply.

2 Literature Review of Indirect Solar Dryer

Sharma et al. [9] successfully investigated the performance of indirect type fruit and vegetable solar dryer. Maximum SAH efficiency was recorded in the range of 45–65%. Two air flow system increases the temperature up to 10–27 °C.

El-Sebaii et al. [10] designed, developed and experimentally investigated a natural convection ITSD for grapes, apples, peas and vegetables, etc. SAH is capable to insert various storage materials. Dryer with and without storage material reduces the desired moisture content in 72 h and 60 h, respectively. Pre-chemical treatment results, approximately 8 h decrease in drying time.

Sreekumar et al. [11] successfully developed an indirect type solar dryer as shown in Fig. 3. It is found that 95% of initial moisture content is reduced to 5% in 11 h and 6 h by open sun drying and indirect type solar drying, respectively.

Montero et al. [12] designed, constructed and installed a hybrid system for the performance analysis of drying agro industrial by-product. Experiments were

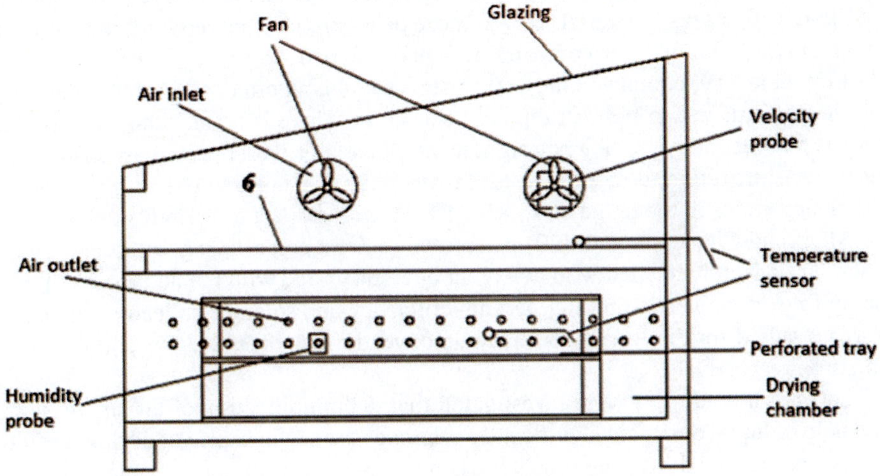

Fig. 3 Indirect type of solar dryer as given by Montero et al. [12]

conducted at a flow rate in between 0.05 and 0.6 m^3/s. Performance improvement was also recorded for indirect solar dryer as compared to direct type solar dryer.

Sebaii and Shalaby [13] experimentally analyzed forced convection ITSD for drying thymus and mint. Initial moisture content of 95% (thymus) and 85% (mint) reduced to desired equilibrium moisture content in 34 h and 5 h, respectively. Midilli and kucuk model is found convenient but page and modified page model is best suited for thin layer drying.

Lingayat et al. [14] analyzed the performance of ITSD for drying banana. V-corrugated absorber plate was used in solar air heater. SAC efficiency and drying efficiency was measured as 31.50% and 22.38%, respectively. They concluded temperature that humidity of air and air velocity are important parameters for performance improvement.

Essalhi et al. [15] compared experimental and theoretical analysis of ITSD and open sun dryer for drying grapes. Final moisture content of 20.2% on wet basis is achieved in 120 h and 201 h by ITSD and open sun dryer, respectively. Midilli et al. model is found best suited for drying grapes.

Sözen et al. [16] successfully developed two designs of SAH, i.e., hollow tube type SAH and heater with iron mesh in the air flow area. Experiment was conducted at three different air flow rates 0.014, 0.011 and 0.009 kg/s. Thermal efficiency of SAH increases 11% by the utilization of iron mesh. Maximum 50.85% solar dryer efficiency was recorded for the dryer in-build with this iron mesh SAH.

Lingayat et al. [17] analyzed drying kinetics and performance parameters of ITSD for banana drying. Initial moisture content of tomato and brinjal reduces from 15.667 to 0.803 kg/kg of db and 10.11–0.498 kg/kg of db. SAC thermal and drying efficiency was calculated 59.09% and 31.4% for tomato. The same was recorded for brinjal as 58.42% and 25.16%, respectively.

Yadavand Chandramohan [18] investigated the effect of fins on the performance of ITSD with storage material. 55.2% more heat gain is observed in case of finned indirect type solar dryer as compared to without finned.

Etim et al. [19] conducted the experiments to find out effect of air flow inlet area on the performance of indirect type of solar dryer. There is no any change is outlet area is provided. Dryer efficiency significantly increases with the increase in air inlet area. Moisture ratio was reduced to 12% from 68.97% in just 9 to 16 h drying. Drying efficiency varies in the range of 13.85–31.84% with respect to various inlet areas.

Mugi and Chandramohan [20] performed indirect solar drying with the help of solar air collector attached with divergent duct integrated with fan and removed this setup for free convection drying. Drying efficiency and solar air collector efficiency was measured for the forced and natural convection 74.98%, 24.95% and 61.49%, 20.13%, respectively.

Sabareesh et al. [21] were investigated that dehumidification of air stream with the help of liquid desiccant significantly improves the performance of drying. Liquid desiccant reduced the drying time 9 h when compared to without desiccant and 13 h when compared to open sun drying.

Salve and Fulambarkar [22] conducted experiments to analyze the combine effect of phase change material, absorber plate coating and mass flow rate. They have recorded maximum temperature and efficiency 96 °C and 33%, respectively.

3 Research Gap and Future Recommendation

Based on the present literature review, the following points of research gaps have been identified for future research works.

1. A north wall reflector is a novel approach to increase the efficiency of indirect type solar dryer.
2. Various techniques like use of phase change material, integration of PV module, different rough surfaces of solar air collector absorber can improve the performance of Indirect type solar dryer.
3. Very few works related to analytical analysis have been conducted. All these analytical results can be optimized to get better output in the term of efficiency.
4. Exergy analysis of indirect type solar dryer is very limited. Also little work related to reduction and elimination of exergy losses has been done.
5. Double pass, double flow techniques can be implemented with indirect type of solar dryer.

4 Conclusion

Various Indirect type solar dryer are successfully reviewed in this paper. It has been observed that ITSD is basically consist of two parts SAH, and drying chamber and connecting ducts. Operating parameters greatly affect the performance of indirect type solar dryer. Air flow rate and mass flow rate significantly influences the performance of indirect type solar dryer. Operational time for solar dryer can be extended by the application of various phase change storage material. Chemical pretreatment and slicing of product also reduces the drying time up to certain level. Various techniques like iron mesh, finned SAH heater, roughened absorber plate surfaces by various means, i.e., coating, ribs, etc. increase the percentage of heat gain. This present review article may be very helpful to those who are working in the area of solar dryers.

References

1. Sukhatme SP, Nayak JK (1996) Solar energy-principles of thermal collection and storage. Tata McGraw Hill, New Delhi
2. Tiwari GN (2004) Solar energy: fundamentals, design, modelling and applications. Narosa, New Delhi

3. Mustayen AGMB, Mekhilef S, Saidur R (2014) Performance study of different solar dryers: a review. Renew Sustain Energy Rev 34:463–470
4. Prakash O, Kumar A, Sharaf-Eldeen YI (2016) Review on Indian solar drying status. Curr Sustain/Renew Energy Rep 3(3):113–120
5. Chauhan YB, Rathod PP (2020) A comprehensive review of the solar dryer. Int J Ambient Energy 41(3):348–367
6. Pradeep K, Dheerandra S (2020) Advanced technologies and performance investigations of solar dryers: a review. Renew Energy Focus 148–158
7. Prakash O, Kumar A (2020) Solar drying systems. CRC Press
8. Lingayat AB, Chandramohan VP, Raju VRK, Meda V (2020) A review on indirect type solar dryers for agricultural crops—Dryer setup, its performance, energy storage and important highlights. Appl Energy 258:114005
9. Sharma VK, Colangelo A, Spagna G (1993) Experimental performance of an indirect type solar fruit and vegetable dryer. Energy Convers Manag 34(4):293–308
10. El-Sebaii AA, Aboul-Enein S, Ramadan MRI, El-Gohary HG (2002) Experimental investigation of an indirect type natural convection solar dryer. Energy Convers Manag 43(16):2251–2266
11. Sreekumar A, Manikantan PE, Vijayakumar KP (2008) Performance of indirect solar cabinet dryer. Energy Convers Manag 49(6):1388–1395
12. Montero I, Blanco J, Miranda T, Rojas S, Celma AR (2010) Design, construction and performance testing of a solar dryer for agroindustrial by-products. Energy Convers Manag 51(7):1510–1521
13. El-Sebaii AA, Shalaby SM (2013) Experimental investigation of an indirect-mode forced convection solar dryer for drying thymus and mint. Energy Convers Manag 74:109–116
14. Lingayat A, Chandramohan VP, Raju VRK (2017) Design, development and performance of indirect type solar dryer for banana drying. Energy Procedia 109:409–416
15. Essalhi H, Benchrifa M, Tadili R, Bargach MN (2018) Experimental and theoretical analysis of drying grapes under an indirect solar dryer and in open sun. Innov Food Sci Emerg Technol 49:58–64
16. Sözen A, Şirin C, Khanlari A, Tuncer AD, Gürbüz EY (2020) Thermal performance enhancement of tube-type alternative indirect solar dryer with iron mesh modification. Sol Energy 207:1269–1281
17. Lingayat A, Chandramohan VP, Raju VRK, Suresh S (2020) Drying kinetics of tomato (Solanumlycopersicum) and Brinjal (Solanummelongena) using an indirect type solar dryer and performance parameters of dryer. Heat and Mass Transfer 1–20
18. Yadav S, Chandramohan VP (2020) Performance comparison of thermal energy storage system for indirect solar dryer with and without finned copper tube. Sustain Energy Technol Assess 37:100609
19. Etim PJ, Eke AB, Simonyan KJ (2020) Design and development of an active indirect solar dryer for cooking banana. Sci African 8:e00463
20. Mugi VR, Chandramohan VP (2021) Energy and exergy analysis of forced and natural convection indirect solar dryers: estimation of exergy inflow, outflow, losses, exergy efficiencies and sustainability indicators from drying experiments. J Clean Prod 282:124421
21. Sabareesh V, Milan KJ, Muraleedharan C, Rohinikumar B (2021) Improved solar drying performance by ultrasonic desiccant dehumidification in indirect forced convection solar drying of ginger with phase change material. Renew Energy 169:1280–1293
22. Salve S, Fulambarkar AM (2021) A solar dryer for drying green chili in a forced convection for increasing the moisture removing rate. Mater Today: Proc

A Comprehensive Review on Cold-Formed Steel Building Components

Kaminee Rathore, M. K. Gupta, and Manoj Verma

Abstract This review paper summarizes the design method and the behaviour of Cold-formed steel (CFS) building components. On the basis of literature review we can investigate the behaviour of CFS Building components such as beam, column, light gauge stud walls in fire consideration and zero temperature. The fire resistance capacity of CFS is implicit which may limit its application. An excellent knowledge about its mechanical properties is indispensable for fire design purposes. CFS sections are light in weight and its construction is also easy, therefore CFS is progressively used in the construction industry. Rolling, pressing, stamping, bending, and other cold-working procedure are carried out at room temperature to form CFS products. Columns, pillars/beams, joists, studs, floor decking, built-up sections, and other structural and non-structural products are made from CFS sections as thin gauge sheets in manufacturing.

Keywords Cold-formed steel · Beam · Column · building component

1 Introduction

Steel's evolution as a construction material, as well as its corresponding manufacturing industry, has played a key role in the advancement of the industrialized world, assisting in the development of our new way of life. Construction of railways, bridges, mine openings, factories for goods fabrication, and electricity generation and transmission would never have advanced to the extent we have today if steel had not been

K. Rathore (✉) · M. Verma
Department of Civil Engineering, University Teaching Department, CSVTU, Bhilai, CG 491107, India
e-mail: pooja.rathore.p@gmail.com

M. K. Gupta
Department of Civil Engineering, Bhilai Institute of Technology, Durg, CG 491001, India

K. Rathore
Department of Mechanical Engineering, University Teaching Department, CSVTU, Bhilai, CG 491107, India

A. Maurya et al. (eds.), *Recent Trends in Mechanical Engineering*, Lecture Notes in Mechanical Engineering, https://doi.org/10.1007/978-981-19-7709-1_46

invented. In the beginning of the twentieth century America was producing a greater amount of steel yearly, than Britain Germany mixed and run the direction in the exportation of hot-rolled steel (HRS) sections, this section is known as "Carnegie beams", these sections are rolled by the grey method. The section we get known as wide flanged beams (WFB) in the USA and, In Britain, Australia, and South Africa these Beams are known as the universal beam. Earlier steel in the building material in reference to which other building materials were judged, but in twentieth century reinforced concrete soon become the major competitor to the steel [1].

1.1 Cold-Formed Steel (CFS)

Cold-formed members are currently widely used in construction of building, bridge construction, transmission towers, storage racks, railway coaches, drainage facilities, highway products, car bodies, and in types of equipment. These CFS members are shown in Fig. 1. The CFS sections are cold-formed (CF) from low allow steel sheet (carbon), plate, steel sheet and flat bar. Cold-rolling machines are used to shape these sections. Bending brake and push brake operations are also used to build these CFS sections. Cold-formed members with thicknesses ranging from 0.378 mm to 6.35 mm, as well as members made from steel plates and bars as thick as 25.4 mm, can all be CF into structural forms. The use of CFS members as structural materials in all forms of building construction started in the 1850s in the United States and Great Britain. But in United State until the 1940s, these steel members were not broadly used in building construction. Presently, CFS members are broadly used as a building material worldwide [2].

Fig. 1 Various shape of CFS sections [2]

1.2 *Advantages of CFS Members*

Due to the extreme shorter project duration, reduced insurance costs, predictability and high accuracy of steel parts, and improved design ability, CFS structural members have a proven track record of delivering cost-effective benefit over the entire building construction period. CFS's long-term durability, as well as its resilience to rust, fungus, and vermin, give it a lifespan that few other construction materials can match. CFS framing satisfies all broad green building norms with optimum sustainability criteria. CFS framing is highly resilient building material available because it is resistant to extreme environmental and earthquake loads, ballistic penetration and blast menace [3]. In building construction, the CFS Structural members are generally gives the following advantages:

- Cold-formed light (CFL) sections can be built with comparatively short periods and light loads because HRS sections are thicker than CFS sections.
- Cold-forming operations allow for the formation of unusual sectional compositions at a low expense, resulting in desirable Strength/Weight ratios (s/w).
- CFS has the ability to manufacture nestable parts, which makes for more lightweight packing and delivery.
- Decks and load-carrying panels can provide a valuable surface for roof, ground, and wall construction, but in some circumstances, load-carrying panels and decks can also provide sealed cells for electrical and other conduits.
- Decks and load-bearing panels are designed to sustain loads that are normal to their surfaces. The load-carrying panels and decks will serve as a shear membrane to resist force in their own planes if they are properly coupled to each other and to the linked members.

As compared to others material such as concrete and timber, the following advantages can be realized for CFS structural members (Fig. 2).

2 Classification of CFS Sections and Their Behaviour

The CF structural steel members can be divided into two categories (Fig. 3).

2.1 *Individual Structural Framing Members (ISFM)*

CFS sections are often used as structural framing in steel structures, as seen in Fig. 4. T-sections, channels sections (C-sections), angle sections, hat sections, Z-sections, I-sections, and tubular members are all common structural framing shapes. In previous studies it has been seen that the sigma sections (ε) acquires various advantages such as less weight, larger torsional rigidity, high load-bearing capacity, and shorter blank

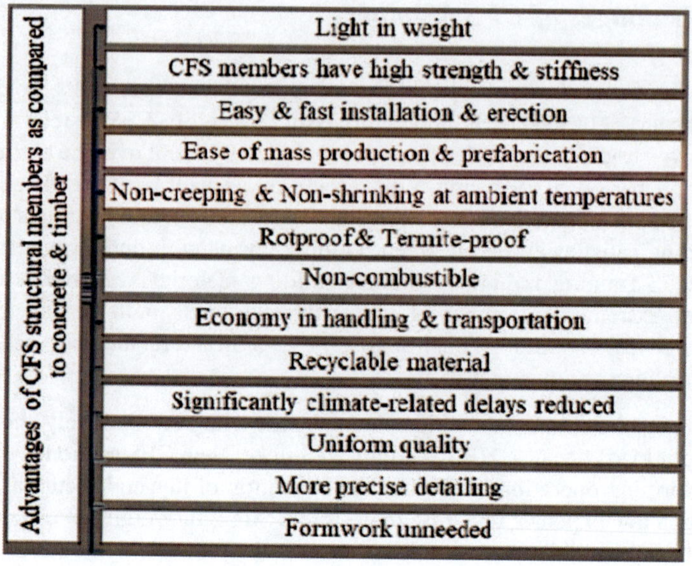

Fig. 2 Advantages of CFS members over concrete and timber [3]

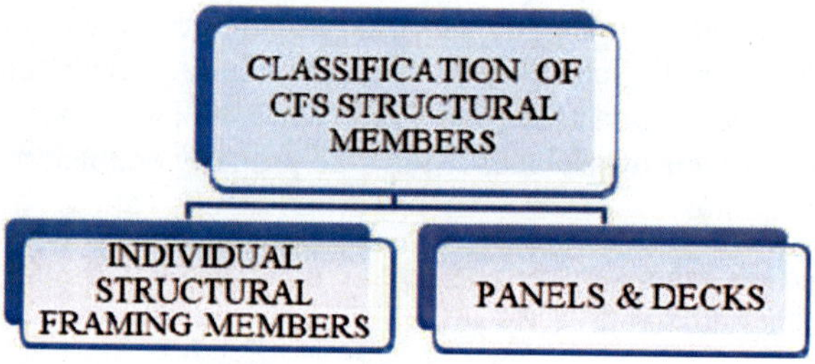

Fig. 3 Classification of CFS structural members [4]

size. For different cases, the different size of CF individual framing members are used. These sizes are listed below.

- In general, the depth of CF individual structural framing members range from 2 to 12 inch (51–305 mm) and thickness of members range from 0.048 to 1/4 inch (1.2–6.4 mm).
- CFS plate sections in thickness of up to 3/4 or 1 inch (19 or 25 mm) have been used in transmission towers, highway sign support structure, and steel plate structures.
- In some cases, the depth and thickness of actual structural framing members in transportation and building construction can be up to 18 inch (457 mm) and 12 inch (13 mm), respectively, or thicker.

Individual framing members play an important role in structural design by bearing weight, supplying stiffness, and providing structural strength. Building up to 6 stories

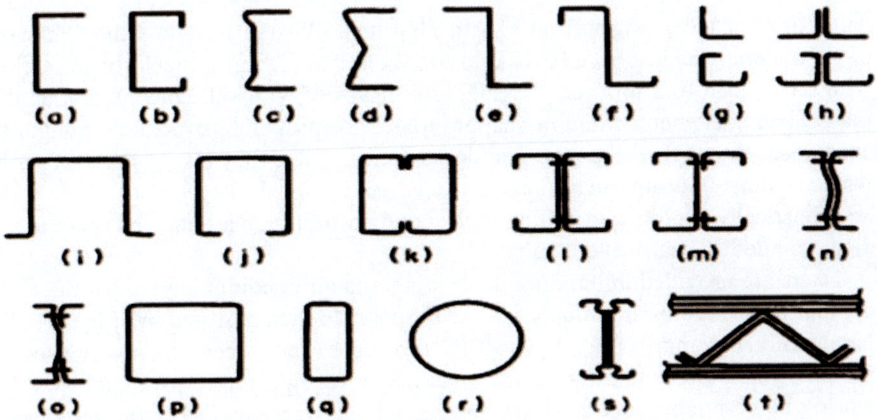

Fig. 4 CFS sections used in structural framing [4]

in height these sections can also be used as main framing members. The main framing of high rise multistory building is typically of heavy HRS and the secondary components may be of CFS members such as decks, panels, and steel joist. The hard HRS forms and the CFS sections balance each other in this case. Space racks, storage shelves, arches, and chord web members of open platform steel joists are all constructed from CFS parts [4].

2.2 *Panels and Decks*

Second category of CFS sections as shown in Fig. 5. In general, these steel sections are used for floor decks, siding material, roof deck, wall panel and bridge form. Some of the deepest panels and decks are CF with the web stiffeners. In general, the

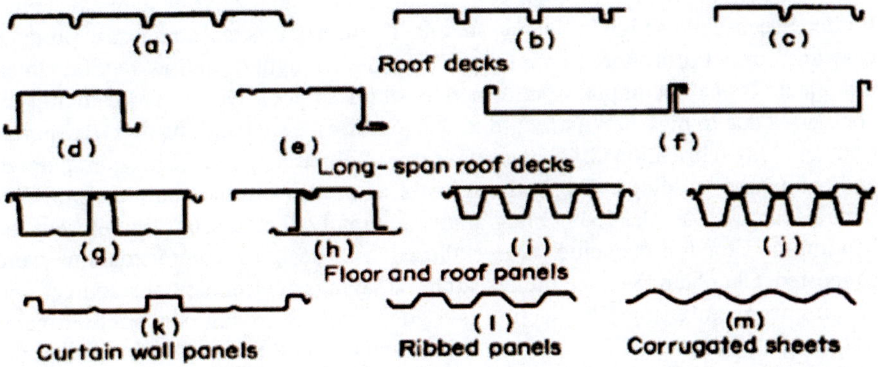

Fig. 5 Decks, panels, and corrugated sheets [4]

depth of CF panels ranges from $1^1/_2$ to $7^1/_2$ inch. (38–191 mm) and the thickness of CF materials ranges from 0.018 to 0.075 inch (0.5–1.9 mm). Steel ribbed section with 0.012 inch (0.3 mm) can be suggested for roof and wall construction as the load-carrying element would be inappropriate. For providing structural strength to carry loads the CFS steel panels and decks are used, and they also give a surface in which roofing, flooring can be done. The CFS panels and decks are also used to form an acoustically conditioned ceiling with sound absorption material, and spacing are also provide for electrical conduits.

Ducts are provided in building for heating and air conditioning so that the cells cellular panels are used as ducts for this purpose. In the construction of hyperbolic paraboloid roof and in folded plate CFS roof decks have been effectively used in the past years. Steel structure using steel decks for hyperbolic paraboloids is the world's biggest light gauge steel structure, and it is designed by LavZetlin Associates. Corrugated sheets are also used to make wall panels, roofs, and sewage systems. Corrugated sheets have been seen to be effective in drainage systems and the arched roofs of under-ground shelters. In general, the pitch of corrugations ranges from $1^1/_4$ to 3 inch (32–76 mm), and the depth varies from 1/4 to 1 inch (6.4–25 mm). The thickness of corrugated CFS sheets generally ranges from 0.0135 to 0.164 inch (0.3–4.2 mm). But corrugations with pitch and depth up to 6 inch (152 mm) and 2 inch (51 mm), respectively, are also available [4].

3 Literature Review

Haidarali and Nethercot [5] studied CFS member have been extensively used as a secondary steel work and main steel frames in the industry of steel construction, and also used as a primary structure. They concluded that the buckling behaviour of CFS lipped (edge stiffened) sections under pure bending was accomplished with the finite element method (FEM). Shu et al. [6] represented that, for corrosion resistance, good appearance, and easy maintenance, stainless steel sections are mostly used in recent year in structural applications, but as a structural material the use of stainless steel is limited because of its high initial investment. The aim of this research was to produce easy and correct procedure to predict the ultimate strength of hollow section which is subjected to axial compression formed by stainless steel. Wan and Mahendran [7] concluded due to high S/W (strength to weight ratio), easy handling and transportation, and easy fabrication CFS beams are extensively used in industrial, commercial, and residential buildings. The study of buckling modes and buckling loads of CFS parts began with an elastic buckling analysis. Finally, the relationship between the bending and torsion capacities was explored, and design recommendations were suggested. Ghannam [8] done the research for savings environmental resources and keeping it green. Using CFS section is a good ways in saving construction material as compared to the HRS sections. Manikandan and Pradeep [9] studied that the compressive resistance of CFS sections may be governed by local or overall buckling. A new CFS section is selected in this study. In general, there are three basic types of buckling occur in thin-walled steel column such as distortion, overall, and local. Various experiments have been performed in the behaviour of CFS sections with V, U, and corrugated form intermediate stiffeners, but it has been found that the effects of corrugated shape stiffeners are contradictory in the behaviour of CFS sections. Chandrikka et al. [10] studied that CFS thin sheet are widely used in building industry in the form of purlins, floor decking, and roof sheeting. In this paper it has to be seen that the different CFS sections used as a beam and ultimate load with respect to transverse stiffeners and inclined stiffeners which is provided in flange. Naganathan et al. [11] studied that for increasing the strength of CFS members, carbon fibre reinforced polymer (CFRP) hold various benefits as compared to the other methods like welding and bolting. CFRP is used as a strengthening material, and it provide good appearance, better resistance to corrosion, high stiffness, adhesive bond, and high strength to weight ratio. Taufiq and Lawson [12] concluded concrete filled tubular columns have high compression and bending resistance and it has also good fire resistance and energy absorption. As a consequence, these columns are most widely used in the design of houses and bridges. They also exposed the behaviour of perforated C-sections and offered due to perforations the shear- bond strength in the web of CFS C-section was over 1.2 N/mm^2. Tiago et al. [13] studied that to increase the fire resistance capacity of CFS built-up column, plasterboard hollow encasement is used as an alternative, and the time of fire resistance test is also increased upto 90 min so that the fire resistance of column is increased to nearly 900% with the use of plasterboard hollow encasement. Krishanu et al. [14] studied and compare the three current guidelines of American Iron and Steel Institute (AISI), Australian and New Zealand Standards (AS/NZS) and Eurocode (EN 1993-1-3), and they concluded that modifications should be incorporated to the current design guidelines of AISI (2016) and AS/NZS (2018) to determine the axial capacity of CFS channel sections under overall buckling because these guidelines are conservative.

After the study of above literature, it is found that researchers have applied different CFS sections as a building components for various purposes as per the building structure. The current study uses overall buckling of thin-walled CFS columns, and stiffeners are used to prevent column from buckling. It has also been shown that stiffeners can enhance the moment capacity of a beam by 15%. Fire resistance capacity of column can also enhance by using plasterboard hollow encasement.

4 Research Gap and Recommendation for Future Work

- Buckling analysis for different CFS sections used as a column shall be made for further work
- Provision shall be made for CFS in design steel code IS: 801–1975
- A numerical study is required to extend this research further to understand the fire resistance capacity of CFS sections.

5 Conclusion

This paper presents the behaviour and uses of different CFS sections (individual sections used as a column and beam, and panels and decks) which are formed by cold-working procedure such as rolling, pressing, stamping, bending. These procedures are carried out at room temperature. Different research article has been reviewed for this purpose, and it has been seen that the thin-walled CFS columns are governed by overall buckling, therefore provision of stiffeners in web prevents the column from buckling, and it has also seen that 15% moment capacity of beam can be increased by use of stiffeners, and at the same time fire resistance capacity of CFS built-up column can also increase by use of plasterboard hollow encasement as a fire line around the columns. CFS sections has high compressive strength as compared to the HRS sections.

References

1. Ball, P. 2016. *A short history of steel.* http://www.theheritageportal.co.za.
2. Yu, w. (1999) Cold-formed steel structures structural engineering handbook. CRC Press LLC, Rolla, MO
3. BuildSteel Powered by the steel framing industry association. https://buildsteel.org.
4. Yu WW (2000) Cold-formed, 3rd edn. Wiley, Rolla, Missouri
5. Haidarali MR, Nethercot DA (2012) Local and distortional buckling of cold-formed steel beams with edge-stiffened flanges. J Constr Steel Res 73:31–42
6. Shu G, Zheng B, Shen X (2013) Experimental and theoretical study on the behavior of cold-formed stainless steel stub columns. Int J Steel Struct 13(1):141–153
7. Wan HX, Mahendran M (2016) Buckling behaviour of cold-formed steel beams under bending and torsion. In: International specialty conference on cold formed steel structures, pp 219–233
8. Ghannam M (2019) Bending moment capacity of cold-formed steel built-up beams. Int J Steel Struct 19(2):660–671
9. Manikandan P, Pradeep T (2018) Effective cross section of cold formed steel column under axial compression. J Instit Eng (India): Ser A 99(2):245–255
10. Chandrikka V, Shanmugavel S, Sivasurya B, Lakshmi BS (2019) Analytical and experimental investigation of cold formed steel sections under bending. Int J Sci Eng Sci 3(4):5–10
11. Naganathan S, Chakravarthy HN, Anuar NA, Kalavagunta S, Mustapha KNB (2019) Behaviour of cold formed steel built-up channel columns strengthened using CFRP. Int J Steel Struct 1–10
12. Taufiq H, Lawson RM (2020) Composite columns using perforated cold formed steel sections. J Constr Steel Res 167:105935
13. Pires TA, do Rêgo Silva JJ, dos Santos MM, Costa LM (2021) Fire resistance of built-up cold-formed steel columns. J Construct Steel Res 177:106456
14. Roy K, Ting TCH, Lau HH, Masood R, Alyousef R, Alabduljabbar H, Alaskar A, Alrshoudi F, Lim JB (2021) Cold-formed steel lipped channel section columns undergoing local-overall buckling interaction. Int J Steel Struct 21(2):408–429

A Review of Recent Advancement in Solar Collector Systems for Water Heating

Yogesh Kumar, Manoj Verma, and Harish Kumar Ghritlahre

Abstract Among all the renewable sources of energy, sun is a valuable source of energy. Solar energy is utilized for various purposes, one of which is water heating via solar collectors in domestic and industry related areas. In contrast to other solar energy applications, solar collector systems for water heating require low maintenance and operating costs. They are broadly classified as active and passive solar water heating systems operating in either direct or indirect mode. This paper reviews the recent advancement and improvements in solar water heaters with different kinds of solar collectors, including both concentrating and non-concentrating types. Some previous works have been studied and summarized which deal with improving the efficiency of solar water heating system (SWHS) by changes in collector design with numerous enhancement techniques of heat transfer, leading to choosing the best option from among them for improving heat transfer. In addition, the research gap and the proposed potential improvements for future work have been given in brief.

Keywords Solar energy · Solar water heater · Flat plate solar collector · Thermal efficiency · Collector storage

1 Introduction

Energy is one of the reasons, for which constant research work is being performed by scientists and researchers from all over the world with the aim of saving it. With the increasing population of the world and increasing demand for energy day by day, the consumption of energy on a global scale has become a matter of concern. Natural gas, oil and coal have been found appropriate to use for fulfilling individual energy requirements but the inadequate supply of theses fuels turns out to be the

Y. Kumar (✉)
Department of Mechanical Engineering, University Teaching Department, CSVTU, Bhilai, Chhattisgarh 491107, India
e-mail: ykverma2016@gmail.com

M. Verma · H. K. Ghritlahre
Energy and Environmental Engineering, University Teaching Department, CSVTU, Bhilai, Chhattisgarh 491107, India

A. Maurya et al. (eds.), *Recent Trends in Mechanical Engineering*, Lecture Notes in Mechanical Engineering, https://doi.org/10.1007/978-981-19-7709-1_47

major deficiency. There are a variety of alternative renewable energy (RE) sources which can be established with replacement of conventional ones. One of the most significant advantages of RE is no carbon dioxide emissions. Among all RE sources, the energy from the sun is one of the promising sources, as well as globally available and emission free. Solar energy's various uses have been widely discovered, including water heating, air conditioning, drying of grains, light applications and food cooking, etc. [1]. Water heating accounts for about 20% of overall energy usage in an average household. Using solar energy is inexpensive, pollution-free and minimizes power expenses from coal, electrical energy, or other sources, that is particularly beneficial to homeowners who use a lot of hot water regularly [2]. Nowadays water heating is mainly done by using conventional fuels, resulting in ecological effluence and environment alteration due to greenhouse gases (GHG) productions. If the global consumption of the conventional (fossil) fuels continues at such a rate, then earth's fossil fuel asset will be depleted by 2050, and world energy need will be nearly 30–46 TW till 2050 and 2100, correspondingly. Water heating by solar energy is one of the globally acceptable applications due to its easy working principle. The solar water heater is a common device that harnesses solar energy and can be used to replace an electric water heater. This device heats water throughout the day and is usually mounted where sunlight is available. Shielded storage tank is used for storing heated water to be made available for morning uses in homes. SWH is not only a safe, simple and reliable technology, but also reasonable in terms of costs. Water heating systems does not rely on fossil fuels and instead uses solar energy to heat stored water. As a result, it saves money, which is a significant benefit of solar heating systems. Because solar energy is free, no charges from electrical utilities are required. One downside is maintenance and corrosion however most systems do not necessitate much attention. Apart from this it also reduces the CO_2 emission footprint [3–5].

This paper examines the different solar collectors used for water heating purpose. The recent advancements are summarized in the research gaps which will help future researchers to identify relevant areas of research. The study includes an overview of solar water heating systems, both active and passive types, essential components of the system and the most recent solar water heater research and advancements.

2 Solar Water Heating System (SWHS)

When heating medium (such as water or other heat transfer fluid (HTF)) is heated directly and no heat exchanging device is involved then it is known as a direct heating system. When heating medium (hydrocarbon oil, nanofluid, etc.) is heated and transfer of heat involves an exchanging device known as heat exchanger, then it is known as an indirect type heating system. These systems are further divided in two types.

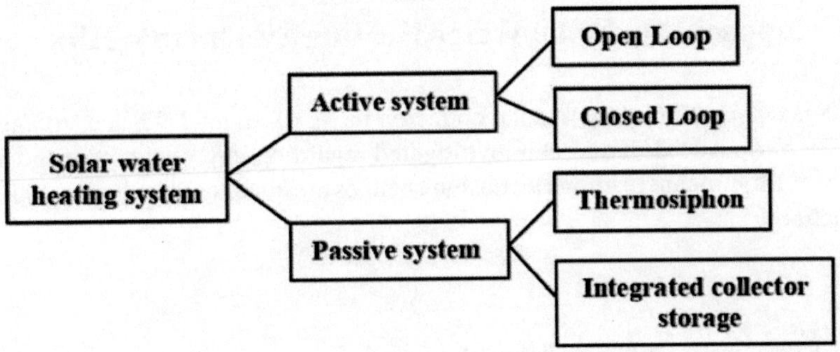

Fig. 1 Types of solar water heating system [6]

2.1 Active Systems

A pump (for circulating heating medium, like water or HTF through collectors), controlling valves and regulators are used in this system. Forced circulation system is another name of active systems. Heating medium gets heated through solar collector, storing units hold heated energy before it is required and then distribution elements distribute the heated energy through medium for end users in a regulated routine in this system. It is again classified in two ways (Fig. 1):

- ***Open-loop (Direct) Active System***
  Open-loop or direct type active system heats the water directly through a solar collector which is pumped to the tank. No heat exchanging devices are used between solar collector and storage tank.
- ***Closed-loop (Indirect) Active System***
  In close loop or indirect type active system, HTF is heated through a solar collector and pumped into the storage tank where heat exchanging devices are used between solar collector and storage tank to transport the heat from HTF to tank water.

2.2 Passive Systems

The concept of this technique is simple: natural convection circulates heating medium in the middle of the solar collector and overhead water tank. After heating of water, its density falls and the lightened water goes to the top of the collector, ready for holding in the tank. When water in the lower part of tank gets cooled, it flows back in to the collector. Thermosiphon method is the best model of passive system.

- Thermosiphon system
- Integrated collector storage.

3 Components of Solar Water Heating System (SWHS)

SWHS is mainly comprised of solar collector, heat exchanger, HTF and storage tanks (Fig. 2). Various HTF have been investigated regarding improvement of the SWHS efficiency. Significant readings on design changes are concluded in the corresponding subsections.

3.1 Flat Plate Solar Collector (FPSC)

A FPSC can be considered as the heart of SWHS and is generally employed on low solar temperature applications. It includes an absorber plate (selectively coated), a transparent glass protection to decrease heat losses from upper side of the absorber plate, HTF, insulation for minimizing heat losses and finally a defensive cover for protecting its components from moisture and dust as shown in Fig. 3.

3.2 Energy Analysis of Collector and Performance Study of SWHS

Energy gained

Useful gained energy (Q_u) from collector is given as:

$$\dot{Q_u} = \dot{m} .C_p.\Delta T = \dot{m} .C_p.(T_{Coll,out} - T_{Coll,in}) \tag{1}$$

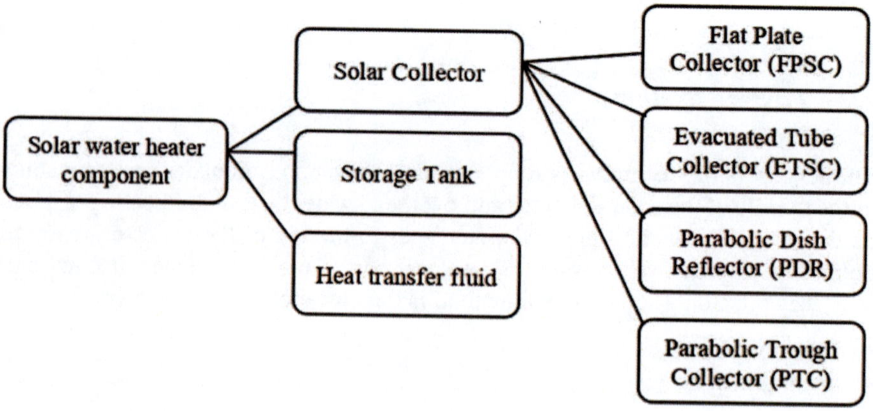

Fig. 2 Components of solar water heating system [6]

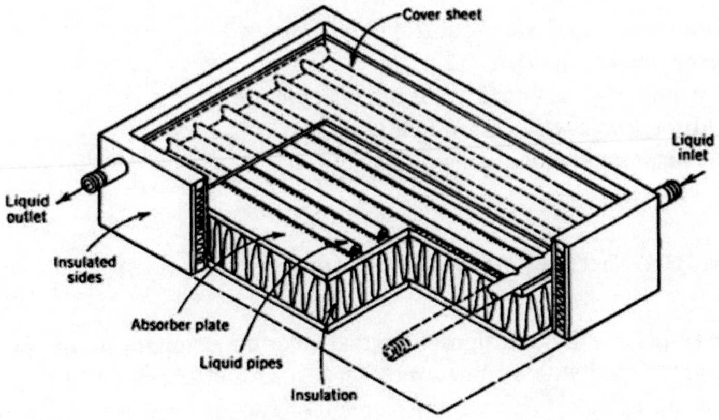

Fig. 3 A typical liquid flat plate collector system [7]

If intensity of sun radiation (G_t) is incident on collector's aperture plane area (A_C) then total radiation acknowledged by collector (Q_{in}) can be given as:

$$Q_{in} = A_c.G_i \tag{2}$$

Collector efficiency (η_{coll})

Performance of SWH is evaluated by, its efficiency, which is calculated by the ratio between useful heat energy attained (Q_u) and heat energy in collector surface:

$$\eta_{Collector} = \frac{Q_u}{A_c.G_i} = \frac{\dot{m}.C_p.(T_{Coll,out} - T_{Coll,in})}{A_c.G_i} \tag{3}$$

Collector heat removal factor (HRF) (F_R)

HRF (F_R) is defined by the following equation:

$$F_R = \frac{\text{Factual useful energy attained at collector surface}}{\text{Maximum available useful energy attained}} \tag{4}$$

(If whole collector will be at the heating medium inlet temperature)

$$F_R = \frac{\dot{m}.C_p.(T_{Coll,out} - T_{Coll,in})}{A_c[G_i.\tau.\alpha - U_L(T_{Coll,in} - T_a)]} \tag{5}$$

where:

Q_{in} Intensity of sun radiation acknowledged by solar collector W;
C_p Specific heat of HTF, kg/s;
$T_{coll,in}$ Temperature of entering water on collector (°C);

$T_{coll,out}$ Temperature of water out from collector (°C);
T_a Ambient temperature (°C);
U_L Overall heat loss coefficient of collector W/m^2;
α Absorption coefficient of plate.
τ Transmission efficiency of glazing.

4 Literature Survey of Solar Water Heating System

Yassen et al. [8] modified design of normal solar collector for household use into an incorporated kind with a ribbed receiver face. Another benefit of this change is its use as a storing reservoir as a replacement of copper tube. A larger storage tank with minimum area was obtained. Examined with different MFR, i.e., 0.0.005 kg/s, 0.0091 kg/s and 0.013 kg/s, it attained regular thermal efficiency of 59%, 65% and 67%, respectively. Author also observed losses at night and suggested proper insulation for avoiding the same.

Visa et al. [9] developed a triangle flat plate SWH with examined area 0.083 m^2 of absorber plate, provided hollow space at base side to circulate water in the middle of top side receiver and base side plate. In this design riser tube was not considered. Selective coatings were used of three individual colors black, green and orange. After examination it was found that black color coating had higher efficiency of about 55%.

Another design modification was performed by El-Assal et al. [10] using side reflector and gave suggestion about left and right attached reflectors tilt angle. The author proposed for left, during cold and summer seasons, 38° and 68° angles and for right 43° and 74.5°, respectively. Improvement in efficiency and exit temperature attained was 58% and 12 °C higher.

Shadow effect on incident solar energy was examined by Farhadi and Taki [11] and resulted in showing that tilt angle, latitude, collector length, collector height and width are the major factors which affect the incident solar energy. They concluded that reducing shadow effect increases the width of the flat plate SWH; provide length or width 70 times of its height for minimizing shadow.

Singh et al. [12] designed and experimented with changing tilt angle and flow rates. Experiment results show that efficiency increased with increasing flow rate at certain times then started decreasing and efficiency decreased with increasing wind speed. Author also stated about factors which affected the performance of solar collector such as absorber plate (it should use materials which can absorb maximum radiation), emissivity of glass cover and its number used, tilt angle and weather situation.

Isravel et al. [13] successfully performed experiment for performance enhancement of SWH with parabolic trough collector (PTC). The experimental setup contains copper (Cu) receiver tube, stainless steel (SS) reflector and structure for support. This experiment was conducted with modification on ring attached twisted tube (RATT) by improved twist ratio and center cut of aluminum twisted tapes and resulted in 24% efficiency improvement than normal twisted tape and 5% on RATT.

Weerasekera et al. [14] experimented with prototype of regularly fabricated PTC-SWH for domestic water heating applications. Experiment was conducted with two types of evacuated receiver tubes (Cu and SS) and two working fluids {diatherm (Therminol–VP 1) and water}. Experiment concluded that diatherm provided 51% greater efficiency than water, responding time of diatherm was low (maximum response rate recorded 0.18 ^{0} C/min) on temperature increase and suitable operation done with low mass flow rate.

Tabassum et al. [15] designed, developed and evaluated PTC-SWH with three different reflection materials–aluminum sheet, aluminum foil and mirror film. Experiment result shows that mirror film has potential to provide better hot water outlet compare to the other two reflectors and average recorded efficiency during experiment was 48%.

Bhakta et al. [16] conducted experiment for performance improvement of SWH using cylindrical type parabolic concentrated collector with nail twisted tube (NTT) and Cu receiver tube for NTT pitch ratios such as 4.787, 6.914 and 9.042. Experiment resulted in highest useful heat gain for twist ratio 4.787. They concluded that nail twist pitch ratio was the key factor to affect and enhance system performance.

Aramesh et al. [17] Comprehensive reviewed of recent experimental work on the development of solar cooking technology were reported and different designs and configurations of solar cookers were compared for their performance. Different solar collectors such as direct type FPSC and direct and indirect type PTSC were discussed. By investigating the performance of different design solar cookers, author concluded that using parabolic concentrating collector have the highest efficiency.

Sathe and Dhoble [18] reviewed recent developments in photovoltaic thermal techniques (PVT) and described the numerical and experimental work done by various researchers on conventional air and water-based PVT systems and typical building integrated PVT systems based on some novel technologies like PCM, heat pipe and nanofluids to understand overall development in PVT technology.

Sadhasivam et al. [19] conducted a numerical investigation where heat exchanger test set up was configured as a closed circle framework, comprising of a test section having rectangular cross-Sect. (200 cm^2) with length and width of 20 mm and 10 mm, respectively, and was produced utilizing copper sheet (1 mm thickness); and the overall length were 1000 mm, a storage tank (14 L), a pump, a detour line, a water cooler and a flow meter. Investigations were carried out using computational fluid dynamics (CFD).

5 Research Gap and Future Work

Based on the literature review, some research gaps have been identified for further research work:

- Optimization and modification of collector design shows enhancement in efficiency by reducing the heat losses and pressure drop.

- Coating is the key factors for efficiency increment. Further work can be done to find novel coating materials for improvement.
- Reflective material enhances the water heater performance, so can work with novel reflective materials for further enhancement.
- More cooling for PV panel ensures its high electrical efficiency.
- Nanofluid-based PVT system exhibited some good results.
- Use of PTSC coupled with FPSC/ETSC collector can further work for domestic hot water heating applications.
- Phase change materials can be used efficiently for thermal management of PV system.

6 Conclusion

This paper effectively reviews a variety of solar water heating collectors. It has been observed that SWHS mainly consists of absorber plates, (frequently prepared of copper, steel or aluminum), transparent glazing cover, working fluid, circulating pipe/channel, insulation, component holding frame and storage tank. Modification in collector design (collector structure, material and absorber design), various types of coating for absorber, use of polymer material, enhancing device, various PCM employment, nanofluids greatly enhance the efficiency of solar collector and solar water heater. The identified research gaps are the outcome of the paper. Future researchers will be able to find appropriate research fields based on these research gaps.

References

1. Sukhatme, S.P. and Nayak, J.K., 2017. Solar energy. McGraw-Hill Education.
2. Tiwari, G.N., 2002. Solar energy: fundamentals, design, modelling and applications. Alpha Science Int'l Ltd.
3. International renewable energy agency, 2018. Available at https://www.irena.org
4. Bazri S, Badruddin IA, Naghavi MS, Seng OK, Wongwises S (2019) An analytical and comparative study of the charging and discharging processes in a latent heat thermal storage tank for solar water heater system. Sol Energy 185:424–438
5. Verma M., 2020.Wind Farm Repowering Using WAsP Software – An Approach for Reducing CO_2 Emissions in the Environment, In: Hashmi S, Choudhury I A (eds) Encyclopedia of Renewable and Sustainable Materials, vol 3. Elsevier, pp 844–859.
6. Jamar AMZAA, Majid ZAA, Azmi WH, Norhafana M, Razak AA (2016) A review of water heating system for solar energy applications. Int Commun Heat Mass Transfer 76:178–187
7. Struckmann, F., 2008. Analysis of a flat-plate solar collector. Heat and Mass Transport, Project Report, 2008MVK160.
8. Yassen TA, Mokhlif ND, Eleiwi MA (2019) Performance investigation of an integrated solar water heater with corrugated absorber surface for domestic use. Renewable Energy 138:852–860
9. Visa I, Moldovan M, Duta A (2019) Novel triangle flat plate solar thermal collector for facades integration. Renewable Energy 143:252–262

10. El-Assal B, Irshad K, Ali A (2020) Effect of Side Reflectors on the Performance of Flat Plate Solar Collector: A Case Study for Asir Region, Saudi Arabia. Arab J Sci Eng 45(2):1035–1050
11. Farhadi R, Taki M (2020) The energy gain reduction due to shadow inside a flat-plate solar collector. Renewable Energy 147:730–740
12. Singh S, Kumar A, Yadav A (2020) Experimental Investigation of Thermal Performance Evaluation of Solar Flat Plate Collector. Materials Today: Proceedings 24:1533–1540
13. Isravel RS, Raja M, Saravanan S, Vijayan V (2020) Thermal augmentation in parabolic trough collector solar water heater using rings attached twisted tapes. Materials Today: proceedings 21:127–129
14. Weerasekera ND, Abdulla AI, Shingdon DR, Cheruiyot K (2019) Feasibility Study of Parabolic Trough Collectors for Residential Water Heating. SSRG International Journal of Mechanical Engineering 6(12):1–6
15. Tabassum S, Mashudur LSMSB, Khanam RSCDM (2019) Design and analysis of parabolic trough solar water heating system. J Architect Environ Struct Eng Res 2(3):1–6
16. Bhakta AK, Panday NK, Singh SN (2018) Performance study of a cylindrical parabolic concentrating solar water heater with nail type twisted tape inserts in the copper absorber tube. Energies 11(1):204
17. Aramesh M, Ghalebani M, Kasaeian A, Zamani H, Lorenzini G, Mahian O, Wongwises S (2019) A review of recent advances in solar cooking technology. Renewable Energy 140:419–435
18. Sathe TM, Dhoble AS (2017) A review on recent advancements in photovoltaic thermal techniques. Renew Sustain Energy Rev 76:645–672
19. Sadhasivam C, Thirumalai R, Kua-anan T (2018) Experimental Measurement and Computational Investigation on the Effect of Inlet Winglet Turbulators on the Heat Exchange Behaviour of Radiator Tubes. Journal of Applied Fluid Mechanics 11:135–139

Feasibility Study of Adsorption Refrigeration System for Air Conditioning System

Vaibhav Kr. Singh, Anirban Sur, and P. V. Bhale

Abstract Extensively use of air conditioning units is an important factor for global warming. Whether it is a building, industries or vehicle, air conditioning becomes very essential nowadays. For vehicles, as the air conditioning system (HVAC) directly takes power from the engine's main drive, it increases specific fuel consumption, which leads to environmental effects like global warming and ozone layer depreciation. The average cooling requirement of the normal passenger car is about 2.5 kW. So instead of using fuel to meet this cooling demand, we can use low-grade energy which is available on the engine exhaust/engine cooling loop. The thermal efficiency of most of the vehicles is 30% and the rest 70% goes ambient in the form of waste heat. This waste heat can work as an energy source for the vehicle HVAC system running by vapor adsorption refrigeration system (VARS). VARS uses low-grade heat energy to produce a cooling effect. As a working pair silica gel water has been proposed, which is chosen based on the required criteria. The system consists of a two-bed generator to get the continuous cooling requirement for the vehicle. In this work, thermodynamic analysis and dynamic analysis of the system have been carried out. From the numerical analysis, influencing parameter has identified. The parametric study has been performed using the numerical model using Simulink to analyze the performance of the system by varying different system parameters. For standard conditions (T_3 = 80 °C, T_1= T_c=35 °C and T_e=7 °C), the COP, SCP and cooling effect of the proposed system observed 0.56, 319.45 W/kg and 3.67 Kw, respectively.

V. Kr. Singh · A. Sur (✉)
SV National Institute of Technology, Surat, India
e-mail: anirbansur26@gmail.com

P. V. Bhale
Symbiosis International (Deemed University), Symbiosis Institute of Technology, Pune, Maharashtra, India

A. Maurya et al. (eds.), *Recent Trends in Mechanical Engineering*, Lecture Notes in Mechanical Engineering, https://doi.org/10.1007/978-981-19-7709-1_48

1 Introduction

Burning fossil fuel and cutting the forest for human needs, increases the greenhouse effect (GHE) which leads to global warming and also increases the global average temperature of the earth. The automobile is one of the major sectors for pollution and GHE. In a passenger car, a vapor compression refrigeration system is used for comfort cooling of the passenger cabin. To meet the cooling demand of a passenger car, the compressor of the VCR system takes power from the engine drive. By including the extra load of the VCR system, fuel consumption of the engine increases. Running the compressor of the cooling system adds extra consumption of the fuel. Which leads to pollution also. The thermal efficiency of an engine of the automobile is about 30–35% and the rest of the energy goes as waste. From the rest of 65–70%, near about 20–25% goes into the cooling circuit. This energy can be used to run the vapor adsorption refrigeration (VAR) system for the cooling application of the vehicle cabin. And the VAR system can be a potential replacement for the vapor compression system (VCR). VAR system can reduce the amount of fuel consumption by 12–17% [1] followed by pollution. In 1987, a researcher, meunier et al. [2] had studied the intermittent adsorption cycle with activated carbon and methanol as a working pair. The result shows that the effect of evaporator temperature on the COP and specific cooling capacity (SCC) is very high. Saha et al. [3] had analyzed the three-stage adsorption chiller with the low-temperature heat source. As the adsorption cycle is dynamic in nature, the author developed a dynamic mathematical model. A modified form of the Freundlich equation had been developed by them. Suzuki et al. [4] proposed that the waste heat from the engine can be used for cooling the cabin and replacing the VCR system, and this study put the foundation to use the VAR system in an automobile. In the year 2000, Zhang et al. [5] developed an intermittent vapor adsorption refrigeration system with zeolite 13X/water as a working pair for automobile cooling. Critoph et al. [6] had designed a compact sorption bed using a plate heat exchanger for automobiles. A study [7] investigated the effect of the VAR on the performance of the engine exhaust emission. Grain size and no layer of silica gel in the adsorber bed also play a very important role. A researcher [8] had studied this effect on the dynamic behavior of the adsorption chiller. In multiple study, a researcher [9–12] had done the numerical analysis of the VAR system with working pair of Activated carbon-methanol and analyze the performance with varying parameters. After studying all above-mentioned research papers, a novel silica gel water adsorption refrigeration paper has been proposed here for vehicle. The main aim of this study is to develop the mathematical model of the VAR system with the context of a passenger vehicle.

2 Mathematical Model

Assumptions: To develop the thermodynamic model of various processes of adsorption refrigeration system, the following considerations have been taken:

1. The Uniform size of the adsorbent (silica gel) is filled throughout the bed.
2. There is no temperature difference between adsorbate gas (water vapor) and adsorbent bed.
3. The refrigerant (water vapor) behaves as the ideal gas.
4. The heat of adsorption and desorption is taken constant and equal.
5. Both of the isosteric processes is taken as a constant volume process.
6. No desorption process takes place before the bed pressure reaches to condenser pressure. Schematic of double bed adsorption system used for analysis is shown in Fig. 1.

Isotherm: One of the most common models used to determine the equilibrium uptake of RD silica gel/water is the modified Freundlich model [3]. It takes into account the partial pressure of the adsorber bed and its corresponding heat exchanger.

$$
\begin{aligned}
x^*(P, T) &= A(T_{ads})\left[\frac{P_{sat}(T_{ref})}{P_{sat}(T_{ads})}\right]^{B(T_{ads})} \\
A(T_{ads}) &= A_0 + A_1 * T_{ads} + A_2 * T_{ads}^2 + A_3 * T_{ads}^3 \\
B(T_{ads}) &= B_0 + B_1 * T_{ads} + B_2 * T_{ads}^2 + B_3 * T_{ads}^3
\end{aligned}
\tag{1}
$$

P_{sat} (T_{ads}) is Saturation vapor pressure corresponding to adsorber/desorber bed and P_{sat} (T_{ref}) is Saturation vapor pressure of the refrigerant, corresponding to the evaporator/condenser. The value of the constant is given in the Table 1. The saturation

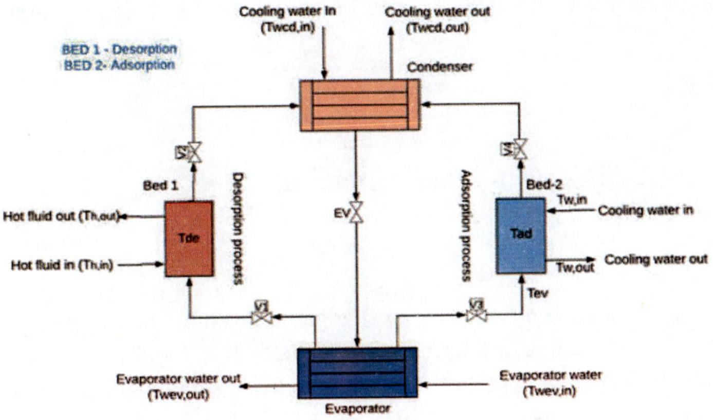

Fig. 1 Schematic of a two-bed adsorption system

Table 1 Value of the constants

Constant	Value	Unit	Constant	Value	Unit
D_{so}	2.54 * 10^{-4}	m^2/s	R_p	1.7*10^{-4}	m
E_a	4.2 * 10^4	J/mol	R	8314	J/mole K
a_0	−6.5314	Kg/kg	b_0	−15.587	–
a_1	0.72452e−1	Kg/kg * K	b_1	0.15915	1/K
a_2	−0.23951e−3	Kg/kg * K^2	b_2	−0.52612e−3	1/K^2
a_3	0.25493e−6	Kg/kg * K^3	b_3	0.5329e−6	1/K^3

pressure of the water vapor at a temperature can be determined by using the Eq. (2). Here temperature (T) is in Kelvin.

$$P_{sat}(T) = 133.32\exp\left(18.30 - \frac{3820}{T - 46.1}\right) \tag{2}$$

Kinetics of the adsorbent:

The adsorption and desorption rate is calculated by the linear driving force (LDF) kinetic equation. According to kinetics, for silica gel and water adsorption, the rate of adsorption is controlled by surface diffusion [13]. The rate of adsorption will be positive and the rate of desorption will be negative. The rate of adsorption for the silica gel and water can be express by the following equation:

$$\frac{dx}{dt} = Ks\left(x^*(P, T) - x\right) \tag{3}$$

where K_s the effective mass transfer coefficient inside the pores is written:

$$Ks = 15\left(\frac{D_s}{R_p^2}\right) \tag{4}$$

The effective diffusivity is defined as follows:

$$Ds = D_{so}e^{-\frac{E_a}{RT}} \tag{5}$$

The energy balance of all the heat exchanger is given by the following equations:

Adsorber: In the Eq. 6, first term in the right side is amount of heat generation by the adsorption process and 2nd term is amount of heat needed to raise the temperature of refrigerant (sensible heating) and 3rd term is amount of heat taken away by cooling water.

$$(M_{ad}C_{ad} + M_aC_a + M_axC_{pr,v})\left(\frac{dT_{ad}}{dt}\right) = M_a\Delta H_{ads}\frac{dx}{dt} + M_aC_{pr,v}\frac{dx}{dt}(T_{ev} - T_{ad}) + \dot{m}_{wad}C_{pw}(T_{w,in} - T_{w,out})(kW) \quad (6)$$

$$T_{w,out} = T_{ad} + (T_{w,in} - T_{ad})exp\left(-\frac{U_{ad}A_{Ad}}{m_{wad}C_{pw}}\right) \quad (7)$$

Desorber: In energy balance Eq. 8 of the desorber bed, 1st term is heat required for the desorption process, and 2nd term is heat given by the hot fluid to the bed.

$$(M_{de}C_{de} + M_aC_a + M_axC_{pr,v})\left(\frac{dT_{de}}{dt}\right) = M_a\Delta H_{des}\frac{dx}{dt} + \dot{m}_hC_{ph}(T_{h,in} - T_{h,out})(kW) \quad (8)$$

$$T_{h,out} = T_{de} + (T_{h,in} - T_{de})exp\left(-\frac{U_{de}A_{de}}{m_hC_{ph}}\right) \quad (9)$$

Condenser: First term in Eq. 10, amount of latent heat released from the vapor, 2nd term is sensible heat release due to a reduction in temperature from desorber to condenser temperature, and 3rd term is amount of heat taken away by cooling water.

$$(M_{cd}C_{pcd})\left(\frac{dT_{cd}}{dt}\right) = -M_aL_v\frac{dx_{des}}{dt} - M_a\frac{dx_{des}}{dt}C_{pr,v}(T_{de} - T_{cd}) + \dot{m}_{wcd}C_{pw}(T_{wcd,in} - T_{wcd,out}) \quad (10)$$

$$T_{wcd,out} = T_{cd} + (T_{wcd,in} - T_{cd})exp\left(-\frac{U_{cd}A_{cd}}{m_{wcd}C_{pw}}\right) \quad (11)$$

Evaporator: First term in the energy balance is the temperature drop of the refrigerant to the evaporator temperature and the second term is the latent heat of vaporization of the liquid refrigerant and the third term is the amount of heat given to evaporator by the water.

$$(M_{ev}C_{ev} + m_{r,ev}C_{pr,l})\left(\frac{dT_{ev}}{dt}\right) = -M_a\frac{dx_{des}}{dt}C_{pr,l}(T_{cd} - T_{ev}) - M_aL_v\frac{dx_{ads}}{dt} + \dot{m}_{wev}C_{p,w}(T_{wev,in} - T_{wev,out}) \quad (12)$$

$$T_{wev,out} = T_{ev} + (T_{wev,in} - T_{ev})exp\left(-\frac{U_{ev}A_{ev}}{m_{wev}C_{p,w}}\right) \quad (13)$$

Mass balance of the evaporator: In the evaporator section, we have an inlet from the condenser and outlet to the adsorber column. So mass balance can be expressed by neglecting the gas phase with this equation

$$\frac{dm_{r,ev}}{dt} = -M_a\left(\frac{dx_{ads}}{dt} + \frac{dx_{des}}{dt}\right) \tag{14}$$

System performance equations: The COP value of the system is defined by the following equation

$$\mathrm{COP} = \frac{\mathrm{Q_{ev}}}{\mathrm{Q_{de}}} \tag{15}$$

where

$$Q_{ev} = \frac{\int_0^{t_{cycle}} m_{wev} C_{pw}\left(T_{ev,in} - T_{ev,out}\right)dt}{t_{cycle}},$$

$$Q_{de} = \frac{\int_0^{t_{cycle}} m_h C_{ph}\left(T_{h,in} - T_{h,out}\right)dt}{t_{cycle}}$$

Specific Cooling Power: The amount of cooling we get from per kg of the refrigerant is called specific cooling power (SCP).

$$\mathrm{SCP} = \frac{\mathrm{Q_{ev}}}{M_a} \tag{16}$$

3 Result and Discussion

Certain performance parameters have been identified to check the performance of the adsorption system. The standard parameters which have been used throughout the study are given in Table 2. The properties of the adsorber/desorber bed, condenser and evaporator have been taken from [14].

Optimum cycle time: One complete cycle time is the time taken by a bed to undergone desorption and adsorption process and ready for desorption. The average cooling requirement of an automobile air conditioning ranges from 2.5 to 3 kW. For the designed system and standard conditions with T_1= 35 °C, T_c= 35 °C, T_e= 7 °C and T_h= 80 °C simulation has been performed and a graph is plotted between adsorption/desorption time and Q_{ev} and COP. From Fig. 2, it can be seen that, by increasing the cycle time, the average cooling effect decreases. For short cycle time, average cooling time is high and COP is less. As cycle time increases, the system moves toward equilibrium, and the cooling effect decreases. So according to the requirement for automobiles, it has been decided that adsorption/desorption time 300 s will be suitable. As shown in Fig. 3, the temperature profile of the VAR system with a double bed has been shown. In the process initially, from 0 s, Bed 1 is undergoing in desorption process while Bed 2 is adsorption. After 300 s when the bed has been switched, adsorption starts in Bed 1 and desorption in bed 2. This is a

Table 2 Standard parameter used for analysis

Parameters	Values	Parameters	Values
M_a	11.5 kg	C_a	0.924 kJ/kgK
M_{ad}	5 kg	$C_{ads/des/ev/cd}$	0.386 kJ/kgK
M_{cd}	6 kg	$C_{p,hw/cw}$	4.18 kJ/kgK
M_{ev}	4 kg	C_{prv}	1.85 kJ/kgK
$(U{*}A)_a$	3942.29 W/K	m_h	0.51 kg/s
$(U{*}A)_{de}$	4241.38 W/K	m_c (ads + cd)	0.89 kg/s
$(U{*}A)_{cd}$	15,349 W/K	m_{evw}	0.27 kg/s
$(U{*}A)_{ev}$	4884 W/K	L_v	2500 kJ/kg
T_1	35 °C	T_3	80 °C
T_c	35 °C	T_e	7 °C

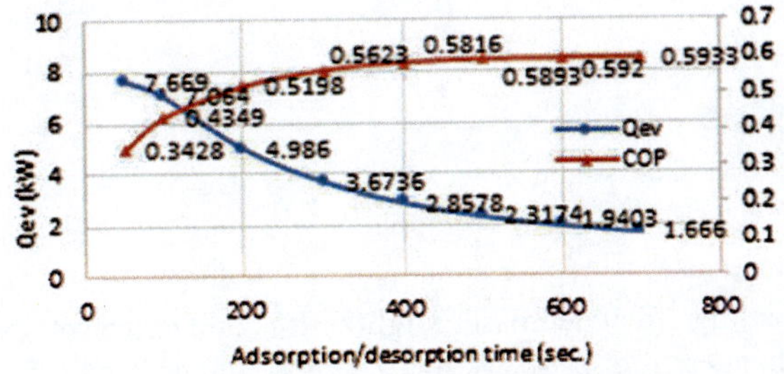

Fig. 2 Variation of cooling effect and COP with a cycle time

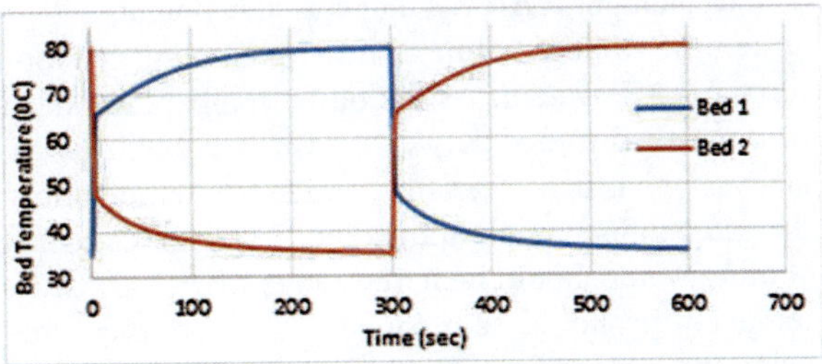

Fig. 3 Temperature profile of the be

complete temperature profile of the bed for one cycle. The longer time it will take to attain the equilibrium state, the more amount of cooling we will get.

Effect of hot water temperature: The maximum temperature of the cycle or VAR system has been decided by the temperature of the hot water flowing to heat the bed for the desorption process. During the thermodynamic analysis, it has been seen that by increasing the maximum temperature of the cycle, performance increases. In Fig. 4, the hot water temperature has been changed from 70 °C to 95 °C (Fig. 5).

By increasing the temperature, heat transfer will take place very quickly in the bed and then approaches equilibrium. It can be noticed from the plot that the initially cooling effect also increases by increasing the water temperature but after 90 °C, the

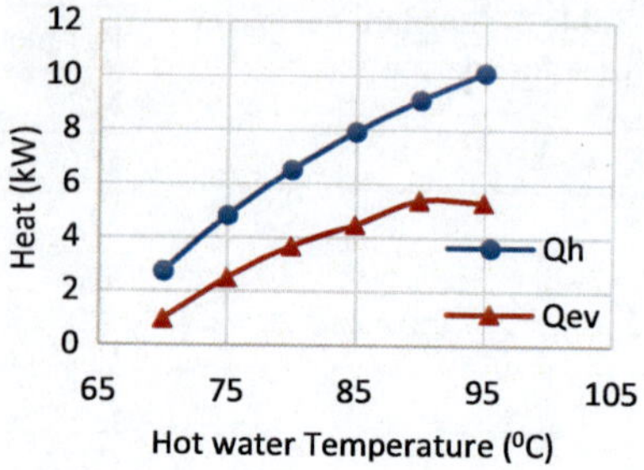

Fig. 4 Heat given and cooling effect with hot water

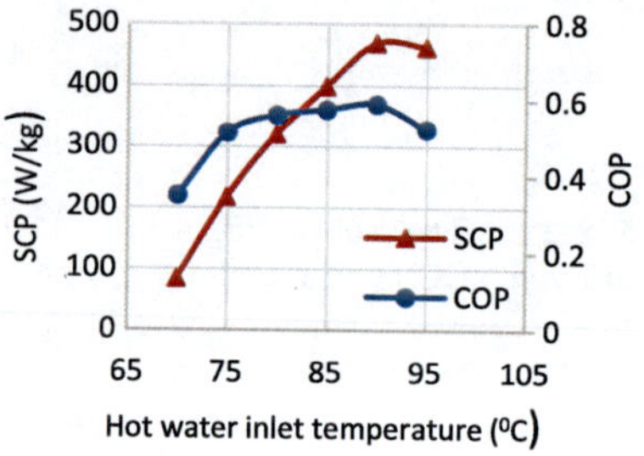

Fig.5 SCP and COP variation with hot water

cooling effect decreases slightly. It is due to irreversibility in the system. Heat input changes from 2.74 kW to 10.13 kW and cooling effect from 0.965 kW to 5.31 kW by increasing the temperature from 70 °C to 95 °C.

Conclusion: VAR system has the potential to replace the conventional VCR system of an automobile. And engine coolant loop has enough energy that it can run the VAR system. As the outlet temperature of the coolant from the engine can be controlled to 368 K [4]. And it is enough to drive the VAR system. From the analysis it has been seen that, condenser temperature and initial bed temperature are the most influencing parameters. For both, condenser and initial bed temperature, cooling water should be in the range of 30–40 °C. It is advisable that the condenser temperature should be below 40 °C. From the numerical analysis, it is shown that using the VAR system has the potential to meet the cooling demand. To implement the VAR system into automobiles, few challenges are there like weight and volume of the system. If the heat transfer area of the adsorber/desorber bed can be increased with the minimum weight of the system then it would be much preferable. To control the valve opening and closing, some kind of automated system must be used. Still, there is a lot of work scope in this field. As any commercial VAR system for automobile is not in the market.

References

1. Lambert MA, Jones BJ (2006) Automotive adsorption air conditioner powered by exhaust heat. Part 1: Conceptual and embodiment design. Proc Inst Mech Eng Part D J Automob Eng 220(7):959–972. https://doi.org/10.1243/09544070JAUTO221
2. Meunier N, Douss F (1987) Effect of operating temperatures on the coefficient of performance of aqua- ammonia refrigerating systems. ASHRAE Trans 77(pt 1):163–170
3. Saha BB, Akisawa A, Kashiwagi T (1997) Silica gel water advanced adsorption refrigeration cycle. Energy 22(4):437–447. https://doi.org/10.1016/S0360-5442(96)00102-8
4. Suzuki M (1993) Application of adsorption cooling system to automobiles.pdf., pp 335–340
5. Zhang LZ (2000) Design and testing of an automobile waste heat adsorption cooling system. Appl Therm Eng 20(1):103–114. https://doi.org/10.1016/S1359-4311(99)00009-5
6. Critoph RE, Metcalf SJ, Tamainot-Telto Z (2010) Proof of concept car adsorption air-conditioning system using a compact sorption reactor. Heat Transf Eng 31(11):950–956. https://doi.org/10.1080/01457631003604459
7. Manzela AA, Hanriot SM, Cabezas-Gómez L, Sodré JR (2010) Using engine exhaust gas as energy source for an absorption refrigeration system. Appl Energy 87(4):1141–1148. https://doi.org/10.1016/j.apenergy.2009.07.018
8. Chakraborty A, Saha BB, Aristov YI (2014) Dynamic behaviors of adsorption chiller: effects of the silica gel grain size and layers. Energy 78:304–312. https://doi.org/10.1016/j.energy.2014.10.015
9. Sur A, Das RK (2017) Experimental investigation on waste heat driven activated carbon-methanol adsorption cooling system, pp 2735–2746. https://doi.org/10.1007/s40430-017-0792-y.
10. Sur A (2015) Numerical modeling and thermal analysis of an adsorption refrigeration system. vol. 23(4):1–11. https://doi.org/10.1142/S2010132515500339
11. Sur A, Das RK, Sah RP, Influence of initial bed temperature on bed performance of an research scholar, Department of mechanical engineering, Indian School of Mines Dhanbad (India), pp 1–13
12. Sur R, Das A (2013) Thermodynamics analysis of adsorption refrigeration system. In: STME
13. Sakoda A, Suzuki M (1984) Fundamental study on solar powered adsorption cooling system. J Chem Eng Jpn. https://doi.org/10.1252/jcej.17.52
14. Alam Akira Akahira K (2004) Mass recovery adsorption refrigeration cycle improving cooling capacity.pdf. Int J Refrig 27:225–234. https://doi.org/10.1016/j.ijrefrig.2003.10.004.

Design and Development of an IoT-Based Gas Monitoring System for Underground Coal Mines

Abhishek Kumar Tripathi, Mangalpady Aruna, N. R. N. V. Gowripathi Rao, and Shashwati Ray

Abstract Safety in underground coal mines is a major challenge whenever the mine comprises of toxic gases. The risk of the presence of gas influences the overall productivity of the mines, which is a subject of concern to the mining industry. Therefore, there is a need for real-time monitoring of underground mine environment, so that the miners can be safeguarded in case of presence of toxic gases. In this paper, an attempt was made to evolve and validate an Internet of Things (IoT)-based gas monitoring system for monitoring underground coal mines environment, which includes multiple sensors for real-time measurement of different gases. The developed IoT-based gas monitoring system was tested and validated in the laboratory, under the controlled environmental conditions, for the measurement of carbon dioxide (CO_2), carbon monoxide (CO) and methane (CH_4) gases. Further, the test results were compared with the readings obtained by the digital multi-gas detector, which confirmed that the developed real-time gas monitoring system yields a good result.

Keywords Internet of Things · Miners · Sensors · Real-time monitoring

1 Introduction

The atmospheric air is a mixture of several gases and its composition is practically constant from the whole surface of the earth up to an altitude of at least 25 km from the sea level. In general, the major gases which present in the atmosphere are

A. K. Tripathi (✉)
Department of Mining Engineering, Aditya Engineering College, Surampalem, A.P, India
e-mail: abhishekkumar@aec.edu.in

M. Aruna
Department of Mining Engineering, National Institute of Technology, Surathkal, India

N. R. N. V. Gowripathi Rao
Department of Agricultural Engineering, Aditya Engineering College, Surampalem, A.P, India

S. Ray
Department of Electrical Engineering, Bhilai Institute of Technology, Durg, C.G, India

A. Maurya et al. (eds.), *Recent Trends in Mechanical Engineering*, Lecture Notes in Mechanical Engineering, https://doi.org/10.1007/978-981-19-7709-1_49

oxygen, nitrogen, carbon dioxide [1]. In addition to these major gasses, in underground coal mines, the other gasses such as carbon monoxide, methane, sodium hydroxide, hydrogen, sulphur dioxide are also present. The methane (CH_4) is one of the harmful gases, emanating from the strata and entrapped in the coal seam [2]. Further, the gasses like carbon monoxide (CO) and carbon dioxide (CO_2) are treated as the key toxic gasses. Carbon monoxide (CO) which is also known as white damp and it produces due to the insufficient supply of oxygen. When a person inhales CO it reacts with haemoglobin and forms carboxy haemoglobin. By determining the percentage of CO, one of the important risk rating, such as Grahams Ratio (CO produced/O_2 consumed) could be determined. Carbon dioxide is a toxic gas which is present underground coal mines. Whenever the percentage of CO and CO_2 increases, in mines environment, the percentage of O_2 reduces. This reduction of oxygen causes improper respiration to the miners [3].

Any changes in the permissible limit of these gases may harm the miners those are working in that atmosphere. Due to the continuous operation of heavy machinery, the level of oxygen decreases, which creates the unconformable condition to the underground miners. Therefore, there is a need to supply fresh air to the underground atmosphere so that the oxygen level can be adjusted to the permissible level [4].

Hence, before the installation of a ventilation fan, it is very much necessary to measure the percentage of various harmful gases present in the underground mines. There are various techniques and instruments are available to detect the presence of above said gasses. CH_4 can be detected using Flame Safety Lamp, M. S. A D-6 Methanometer, Automatic Fire Damp Detector, Methane Monitor (MEMACS–I), Riken Interferometer, etc. [5]. The level of CO can be measured using various devices, such as CO detector tubes, Hoopcalite tubes, Hoolamite tubes and multi-gas detector. Further, the CO_2 can be detected using detector tube and multi-gas detector. The main drawback of these gas measuring devices is that the person has to enter into the hazardous underground mine environment for detecting the percentage of gases. The other drawback is the recorded data by the instrument cannot be stored in detector device [6].

The continuous gas monitoring technique, such as tube bundle system (TBS) operation fails if the tubes get damage. Further, this technique fails to notify a danger to the executives of the mine [7]. Thus, the TBS technique also does not seem to be an alternate to measure gas and ventilation airflow as per statutory. Therefore, there is a need for an appropriate technique that can perform continuous monitoring of gases and store the data for assessing the future risk of action.

The use of Internet of things (IoT) technology in coal mine gas monitoring system helps in notifying the danger to the safety personal and also it is capable of taking counter measures against the gases when it exceeds their permissible limit. The IoT technology is an emerging concept that has a sufficient potential to execute anything in virtual environment. It gives the flexibility to facilitate an automatic alarming signal both in underground and surface of a mine. The application of IoT technology makes the things easy and improve the quality of doing work. IoT technology is nothing but the web connectivity of physical devices of the everyday life. It basically consists of the electronics devices, Internet connectivity and other forms of hardware, like sensors [8].

2 Literature Review

In order to make the underground gas monitoring system from preventive to predictive measures, which can improve the fast-decision-making process, the adoption of Internet of Things Technology can play a vital role in enhancing the real-time gas monitoring system. A study conducted by Yinghua et al. (2012) demonstrated that the interconnection of IoT to the radio frequency identification (RFID), inferred sensors, GPS, laser scanners and other sensing devices can help in exchanging the information and communication (which is achieved by intelligent identification) to the supervisor and management, so that the faulty location can be tracked. Further, they proposed the wide applications of IoT technology in mining industry, such as tracking and supervision of underground personals, monitoring and creation of warning of environment, monitoring and warning of devices, supervision of management and many other ways [9]. The introduction of IoT in mines can be achieved by three possible ways, such as open platform communication server, data access mode and Ethernet access mode. In their study, the automation of the coal mines through IoT was emphasized, that helps the management for gathering the information, digital monitoring of the underground fields and process automation [10].

Based on the previous studies it is understood that the IoT technology is having sufficient potential in predicting the underground mine gasses so that its operation can be shifted from preventive to predictive. This will help the safety of the miners by giving accurate measurement of the underground gasses without any physical damage. Thus, in order to ensure safety in the underground coal mines, it is essential to conduct further studies for the development of sensor-based gas monitoring system. Therefore, in this paper an attempt has been made to design and develop an IoT-based gas monitoring system.

3 System Architecture of the Proposed Gas Monitoring System

To monitor the underground mine gasses an automatic IoT-based gas monitoring system was designed and developed, and the experiment was carried out to see its efficiency. The developed monitoring system consists of four units, namely suction unit, analysing unit, transmitting unit and auto control unit. Suction unit intake the sample air with the help of vacuum pump so that the sample air can be sent to the analysing unit for sensing the percentage of gasses in the sample air. The transmitting unit consists of Wi-Fi and GSM module. The auto control unit is directly connected

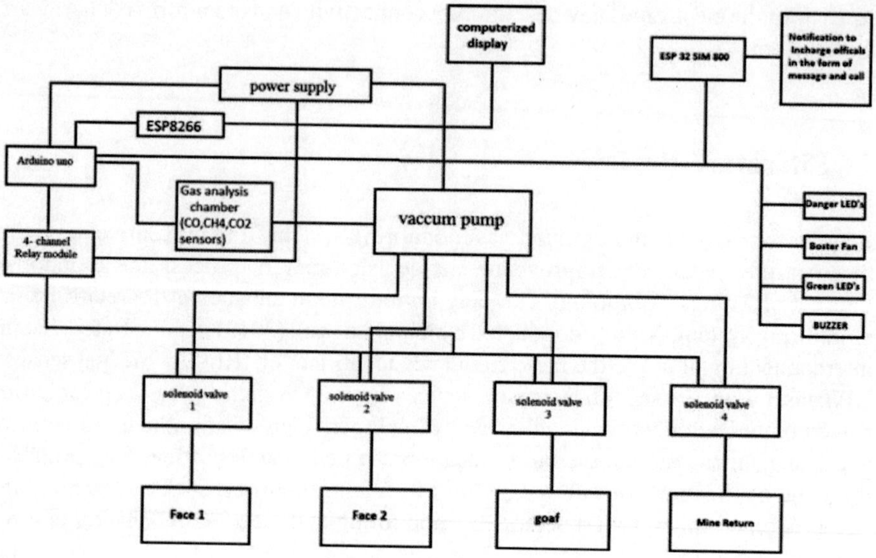

Fig. 1 Block diagram of the gas monitoring system

with Arduino, which automatically starts and perform its work to control the gases whenever they are exceeding their pre-set value. The flow chart or block diagram of the entire gas monitoring system is presented in the Fig. 1 and it show the sequence of operation of all the four units.

4 Development of Gas Monitoring System

This section briefly explains the constructional details of four main parts of the gas monitoring systems, such as suction unit, analysing unit, transmitting unit and auto control unit. Suction unit collects the sample air through the sample pipes using vacuum pump, which is connected to a relay unit. The sample pipes are operated by 12 V DC power. The collected air samples are analysed by the analysing unit to detect the presence of gasses and it activate warning signal if the gas concentration exceeds the permissible limit. The analysing chamber consist of three gas sensors, such as MQ2, MQ7 and MQ135 for analysing methane, carbon monoxide and carbon dioxide gases, respectively.

The transmission unit consists of node MCU (i.e. microcontroller unit). This is used for data transmission from Arduino to Thing Speak through wireless network so that the analysed data can be transferred to the respective authorities for taking appropriate action against any danger. Figure 2 represents the electrical circuit diagram of the gas analysis system. The constructional view of the automatic gas analysing system is presented in Fig. 3.

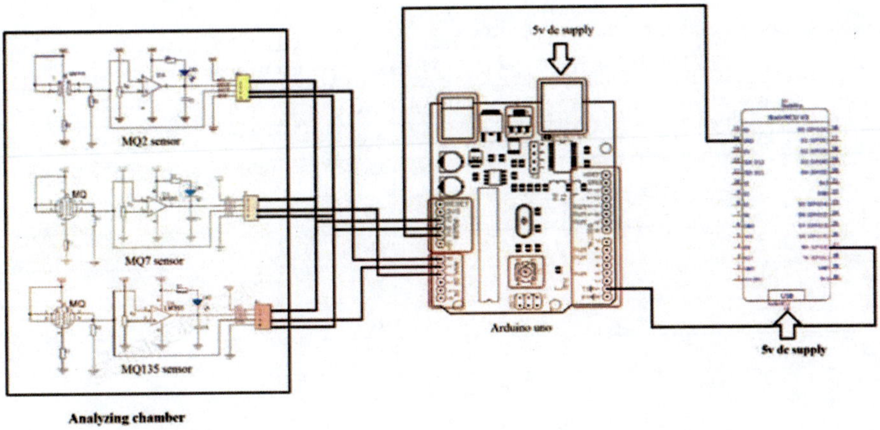

Fig. 2 Circuit diagram of analysis unit

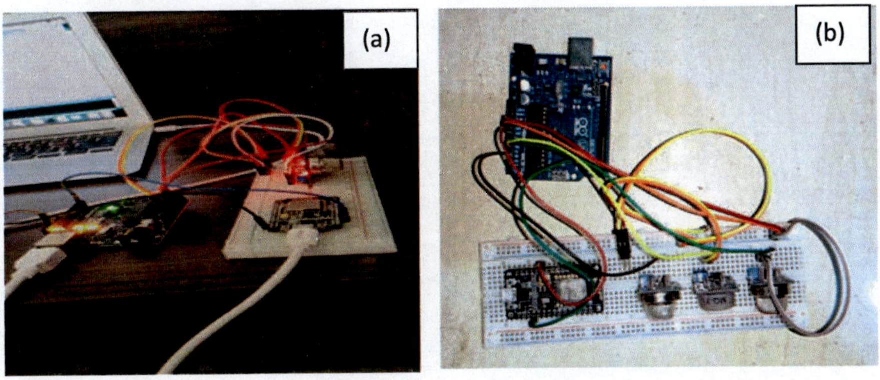

Fig. 3 Photographic view of **a** Analysing and **b** Transmitting unit

5 Results and Discussion

This section is nothing but the experimental analysis of the developed IOT based gas monitoring system. The developed gas monitoring system was tested in a laboratory, in a closed chamber, under the controlled environmental conditions. As stated earlier, MQ2, MQ7, MQ135 sensors were used for detection of methane, carbon monoxide and carbon dioxide gases. For methane gas, readily available liquid petroleum gas (LPG) was utilized and diesel engines was used to generate carbon monoxide and carbon dioxide gases. The concentration of above said gases were also measured using digital multi-gas detector (Model: MSA ALTAIR 4XR). The readings obtained by the gas monitoring system and multi-gas detector are presented in Tables 1 and 2, for their comparison.

Table 1 Comparison of readings of methane gas

S. No	As obtained by monitoring system (PPM)	As obtained by multi-gas detector (PPM)
1	196	190
2	203	205
3	199	201
4	197	192
5	183	181
6	183	180
7	182	181
8	180	179

Table 2 Comparison of readings of carbon monoxide and carbon dioxide

Sl. No	Carbon Monoxide	Carbon Dioxide		
	As obtained by monitoring system (PPM)	As obtained by multi-gas detector (PPM)	As obtained by monitoring system (PPM)	As obtained by multi-gas detector (PPM)
1	359	360	103	105
2	368	370	104	106
3	367	369	106	106
4	374	375	108	109
5	117	114	110	111
6	112	110	114	113
7	106	108	116	118
8	103	104	118	120

As observed in Tables 1 and 2, the gas concentrations recorded by the gas monitoring system is very close to that of readings obtained by the multi-gas detector unit. Hence, it can be concluded that the developed gas monitoring system is pretty reliable and can be deployed in underground coal mines for monitoring the mine environment.

6 Conclusions

In underground coal mines miners are exposed to high-risk environment where they supposed to work under the influence of many hazardous gasses. The early detection of these gases is a major challenge in underground coal mines. In this paper, an attempt was made to design and develop an Internet of Things (IOT)-based gas monitoring system for monitoring of underground coal mines environment, which was validated using conventional digital multi-gas detector. The developed system consists of multiple gas sensors for real-time measurement of concentration of different gases and these recorded readings can be transmitted through Wi-Fi to the respective mine officials, so that they can take proper precautions in case of gases exceeding their threshold limits.

For all the three gases, eight observations were taken with the gas monitoring system as well as with the multi-gas detector, under the same controlled environmental conditions. The results of both the measuring systems are well compared. This study demonstrated that the developed gas monitoring system works suitably with very minute deviation in its results and hence can be used in underground coal mines for continuous monitoring of its environment.

References

1. Johnson FS (1970) The oxygen and carbon dioxide balance in the earth's atmosphere. In: Global effects of environmental pollution. Springer, Dordrecht, pp 4–11
2. Singh H, Mallick J (2015) Utilization of ventilation air methane in Indian coal mines: prospects and challenges. Procedia Earth Planet Sci 11:56–62
3. Baldocchi D, Penuelas J (2019) The physics and ecology of mining carbon dioxide from the atmosphere by ecosystems. Glob Change Biol 25(4):1191–1197
4. Toraño J, Torno S, Menendez M, Gent M, Velasco J (2009) Models of methane behaviour in auxiliary ventilation of underground coal mining. Int J Coal Geol 80(1):35–43
5. Tong L, Liu L, Qiu Y, Liu S (2013) Tunneling in abandoned coal mine areas: Problems, impacts and protection measures. Tunn Undergr Space Technol 38:409–422
6. Liang Y, Zhang J, Wang L, Luo H, Ren T (2019) Forecasting spontaneous combustion of coal in underground coal mines by index gases: a review. J Loss Prev Process Ind 57:208–222
7. Zipf RK Jr, Marchewka W, Mohamed K, Addis J, Karnack F (2013) Tube bundle system: for monitoring of coal mine atmosphere. Min Eng 65(5):57
8. Mishra PK, Kumar S, Kumar M, Kumar J (2019) IoT based multimode sensing platform for underground coal mines. Wireless Pers Commun 108(2):1227–1242
9. Yinghua Z, Guanghua F, Zhigang Z, Zhian H, Hongchen L, Jixing Y (2012) Discussion on application of IOT technology in coal mine safety supervision. Procedia Eng 43:233–237
10. Sharma A, Gupta S, Sharma A (2017) IoT in mining: a review. Int J Electron, Electr Comput Syst 6(2):84–91

Optimization of Process Parameters of Friction Stir Welding for Joining of High Strength Aluminum Alloy

Md. Parwez Alam, Deepak Kumar, and A. N. Sinha

Abstract The latest generation of Al-Li alloys is considered a suitable material for several production applications. Its high strength-to-weight ratio offer fuel-saving and payload in aerospace. Friction stir welding is assumed as a suitable fabrication process. It is a broadly flourishing joining technique due to economical, eco-friendly, energy-efficient. Sound weld joint achieves only when proper process parameters were applied. In this paper, input parameters were optimized using the Taguchi method based on Taguchi's L9 orthogonal array. Experiments have been carried out based on three process parameters with three-level. Mathematical models were developed to draw a relationship between process parameters and response (tensile strength of the welded joint) and found a model accuracy 99.8%. The improvement in tensile strength on the optimum condition was found 5.32%. Confirmation tests have been carried out, and found the confirmatory experimental results show a good agreement with predicted outcomes.

Keywords Friction stir welding · Al-Li alloy · Taguchi method · ANOVA

1 Introduction

High-strength aluminum alloys are broadly flourishing materials in the manufacturing industries. Its wide variety and concerned inherent properties allow using from kitchen utensils to aerospace industries. Numerous scientists stimulate research on the high quality of the material. The latest generation Aluminum–lithium alloy emerges as futuristic material. The lesser amount of Lithium changes the significant properties of the alloy [1]. These alloys often use in the aviation industry due to their high strength to light weight ratio [2]. The abovementioned materials successfully join by the friction stir welding (FSW) method. This process permits to joining of similar, dissimilar equal, and unequal thickness of the plate. The automated robot successfully operated the friction stir welding process. Consequently, the process

Md. Parwez Alam (✉) · D. Kumar · A. N. Sinha
Mechanical Engineering Department, NIT Patna, Patna 800005, India
e-mail: parwez.me13@nitp.ac.in

A. Maurya et al. (eds.), *Recent Trends in Mechanical Engineering*, Lecture Notes in Mechanical Engineering, https://doi.org/10.1007/978-981-19-7709-1_50

window of the joint configuration increased. The various joints arrangements like lap joints, butt joints, T joints, and even fillet joints can easily weld.

The performance of the FSW depends on several process parameters. The proper combination yields the best joint strength. Numerous researchers worked on the experimental and numerical analysis and reported that the rotational speed (RS), welding speed (WS), tool profile, the tilt angle of the tool is the vital influencing process parameter.

Previous researchers reported that the weld joint was mostly affected by the rotational speed, welding speed, and tool geometry [3, 4]. Four different RS from 600–1200 rpm applied by Chen et al. [5], and the author observed the highest tensile strength 350.8 MPa at 800 rpm. A similar trend of the result was also reported by Mao et al. [6]. The speed of the rotating tool along the weld line is termed welding speed. It is responsible for heat input per unit length. Ahmed and Saha [7] found that welding speed is the second most influencing parameter, followed by rotational speed. Considerably the less study was found on the number of passes [8]. Some researchers applied the different ratios of rotational speed to welding speed [6]. Rao et al. reported that the weld quality is mostly affected by the ratio of RS to WS (ω/v) [9].

Many research works have been published on the optimization of process parameters of friction stir welding [10–12]. It reduces the number of experiments accordingly to experimental cost. Medhi et al. [13] evaluated the optimal parameter for FSW by carried out a time-consuming trial and error technique. They applied six different RS and WS and found maximum joint strength 142.32 MPa for the parameters 1693 rpm rotational speed and 2.72 mm/s welding speed. Based on the literature report, it can be demonstrated that the optimum parameter plays a significant role in successful welding.

Irrespective of the abundant industrial application of joints of the latest generation of Al–Li alloy, there exist very few literatures that try to compute the optimal process parameters of FSW. This paper tries to bridge the gap as identified by developing the mathematical modeling and optimizing by Taguchi method. In the present paper, three process parameters, RS of the tool, WS, and number of passes of the FSW tool, were considered to study the optimum tensile strength of FSW of AA2099 T8. The flow chart of the present analysis is shown in Fig. 1.

2 Experimental Procedure

The friction stir welding was performed on the third generation of Al-Li alloy 2099 T8 plate of dimension 130 × 120 × 5 mm by the CNC vertical milling machine. Before welding, workpiece surfaces were gently finished to avoid the gap between them. The experimental set-up and FSW tool are shown in Fig. 2. To measure joint strength, specimen were prepared as per ASTM: E8M04 guideline. The tensile tests were conducted on UTM, model Zwick/Roell Z250.

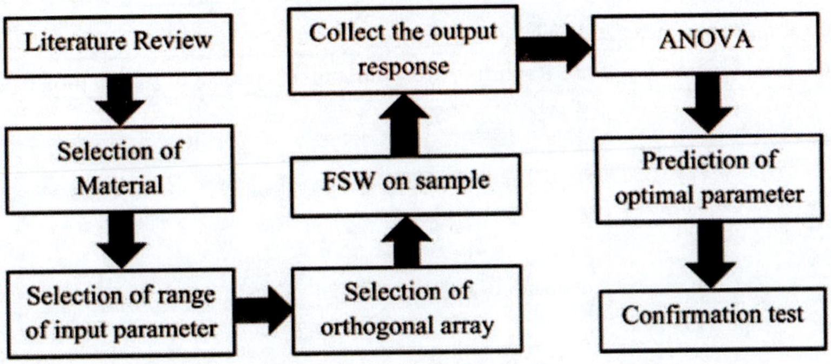

Fig. 1 Research methodology of the present analysis

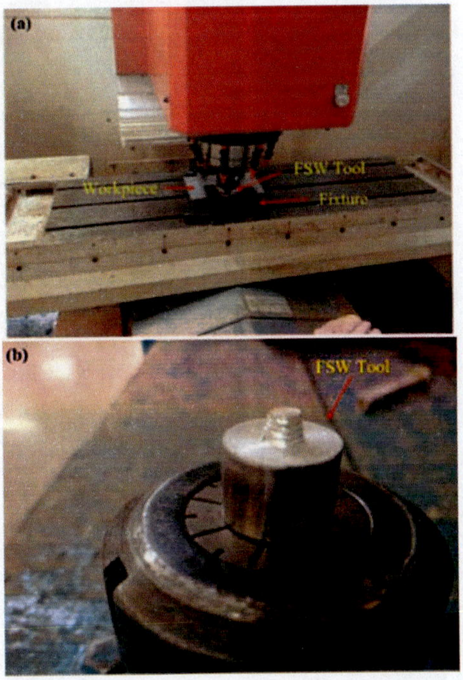

Fig. 2 Experimental set-up and FSW tool

The input parameter, namely, RS, WS, and the number of passes of the tool, are listed in Table 1. In the present study, three variables and three levels of input parameters are selected as per the L9 orthogonal array.

The experimental analysis and the regression fit values are shown in Table 2. The average percentage error was found at 0.321%. Based on the experimental output, the mathematical model of tensile strength is developed. It is given in Eq. (1).

Table 1 Process parameter and their values

Selected level for each factor	RS (rpm)	WS (mm/min)	Number of pass of the FSW tool
Level 1	1100	60	1
Level 2	1300	75	2
Level 3	1500	90	3

Table 2 Analysis of experimental and regression model fit values

Sl No	Input parameter			Analysis		
	RS	WS	No. of passes	Exp. (TS)	Reg. (TS)	% Error
1	1100	60	1	247	247	0
2	1100	75	2	284	282.67	0.47
3	1100	90	3	259	260.33	0.52
4	1300	60	2	305	306.33	0.43
5	1300	75	3	318	318	0
6	1300	90	1	302	300.67	0.44
7	1500	60	3	252	250.67	0.53
8	1500	75	1	266	267.33	0.5
9	1500	90	2	265	265	0

$$\begin{aligned}&\text{The mathematical equation of tensile strength} = -2167 + 3.028 \times \omega \\ &+12.6 \times v + 47 \times n - 0.001154 \times \omega^2 - 0.0785 \times v^2 - 11.33 \times n^2 \\ &-0.000444 \times \omega \times v\end{aligned} \quad (1)$$

where ω, v and n is represented rotational speed, welding speed, and number of passes, respectively. It was observed from the mathematical model, the R square and R square Adj for the tensile strength is 99.80 and 98.37%, respectively. It is worth noting that a quadratic model was used to establish the relationship between the input variables and the joint strength.

3 Results and Discussion

In the present study, experiments were conducted as per the Taguchi method given in Table 2. The tensile strength of the specimen is the main characteristic considered for describing the quality of welded joints. The mean and signal to noise (S/N) ratio can be calculated to find the response's influencing parameter. Here, the signal indicates desirable output, and noise represents undesirable output. In this study, tensile strength was a desirable characteristic. It should be maximum. Therefore, the

"Larger is better" concept was adopted. The values of the signal to noise ratio (η) were computed by using the Eq. (2)

$$\eta = -10\log_{10}\frac{1}{n}\sum_{i=1}^{1}\frac{1}{y_i^2} \tag{2}$$

where η is signal to noise ratio, n is the number of experiments; i is the experiment number, and y is the response value of experiment number i. The output results were then converted into means of tensile strength and corresponding S/N ratio. In signal to noise ratio, the average responses indicated by the signal and the average deviations from the sensitiveness in experimental results displayed by the noises. The Mean and S/N ratio are calculated and depicted in Table 3.

In the present analysis, the MINITAB software is used to compute S/N ratio and Analysis of variance (ANOVA). ANOVA results reveal the impact of each input parameter on the tensile strength. Based on obtained results, the optimum level setting is $RS_2WS_2NoP_2$, i.e., RS at 1300 r/min, WS at 75 mm/min, and a double pass. ANOVA table is depicted in Table 4.

It can be seen from the ANOVA table the RS is the key influencing parameter. The contribution of RS, WS, and NoP of the FSW tool was observed 81, 13, and 5%, respectively. The percentage contribution of the input parameter is shown in Fig. 3 and Table 4.

Table 3 Means and signal to noise ratio for tensile strength

Level	Means			Signal to noise ratio		
	Rotational speed	Weld Speed	Number of pass	Rotational speed	Weld Speed	Number of pass
1	263.3	268	271.7	48.40	48.52	48.65
2	**308.3**	**289.3**	**284.7**	**49.78**	**49.20**	**49.07**
3	261.0	275.3	276.3	48.33	48.78	48.78
Delta	47.3	21.3	13	1.45	0.68	0.42
Rank	1	2	3	1	2	3

Table 4 ANOVA of tensile strength

Input parameter	Signal to noise ratio						
	DF	Seq SS	Adj SS	Adj MS	F	P	% of Contribution (%)
RS	2	4270.89	4270.89	2135.44	300.3	0.003	81.35
WS	2	704.89	704.89	352.44	49.56	0.020	13.43
NoP	2	260.22	260.22	130.11	18.30	0.052	4.96
Error	2	14.22	14.22	7.11			0.27
Total	8	5250.22					100

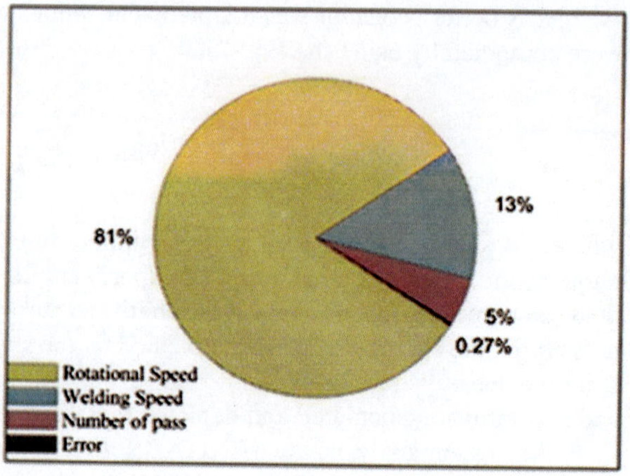

Fig. 3 Percentage contribution of process parameter

It is clearly revealed from Fig. 3 rotational speeds are the significant influencing process parameter, and it has 81% contribution in tensile strength. It is reported that heat input increases with increasing rotational speed resulting in proper material mixing. Further increasing of RS reduces the joint strength. It is due to heat input increases by increasing RS. At high temperature the friction coefficient between tool and workpiece decreases.

Heat input per unit length during friction stir welding is related to welding speed. In this study, the second level welding speed of 75 mm/min yields the maximum tensile strength. Lower welding speed produces excessive heat. It causes grain growth, defects, and reduces the strength of the joint. On the other side, a lack of the mixing of the material was observed at high welding speed. It can be demonstrated that insufficient heat responsible for inappropriate material mixing during the welding and fails to fill the gap behind the tool, whereas, at high heat input, excessive flashes were found. Better refinement of weld joint can be achieved by increasing the number of passes of the tool [8].

3.1 Prediction Value of Tensile Strength and Confirmation Test

The predictive value of the tensile test is calculated at the parameter 1300 rpm, 75 mm/min, and double pass and found 327.22 MPa. From the Taguchi analysis, it can be seen that the second level is the optimum parameter. This condition is not present in Table 3. A confirmation test was performed on the parameter RS 1300 rpm, WS 75 mm/min, and double pass and found 323 MPa joint strength of the welded specimen. It was found the predicted results of the model and experimental results were followed in good agreement with the given parameters. The comparative study of tensile strength is shown in Table 5.

Table 5 Improvement in tensile strength

	Input setting	Output result
Experimental value	Initial setting ($\omega_2 v_2 n_3$)	318 MPa
	Optimum setting ($\omega_2 v_2 n_2$)	323 MPa
	Improvement in percentage	5.32%
Predicted value	Optimum setting ($\omega_2 v_2 n_2$)	327.22 MPa
	Error in percentage	1.29%

Contour plots explaining the relation between the tensile strength and two control variables. Figure 4 shows the contour plots between the process parameters and tensile strength. Figure 4a indicates the second level of RS, and the second level of WS yields the maximum value of tensile strength. Figure 4b shows that the maximum value of tensile strength could be attained at the double pass of the tool and the second level of RS. Figure 4c shows that the maximum tensile strength at the second level of WS and double pass of the tool. The study observed that the sound weld joint was achieved by a proper combination of the input parameter.

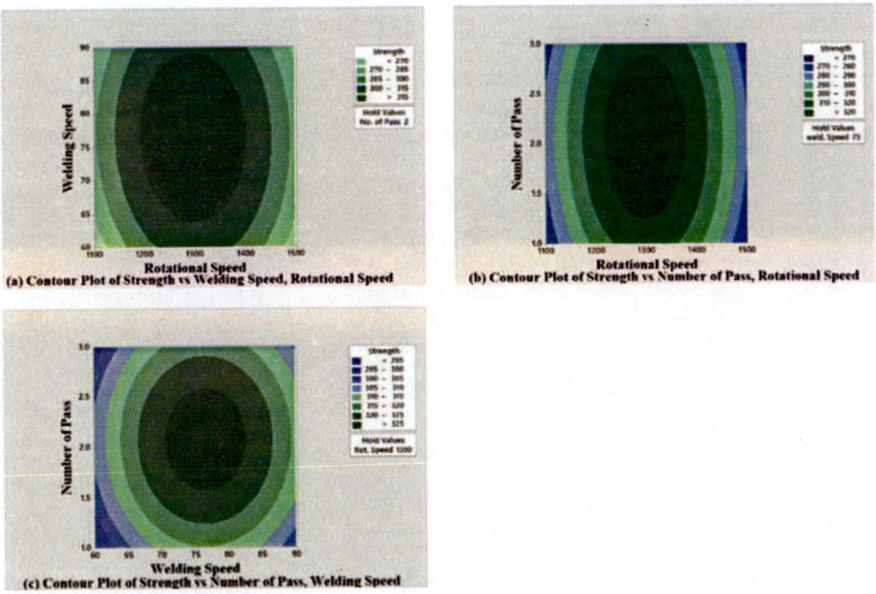

Fig. 4 Contour plot for tensile strength **a** Welding speed versus Rotational speed **b** Number of pass versus Rotational speed **c** Number of pass versus Welding speed

4 Conclusion

In the present analysis, process parameters for AA 2099 T8 were optimized by the Taguchi method based on Taguchi's L9 orthogonal array. Input process parameters have been taken as rotational speed, welding speed, and the number of passes of the FSW tool. The effect and contribution of each process parameter on the tensile strength were studied. The important conclusions from the present analysis are summarized as follow:

- A mathematical model is developed to speculate the tensile strength of FSW of AA 2099 T8 with a model accuracy 99.8%.
- The confirmatory experimental results show a good agreement with predicted outcomes
- It was found that RS was the most significant parameter, followed by welding speed and number of pass, respectively.

References

1. Alam MP, Sinha AN (2019) Fabrication of third generation Al–Li alloy by friction stir welding: a review. Sādhanā 44(6):1–13
2. Alam MdP, Sinha AN (2019) Effect of heat-assisting backing plate in friction stir welding of high strength Al–Li alloy. Energy Sources, Part A: Recover, Util, Environ Eff 1–12 (2019)
3. Kumar S, Sethi D, Choudhury S, Roy BS, Saha SC (2020) An experimental investigation to the influence of traverse speed on microstructure and mechanical properties of friction stir welded AA2050-T84 Al–Cu–Li alloy plates. Mater Today: Proc 26:2062–2068
4. Banik A et al (2021) Determination of best tool geometry for friction stir welding of AA 6061-T6 using hybrid PCA-TOPSIS optimization method. Measurement 173:108573
5. Chen H, Fu L, Liang P (2017) Microstructure, texture and mechanical properties of friction stir welded butt joints of 2A97 AlLi alloy ultra-thin sheets. J Alloy Compd 692:155–169
6. Mao Y, Ke L, Liu F, Huang C, Chen Y, Liu Q (2015) Effect of welding parameters on microstructure and mechanical properties of friction stir welded joints of 2060 aluminum lithium alloy. Int J Adv Manuf Technol 81(5):1419–1431
7. Ahmed S, Saha P (2020) Selection of optimal process parameters and assessment of its effect in micro-friction stir welding of AA6061-T6 sheets. Int J Adv Manuf Technol 106(7):3045–3061
8. de Lima Lessa CR, Landell RM, Bergmann L, dos Santos JF, Kwietniewski CEF, Reguly A, Klusemann B (2020) Two-pass friction stir welding of cladded API X65. Procedia Manuf 47:1010–1015
9. Rao J, Payton EJ, Somsen C, Neuking K, Eggeler G, Kostka A, dos Santos JF (2010) Where does the lithium go?–a study of the precipitates in the stir zone of a friction stir weld in a li-containing 2xxx Series Al Alloy. Adv Eng Mater 12(4):298–303
10. Koilraj M, Sundareswaran V, Vijayan S, Rao SK (2012) Friction stir welding of dissimilar aluminum alloys AA2219 to AA5083–optimization of process parameters using Taguchi technique. Mater Des 42:1–7
11. Wahid MA, Masood S, Khan ZA, Siddiquee AN, Badruddin IA, Algahtani A (2020) A simulation-based study on the effect of underwater friction stir welding process parameters using different evolutionary optimization algorithms. Proc Inst Mech Eng C J Mech Eng Sci 234(2):643–657

12. Mirabzadeh R, Parvaneh V, Ehsani A (2021) Estimating and optimizing the flexural strength of bonding welded polypropylene sheets by friction-stir welding method. Proc Inst Mech Eng, Part L: J Mater: Des Appl 235(1):73–86
13. Medhi T et al (2021) An intelligent multi-objective framework for optimizing friction-stir welding process parameters. Appl Soft Comput 104:107190

Effect of Through Slots and Dwarf Finned Heat Sinks on Electronic Cooling Load

Rajshekhar V. Unni and M. Sreedhar Babu

Abstract The objective of the paper is to investigate the effect of fin geometrical parameters of a heat sink, on the rate of heat dissipation for effective cooling purpose. The cooling rate is of paramount interest, especially in electronic systems accompanied with heating elements. Experiments were conducted on solid (*S*), through slot (*TS*) and dwarf (*D*) type of heat sink variants considering 9, 6, 5 and 3 fin configurations. Steady and natural convection mode was considered with a heat supply of 5 to 35 W range. Experimental analysis revealed that, heat sink (*HS*) with through slots and dwarf fins enhanced the rate of heat dissipation from the base surface of a heat sink. Maximum rate of heat dissipation (thus cooling) was exhibited by a heat sink with 6 fins *TS*, owing to the twin effect of bidirectional and bulk motion of air molecules along the fin surfaces as compared to the solid and dwarf type of heat sinks. Besides this, the fins with through slot reflected 41.7% material saving as compared to solid type of heat sink.

Keywords Heat sink · Heat transfer coefficient · Fins spacing · Through slot · Dwarf · Cooling · Rate of heat dissipation · Natural convection

1 Introduction

Electronic components generate excess heat while operating, due to electrical resistance offered by the elements. If not treated, overheating of the system may lead to technical breakdown and eventually a failure. Thus, thermal management is significant in offering thermal solutions. One such solution is to treat the overheating component either by free or forced convection. However, forced convection heat transfer is complex, requires more space and also dependent on external air flow (fan),

R. V. Unni (✉) · M. Sreedhar Babu
Department of Mechanical Engineering, Jain College of Engineering, Belagavi, Karnataka 590014, India
e-mail: rajshekhar.rvu@jainbgm.in

M. Sreedhar Babu
e-mail: sreedhar@jainbgm.in

A. Maurya et al. (eds.), *Recent Trends in Mechanical Engineering*, Lecture Notes in Mechanical Engineering, https://doi.org/10.1007/978-981-19-7709-1_51

whereas free or natural convection is safe, simple and also cost effective. In electronic applications like LED television set, stabilizer and amplifiers, free convection mode is generally employed.

Heat sink (HS) is a component which facilitates the flow of heat from a hot device. Besides this, HS saves space and thus proves to be economical. In many electronics gadgets, variety of HSs are adopted for cooling of electronic components such as PCB, LED, transformers and transistors. In most of the applications, natural convection mode is generally preferred due to the advantages offered. HS geometry or fin arrays providing the best possible cooling rate becomes substantial. Hence, the optimum geometry of HS facilitating the maximum cooling rate is an area to explore and contribute. Apart from geometry, HS material having good thermal conductivity is also preferred for better heat dissipation. Among the high thermal conductivity materials, aluminum is an ideal candidate as it has an admirable thermal conductivity and also cost effective. In the following section, literature with respect to heat sink geometry is presented.

2 Literature Review

Starner et al. [1] investigated experimentally rectangular fin arrays. Four fin arrays were studied under natural convection heat transfer with base horizontal, vertical, 45° and average convection coefficients were found. During experiments, fin length, fin width and fin thickness were fixed, but fin height (0.25–1 inch) and fin spacing (0.25–0.313 inch) were varied and 14 and 17 number of fins used. Results indicated using optimum fin spacing enhances heat transfer. Harahap et al. [2] investigated experimentally the vertical fins horizontal base HS. Two sets of configurations were studied to understand the effect of the fin length on average heat transfer coefficient (HTC). It was concluded that, for shorter fin length, HTC was superior compared longer fin length. Jones and Smith [3] studied vertical fins on a horizontal base under natural convection at steady state. The results showed, fin spacing has a greater influence on convection coefficient and hence heat dissipation too. Leung et al. [4] experimentally investigated three experimental cases, namely vertical fins, base horizontal and vertical base and horizontal fins vertical base. Experiments were conducted for the same geometry of HS and excess temperature, for natural convection. Heat transfer by vertical fins with base horizontal was maximum and horizontal fins base vertical was minimum. Optimum fin spacing was found for all the cases. Leung et al. [5] experimentally investigated two cases, i.e., fin base vertical and horizontal. It was inferred that with increase in fin length, optimum fin spacing has increased. Leung et al. [6] investigated experimentally HS (vertical fins with vertical base) under free convection mode. The experiments were conducted for five fin thickness (1, 3, 6, 9 and 19) and two base temperatures. The optimal fin thickness was noted 3 $\pm$ 0.5 mm, for spacing of the fins less than 20 mm, also the optimal fin thickness decreased with decrease in base temperature. Leung et al. [7] investigated experimentally rectangular fins with rectangular base. Introduction of short fins enhanced heat dissipation rate from the surface. Yuncu et al. [8] investigated experimentally rate of heat transfer from rectangular fin arrays on a horizontal base under natural convection. The influence of the surface to ambient air temperature difference was not significant on optimum fin spacing, but as the fin height was increased, optimum fin spacing was found decreasing for the maximum rate of dissipation. Yuncu et al. [9] investigated experimentally rectangular fin arrays under free convection. The experimental results of present work were compared with horizontal base higher rate heat transfer which was observed for same geometry and other experimental conditions. Results indicated, thermal performance of heat sinks is influenced by fin spacing. For a given temperature difference ($T_{base} - T_{ambient}$), for each height of fin considered, there is optimal fin spacing. Yuncu et al. [10] investigated experimentally steady state heat transfer from rectangular fins with vertical base. A total of 30 fin configurations were investigated. From the results, it was inferred that, for a given excess difference and fin height, there is an optimum fin spacing for which the heat transfer is maximum. The present work and previous work of the researcher were pooled, and optimum fin spacing was found between 6.1 and 11.9 mm. Umesh et al. [11] experimentally investigated fin arrays under free convection with and without perforation for 0–90° inclinations. 32% increase in HTC was noted and also 30% material saving. Umesh et al. [12] investigated fin arrays under free convection, and the results of solid fin array block and dwarf fins were compared. Dwarf fin heat sink showed 28% increase in HTC and 25% saving in material as compared to solid fin heat sink. Haghighi et al. [13] experimentally investigated the performance of plate fins and plate cubic pin-fins HS. From the results, it can be inferred that, plate cubic pin-fins have exhibited improved heat transfer, compared to plate fins. Shahrubudin et al. [14] reviewed 3D printing technology. With the help of 3D printing technology, even metal alloys can be printed by the help of material deposition. Hussein [15] reviewed investigations attempted enhancing the HS thermal performance. The various enhancement techniques used for optimizing the hydrothermal design, also the way in which the HS thermal performance is affected by orientation, shapes, perforation, slot, interruption and space between fins and their arrangement under free and forced convection condition were reviewed. Aziz et al. [16] experimentally investigated wavy fins having 3 amplitudes, and it was concluded from the results that, there is heat transfer enhancement with wavy fins than compared the rectangular fins. Anilkumar et al. [17] experimentally investigated the influence of slitted rectangular on fins heat transfer. It was inferred that the use of slitted rectangular fins improved the heat transfer by 58%. Cheng-Hung et al. [18] investigated numerically and experimentally a 3D optimum pin fin heat sink (PFHS) under natural convection. Results indicate, the tapered pin heat sink with a 66° angle has least thermal resistance among all designs, and its thermal resistance is 18.7% smaller than that of a traditional PFHS, showing the validity of this design algorithm in estimating the optimal variables of the natural convection PFHSs. Rajshekhar et al. [19, 20] experimentally investigated the effect of perforations on fins. Solid and perforated HS were also compared. From the results, it was found that, use of perforations enhances the HTC and also saves material. In subsequent work by the authors [20], solid HS was compared with modified HS. Results indicated an increase in heat dissipation with saving of material.

It is evident from above discussion that the modifications in fins will improve rate of heat dissipation. In the present work, modified heat sinks have been studied to check the feasibility of promoting higher heat transfer rates.

3 Materials and Methodology

In the present work, aluminum alloy (6082) was selected as a HS material. The properties of Al (6082) are listed in Table 1. Schematic view of an experimental setup is as shown in Fig. 1a. Current and voltages were regulated using calibrated electrical meters. For temperature measurement, K-type thermocouples were used.

3.1 The Heat Sink and the Heating Coil

Figure 1b represents the symmetric HS assembly. More related to constructional features are discussed by author elsewhere [19]. Figure 1a represents the details of the experimental measurements. Figure 2 represents the notations used. The uncertainties

Table 1 Properties of aluminum alloy 6082

Material property	Units	Property value
Density	Kg/m^3	2680
Melting point	°C	555
Modulus of elasticity	GPa	70
Electrical resistivity	Ωm	0.038×10^{-6}
Thermal conductivity	W/m K	180
Thermal expansion	1/K	24×10^{-6}

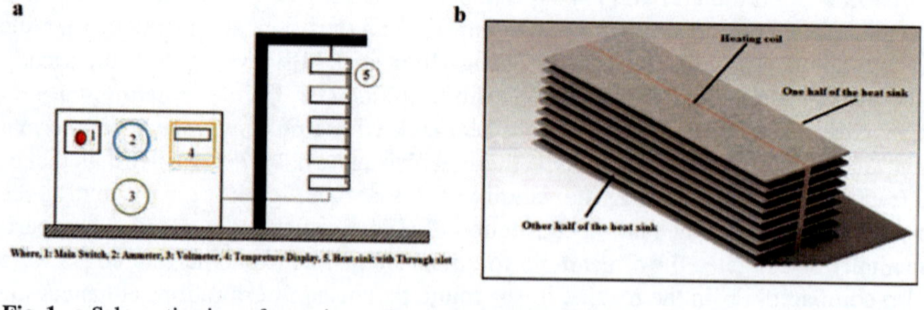

Fig. 1 **a** Schematic view of experimental test-rig, **b** Symmetric HS assembly

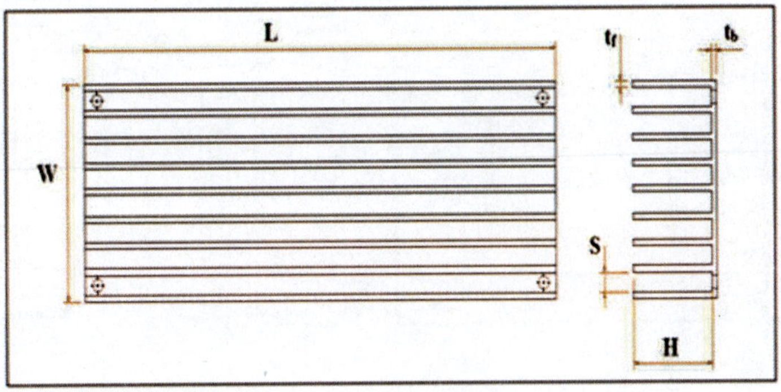

Fig. 2 Notations of a heat sink

associated with experimental measurements are well within 3.5% through Kline et al. method [21].

4 Results and Discussion

4.1 Temperature History for Steady State Analysis and Experiments with 9, 6, 5 and 3 Fin HS Configuration

Surface temperatures from T_1, T_2, T_3, T_4 *and* T_{amb} were noted at regular time intervals to ensure a steady state condition. The test-rig achieved steady state condition almost after 2–3 h. The thermocouples were located as shown in Fig. 2. At steady state, for the supplied heat input, the (HTC) was calculated accordingly [11]. More related to configurations discussed by author elsewhere [20]. Table 2 represents the details of the configurations used in the experiments. Note that only the fin spacing *s* is varied by reducing number of fins to obtain different configurations. Figure 3a, b, c represents the details of 9 fins HS configurations.

From Fig. 4a, b, c, it is seen that, excess temperature keeps on increasing with increase in heat input. Among 9, 6 and 5 fin configurations considered, modified

Table 2 Experimental details and configuration

Experiment with solid fins HS	Experiment with through slot fins HS	Experiment with dwarf fins HS
9 fins HS	9 fins HS	9 fins HS
6 fins HS	6 fins HS	6 Fins Hs
5 fins HS	5 fins HS	5 fins HS
–	3 fins HS	3 fins HS

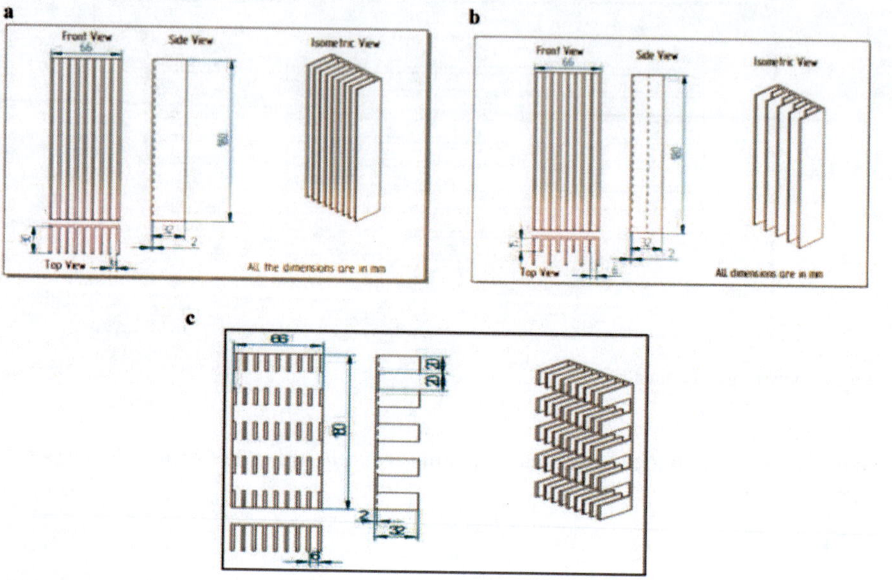

Fig. 3 **a** Fin configuration geometry for 9 fins HS, **b** Fin configuration geometry for 9 fins dwarf HS, **c** Fin configuration geometry for 9 fins dwarf HS

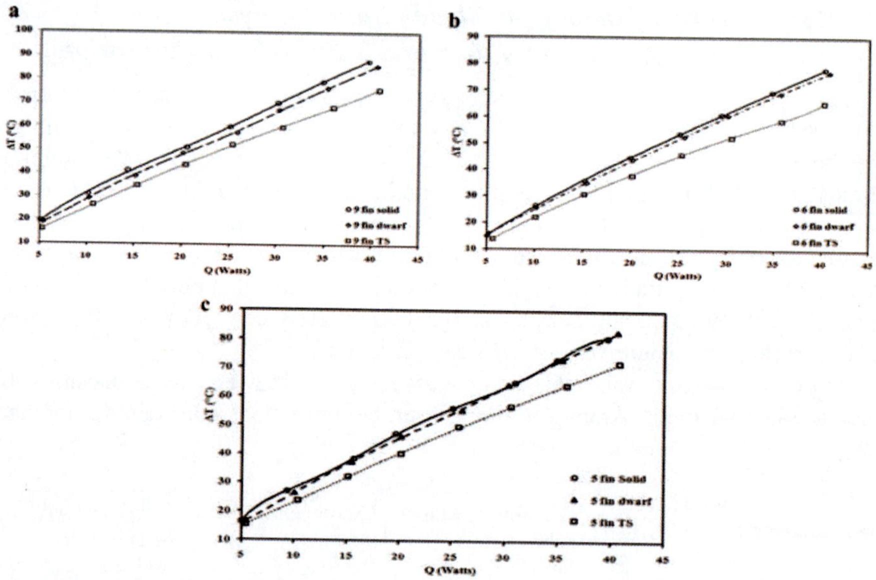

Fig. 4 **a** Variation of base temperature difference for 9 fins configurations, **b** variation of base temperature difference for 6 fins configurations, **c** variation of base temperature difference for 5 fins configurations

HS gave better cooling effect and solid HS gave the least cooling effect. Among the modified HS, TS configurations gave the highest cooling effect. The amount air flow in modified HS configurations is more in comparison with solid HS configurations. Cooling effect is better in case of modified HS.

From Fig. 5a, it is seen that, the cooling effect for 6 fins solid HS configurations is highest, whereas for 9 fins HS, it is least and for 5 fins HS it is moderate. From Fig. 5b, c, it is observed that, the cooling effect for 6 fins modified HS configurations is highest, whereas for 3 fins HS, it is least, and for 9, 5 fins HS, it is moderate. It must be noted that 3 fins HS data is available only for modified HS configurations. From Fig. 5a, b, c, it is observed that, 6 fins solid HS or 6 fins modified HS are promoting optimum cooling among the configurations considered.

From Fig. 6a, b it is observed that, for 6 fins (10.8 mm spacing) modified HS, the heat dissipation rate is highest and the rate of heat dissipation peaks at 6 fins HS configuration and decreases on either side, i.e., 9 fins (6 mm spacing) and 5 fins (14 mm spacing) HS configurations. The heat dissipation rate for 3 fins (30 mm spacing) modified HS is the least. It is inferred from the figures that whether the HS is solid or modified, for a given fin length, fin height and fin thickness, as the fin spacing increases, the rate of cooling reaches toward an optimum value and then decreases, therefore, for the present set of experiments conducted, the optimum fin spacing is at 10.8 mm. In case of HS with 10.8 mm spacing, the influence of conduction heat transfer and convection heat transfer is optimum.

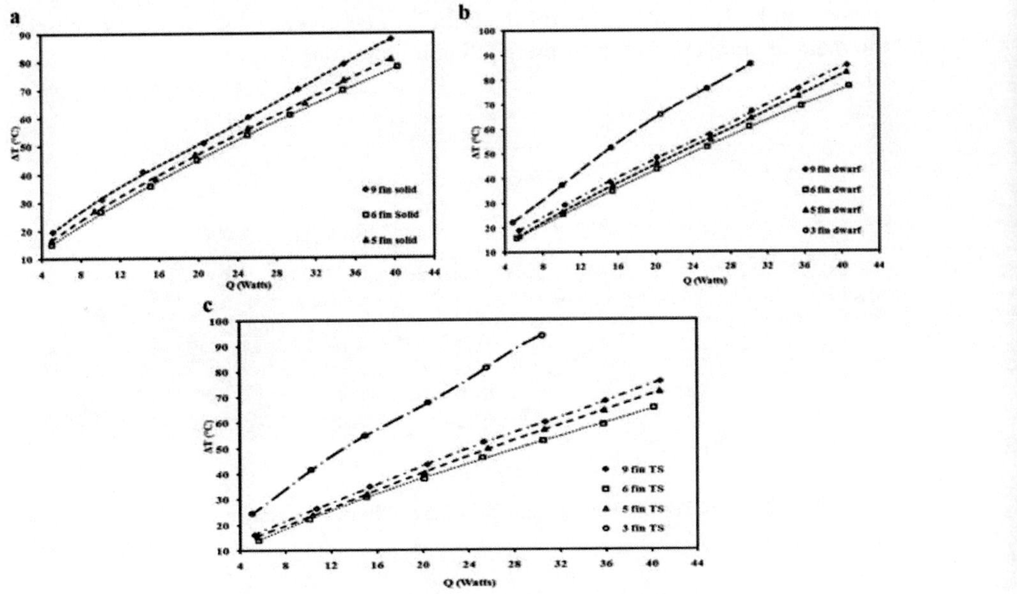

Fig. 5 **a** Variation of base temperature difference for different fin configurations (solid), **b** variation of base temperature difference for different fin configurations (dwarf), **c** Variation of base temperature difference for different fin configurations (through slot)

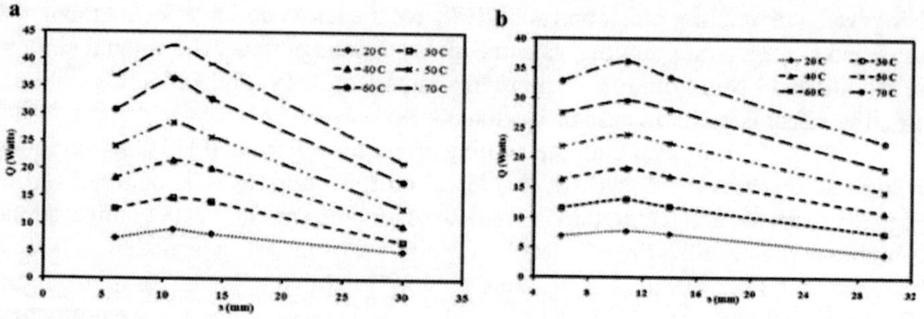

Fig. 6 **a** Variation of heat dissipation rate for different fin configurations (dwarf), **b** Variation of heat dissipation rate for different fin configurations (through slot)

From Fig. 7, it is clear that, at 50 °C, the heat dissipation from 6 fins TS HS was the highest, the heat dissipation from 6 fins dwarf HS was moderate and the heat dissipation from 6 fins solid HS was the least. The 6 fins TS HS was found to be the best configuration among the selected configuration of experiments. In comparison with solid HS, TS and dwarf HS promoted cooling process. For 3 fins modified HS, it can be observed that, the heat dissipation from TS configuration is least, as material available for conduction heat transfer is minimum. The heat conducted from the base is not optimum hence total heat dissipated by conduction and convection mode is less. In Table 3, the details of percentage material (mass) saved and percentage increase in HTC are presented for the 6 fins HS configurations. Figure 8 represents the mass of comparison for different HS configurations.

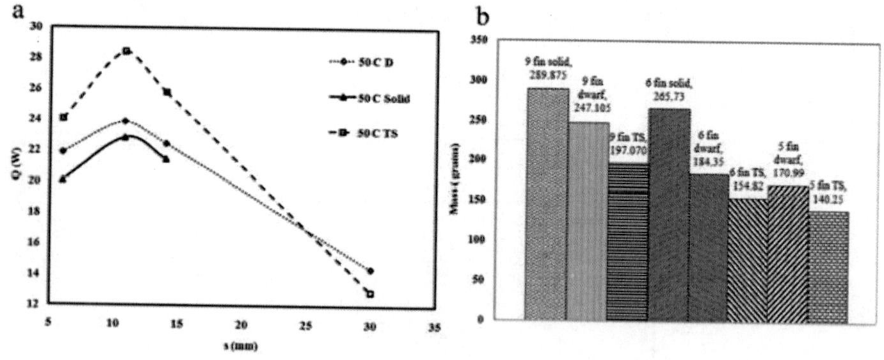

Fig. 7 **a** Variation of heat dissipation rate for different fin configurations at 50 °C **b** Comparison of heat sink mass

Table 3 Details of material saved and increase in HTC

Fin array configuration	% Material saved	% Increase in HTC
6 fins solid HS	0	0
6 fins through slot	41.74	81.68%
6 fins dwarf HS	30.625	32.4%

5 Conclusion

Experiments were conducted considering several heat sink configurations to quantify the scope of heat dissipation for improved cooling process. It has been inferred that, heat sinks with modifications such as through slot and dwarf type yielded higher amount of heat dissipation rates as compared to solid heat sink. The maximum heat dissipation rate amounting to 26.26% was noted in case of through slot heat sink with 6 fin configuration as compared to solid heat sink. The dwarf type of heat sink with 6 fin configuration exhibited 9.74% of heat dissipation over solid heat sink. Among the various configurations tested, 6 fin through slot heat sink exhibited higher rate of heat dissipation and thus maximum cooling effect. Material savings of 41.74% in through slot and 30.62% in dwarf over solid heat sink were obtained.

Acknowledgements Authors are thankful to Principal and Director of Jain College of Engineering, Belagavi, and Dean of Research for support and motivation through the research work.

References

1. Starner KE, McManus HN (1963) An experimental investigation of free convection heat transfer from rectangular fin arrays. J Heat Transfer 85:273–278
2. Harahap F, McManus HN (1967) Natural convection heat transfer from horizontal rectangular fin arrays. Trans ASME, Ser C, J Heat Transfer 89:32–38
3. Jones CD, Smith LF (1970) Optimum arrangement of rectangular fins on horizontal surfaces for free convection heat transfer. J Heat Transfer 92:6–10
4. Leung CW, Probert SD, Shilston MJ (1985) Heat exchanger design: thermal performances of rectangular fins protruding from vertical or horizontal rectangular bases. Appl Energy 19:287–300
5. Leung CW, Probert SD, Shilston MJ (1986) Heat-transfer performances of vertical rectangular fins protruding from rectangular bases: effect of fin length. Appl Energy 22:313–318
6. Leung CW, Probert SD (1987) Heat-exchanger design: optimal uniform thickness of vertical rectangular fins protruding perpendicularly outwards, at uniform separations, from a vertical rectangular base. Appl Energy 26:111–118
7. Leung CW, Probert SD (1989) Thermal effectiveness of short-protrusion rectangular, heat-exchanger fins. Appl Energy 34:1–8
8. Yüncü H, Anbar G (1998) An experimental investigation on performance of rectangular fins on a horizontal base in free convection heat transfer. Heat Mass Transfer 33:507–514
9. Yuncu H, Guvenc A (2001) An experimental investigation on the performance of rectangular fins on a vertical base in free convection heat transfer. Heat Mass Transfer 37:409–416
10. Yazicioğlu B, Yüncü H (2007) Optimum fin spacing of rectangular fins on a vertical base in free convection heat transfer. Heat Mass Transfer 44:11–21

11. Awasarmol UV, Pise AT (2015) An experimental investigation of natural convection heat transfer enhancement from perforated rectangular fins array at different inclinations. Exp Thermal Fluid Sci 68:145–154
12. Awasarmol UV, Pise AT (2017) Experimental study of heat transfer enhancements from array of alternate rectangular dwarf fins at different inclinations. J Inst Eng India Ser. https://doi.org/10.1007/s40032-017-0383-9
13. SadrabadiHaghighi HR, Goshayeshi, Safaei MR (2018) Natural convection heat transfer enhancement in new designs of plate-fin based heat sinks. Int J Heat Mass Transfer 125:640–647
14. Shahrubudin N, Leea TC, Ramlan R (2019) An overview on 3d printing technology: technological, materials, and applications. Procedia Manuf 35:1286–1296
15. Dhaiban HT, Hussein MA (2020) The optimal design of heat sinks: a review. J Appl Comput Mech 6(4). https://doi.org/10.22055/JACM.2019.14852
16. HakanAltun A, Ziylan O (2019) Experimental investigation of the effects of horizontally oriented vertical sinusoidal wavy fins on heat transfer performance in case of natural convection. Int J Heat Mass Transfer 139:425–431
17. Sathe A, Sanap S (2020) Experimental analysis of effect of slitted rectangular fins on heat sink under natural convection heat transfer. Int J Ambient Energy. https://doi.org/10.1080/01430750.2020.1778083
18. Huang C-H, Wu Y-T (2021) An optimum design for a natural convection pin fin array with orientation consideration. Appl Thermal Eng 188:116633
19. Rajshekhar VU, Metikela SB (2021) Experimental Investigation of heat sinks with and without perforation—addressing toward higher cooling rates and optimum material. In: Recent advances in mechanical infrastructure. Lecture notes in intelligent transportation and infrastructure. Springer, Singapore. https://doi.org/10.1007/978-981-33-4176-0_23
20. Rajshekhar VU, Metikela SB (2021) Experimental investigation of electronic cooling rates considering perforated and dwarf based heat sinks. In: 3rd international conference on advances in mechanical engineering and nanotechnology. Manipal University Jaipur, India
21. Kline SJ, McClintock FA (1953) Describing uncertainties in single sample experiments. Mech Eng 3–8

A Framework for Leagile Production System Using Analytical Hierarchical Process

Soumil Mukherjee, Vinay Kumar Bohra, and Vikram Sharma

Abstract The manufacturing industry in several parts of the world finds it difficult to survive over the last financial year. Companies are now looking forward to new survival strategies to enhance productivity and profits. Leagile manufacturing will be one such strategy. Leagile manufacturing combines the concepts of lean production systems and agile production systems. Lean production aims to improve efficiency by eliminating waste while working with minimal resources at disposal. The agile production systems aim to quickly capture new and changing customers' requirements using a flexible manufacturing setup. This paper proposes the use of leagile production as a competitive strategy for the manufacturing industry. The research uses Analytical Hierarchy Process (AHP) to propose a conceptual leagile implementation framework following a practitioners' perspective.

Keywords AHP · Leagile · Manufacturing

1 Introduction

The manufacturing industry comprises design, production of components, sub-assemblies, and assemblies using raw materials, quality assurance, and timely delivery of goods to the customers. The manufacturing sector is one of the significant sources of revenue for most of the economies of developed and developing countries. Hence, this sector is often encouraged by the governments of these countries. The ongoing pandemic that has prevailed throughout the last year has had a severe impact on the growth of the manufacturing industry world over. The industry is compelled to change its competitive strategy. According to a report by Industrial automation

S. Mukherjee (✉) · V. K. Bohra · V. Sharma
The LNM Institute of Information Technology (LNMIIT), Jaipur, Rajasthan 302006, India
e-mail: 15pcs003@lnmiit.ac.in

V. K. Bohra
e-mail: 17ume064@lnmiit.ac.in

V. Sharma
e-mail: vikram.sharma@lnmiit.ac.in

A. Maurya et al. (eds.), *Recent Trends in Mechanical Engineering*, Lecture Notes in Mechanical Engineering, https://doi.org/10.1007/978-981-19-7709-1_52

India [8], the component manufacturing industry belonging to automotive, aerospace, electronics, and machine tools has been hit hard due to the economic meltdown. The component manufacturing sector is heavily hit by the impact of the virus and shut production and factory operations in the automotive, electronics, aerospace verticals, among others. There has been a fall in demand and severe disruptions in the supply chain. A report on a survey conducted by Pricewatercoopers [11] brings out various challenges the manufacturing industry faces because of the COVID-19 pandemic. It has been indicated that 64 percent of respondents feel there is a global recession, 41 percent indicate a reduction in productivity, 40 percent indicate reduced consumption, and 23% indicate supply chain disruptions.

Emerging technologies have made processes smoother, more reliable, and more accessible. Integrating the latest technology in the supply chain has become essential to eliminate redundancy [16]. Decisions that need support methods are difficult by definition, and therefore, complex to model. A trade-off between perfect modeling and usability of the model should be achieved [13, 16]. As the businesses are now looking forward to changing their survival cum competitive strategy, leagile production can become a means to improve productivity. In most of the industrial sectors, the demand volatility is constantly rising. The ever-changing market ensures that demand in almost every sector has become more volatile than it was earlier. Reacting quickly to changes has become an essential characteristic for any commercial supply chain [4]. Leagile combines concepts of lean and agile production systems [14]. While lean production aims to improve efficiency by eliminating waste when working with the minimal resources at disposal [17, 18], agile production systems aim to quickly capture new and changing customers' requirements by using flexible manufacturing setups. Lean production uses techniques such as value stream mapping and just in time [10, 15]. Both lean and agile concepts are separated by a strategic point known as the decoupling point. Upstream of which there is a level schedule and downstream, and the system is agile to meet the volatile needs of customers. Research by Virmani et al. [16] claims that leagile production systems are among the most dominant research areas in operations management in the recent past. It involves the advantages of both lean and agile systems [17]. Balakrishnan et al. [3] present a leagile manufacturing paradigm for a pumps manufacturing unit. Raj et al. [12] present the performance metrics of a leagile supply chain and develop a leagility index.

This study developed a conceptual model for leagile production using the Analytical Hierarchy Process (AHP) that T. L. Saaty developed in 1971–1975 [7]. AHP is one of the leading mathematical models used for multi-criteria decision-making, planning, resource allocation, and in-conflict resolution [5, 9]. This is a method in which both physical (tangible) and psychological (intangible) measures are considered [14]. In this process, a hierarchic or a network structure is used to represent the problem. It then analyzes and prioritizes different criteria by 'paired comparison' to derive a 'judgment matrix'. Further applying the technique to a 'judgment matrix' to obtain a 'priority matrix', which gives us the most and least influential criterion. These matrices are positive and reciprocal.

2 Framework of Enablers to Leagile Production System Using AHP Methodology

AHP approach achieves pairwise comparisons among factors or criteria to prioritize them at each hierarchy level using the eigenvalue calculation [6]. AHP structures a decision problem into a hierarchy with a goal, decision criteria, and alternatives [1]. Analytical Hierarchy Process (AHP) is a mathematical tool of problem-solving used among various levels of management authorities. The technique is used to rank the criterion and alternatives for any given problem at hand. The steps in performing an AHP for a problem statement are as follows.

The first step is to identify the problem statement that needs to be resolved. This problem statement goes at the top of the hierarchy tree. The criteria are kept at the second level in the hierarchy, and the alternatives are kept at level three. Each alternative has its value of criteria associated with the problem. The second step is to create a pairwise comparison matrix of all the criteria in the second level. This pairwise matrix gives the relative importance of each attribute with respect to the other attributes relevant in the problem statement. The scale used for the ranking of attributes of pairwise comparison is as under (Table 1).

The third step is to create a normalized pairwise matrix which is calculated by dividing all the column elements by the sum of that column. The fourth step involves calculating the average of all the elements of each row called criteria weight. Each criteria weight is then multiplied with its respective column and a new matrix is received. After the previous multiplication, all the row elements are added to derive a weighted sum. The weighted sum value for each row is divided by the criteria weight for that row. The ratios obtained in the last step are averaged to yield a value called Lambda max. Lambda max is the maximum eigenvalue of the matrix and it is needed to calculate Consistency Index (CI). Lambda max is determined by dividing all the elements of the weighted sum matrices by priority vector for each criterion and then averaging these values [2]. Consistency Index is calculated using the following formula:

$$\text{ConsistencyIndex(C.I.)} = (\text{LambdaMax.} - n) / (n - 1) \tag{1}$$

Table 1 Attribute ranking scale of pairwise comparison

Attribute	Relative importance
1	Equal importance
3	Moderate importance
5	Strong importance
7	Very strong importance
9	Extreme importance
2, 4, 6, 8	Intermediate values
1/3, 1/5, 1/7, 1/9	Values for inverse comparison

Table 2 Random index values

n	Random index (RI)
1	0
2	0
3	0.58
4	0.90
5	1.12
6	1.24
7	1.32
8	1.41
9	1.45
10	1.49

In the last step, Consistency Ratio is calculated as follows:

$$\text{ConsistencyRatio(C.R.)} = \text{ConsistencyIndex(C.I)} / \text{RandomIndex(R.I)} \quad (2)$$

where the Random Index is the Consistency Index of a randomly generated pairwise matrix. The Consistency Ratio tells the decision maker how consistent the decisions have been when making pairwise comparisons. The CR value is inversely proportional to consistent decision-making [2] (Table 2).

The enablers of leagile production system were studied from the extant literature and validated by the practitioners. The table used for AHP analysis in this research is given below (Table 3).

The framework of enablers for achieving a leagile production system is proposed as follows (Figs. 1 and 2).

The data for AHP analysis was collected using Delphi method, where in researchers submit a questionnaire to a panel of experts and multiple rounds are taken into consideration where experts are permitted to adjust their response in the subsequent round based on answers from groups of panelists. The questionnaire was sent out to 63 industry practitioners. Two thirds of the total respondents were employed as production supervisors or managers. A third of the respondents held operations and lean manager positions during the study. The major sectors where the respondents belong to can be broadly classified into automotive, metal forming, packaging, and retail industries. The median working tenure of the respondent group was 12 years (Tables 4 and 5).

Table 3 Enablers of leagile production system

Enabler	Reviewed by
Involvement of suppliers The proper quality and quantity of deliveries Spontaneous provision of raw materials Supplier preparation	Carvalho (2011) Vonderembse (2006), Carvalho (2014) Bowen (2011)
Contribution from top management Employee coaching and mentoring for performance development Transparency in exchanging details Commitment to green production	Hong et al. (2012), Wang et al. (2013) Lee (2002), Cicculло F (2017) Bowen (2011)
Customers Engagement Feedback on quantity, cost, and quality Customer recommendation environment Customer recommendation implementation Encouraging the client to turn from their hand to a green chain	Agarwal (2007) Mittal (2017) Vonderembse (2006) Gimenez (2012), Caniels (2013)
Integration of IT and optimization of capital Department communication to eliminate wastage Efficient supply chain and its quantification Retention of minimum inventory level Including natural resource use	Mittal (2017) Gunasekran (2001) Melton (2005) Mittal (2017)
Regulation of the product life cycle Material recycling Proper waste management Implementation of reverse logistics in the chain	Sarkis (2003) Zhu, Sarkis (2004) Sarkis(2003)

Based on criteria weight, the factors can be ranked in the current order:

1. Contribution from top management
2. Customers engagement
3. Integration of IT and optimization of capital
4. Regulation of product life cycle
5. Involvement of supplier.

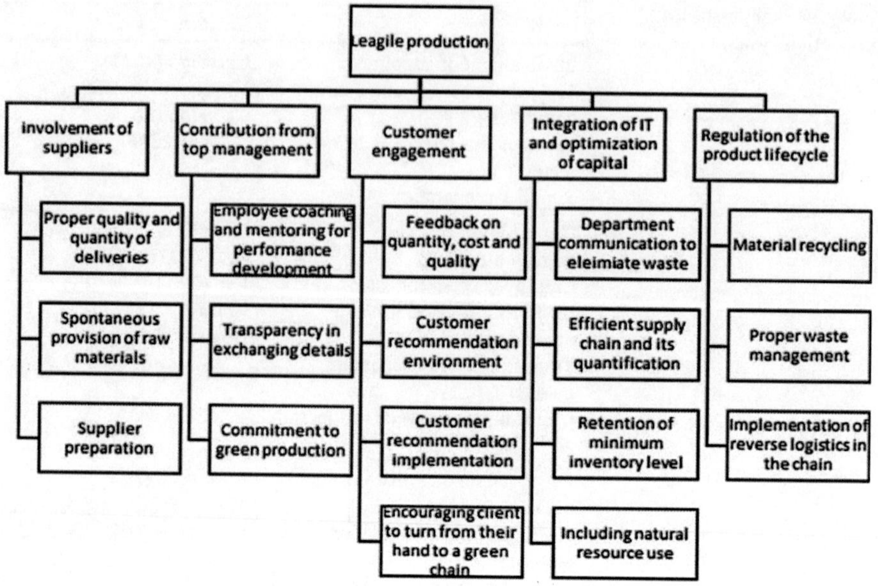

Fig. 1 AHP-based framework for enablers of leagile production system

Lambda max for inconsistent pairwise comparison matrix is always greater than n, and the difference between the root and n is equal to the sum of the remaining eigenvalues. The quotient of this difference divided by (n–1) is defined as the consistency index (CI), which is the index of the consistency of judgments across all pairwise comparisons [4]. To check the consistency (Table 6).

Ratio of weighted sum/criteria weight (Table 7).

1. Involvement of supplier = 4.24
2. Top contribution from management = 5.27
3. Customers engagement = 5.33
4. Integration of IT and optimization of capital = 5.33
5. Regulation of product life cycle = 5.33

$$\text{LambdaMaximum} = 5.132$$

$$\text{ConsistencyIndex} = (5.132 - 5)/(5 - 1) = 0.033$$

$$\text{ConsistencyRatio} = \text{ConsistencyIndex}/\text{RandomIndex}$$

As we have N = 5, R.I. = 1.12.

C.R. = 0.033/1.12 = 0.029 < **0.1**

The matrix can be assumed to be consistent since the Consistency Ratio is less than 0.1

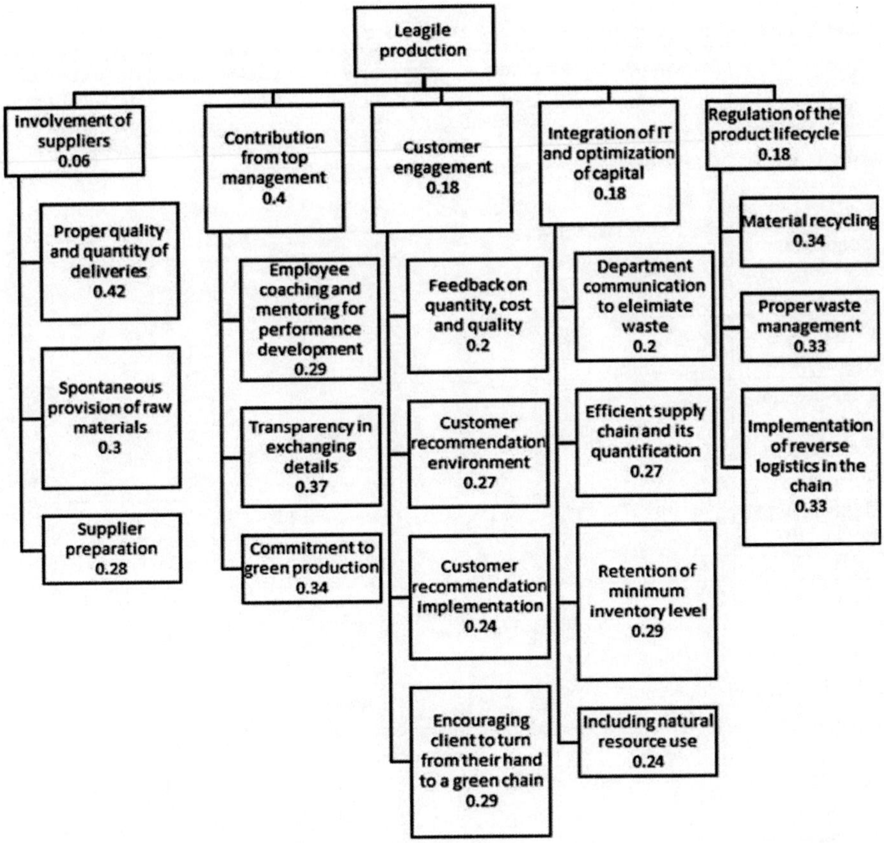

Fig. 2 Weighting of enablers based on experts responses

3 Inference and Discussion

A framework to prioritize the implementation of enablers in the leagile production system was developed using the AHP methodology. Having applied the AHP model to the respondent dataset and interpreting the results, since the Consistency Ratio has a value less than 0.1, it makes the proposed methodology acceptable. It can be inferred that the attributes (data and parameters) used are consistent. After enumerating the percentage weightage of the primary enablers of the leagile implementation strategy, the contribution of top management comes out as the most significant enabler. This model also tells us that supplier relations and continuous improvement enablers can be considered essential sub-factors of any agile supply chain. The primary importance of continuous enablers could be understood from the fact that agility in the supply chain is a continuous process. Changes should be made as and when necessary. This might mean using the right and latest technologies, implementing, and improving proper techniques for human resource management, cutting down tasks

Table 4 Pairwise comparison matrix

Criteria	Involvement of supplier	Contribution from top management	Customer engagement	Integration of IT and optimization of capital	Regulation of product life cycle
Involvement of supplier	1	1/7	1/5	1/5	1/5
Contribution from top management	7	1	3	3	3
Customer engagement	5	1/3	1	1	1
Integration of IT and optimization of capital	5	1/3	1	1	1
Regulation of product life cycle	5	1/3	1	1	1
Sum	23	2.13	6.2	6.2	6.2

Table 5 Normalized matrix

Criteria	Involvement of supplier	Contribution from top management	Customer engagement	Integration of IT and optimization of capital	Regulation of product life cycle	Criteria weight
Involvement of supplier	0.04	0.08	0.06	0.06	0.06	0.06
Contribution from top management	0.30	0.44	0.42	0.42	0.42	0.42
Customer engagement	0.21	0.16	0.19	0.17	0.18	0.18
Integration of IT and optimization of capital	0.21	0.16	0.19	0.17	0.18	0.18
Regulation of product life cycle	0.21	0.16	0.19	0.17	0.18	0.18

that can be avoided, and improving effectiveness and efficiency. The involvement of suppliers is an important factor and determines the speed and availability of the resources. Customer engagement, technology integration, and capital optimization when necessary are vital aspects for agility. Regulation of the product life cycle

Table 6 Consistency check

Criteria	Involvement of supplier	Contribution from top management	Customer engagement	Integration of IT and optimization of capital	Regulation of product life cycle	Weighted Sum Value
Criteria weights	**0.06**	**0.40**	**0.18**	**0.18**	**0.18**	
Involvement of supplier	0.05	0.06	0.034	0.034	0.034	**0.212**
Contribution from top management	0.35	0.44	0.51	0.51	0.51	**2.32**
Customer engagement	0.25	0.147	0.17	0.17	0.17	**0.907**
Integration of IT and optimization of capital	0.25	0.147	0.17	0.17	0.17	**0.907**
Regulation of product life cycle	0.25	0.147	0.17	0.17	0.17	**0.907**

Table 7 Consistency ratio

n	Random Index (RI)
1	0
2	0
3	0.58
4	0.90
5	1.12

leads to enabling new strategies, new technologies, adding or cutting down process steps, but the process can lead to complexities that would contradict agility. Reducing delays and avoiding wastage of resources is essential for any agile supply chain and can facilitate the transition to being leagile.

4 Conclusion

The manufacturing industry has been severely impacted by the COVID-19 pandemic worldwide, affecting the livelihood of many. The industry now needs to look forward to a more resilient survival strategy. This research proposes leagile production as the survival and competitive strategy in the changing world order. In this study, enablers of leagile manufacturing were studied from the extant literature and validated by industry practitioners. The study bolsters the findings of several previous researchers that claim top management commitment as one of the most critical factors for the success of any improvement strategy. As a future scope, the study can be taken forward using more advanced multi-criteria decision-making algorithms, statistical analysis of data collected through surveys, and confirmatory analysis.

References

1. Acharya V, Sharma SK, Gupta SK (2018) Analyzing the factors in industrial automation using analytic hierarchy process. Comput Electr Eng 71:877–886
2. Aupetit B, Genest C (1993) On some useful properties of the Perron eigenvalue of a positive reciprocal matrix in the context of the analytic hierarchy process. Eur J Oper Res 70:263–268
3. Balakrishnan K, Devadasan SR, Thilak VMM, Soundaram DS (2020) Leagile manufacturing paradigm in the production of jet pump-an implementation experi-ence. Int J Prod Qual Manag 30(4):462–487
4. Barzilai J, Lootsma FA (1997) Power relation and group aggregation in the multiplicative AHP and SMART. J Multi-Criteria Decis Anal 6:155–165
5. Dos Santos PH, Neves SM, Sant'Anna, D. O., de Oliveira, C. H., & Carvalho, H. D. (2019) The analytic hierarchy process supporting decision making for sustainable development: an overview of applications. J Clean Prod 212:119–138
6. Hilorme T, Tkach K, Dorenskyi O, Katerna O, Durmanov A (2019) Decision making model of introducing energy-saving technologies based on the analytic hierarchy pro-cess. J Manag Inf Decis Sci 22(4):489–494
7. Ho W, Ma X (2018) The state-of-the-art integrations and applications of the analytic hi-erarchy process. Eur J Oper Res 267(2):399–414
8. Industrial Automation India (2020) Impact of Covid 19 on manufacturing industries. https://www.industrialautomationindia.in/industryitm/10187/Impact-of-COVID-19-on-the-Manufacturing-Industries/industry. Accessed 2nd Jan 2021
9. Ishizaka A, Labib A (2009) Analytic hierarchy process and expert choice: benefits and limitations. OR Insight
10. Power DJ, Sohal AS (2000) An empirical study of human resource management strategies and practices in Australian just-in-time environments. Int J Oper Prod Manag 932–958
11. PwC (2020) COVID-19: What it means for industrial manufacturing, https://www.pwc.com/us/en/library/covid-19/coronavirus-impacts-industrial-manufacturing.html. Accessed 3 Jan 2021
12. Raj SA, Jayakrishna K, Vimal KEK (2018) Modelling the metrics of leagile supply chain and leagility evaluation. Int J Agile Syst Manag 11(2):179–202
13. Sharma, V., Dixit, A. R., & Qadri, M. A. (2016). Modeling lean implementation for manufacturing sector. Journal of Modelling in Management.
14. Sharma V, Dixit AR, Qadri MA (2015) Impact of lean practices on performance measures in context to Indian machine tool industry. J Manuf Technol Manag
15. Sharma V, Dixit AR, Qadri MA (2014) Analysis of barriers to lean implementation in machine tool sector. Int J Lean Think 5(1):5–25
16. Virmani N, Sharma V (2019) Prioritization and Assessment of leagile manufacturing enablers using Interpretive Structural Modeling (ISM) approach. Euro J Ind Eng 13(6):701–722
17. Virmani N, Saha R, Sahai R (2018) Leagile manufacturing: a review paper. Int J Prod Qual Manag 23(3):385–421
18. Virmani N, Saha R, Sahai R (2018) Social implications of leagile manufacturing system: TISM approach. Int J Prod Qual Manag 23(4):423–445

Various Concepts on Variable Inertia Flywheel in Rotating System

Dharmendra Kumar and Anil C. Mahato

Abstract Variable inertia flywheel is an innovative approach for storing energy in a rotating system. It may replace the constant inertia flywheel effectively from the conventional rotating system. The variable inertia flywheel has less weight, and it has a great potential to adjust the moment of inertia according to the load of the system. The rotating system with variable inertia flywheel possesses less weight, more flexible and compact. Besides, the variable inertia flywheel has a significant role to reduce the vibration of the system. This paper presents various methods to obtain variable inertia flywheel. Additionally, a comparative study on the influence of various vehicle suspension parameters using variable inertia flywheel and a constant inertia flywheel is addressed.

Keywords Rotating system · Variable inertia flywheel · Constant inertia flywheel · Vibration reduction

1 Introduction

Variable inertia flywheel (VIF) is importance equipment in the fields of energy storage and power control strategies in rotating system [1]. The working principle of the VIFs is simple and it is based on the centrifugal force which is developed during rotation of the VIF. The moment of inertia of the conventional flywheel is constant throughout the entire operation, while the same is variable in case of VIF, according to the speed of the rotating machine [2, 3]. To control the speed fluctuation of a rotating system, a heavy weight Constant Inertial Flywheel (CIF) is attached on it [4, 5]. The heavy weight CIF makes the whole system bulky as a result, it increases the start-up torque requirement for the machine. Moreover, it makes more difficulties to transport the rotating system and increase the overall cost of the system [6]. The stated drawbacks of the rotating system are eliminated using VIF instead of CIF. The VIF worked on the principle of centrifugal force and it contains different components such as

D. Kumar (✉) · A. C. Mahato
Department of Mechanical Engineering, Birla Institute of Technology, Mesra, Jharkhand 835215, India
e-mail: phdme10053.19@bitmesra.ac.in

A. Maurya et al. (eds.), *Recent Trends in Mechanical Engineering*, Lecture Notes in Mechanical Engineering, https://doi.org/10.1007/978-981-19-7709-1_53

rotating mass elements (movable blocks) that are moved radially outward direction, compression spring, rim and others. When the speed of the flywheel is increased, the radius of rotation of the movable blocks is increased as it slides into the sliding slot and moves away from the center of rotation and vice versa [7]. The compression springs are compressed during the travel of the moving blocks and develop a resistive force on it. The developed resistive force helps to return the moving blocks when the speed of the machine is decreased. As the positions of the movable blocks are changed with the variation of the speed, the radius of gyration is also changed. As a result, the moment of inertia is varied and it is directly proportional to the square of the radius of gyration. Therefore, when the VIF is used in rotating system, the moment of inertia is automatically varied according to the change of the speed of the system [8]. Due to adjustable moment of inertia, the VIFs are more flexible, provide smooth operation, generate less vibration, more economical and effectively utilize the available energy [9]. Based on the driving mechanism, VIFs are classified as active and passive type VIF. The passive VIF is popular due to its simple structure, ease of handling and independent from driving mechanism, while active VIF needs driving mechanism for performing their desired functions [10]. In addition, based on the design, it can be classified as Moveable Mass Block (MMB) VIF and fluidic VIF [1]. The design of MMB-VIF has been varied according to the loading and unloading conditions of the system. In [11], Yang et al. designed a VIF with three moveable mass blocks, and in [12, 13], authors designed the same, considering four movable mass blocks. Another literature, i.e., [14], Zhang et al. used eight numbers of movable mass blocks to design the VIF. Above authors changed their respective design according to the loading conditions of the system. The application of the VIF in various engineering fields is an important aspect and presently, its performance found significant satisfactory in various fields such as centrifugal pendulum [15], hybrid power system [16], diesel generator [14, 17], automobile [18], punching press, energy wave converter system [13], vibration harvesting device [8, 12, 19–22], energy storage system [23] and microgrid application [24]. In automobile, a band variable inertia flywheel has been used in urban transit bus and it was found that 30% energy saving [18]. However, the application of the VIF is still limited due to its complex structure, maintenance difficulties, leak proof condition and fewer life cycles of its components [10].

This paper presents an overview on the various designs of VIF and its application in various engineering field. Also, various benefits of the VIF over the CIF are presented briefly.

2 Two-Terminal Mass System

In [25], Li et al. presented a two-terminal mass system with a combination of a flywheel and screw transmission. Another two-terminal mass system, which is a combination of an inerter and rack-gear transmission, is developed by Smith and Wang in [26]. The schematic diagram of the two-terminal mass system is shown

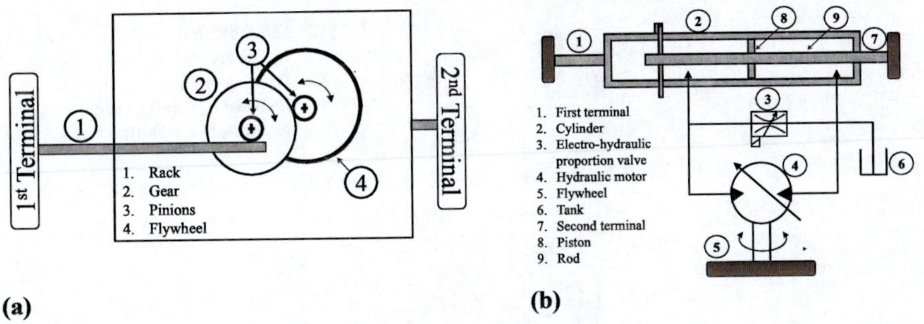

Fig. 1 **a** Two-terminal mass system with an inerter and rack-gear transmission **b** Two-terminal mass system with a flywheel used in power hydraulic system

in Fig. 1a. Additionally, Li et al. present another concept on a two-terminal mass system where a flywheel is incorporated into the power hydraulic system [27], which is shown in Fig. 1b. The two-terminal mass system may effectively used in vehicle suspension for vibration absorption [28] and it is also known as a constant moment of inertia suspension system (CMI suspension). It is found that the performance of CMI suspension is better as comparison to the conventional suspension system. However, some limitations of this system still to be addressed. To overcome its limitations, one novel two-terminal variable moment of inertia hydraulic-flywheel system (VMI system) has been introduced [12, 29]. The performance of the VMI suspension system is examined by considering the ride comfort, tire grip and suspension deflection. It is found that the performance of the suspension system is improved by incorporating the VMI flywheel into the suspension system.

2.1 Various Designs of VIF

To obtain variable inertia, various concepts have been adopted previously to design the VIF. The most important and significant techniques that are used to design the VIF are moveable mass block (MMB)-based VIF [1], fluidic VIF [30] and a combination of both [31]. The working principle of the MMB-VIF is based on centrifugal force and the main components of the stated system are compression springs, moveable masses, rim and other components. The parameters that affect the performance of the MMB-VIF are spring stiffness, weight and size of the movable mass block and their composition. The forces involved in MMB-VIF are inertia force, spring force, friction force, Coriolis force and centrifugal force, while in CIF, mainly inertia force is dominating. The number of movable blocks can be increased to meet the required demand and working conditions. In [11], Yang Tie-Her discussed a concept on VIF which is based on three movable blocks is shown in Fig. 2. All blocks are placed on the sliding rods and each sliding rods makes an angle of 120° with each other. When the speed of the flywheel increased/decreased, the compression springs are

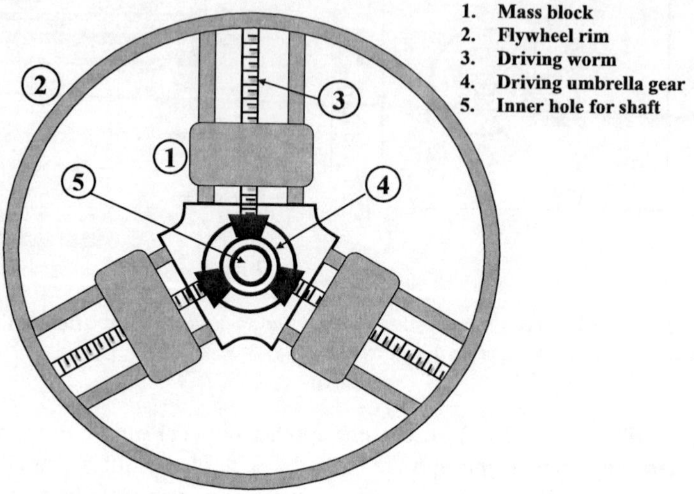

Fig. 2 Variable inertia flywheel with three movable mass blocks

being compression/expansion stage to provide the spring forces to the sliders [11]. The stated design of the MMB-VIF is not widely used due to the low range of changing capacity of the moment of inertia and for the possibility of developing a huge unbalanced force into the system. Another similar concept but slightly different design of MMB-VIF is presented by various authors [12, 29]. All these designs contain four MMB instead of three mass blocks and these are traveled into the slots. However, the concept of the working principle is similar to the previous design. The schematic diagram of the four MMB-based flywheel design is shown in Fig. 3. Moreover, in [32], authors presented another design of VIF which is based on four MMB is shown in Fig. 4. This design has been modified from previous similar designs by considering single spring in each slot. However, the principle of operation is similar.

The flywheel is mostly used as an energy storage device and vibration harvesting device. To reduce the vibration, various shock harvesting devices are used. Presently, most shock harvesting devices harvest the generated oscillation movement by converting it into heat energy. The heat loss increases the fuel consumption of the vehicle. To overcome the issue, a flywheel is used along with other shock harvesting devices into the vehicle suspension. The flywheel converts the oscillating suspension movement into rotary motion, and developed rotary motion can be utilized for performing other auxiliary functions of the vehicle. In [12], Xu et al. show that if MMB-VIF is used in vehicle suspension systems instead of CIF, the function of harvesting the oscillating motion is achieved effectively and efficiently.

Many authors used MMB-VIF in diesel generator objective are to reduce the power fluctuation under variable loading and disturbance conditions [14, 17]. When the diesel generator is connected with MMB-VIF, the stability of the system is better than the system whenever the same generator is connected with CIF. In [14], Zhang

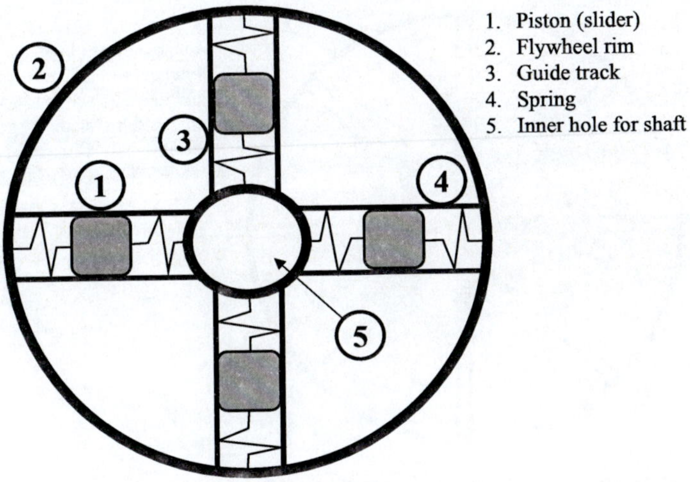

Fig. 3 Variable inertia flywheel with four movable mass blocks

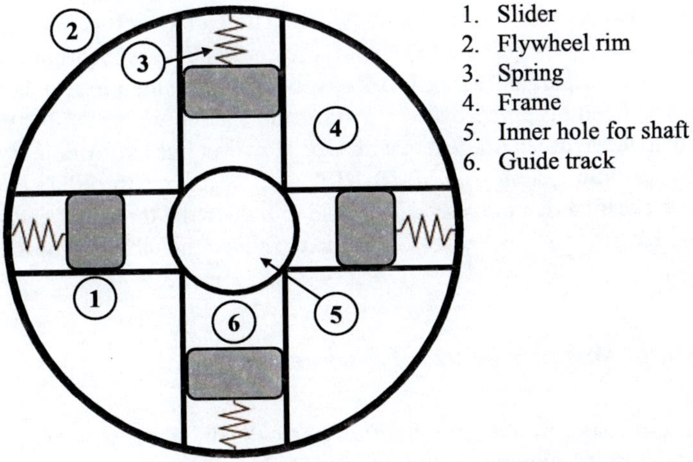

Fig. 4 Variable inertia flywheel with four movable mass blocks and single spring in each slot

et al. modified the design of VIF with eight MMB and used a closed-loop frequency response characteristics curve for differentiating the performance of MMB-VIF and CIF in diesel generators. The modified VIF with eight MMBs is shown in Fig. 5. In the closed-loop characteristics curve, MMB-VIF decreases the magnitude of the peak value and bandwidth. The decrement of maximum magnitude is the indication of smooth response of MMB-VIF and bandwidth is the performance under defined disturbance. The MMB-VIF can be used in a power hydraulic system to reduce the speed fluctuation of the system [33]. In [34], authors presented another literature

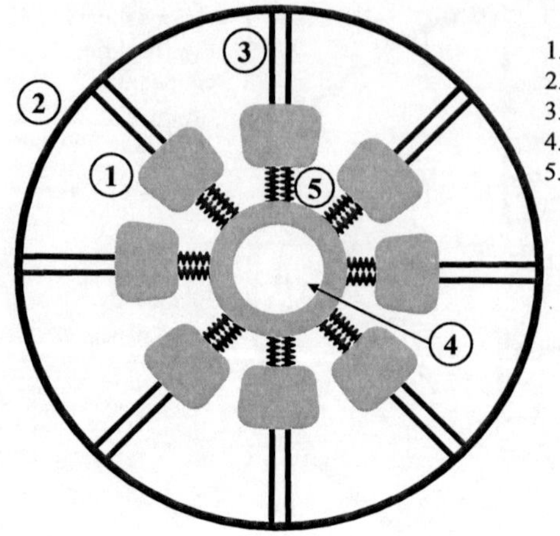

Fig. 5 Variable inertia flywheel with eight movable mass blocks

based on a hydraulic system with MMB-VIF. That literature is based on the comparative study between CIF, VIF and soft switch. The soft switch is a device used to reduce the throttling energy loss [35]. All these equipment are incorporated into the hydraulic system, separately as well collectively to identify their role in the hydraulic system. It has been observed that the hydrostatic system with MMB-VIF has less vibration than the hydrostatic system with CIF. However, the hydrostatic system with CIF and hydrostatic system with MMB-VIF has almost same throttling energy loss. The authors clarified that the VIF has no role to reduce the throttling energy loss of the system. However, it has a significant role to reduce the vibration of the system.

2.2 *Fluidic Variable Inertia Flywheel*

To obtain a variable moment of inertia, when the flywheel design is based on liquid contain on it, is known as fluidic variable inertia flywheel. Previously, many researchers presented various design concepts on fluidic variable inertia flywheel. In [36], Maxwell used a hollow structure to design the fluidic variable inertia flywheel. According to that design, energy is stored by accumulating fluid in definite space, whereas energy is supplied by traveling fluid to another definite space of the hollow structure. Therefore, the moment of inertia of the flywheel changes by changing the position of the fluid. However, the fluidic VIF is not performed well when treated as a high energy storage system. In [37], author presented an innovative arrangement on fluidic variable inertia flywheel which is shown in Fig. 6. In this concept, the fluid enters into the flywheel from the tank and comes out from the cylinder. The fluid flow is controlled by the piston movement and the spring force. The proposed design

Fig. 6 Fluidic variable inertia flywheel

1. Load
2. Fluidic flywheel chamber
3. Atmospheric pressure variation
4. Piston
5. Pressure distribution of the liquid
6. Spring
7. Pipe used for liquid supply
8. Reservoir

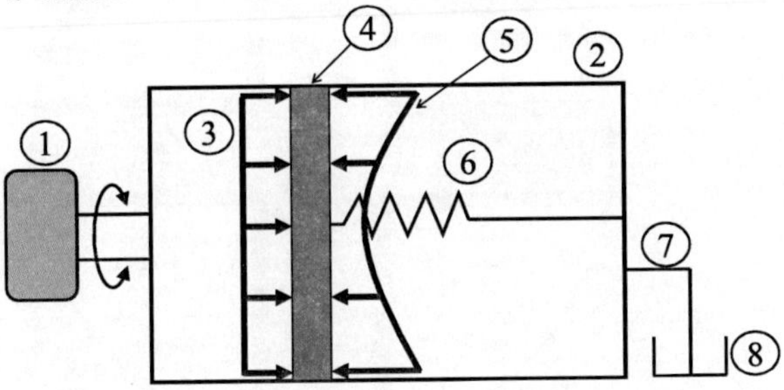

of the flywheel can absorb and release the desired energy without any fluctuation in angular velocity (up to 2% coefficient of fluctuation).

Dugas [38] discussed another innovative approach on fluidic VIF, in which the liquid chamber contains holes for liquid inhaling and air exhaling. Also, some one-way restriction valves are placed into it. The function of a one-way restriction valve is to allow the fluid flow in such a way that the direction of flow is just opposite to the direction of rotation of the flywheel body. When the system worked with increased load, the flywheel supplies the required surplus amount of energy, and hence, the rotary motion of the flywheel is slow down. Hence, energy stored in the fluid is supplied to the flywheel to meet the desired energy demand in the system.

2.3 *Hybrid Variable Inertia Flywheel*

In hybrid VIF, the inertia of the flywheel is adjusted mechanically as well as by fluid force [39]. In [31], authors used magneto-rheological fluid into the novel magneto-rheological variable inertia flywheel (MR-VIF). This flywheel consists of an aluminum frame and four magneto-rheological dampers which traveled through the four identical slots as shown in Fig. 7. Each damper is made of a cylinder, piston, MRF and springs. The effect of rotational speed and excitation current applied to MR damper on the moment of inertia is estimated and the performance of the proposed flywheel is validated experimentally. The additional damping force due to the MRF, the adaptive adjusting of the moment of inertia can be reached up to 27%. In addition, the authors show that the magneto-rheological damping force increases with the increase of electric current or vice versa.

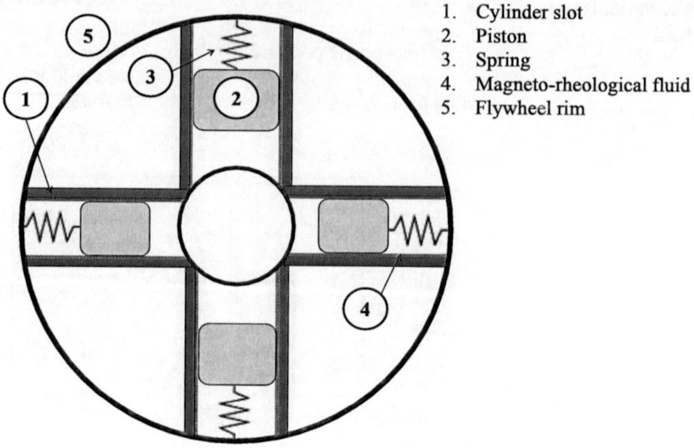

Fig. 7 Magneto-rheological variable inertia flywheel (MR-VIF)

3 Conclusions

This study reviewed various concepts of variable inertia flywheel to obtain an adjustable moment of inertia. It is found that there three ways the variable inertia flywheel can be designed and these are moveable mass block variable inertia flywheel (MMB-VIF), fluidic variable inertia flywheel and magneto-rheological variable inertia flywheel (MR-VIF). The following highlighted points are obtained from the present study.

- The variable inertia flywheels are suitable whenever variable loading and unloading are involved in the system.
- The MMB-VIF is feasible but the fluidic variable inertia flywheel is not commercialized yet due to its complex structure. It requires future work in numerous areas.
- The variable inertia flywheel not only reduces the weight of the machine, but it also provides various benefits such as reduction of vibration and improve system stability.

References

1. Ullman DG (1978) A variable inertia flywheel as an energy storage system, Doctoral dissertation, The Ohio State University
2. Bolund B, Bernhoff H, Leijon M (2007) Flywheel energy and power storage systems. Renew Sustain Energy Rev 11(2):235–258
3. Akram W, Bankar VH (2017) Design and analysis of various shapes of flywheel. Int J Sci Res Sci Technol 3(1):408–417

4. Arani AK, Karami H, Gharehpetian GB, Hejazi MSA (2017) Review of flywheel energy storage systems structures and applications in power systems and microgrids. Renew Sustain Energy Rev 69:9–18
5. Pullen KR (2019) The status and future of flywheel energy storage. Joule 3(6):1394–1399
6. Faraji F, Majazi A, Al-Haddad K (2017) A comprehensive review of flywheel energy storage system technology. Renew Sustain Energy Rev 67:477–490
7. Moosavi-Rad H (1988) The application of a band variable-inertia flywheel to an urban transit bus, Doctoral thesis, Oregon State University
8. Jalili N, Fallahi B, Kusculuoglu ZK (2001) A new approach to semi-active vibration suppression using adjustable inertia absorbers. Int J Model Simul 21(2):148–154
9. Dreidy M, Mokhlis H, Mekhilef S (2017) Inertia response and frequency control techniques for renewable energy sources: a review. Renew Sustain Energy Rev 69:144–155
10. Yamazaki M (2005) Variable mass flywheel mechanism. YGK Co Ltd., U.S. Patent US6915720B2
11. Yang T (1992) The principles and structure of variable inertia flywheels. Patent EP, 508790
12. Xu T, Liang M, Li C, Yang S (2015) Design and analysis of a shock absorber with variable moment of inertia for passive vehicle suspensions. J Sound Vib 355:66–85
13. Li Q, Li X, Mi J, Jiang B, Chen S, Zuo L (2020) A tunable wave energy converter using variable inertia flywheel. IEEE Trans Sustain Energy
14. Zhang Y, Zhang X, Qian T, Hu R (2020) Modeling and simulation of a passive variable inertia flywheel for diesel generator. Energy Rep 6:58–68
15. Ishida Y, Inoue T, Fukami T, Ueda M (2009) Torsional vibration suppression by roller type centrifugal vibration absorbers. J Vib Acoustics 131(5)
16. Su HK, Liu T (2010) Design and analysis of hybrid power systems with variable inertia flywheel. World Electr Veh J 4(3):452–459
17. Yuan LG, Zeng FM, Xing GX (2010) Research on the design and control strategy of variable inertia flywheel in diesel generator unit under pulsed load. In: 2010 international conference on computing, control and industrial engineering, vol 1, pp 187–189. IEEE
18. Elliott CM, Mintah B, Lapen DA (2009) Variable inertia flywheel. Caterpillar Inc, U.S. Patent Application 12/216,123
19. Megahed SM, Abd El-Razik AK (2010) Vibration control of two degrees of freedom system using variable inertia vibration absorbers: modeling and simulation. J Sound Vib 329(23):4841–4865
20. Li C, Liang M, Wang Y, Dong Y (2012) Vibration suppression using two-terminal flywheel. Part I: modeling and characterization. J Vib Control 18(8):1096–1105
21. Jauch C (2015) A flywheel in a wind turbine rotor for inertia control. Wind Energy 18(9):1645–1656
22. Pfabe M, Woernle C (2016) Reducing torsional vibrations by means of a kinematically driven flywheel—Theory and experiment. Mech Mach Theory 102:217–228
23. Li C, Liang M, Wang Y, Dong Y (2012) Vibration suppression using two-terminal flywheel. Part II: application to vehicle passive suspension. J Vib Control 18(9):1353–1365
24. Braid J (2014) Conceptual design of a liquid-based variable inertia flywheel for microgrid applications. In: 2014 IEEE international energy conference (ENERGYCON), pp 1291–1296
25. Li C, Wang S, Kang L, Lei S, Yu Q (2010) Two-terminal manipulation of masses: application to vibration isolation of passive suspensions. J Vibroeng 12(2):143–255
26. Smith MC, Wang F (2004) Performance benefits in passive vehicle suspensions employing inerters. Veh Syst Dyn 42(4):235–257
27. Li C, Liang M, Yu Q (2011) Multi-objective optimization of vehicle passive suspension with a two-terminal mass using chebyshev goal programming. World Acad Sci, Eng Technol 52(126):633–638
28. Li C, Liang M (2011) Characterization and modeling of a novel electro-hydraulic variable two-terminal mass device. Smart Mater Struct 20(2):025004
29. Xu T (2013) Design and analysis of a shock absorber with a variable moment of inertia flywheel for passive vehicle suspension, Doctoral dissertation, University of Ottawa

30. Burstall OW (2005) Variable inertia flywheel. Perkins Engines Co Ltd, U.S. Patent 6,883,399
31. Dong X, Xi J, Chen P, Li W (2018) Magneto-rheological variable inertia flywheel. Smart Mater Struct 27(11):115015
32. Yang S, Xu T, Li C, Liang M, Baddour N (2016) Design, modeling and testing of a two-terminal mass device with a variable inertia flywheel. J Mech Design 138(9)
33. Kushwaha P, Ghoshal SK, Dasgupta K (2020) Dynamic analysis of a hydraulic motor drive with variable inertia flywheel. Proc Inst Mech Eng, Part I: J Syst Control Eng 234(6):734–747
34. Mahato AC, Ghoshal SK, Samantaray AK (2019) Influence of variable inertia flywheel and soft switching on a power hydraulic system. SN Appl Sci 1(6):1–13
35. Mahato AC, Ghoshal SK, Samantaray AK (2017) Influence of locking and passive soft switching on hydraulic circuit efficiency. SIMULATION 93(3):237–249
36. Maxwell TJ (1982) Hydraulic flywheel. U.S. Patent 4,335,627
37. Van de Ven J (2009) Fluidic variable inertia flywheel. In: 7th international energy conversion engineering conference, p 4501
38. Dugas PJ (2011) Variable inertia flywheel. U.S. Patent Application 13/136,064
39. Matsuoka T (2011) Vibration suppression device having variable inertia mass by MR-fluid. In: International design engineering technical conferences and computers and information in engineering conference, vol 54785, pp 1181–1185

Parametric Study of a Revolving Piezoelectric Tapered-Bimorph Beam Subjected to Pulsating Axial Load Considering Various Boundary Conditions

Rakesh Ranjan Chand and Amit Tyagi

Abstract Designing any rotational vibration energy harvester requires a comprehensive study of rotating sandwiched structures with piezoelectric layers and under external loading. In this article, the parametric investigation of a rotating piezoelectric bidirectional-tapering-bimorph beam comprises both width and thickness tapering central host and piezoelectric patches on its surfaces, under axial pulsating load is done. The mathematical modeling of the system is done using Hamilton's equation. The effect of four boundary arrangements and different piezoelectric patch thicknesses, taper parameters, and rotational frequencies on the system's response is studied with the help of parametric instability regions and static load graphs using the MATLAB program. The results demonstrate that an increase in the thickness taper parameter increases resonant frequencies of the structure significantly compared to the marginal rise with the width counterpart. The resonant frequencies decreased with an increase in the piezoelectric patch thickness up to a particular value; after that, the frequencies are increased. The pinned–pinned system provides the lowest first resonant frequency for any set of operating and system parameters; however, practical implementation of this system in the energy harvesting devices will be complicated.

Keywords Rotational vibrating system · Tapered-bimorph · Piezoelectric-composite · Static and dynamic loading · Parametric analysis

1 Introduction

The design of rotational VEHs for the condition assessment of rotary machine parts like turbines, propellers, and Tire Pressure Monitoring System (TPMS) requires widespread analysis of rotating piezoelectric sandwich structures under external load.

R. R. Chand (✉) · A. Tyagi
Mechanical Engineering Department, IIT (BHU), Varanasi 221005, India
e-mail: rakeshrchand.rs.mec17@itbhu.ac.in

A. Tyagi
e-mail: atyagi.mec@iitbhu.ac.in

A. Maurya et al. (eds.), *Recent Trends in Mechanical Engineering*, Lecture Notes in Mechanical Engineering, https://doi.org/10.1007/978-981-19-7709-1_54

If any of the inertia, stiffness, loading, or damping terms of the system's motion equations is periodic, the system vibrates vigorously even under the critical level of loading Nayfeh and Mook [8]. Therefore, the parametric analysis of such systems is of much interest to obtain the critical configuration parameters. Nelson [9] and Gmur and Rodrigues [5] presented finite element methods to get critical speeds of revolving tapered circular shafts, using the Timoshenko beam concept. Ko [6] investigated the flexural performance of a revolving tapered-sandwich structure. A computational approach was reported to examine the free-vibration of spinning Euler beams with several end conditions by Bauer [1] and found that the modal frequencies either increase or decrease linearly with an increase in the revolving speed. The mode shapes of revolving structure with rectangular and circular cross sections were acquired by utilizing the Wittrick–Williams procedure by Lee [7] to obtain the stability of revolving beams of irregular cross section with different end situations. Chand et al. [2, 3] explored the parametric instability of exponentially and parabolic varying rotary beams under axial dynamic load. It may be noticed that the parametric study of the PZT-coupled tapering-bimorph rotating system under dynamic axial loading and various boundaries has not been reported yet. This paper presents the parametric study of an axially loaded rotating bidirectional-tapered-bimorph piezoelectric beam with a linearly converging width and thickness system with C–F, C–P, C–C, and P–P arrangements.

2 Mathematical Modeling

A bidirectional-tapered-bimorph piezoelectric beam with linearly converging width and thickness revolving about Z'-axis with a hub of the radius R_0 is depicted in Fig. 1.

The host beam's thickness is also converging from one to another end, while the piezoelectric patch thickness remains constant throughout the span. The beam is under external force $f(t) = f_s + f_d \cos(\omega t)$ at the centroid $x' = R_0 + l$, where l is the span of the bimorph-tapered structure, ω_d is the excitation frequency, f_d is the maximum amplitude of the transient load, and f_s is the static part, as portrayed in Fig. 1b. As shown in Fig. 1c, t_p is the PZT patches, h and $h(x)$ are the substrate thicknesses at the hub side and at any standard section $x' = R_0 + x$, correspondingly, w and $w(x)$ represent the widths, and $A_p(x)$, $A_h(x)$ are the cross-sectional area of the PZT patch and the substrate beam, correspondingly at $x' = R_0 + x$.

Considering the transverse displacement $\delta(x, t)$ is small, the total potential energy (E_P), kinetic energy (E_T), work done by the axial load (W_e), and the resultant centrifugal force (W_c) expressions for the tapered system are adopted from Chand and Tyagi [4]. Again considering, E_h and E_p, $I_h(x)$ and $I_p(x)$, ρ_h and ρ_p denote Young's modulus, area moment of inertia, and density of the host beam and the PZT patches, whereas N represents the rotary speed of the structure. The terms with h and p subscripts represent the substrate and the PZT patches, respectively. The energy expressions can be written in terms of elemental displacement, $\{q\} = \{q_1 \ldots\ldots q_n\}^T$ as follows;

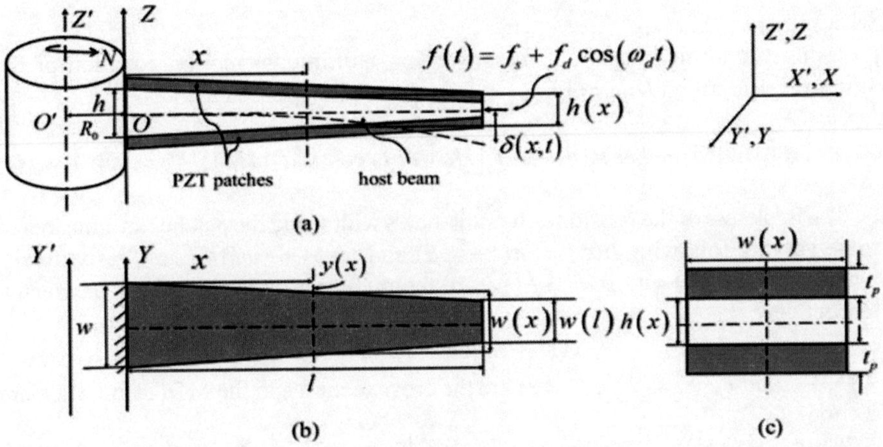

Fig. 1 Representation of the rotating bidirectional-tapered-bimorph model having a linearly varying cross section and fixed on a hub under the influence of axial load **a** front, **b** top, and **c** side view

$$E_P = \frac{1}{2}\{q\}^T[K_u]\{q\} \tag{1}$$

$$E_T = \frac{1}{2}\{\dot{q}\}^T[M]\{\dot{q}\} \tag{2}$$

$$W = \frac{1}{2}\{q\}^T f(t)[H]\{q\} + \{q\}^T[K_c]\{q\} \tag{3}$$

Considering, $W = W_e + W_c, \phi = {}^x/_l, \lambda = {}^\delta/_l$ and $m(\phi)$ as the mass distribution, the matrices above are expressed as:

$$[M_{ij}] = \int_0^l \{m_h(\phi)\lambda_i(\phi)\lambda_j(\phi)\}d\phi + 2\int_0^l \{m_p(\phi)\lambda_i(\phi)\lambda_j(\phi)\}d\phi \tag{4}$$

$$[K_{u_{ij}}] = \int_0^l \left\{E_h I_h(\phi)\lambda_i''(\phi)\lambda_j''(\phi)\right\}d\phi + 2\int_0^l \left\{E_p I_p(\phi)\lambda_i''(\phi)\lambda_j''(\phi)\right\}d\phi \tag{5}$$

$$[K_{c_{ij}}] = \int_0^l \{m_h(\phi)N^2 R_0 \lambda_i(\phi)\lambda_j(\phi)\}d\phi \tag{6}$$

$$[H_{ij}] = \int_o^l \left\{\lambda_i'(\phi)\lambda_j'(\phi)\right\}d\phi \tag{7}$$

where i, $j = 1,2,3,\ldots,N$.

Using Hamilton's equation, the governing multimodal motion equation of the system considering $\left[K_{ij}\right] = \left[K_{u_{ij}}\right] + \left[K_{c_{ij}}\right]$ can be expressed as;

$$[M_{ij}]\{\ddot{q}\} + [K_{ij}]\{q\} - \{f_s[H_{ij}] - f_d\cos(\omega_d t)[H_{ij}]\}\{q\} = \{0\} \tag{8}$$

The thickness of the substrate, the substrate's width, and the patches are considered to be varying following $h(\phi) = h(1 - \phi\beta)$ and $w(\phi) = w(1 - \phi\alpha)$, respectively, where $\alpha = 1/\left(\frac{l}{w}\right)$ and $\beta = 1/\left(\frac{l}{h}\right)$ are the thickness and width taper parameter, respectively.

Similarly, $A_h(\phi) = A_h(1 - \phi\alpha)(1 - \phi\beta), A_p(\phi) = A_p(1 - \phi\alpha), I_h(\phi) = \frac{w(\phi)h(\phi)^3}{12}$, and $I_p(\phi) = I_p(1 - \phi\alpha)$ are the cross section and the MOI at any standard section.

The mode shapes are obtained using Galerkin's discretization by assuming a series solution as;

$$\lambda(\phi, t) = \sum_{r=1}^{\infty} \lambda_r(\eta)\varphi_r(t) \tag{9}$$

where $\lambda_r(\phi)$ is a coordinate function and $\varphi_r(t)$ is a function of time. The coordinate functions for the C–P, C–C, C–P, and P–P systems are taken from Chand and Tyagi [4]. Substituting Eq. (9) and considering the modal matrix $[T]$ equivalent to $[M]^{-1}[K]$, Eq. (8) can be rewritten as;

$$\{\ddot{q}\} + \left[\omega_n^2\right]\{q\} + f_d\cos(\omega_d t)[B]\{q\} = \{0\} \tag{10}$$

where $\left[\omega_n^2\right] = [M]^{-1}[K]$ and $[B] = -[T]^{-1}[M]^{-1}[K][T]$ and now Eq. (10) can be written as

$$\ddot{q}_n + \omega_n^2 q_n + f_d\cos(\omega_d t)\sum_{m=1}^{R} b_{mn}q_n = 0 \tag{11}$$

Equation (11) describes R coupled Hill's equations with complex parts, which are $\omega_n = \omega_{n,R} + j\omega_{n,I}$ and $b_{n,m} = b_{nm,R} + jb_{nm,I}$. Saito-Otami formulation [2] for the undamped case ($\omega_d = 2\omega_n$, $n > 1$) is employed to obtain primary instability areas.

Now substituting $\{\ddot{q}\} = \{0\}$ and $f_d = 0$ in Eq. (8) gives an eigenvalue problem;

$$[K]^{-1}[H]\{q\} = \frac{1}{f_s}\{q\} \tag{12}$$

The real part of the reciprocal of $[K]^{-1}[H]$ gives Euler's buckling loads.

3 Numerical Results and Discussions

A MATLAB code is formulated to obtain the system's parametric instability areas. The properties and dimensions of the system used in the simulation are listed in Table 1.

Figure 2 depicts the shrinkage and movement of the instability zones toward higher frequency values for the increase in speed from $N = 1$ to 50rps, for all the boundary conditions. This is due to the surge in the centrifugal forces, which diminishes the external load's consequence.

The effect of the width taper parameters $\alpha = 0.2 - 0.8$ on the instability regions is inspected for all the boundary arrangements and depicted in Fig. 3. It demonstrates that an increase in width taper parameter increases the structure's flexural stiffness, consequently shifting the instability areas to a higher frequency. A similar effect is observed in the case of an upsurge in the thickness taper parameter $\beta = 0.1 - 0.8$(figure not given), but the system response is higher than the previous case because of the more substantial influence of thickness variation on the stiffness of the system. This trend in the variation of resonance frequencies agrees with results presented by Chand et al. [3] and Pradhan and Dash [10].

Table 1 Geometric dimensions and physical properties of the proposed system

Substrate beam		PZT-850 patch	
l, w, h, R_0(mm)	60, 50,0.3–1, 50	l, w, t_p(mm)	60, 50,0.2
ρ_h, E_h	2.7 g/cm^3, 4.5GPa	ρ_p, E_p	7.6 g/cm^3, 63GPa

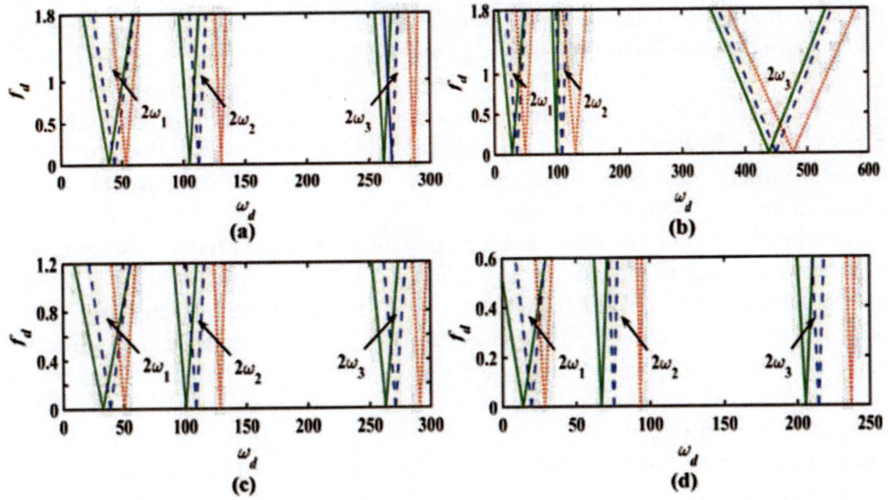

Fig. 2 Parametric instability plot with, $N = 1$ (full line), 25 (dashed), and 50 rps (dotted) for **a** C–C, **b** C–F, **c** C–P, and **d** P–P arrangements

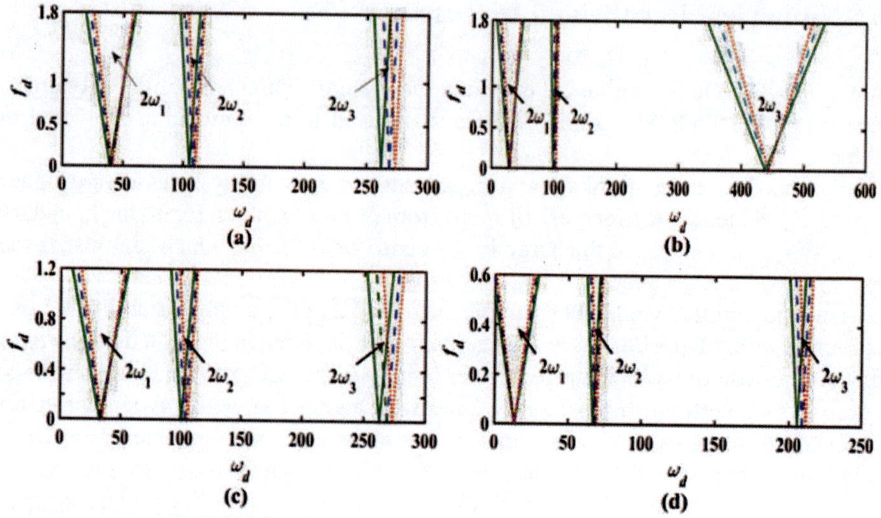

Fig. 3 Parametric instability plot with $\alpha = 0.2$ (full line), 0.5 (dashed), and 0.8 (dotted) for **a** C–C, **b** C–F, **c** C–P, and **d** P–P arrangements

A leftward shift in the system's modal frequencies with an upsurge in the PZT patches' thickness is observed (figure not given). This is due to the increase in the system's mass, which causes the decline in rigidity. It can be noticed that the P–P system provides the lowest first resonant frequency for any set of operating and system parameters.

The instability regions acquired are verified by solving Eq. (10) using the Runge–Kutta method and plotting the amplitude versus time curves. The analysis is carried out for all boundary conditions and all three modes. The graph for the first mode of the C–F system is given in Fig. 4. When $\omega_d = 13.8$ (inside of the instability zone) is chosen, the displacement amplitude went on swelling (unstable), given in Fig. 4a. Figure 4b displays the decreasing amplitude (stable), for which $\omega_d = 14.5$ is chosen from outside the instability zone. These curves validated the instability area formulations proposed in this article.

The system's response under static loading is inspected for various rotating speeds, both the taper parameters and the PZT thicknesses. The static load factor decreases with a surge in the PZT patch thickness, but it increases with an increase in the thickness (figure not shown) after a certain thickness. This is because, up to a certain patch thickness level, the increase in the system's stiffness is less than the increase in the system's mass, but after that, any rise in the thickness of the PZT patches leads to a more rigid structure. As depicted in Fig. 5a, b, with an increase in both the thickness and width taper parameters, the static load upsurges because of the rise in the structure's flexural rigidity. But, the stability of the structure is more responsive toward the variation in thickness taper than that of the width. This is due to the

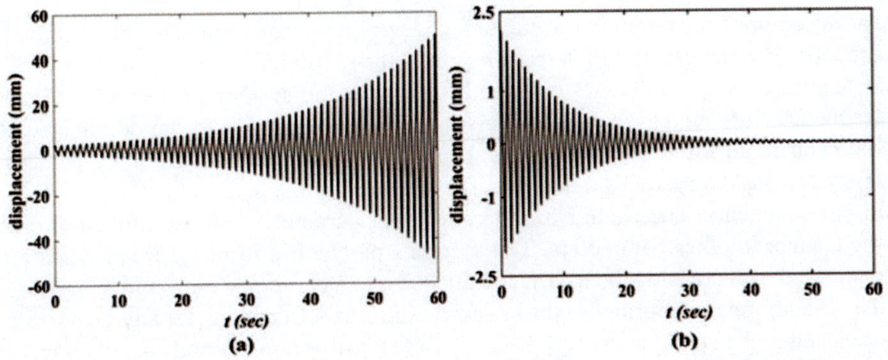

Fig. 4 Amplitude versus time curve for C–F arrangement with **a** $\omega_d = 13.8$ (inside the instability zone) **b** $\omega_d = 14.5$(outside the instability zone)

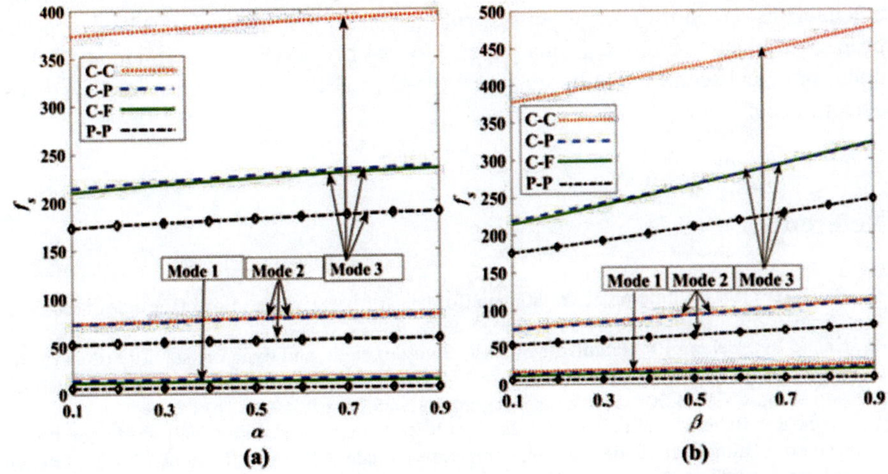

Fig. 5 Buckling load plot for C–F, C–P, C–C, and P–P arrangements with **a** $\beta = 0.1$, $t_p = 0.2$, $N = 1$, $\alpha = 0.1 - 0.9$ **b** $\alpha = 0.2$, $t_p = 0.2$, $N = 1$, $\beta = 0.1 - 0.9$

greater influence of the thickness taper parameter on the moment of inertia of the tapered-bimorph system.

4 Conclusions

Parametric analysis of a novel revolving bidirectional-tapering-bimorph system consists of two PZT patches and a central host, subjected to axial dynamic and static load, and the effect of various configurations and operating parameters with four boundary conditions is presented in this paper. A multimodal motion equation for the composite system is formulated via Hamilton's equation. The effect of both the taper parameters, the PZT thickness, and the revolving speed on the instability is analyzed using MATLAB code and found that the system's principal resonant frequencies are increased with the driving speed increase due to the dominance of the centrifugal force over the external load. The surge in both the taper parameters increases the system's flexural stiffness, consequently stabilizing the system for all arrangements. The change in thickness taper parameter is most influential than the change in other parameters. The system's parametric resonant frequencies are decreased by increasing the PZT patch thickness to a particular value and then rising. The P-P arrangement provides the lowest first resonant frequency for any set of operating and system parameters. The Runge–Kutta fourth-order method is used for result confirmation, which verified the results. It is also concluded that a rise in revolving speed and both the taper parameters raises the static load factor but declines with an upsurge in the thickness of the PZT patches up to a certain level. This research can be applied for low-frequency RVEH for real-time condition assessment of rotating mechanisms like turbines, propellers, robotic manipulators, and TPMS by opting proper parameters offered in this paper. However, extensive experimental studies under practical scenarios to be go together with to evaluate the presented system's performance.

References

1. Bauer HF (1980) Vibration of a rotating uniform beam, part I: orientation in the axis of rotation. J Sound Vib. https://doi.org/10.1016/0022-460X(80)90651-3
2. Chand RR, Behera PK, Pradhan M (2019) Study of static and dynamic stability of an exponentially tapered revolving beam exposed to a variable temperature grade under axial loading. Int J Acoust Vib 24:504–510. https://doi.org/10.20855/ijav.2019.24.31357
3. Chand RR, Behera PK, Pradhan M, Dash PR (2019) Parametric stability analysis of a parabolic-tapered rotating beam under variable temperature grade. J Vib Eng Technol 7:23–31. https://doi.org/10.1007/s42417-018-0071-y
4. Chand RR, Tyagi A (2021) Parametric analysis of a rotational piezoelectric-coupled tapered-bimorph structure with various boundary conditions under transient axial loading. J Vib Eng Technol. https://doi.org/10.1007/s42417-020-00272-9
5. Gmür TC, Rodrigues JD (1991) Shaft finite elements for rotor dynamics analysis. J Vib Acoust Trans ASME 10(1115/1):2930212
6. Ko CL (1989) Flexural behavior of a rotating sandwich tapered beam. AIAA J 10(2514/3):10120
7. Lee HP (1996) Dynamic stability of spinning beams of unsymmetrical cross-section with distinct end conditions. J Sound Vib. https://doi.org/10.1006/jsvi.1996.0013
8. Nayfeh AH, Mook DT (1995) Nonlinear Oscillations
9. Nelson HD (1980) A finite rotating shaft element using Timoshenko beam theory. J Mech Des Trans ASME 10(1115/1):3254824
10. Pradhan M, Dash PR (2016) Stability of an asymmetric tapered sandwich beam resting on a variable Pasternak foundation subjected to a pulsating axial load with thermal gradient. Compos Struct. https://doi.org/10.1016/j.compstruct.2016.01.010

Numerical Analysis of Buoyant Balloon for Airborne Wind Turbines

R. S. Jegan Vishnu and Beena D. Baloni

Abstract In India, nearly 45 TWh of wind power has been harnessed over a decade using low to medium altitude winds. Winds at higher altitudes remain stable and have higher velocity; it can be harnessed by the use of turbines placed in buoyant filled balloons at higher altitudes. Studies suggest the usage of the convergent-divergent shape of the balloon can increase the velocity of air at the minimum area where the turbine can be placed. The buoyant balloon is designed in the shape of an airfoil. To reduce the effect of drag, different NACA 4-digit airfoils are taken for analysis in XFLR5 by varying thickness from 10 to 30% of the chord length, the maximum camber of 1–3% of chord, and maximum camber position in tenths of chord was varied from 1 to 9% with an increment of 1%. The results indicate that NACA 1730 shows the least coefficient of drag about 8.345E–03 at 0° angle of attack. The preliminary calculation has shown the volume of 19.16 m^3 of hydrogen gas is required to make the entire Airborne Wind Turbine (AWT) float in the air. The length of the airfoil to accommodate the volume is found to be 2.82 m. Numerical analysis of the balloon model is done using ANSYS FLUENT.

Keywords Airborne wind turbines · XFLR5 · Buoyant balloon · High-altitude wind rnergy. National Advisory Committee for Aeronautics (NACA)

Abbreviations

AWT	Airborne Wind Turbines
C_D	Coefficient of drag
NREL	National Renewable Energy Laboratory
AOA	Angle of attack

R. S. Jegan Vishnu (✉) · B. D. Baloni
Department of Mechanical Engineering, Sardar Vallabhbhai National Institute of Technology (SVNIT), Surat, India
e-mail: p19tm009@med.svnit.ac.in

A. Maurya et al. (eds.), *Recent Trends in Mechanical Engineering*, Lecture Notes in Mechanical Engineering, https://doi.org/10.1007/978-981-19-7709-1_55

1 Introduction

At present, the maximum amount of energy is being obtained from non-renewable energy resources. The deterioration of these resources at faster rates has a direct impact on the atmosphere. The researches show that we must employ other sources of energy forms for our day-to-day usage. In the past decade, India has harnessed nearly 45 TWh of power from wind. The harnessing of high-altitude wind energy was first proposed by Loyd in 1980 [1], he proposed that wind turbines tied to kites can generate nearly 5 times higher power than those installed in the ground. The windmills used for power generation can harness winds at altitudes of 90 m, and Marvel et. al [2] suggested that the higher altitude winds are more stable and the speed is nearly 10 times powerful compared to lower altitude winds. Cherbuni et al. [3] classified the Airborne Wind Turbines (AWT) as Fly Gen (FG) and Ground Gen (GG) based on the nature of power generation in air and ground, respectively. FG is further classified based on the nature of wind as a crosswind and non-crosswind system, while GG is classified based on their moment in the ground as a fixed and movable ground system. The present work is based on the non-crosswind Fly Gen type of AWT. Roshan et al. [4] studied that proper enclosure of wind turbines increases wind velocity and thereby more power could be generated. Two types of ducts, namely, simple and stepped duct were used. Placing the turbine at the inlet of the stepped duct increased power production by approximately 16%. Grassman et al. [5] studied the effect of shroud design that keeping the turbine at the throat of the shroud can increase the power production by a factor of 5. Hansen [6] reported the extraction of power greater than the Betz limit could be achieved by placing the turbine in a shrouded duct. Criteria for designing shrouds were given by Samson and Katebi [7] using shroud of three different area ratios. The selection of gas to be filled in a shroud for making it buoyant was discussed. Hydrogen and helium were proposed to be used. Zefreh [8] devised the points to be considered for designing the rotor and performed unsteady simulation using the NREL Phase IV rotor. The rotor placed in the shell shown a maximum increase of 17% torque compared to the rotor placed without a shell. Suri et al. [9] gave the performance of airfoil-shaped aerostat. The turbine was modeled as an actuator disk, while the outer shell was an airfoil-shaped shell to provide lift. The lift-to-drag ratio increased with an increase in the camber of the airfoil. The aerodynamic shape of the shell augmented the flow, thereby increased performance. Saleem and Kim [10] studied the performance of airfoil-shaped buoyant shells using three different airfoils NACA 5415, NACA 5425, and NACA 9415. The position of the camber near the chord and thickness of the airfoil both was responsible for an increase in rotor thrust coefficient. The shell with higher camber and thickness exhibited a maximum power coefficient. The shell with higher camber shown a higher duct thrust coefficient. Kim and Ali [11] studied the rotor for airborne wind turbines and made a comparative study with Blade Element Momentum Theory (BEM) and CFD. It was seen that placing the rotor in the diffuser typed shell improved the thrust coefficient by 23% and the power coefficient by 21%. They suggested that redesigning the diffuser shell in the shape of an airfoil can effectively improve the overall power coefficient. Kashyap et al. [12] made the estimation study for calculating the volume of balloon required for making the total system weight remain airborne. Morgado et al. [13] compared the predictions of CFD and XFLR5 and declared that XFLR5, which uses XFOIL's analytical code as its base, is the best tool for analyzing the airfoils. Research works done till now focused mainly on the performance of wind turbine when placed inside a shroud, and there is the scope of research in selecting the balloon shape for minimizing the drag, which results in enhancement of AWT's performance. Therefore, the selection of airfoil for balloon profile by XFLR5 and its CFD analysis with ANSYS FLUENT is discussed in the present paper.

2 Selection of Airfoil with XFLR5

The total weight of the system as calculated by Kashyap et al. [12] is 20.863 kg, and the corresponding volume of hydrogen required is 19.13 m^3 for generating power of 5 KW. Among the various geometrical shapes, airfoil produces lesser C_D and hence it is chosen. Minimum drag and maximum volume are the two important parameters for the selection of airfoil for the balloon. NACA 4-digit airfoils are considered for the analysis, because of their convergent-divergent shape. The first digit describes the maximum camber as a percentage of the chord. The second digit specifies the position of the maximum camber from the airfoil leading edge in tenths of a percent of the chord. The last two digits specify the maximum thickness of the airfoil in terms of the percentage of the chord. 6 different Reynolds number are considered for velocity range of 12–22 m/s with an increment of 2 m/s, and the analysis is carried out for each airfoil with the angle of attack (AOA) varying from-10° to 10° with an increment of 1°. Velocity increment increases Reynolds number, thereby decreasing the skin friction and drag coefficient. The trend of decrement in C_D is observed when Reynold's number is varied. When the thickness of the airfoil increases, the flow tends to be separated from the airfoil surface because of which drag increases. This is checked with the XFLR5 software by selecting different airfoils and to give a clear insight of the result, C_D corresponding to the Reynold number with the velocity of 12 m/s is given in Table 1.

It can be inferred that with an increase in thickness of the airfoil, C_D keeps increasing. The minimum drag criteria restrict to go further and hence NACA0030 is chosen for further analysis though it shows greater drag compared to NACA 0020 and NACA 0010, as it can satisfy the needs to accommodate the required volume of

Table 1 Coefficient of drag for airfoils with varying thickness

NACA	C_D (xE–03) at 0° Angle of attack (AOA) for NACA 00X0 airfoils				
	0010	0020	0030	0040	0050
C_D	4.64	6.74	8.75	12.03	19.19

hydrogen. Further, airfoils are analyzed with variation in the position of maximum camber (2nd digit in case of NACA 4-digit airfoils) from 1 to 9 keeping the maximum camber as a percentage of the chord (1st digit in case of NACA 4-digit airfoils) as constant. The maximum camber is varied from 1 to 3% of the chord, as the divergent portion of the airfoil needs to be larger enough to stabilize the flow past the rotor. The airfoils are divided into a smaller number of panels (smaller grids for calculation of flow over the airfoil). The average coefficient of drag for NACA 1X30, 2X30, and 3X30 airfoils for the velocity of 12 m/s is given in Table 2. From the above table, NACA1730 shows a lesser average C_D (i.e., 8.345) among NACA 1X30 airfoils. Similarly, NACA 2530 and NACA 3530 airfoils show lower average C_D compared to other NACA 2X30 and NACA 3X30 airfoils. Since NACA 1730 shown the least drag compared to all other airfoils, it is chosen for further studies. C_D versus AOA for three fewer drag airfoils with maximum camber variation is given in Fig. 1. The throat area is maintained as 3 m to accommodate the rotor of AWT [12].

Table 2 Average CD for airfoils at 12 m/s and Re = 2,173,410

NACA	C_D for NACA airfoils at 0° AOA (xE-03)								
	1130	1230	1330	1430	1530	1630	1730	1830	1930
C_D	8.425	8.465	8.452	8.448	8.403	8.385	**8.345**	8.352	8.377
NACA	2130	2230	2330	2430	2530	2630	2730	2830	2930
C_D	8.542	8.593	8.57	8.533	**8.395**	8.475	8.655	8.805	8.905
NACA	3130	3230	3330	3430	3530	3630	3730	3830	3930
C_D	8.835	9.00	9.037	8.802	**8.605**	8.828	9.232	9.593	9.903

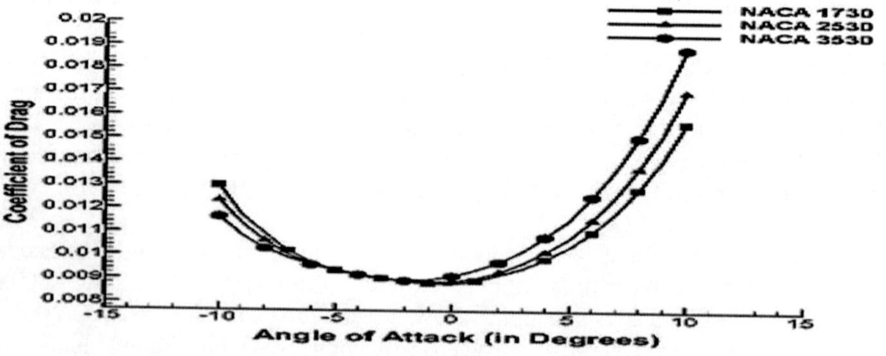

Fig. 1 Coefficient of drag versus angle of attack for airfoils with minimum drag

3 Numerical Simulation

The numerical simulation of the NACA 1730 airfoil is carried out with ANSYS FLUENT to get an idea about flow behavior within and outside the periphery of the AWT balloon. The airfoil coordinates from XFLR5 are imported to Solidworks for modeling. To compensate for the volume of 19.13 m^3 of hydrogen gas, the length of the airfoil was estimated to be 2.82 m. The Solidworks model of the balloon is shown in Fig. 2. The domain size of 25C × 15C × 15C is considered. The model is meshed using ANSYS ICEM CFD. Minimum orthogonality of 0.6 and skewness of 0.4 is maintained. Steady-state analysis is performed over the balloon using a pressure-based solver, as the velocity remains very low. K-epsilon turbulence model with standard wall function is used. Wall y+ of 50 [14] is maintained near the balloon region. The mesh of the model is shown in Fig. 3. To perform the grid independence check, three mesh sizes are used and the respective C_D is checked and given in Table 3. The properties of air at 150 m altitude are calculated using the formulae given below. Table 4 represents the air properties at 150 m altitude calculated using the mentioned standard equations, which are used as boundary conditions for the analysis. Convergence criteria are maintained as 10^{-6}.

$$T = T_o - \lambda h \tag{1}$$

$$P = P_o\left(\frac{T}{T_o}\right)^{\frac{g}{\lambda R}} \tag{2}$$

$$\rho = \rho_o\left(\frac{T}{T_o}\right)^{\frac{g}{\lambda R}-1} \tag{3}$$

Fig. 2 Balloon model

Fig. 3 Mesh of the model

Table 3 Grid independence study

Element numbers	C_D
354,502	0.00868
435,112	0.00839
525,322	0.00834

Table 4 Air properties at 150 m altitude

Property	@150 m altitude
Abs. pressure (Pa)	99,535.96
Temperature (K)	287.025
Density (Kg/m^3)	1.207
Dynamic viscosity (Kg/ms)	1.7839E-05

$$\mu = 1.458 * 10^{-6}\left[\frac{T^{1.5}}{T + 110.4}\right] \quad (4)$$

where $T_o = 288.15K$; $P_o = 101325Pa$; $\rho_o = 1.225\frac{Kg}{m^3}$; $R = 287\frac{J}{Kg.K}$; $\lambda = 0.0065K^{-1}$

4 Results and Discussion

The numerical simulation of NACA 1730 airfoil, as balloon shape of AWT, is carried out with the ANSYS FLUENT, and results for the same are analyzed. C_D of the balloon from CFD is found to be 0.00834. Since XFLR5 uses XFOIL's analytical code [15] as its base, the result obtained by CFD is compared and validated by the results of XFLR5. C_D obtained from XFLR5 is 0.008345 which indicates a 0.05%

variation in the value of the coefficient of drag from CFD one. To get the idea about flow behavior within the fluid domain, the pressure and velocity contour are plotted and represented in Figs. 4 and 5, respectively.

Comparing Figs. 4 and 5, it is seen that increase in velocity results in a decrement in pressure. For an airfoil, low-pressure acts on the top surface, and high-pressure

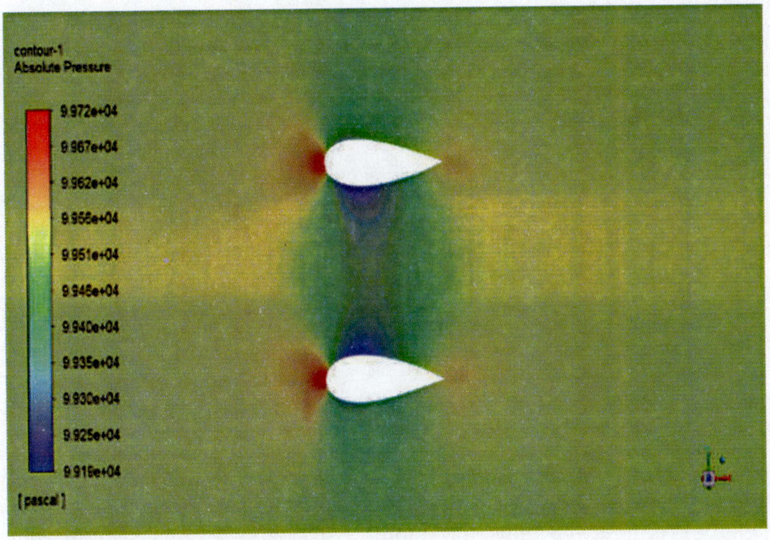

Fig. 4 Pressure contour at symmetry plane

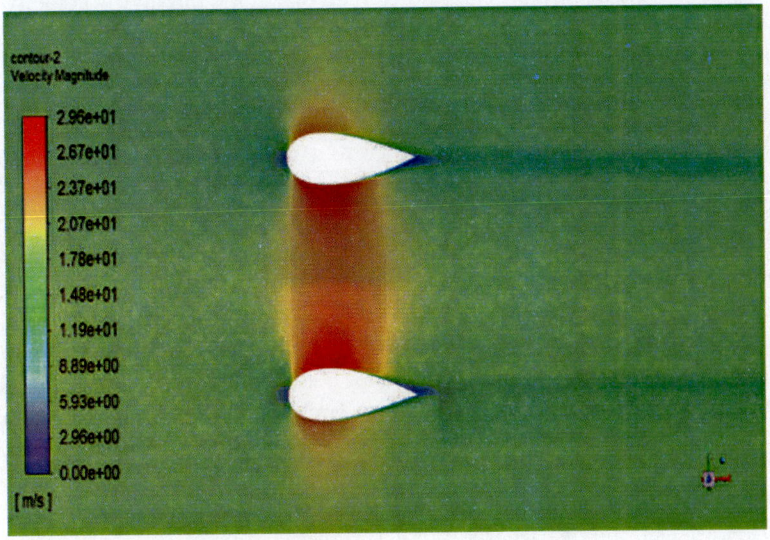

Fig. 5 Velocity contour at symmetry plane

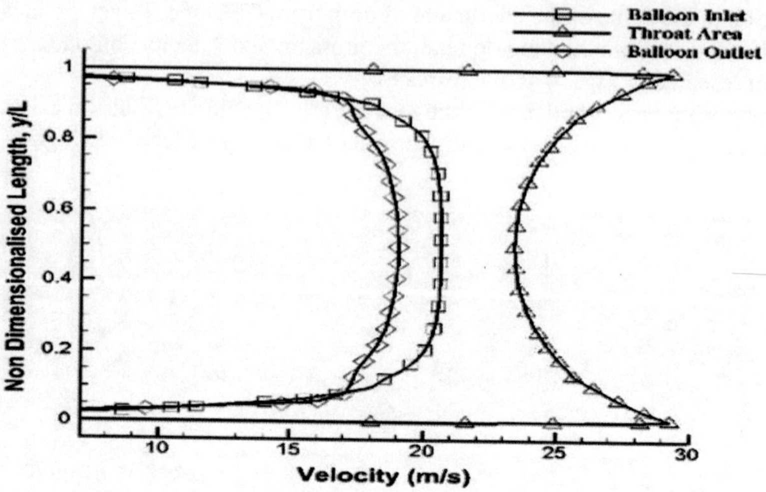

Fig. 6 Velocity variation at inlet, throat, and outlet of balloon

acts on the bottom surface, which results in lift generation. The same effect is seen, the inner surface of the balloon has lower pressure compared to the outer surface as shown in Fig. 4. At the leading edge of the airfoil, the velocity becomes zero, which results in higher-pressure formation at the stagnation point. Figure 6 shows the velocity profile at the inlet, the outlet of the balloon cross section, and at the throat area of the balloon where the rotor is proposed to be placed. Because of the reduction in the area near the airfoil, an increment in velocity of 28% is seen near the throat region of the balloon. It can also be seen that at the throat, the velocity varies near the surface because of the wall and then decreases and matches with the free stream after the formation of the boundary layer. The velocity profile at the inlet and outlet of the balloon remains parabolic.

5 Conclusion

C_D predicted by XFLR5 and FLUENT is equal and found to be 0.00834. The pressure and velocity contour show the predominance of drag force compared to the lift force. This work suggests the performance of static balloon-shaped shell produces drag similar to the airfoil even though balloon area creates more drag, and the C_D lies in the acceptable range for airfoils. Researches so far shown the increment in power coefficient when the rotor is placed inside the diffuser. The buoyancy effect has not been considered throughout the simulation. Authors suggest incorporation of unsteadiness along with the control surfaces on the balloon to depict the exact scenario of the shroud of Airborne Wind Turbines.

References

1. Loyd ML (1980) Crosswind kite power for large-scale wind power production. Energy 4(3)
2. Marvel K, Kravitz B, Caldeira K (2012) Geophysical limits to global wind power. Nat Climate Change 3:118–121
3. Cherubini A, Papini A, Vertechy R, Fontana M (2015) Airborne wind energy systems: a review of the technologies. Renew Sustain Energy Rev (51):1461–1476
4. Roshan SZA, Rad S, Manouchehr RANS (2015) Simulations of the stepped duct effect on the performance of ducted wind turbine. Wind Eng Indus Aerodyn (145):270–299
5. Grassmann H, Bet F, Ceschia M, Ganis ML (2004) On the physics of partially static turbines. Renew Energy 4:491–499
6. Hansen M (2008) Aerodynamics of wind turbines, 2nd edn. New York
7. Samson J, Katebi R (2014) Shroud design criteria for a lighter than air wind energy system. Physics (524)
8. Zefreh MA (2016) Design and CFD analysis of airborne wind turbines for boats and ships. Aerosp Sci
9. Suri D (2019) Lighter-than-air wind energy systems: stability analysis. Mech Mater Renew Energy
10. Saleem A, Kim MH (2018) Aerodynamic analysis of an airborne wind turbine with three different aerofoil-based buoyant shells using steady RANS simulations. Energy Convers Manag (177):233–248
11. Kim M-H, Ali QS (2021) Design and Performance analysis of an airborne wind turbine for high-altitude energy harvesting. Energy
12. Kashyap N, Rao S, Charan KS, Reddy A, Shetty R, Verma N, Baloni B (2020) Optimized balloon model for high altitude airborne wind turbine. In: 1st Asia Pacific conference on sustainable development of energy, water, and environment system. Australia
13. Morgado J, Vizinho R, Silvestre MAR, Páscoa JC (2016) XFOIL versus CFD performance predictions for high lift low Reynolds number airfoils. Aerosp Sci Technol 52:207–214
14. ANSYS Learning Forum. https://forum.ansys.com/discussion/1522/near-wall-treatment. Last accessed 05/2018
15. XFoil. https://web.mit.edu/drela/Public/web/xfoil/xfoil_doc.txt

Numerical Investigation of Heat Transfer Characteristics of Pin–Fin Array for Natural Convection

Siddhartha Kumar Singh and Vandana Agrawal

Abstract Natural convection through fins is utilized for thermal management of various engineering devices by dissipating the excess heat. The prominent influencing parameters for fins are its size and spacing. This paper is aimed at numerically investigating the heat transfer augmentation using pin fins in a rectangular cavity filled with air. Simulations are done to study the effect of prominent influencing parameters, namely fin spacing, fin height and Rayleigh number on heat transfer. It is expected that the present work will help in the design of fins.

1 Introduction

Natural convection in enclosed spaces is widely encountered in numerous engineering applications like in dissipation of heat in electronic devices, nuclear reactors, photovoltaic collectors, etc. It has distinct advantages in terms of less to no maintenance, durability and no power consumption. However, for natural convection dominant regime, the heat dissipating sinks consume up to 40% of total system volume which may be unacceptable in the contemporary compactness scenario. This has attracted the attention of engineers toward the use of fins.

Various experimental tests and numerical simulations have been conducted by researchers to analyze transfer of heat by natural convection for different types of fins. Pin fins of round shape was found to be having high potential of transfer of heat and lower sensitivity with the flow pattern of air [1, 2]. Sunderland and Zografos [3, 4] have done the analysis for the orientation of fins and found that orientation is not a predominant factor influencing the performance of system. However, their findings were conflicting with the results given by Vemuri and Sparrow [5, 6]. Taji et al. [7] experimentally analyzed the effect of longitudinal spacing of fins on performance of fin arrays. Rasouli and Narayanan [8] carried out analysis for inclined pin fins of square shape in staggered arrangement to get the relation of variations in geometrical

S. K. Singh (✉) · V. Agrawal
Mechanical Engineering Department, Motilal Nehru National Institute of Technology Allahabad, Prayagraj, India
e-mail: singhsiddharthakumar@gmail.com

A. Maurya et al. (eds.), *Recent Trends in Mechanical Engineering*, Lecture Notes in Mechanical Engineering, https://doi.org/10.1007/978-981-19-7709-1_56

parameters with thermofluidic properties. Micheli et al. [9] performed a comparison between plate-type and pin-type micro-fins in natural convection environment and showed that the performance of pin fins is better in comparison to plate fins. Nada et al. [10] studied natural convection using rectangular fins in annular spaces. Mehdi et al. [11] experimentally investigated tube exchangers with square fins. Liu and Zhang [12] and Corcione et al. [13] have done numerical study for plates in natural convection. Wang et al. [14] carried out a numerical study for a heated cylinder placed vertically at various locations inside the cavity. A study of natural convection in enclosed domains with differentially heated walls was carried out by Fabregat et al. [15] to investigate laminar and turbulent hat transfer.

As discussed above, lot of work has been done for the study of heat transfer for various shapes of fins [1, 2, 9–12], their orientation [3–6, 8], boundary conditions [10, 12–14], flow conditions [15], etc. Also, studies [1, 2, 9] show that the performance of pin fins with round shape is better than other shapes of fins. However, very few literatures are available regarding thermal characteristics of pin fins in an enclosed cavity of rectangular shape of constant aspect ratio with prevailing conditions pertaining to those of natural convection despite having large number of applications like in biomedical devices, turbine airfoil cooling, PCB cooling, etc. [16, 17]. Although some literatures have done the study for a wide range of Rayleigh number (*Ra*) [17], heat transfer study in the literature for the laminar zone in the *Ra* range of 700,000–940,000 with close proximate values was not found. The present study has been carried out for the investigation of heat transfer in the laminar zone to fulfill the above-mentioned gap. As in the present work, variation of Nusselt number (*Nu*) with fin spacing (X), fin height (H) and *Ra* has been computed, so it is expected that the present study will help in the pin–fin design for the devices [16, 17] working in laminar zone with better heat dissipation leading to the reduction in volume and cost. In the present work, numerical simulation has been done for the above-mentioned computation. Also, initially, the empirical correlations given by Holland's et al. [18] are used for the validation of the numerical model.

2 Numerical Model Formation

The computational domain for the present work is shown in Fig. 1. The pin–fin diameter (D) considered in the work is 10 mm [19]. H has been considered in the range of 15 mm–40 mm to investigate heat transfer rate for fin height-to-channel height ratios (H/H_c) in the range of 0.25–0.67 [20]. Further, X is considered in the range of 20 mm–80 mm as the optimum values of heat transfer rate were achieved [3, 4] when the ratio of D/X is around 1/3 for a base plate size of 203 mm × 203 mm. In order for the end wall effects to be minimum, a sufficiently large rectangular cavity (Fig. 1) with air is considered. The surrounding walls of the enclosure are assumed adiabatic, and heat exchange takes place between the top and bottom surfaces only. The material of pin fins is taken to be pure aluminum being the lightweight and having high thermal conductivity. Numerical simulation is carried out only for the

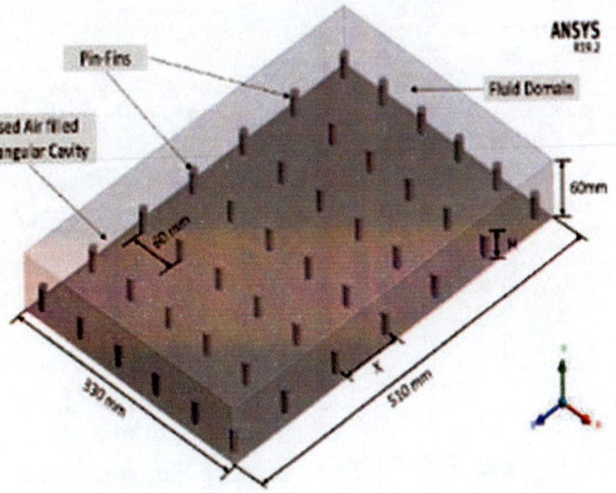

Fig. 1 Schematic diagram of rectangular enclosure

quarter portion of the domain using ANSYS Fluent 19.2 commercial software by considering symmetrical boundary condition (Fig. 2). Numerical model for simulation has been formed by solving the governing equations for conservation of mass (Eq. (1)), momentum (Eq. (2)) and energy (Eqs. (4) and (5)) by considering the flow to be laminar as the Rayleigh number (Eq. (6)) used in the present analysis is less than 10^9. Equations (1)–(6) are given below.

$$\frac{\partial \rho}{\partial t} + \frac{\partial}{\partial x_i}(\rho u_i) = 0 \tag{1}$$

$$\frac{\mathrm{d}}{\mathrm{d}t}(\rho u_i) = -\frac{\partial p}{\partial x_i} + \frac{\partial}{\partial x_j}\left[\mu\left(\frac{\partial u_i}{\partial x_j} + \frac{\partial u_j}{\partial x_i}\right)\right] + (\rho - \rho_\infty)g_i \tag{2}$$

Fig. 2 Boundary conditions

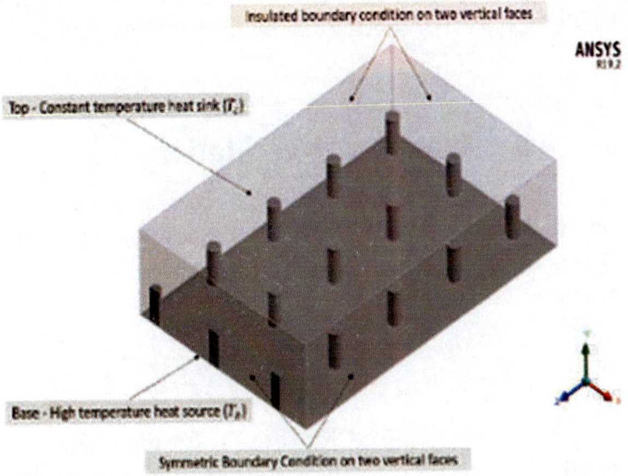

where

$$p = p_s + \rho_\infty g z \tag{3}$$

$$\frac{\mathrm{d}}{\mathrm{d}t}(\rho c_p T) = \frac{\partial}{\partial x_i}\left(k\frac{\partial T}{\partial x_i}\right) \tag{4}$$

$$\frac{\partial}{\partial x_i}\left(\frac{\partial T}{\partial x_i}\right) = 0 \tag{5}$$

$$Ra = gL^3\beta Pr(T_b - T_c)/\gamma^2 \tag{6}$$

In the above equations, ρ and ρ_∞ are the density of fluid near to the heated surface and away from the surface, respectively, μ is the dynamic viscosity, g is acceleration due to gravity, t is time and p_s is the stagnation pressure. Further, u_i and u_j are the velocity of fluid in x_i and x_j directions, respectively, c_p is specific heat, k is thermal conductivity of fluid and T is temperature. In Eq. (6), Ra is given for an enclosure of height L, hot-base surface temperature T_b and cold-top surface temperature T_c. Here, Pr is the Prandtl number of the fluid, γ is the kinematic viscosity of the fluid, g is the acceleration due to gravity and β is the coefficient of volume expansion.

Numerical model formed in this way is used for the simulation for the computation of Nu. Nu (Eq. (7)) provides convection heat transfer coefficient (h) for the fluid with thermal conductivity K_a which is used to compute the heat transfer by convection (Q_{cv}) (Eq. (8)) taking place over the area A_{cv}.

$$Nu = hL/K_a \tag{7}$$

$$Q_{cv} = h(T_b - T_t)A_{cv} \tag{8}$$

The obtained numerical model is validated (Sect. 3) before the computation work.

3 Numerical Model Validation

Empirical correlation (Eq. (9)) obtained from experimental data by Holland's et al. [19] is used for the validation of numerical model.

$$Nu = 1 + 1.44\left[1 - \frac{1708}{Ra}\right] + \left[\frac{Ra^{1/3}}{18} - 1\right] (Ra < 10^8) \tag{9}$$

Nu is plotted against Ra (Fig. 3) for the results obtained from the simulation (s) and the empirical correlation (e). Percentage deviation for s is computed by Eq. (10).

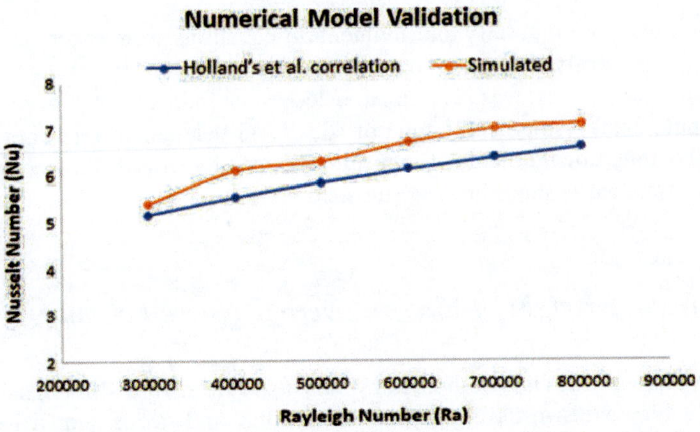

Fig. 3 Validation of numerical model

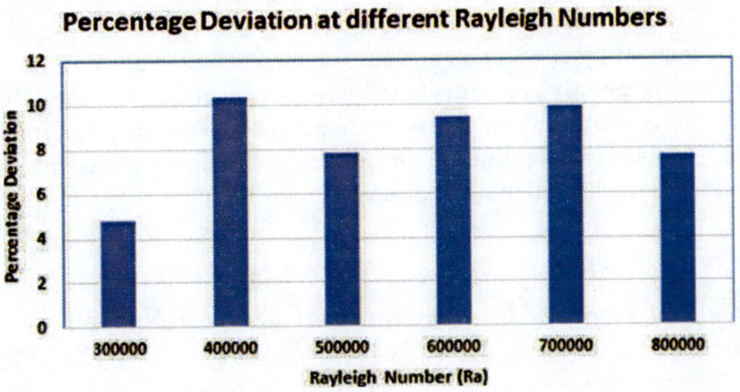

Fig. 4 Percentage deviation

$$\text{Percentage deviation} = \frac{s - e}{e} * 100 \qquad (10)$$

It is evident from Figs. 3 and 4 that both the results are in agreement with each other and the maximum deviation is about 10%. The validated numerical model so obtained is used for finding the effect of H, X and *Ra* on *Nu* discussed in Sect. 4.

4 Results and Discussions

Effect of H, X and *Ra* on *Nu* is found out using convective terms in Navier–Stokes equation, which were discretized using SIMPLE algorithm [21, 22] using upwind scheme of second order for achieving high degree of accuracy. The convergence criteria for energy, continuity and momentum equations were taken as 10^{-7}, 10^{-5} and 10^{-5}, respectively. Transient analysis is carried out till a steady state is reached. Mesh independency test was performed to access the independence of mesh size on the computational results, and a total of 43, 61, 489 elements were considered for the final computational domain. The results obtained by simulation work on each of them are discussed in the following sub-sections.

4.1 Effect of Height of Fins on Average Nusselt Number

In the present work , Nu is computed by varying H from 15 to 40 mm as described in Sect. 2 for sixteen different combinations of X and Ra as shown in Fig. 5a–d. It can be seen that each curve is moving in the upward direction continuously which shows that for the conditions mentioned in Sect. 2 for the present work, Nu always increases by increasing H. The behavior of the curves can be explained by "Newton's law of cooling" which gives a direct proportional relationship between rate of heat transfer by convection and surface area of heat transfer. The increase in height of the fins directly conforms to increase in area effectively available for heat transfer, resulting in improvement in performance of the system. Also, for the present conditions, increase in height of the fins results in formation of discrete convection cells, resulting in increased transfer of heat.

4.2 Effect of Longitudinal Spacing of Fins on Nusselt Number

The study of effect of X on Nu has been done by considering twelve different combinations of H and Ra as given in Fig. 6a–c. It can be seen that all the plotted curves initially move in the upward direction reach to a peak value and then move in the downward direction. This behavior of the curves can be explained with the combination of two factors, as the value of X reduces, the number of fins per unit area increases, resulting in increase in total area available for heat transfer which leads to increase in heat transfer rate; however, at the same time, hindrance for the convection cells to move between adjacent fins also increases which leads to reduced rate of heat transfer. Increasing trend of Nu continues up to the point where the former factor dominates over the later. The peak value obtained for these curves is at around X = 40 mm or X/D $\approx$ 4. Once the hindrance factor dominates, the Nu decreases with X. So, for the conditions given in Sect. 2, the optimum value of X/D can be considered around 4. However, the position of curves shifts according to the value of H and Ra.

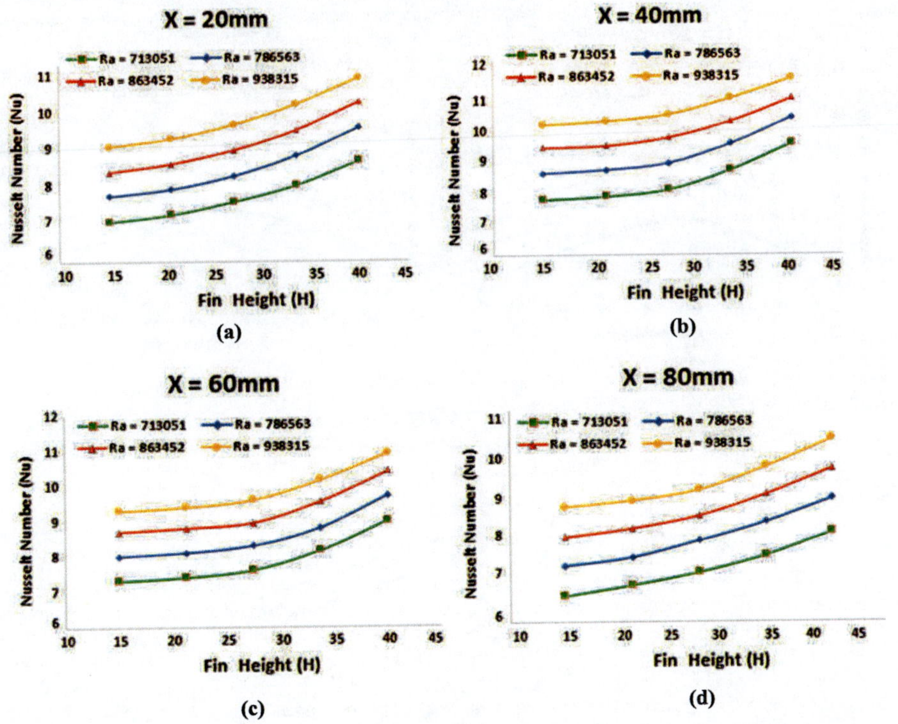

Fig. 5 Variation of Nusselt number with fin height for constant fin spacing: **a** X = 20 mm; **b** X = 40 mm; **c** X = 60 mm; **d** X = 80 mm

4.3 *Influence of Rayleigh Number on Nusselt Number*

The effect of *Ra* on *Nu* was studied by using the various combinations of H and X. Figure 7 shows the result of it by plotting the curve for each of the combination with different colors. The common behavior in all these curves is observed that all of them are moving in the upward direction continuously which shows that increase in *Ra* always leads to increase in *Nu*. This behavior of the curves is due to the increase in force of buoyancy, resulting in improvement in the intensity of flow as *Ra* increases. Curve for bare plate can be seen at the lowest position as heat transfer with fins is always better than with no fin condition. Further, the position of curves from lowest to highest for various (H, X) values can be seen as (15, 80), (15, 20), (27.5, 80), (15, 60), (27.5, 20), (27.5, 60), (15, 40), (27.5, 40), (40, 80), (40, 20), (40, 60) and (40, 40). It shows that under the given conditions, fins with H = 40 mm and X = 40 mm gives the maximum heat transfer rate. Also, the effect of fin spacing is seen to be dominating over the effect of fin height as the position of curve with H = 15 mm and X = 40 mm is higher than the curves with H = 27.5 mm having X as 20, 40 and 80 mm. Also, curve with H = 27.5 mm and X = 40 mm is very close to the curve with H = 40 mm and X = 80 mm. It shows that only increasing the fin height may

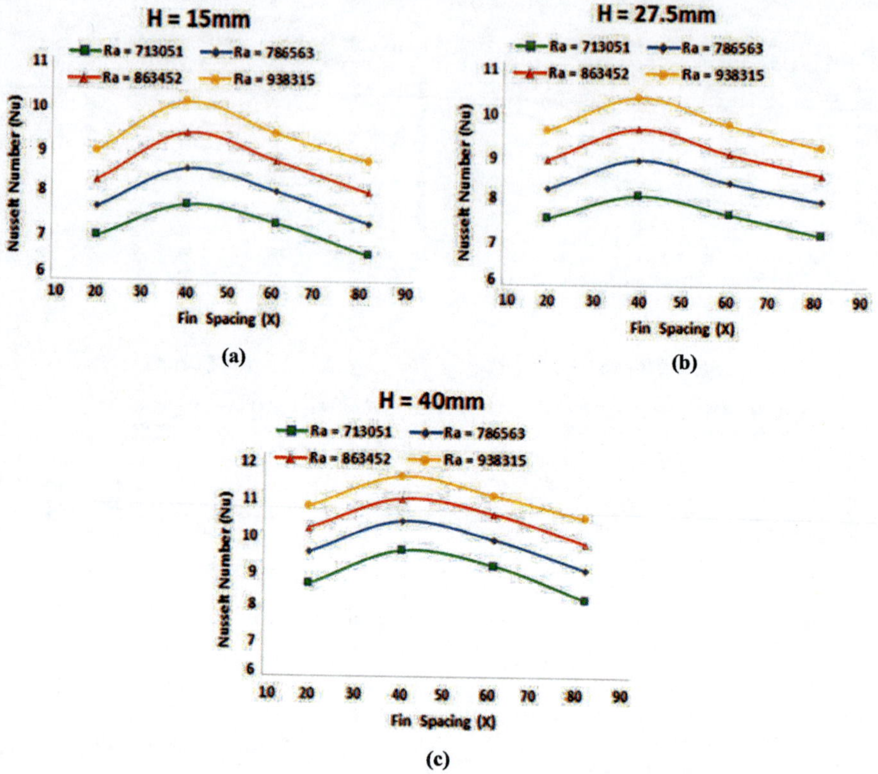

Fig. 6 Variation of Nusselt number with spacing of fins for constant height of fins: **a** H = 15 mm; **b** H = 27.5 mm; **c** H = 40 mm

not affect the heat transfer rate much, in fact sometimes the fins with lesser height with optimum fin spacing may perform better. Use of fins with lesser height with optimum spacing will consume less volume, and also, the requirement of material for manufacturing it will be reduced.

5 Conclusion

Natural convection with pin–fin array in a rectangular cavity filled with air is studied numerically to find the effect of fin height, longitudinal fin spacing and Rayleigh number on Nusselt number. Based on the results obtained, conclusions can be given as: (a) increase in fin height was continuously increasing heat transfer. (b) As the fin spacing was increased, Nusselt number initially increased, reached to a maximum value, and then was continuously reducing. Hence, a proper fin spacing for a given condition can be selected to obtain maximum heat transfer. (c) As Rayleigh number

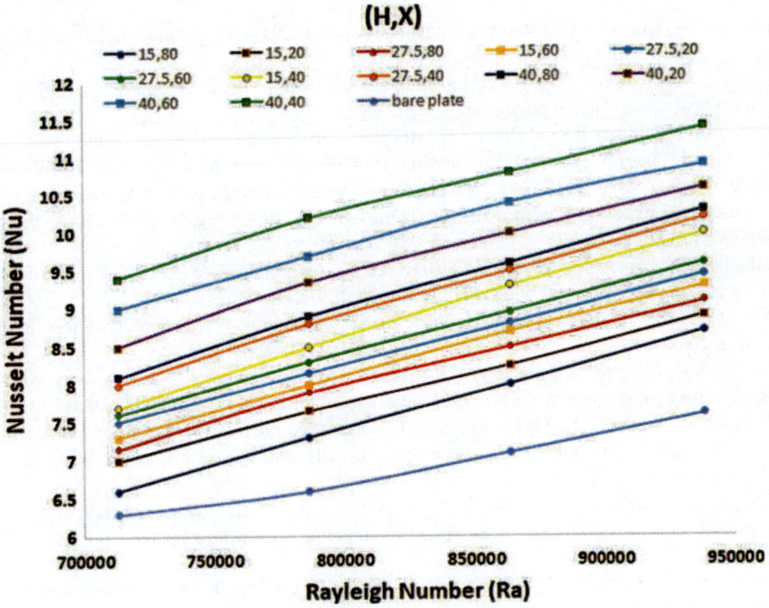

Fig. 7 Variation of Nusselt number with Rayleigh number

was increased, Nusselt number was also increased; however, the effect of various combinations of fin height and fin spacing was also observed at the same time. Fins with lesser height with proper spacing was found to be giving higher Nusselt number for the same Rayleigh number. It shows that for the required rate of heat transfer, fins with smaller height with proper spacing may be used. It will lead to reduction in volume consumed by fins, reduction in material required by them and ultimately reducing the cost and space.

References

1. Peng Y (1984) Heat transfer and friction loss characteristics of pin fin cooling configurations. J Eng Gas Turbines Power 106:246–251
2. Brigham BA, VanFossen GJ (1984) Length to diameter ratio and row number effects in short pin fin heat transfer. J Eng Gas Turbines Power 106:241–245
3. Zografos AI, Sunderland JE (1990) Natural convection from pin fin arrays. Exp Thermal Fluid Sci 3(4):440–449
4. Zografos AI, Sunderland JE (1990) Numerical simulation of natural convection from pin-fin arrays. ASME HTD 157:55–66
5. Sparrow EM, Vemuri SB (1985) Natural convection/radiation heat transfer from highly populated pin fin arrays. J Heat Transfer 107(1):190–197
6. Sparrow EM, Vemuri SB (1986) Orientation effects on natural convection/radiation heat transfer from pin-fin arrays. Int J Heat Mass Transf 29(3):359–368

7. Chavan SG, Bhosale SY, Deshpande HN (2018) CFD analysis of convective heat transfer from inverted trapezoidal plate fin array. Int Sci Technol J 7(3):189–198
8. Erfan R, Narayanan V (2016) Single-phase cryogenic flow and heat transfer through microscale pin fin heat sinks. Heat Transfer Eng 37(11):994–1011
9. Leonardo M, Reddy KS, Mallick TK (2016) Experimental comparison of micro-scaled plate-fins and pin-fins under natural convection. Int Commun Heat Mass Transfer 75:59–66
10. Nada SA, Said MA (2019) Effects of fins geometries, arrangements, dimensions and numbers on natural convection heat transfer characteristics in finned-horizontal annulus. Int J Therm Sci 137:121–137
11. Mehdi K, Yaghoubi M, Keyhani A (2018) Experimental study of natural convection from an array of square fins. Exp Thermal Fluid Sci 93:409–418
12. Zhang X, Liu D (2010) Optimum geometric arrangement of vertical rectangular fin arrays in natural convection. Energy Convers Manage 51(12):2449–2456
13. Massimo C, Cretara L, Quintino A, Spena VA (2021) Dimensionless correlations for natural convection heat transfer from an enclosed horizontal heated plate. Heat Transfer Eng 1–18
14. Zhiheng W, Wang T, Xi G, Huang Z (2021) Periodic unsteady natural convection in square enclosure induced by inner circular cylinder with different vertical locations. Int Commun Heat Mass Transfer 124:1–8
15. Alexandre E, Pallarès J (2020) Heat transfer and boundary layer analyses of laminar and turbulent natural convection in a cubical cavity with differently heated opposed walls. Int J Heat Mass Transf 151:1–12
16. Chiang H, Kleinstreuer C (1991) Analysis of passive cooling in a vertical finite channel using a falling liquid film and buoyancy-induced gas-vapor flow. Int J Heat Mass Transf 34(9):2339–2349
17. Das D, Roy M, Basak T (2017) Studies on natural convection within enclosures of various (non-square) shapes–a review. Int J Heat Mass Transf 106:356–406
18. Hollands KGT, Raithby GD, Konicek L (1975) Correlation equations for free convection heat transfer in horizontal layers of air and water. Int J Heat Mass Transf 18(7–8):879–884
19. Fisher TS, Torrance KE (1998) Free Convection Limits for Pin-Fin Cooling. J Heat Transfer 120(3):633–640
20. Dogan M, Sivrioglu M (2009) Experimental investigation of mixed convection heat transfer from longitudinal fins in a horizontal rectangular channel: In natural convection dominated flow regimes. Energy Convers Manage 50(10):2513–2521
21. Patankar SV (2018) Numerical heat transfer and fluid flow. Taylor and Francis. ISBN 9781315275130
22. Ferziger JH, Perić M (2002) Computational methods for fluid dynamics. Springer, Berlin. ISBN 3-540-42074-6

Study of Traction Forces at Elevated Temperatures During Micro-Scratch Tests on 45S5 Bioglass

Jitendra Kumar Singh, U. S. Rao, and Ram Pyare

Abstract Modern machining processes are now focused on machinability aspects of brittle and hard to machine materials. As per concerns raised in the field of machinability of brittle materials, the prime obstacle in the process is their hardness. Such carping issues can be suppressed through a bit increase in softness of the brittle materials. Therefore, heating of such material can raise the bar for the softness in such materials so as to their machinability. But the localized heating leads to a higher temperature gradient between heated and non-heated regions of the work material so as the thermal stresses in it which results into micro- as well as macro-cracks. Such deplorable issues can be solved through bulk heating of the work materials. Apparently, the comparisons of their softness can be examined through the scratch tests performed at different temperatures. In the view of above elucidation, 45S5 bioglass samples have been used for micro scratch tests as well as a portable heating setup is used to heat those samples. The temperature values are kept between room temperature (27 °C) and 420 °C during the performed tests. Subsequently, traction force, coefficient of friction during the scratch tests and scratch images are compared as the elucidated outcome softness of the material. It is found that there is a significant reduction in traction forces and coefficient of frictions during tests with rise in the sample temperature.

Keywords Bioglass · Brittleness · Machinability · Ductility · Traction forces · Coefficient of friction · Micro-scratch · Elevated temperature scratch test

J. K. Singh · U. S. Rao (✉)
Department of Mechanical Engineering, IIT BHU, Varanasi, UP 221005, India
e-mail: sruppu.mec@iitbhu.ac.in

R. Pyare
Department of Ceramic Engineering, IIT BHU, Varanasi, UP 221005, India

A. Maurya et al. (eds.), *Recent Trends in Mechanical Engineering*, Lecture Notes in Mechanical Engineering, https://doi.org/10.1007/978-981-19-7709-1_57

1 Introduction

Ceramic materials are the new age of the materials and manufacturing processes because they have better material properties as compare to metals. For example, many of the ceramics are light in weight with very good mechanical properties like hardness, toughness, etc., and some of the ceramics possess great thermal insulation. Apparently, some ceramics show a great biocompatibility as implant materials. Therefore, the modern age is now focused to manufacture and process ceramic materials in the view of extensive uses of them like electronics parts, semiconductors, insulators, high-quality cutting tools, aesthetic items, wear resistant applications, utensils, biological implants, space craft parts, air craft parts, refractories, etc. Apart from these alluring uses of ceramics, they involve complex manufacturing processes due to their high hardness and brittleness. Subsequently, they possess poor machinability due to their high hardness and brittleness because of the fact that brittleness leads to poor surface finish, while hardness leads to low machinability. Such deplorable issues can be suppressed by softening of the materials. Therefore, heating of such material can raise the softness in such materials [1] so as to their machinability.

Laser sources are widely used nowadays to assist the machining processes which is a hybrid method that uses a laser for heating the workpiece prior to material removal with a conventional cutting tool. The yield strength of a brittle material decreases to below the fracture strength reducing the brittleness of the material at the elevated temperature. Also at elevated temperatures, same phenomenon happens for the tough and ductile materials, thus reducing surface roughness and cutting forces as well as tool wear [2]. It is also stablished that with the softening of glassy phase material, the material removal can be achieved through a combination brittle fracture and plastic deformation [3]. The application orientations of laser sources have also been the other area of research interest. Thus, a double-ramp laser source for LAM on workpiece is developed to prelude thermal fracture of the workpiece due to low thermal diffusivity, fracture toughness and tensile strength than silicon nitride [4]. Subsequently, laser-assisted machining (LAM) has been identified an important area of research to improvise the benefits of LAM and it becomes necessary to understand the interaction effects of process parameters during LAM on different workpiece material and to develop the optimum levels of process parameters to achieve lower cutting force, surface finish and low machining cost [5]. In continuation of that, the optimal conditions of machining process parameters are established within the test matrix [3, 6–8]. Laser is an extremely localized thermal process because of the fact that during laser beam and work material interaction in a small portion of the work material, the incident energy beam is absorbed and high temperature is developed in the region of the beam spot, resulting in softening, local yielding, melting, burning or evaporation [9–11]. It is important to note here the fact that the applications of laser sources are limited to localized heating only on the workpieces. It may lead to non-uniform heat affected zones (HAZ) so as to larger heat gradient in workpieces between the heated and non-heated zones which leads to higher thermal stresses as shown in Fig. 1. Most of the recent researches on LAM have been largely focused

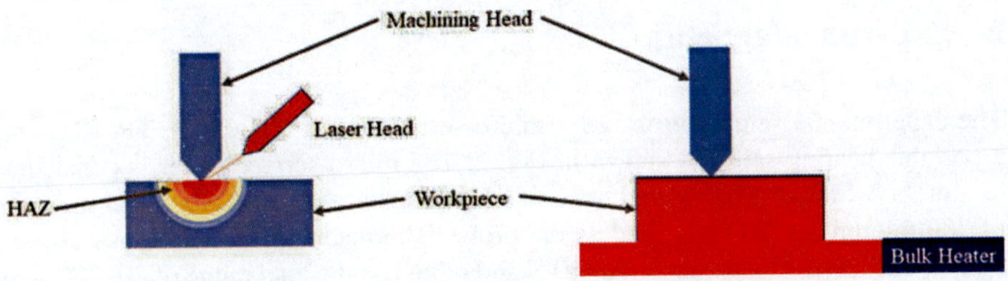

Fig. 1 Localized heating versus bulk heating

on laser-assisted turning. However, other machining processes like milling, drilling and grinding play a vital role in production systems.

In parallel, researchers are concerned to find biocompatible materials suitable for human anatomy applications. Therefore, it gained significant attraction in the recent research trends. For instance, bioglass is found to be a better substitute for human body implants as preferred over stainless steels or zirconia coated steels [12]. In fact, bioactive glass as implants for human body continues to be at the leading material to the research in the medical field [13]. Due to the brittleness and poor load bearing strength of the bioglass and bioactive ceramics [14], they have poor machinability which restrict the use of it as bulk implants. This draws the attention of research for the machinability aspects of such materials.

In the view of above elucidation, the scratch tests can be performed to compare the traction forces and coefficient of friction during scratch tests. A study has shown that strain, strain rate and temperature are some of the important factors in explaining the material deformation mechanism during scratching [15]. In support of that, the scratch tests were performed and evaluated that the scratch resistance of nickel-nano-SiC composite was found to be higher than the nickel-micro-SiC composite as indicated by the higher scratch load showing better adhesion of the deposits [16].

Subsequently, an elevated temperature scratch experiment was carried out on materials for high temperature applications, namely an austenite, a ferritic metal matrix composite and a Ni-based metal matrix composite. The influence of load and temperature on the wear behavior was studied and it is found that due to their inhomogeneous microstructure, multiphase materials have instable scratch behavior [17]. Apparently, the scratch tests are found to be a great measure to elucidate the machinability aspects of a material. Therefore, it has been often used to run a pilot study about machinability aspect of different hard to machine materials [18–23].

In this research work, it is intended to perform an elevated temperature scratch test on 45S5 bioglass samples. In addition of that, the authors have preferred a bulk heating method over localized heating because it provides more uniform HAZ and lower heat gradient in workpieces which leads to lower thermal stresses, which makes it suitable to all machining processes as shown in Fig. 1. In the purpose of that, a multipurpose portable heating setup have been used which is capable of holding as well as heating the samples of size $23 \times 21 \times 10$ mm^3. This setup can raise the sample temperature from room temperature up to 420 °C [24].

2 Experimental Setup

The experimental setup consists of a micro-scratch tester (DUCOM, TR-101) and a portable heating setup as shown in Fig. 2. The micro-scratch tester is consisting of a probe which is directly mounted on a load sensor. A C-type Rockwell indenter of 200-micron tip radius is used as the probe. The machine has two types of load sensor, one has the load range of 0–20 N and other has the load range of 20–200 N as well as a ball indenter, used as the probe. There are two types of sliders, the X-axis slider and Y-axis lower drive. The purpose of X-axis slider is to provide motion for scratch or ball sliding, whereas the Y-axis lower drive is used to move the sample in perpendicular direction to the scratch direction to reorient the probe for new scratch in XY-plane. Subsequently, the machine is consisting a mounting block which holds the load sensor and connects it to the Z-axis carriage. The function of Z-axis carriage is to move probe toward sample surface and apply required load for scratch tests. All drives for X, Y and Z axes along with the load sensors are connected to data acquisition system (DAQ). Apparently, they are operated through a computer connected to the DAQ. The 45S5 bioglass samples are mounted on portable heating setup with the help of a sample holder. Here, the sample holder can hold a sample of size $23 \times 21 \times 10\ mm^3$. Further, the assembly of sample and the portable heating setup is mounted on clamping vice of micro-scratch tester, and the electrical connections are made between voltage variac and the portable heating setup (Fig. 2). The voltage variac will work as the heat controller for the portable heating setup and it will be operated on 220 V AC current[24]. Furthermore, the temperature of the sample is observed with the help of an infrared thermometer (Fluke 572–2).

3 Experimental Details

3.1 Sample Preparation

45S5 bioglass samples are prepared with the help of melting process. Total 6 samples are cut in $23 \times 21 \times 10\ mm^3$ size and polished with the help of emery sheets of grit size 220, 320, 1000, 1500 and 2000 subsequently. Hence, the average surface roughness is kept between a very close range from 0.1164 μm to 0.1478 μm to maintain the scratch behaviors on every sample as given in Table 1. The average surface roughness is the average of surface roughness taken along sample length 23 mm and sample width 21 mm. Mitutoyo SV-2100 has been used for surface roughness measurement. While measurement, the sampling length is kept 0.8 mm and the evaluation length is 4 mm as well as the traverse length of probe is kept 4.8 mm. Subsequently, Fig. 3 shows the typical roughness plot along the length for first sample.

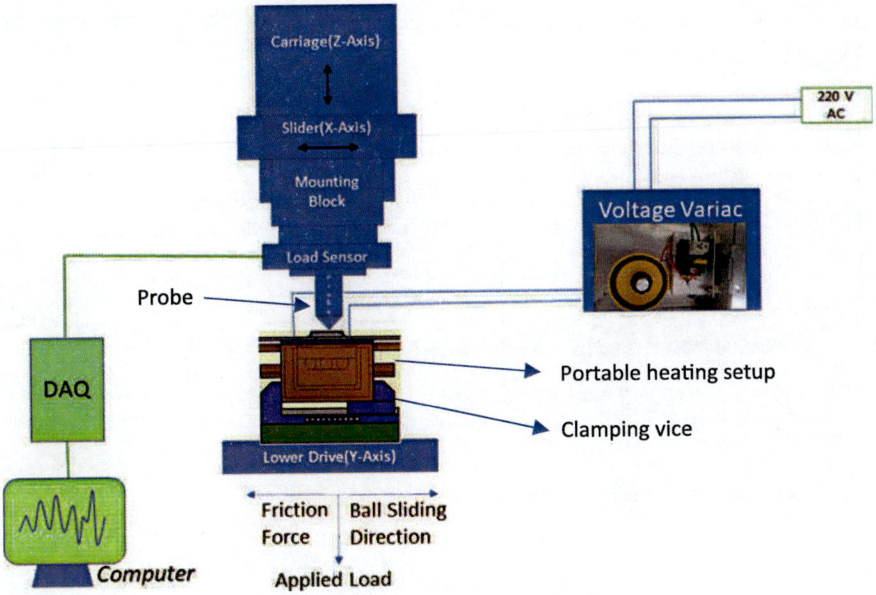

Fig. 2 Schematic diagram of elevated temperature scratch testing setup

Table 1 Surface roughness of polished samples

Sample no	Along length (μm)	Along width (μm)	Average roughness (μm)
1	0.1274	0.1681	0.1478
2	0.1317	0.1504	0.1411
3	0.127	0.1057	0.1164
5	0.1443	0.1479	0.1461
6	0.1836	0.1032	0.1434

3.2 Operating Parameters

Table 2 gives the operating parameters for the elucidated experiments, where the four average temperature values selected to perform scratch tests are 27, 210, 310 and 420 °C. The scratch tester is capable of performing scratch tests at constant load as well as at ramp load. Although, the machine is having two load cells, one is of 0–20 N and the other one is of 20–200 N, the 20–200 N load cell is selected for the experiments. Additionally, the ramp load is selected to perform the experiments with the loading rate of 0.2 N/mm, while the stoke length is kept as 5 mm. It is in such a way that the starting load will rise by 1 N at the end of the scratch of 5 mm length. There are five starting loads selected as 20, 21, 22, 23 and 24 N, so that the total load range between 20 and 25 N will be covered during tests. Apparently, the

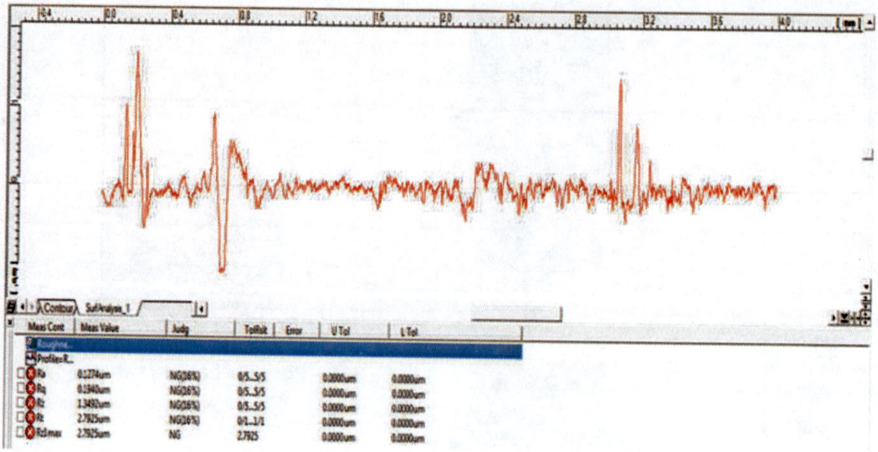

Fig. 3 A typical roughness plot along the length for first sample

Table 2 Operating parameters during scratch test

Temp. (°C)	Load type	Load range (N)	Start load (N)	Loading rate (N/mm)	Stroke (mm)	Scratch speed (mm/sec)	Scratch offset (mm)
27	Ramp	20–200	20	0.2	5	0.5	1
210			21			1	
420			22			2	
			23				
			24				

offsets between any two scratches are kept constant at 1 mm. Three scratch speeds are selected at every combination to perform the experiments. These scratch speeds are 0.5, 1s and 2 mm/s.

3.3 *Experimental Procedure*

The sample is mounted on the portable heating setup and this assembly is mounted on scratch tester as discussed previously. Figure 4 shows the image of experimentation, whereas, Fig. 5 shows the workpiece image in the experimental condition.

It was observed that there is a lot of thermal cracking at fast heating rate due to sudden high temperature gradient between the sample's heated and non-heated zones. This high temperature gradient induces high thermal stresses. Therefore, the heating rate of the sample is kept less than 10 °C/min to avoid thermal cracking in the samples. The desired temperature is achieved with help of the heat controller and the

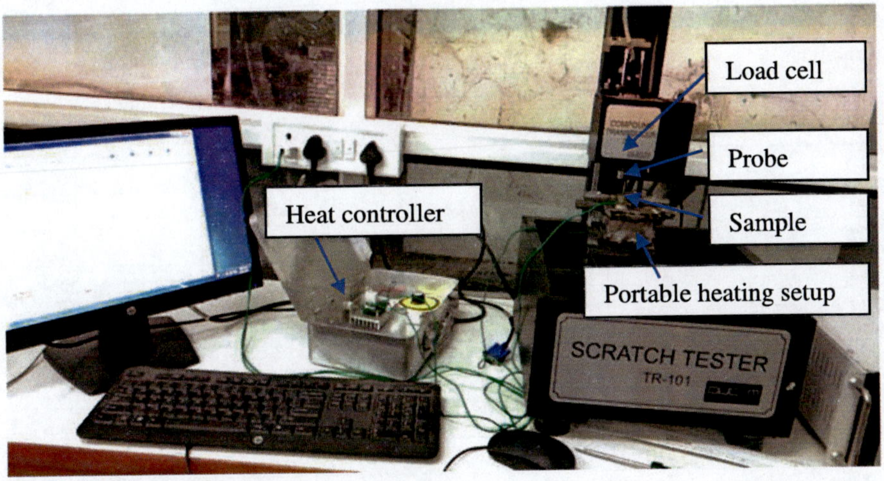

Fig. 4 Image of experimentation

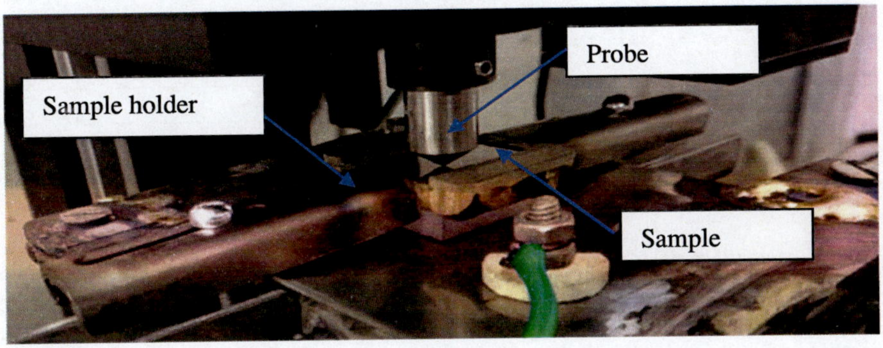

Fig. 5 Workpiece image in experimental condition

steady state is observed for 10 min as the minimum time span. This ensures the least temperature variation during the scratch tests at desired temperatures. Additionally, the continuous reading of the temperature has been observed during scratch tests and made sure that the temperature is in the range of ±5 °C from the desired temperature value. In total, a set of 45 scratch conditions are performed and allocated a specific number as shown in Table 3. Each of the scratch conditions is repeated for three times, and the best run was selected on the basis of consistent results for further analysis.

Table 3 Table of experiments indicating different scratch numbers during a set of 45 scratch conditions

Start load (N)	Scratch speed (mm/s)	Scratch numbers according to the different temperatures					
		Sample No	27 °C	Sample No	210 °C	Sample No	420°C
20	2	1	1.1	3	3.1	5	5.1
21	2	1	1.2	3	3.2	5	5.2
22	2	1	1.3	3	3.3	5	5.3
23	2	1	1.4	3	3.4	5	5.4
24	2	1	1.5	3	3.5	5	5.5
20	1	1	1.6	3	3.6	5	5.6
21	1	1	1.7	3	3.7	5	5.7
22	1	1	1.8	3	3.8	5	5.8
23	1	2	2.1	4	4.1	6	6.1
24	1	2	2.2	4	4.2	6	6.2
20	0.5	2	2.3	4	4.3	6	6.3
21	0.5	2	2.4	4	4.4	6	6.4
22	0.5	2	2.5	4	4.5	6	6.5
23	0.5	2	2.6	4	4.6	6	8.6
24	0.5	2	2.7	4	4.7	6	6.7

4 Results and Discussion

Scratch tests were performed as shown in Table 3. After performing the mentioned tests, the graphs of different starting loads are clubbed together to reduce the sample numbers of graphs so that the complete range of 20–25 N of starting normal load is plotted on a single graph with reference to different temperatures as shown in Fig. 6. It shows the comparative plots for traction force and coefficient of friction at different temperatures (27, 210 and 420 °C) during scratch tests, whereas Fig. 6a shows the plot for traction force at scratch speed of 2 mm/s, Fig. 6b shows the plot for traction force at scratch speed of 1 mm/s, Fig. 6c shows the plot for traction force at scratch speed of 0.5 mm/s, Fig. 6d shows the plot for coefficient of friction at scratch speed of 2 mm/s, Fig. 6e shows the plot for coefficient of friction at scratch speed of 1 mm/s and Fig. 6f shows the plot for coefficient of friction at scratch speed of 0.5 mm/s.

It is continuously observed from Fig. 6 that the traction forces as well as coefficient of friction during scratch tests are decreasing with the increase in sample temperature. This phenomenon repeats even at different scratch speeds as shown in Fig. 6. This decrease in traction forces and coefficient of frictions can lead to better machinability of hard to machine materials.

Figure 7 shows the three typical SEM images Fig. 7a, b, c of scratch sets 1.3, 3.3 and 5.3, respectively, which are at 27, 210 and 420 °C of surface temperatures, respectively. The scratch speed during these scratches is 2 mm/s and the load range

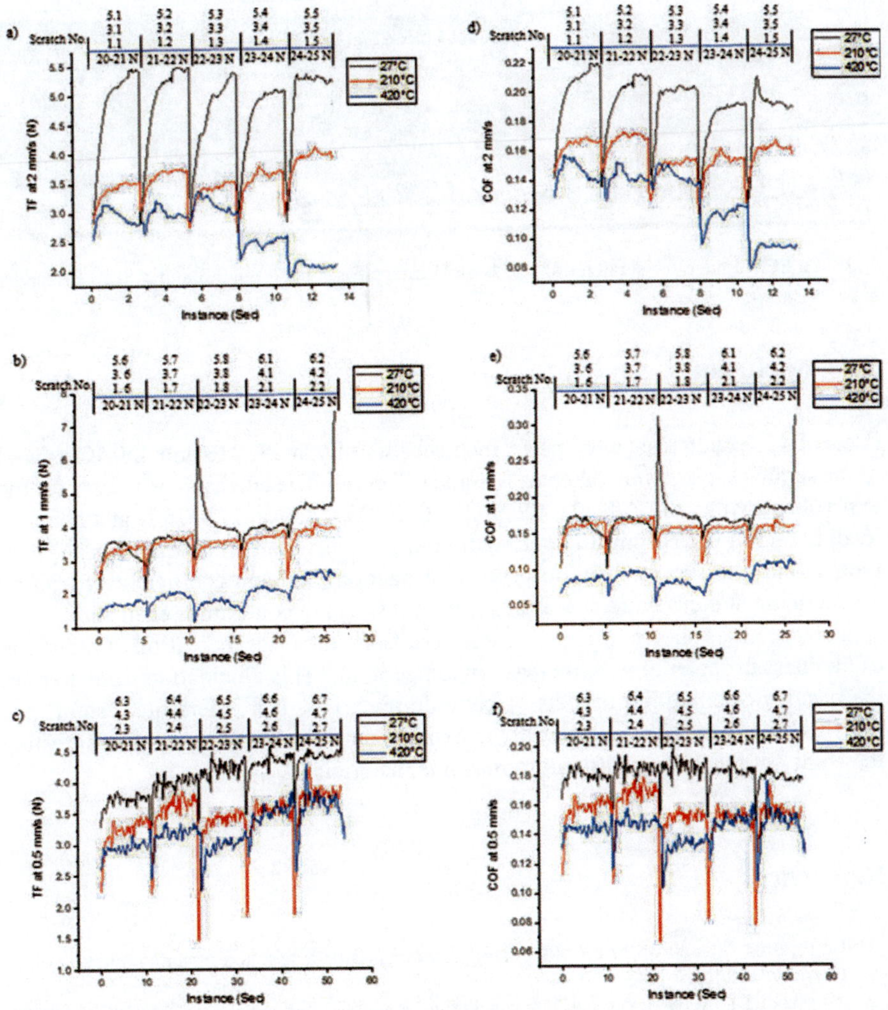

Fig. 6 Comparative plots for traction force and coefficient of friction at different temperatures (27, 210 and 420 °C) during scratch tests-**a** Plot for traction force at scratch speed of 2 mm/s, **b** Plot for traction force at scratch speed of 1 mm/s, **c** Plot for traction force at scratch speed of 0.5 mm/s, **d** Plot for coefficient of friction at scratch speed of 2 mm/s, **e** Plot for coefficient of friction at scratch speed of 1 mm/s, **f** Plot for coefficient of friction at scratch speed of 0.5 mm/s

was 22–23 N. This comparison is a positive indication of reduction in traction forces and coefficient of friction during the scratch. The images indicate that the lateral cracks are reducing with the rise in temperature but median and radial cracks may be increasing. Overall, the reduction in traction force and coefficient of friction may lead to better machinability.

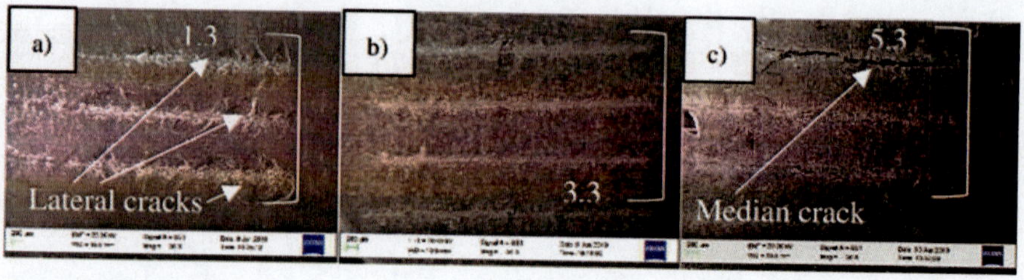

Fig. 7 **a** SEM image for scratch no. 1.3, **b** SEM image for scratch no. 3.3 and **c** SEM image for scratch no. 5.3

5 Conclusion

A set of 45 scratch tests were performed successfully at 27, 210 and 420 °C as well as the scratch speeds were taken as 2, 1 and 0.5 mm/s. The normal load ranges for the experiments were 20–21 N, 21–22 N, 22–23 N, 23–24 N and 24–25 N at a constant loading rate of 0.2 N/mm. In continuation of that, different comparative graphs are plotted and it is successfully observed that traction load and coefficient of friction are reducing at elevated temperatures. These results are continuous even at different scratch speeds as shown in Fig. 6. These results are the clear indication of reduction of hardness of material with increase in temperature. This elucidation leads to raise the bar of machinability aspects of ceramic materials. It is, therefore, established that this process can be a great utility in existing machining processes and for testing feasibility of machining difficult to machine materials.

References

1. Le Bourhis E, Metayer D (2000) Indentation of glass as a function of temperature. J Non-Crystalline Solids 34–38
2. Manshadi ST (2009) Laser assisted machining of inconel 718 superalloy. McGill University Montréal.
3. Lei S, Shin YC, Incropera FP (2001) Experimental investigation of thermo-mechanical characteristics in laser-assisted machining of silicon nitride ceramics. J Manuf Sci Eng 123(4):639–646
4. Rebro PA, Shin YC, Incropera FP (2004) Design of operating conditions for crackfree laser-assisted machining of mullite. Int J Mach Tools Manuf 44(7–8):677–694
5. Jeon Y, Park HW, Lee CM (2013) Current research trends in external energy assisted machining. Int J Precis Eng Manuf 14(2):337–342
6. Lei S, Shin YC, Incropera FP (2000) Deformation mechanisms and constitutive modeling for silicon nitride undergoing laser-assisted machining. Int J Mach Tools Manuf 40(15):2213–2233
7. Rebro PA, Shin YC, Incropera FP (2002) Laser-assisted machining of reaction sintered mullite ceramics. J Manuf Sci Eng 124(4):875–885
8. Chang C-W, Kuo C-P (2007) Evaluation of surface roughness in laser-assisted machining of aluminum oxide ceramics with Taguchi method. Int J Mach Tools Manuf 47(1):141–147
9. George C (1991) Laser machining—theory and practice. Springer, New York
10. Chryssolouris G et al (1988) Theoretical aspects of a laser machine tool

11. Haferkamp H, Seebaum D (1994) Material removal on tool steel using high power CO_2-lasers. In: Proceedings of the conference on laser assisted net shape engineering (LANE′ 94)., Meisenbach, Bamberg
12. Hench LL et al (1971) Bonding mechanisms at the interface of ceramic prosthetic materials. J Biomed Mater Res 5(6):117–141
13. Hench LL (2013) Chronology of bioactive glass development and clinical applications
14. Thompson ID, Hench LL (1998) Mechanical properties of bioactive glasses, glass-ceramics and composites. Proc Inst Mech Eng [H] 212(2):127–136
15. Briscoe BJ, Pelillo E, Sinha SK (1996) Scratch hardness and deformation maps for polycarbonate and polyethylene. Polym Eng Sci 36(24):2996–3005
16. Narasimman P, Pushpavanam M, Periasamy V (2012) Wear and scratch resistance characteristics of electrodeposited nickel-nano and micro SiC composites. Wear 292:197–206
17. Varga M et al (2017) Study of wear mechanisms at high temperature scratch testing. Wear 388:112–118
18. Pazhanivel B, Kumar TP, Sozhan G (2015) Machinability and scratch wear resistance of carbon-coated WC inserts. Mater Sci Eng, B 193:146–152
19. Wang N et al (2019) Research on the machinability of A-plane sapphire under diamond wire sawing in different sawing directions. Ceram Int 45(8):10310–10320
20. Zhang Z et al (2010) Nanoscale machinability and subsurface damage machined by CMP of soft-brittle CdZnTe crystals. Int J Adv Manuf Technol 47(9–12):1105–1112
21. Zhang S et al (2015) Structural optimisation and synthesis of multilayers and nanocomposite AlCrTiSiN coatings for excellent machinability. Surf Coat Technol 277:23–29
22. Tseng AA et al (2009) Scratching properties of nickel-iron thin film and silicon using atomic force microscopy. J Appl Phys 106(4):044314
23. Da Silva L et al (2020) Critical assessment of compacted graphite cast iron machinability in the milling process. J Manuf Process 56:63–74
24. Singh JK, Rao US, Pyare R (2020) Design and development of a multipurpose portable heating setup for 45S5 bioglass and other ceramic samples. Rev Sci Instrum 91(10):105108

A Study on Green Manufacturing Research from 2010 to 2020: A Bibliometric Research

Utkal Khandelwal, Avnish Sharma, and Aneesya Panicker

Abstract Shifting trend of consumer, businesses and government compliances toward environment-friendly products and processes are now shifting the industries toward sustainable manufacturing process better known as "green manufacturing" (GM). Green manufacturing can be elucidated as the utilization of high-efficient materials in manufacturing that reduces harmful environmental impact. It confines critical manufacturing concerns such as complying with environmental laws and regulations, natural resource conservation, controlling toxic substance and the management of waste. This trend has led to increasing interest of researchers in green manufacturing. In order to facilitate those researchers, this research presents the bibliometric analysis of the research on green manufacturing from the year 2010–2020. The Scopus database was used to determine year-wise publications, most cited papers, most prolific authors, countries and institutions. Result of bibliometric analysis shows that research interest is continuously increasing as number of publication is increasing after 2015 continuously. This analysis provides a direction to those who are entering the field of green manufacturing research, providing information on which journals to consult, which authors, institutions and countries are most eminent, and keywords which was frequently used in the green manufacturing researches.

Keywords Green Manufacturing (GM) · Environmental impact · Green initiatives and practices · Sustainability · Review

1 Introduction

In last few decades, business has taken a strain through scarcity of resources leading to reduced cost competitiveness [7]. This has led to firms seeking opportunities to develop business models based on circular economy principles inspired by green practices for sustainable development [1]. Industries are now shifting to adopt GM because of meeting regulative compliances, catering the shifting demand of green

U. Khandelwal (✉) · A. Sharma · A. Panicker
Institute of Business Management, GLA University, Mathura, U.P 281406, India
e-mail: utkal.khandelwal@gla.ac.in

A. Maurya et al. (eds.), *Recent Trends in Mechanical Engineering*, Lecture Notes in Mechanical Engineering, https://doi.org/10.1007/978-981-19-7709-1_58

products and ensuring the sustainability in their businesses. GM can be described as process of using the correct and high-efficient materials in the production process to diminish the harmful environmental impact [8]. This may involve "the conscious integration of environmental management initiatives during a product's life cycle and cover critical manufacturing issues such as the design of green products with reusable and recyclable content, pollution control and environmental protection, environmental regulatory compliance and waste management", just to mention a few [11]. Rapid shifting of industries from conventional manufacturing to green manufacturing attracts researchers for conducting various researches on GM in last two decades, thus expediting researches on GM.

2 Concept of Green Manufacturing

Practicing GM is the need of the hour for healthy and happy survival and sustainability of industries and human societies [4]. For long-term sustainability, it is essential to establish an efficient global manufacturing standard, which ensures that every company willing to compete globally must produce products which comply with the green regulations [5]. GM usually involves the renovation of industrial operations in three ways—use of green energy, development and sale of green products and adopting green processes for different business operations [12]. According to the latest global survey by the Boston Consulting Group (BCG), 92% of the organizations were found to be engaged in adoption of green initiatives to minimize the adverse environmental impact, thereby benefiting through cost savings, brand enrichment, superior regulatory transactions and building superior investors interest [9, 10]. In the power sector, green initiatives have the major impact on dropping CO_2 emissions' standards followed by industrial and transportation sector [2, 13]. In global context, the European Union (EU) issued directions to member states for implementation of GM. Germany started with "Green movement", and factories in the USA were also practicing different GM philosophies for dropping the use of energy and recycling water for implementation of ISO 14001 and reprocess packaging [3]. This research is novel as it gives the glimpse of latest research articles published in the field of GM.

3 Research Method and the Bibliometric Indicators

This study has adopted a systematic review approach [14] and collected peer-reviewed journal articles to attain the research objectives. The GM articles which were written in English language including British and American were searched and accessed from Scopus database, as it has wide coverage of quality journals. Although choice of articles written in English language only may be considered as a limitation of this study, the keyword "green manufacturing" was used for searching the relevant research articles. More than 3071 research documents were examined and reviewed for collecting the relevant research information. The authors analyzed diverse titles, abstracts, keywords, research frameworks and different headings and sub-headings to reach the research outcome. For analysis purpose, bibliometric analysis was used, and the studies from 2010 to 2020 on GM were incorporated. The different units of analysis include—year-wise number of research on GM from 2010 to 2020, most prolific countries, institutions, authors, journals having maximum publication, most cited research articles on GM, etc.

4 Results and Findings

4.1 Publication Year

Figure 1 indicates that research on GM always catches the attention of researchers. The amount of research conducted on GM was comparatively less in the beginning of 2011, but subsequently increased from 2018 to 2020. The maximum publications on GM were done in the year 2020, i.e., 1732.

4.2 Most Prolific Institutions

Figure 2 demonstrates list of worldwide institutions that have published more than 25 researches on GM. The Ministry of Education, China, is one which publishes maximum research on GM processes and activities, i.e., 128, whereas other academic institutions and universities have published a reasonable research due to several academic and administrative reasons.

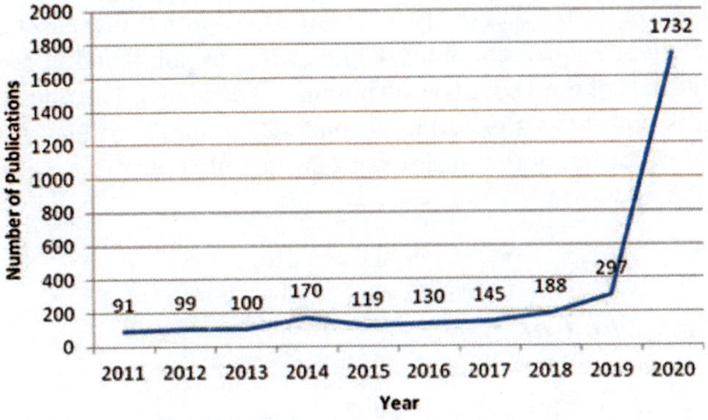

Fig. 1 Scopus publication on green manufacturing

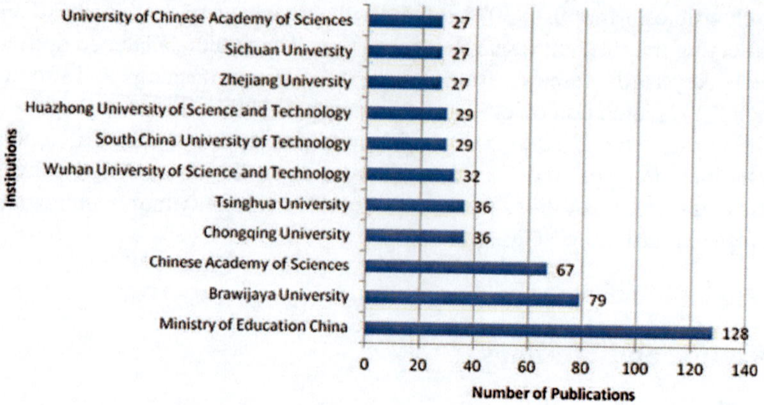

Fig. 2 Most productive institutions' publications wise

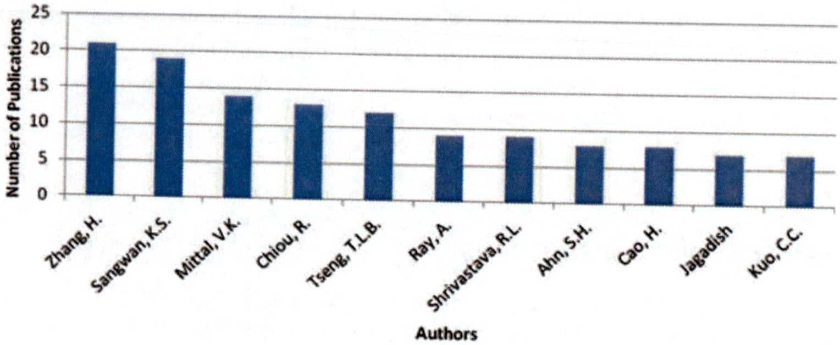

Fig. 3 Authors who have published on green manufacturing

4.3 Most Prolific Authors

Scopus database indicates that the most prolific GM author is Hua Zhang of Wuhan University of Science and Technology, China, who has published 21 research articles in different journals of repute with good number of citations. The figure indicates that only 11 authors out of the 159 had published 127 research articles with an average of 3–6 articles on different dimensions of GM and also received a greater attention of scholars (Fig. 3).

4.4 Most Prolific Countries

Figure 4 indicates China having maximum publications, i.e., 1240, followed by India with 339 publications, the USA, Indonesia, the UK, South Korea, Malaysia, Germany, etc.

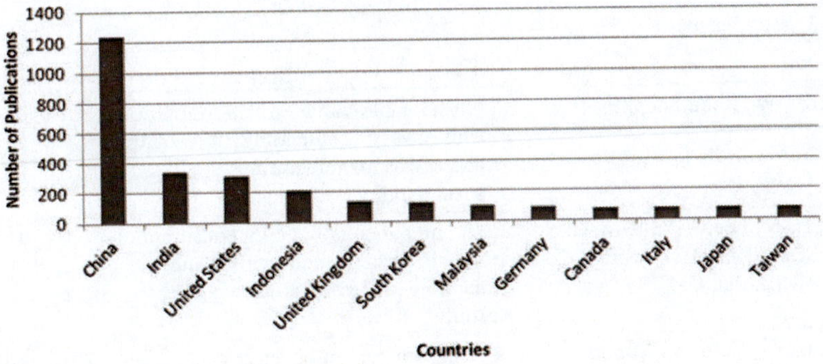

Fig. 4 Countries where green manufacturing research has been published

4.5 Most Cited Papers

Table 1 shows the most cited research titles on "Green manufacturing" with authors' name and year of publications. The number of citations in an article is the most significant indicator to decide an article's overall quality [6]. Titirici, White, Brun, Budarin, Su, Del Monte, Clark and MacLachlan, published in the year 2015, have received maximum citations, i.e., 634, and title a resource-based view of green supply management published in 2011 and authored by Gavronski, Klassen, Vachon and Nascimento has received minimum citations in comparison, i.e., 148. All articles which received greater attentions of research scholars have cited more than 100 times.

Table 1 Authors with title, publication year, and citations

Authors	Title	Year	Cited by
Titirici, M.-M., White, R. J., Brun, N., Budarin, V. L., Su, D. S., Del Monte, F., Clark, J. H., MacLachlan, M. J	"Sustainable carbon materials"	2015	634
Luo, H., Du, B., Huang, G. Q., Chen, H., Li, X	"Hybrid flow shop scheduling considering machine electricity consumption cost"	2013	231
Li, L	"China's manufacturing locus in 2025: With a comparison of "Made-in-China 2025" and "Industry 4.0""	2018	210
Deif, A. M	"A system model for green manufacturing"	2011	210
Zhang, Y. N., Cao, X., Larose, S., Wanjara, P	"Review of tools for friction stir welding and processing"	2012	198

(continued)

Table 1 (continued)

Authors	Title	Year	Cited by
Faulkner, W., Badurdeen, F	"Sustainable Value Stream Mapping (Sus-VSM): Methodology to visualize and assess manufacturing sustainability performance"	2014	190
Yoon, H.-S., Lee, J.-Y., Kim, H.-S., Kim, M.-S., Kim, E.-S., Shin, Y.-J., Chu, W.-S., Ahn, S.-H	"A comparison of energy consumption in bulk forming, subtractive, and additive processes: Review and case study"	2014	165
Behrendt, T., Zein, A., Min, S	"Development of an energy consumption monitoring procedure for machine tools"	2012	164
Shrouf, F., Miragliotta, G	"Energy management based on Internet of Things: Practices and framework for adoption in production management"	2015	158
Govindan, K., Diabat, A., Madan Shankar, K	"Analyzing the drivers of green manufacturing with fuzzy approach"	2015	153
Gavronski, I., Klassen, R. D., Vachon, S., Nascimento, L. F. M. D	"A resource-based view of green supply management"	2011	148

4.6 Journals with Highest Publications

Figure 5 presents that the list of journals published most research articles on GM. According to the Scopus official website, these journals and proceedings have a greater research influence. It shows that IOP Conference Series Earth and Environmental Science has published 157 research documents on GM, which is higher as compared to other journals. The Journal of Cleaner Production ranks second with 128 research publications. According to the Scopus database, 148 journals have published research on green manufacturing.

4.7 Important Keywords in the Publications of Green Manufacturing

Keywords' occurrence indicates the effect on academics and other fields of green manufacturing. The green manufacturing articles, widely searched in other journals, are given in Table 2. This table shows those keywords which occurred more than 100 times during last ten years. The keyword which was most widely searched was green manufacturing which was searched 2842 times in the last ten years.

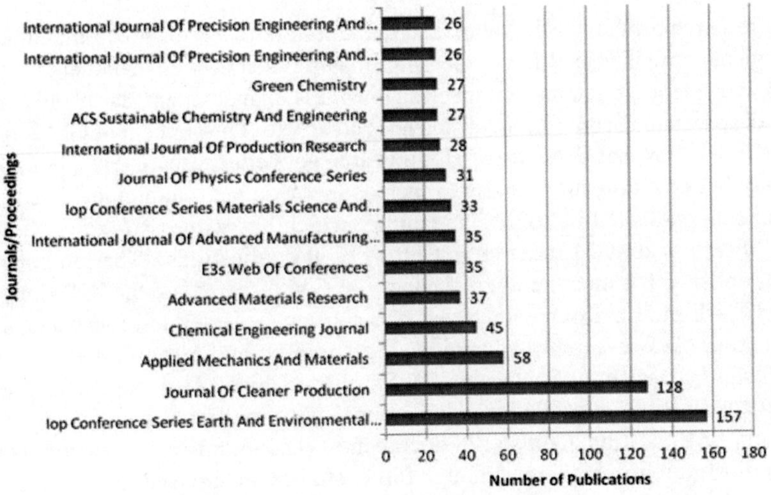

Fig. 5 List of journals published research on green manufacturing

Table 2 Keywords occurrence in last five year publications of green manufacturing

#	Keyword	Occurrence	#	Keyword	Occurrence
1	Green manufacturing	2842	12	Carbon	128
2	Manufacture	498	13	Costs	127
3	Sustainable development	464	14	Efficiency	120
4	Industrial research	223	15	Fabrication	117
5	Energy utilization	220	16	Decision-making	115
6	Energy efficiency	160	17	Manufacturing industries	110
7	Article	153	18	Environmental protection	108
8	Environmental impact	141	19	Supply chains	108
9	Scanning electron microscopy	132	20	Energy conservation	107
10	Sustainability	132	21	Environmental management	103
11	Life cycle	131	22	Manufacturing process	101

5 Discussion and Conclusion

The bibliometric analysis presented in this article offers direction to those who are planning to enter in the field of GM, guiding toward selection and consultation to relevant journals and eminent authors. The bibliometric analysis of 3071 GM research documents collected from the Scopus database shows that 11 are the most cited research articles in the field. The maximum numbers of green manufacturing documents are published in engineering, material science, environmental Science, computer science, chemical engineering, business, management, earth and planetary sciences, decision sciences, social sciences areas, etc. The concept of GM originated in the late 1980s, but only after 2015, this area got better researchers' attention, and the number of publications starts to increase year on year, attaining 1732 research documents published in 2020. The country which has witnessed the good number of publications on GM research is China (1240 publications). The journals which have published the most number of papers on GM research are the IOP Conference Series Earth and Environmental Science as conference series and Journal of Cleaner Production (285 research documents). Most prolific author on GM is Hua Zhang of Wuhan University of Science and Technology, China (21 documents). The most productive institution/affiliation publications' wise on green manufacturing is the Ministry of Education, China (128 documents). The findings of this study could be useful for future researchers, although future studies are advised to include research articles that belong to the other databases such as Web of Science, IEEE Xplore, ScienceDirect, Directory of Open Access Journals, JSTOR, SCI, ESCI and journals with high impact. Future research scholars are advised to understand the role of different government and private agencies in strategic planning and execution of GM practices across the organizations. They are recommended to investigate numerous emerging issues and challenges in the field of green manufacturing across the nation and globe.

References

1. Bag S, Wood LC, Xu L, Dhamija P, Kayikci Y (2020) Big data analytics as an operational excellence approach to enhance sustainable supply chain performance. Resour Conserv Recycl 153:104559
2. Bhattacharya A, Jain R, Chaudhry A (2011) Green manufacturing: Energy, products and processes. The Boston Consulting Group-Confederation of Indian Industry. Accessed from https://www.cii.in/webcms/Upload/BCG-CII%20Green%20Mfg%20Report.pdf
3. Brown AS (2009) The many shades of GREEN. JO Mech EngCIME (ASME) 131(1):22–29. ISSN: 00256501
4. Bylinsky G (1995) Manufacturing for reuse, Fortune 500 Current Issue, 131(2):102–112. Accessed from http://money.cnn.com/magazines/fortune/fortune
5. Durham, (2002) Environmental benign manufacturing: current practices and future trends. JOM 54(5):34–37
6. Duque Oliva EJ, CerveraTaulet A, Rodríguez Romero C (2006) A bibliometric analysis of models measuring the concept of perceived quality in providing internet service. Innovar 16(28):223–243
7. Feger ALR (2014) Creating cross-functional strategic consensus in manufacturing facilities. Int J Oper Prod Manag 34(7):941–970
8. Ghazilla R, Sakundarini N, Abdul-Rashid S, Ayub N, Olugu E, Musa S (2015) Drivers and barriers analysis for green manufacturing practices in Malaysian SMEs: a preliminary findings. Procedia CIRP 26:658–663
9. Gungor A, Gupta SM (1999) Issues in environmentally conscious manufacturing and product recovery: a survey. Comput Ind Eng 36(4):811–853

10. Gutowski T (2002) Environmentally benign manufacturing and ecomaterials; product induced mater flows. Mater Trans 43(3):282–284
11. Rehman MA, Seth D, Shrivastava RL (2016) Impact of green manufacturing practices on organisational performance in Indian context: an empirical study. J Clean Prod 137:427–448
12. Shrivastava SK (2003) How green are Indian firms. JO Prod 44(2):294–302
13. Thanki S, Thakkar JJ (2020) An investigation on lean-green performance of Indian manufacturing SMEs. Int J Prod Perform Manag 69(3):489–517
14. Tranfield D, Denyer D, Smart P (2003) Towards a methodology for developing evidence-informed management knowledge by means of systematic review. Br J Manag 14:207–222. https://doi.org/10.1111/1467-8551.00375

Effect of Variation in Layer Thickness on Tensile Strength of a 3D-Printed Object Produced from PLA Filament

Samriddhya Ray Chowshury and Rituparna Biswas

Abstract Three-dimensional printing is a rapid prototyping (RP) technology where layer-by-layer material addition is done in order to get the desired dimensional accuracy at a much faster rate. The paper looks into a very important property of a material, which is its tensile strength. In this regard, the thickness of each layer plays an important role for both conforming dimensional accuracy and to meet required mechanical properties. In this study, the effect of varying layer thickness for each specimen on the tensile strength of 3D-printed object, keeping every other parameter unaltered, has been experimented and analyzed. For this aim, three different printing layer thicknesses, i.e., 0.2, 0.25 and 0.3 mm, have been printed in three specimens of polylactic acid (PLA). The tensile strength of each specimen has been measured by a typical setup that includes a dial gauge and jig setup, which concluded about the applied load that the specimens can withstand before tearing. From the experimental outcomes and the geometrical design of the specimens, the tensile strength of each specimen has been determined. From the study, it is analyzed that ability of the PLA specimens to withstand the load gets significantly high with the lesser layer thickness.

Keywords Additive Manufacturing (AM) · Three-Dimensional Printing (3D printing) · Fused Deposition Modeling (FDM) · Polylactic Acid (PLA) · Tensile strength

---

S. R. Chowshury (✉) · R. Biswas
Department of Mechanical Engineering, Budge Budge Institute of Technology, Nishchintapur, Budge Budge, West Bengal 700137, India
e-mail: samriddhya@gmail.com

A. Maurya et al. (eds.), *Recent Trends in Mechanical Engineering*, Lecture Notes in Mechanical Engineering, https://doi.org/10.1007/978-981-19-7709-1_59

1 Introduction

Three-dimensional printing as an additive manufacturing technologies has gained prominence in the recent past and is increasingly becoming important not only as a prototype manufacturing technology, but also as a mass manufacturing industry.

Ayrilmis [1] had studied the effect on surface integrity of a 3D-printed object with a varying layer thickness. He experimented with wood flour or PLA filament which was used for printing specimens with four different layer thicknesses, i.e., 0.05, 0.1, 0.2 and 0.3 mm. It is evidenced that the surface smoothness got improved with reducing layer thickness.

For identifying and understanding the role of tensile strength in a 3D-printed object, it was reviewed that Haoa et al. [2] had studied that the 3D-printed specimens were prepared with continuous carbon fiber-reinforced thermosetting composites, and then, their mechanical properties were characterized. In his experiment, the 3D printing platform was firstly fabricated to prepare the composite lamina and grid with FDM. The mechanical properties of the composites were characterized. It was observed that the mechanical properties of the printed thermosetting composites were superior to that of similarly printed thermoplastic composites and printed fiber-reinforced composites. Khabia and Jain [3] demonstrated the effect of layer thickness variation on tensile strength of specimens printed with Z-ABS filament. They concluded that for test specimens printed with thickness parallel to build platform, change in tensile strength remains unaltered.

Rao et al. [4] experimented with varying printing temperature, layer thickness and filament pattern and projected the effect on tensile strength of the printed specimens. The experimentation was framed for full factorial method with three stage variations in each parameter and analyzed with ANOVA. Through optimization technique, it was indicated that both the main effects and interactions between the factors can be estimated as a part of this ANOVA test. In the experimental investigation by Vălean et al. [5], it was testified that spatial printing direction (0, 45, 90°) has less influence on the Young's modulus and higher influence on tensile strength, whereas the numbers of layer in printing leads to significant reduction in both Young's modulus and tensile strength. Wagari Gebisa and Lemu [4], in their paper had designed their experiments for full factorial method using five parameters (air gap, raster angle, raster width, contour width and contour number) of printing with ULTEM 9085 polymeric material filament. It this study, it has been observed that except from raster angle, other parameters are playing insignificant role in influencing the tensile strength of the material. Yao et al. [5] had measured and analyzed theoretically the effect of printing angles, i.e., 0, 15, 30, 45, 60, 75, 90° and layer thickness, i.e., 0.1, 0.2, 0.3 mm on the tensile strength of 3D-printed object. It was concluded through analysis that both the parameters play significant role in tensile strength of the 3D-printed object.

2 Experimentation and Determination

2.1 PL Filament and Specimen Preparation

The objective of this study is to test the tensile strength of 3D-printed objects with varied layer thicknesses. The experiment looks into a very important characteristic of material, which is strength. The tensile strength is measured by this experiment, and the result displays the perfect parametric condition of getting the best strength out from the specimen. The specimens are designed in computer-aided design (CAD) software and then printed in a 3D printer. We will be conducting two experiments where two parameters are varied. For this experiment, the layer thickness is varied for the creation of three similar specimens. The specimens are tested out in a jig, and for each of the specimen, we find the ultimate tensile strength. Studying the results of our experiment, we can have a more clear idea about the conditions at which we will get the best print quality. Also, the manner of breakage of each single specimen and the pattern of breakings overall displayed by the specimens will also help us state the properties of the material (Fig. 1).

2.2 Experimental Setup

The expected possible range of tensile strength has been studied, and with the alignment of the required range of load to meet, the permissible tensile failure specimen as well as the tensile load testing setup has been fabricated. For this purpose, a jig is made to test the tensile strength of the three-dimensional-printed specimens. The jig has a capability to test tensile strength up to 100 kg-wt. The specimen is designed with proper calculations, which helps to get a rough idea of the experimental results. The specimen is then designed in computer-aided design (CAD) software, and further improvements are made if necessary. The specimen is then transferred into StereoLithography File (.STL) and further decrypted to G-codes (machine level language) which can be understood by three-dimensional printers. The prints are made with PLA material and printed in three different layer thicknesses starting from 0.2 mm to 0.25 mm and then to 0.3 mm. All the printed parts are checked to make sure that all parts are free of defects. Then, all three prints are put to test, and all the desired results are noted down in required tabular format. The test is concluded. The tensile strength is calculated (Fig. 2).

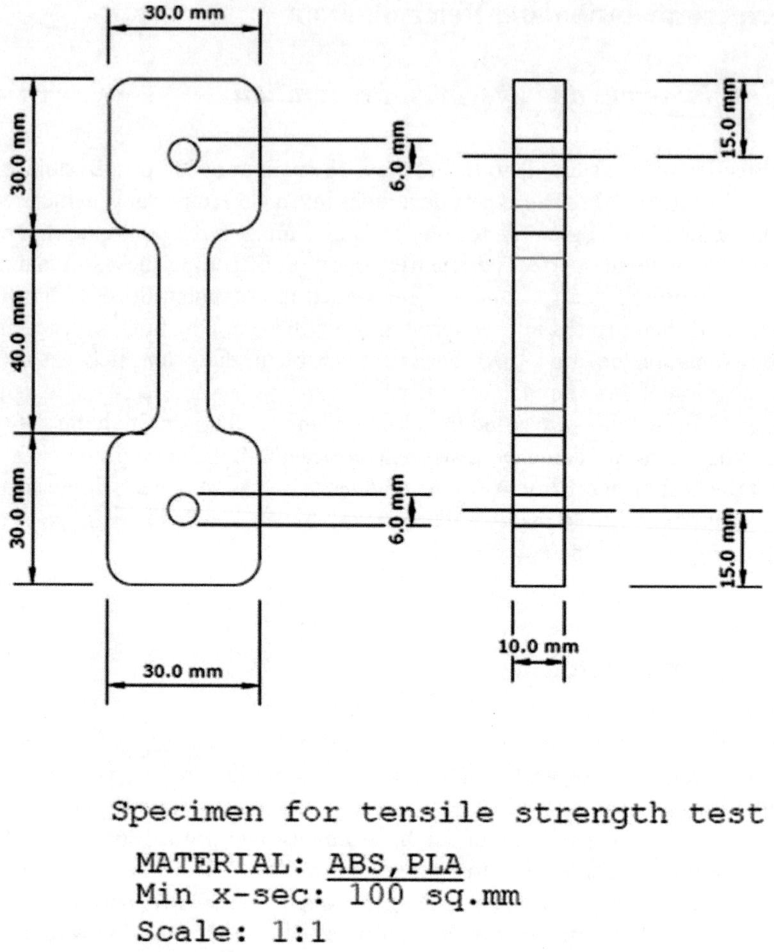

Fig. 1 Specimen for the tensile strength test

2.3 Determination of Permissible Tensile Load

The determination of maximum permissible tensile load of the specimens has been carried out with the help of the prepared experimental setup. The experimental observations are as follows (Table 1 and Fig. 3).

2.4 Results and Discussions

During the experimentations, few diversified observations are made in Fig. 3.

Fig. 2 Experimental setup for load determination for specimen with varying layer thickness

Table 1 Experimental data and observations

Specimen no	Specimens	Breaking load (kg-wt)	Observations	Probable cause of failure
1	0.20 mm	58.5	PLA showed some ductility, and cup-cone formation on the time of breaking is clearly visible	The individual fiber of the printed specimen failed under the load under normal condition
2	0.25 mm	47	The broken part is not homogeneous in nature, and all the fibers are not broken samely	The layers of the specimen failed before the fiber failed
3	0.30 mm	48.5	Same as observation number 2, the layers tend to chip out rather than fiber failure	Increase in thickness of the layer may result to decrease in adhesion of layers which causes layers to separate before the ultimate load the fibers can take

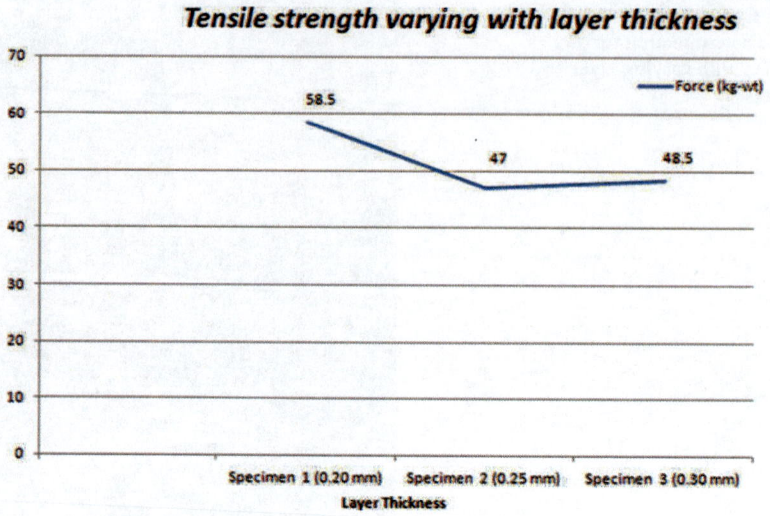

Fig. 3 Load determination for specimen with varying layer thickness

The specimens are broken in different ways which have been shown in Fig. 4. On the basis of the breaking pattern, the observations have been made.

On the basis of the experiments, it is very evident that Specimen 1 showed some ductility and cup-cone formation at the time of breaking, which is clearly visible (Fig. 4a), Specimen 2 shows that the broken parts are not homogeneous in nature and all the fibers are not broken in the same way (Fig. 4b), and in Specimen 3, we observe a similarity with the previous observation, and the layers tend to chip out rather than fiber failure (Fig. 4c).

3 Conclusion

The result and observation on the present study on 3D printing showed that layer thickness plays a prominent role in improving the tensile strength of the product. From this study, this can be concluded that the lesser layer thickness is always good for fusing of the consecutive layered material more homogeneously, but a very big difference in layer thickness may result in early failure.

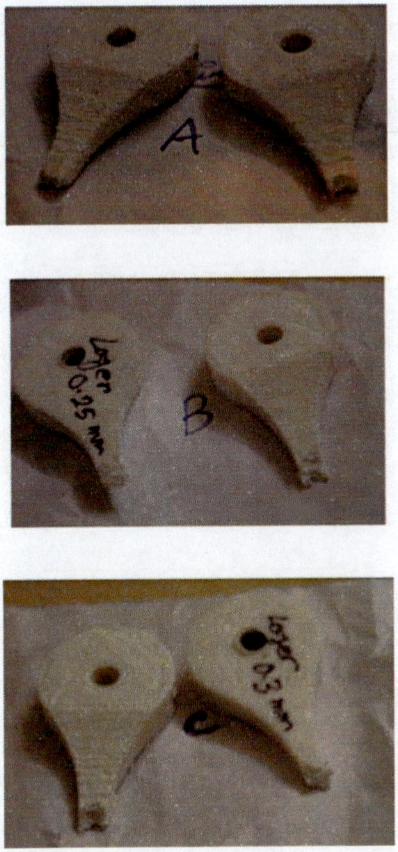

(a) Specimen A

(b) Specimen B

(c) Specimen C

Fig. 4 Pattern of breaking during the load determination for specimen A (**a**), specimen B (**b**) and specimen C (**c**)

References

1. Ayrilmis N (2018) Effect of layer thickness on surface properties of 3D printed materials produced from wood flour/PLA filament. Polym Test 71(2018):163–166
2. Haoa W et al (2018) Preparation and characterization of 3D printed continuous carbon fiber reinforced thermosetting composites. Polym Test 65(2018):29–34
3. Khabia S, Jain KK (2020) Influence of change in layer thickness on mechanical properties of components 3D printed on Zortrax M 200 FDM printer with Z-ABS filament material and Accucraft i250+ FDM printer with low cost ABS filament material. Mater Today: Proc https://doi.org/10.1016/j.matpr.2020.02.268
4. Rao VDP et al (2019) Effect of fused deposition modelling (FDM) process parameters on tensile strength of carbon fibre PLA. Mater Today: Proc 18(2019):2012–2018
5. Vălean C et al (2020) Effect of manufacturing parameters on tensile properties of FDM printed specimens. Procedia Struct Integrity 26(2020):313–320
6. Wagari Gebisa A, Lemu H (2019) Influence of 3D printing FDM process parameters on the tensile property of ULTEM 9085. Procedia Manuf 00:331–338

7. Yao T et al (2020) (2020) Tensile failure strength and separation angle of FDM 3D printing PLA material: experimental and theoretical analyses. Compos B 188:107894
8. Zhao Y, Chen Y et al (2019) (2019) Novel mechanical models of tensile strength and elastic property of FDM AM PLA materials: experimental and theoretical analyses. Mater Des 181:108089

Selection of Total Quality Management Implementation Success Factors for Vocational Education Institutes Using Analytic Hierarchy Process

Arish Ibrahim

Abstract Vocational education institutes provide skill-based courses that prepare students for the career in wide range of occupational fields, services and livelihood. The job market and community expect quality training and competent students from these institutes meet the demand of skilled labor. Adaption of total quality management principles that focused on continuous improvement will support the quality assurance at all levels, and the implementation depends upon various success factors. This paper focused on identifying the success factors relevant to vocational education, and these factors are ranked according to their importance. Analytic hierarchy process is applied for the ranking of ten identified success factors, and consistency check confirmed the evaluation results. The results demonstrated that senior management commitment is the most important success factor and adaptability to disruptions due to pandemics is also identified as another major success factor for the total quality management implementation.

Keywords Total quality management · Vocational education · Analytical hierarchy process · Success factors

1 Introduction

Total quality management approach involves every member in the organization for achieving long-term success through continuous involvement and customer satisfaction. It involves various tools and techniques that applied to put quality improvement in action by promoting culture of quality and quality assurance. Through this approach, the organization can cope with the changing needs and demands of customer and design products and services that meet the expectations. The implementation always requires the mindset of change for quality and well-designed planning. Top management will decide the right method of implementation and the degree of success depends upon the management commitment and employee involvement.

A. Ibrahim (✉)
BITS Pilani Dubai Campus, Dubai, UAE
e-mail: arishibrahim@outlook.com; p20200003@dubai.bits-pilani.ac.in

A. Maurya et al. (eds.), *Recent Trends in Mechanical Engineering*, Lecture Notes in Mechanical Engineering, https://doi.org/10.1007/978-981-19-7709-1_60

Implementation starts with management commitment and educates them on basic principles and methods of total quality management. An assessment of current status of organization and level of customer satisfaction is required to develop a master plan. A steering committee can support the master plan and clarifies the vision, mission and values. Then, critical success factors are identified along with the customer groups to focus and continuous monitoring of these factors for developing an improvement plan. Establishment of quality improvement teams and quality training program backed up the entire process. The changes will be integrated and standardized for improvement in daily process and continuous evaluation of the progress against plan for any adjustment required. Total quality management is widely applied in industrial setting at the beginning and gradually got attention in the service industry also. The movement for total quality management in education sector is of more recent origin as it is considered as service sector. In education sector, learners and parents are the primary customers, while employers and government constitute the secondary external customers. The vital team of teachers and supporting staff falls into the internal customers' category. Meeting all the needs of these customers beyond their expectations is the main driving force for the success of total quality management implementation in education sector. The educational institutes are accountable for demonstrating high standards and commitment to community in all of their activities; especially, vocational education sector provides skilled labor and their quality should be high as required by the industry standards. Total quality management supports this accountability and responsibility by continuous improvement and quality assurance even from the basic activity of student's admission till the end stage of work placement. Teaching and learning processes are designed to meet the quality standards as per the industry expectations and infrastructure supports this effort. To achieve the objective of quality, the implementation of total quality management in organization needs to be carried out carefully and well planned. There are so many factors that make this implementation a success one or not, and those factors are known as critical success factors. These critical success factors need to address as per their importance and make sure that actions are taken to accomplish those factors. The success factors are those managed areas that must be given special attention to achieve the objective of high performance. Success factors can be internal indicators such as teacher's professional development and can be external indicator such as customer satisfaction. This study focused on identification of success factors for the implementation of total quality management in vocational education institute. These identified factors will be ranked and compared based on the importance using analytic hierarchy process.

2 Literature Review

Due to the growing need of high-quality skilled worker, technical education institutes are also started focusing on quality management and continuous improvement practices. A recent study by Hu et al. [3] about the quality management of vocational skill identification evidenced that the customer satisfaction level of vocational skill

Table 1 Total quality management success factors with their references

S. No	References	Success factors
1	Salleh et al. [10]	Management commitment Total customer satisfaction Employee involvement Continuous improvement Training and development Communication Teamwork
2	Sfakianaki (2018)	Leadership Student focus Continuous improvement Process control and involvement Education and training Measurement and evaluation Change management
3	Janette et al. [4]	People empowerment Continuous improvement Leadership commitment Stakeholder satisfaction
4	Anil et al. (2012)	Infrastructure Training and placement Administration Development

development institute increased after the implementation of the quality management system. Various literature works are also published on the implementation of the total quality management in education sector among different countries. In a higher education context, Kwarteng [6] identified service quality dimensions that can be measured from external customer's perspective. He recommended the importance of management responsibility to listen to the voice of the external customers regularly, and management should address the service gaps identified by the external customers. In Indonesia, the government is focusing on enhancing the quality of vocational schools through an approach of transforming vocational schools into vocational school-based enterprises. Khurniawan et al. [5] from Indonesia analyzed the factors in this approach along with total quality management and identified that total quality management has a significant positive effect on increasing school governance. A study conducted among Saudi Arabia university students for examining the influence of service quality on student's satisfaction by Sohail et al. [11] revealed that four of the five dimensions of service quality, namely, tangibility, reliability, responsiveness and assurance had a significant effect on student's satisfaction (Table 1).

The fifth factor empathy was not affecting or contributing to student's satisfaction. Hota et al. [2] examined the factors that affect the quality of resources and efficiency of the institutions in public and private management institutions in Odisha, India, and the results have shown that a correspondence number is high and positive among internal and external stakeholder's relationship. Furthermore, the support structure is an element of TQM, which directly affects the quality of the resources of the educational institutions. To support this study, an extended literature review was also carried out to gather the various success factors that affect the successful implementation of total quality management. Most of the studies or literature is found in higher education sectors, and still, there is a research gap existing in the field of vocational education. Bouranta et al. [1] conducted a systematic literature review on the studies on quality management in primary and secondary schools and presented the finding the highest number of studies on quality management in primary and secondary education. As there are no adequate studies conducted in vocational education sector, so the previous works done in the field of higher education is reviewed for listing the success factors. One of the selected study by Salleh [10] identified management commitment, total customer satisfaction, employee involvement, continuous improvement, training and development, communication and team work as total quality management implementation success factors.

We can observe that some of the factors are repeated in most of the works such as management commitment, employee involvement, customer satisfaction, etc., and those factors are varied based on the institution or method applied. In this study, the common and basic factors are considered along with the factors relevant to vocational education sector as per the recent trends and development.

3 Analytic Hierarchy Process for Selecting Success Factors

Analytic hierarchy process was developed by Saaty [8] to solve the complex decision-making problems and useful when the decision-maker is unable to generate a utility function. Analytic hierarchy process makes assessments, prioritization and selection among available options and theory of measurement through pairwise comparisons and relies on the judgment of experts to derive priority scales. Laosirihongthong et al. [7] in their study applied analytic hierarchy process that allows them to define the success factors in a hierarchical structure. Analytic hierarchy process is selected for this study over other modeling and decision-making tools, the main reason is that analytic hierarchy process records both subjective and objective evaluations along with the capability of checking consistency of the decision-maker evaluations. Analytic hierarchy process is also having less complexity compared to other multi-criteria decision-making methods, and this approach is used for similar works related to the identification of total quality management implementation success factors.

The analytic hierarchy process starts with the problem identification and construction of hierarchy structure to identify the set of criteria and alternatives that influences the objective. The relative importance of success factors is then determined by generating pairwise comparison matrix using Saaty scale for each success factor to indicate its dominance to other factors (Table 2).

The next step is to develop a normalized comparison matrix by dividing each entry in column of pairwise comparison matrix by sum of all the entries in the same column. Then, priority weights for each success factor are calculated by adding

Table 2 Saaty scale

Importance	Priority
1	Equal importance
3	Moderate importance
5	Strong importance
7	Very strong importance
9	Extreme important
2, 4, 6, 8	Intermediate values between two adjacent judgments

and averaging each row of the normalized values. Finally, for the confirmation of robustness of result, a consistency check will be conducted. An index of consistency is calculated for checking numerical and transitive consistency. Formula CI = (Eigen vector_{max}–n)/(n–1), where n is the size of the matrix (Table 3).

Step 1—Defining the problem and hierarchy structure:

The most important success factors are identified using literature review and based on studies on recent developments in vocational education field (Table 4).

For this, initially relative weight is calculated by multiplying pairwise comparison matrix and priority weights. Then, Eigen vector is calculated by dividing relative weight to the corresponding priority weight and noted the maximum Eigen vector to get the CI value. After calculating consistency index, based on the appropriate random consistency index (RCI) data by Saaty [9], calculate the consistency ratio using formula CR = CI/RCI. Average random consistency index (R.I.) is derived from a sample of size 500, of a randomly generated reciprocal matrix using the scale

Table 3 Random consistency index

n	1	2	3	4	5	6	7	8	9	10
RI	0	0	0.58	0.9	1.12	1.24	1.32	1.41	1.45	1.49

Table 4 Success factors

S. No	Success factor	Code
1	Senior management Commitment	SMC
2	Employee involvement and teamwork	EI
3	Students performance and work placement	SPWP
4	Industry partnership and Entrepreneurship	IPE
5	Teaching and learning	TL
6	Competency tests and assessment	CTA
7	Training and development	TD
8	Infrastructure and technology	IAT
9	Curriculum and standards	CS
10	Adaptability to disruptions	ATD

1/9, 1/8,..., 1, 0.8, 9 to see if it is about 0.10 or less. If consistency ratio is less than 10%, we can confirm that our judgements are valid, and the evaluations are acceptable. According to Saaty [9], the priority of consistency to obtain a coherent explanation of a set of facts must differ by an order of magnitude from the priority of inconsistency which is an error in the measurement of consistency. Thus, on a scale from 0 to 1, inconsistency should not exceed 0.10 by very much.

The last factor adaptability to disruption is added as per the recent relevance in continuing education through online or blended mode during disruptions such as outbreak of COVID-19.

Step 2—Pairwise comparison matrix generation (Table 5):

Step 3—Normalization of criteria pairwise comparison matrix (Table 6):

Table 5 Pairwise comparison matrix

	SMC	EI	TD	CS	TL	CTA	SPWP	IAT	IPE	ATD
SMC	1	3	2	1	1	1	1	2	2	2
EI	0.33	1	2	1	1	1	2	2	2	1
TD	0.50	0.50	1	2	3	1	1	2	1	2
CS	1	1	0.50	1	2	2	2	2	1	2
TL	1	1	0.33	0.50	1	2	2	2	2	2
CTA	1	1	1	0.50	0.50	1	1	2	2	2
SPWP	1	0.50	1	0.50	0.50	1	1	2	2	1
IAT	0.50	0.50	0.50	0.50	0.50	0.50	0.50	1	2	2
IPE	0.50	0.50	1	1	0.50	0.50	0.50	0.50	1	1
ATD	0.50	1	0.50	0.50	0.50	0.50	1	0.50	1	1
SUM	7.33	10	9.83	8.50	10.50	10.50	12	16	16	16

Table 6 Normalized matrix

	SMC	EI	TD	CS	TL	CTA	SPWP	IAT	IPE	ATD	Sum	Priority weights
SMC	0.14	0.30	0.20	0.12	0.10	0.10	0.08	0.13	0.13	0.13	1.41	0.14
EI	0.05	0.10	0.20	0.12	0.10	0.10	0.17	0.13	0.13	0.06	1.14	0.11
TD	0.07	0.05	0.10	0.24	0.29	0.10	0.08	0.13	0.06	0.13	2.54	0.12
CS	0.14	0.10	0.05	0.12	0.19	0.19	0.17	0.13	0.06	0.13	1.26	0.13
TL	0.14	0.10	0.03	0.06	0.10	0.19	0.17	0.13	0.13	0.13	1.16	0.12
CTA	0.14	0.10	0.10	0.06	0.05	0.10	0.08	0.13	0.13	0.13	2.42	0.10
SPWP	0.14	0.05	0.10	0.06	0.05	0.10	0.08	0.13	0.13	0.06	0.89	0.09
IAT	0.07	0.05	0.05	0.06	0.05	0.05	0.04	0.06	0.13	0.13	0.68	0.07
IPE	0.07	0.05	0.10	0.12	0.05	0.05	0.04	0.03	0.06	0.06	1.56	0.06
ATD	0.07	0.10	0.05	0.06	0.05	0.05	0.08	0.03	0.06	0.06	0.61	0.06
SUM	1	1	1	1	1	1	1	1	1	1		

Table 7 Relative weight calculation

1	3	2	1	1	1	1	2	2	2		0.14		1.54
0.33	1	2	1	1	1	2	2	2	1		0.11		1.25
0.50	0.50	1	2	3	1	1	2	1	2		0.12		1.36
1	1	0.50	1	2	2	2	2	1	2		0.13		1.37
1	1	0.33	0.50	1	2	2	2	2	2	X	0.12	=	1.24
1	1	1	0.50	0.50	1	1	2	2	2		0.10		1.07
1	0.50	1	0.50	0.50	1	1	2	2	1		0.09		0.95
0.50	0.50	0.50	0.50	0.50	0.50	0.50	1	2	2		0.07		0.72
0.50	0.50	1	1	0.50	0.50	0.50	0.50	1	1		0.06		0.69
0.50	1	0.50	0.50	0.50	0.50	1	0.50	1	1		0.06		0.66

Table 8 Eigen vector calculation

Success factors	Relative weight	Priority weights	Eigen vector
SMC	1.54	0.14	10.97
EI	1.25	0.11	10.99
TD	1.36	0.12	11.04
CS	1.37	0.13	10.84
TL	1.24	0.12	10.68
CTA	1.07	0.10	10.73
SPWP	0.95	0.09	10.76
IAT	0.72	0.07	10.64
IPE	0.69	0.06	10.89
ATD	0.66	0.06	10.83

Step 4—Consistency check (Tables 7 and 8):

CI = (Eigen $vector_{max}$-n)/(n-1) = (11.04–10)/(10–1) = 0.12.

From random consistency index table, RCI = 1.49.

Consistency ratio CR = CI/RCI = 0.12/1.49 = 0.08.

The evaluations in the study are consistent and acceptable as the consistency ratio is less than 0.10.

Step 5—Ranking of success factors.

Based on the priority weights calculated the success factors are ranked in Table 9.

Table 9 Ranking of success factors

Rank	Success factor
1	Senior management Commitment
2	Curriculum and standards
3	Training and development
4	Teaching and learning
5	Employee involvement and teamwork
6	Competency tests and assessment
7	Students performance and work placement
8	Infrastructure and technology
9	Industry partnership and entrepreneurship
10	Adaptability to disruptions

4 Results

From the analytic hierarchy process, the most important success factor for total quality management implementation identified is the commitment of institute management toward the quality objective. Development of competent learners who met all the industry standards needs a well-designed and updated curriculum, and this element made it as the second most important success factor. Teaching and learning process is a fundamental success factor which depends on the qualification and professional development of teachers. Teamwork is crucial as interdepartmental collaboration facilitates effective training and learning. All the competency-based tests both summative and formative assessments help to track the progress of students, and the achievement of competency will open the door of work placement for the successful students. Well-equipped workshops, internet and digital tools enhance the teaching and learning process, and industry partnership supports the career-based training for the students. The last identified factor is adaptability of change which is backed up by all the other factors that enable the institute to migrate suddenly to online learning mode during the period of pandemic.

5 Conclusion

Total quality management enables continuous improvement, quality service and customer satisfaction in industrial and service sectors. Education sectors already started implementing total quality management strategy for delivering high-quality education to students. A wide study is carried about the implementation in higher education, but there is a research gap exists in the case of vocational education sector. As vocational educational institutes develop skilled labor to the economy, there is no compromise in quality training and development of graduated students.

In this study, a detailed evaluation of total quality management implementation success factors in vocational education institutes is carried using analytical hierarchy process. This study successfully identified and ranked the important success factors and pinpointed the main ten success factors. The most important success factor is management commitment followed by curriculum, standards, training and development. We can find that academic, non-academic and facilities factors are listed in the success factors that highlight the importance of collaboration of all stakeholders which is very important for the successful implementation of total quality management. This study implicates that the identification and ranking based on importance will help the institutes to focus on the critical factors and make the effort of achieving quality more effective. As a future scope of work, more advanced analysis and decision-making tools also can be involved for the success factor ranking and identification. A framework can be generated based on the success factors and needs to be validated with an implementation case study.

References

1. Bouranta N, Psomas E, Antony J (2020) Findings of quality management studies in primary and secondary education: a systematic literature review. TQM J 33(3):729–769
2. Hota P, Nayak B, Sarangi P (2020) Integration of total quality management principles to enhance quality education in management institutions of Odisha. Mater Today: Proc
3. Hu P, He Y, Yu L (2021) Quality management of vocational skill identification for higher vocational education. 517(SSCHD 2020):490–493
4. Janette R, Madonna V, Nunilon A (2018) TQM paradigm for higher education in the Philippines. Qual Assur Educ 26(1):101–114
5. Khurniawan AW, Sailah I, Muljono P, Indriyanto B, Maarif MS (2021) Strategy for improving the effectiveness of management vocational school-based enterprise in Indonesia. Int J Educ Practice 9(1):37–48
6. Kwarteng AJ (2021) An assessment of outcome criteria associated with the implementation of TQM in a higher education institution in Ghana. Cogent Educ 8(1)
7. Laosirihongthong T, Rahman SU, Saykhun K (2006) Critical success factors of six sigma implementation: an analytic hierarchy process-based study. Int J Innov Technol Manag 3(3):303–319
8. Saaty TL (1977) A scaling method for priorities in hierarchical structures. J Math Psychol 15(3):234–281
9. Saaty RW (1987) The analytic hierarchy process-what it is and how it is used. Math Model 9(3–5):161–176
10. Salleh NM, Zakuan N, Ariff MSM, Bahari AZ, Chin TA, Sulaiman Z, Yatim SM, Awang SR, Saman MZM (2018) Critical success factors of total quality management implementation in higher education institution: UTM case study. AIP Conf Proc 2044(1):20007
11. Sohail MS (2021) Student's perceptions of service quality in Saudi universities: the SERVPERF model. Learn Teach High Educ: Gulf Perspect 17(1):54–66

Determination of Pressure and Force Coefficient of Bridges Due to Wind: Practical Investigation and Simulation Using LS-DYNA

Indrani Chattopadhyay, Poulomee Roy, and Somnath Karmakar

Abstract The structural failure in the bridge decks often causes aerodynamic instability of the structures or discomfort to the travellers. This structural deficiency may also lead to a catastrophic failure or sudden collapse due to aeroelastic instability of the bridge deck. To understand the fluid–structure mechanism, computation fluid dynamics (CFD) is the best feasible alternative compared to wind tunnel testing which is expensive and time-consuming and required full-scale physical experiment. In this paper, one High Level Bridge in India of 46 m span of prestressed concrete single box girder bridge deck at mid span, located near Durgapur, West Bengal, has been studied with different geometric features and varying the distance between two single bridges' decks in parallel case. Further, it has been analysed using incompressible computational fluid dynamics (ICFD) solver using LS-DYNA (R-7.1–2.16) version to evaluate various fluid parameters of the bridge deck under the influence of wind such as pressure contour region and fluid velocity region surrounding the bridge decks in single and parallel bridge decks with various gaps, displacements in X direction and Y direction, drag forces and lift forces. From the simulation, it is found that variation of pressure continues overall on the structure and maximum pressure accumulates at the sharp edge corner of the geometry.

Keywords Force coefficient · Pressure · LS-DYNA · Drag force · Parallel bridge deck · Lift force

1 Introduction

An infinite number of aerodynamic instabilities like flutter, buffeting, galloping and various modes of vibrations are shown in slender structural element when come in contact with a fluid flow due to the fluid–structure interaction mechanism. The aim of the study is to explore the different pressure and force coefficients of bridge decks such as the pressure and velocity profile, X and Y displacement, lift force and

---

I. Chattopadhyay (✉) · P. Roy · S. Karmakar
Department of Civil Engineering, NIT, Durgapur, India
e-mail: indranichattopadhyay111@gmail.com

A. Maurya et al. (eds.), *Recent Trends in Mechanical Engineering*, Lecture Notes in Mechanical Engineering, https://doi.org/10.1007/978-981-19-7709-1_61

drag force of bridge decks under wind induced vibration. The High Level Bridge of 46 m span near Durgapur of the PSC single box girder bridge deck is analysed with different geometric features and provisions of bridge decks (in single and parallel positions) using the LS-DYNA software in ICFD solver. When the wind is getting obstructed by a bluff body, i.e. a structure with an arbitrary shape, it sheds some amount of vortices in a particular pattern (periodic or non-periodic), particularly in a low wind speed. As a result, these shedding vortices impart some frequency in the adjacent zone of structure. After a certain time, this vortex shedding frequency (f_s) approaches to the natural frequency of the structure (f_n) at a certain wind speed and produces large amplitudes of wind-induced oscillation; this phenomena is defined as "locked-in" and the region where it takes place as "lock-in" region [1]. It is found that the slender structures with a large aspect ratio, i.e. a huge difference between the length and width, mostly experience the aerodynamic instability accompanying with multi-mode vibration [2]. If the bridge deck consists of multiple box cells, the number of "lock-in" regions are more than 1 which indicates significant nonlinearity in the amplitude of the wind-induced vibration [3]. A study has enlightened the abrupt change in pressure contour region alternatively during the wind-induced mechanisms in long-span bridges when the vortices were developed in two sides of the bridge deck sections [4]. For analysing the distributed aerodynamic and aeroelastic pressure surrounding a twin-box bridge decks, the proper orthogonal decomposition (POD) method based on pressure modes' variation is found more accurate. The cable-stayed twin-box deck bridge of Stonecutters was selected to examine the effect of Typhoon Hato using POD method [5]. The geometry of bridge decks' shape also plays important role to mitigate or exaggerate the aerodynamic instability. Providing triangular fairings in deck design helps to mitigate the large amplitude. To avoid aeroinstability, the sharp edge of fairings separates the boundary layer of the flowing fluid [6]. Parallel bridge deck concepts are very common in developed cities to control the traffic density. The distance provided between two bridge decks can severely affect the bridge structure due to aerodynamic instability. Two neighbouring Haihe bridges with semi- and full-closed bridge decks are examined using CFD techniques to measure the influence of aerodynamic interference between the two bridge decks, and the results show that the influence of wind-induced vibration varies with the shapes of windward and leeward decks [7]. Investigation explores that vortex shedding occurs in between the parallel bridge decks depending upon the gaps, and thus, it can fluctuate different aerodynamic properties [8]. It is observed that drag force is proportional to the cross-sectional area projected in the direction of fluid flow [9]. Drag is influenced by geometric shape, viscosity, compressibility texture of the surface area and lift which causes induced drag [10]. Rain–wind-induced vibration may also be a threat to the bridge engineers. The presence of rain and wind together can produce large amplitude resulting in aerodynamic damping [11]. A new stochastic approach based on spring-suspended sectional model (SSSM) is proposed to measure the aerodynamic forces and fluctuating wind speeds around the deck cross-section [12]. An experiment is also conducted over a freely vibrating cylinder exposed to currents and placed in parallel directions to the cylinder axis to observe the proximity of boundary effects [13]. Further studies also enlighten the mechanism of the structure and linear stability of two-dimensional hollow vortex equilibria. It is found from the study that one of the shape distortion modes of a circular hollow vortex has the same frequency as one of the modes is corresponding to the displacement of the vortex centroid [14].

In this paper firstly, the original bridge deck structure at mid span of the High Level Bridge near Durgapur has been studied through simulation without any further modification in geometry. Secondly, the internal box from the deck is removed and analysed. Thirdly, the crash barrier from the bridge deck cross-section is removed and wind effect on the model is observed. Lastly, the corners of the deck are removed with a judiciously taken curvature as fillet to check the intensity of the wind induction and its effect on different dynamic characteristics of the bridge deck. Apart from the single bridge deck modelling, parallel provision of bridge deck with a distance of 2 m and 4.5 m (as per cl. 2.5.1_table 2.2 of IRC: SP: 87–2013) between the two bridge decks is also modelled as depicted in Fig. 1.

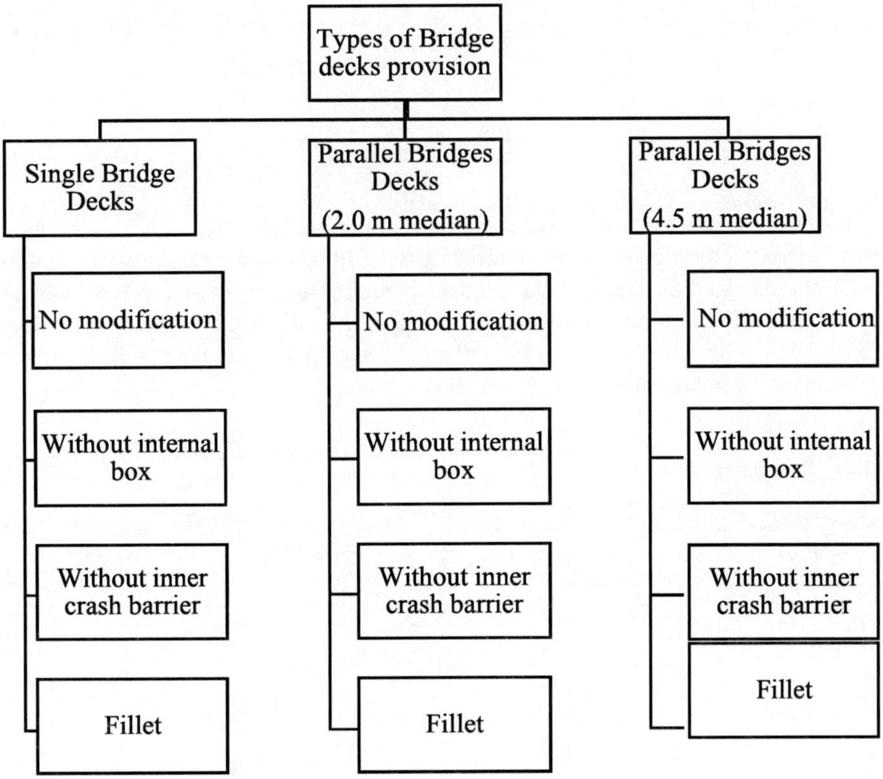

Fig. 1 Types of bridge decks using for simulation in LS-DYNA

2 Methodology

Computational methods in unsteady wind loading can be systematically applied to measure the aerodynamics threat to the structure. Basically, computational fluid dynamics can be defined as the blending of numerical analysis and algorithm to analyse fluid flow problems. In this paper, the LS-DYNA (R-7.1–2.16) version is used to analyse the force and pressure coefficient of bridges using "lsdyna_smp_d_R901_winx64_ifort131" incompressible fluid (ICFD) solver, an implicit solver based on the Navier–Stokes equation. Incompressible fluid refers to the fluid flow when the particles will maintain constant density and the Mach number is lower than 0.3. In simulation process, air has been taken as incompressible fluid considering that the flow velocity will be less than 100 m/s, i.e. low subsonic flow. Traditionally, the Navier–Stokes equation represents in Eulerian conventional form combining the continuity equation with the conservation of momentum and mass for incompressible Newtonian fluids as follows:

$$\rho\left(\frac{du_i}{dt}+u_j\frac{\partial u_i}{\partial x_j}\right)=\frac{\partial\sigma_{i,j}}{\partial x_j}+\rho f_i \text{ in } \Omega$$

$$\frac{\partial u_i}{\partial x_i}=0 \text{ in } \Omega \tag{1}$$

In the application of CFD tool, assigning proper boundary condition to the fluid zones is very important to avoid incorrect results. In this study, the boundary conditions are assigned to simulate the model at inlet boundary with varying velocity boundary and outlet boundary with specific pressure condition. Free-slip boundary defines where fluid has zero velocity and non-slip boundary, i.e. the obstruction wall or element, as represented in the block diagram of Fig. 2.

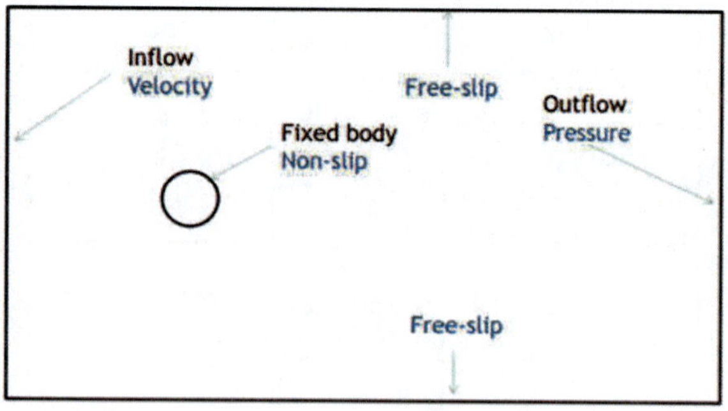

Fig. 2 Boundary conditions applied in modelling

3 Validation

In this paper, the model which is used for the validation is taken from the Journal of Wind Engineering and Industrial Aerodynamics, "Numerical analysis on the difference of drag force coefficients of bridge deck sections between the global force and pressure distribution methods" by Yan Han, Hao Chen, C.S.Cai, GujioXu, Lian Shen and Peng Hu. This paper shows the generated pressure contour using ICFD tools in ANSYS CFX 14.5 with three different geometrical configurations of the cross-section of Sutong Bridge, China.

The three models are simulated using free-slip and non-slip boundary conditions. To ensure the development of turbulent wake from the inlet, the distance from the inlet to the leading edge of the section and the distance from the outlet to the rear edge of the section are set as 6B and 12B, respectively, with the aspect ratio 10 of the bridge cross-section. The top wall and bottom wall are set as free-slip wall boundary conditions. The no-slip wall boundary conditions are applied on the surfaces of the sections. The wind velocity vector and turbulence intensity can be specified at the inlet. Turbulence intensities are set as default, and the value of intensity is 5%. The pressure outlet is set as the outlet boundary condition, and the relative pressure is zero. The symmetry boundary conditions are employed on the two ends of the models.

In LS-DYNA, the model is incorporated with the same data and boundary conditions given in reference model discussed above. In Fig. 3, one comparative study regarding pressure profile region between the ANSYS CFX 14.5 and LS-DYNA (R-7.1–2.16) version is provided. It shows the pressure contour region that is generated in the surrounding surface and corner; below, the deck portions are same as the pressure profile generated in LS-DYNA ICFD tools. In both cases, the pressure contour region is approximately same.

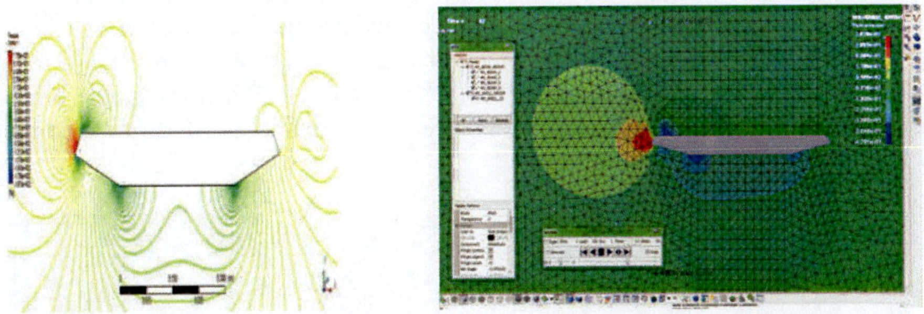

Fig. 3 Pressure contour of the Sutong bridge cross-section in ANSYS CFX 14.5 and LSDYNA

4 Results and Discussions

A 2D model has been prepared, and simulation is carried out with help of collected data from actual construction site. This model has been simulated on the basis of following parameters such as pressure contour region, fluid velocity region, X displacement, Y displacement, lift force and drag force.

(a) A variety of pressure contour region is formed on the surface region of the geometry of bridge deck when the wind flows through the bridge deck. In this study, the different pressure contour patterns of bridge deck have been briefly introduced. Figure 4(i) shows that the sharp edge corner holds the accumulation of maximum pressure. When the crash barrier is removed, two negative suction pressure zones are appeared on the bridge deck. Although variation of pressures are prominent in all the different models of the bridge deck with the wind propagations [4].

(b) The fluid velocity surrounding the bridge deck is also played an important role to analyse the wind induced motion. Figure 4(ii) shows that at the leeward side, there is no velocity due to the obstruction of the deck itself. Increasing the depth of the deck, stagnant velocity zone, i.e. zero velocity zone, surrounding the deck can be formed. The curvature provided at the crash barrier enhances the velocity.

(c) While representing the X displacement, the three different cases, i.e. no modification of bridge deck, removing internal box cell of the deck and without crash barrier, are merged together in a single graph to show the fluctuations of the displacements due to wind effect as depicted in Fig. 5. The bridge deck with judiciously taken fillet at the corners produces the highest displacement value.

(d) The result of the numerical simulations is represented in Fig. 6, i.e. variation of Y displacements for the single bridge deck placed in parallel position with 2 m gap between the decks. The study shows that the displacement of bridge deck cross-section increases with the distance between the gaps of the bridge deck.

(e) Lift force can be defined as the aerodynamic force acting as a resistance force perpendicular to the oncoming fluid flow direction. When the single bridge decks are placed parallel with a gap of 4.5 m between the deck shows an abrupt

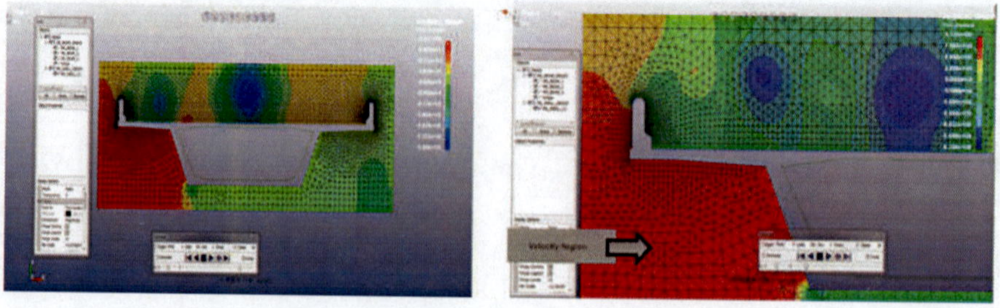

Fig. 4 **i** Pressure contour region of bridge deck cross-section with no modification. **ii** Analysis of fluid velocity region of the bridge deck cross-section with no modification

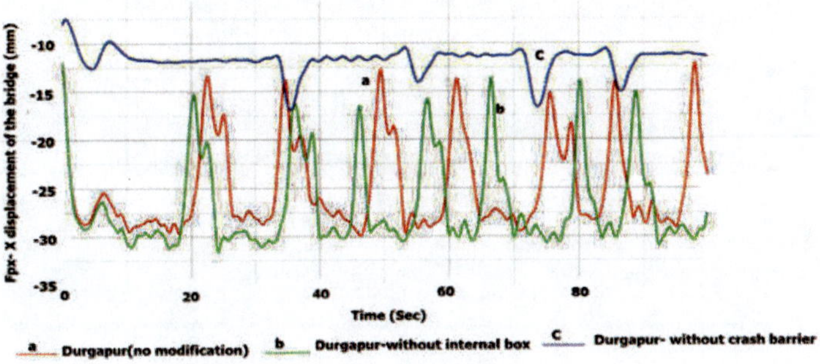

Fig. 5 X displacement of the bridge deck cross-section

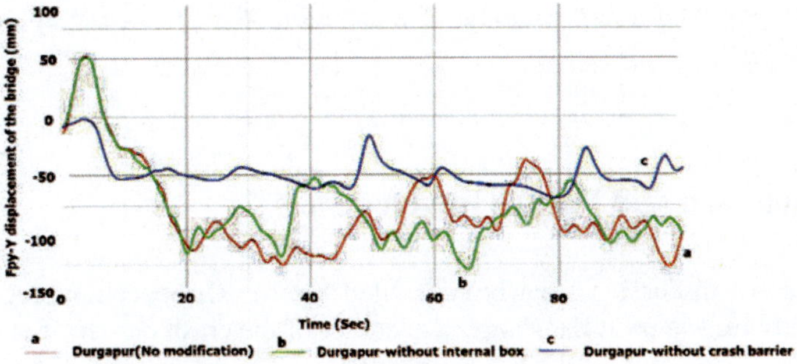

Fig. 6 Y Displacement of bridge deck cross-section of 2 m gap

fluctuation in the graph of lift forces which can cause of a catastrophic failure of the bridge decks [7] (Fig. 7).

(f) Drag force increases with the projected area. The projected area will come in contact with the wind; it will create a huge amount of drag force [9]. In Fig. 8,

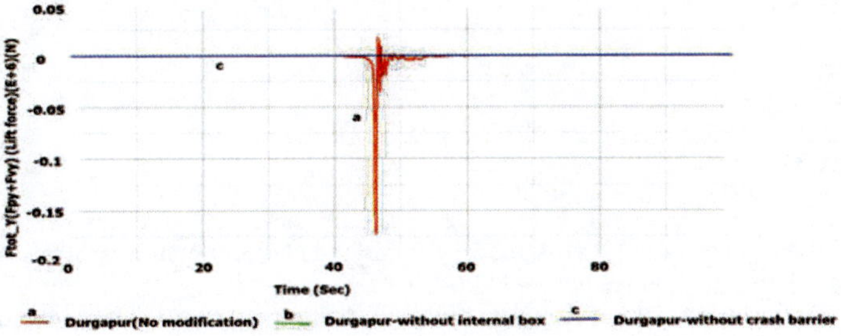

Fig. 7 Lift force analysis of the bridge deck cross-section of 4.5 m gap

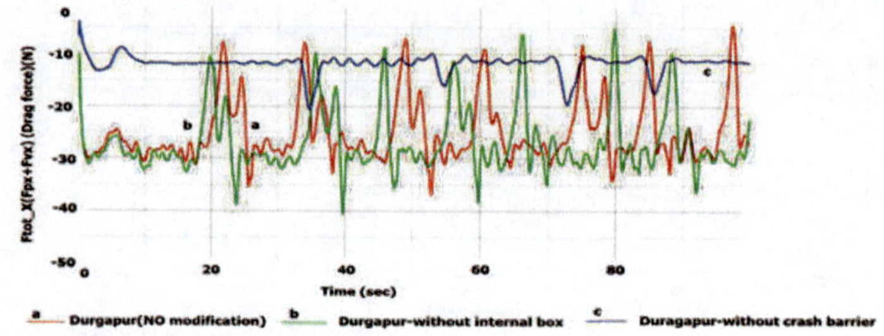

Fig. 8 Drag force analysis of the bridge deck cross-section

the parallel bridge deck with a 4.5 m gap is found not at all suitable due to the development of a large amount of fluctuation of drag force at approximately 47 s.

5 Conclusion and Future Prospective

From the above discussion, it can be concluded that the velocity contour region shows significantly less value at the sharp edge corner of the crash barrier. It is noted that the windward side of the deck has a higher value of pressure zone compared to the leeward side. Fillet shape corner enhances the effect of the wind induction on the bridge decks. So from the above discussion, it can be concluded that the application of this CFD tool may serve as a first-stage design tool before finalised the design of a structural element like bridge deck. For better improvement, 3D modelling features can be incorporated in this ICFD solver to make it more feasible to use in the analysis of wind engineering and aerodynamics stability of structures.

Acknowledgements The authors are grateful to Dr. Debashis Sarkar, Associate Professor and HOD of Mechanical Engineering Department of Asansol Engineering College to extend his cooperation and guidance to write this paper.

References

1. Mashnad M, Jones NP (2014) A model for vortex-induced vibration analysis of long-span bridges. J Wind Eng Ind Aerodyn 134:96–108
2. Wu X, Ge F, Hong Y (2012) A review of recent studies on vortex-induced vibrations of long slender cylinders. J Fluids Struct 28:292–308
3. Wu T, Kareem A (2012) An overview of vortex-induced vibration (VIV) of bridge decks. Front Struct Civ Eng 6(4):335–347
4. Zhou L, Ge Y (2008) Wind tunnel test for vortex-induced vibration of vehicle-bridge system section model. J Braz Soc Mech Sci Eng 30(2):110–117

5. Xu Y-L et al (2019) Buffeting-induced stress analysis of long-span twin-box-beck bridges based on POD pressure modes. J Wind Eng Ind Aerodyn 188:397–409
6. Yoshimura T et al (2001) Prediction of vortex-induced oscillations of a bridge girder with span-wise varying geometry. J Wind Eng Ind Aerodyn 89(14–15):1717–1728
7. Meng X, Zhu L, Guo Z (2011) Aerodynamic interference effects and mitigation measures on vortex-induced vibrations of two adjacent cable-stayed bridges. Front Archit Civil Eng China 5(4):510–517
8. Li H et al (2011) Investigation of vortex-induced vibration of a suspension bridge with two separated steel box girders based on field measurements. Eng Struct 33(6):1894–1907
9. Pavlovskaia E et al (2016) Multi-modes approach to modelling of vortex-induced vibration. Int J Non-Linear Mech 80:40–51
10. Han Y et al (2016) Numerical analysis on the difference of drag force coefficients of bridge deck sections between the global force and pressure distribution methods. J Wind Eng Ind Aerodyn 159:65–79
11. Hua J, Zuo D (2019) Evaluation of aerodynamic damping in full-scale rain-wind-induced stay cable vibration. J Wind Eng Ind Aerodyn 191:215–226
12. Yan L, He XH, Flay RG (2019) Experimental determination of aerodynamic admittance functions of a bridge deck considering oscillation effect. J Wind Eng Ind Aerodyn 190:83–97
13. de Oliveira Barbosa JM, Qu Y, Metrikine AV, Lourens E-M (2017) Vortex-induced vibrations of a freely vibrating cylinder near a plane boundary: experimental investigation and theoretical modelling. J Fluids Struct 69:382–401
14. Llewellyn Smith S, Crowdy D Structure and stability of hollow vortex equilibria. Under consideration for publication in J Fluid Mech

Investigation of the Aerodynamic Phenomenon on a Box-Cell Highway Bridge with Varying Wind Attack Angles Using Incompressible Computational Fluid Dynamics

Poulomee Roy, Indrani Chattopadhyay, and Somnath Karmakar

Abstract The design method of long-span structure has been revolutionized in last decades, and aerodynamic and aeroelastic phenomenon has been increasingly prioritized to avoid cataclysm due to oscillation of bridge because of wind. The structures like long-span bridge show higher sensitivity toward wind turbulence even leading to disastrous collapse and (or) displacement of structure. Therefore, investigation of wind phenomenon plays crucial role in design of slender structures. Objective of the project is to obtain aerodynamic phenomenon, e.g., drag force, lift force and frictional force coefficient of a bridge at Durgapur, West Bengal, India, using incompressible computational fluid dynamics (ICFD) solver of finite element software LS-DYNA. The purpose is to analyze effect of axial vortex in the direction of span in 3D case through CFD simulation instead of experiment (wind tunnel test). CFD is an effective simulation technique with wide range of implementation. In this paper, one High Level Bridge near Durgapur, India, 46 m span, pre-stressed concrete single box girder bridge deck is perused with mid-span geometry in single, parallel bridge deck configuration at various wind attack angles with respect to X-axis (rolling) and Y-axis (yaw), with wind velocity as per clause 6.3, IS875, part III, 2015 in Durgapur. For this purpose, ICFD solver of LS-DYNA (R-4.3.5) is used where various fluid parameters (e.g., density, viscosity, velocity region surrounding the bridge deck) are evaluated. It is observed that the pressure and drag coefficient are reducing at regular interval with reduced wind attack angle. Further study concludes that drag coefficient varies irrespective of direction, i.e., windward or leeward of inclination angle.

Keywords Wind attack angle · Rolling angle · Yaw angle · Drag force coefficient · Lift force coefficient · Parallel bridge deck

P. Roy (✉) · I. Chattopadhyay · S. Karmakar
Department of Civil Engineering, NIT Durgapur, Durgapur, India
e-mail: poulomeekatwa@gmail.com

A. Maurya et al. (eds.), *Recent Trends in Mechanical Engineering*, Lecture Notes in Mechanical Engineering, https://doi.org/10.1007/978-981-19-7709-1_62

1 Introduction

1.1 Overview

Bridges are considered to be one of the most important connecting structures between two distinct geographies having any kind of obstacles in between. The vortex shedding phenomena has been one of the major issues in bridge engineering as it causes catastrophic failure of the slender and flexible bridge components though they are properly designed [1]. Severe aerodynamic response to bridges makes it oscillate causing discomfort to the traffic. The suspension bridges with span of greater length and bridges supported by cables are greatly affected by wind even at moderate velocity, and sometimes, this turns very critical for safety and serviceability of such slender structures [2]. There are still some questions regarding such aerodynamic responses which are still unanswered.

The failure of bridges due to vortex shedding is not a sudden incident as the vibrations caused by wind gradually weaken the structure and lead to fatigue of the bridge structure. After several incident of failure, prediction of the dynamic response of bridges under different wind conditions and finding solutions to this problem has become a popular research topic worldwide [3]. Vortex-induced vibration of bridges is dependent upon multiple parameters, and even, very negligible changes in those parameters can cause severe damage to the structure [4]. One of the major examples of such failure is the failure of Tacoma Narrows Bridge. It collapsed dramatically within just a few months of construction. The objective of the project is to obtain pressure drag force coefficient and plot the trend of varying pressure drag force coefficient in accordance with various wind attack angles of the bridge [5] situated at Durgapur, West Bengal, using ICFD Solver of Finite Element Software LS-DYNA. It is a High Level Bridge (HLB) of 46 m span of PSC single box girder bridge deck at mid-span near Durgapur. The obtained pressure drag force will be used for a comparative study between the 2D and 3D analyses to investigate the effect of axial vortex generated in the direction of span in the 3D case. The velocities will be varied to study the effect of Reynolds number and corresponding drag and lift forces [6].

2 Literature Review

In order to proceed with our current studies, previous research works have been thoroughly reviewed and studied.

Larsen et al. studied the long-span bridges seem to be affected by severe VIV at low wind speed even below 10 m/s [1]. Xu et al. proposed a new method for restricting VIV of long-span bridge deck using tuned mass damper inerter. Even more than 94% of the deck vibration was possible to control using tuned mass damper inerter [2]. Zhang et al. have observed that vortex generators can highly distort the regular span-wise vortices and are accompanied by a span-wise mismatch in the vortex shedding process. Consequently, the regular span-wise vortex shedding is substantially suppressed, resulting in a considerable reduction in vortical strength in the wake [3]. Wang et al. proposed that the active control methods require power inputs, whereas passive VIV control can be achieved by modifying the geometry of the structure without any energy input that makes this method easier to implement [12]. Xing et al. have given a light about the effect of wind fairing angle, aerodynamic coefficient, pressure distribution and velocity profile which are deduced by CFD modeling [5]. The thorough review of these paper has been performed while writing this paper. Apart from that while modeling in CFD, some references have been taken to proceed to a new exploration.

3 Introduction to LS-DYNA Incompressible CFD Solver

3.1 Fluid Mechanics Equation

The present solver utilizes Navier–Stokes equation (incompressible) for simulation purpose. Taking an assumption as Mach number is lower than 0.3, i.e., the fluid is incompressible:

$$\mathrm{M} = \frac{v}{a} < 0.3 \tag{1}$$

where V denotes the velocity of the flow with respect to a fixed object and is similar to the speed of sound in that medium (e.g., air) $v \gg 370km/h$). The differential form of the continuity of flow equation is as follows:

$$\frac{\partial \rho}{\partial t} + \nabla \cdot (\rho \vec{u}) = 0 \tag{2}$$

3.2 Governing Equation

Where

ρ is density of air;
U is wind velocity;
H is bridge deck height;
L is the length (span) of the bridge;
CH (α) is coefficient of global drag force;

FHP and FSS are the pressure drag force and frictional drag force, respectively, that can be derived by integration of the pressure and the shear stresses. The final form of the equation is:

$$\rho\left(\frac{du_i}{dt} + u_j\frac{\partial u_i}{\partial x_j}\right) = \frac{\partial \sigma_{i,j}}{\partial x_j} + \rho f_i \text{ in } \Omega$$

$$\frac{\partial u_i}{\partial x_i} = 0 \text{ in } \Omega \tag{3}$$

4 ICFD Numerical Simulation and Validation

Primarily, Sutong Bridge is subjected to three-dimensional numerical simulation, and the pressure drag coefficient and total drag coefficient are investigated and validated with a value obtained by the wind tunnel test. Furthermore, the effect of Reynolds number is also simulated, and proper quality check is executed. The drag force, pressure and frictional forces can be easily computed through simulation here. The analysis is done for obtaining the variation of the drag force coefficient between the global force and pressure distribution (local). In accordance to body axis coordinate system, F_H denotes total drag force, F_V represents lift force, M indicates pitching moment at unit length of bridge and α signifies the wind attack angle. The global or total drag force, F_H, is computed as the summation of the pressure drag force and the frictional drag force.

$$\mathrm{FH} = 0.5\rho \mathrm{U}^2 \mathrm{C_H}(\alpha)\mathrm{HL} = \mathrm{F_{HP}} + \mathrm{F_{SS}} \tag{4}$$

$$\mathrm{F_{HP}} = \sum_{i=1}^{n} pi \cdot \Delta\mathrm{Li} \cdot \sin(\theta i) \tag{5}$$

$$\mathrm{F_{SS}} \sum_{i=1}^{n} \tau\mathrm{ix} \cdot \Delta\mathrm{Li}) \tag{6}$$

4.1 Geometries, Boundary Condition and MESH

The Sutong Bridge serves as the pathway linking Nantong and Suzhou in Jiangsu province in China separated by Yangtze River which was famed as the longest main span cable-stayed bridge with span of 1088 m. However, for the validation purpose, the length of 1 m and width of 0.585 m are considered as model dimension. The bridge is scaled down to 1:70 in order to reduce the simulation efforts. The aspect ratio of 10 (same as reference bridge deck) and fairing angle of 67° are considered while computation. The distance between inlet and leading edge of the section is considered as 6B, and the distance of outlet and rear of the model is 12B. This values were decided for ensuring the independent inlet condition of the boundary which permits advancement of turbulent wake from inlet. The top and bottom walls were provided at a distance of 10B to restrict blocking ratio to less than 5%. The outlet boundary condition is the pressure outlet for this model (taking an assumption of relative pressure = 0). On both ends of the models, symmetry condition is adopted. AutoMesh is employed to analyze flow separation. Different meshing numbers are taken as per requirement (grid independence).

4.2 *Numerical Model and Simulation Results*

Integration of the pressure drag force distributions on Sutong Bridge cross-section is calculated to scrutinize influence of the wind attack angle on the friction drag force with the (mean) wind speed of 25 m/s, acoustic wave velocity of, i.e., 332 m/s, air density as 1.225 kg/m^3. The trend of coefficient of total and pressure drag forces of the Sutong Bridge section versus the angle of attack of wind is plotted in the graph below. From the following graph of the model, the result and methodology of the solver are validated (Fig. 1).

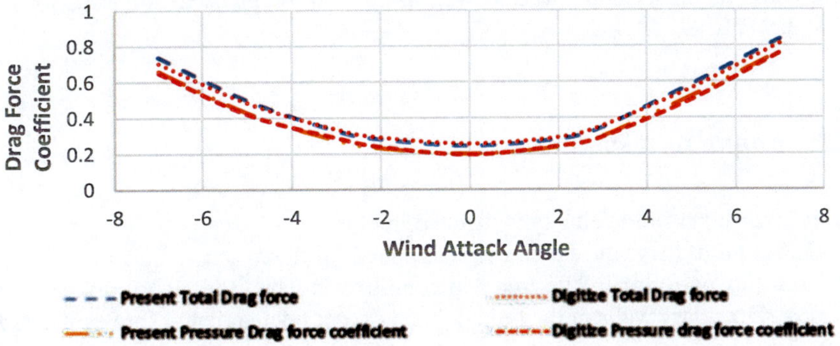

Fig. 1 Aerodynamic coefficients of **a** total drag; **b** pressure drag versus wind angle of attack

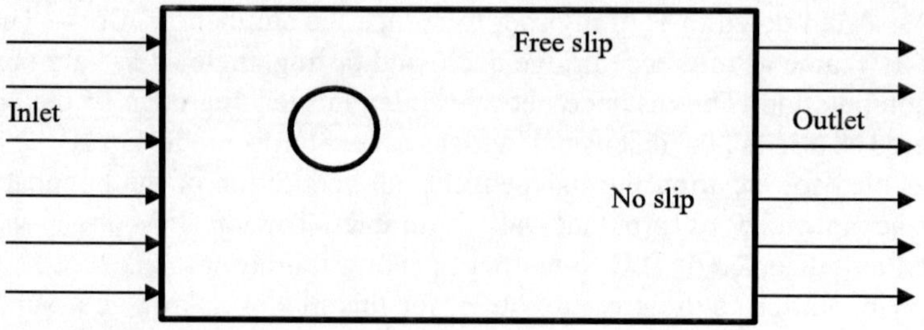

Fig. 2 Boundary condition

5 ICFD Numerical Simulation and Validation

5.1 *Preprocessing*

The initial step is to draw the geometry and assign fluid domain as per requirement. The domain is distributed into smaller sections (mesh generation).

5.2 *Solver*

Once the problem of the physics is identified, then it is the time to assign various values and boundary conditions as well part ids to the physical solver (LS-DYNA (R-4.3.5)) [10]

5.3 *Boundary Condition*

Inlet boundary: fluid velocity is denoted by this.

Outlet boundary: fluid pressure is assigned for this.

Free-slip boundary: The free-slip condition is when the plane surface is frictionless. Boundary walls of the fluid domain except inlet and outlet are explicated as free slip.

Non-slip boundary: This boundary condition is made to the surface of the obstruction which is known as the no-slip condition (Fig. 2).

5.4 *Calculation of Wind Speed*

$$Vz = Vb \times K1 \times K2 \times K3 \times K4 \tag{7}$$

where V_z is the design wind speed at height z in m/s. V_b is the basic wind speed, K1 is the basic probability factor, K2 is the terrain roughness and height factor, K3 is topography factor and K4 is the importance factor for the cyclonic region. As the bridge considered in this this study is situated in Durgapur, West Bengal, the values of above mentioned factors have been found as follows: K1 = 1.0 (from Table 1 of IS 875 part III 2015); K2 = 0.91 (for terrain category 3 and eight up to 10 m, Table 2 of IS 875 part III 2015); K3 = 1.0 (from clause 6.3.3.1 of IS 875 part III 2015); K4 = 1.0 (from clause 6.3.4 of IS 875 part III 2015). The height of the bridge deck from the ground has been found to be 6.825 m.Vb = 47 m/s for Durgapur (from Annex A of IS 875 part III 2015). The modified wind speed after multiplying the basic wind speed with the factors K1, K2, K3, K4 has been reduced to 153.97 kmph or 42.77 m/s. The turbulence intensity variations with height for terrain category 3 (i.e., Durgapur location) have been obtained using the relation given below (clause 6.5 of IS 875 part III 2015).

$$Iz, 3 = Iz, 1 + 3/7(Iz, 4 - Iz, 1) \tag{8}$$

$$Iz, 1 = 0.3507 - 0.0535\, log\ 10(z/z0, 1) \tag{9}$$

$$Iz, 4 = 0.466 - 0.1358\, log\ 10(z/z0, 4) \tag{10}$$

Z0 = equivalent aerodynamic roughness height. The calculations are done considering the above equations for the bridge located near Durgapur having basic wind speed of 47 m/s. The value of turbulence intensity obtained is = 0.26107.

5.5 *Bridge Deck Without Shape Modification*

Two dimensional single deck bridge of span 1 m along with varying wind attack angles (α) as mentioned below are taken into consideration for the simulation. Wind attack angle of $\alpha = 7°, 5°, 3°, 2°$, clockwise and counterclockwise. ∂y is velocity of air, H is height and L is the length of the bridge deck. The mid-span geometry of the bridge has been taken into consideration. Total drag force is the sum of pressure drag force and frictional drag force (Figs. 3 and 4).

$$Cd_T = (Cd)_p + (Cd)_f \tag{11}$$

Table 1 Force coefficient versus attack angle of wind

Wind attack angle	−7°	−5°	−3°	−2°	0°	2°	3°	5°	7°
Pressure drag force (kN)	3192.99	29,947.17	2847.39	2746.89	2599.36	2746.89	2856.64	2948.76	3270.99
Drag force coefficient	0.89055	0.821973	0.750044	0.7441143	0.724967852	0.76611432	0.7967239	0.82241636	0.91228709

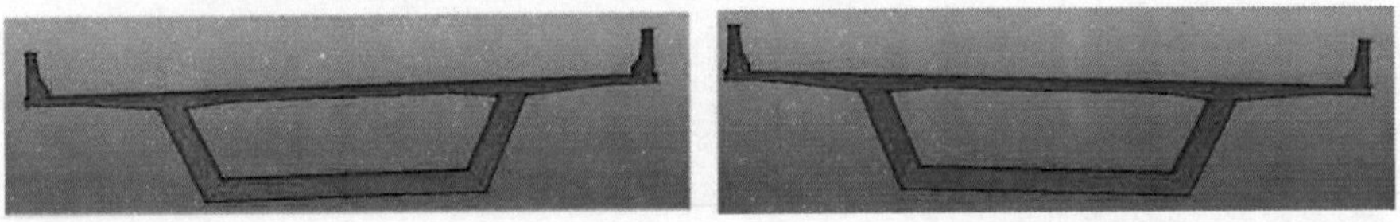

Fig. 3 Bridge deck with no modification subjected to wind attacking angle at 2°

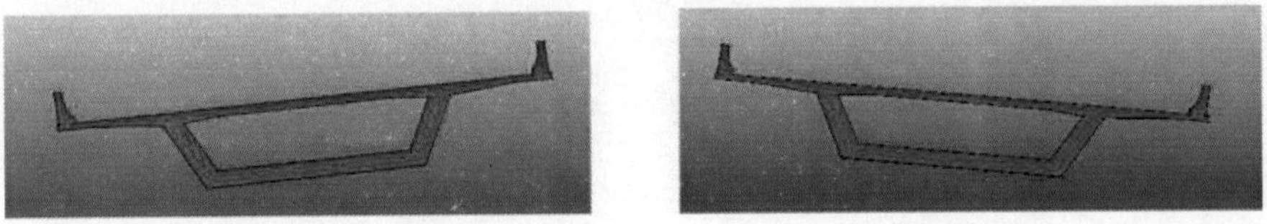

Fig. 4 Bridge deck with no modification subjected to wind attacking **angle** at 7°

6 Result and Discussion

In this chapter, pressure drag force has a significant influence on the total drag force and frictional drag force. When the wind attacking angle is gradually decreasing in magnitude, pressure drag force value shows a declination. Validation of numerical simulation is executed by LS-DYNA, in which materials and the types of elements used are mentioned. Some disparity of values is noted due to variation in mesh size of the elements and some values of particular parameters, whose reference has been taken from other literatures of this genre [10]. The coefficient of total and pressure drag forces is directly proportionate to the absolute value of wind attack angle. With gradual increase or decrease of the wind attack angles, the coefficient of pressure drag increases as shown in the graph, and at 0° angle, this coefficient is minimum (Fig. 5). Frictional drag force contributes minimum to the total drag force when the attacking angle is maximum, and frictional drag force contributes maximum to the total drag force when the wind attack angle is 0° (Fig. 6). Taking density (1.225 kg/m^3), deck height (2.65 m) and velocity (42.77 m/s) constant, we can see the variation of the two variable below.

7 Conclusion and Future Scope

In this chapter, the relation between drag force and wind angle is depicted. The variation of the Reynolds number has a comparatively less significance on the pressure drag force, and it can be ignored. We notice from the graph that the pressure drag coefficient is reducing at a regular interval when the wind attack angle is being reduced in magnitude. The pressure drag coefficient varies irrespective the direction (counterclock or clock) of the angle of inclination. With reduction of angle of inclination of bridge deck by 0.5% (bridge deck also having vertical inclination on

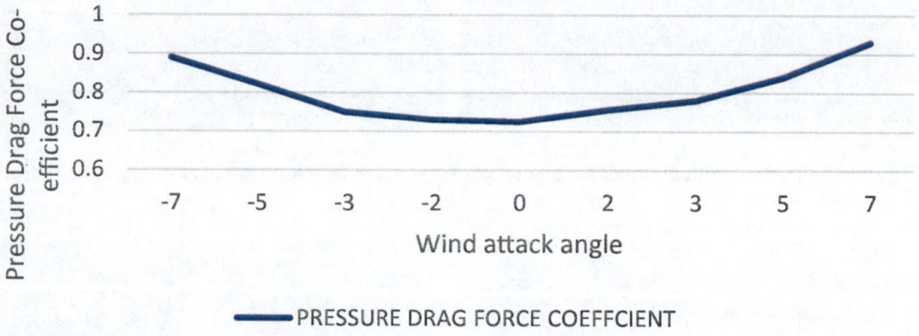

Fig. 5 Pressure drag coefficient versus wind attack angle

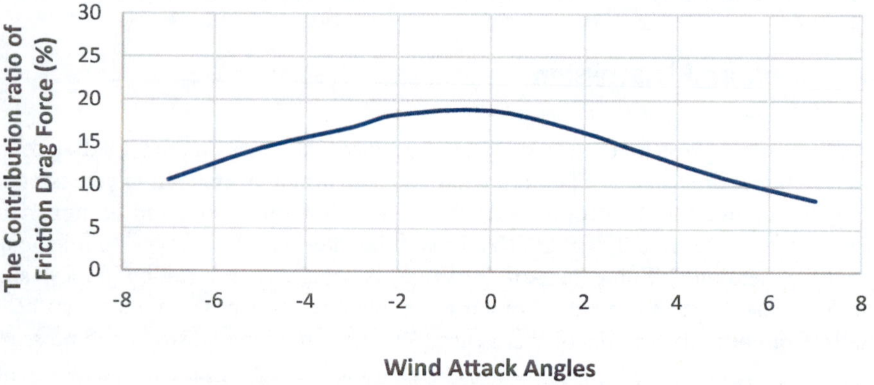

Fig. 6 Contribution of frictional drag force versus wind attack angle

windward side), the value of force coefficient seems to reduce up to 30%. With these data in hand, we can perform a trend analysis model in R-Studio and make a tool which can predict the value of pressure drag coefficient for any angle and velocity all over a demographic region. We can further investigate the contribution of frictional drag force at various wind attack angles.

References

1. Larsen A, Esdahl S, Andersen JE, Vejrum T (2000) Storebaelt suspension bridge– vortex shedding excitation and mitigation by guide vanes. J Fluids Struct 88(2):283–296
2. Zhu Q, Xu YL, Zhu LD, Li H (2018) Vortex-induced vibration analysis of longspan bridges with twin-box decks under non-uniformly distributed turbulent winds. J Wind Eng Ind Aerodyn 172:143–41
3. Zhang Z, Xu F Added mass and damping effects on vibrating bridge decks in still air. J Wind Eng Ind Aerodyn 191:227–238

4. Sarpkaya T (2004) A critical review of the intrinsic nature of vortex-induced vibrations. J Fluids Struct 19:389–447
5. Li Y-L, Chen X-Y, Yu C-J, Togbenou K, Wang B, Zhu L-D (2018) Effects of wind fairing angle on aerodynamic characteristics and dynamic responses of a streamlined trapezoidal box girder. J W.E & I.A 177–69–7
6. Helgedagsrud TA, Bazilevs Y, Mathisen KM, Øiseth OA (2019) ALE-VMS methods for wind-resistant design of long-span bridges. J Wind Eng Ind Aerodyn 191:143–153
7. Leclercq T, Emmanuel L (2018) Vortex-induced vibrations of cylinders bent by the flow. J Fluids Struct 80:77–93
8. Govardhan R, Williamson CHK (2004) Critical mass in vortex-induced vibration of a cylinder 23:17–27
9. Hansen SO (2007) Vortex-induced vibrations of structures (Structural Engineers World Congress, Bangalore, India)
10. Gao X (2016) Assessment of simulating underwater impact using the icfd Solver in lsdyna (Memorial University of Newfoundland Canada)
11. Tophøj L, Hansen SO A point vortex model for aerodynamic derivatives for tandem multi-deck structures
12. Wang J, Jakobsen JB, McTavish S, Larose GL Aerodynamic performance of a grooved cylinder in flow conditions encountered by bridge stay cables in service

Design and Development of Arachis Hypogaea (Peanut) Decortication Machine

M. Sarvesh and B. U. Balappa

Abstract Agriculture plays a significant role in Indian economic output as its contribution to GDP is 19.9%. Oilseeds are commercialized worldwide; hence, mechanization is required. Mechanisms and machines are available for these crops, but not scientifically designed. The focus of this research is toward small land holding farmers (<2 hectare). The reason is that irrigated land is reducing. An attempt has been made for small scale industries, as they can implement to develop and manufacture Arachis hypogaea decortication machine. Literature survey has been made for available machines. With field survey, based on voice of customer, the product specification is built. To fulfill the requirement, a number of concepts were generated. The best feasible concept is selected and supported with design calculations by adopting combination matrix method (CMM). The geometric model of decorticator was developed. The concave clearance of 9.5 mm was considered to be optimal. A blower was designed using back-faced impeller blades and was analyzed for velocity components. The feasible speed of impeller was 720 rpm. The CFD simulation of blower showed that air flow at outlet was 5.45 m/s, exerting a force of 1.97 N for separation of shells without affecting the kernels.

Keywords Agriculture · Compact machine · Computational fluid dynamics · Blower design · Finite element analysis · Concave clearance

1 Introduction

Agriculture is an art and science of cultivating crops for bonafide supply of food for the nation. India is called 'The land of agriculture' as it was practiced way before 2000 B.C [6]. It also ranks first in the world with highest net cropped area. From a survey of worldwide spectrum, India ranks second in farm outputs, and the contribution of agriculture to India's gross domestic product (GDP) is nearly over 25% [5]. Our country's broad-based growth in terms of economy is steadily declining

M. Sarvesh (✉) · B. U. Balappa
Department of Mechanical Engineering, M.S. Ramaiah University of Applied Sciences, Bengaluru, India
e-mail: sarvesh.ronaldo2@gmail.com

A. Maurya et al. (eds.), *Recent Trends in Mechanical Engineering*, Lecture Notes in Mechanical Engineering, https://doi.org/10.1007/978-981-19-7709-1_63

due to instability in climatic conditions and agroecosystem. These conditions also have a tremendous negative effect on small land holding farmers (<2 ha) and marginal land holding farmers (<1 ha) as well [11]. Agricultural products are exported that accounts for 15% of earning and 57% in consumer price index [15]. Oilseeds are a significant category of crops, used for consumption as well as for extraction of oil. It is requisite commodity of trade and is grown under diverse agroclimatic conditions. Peanut is an oil seed that is commercialized worldwide for consumption and extracted for oil as well. India cultivates about 7.74 million hectares of peanuts and produces 7.61 million ton of peanut with a production rate of 991.8 kg per hectare. The domestic consumption rate of peanut in India is 1.162 lakh in metric ton as of the year 2019–2020, with depletion in growth rate of −0.68% compared to 2018–2019 [2]. Peanuts are grown in 11 districts of Karnataka. Haveri is a district that has the highest yield of 2267 kg per hectare, and Dharwad is the second largest producer of peanut with a production rate of 44,010 metric ton [2]. Decorticator is a machine that separates husk and kernels by performing an operation called shelling. The peanut decortication machine is used for post-harvesting of peanuts by crushing the shell to separate kernels. Manual decortication is monotonous and also has low production rate. A power driven mechanism can be implemented to increase production rate with less human effort. Compact power operated decorticators can be developed to solve such issues. Hence, a machine is developed to support small and marginal land holding farmers.

2 Methodology for Development

The design specification for peanut decorticator was arrived through field survey conducted at Hassan district of Karnataka followed by voice of customers involved in peanut cultivation and benchmarking of existing decortication machine. From the field survey, data regarding peanut cultivation and post harvesting techniques were gathered. Comparative benchmarking was performed on existing decortication machines, and specifications were evaluated in order to design a feasible decorticator for small land holding farmers. Conceptual designs were generated and evaluated using selection matrix based on relative significance. The parallel clearance disk mechanism (PCDM) was identified to be the best, as it is compact, easy to operate and consumes less power with high decortication efficiency. An approach called CMM is implemented in calculation and synthesis of machine components.

2.1 Design Consideration

The peanuts get decorticated by undergoing compression and shearing. Concave clearance maintained between PCDM employs the principle of shear force. Hence, some considerations were made while designing Arachis hypogaea decortication machine as follows.

- Requisite materials were incorporated for various components of the machine that offer sufficient strength and stability (i.e., mild steel, aluminum).
- Machine was designed using standard parts, for ease of assembly and availability as per DFMA.

The major aspect is to reduce the cost of manufacturing, ease of assembly and maintenance while satisfying the structural strength and integrity of machine.

2.2 Detailed Design of Machine Components

The primary approach toward design was by manually measuring the dimensions of 20 different sizes of peanut. The variety of peanut used was ICGS-II. The physical properties of peanut and kernel are shown in Table 1.

The Arachis hypogaea decortication machine consists of a hopper, inner drum, outer drum, concave sieve, frame, discharge outlet and a customized blower unit for discharging the shells. The shape of hopper used was a square prism, where appropriate dimension was selected by incorporating CMM. Equation (1) [7] was solved in MATLAB for determining an appropriate volume of hopper for various values of height, inlet area and outlet area.

$$V_h = \frac{h}{3} \times \left[(a_1 + a_2) + (\sqrt{a_1} + a_2) \right] \tag{1}$$

Hence, the hopper had a volume of 0.0511 m^3 to accommodate 35 kg of peanut (considering bulk density) [1]. The inlet area, outlet area and height of hopper were

Table 1 Physical properties of peanut and kernel

Physical properties	Peanut	Kernel
Moisture (%, d.b.)	4.9	4.5
Average length (mm)	36.8	11.21
Average width (mm)	17.2	7.56
Average thickness (mm)	16	6.93
Average volume (mm^3)	9987.8	587.3
Average bulk density (kg/m^3)	635	565
Average solid density (kg/m^3)	1190	905
Average mass (kg)	6.34×10^{-3}	0.33
Angle of response (deg)	25.3–32	17
Coefficient of friction	0.31–0.77	0.23–0.77

Source Akcali et al. (2006), Olajide and Igbeka (2003) [1, 10]

160 mm^2, 40 mm^2 and 260 mm, respectively. The sliding angle of hopper was 21.03^0. The mass of kernel in 35 kg calculated using bulk density of kernel is 28.87 kg [1, 10], and volume occupied after crushing is 0.0454 m^3.

$$V_d = \pi R^2 \times h_d \tag{2}$$

By incorporating CMM and solving Eq. (2) [9], the volume of inner and outer drums was 0.0517 m^3 and 0.0665 m^3. The inner drum was designed for a radius of 220 mm and height of 340 mm. Similarly, outer drum had a radius of 230 mm and height of 400 mm. The inner drum also known as crusher should rotate at 360 rpm, with an angular velocity, ω_{in}, which is 37.7 m/s to enable shearing of shells [7]. The mass of roasted peanut, M_p, between the clearances was 9.398 kg. The crusher requires input power of 495 W and input torque of 13.13 Nm, calculated using Eq. (3) [7, 9]. Considering acceleration due to gravity as 9.81 m/s^2 and r = 0.1.

$$P_{in} = T_{in} \times \omega_{in} = \left[M_p \times 9.81 \times 0.1\right] \times \left[\frac{2\pi N}{60}\right] \tag{3}$$

A motor of 1 HP was used, that had a rotation speed of 1440 rpm. The efficiency of the motor and system was assumed to be 85% and 80%, respectively [3, 7, 13]. V-Belt 'A—section groove type' was used for transmission between motor–decortication unit and decortication unit–blower unit. The diameter of motor pulley, decortication unit pulley and blower pulley were 75 mm, 300 mm and 150 mm, respectively. The center distance between motor pulley and decortication unit was 666 mm and between decortication unit and blower unit was 508.4 mm.

$$P = (T_1 - T_2) \times V \tag{4}$$

$$\frac{T_1}{T_2} = e^{\mu\theta} \tag{5}$$

The belt tensions between the pulleys were calculated using Eqs. (4) and (5) [9]. Hence, tension of 294.38 N and 136.13 N was generated between motor and decortication units, while the belt tension between decortication unit and blower unit was 206.76 N and 134.23 N.

2.3 *Design of Impeller*

Blower is a device that converts driven energy into kinetic energy by a rotating impeller to facilitate the separation of shell [7, 14]. A back-curved vane was used to design the impeller, because losses are reduced due to low radial and tangential components with high pressure raise. The moment of fluid and exerted fluid force are proportional to the rotational speed of impeller. The larger the diameter of an impeller, with a constant speed of rotation, generates higher head [3]. The mass of shell and kernel is 0.76×10^{-3} kg and 5.64×10^{-3} kg [1, 10]. By taking acceleration due to gravity into consideration, the mass is converted into force, i.e., 7.54×10^{-3} N and 55.35×10^{-3} N. A form factor, $\Delta_{sh} = 7$, is the force ratio of kernel and shell. Hence, a minimum force of 52.80×10^{-3} N is required to blow away the shell.

3 Velocity Components of Blower (Analytically)

The blower has no guide vane; hence, the entry of fluid is radial (V_{r1}). Also, the radial component of absolute velocity is same as inlet velocity ($V_1 = V_{r1}$) [14]. The flow of fluid is assumed to be completely guided by the blades. The performance characteristics of the blower involve mass flow rate, specific speed and pressure ratio. Also, the operating conditions like inlet pressure, density of fluid and inlet temperature have an influence on performance [13, 14]. The velocity component of blower was calculated from 300 to 800 rpm. The solution is formulated using MATLAB. Hence, N = 720 rpm was the acceptable speed for blower. The velocity components at 720 rpm were calculated. By substituting the values from 'Table 2' in Eqs. (6), (7), (8), (11), (12), and (13), the values of velocity components were obtained.

$$U1 = (2\pi \times N \times r_b 1)/60 \tag{6}$$

$$(V_1 = V_r 1) = U_1 \times tan(\beta_1) \tag{7}$$

$$[\text{Expected flow rate}] Q = 2\pi \times r_b 1 \times D_{bw} \times V_{r1} \tag{8}$$

$$V_D = Q/A_D \tag{9}$$

$$H = V_D^3/2g \tag{10}$$

Table 2 Design parameters of blower

Blower design parameters	Values
Inlet vane angle, β_1	79°
Outer vane angle, β_2	59°
Internal volute radius, r_{b1}	40 mm
External volute radius, r_{b2}	200 mm
Vane width at suction eye and tapered end, D_{bw}	320 mm

$$U_2 = (2\pi \times N \times r_b 2)/60 \tag{11}$$

$$V_{r2} = Q/(2\pi \times r_b 2 \times Dbw) \tag{12}$$

$$V_{\tau 2} = U_2 - (V_r 2/tan\beta 2) \tag{13}$$

$$V_2 = \sum (V_{\tau 2}^2 + V_{r2}^2) \tag{14}$$

The fluid used in blower is atmospheric air. The average density (ρ_{air}) and absolute viscosity (μ_{air}) of air are 1.2162 kg/m^3 and 18×10^{-6} Ns/m^2 for a temperature range of 0 °C–35 °C [8]. Power input at impeller is calculated using Eq. (15).

$$P_{\text{Impeller}} = \rho g Q H \tag{15}$$

Therefore, an additional input power required at impeller is (P_{Impeller} = 390.0135 W). The specific speed of blower is 19340.

4 Computational Fluid Dynamics of Blower

The computational fluid dynamics (CFD) enables the analysis of fluids using methodology of numerical solution. The static pressure, absolute pressure and velocities of fluid at inlet and outlet of the blower were analyzed. The 2D CFD model of blower was created in HyperMesh, and quad meshing was performed. The analysis of the model was carried out in Ansys Fluent. The moving reference frame (MRF) method was adapted as it is simple, vigorous and efficient for steady-state simulation of rotating machineries [4]. The 2D model was turbulent as Reynolds number (R_e = $1.63 \times 10^5 > 4000$) [3, 8, 14], calculated from Eq. (16).

$$\text{Re} = (\rho \times U1 \times \text{d}_{\text{bi}})/\mu_{\text{air}} \tag{16}$$

The MRF combined with realizable K-epsilon model is used in case of turbulence [14]. The k-epsilon realizable model is more efficient than normal k-epsilon model as it efficiently predicts dissipation rate and boundary layer characteristics [3]. The cell zone conditions at inlet and casing fluid are kept stationary. Motion is provided to impeller fluid at 720 rpm. The velocity magnitude at inlet is specified as 3.0159 m/s, turbulence intensity is 5% and hydraulic diameter is 0.08 m. The spatial discretization is 'Green—Gauss Node Based' as it is more accurate when compared to 'Least Squares Cell Based' and 'Green—Gauss Cell Based' [12].

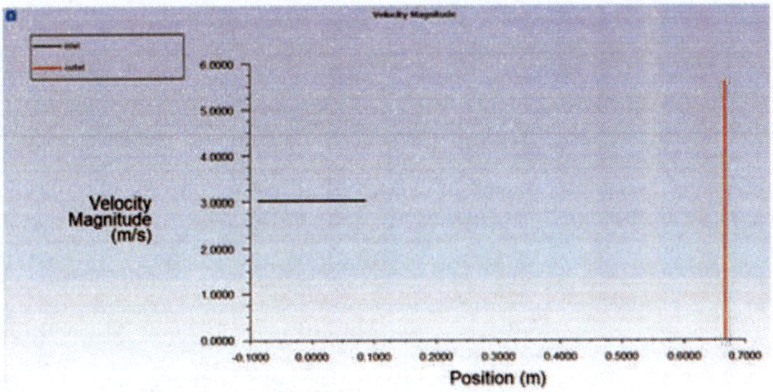

Fig. 1 Velocity at inlet and outlet of blower

5 Results and Discussion

5.1 *CFD Analysis of Velocity Components of Blower*

The magnitude of velocity of air flowing out of the blower is 32.8 m/s. The velocity of air at the impeller blade in X-component is 26.4 m/s, and velocity of air flowing through the outlet is 4.30 m/s. Using the velocity of fluid flowing through the blower outlet, force exerted by the fluid can be calculated. The velocity at X is very significant because flow is linear at outlet. The analytical and simulated values of velocity component are compared in Table 2. The overall error percentage is lesser than 1.

$$\% \text{ of error} = (\mathrm{S} - \mathrm{A})/\mathrm{S} \tag{17}$$

The simulated result of blower velocity at inlet and outlet is shown in Fig. 1, and the outlet velocity of blower is $V_{out} = 5.45$ m/s. Hence, using the value of outlet velocity, the force of air flowing out of blower is calculated (Table 3).

5.2 *Force of Air Exerted by the Blower on Shell*

The force of air (F_{air}) discharging out of blower is calculated using average density of air, area of impact and acceleration [14].

$$F_{air} = \times AD \times V_{out}^2 \tag{18}$$

Hence, $F_{air} = 1.9748$ N calculated using Eq. (18). The calculated value F_{air} is greater than F_{sh}. The air flowing out from the outlet was 37.39 times greater than

Table 3 Comparison of analytical and CFD simulation results

Velocity (m/s)	Analytical result (A)	Simulated result (S)	% of error < 1
Linear velocity	15.0790	14.5000	−0.0675
Radial velocity	15.5150	16.0000	0.0303
Tangential velocity	13.2150	13.5000	0.0211
Relative velocity	22.6710	19.0000	−0.1932
Relative tangential velocity	13.3740	13.5000	0.0093
Overall percentage of error	0.3215%		

S = Simulated result, A = Analytical result

F_{sh}. Hence, the velocity of air produced by the blower is sufficient to push away the peanut shells [16].

6 Product Specification

The product specification of Arachis hypogaea decortication machine is shown in Table 4. The CAD drawings of machine were created according to design for assembly and manufacture (DFMA) by standardization of all components (Fig. 2).

7 Conclusion

The Arachis hypogaea decortication machine, on evaluation for dimension using CMM, showed that all machine components are synthesized according to product specification and dimension. The QFD and benchmarking techniques help in arriving relative significance score for technical attributes. The selected disk-type mechanism is the most feasible design based on evaluation of product specification and concepts. The designed impeller on assessment of blade velocity showed similar results for both analytical solution and simulation performed in Ansys R2 Fluent, with a total error percentage of 0.3214 (<1). Hence, design of blower was successful. The air that flowed at 5.45 m/s at blower outlet was sufficient enough to push away the shells of peanuts without affecting the kernels. Finally, on the basis of analytical calculation

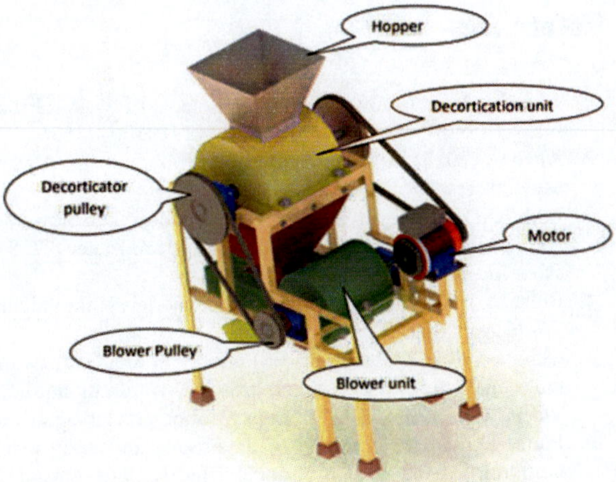

Fig. 2 Arachis hypogaea decortication machine

Table 4 Arachis hypogaea decorticator specification

Components	Specification
Operation type	Power operated (electricity)
Motor capacity	1 HP
Radius of inner drum	220 mm
Radius of outer drum	230 mm
Main shaft diameter	44 mm
Blower shaft diameter	50 mm
Blower pulley diameter	150 mm
Decortication pulley diameter	300 mm
Motor pulley diameter	75 mm
Concave clearance	9.5 mm
Hopper capacity	35 kg
Decortication capacity	80–100 kg/hr

and virtual simulation, the designed Arachis hypogaea decorticator is standardized according to DFMA and is feasible for small land holding farmers.

Acknowledgements The authors like to thank M.S. Ramaiah University of Applied Sciences for providing access to their facilities and support of farmers from Hassan district of Karnataka.

References

1. Akcali I, Ince A, Guzel E (2006) Selected Physical Properties of Peanuts. Int J Food Prop 9:25–37. https://doi.org/10.1080/10942910500471636
2. APEDA (2020) India production of Groundnut. In: Agriexchange.apeda.gov.in. Accessed 11 May 2021
3. Budiman H (2016) Effect of impeller condition on blower performance, case study: impeller of K-2201 in Pt. Pertamina RU III. Appl Mech Mater 836:54–59. https://doi.org/10.4028/www.scientific.net/amm.836.54
4. Bulat M, Bulat P (2013) of Turbulence models in the calculation of supersonic separated flows. World Appl Sci J 27:1263–1266
5. Chinnasamy P, Maheshwari B, Dillon P et al (2018) Estimation of specific yield using water table fluctuations and cropped area in a Hardrock aquifer system of Rajasthan, India. Agric Water Manag 202:146–155. https://doi.org/10.1016/j.agwat.2018.02.016
6. Fuller D, Murphy C (2017) The origins and early dispersal of horsegram (Macrotyloma uniflorum), a major crop of ancient India. Genet Resour Crop Evol 65:285–305. https://doi.org/10.1007/s10722-017-0532-2
7. Khamprapai P, Laohavanich J, Chiawchanwattana C, Yangyuen S (2018) A shelling machine for Inca peanuts. In: Third international conference on engineering science and innovative technology (ESIT). IEEE, pp 1–4
8. Kothandaraman C, Subramanyan S (2014) Heat and mass transfer data book. New Academic Science, London
9. Mahadevan K, Reddy B (2018) Design data handbook for mechanical engineers in SI and Metric Units, 4th edn, pp 3–500
10. Olajide J, Igbeka J (2003) Some physical properties of groundnut kernels. J Food Eng 58:201–204. https://doi.org/10.1016/s0260-8774(02)00323-0
11. Russell R, Holden P, Garvey C et al (2006) Investigation of the phase morphology of bacterial PHA inclusion bodies by contrast variation SANS. Phys B 385–386:859–861. https://doi.org/10.1016/j.physb.2006.05.126
12. Shima E, Kitamura K, Haga T (2013) Green–Gauss/weighted-least-squares hybrid gradient reconstruction for arbitrary polyhedra unstructured grids. AIAA J 51:2740–2747. https://doi.org/10.2514/1.j052095
13. Shukla A, Harsha S (2015) An experimental and FEM modal analysis of cracked and normal steam turbine blade. Mater Today: Proc 2:2056–2063. https://doi.org/10.1016/j.matpr.2015.07.191
14. Son P, Kim J, Byun S, Ahn E (2012) Effects of inlet radius and bell mouth radius on flow rate and sound quality of centrifugal blower. J Mech Sci Technol 26:1531–1538. https://doi.org/10.1007/s12206-012-0311-0
15. Srivastava P, Singh R, Tripathi S, Raghubanshi A (2016) An urgent need for sustainable thinking in agriculture – an Indian scenario. Ecol Ind 67:611–622. https://doi.org/10.1016/j.ecolind.2016.03.015
16. Ince A, Ugurluay S, Güzel E, Özcan T (2009) Mechanical behavior of hulled peanut and its kernel during the shelling process. Res Gate 92:92–99. https://doi.org/ISSN0031-7454

Dynamic Response of Simply Supported Beam Carrying Rotating Unbalance and a Damper with CuO Nanolubricants

Abhijeet G. Chavan and Y. Prasannatha Reddy

Abstract This work investigates the dynamic response of simply supported beam subjected to harmonic excitation by means of rotating unbalance. Passive viscous damper plays crucial role in controlling vibration response of system operating at resonance. Viscous damper containing CuO nanolubricants is used for suppressing the vibrations. Copper oxide (CuO) nanolubricants are prepared by two-stage process, which involves addition of CuO nanoparticles to lubricants and mixing by ultrasonication process for better dispersion stability. Orthogonal array technique is adopted for deciding set of experiments. Experiments are conducted at various speeds and nanoparticle concentrations. RMS acceleration values of the vibrating system are recorded for each experiment. Dynamic performance of the system is compared for various combinations of plain oil, nanolubricants and speed. Results show improved dynamic performance by use of CuO nanolubricants.

Keywords Nanolubricants · Damper · Dynamic response · Simply supported beam

1 Introduction

The use of nanoparticles in various fields of engineering is becoming a common practice. When suspended in base fluid, nanoparticles offer certain advantages. Numerous studies reported enhanced physical properties of base fluid with addition of nanoparticles. After addition in lubricating fluids, certain nanoparticles enhance properties, like viscosity [1–3], thermal conductivity [4–6] of base fluid. Few nanolubricants

A. G. Chavan (✉) · Y. P. Reddy
Mechanical Engineering Department, Sinhgad College of Engineering, Vadgaon, Pune, Maharashtra 411041, India
e-mail: abhijeetchavan101@gmail.com

Y. P. Reddy
e-mail: ypreddy.scoe@sinhgad.edu

A. G. Chavan
MIT School of Engineering, MITADT University, Pune, Maharashtra 412201, India

A. Maurya et al. (eds.), *Recent Trends in Mechanical Engineering*, Lecture Notes in Mechanical Engineering, https://doi.org/10.1007/978-981-19-7709-1_64

show superior friction performance [7, 8] along with negative wear characteristics which are useful while dealing with surfaces having friction.

Although use of passive viscous dampers is common for various structural applications which are subjected to resonance, there are comparatively less studies dealing with damping capabilities of viscous dampers having nanolubricants as working fluid. Nanolubricants can be effectively used in place of normal lubricating oils in viscous dampers for absorbing shocks and vibrations, as there is significant improvement in the viscosity of nanolubricant. Thus, nanolubricants when used in passive viscous damper can offer effective solution for damping the vibrations at resonance.

Yeh et al. [9] assessed the performance of shear thickening fluids, which was made up of nanoparticles and polyethylene glycol, and reported the improved dynamic performance of damper. In the present work, effort is made to assess the damping capability of CuO nanolubricants for a particular application of simply supported beam carrying rotating unbalanced mass.

2 Theory

The model of simply supported beam with added lumped masses subjected to harmonic excitation [10] is provided in Eq. (1).

$$E_i I_i \frac{\partial^4 y_i(x_i, t)}{\partial x_i^4} + P(t)\frac{\partial^2 y_i(x_i, t)}{\partial x_i^2} + \rho_i A_i \frac{\partial^2 y_i(x_i, t)}{\partial t^2} = 0 \tag{1}$$

Here, 'EI' is flexural rigidity of beam, 'ρ' is density and 'A' is cross-sectional area of beam, and 'P' is external harmonic excitation. When simply supported beam is subjected to harmonic excitation, the response of the system leads to resonance in particular situation. When excitation frequency matches with one of the natural frequencies of beam, resonance occurs, this may result in catastrophic failure. For controlling the amplitude at resonance, dampers are provided which absorb the energy and significantly reduce the amplitude of motion at resonance.

Passive viscous dampers are used widely for many engineering applications. The viscosity of lubricating oil is one of the key parameters in deciding the performance of damper. Nanolubricants have superior viscosity as compared to the base lubricants. The present work involves use of copper oxide nanoparticles as an additive in the base lubricant to improve the performance of damper.

3 Methods and Materials

3.1 Nanolubricants

Nanolubricants were prepared by suspending different weight fractions (0.1% and 0.5%) of copper oxide (CuO) nanoparticles in commercially available lubricating oil (15 W-40). Oleic acid was used as surfactant for enhancing the dispersion stability of CuO nanoparticles in base oil [11, 12]. Bath-type ultrasonicator was used for preparing homogenous mixture of CuO nanoparticles in lubricating oil. Prepared nanolubricant shows good dispersion stability up to 15 days, with no visible agglomeration. After this period, ultrasonication process is required for mixing agglomerated particles evenly. Vibratory system involving rotating unbalance is shown in Fig. 1. Beam is made of steel having rectangular cross-section of 25 × 10 mm. Span between the supports measures 1030 mm. One end is kept pinned and other is kept roller for the beam; thus, the condition simply supported beam is achieved. Beam is loaded with an electrical motor which carries two disks along with unbalanced masses. Assembly of electric motor and unbalanced masses exerts harmonic unbalanced force on beam element. Damper is connected at the base of motor. Damper is having piston with three openings in the form of circular holes to control the lubricant flow through slider. An accelerometer is mounted at base of motor to measure the vibration amplitude; tachometer is utilized to measure the speed of motor.

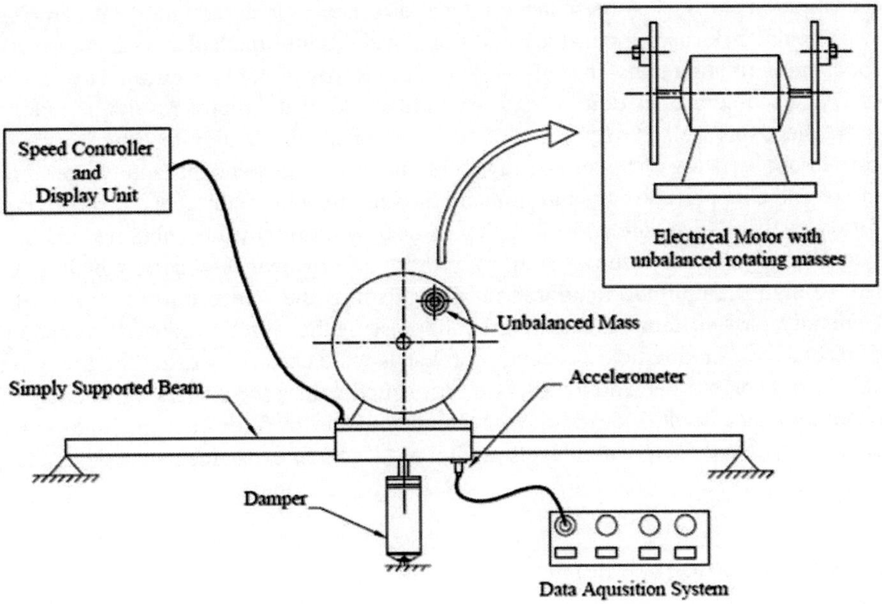

Fig. 1 Schematic of experimental setup

Table 1 Factors and levels selected for experimentation

	Factors			
Levels	Speed (S)	Values (rpm)	Nanoparticle concentration (C)	Values
	S1	282	C1	0%
	S2	565	C2	0.1%
	S3	847	C3	0.5%
	S4	1130		

3.2 *FEA Model*

Finite element model of system is prepared, and modal analysis is carried out by using ANSYS software tool. From modal analysis, first natural frequency of the system is 9.47 Hz. This is verified by experimental value of natural frequency of the system found by using FFT analyzer which is in good agreement with theoretical value.

4 Experiment

A simply supported beam, carrying a rotating unbalance at its mid span, is selected for vibration analysis. A permanent magnet direct current electric motor is carrying two circular disks on its both sides; additional weights are attached to each of the disk for creating rotating unbalance effect. Speed of electric motor is controlled by means of DC speed controller unit. Variation in the speed of the motor results in change in frequency as well as magnitude of rotating unbalance. Design of experiments is carried out by using orthogonal array technique. Two independent factors, speed of motor and nanoparticle concentration are having three and four levels, respectively. Three levels of speed are chosen so that frequency ratio (r) values obtained are 0.5, 1 and 1.5. When the value of frequency ratio is 1, resonance occurs, which gives rise to higher amplitude of vibration. This justifies the selection of speed levels. Generally, passive dampers are used for suppressing the vibration caused on account of resonance. Nanoparticle concentration levels were chosen by using the literature values and some trials which indicates best suited values for dispersion stability of nanolubricants. Table 1 shows the values of levels utilized for both the factors. By considering these factors and levels, a total of 12 sets of experiments are designed.

5 Measurement Scheme

For vibration measurement, IEPE accelerometer was used, and accelerometer is mounted by means of magnet mount. For speed measurement, digital tachometer was used. A sampling frequency of 200 Hz was used for measurement of acceleration, and signal was recorded for 10 s for further processing. From recorded signal, RMS acceleration and maximum acceleration values were calculated.

6 Results

The recorded values of RMS acceleration levels and maximum acceleration levels at various speeds for different oils are tabulated in Table 2. Typical measurement recorded for plain oil at resonance is shown in Fig. 2.

At resonance, i.e., at 565 rpm speed, nanolubricant sample 0.1% CuO shows least value of RMS acceleration, while plain oil shows maximum value of RMS acceleration.

Figure 3a and b shows values of RMS and maximum acceleration for different operating speeds. Nanolubricant sample 0.1% CuO performs well, as it shows

Table 2 RMS acceleration for various speeds and lubricants

Speed (rpm)	RMS acceleration (g)			Maximum acceleration (g)		
	Plain oil	0.1% CuO	0.5% CuO	Plain oil	0.1% CuO	0.5% CuO
280	0.22	0.178	0.214	1.02	0.742	1.358
565	0.314	0.267	0.307	1.553	1.148	1.147
848	0.782	0.673	0.773	3.372	2.34	3.681
1130	1.112	1.09	1.167	3.039	2.953	2.875

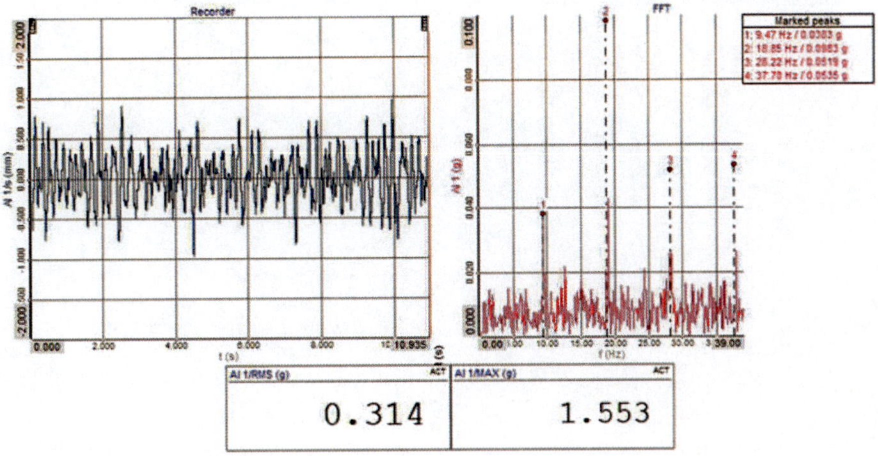

Fig. 2 Typical vibration measurement for plain oil at 565 rpm

least values of RMS acceleration and maximum acceleration almost throughout the operating range.

For comparing the performance of nanolubricants, percentage reduction in RMS acceleration is calculated for 0.1% CuO and 0.5% CuO nanolubricant samples. Figure 4 shows the comparative performance of nanolubricants with respect to plain oil. For all speed values, 0.1% CuO is found to be effective than plain oil.

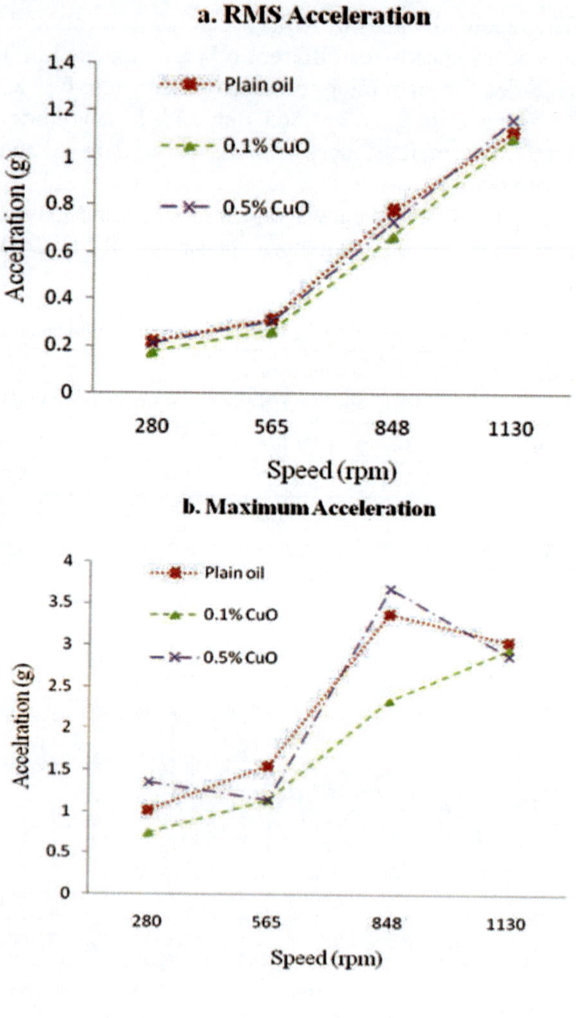

Fig. 3 **a** RMS acceleration at various speeds and lubricants, **b** Maximum acceleration at various speeds and lubricants

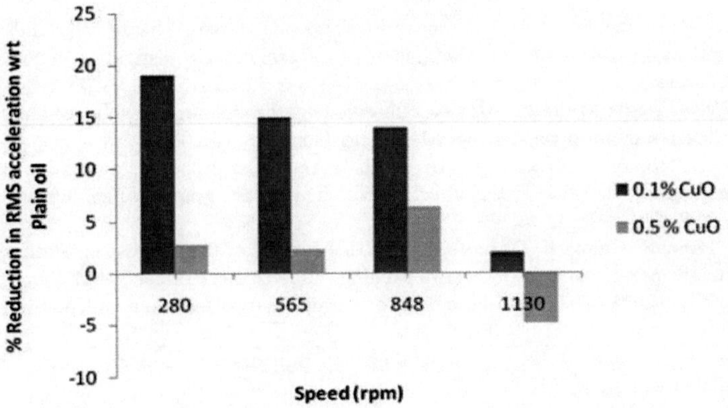

Fig. 4 Percentage reduction in RMS acceleration for nanolubricants in comparison with plain oil

7 Conclusion

Simply supported beam carrying rotating unbalanced mass was coupled with viscous damper, and its dynamic response was recorded for different speeds. RMS and maximum values of acceleration are recorded and compared. Nanolubricant with 0.1% CuO shows superior performance as compared to plain oil and nanolubricant with 0.5% CuO. The RMS acceleration was reduced by 19.1% at 280 rpm as compared to plain oil. At resonance speed, both 0.1% and 0.5% CuO nanolubricants show 14.9% and 2.2% reduction in RMS acceleration, respectively. At higher speed of 1130 rpm, 0.5% CuO shows degraded performance than plain oil.

A small fraction of CuO nanoparticles added in the base lubricant improved the viscosity of lubricant and the performance of viscous damper. Damper utilizing nanolubricants is found to be more effective at resonance than damper utilizing plain lubricant.

References

1. Battez AH, González R, Viesca JL, Fernández JE, Fernández JMD, Machado A, Chou R, Riba J (2008) CuO, ZrO2 and ZnO nanoparticles as antiwear additive in oil lubricants. Wear 265:422–428
2. Peña- L, Taha-Tijerina J, Garza L, Maldonado-Cortés D, Michalczewski R, Lapray C (2015) Effect of CuO and Al2O3 nanoparticle additives on the tribological behavior of fully formulated oils. Wear 332:1256–1261
3. Gulzar M, Masjuki HH, Kalam MA, Varman M, Zulkifli NWM, Mufti RA, Zahid R (2016) Tribological performance of nanoparticles as lubricating oil additives. J Nanoparticle Res
4. Kole M, Dey TK (2013) Enhanced thermophysical properties of copper nanoparticles dispersed in gear oil. Appl Therm Eng 56:45–53
5. Nicoletti R (2014) The importance of the heat capacity of lubricants with nanoparticles in the static behavior of journal bearings. J Tribol 136:044502

6. Asadi A, Pourfattah F (2019) Heat transfer performance of two oil-based nanofluids containing ZnO and MgO nanoparticles; a comparative experimental investigation. Powder Technol 343:296–308
7. Shafi WK, Charoo MS (2021) An overall review on the tribological, thermal and rheological properties of nanolubricants. Tribol Mater Surf Interfaces 15:20–54
8. Ali MKA, Hou X, Abdelkareem MAA (2020) Anti-wear properties evaluation of frictional sliding interfaces in automobile engines lubricated by copper/graphene nanolubricants. Friction 8:905–916
9. Taylor P, Yeh F, Chang K, Chen T, Yu C (2014) Journal of the Chinese institute of engineers the dynamic performance of a shear thickening fluid viscous damper 37–41
10. Sochacki W (2008) The dynamic stability of a simply supported beam with additional discrete elements. J Sound Vib 314:180–193
11. Li D, Xie W, Fang W (2011) Preparation and properties of copper-oil-based nanofluids. Nanoscale Res Lett 6:373
12. Mahbubul IM (2019) 2 - preparation of nanofluid. In: Mahbubul IM (ed) Preparation, characterization, properties and application of nanofluid. William Andrew Publishing, pp 15–45

Effect of Vane Shape on the Performance of the Water Rotor

Vimal Patel, Bheemalingeswara Reddy, Vikram Rathod, and Ravi Patel

Abstract Water rotor blades may have three blades, and each cross-section of the vane is comprising a concave and convex profile and these vanes are extended around the drum between two disks. The main aim of the study is to investigate the water rotor by providing fillet radius at the vane edge. In the current study, one sharp vane edge and four smooth vane edges were analyzed using 2D CFD ANSYS Fluent Solver. The CFD model is validated by the values obtained experimentally published in the open literature. Based on a numerical study, pressure and velocity distributions around the water rotors were analyzed and discussed. The obtain results indicate that sharp vane edge water rotor, i.e., zero fillet radius model gives good performance than any smooth vane edge rotors. Furthermore, the sharp edge water rotor produces C_{Pmax} which is 0.17 at λ value 0.8.

Keywords Water rotor · Hydrokinetic turbine · Renewable energy · Hydraulic machines · Applied energy · Horizontal axis · Savonius turbine

Nomenclature

C_P	Coefficient of power
C_T	Coefficient of torque
C_{Pmax}	Max coefficient of power
D	Diameter of rotor (mm)
d	Diameter of chord (mm)
T	Thickness of blade (mm)
λ	Tip speed ratio
ρ	Density of water (kg/m^3)
θ	Angular displacement (deg)
F_D	Drag force (N)

V. Patel (✉) · B. Reddy · V. Rathod · R. Patel
MED, Sardar Vallabhbhai National Institute of Technology, Surat, Gujarat, India
e-mail: vimal.iitbombay@gmail.com

A. Maurya et al. (eds.), *Recent Trends in Mechanical Engineering*, Lecture Notes in Mechanical Engineering, https://doi.org/10.1007/978-981-19-7709-1_65

F_L	Lift force (N)
H	Height of rotor (mm)
r	Fillet radius (mm)
P	Power output (Watts)
T	Torque (N-m)
U_∞	Inlet velocity of water (m/s)
$y+$	First height cell thickness
ω	Angular velocity (rad/sec)

1 Introduction

Over 71% of the earth's surface covered in slow-moving water. Various types of hydrokinetic turbines have been proposed in the previous works of literature to convert slow-moving water energy into mechanical energy. Hydrokinetic turbines are broadly divided into two groups, i.e., axial flow turbines and cross-flow turbines [1, 2]. Savonius water turbines are simple in construction and self-starting devices and produce power economically without the need of big structures, i.e., penstocks, dams, etc. [3, 4]. Semicircular, Benesh, modified Bach profiles are also investigated for Savonius turbine. The maximum power coefficient found in experimental and numerical studies is 0.21 at λ 0.72 and 0.23 at λ 0.7, respectively, for the Savonius water turbine [5, 6]. Many research works are done on semicircular rotors either numerically or experimentally; hence, in the present study, novel design of a water rotor is proposed which helps drag and lift forces contribute the power generation.

2 Conceptual Discussion

As discussed in the earlier, Savonius conventional water turbines are drag force devices. A new design of the water rotor produces high drag force, and a small lift force helps generate high power compared to semicircular blade profile as shown in Fig. 1a. In the present study, an investigation was carried out by incorporating a fillet radius at the vane edge for a sharp edge rotor. Such that delay of the boundary layer of separation occurs and wake zone size decreases in smooth edge rotors (as shown in Fig. 1b.) which results in small amount of decreasing in drag force, and large amount increase of lift force would produce highest power coefficient compared to sharp vane edge rotor.

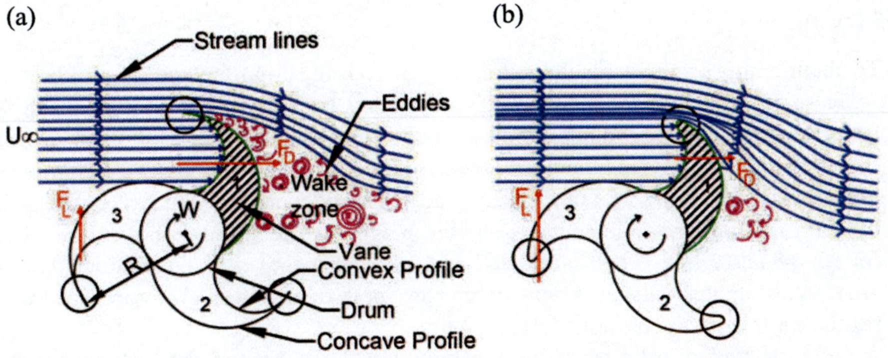

Fig. 1 Conceptual representation of flow over water rotor with **a** sharp vane edge, **b** smooth vane edge

3 Numerical Procedure

CFD is now the most widely used method for resolving a wide range of fluid flow concerns. CFD simulation of a 3D model of the horizontal axis turbine is required when the turbine blade is not symmetrical along the horizontal axis. Because the turbine blades in this study are symmetrical about the horizontal axis and to reduce the computational time, 2D CFD model is used in the present work.

a. ***Methodology Validation of Savonius Turbine***

Geometry

The geometry was designed by Anuj kumar and Saini [6] as shown in Fig. 2. It is conventional Savonius 2-bladed water turbine having that D is 118 mm, d is 60 mm and t is 2 mm. The turbine modeled using AutoCAD software.

To know the optimized dimensions for rectangular domain, simulations are carried out considering various domain sizes. The C_p value becomes almost stable for domain 4000 × 600 which has been selected for the present research.

Fig. 2 Schematic diagram of conventional

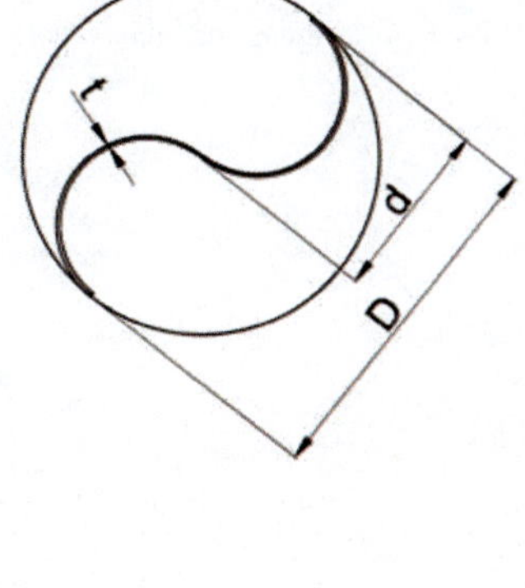

Meshing

To obtain a high accuracy results, generating mesh is the important tool of simulation. After selecting optimized domain, high-qualified unstructured triangular mesh is generated for inner and outer domains using ANSYS Meshing tool as shown in Fig. 3. Particularly, much attention is taken for doing rotating domain because where flow physics is captured. In the present study, grid independence study is performed by varying number of elements as shown in Fig. 5. The simulation results of C_m for 96,448 and 191,828 cells are 0.262 and 0.263, respectively. The obtained values are roughly stable; hence, elements which are more than 96,448 give good accuracy results for this particular domain (Fig. 4).

Inflation tool is used to define the boundary layer around the turbine blades. Inflation layers are 10, and growth rate as 1.2 has been chosen. In the present study to calculate first height, cell thickness y+ value less than 1 is selected [3]. The mesh

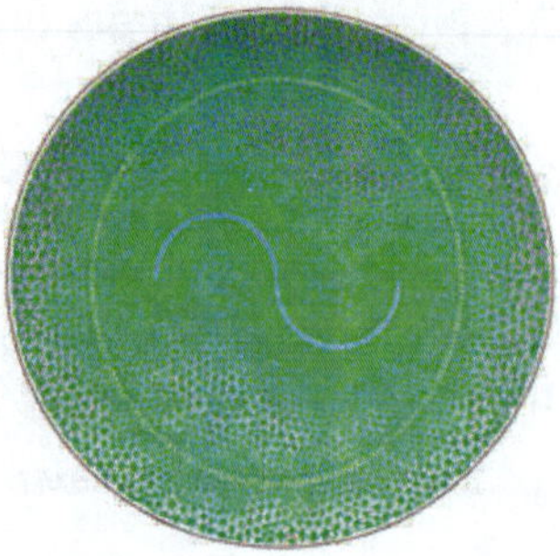

Fig. 3 Mesh generation around the turbine

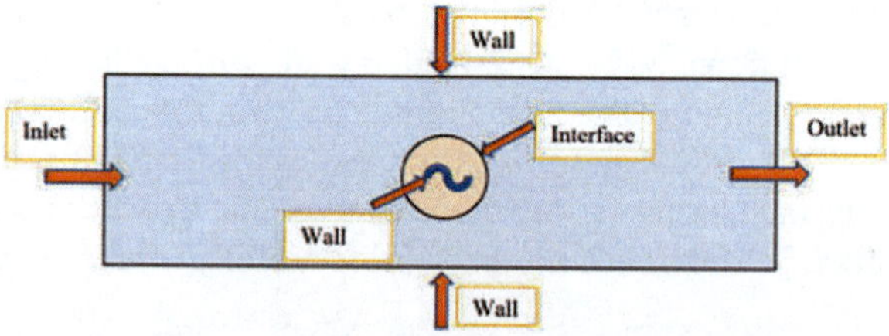

Fig. 4 Two-dimensional computational domain and boundary conditions

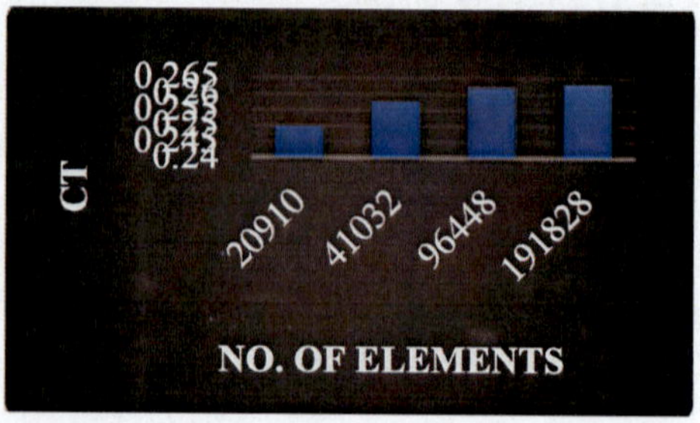

Fig. 5 Grid independence study

Table 1 Values of boundary conditions

Sr. No	Parameter	Value
1	Fluid type	Water
2	Density (kg/m^3)	998.2
3	Inlet water velocity(m/sec)	0.5
4	Outlet pressure (gauge)	0
5	Inner domain	Rotating
6	Outer domain	Stationary

has average value of orthogonal quality, aspect ratio and skewness 0.95, 1.62 and 0.008, respectively, which are acceptable and give good accuracy results.

Analysis

To solve unsteady Reynolds averaged Navier–Stokes equations (*RANS*), the Fluent 16.0 software is used. Pressure-based solver and absolute velocity formulation were set in the solver. A realizable version of k-ε turbulence model gives good results for complex flows with separation flows and strong pressure gradients. Saeed et al. many researchers numerically investigated to optimize the turbulence models by considering four models, i.e., standard k-ε, RNG k-ε, realizable k-ε, and SST of k-ε. Realizable k-ε model values are close to the experimental values. Hence, realizable k-ε turbulence model is chosen in the present study. The boundary conditions are considered as shown in Table 1.

The sliding mesh model (*SMM*) method is used to create interfaces and rotates the rotor at different angular velocities with time. Rotational speed values are given gradually increased from lower values to higher values to observe the flow field and optimize the power output. SIMPLE scheme is selected to couple pressure–velocity algorithm, and for spatial discretization, second-order upwind algorithm is chosen. The convergence criterion for residuals 1×10^{-6} has been taken. The final step in the simulation procedure is to calculate the time step size. Under the present study, the time step size is taken as 3^0/step size, i.e., to rotate one complete rotation of rotor 120-time steps taken.

b. Modeling and parameters to be investigation

The present study was carried out using water rotor on horizontal axis turbine. The research has been done for different vane shapes with variation of fillet radius (*FR*) in mm keeping drum size as constant on edges of rotor as shown in Fig. 7. The number of tested rotors is five models. The dimensions of each rotor are shown in Table 2. The schematic diagram of 3D water rotor and its projections is shown in Fig. 6. In the analysis, free surface is assumed as no effect to satisfy boundary condition symmetry has been taken [6] and gravity 9.81 m/sec^2 has been taken into account.

Table 2 Dimensions of water rotor

Parameter	Model 1	Model 2	Model 3	Model 4	Model 5
Fillet radius (r) mm	0	0.5	1	2	3
Diameter (D) mm	118	117	116	114	112
Height (H) mm	187	185	183	180	177

Fig. 6 Schematic 3D view of water rotor and its projections

a) FR_0 b) FR_0.5 c) FR_1.0 d) FR_2.0 e) FR_3.0

Fig. 7 Two-dimensional models of tested water rotor geometries

4 Results and Discussions

c. ***Validation***

To validate numerical methodology, conventional two-bladed semicircular turbine 2D simulation results are compared with both 3D numerical and experimental investigations which are done by Anuj and Saini [6]. The values of simulation are close to the open literature results as shown in Fig. 8.

d. ***Optimization of vane shape***

Many researchers are tried to improve the power coefficient of water turbines by stage counting, helical angle, aspect ratio, etc., as mentioned in the survey. Under the present investigation, five different models by varying vane shapes were used to

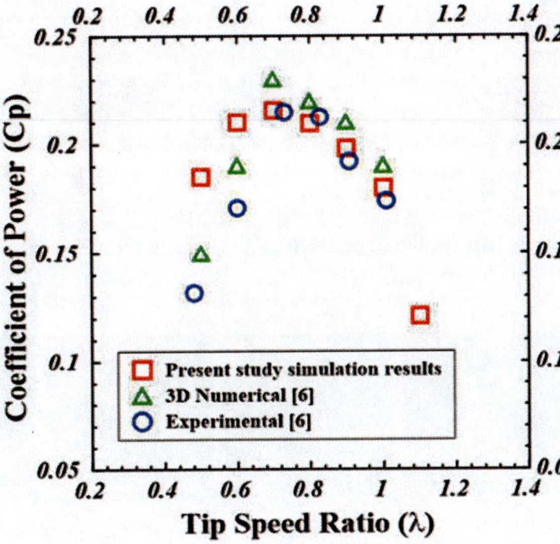

Fig. 8 Validation of C_P with available experimental and 3D numerical values

increase the performance of turbines. Results obtained with the present investigation are explained with pressure and velocity contours.

The given pressure contours are for all five models simulated with λ 0.8 at azimuthal position $\theta = 30°$ as shown in Fig. 9. High pressure (Zone_A) acts on advancing blades, and low pressure (Zone_B) acts on retuning blades. Due to the high amount of pressure difference acts on the blade, profiles drive to the rotation of turbine which produces torque. The wake zone (Zone_C) size is observed for model 1(a) is high compared to all five models. In the present research, investigation carried out with the assumption of reduction of wake zone size leads to increase lift force, and it can increase the coefficient of power for smooth vane edge rotors. The power coefficient (C_P) values are high for model 1 and gradually declining by making smoothening edge. The power coefficient and moment coefficient obtained for varying l range from 0.2 to 1.2 for all five models with different fillet radiuses (*FRs*) as shown in Figs. 11 and 12. The maximum C_P value obtained for model 1, 2, 3, 4, 5 at TSR of 0.8, 0.8, 0.8, 0.8, 0.6 is 0.17, 0.155, 0.146, 0.121, 0.098 respectively. The distribution of velocity for all different vane geometries is shown in Fig. 10. The lower velocity observed downstream of model 1 is a high amount compared to model 5; this represents that more kinetic energy is conserved; hence, model 1 efficiency is high compared to model 5.

The torque of turbine shaft for all different angular velocities calculated using the below relation is:

$$T = \frac{1}{4} C_T \rho {U_\infty}^2 D^2 H \tag{1}$$

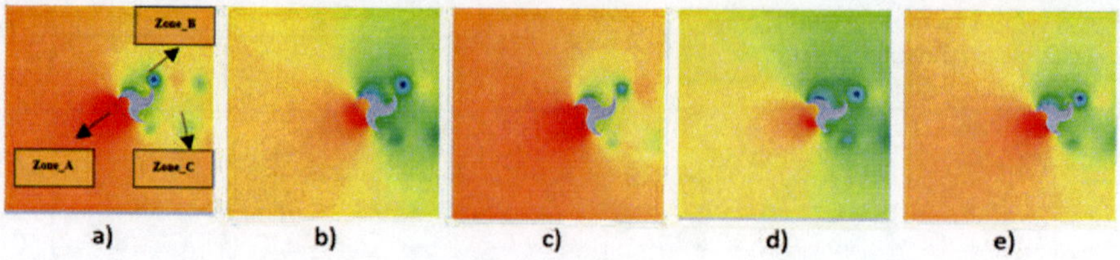

Fig. 9 Distribution of pressure on **a** FR_0.0, **b** FR_0.5, **c** FR_1.0, **d** FR_2.0, **e** FR_3.0 for λ 0.8 at $\theta = 30^0$

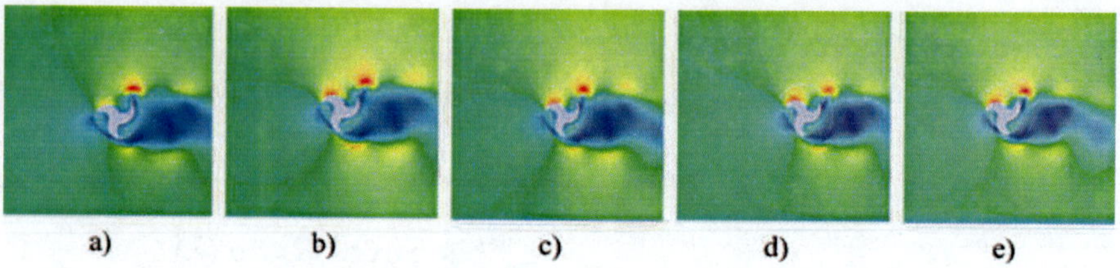

Fig. 10 Distribution of velocity on **a** FR_0.0, **b** FR_0.5, **c** FR_1.0, **d** FR_2.0, **e** FR_3.0 for λ 0.8 at $\theta = 30°$

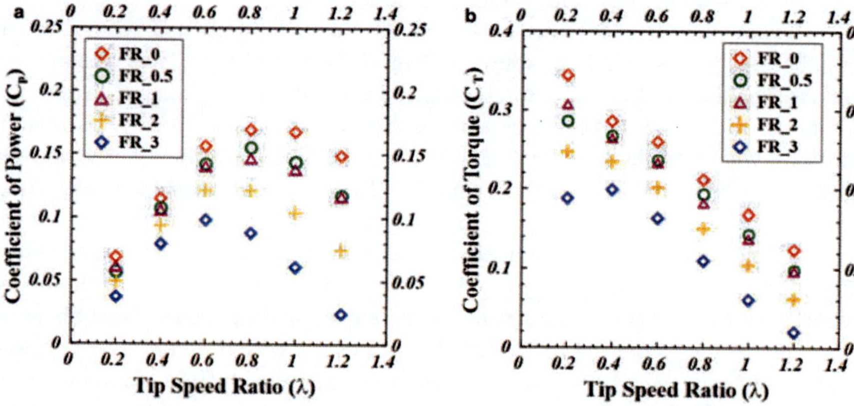

Fig. 11 Performance characteristics of non-dimensional parameters, **a** C_p versus λ, **b** C_m versus λ

The variation of torque at different angular velocities and for all five models is shown in Fig. 12b. The power output from the turbine is also calculated by multiplying the torque and tip speed ratio, and the results for all different geometries are shown in Fig. 12a. The maximum amount of power obtained for model 1 is 0.233 W and gradually decreased from model 1 to model 5 observed.

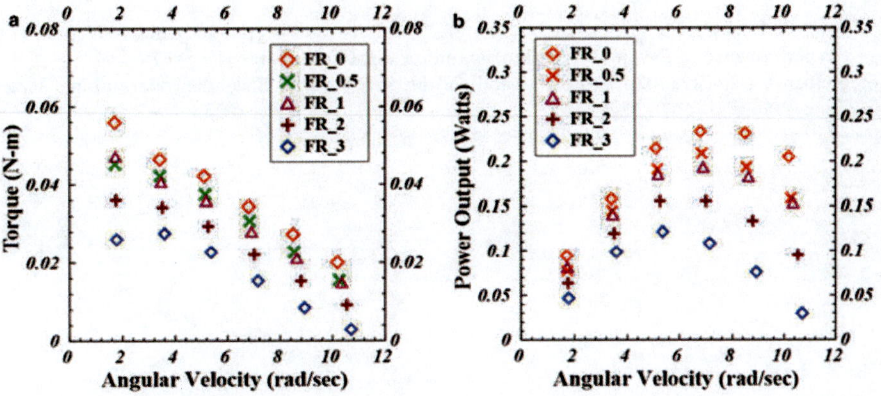

Fig. 12 Variation of **a** power output versus angular velocity, **b** torque versus angular velocity

5 Conclusions

The current study concentrates on the effect of smoothing at the vane edge on the performance of the water rotor. For this purpose, 2D unsteady numerical simulations are performed using k-ε realizable turbulence model. The C_P value of sharp edge and four smooth vane edge rotors are found to be 0.170, 0.155, 0.146, 0.121 and 0.098 at λ value of 0.8, 0.8, 0.8, 0.8 and 0.6, respectively. As compared to smooth vane edge rotors, total pressure and velocity magnitude near the advancing blade profile of sharp vane edge water rotor are higher. Hence, the study concluded that smooth vane edge rotors give low efficiency than sharp vane edge water rotors. Authors suggest that a sharp vane edge water rotor is more feasible to capture free-flowing energy in open channels. Furthermore, the efficiency of the novel design of sharp vane edge of water rotor can increase by twist angle, multistage, aspect ratio, deflector, etc.

References

1. Patel V, Savalia D, Panchal M, Rathod N (2016) Experimental investigations of hydrokinetic axial flow turbine. In: Proceedings of the world congress on engineering, vol 2
2. Kumar A, Saini RP (2015) Investigation on performance of improved savonius rotor: an overview. In: 2015 international conference on recent developments in control, automation and power engineering (RDCAPE), IEEE
3. Patel V, Eldho TI, Prabhu SV (2018) Theoretical study on the prediction of the hydrodynamic performance of a Savonius turbine based on stagnation pressure and impulse momentum principle. Energy Convers Manag 168:545–563
4. Patel V, Eldho TI, Prabhu SV (2017) Experimental investigations on Darrieus straight blade turbine for tidal current application and parametric optimization for hydro farm arrangement. Int J Marine Energy 17:110–135

5. Patel V, Bhat G, Eldho TI, Prabhu SV (20017) Influence of overlap ratio and aspect ratio on the performance of Savonius hydrokinetic turbine. Int J Energy Res 41(6):829–844
6. Kumar A, Saini G (2020) Flow field and performance study of Savonius water turbine. Mater Today: Proc

Design and Aerodynamic Analysis of Small HAWT Using a Novel Computational Approach

Vikash Anand, Deavshish, and Sanatan Kumar

Abstract With the development of human living standard and rapid industrialisation, global consumption of electricity has increased many times. By proper planning, modelling, and design, effective utilisation of renewable energy can be achieved with the aim to fulfil electricity demand. In this context, wind energy is one of the most efficient clean energy that has potential to be the frontrunner. However, designing and control of wind energy system are somewhat difficult because of erratic nature of wind. This paper presents an analytic approach to design small horizontal axis wind turbine (HAWT) blade and investigate its characteristic. Computational code has been developed in MATLAB to find out the parameters of optimal blade shape of three-bladed wind turbine (WT). The performance of WT obtained analytically from this procedure is compared and verified with several other works reported earlier in various articles. The proposed design removes some of the lacuna existing in past works and improves the performance of WT system.

Keywords Wind turbine · Twist · Chord · Airfoil · Lift and drag coefficient · Reynold number

1 Introduction

With the rapid development of the technologies and human living standard, demand for the electricity has increased manifold. Conventional energy sources, i.e., fossil fuels have long been used for the generation of electricity. However, gradual depletion of these sources and increasing environmental concerns have triggered the need of utilising the alternative source of energy, the renewables such as wind, solar, hydro, biomass, geothermal, and few others. Potential of these renewable energy (RE) sources

V. Anand (✉)
Kalinga Institute of Industrial Technology, Bhubaneswar, India
e-mail: vikash.anandfel@kiit.ac.in

Deavshish · S. Kumar
National Institute of Technology, Jamshedpur, India

A. Maurya et al. (eds.), *Recent Trends in Mechanical Engineering*, Lecture Notes in Mechanical Engineering, https://doi.org/10.1007/978-981-19-7709-1_66

have been neglected in the past. From the last few years, they are getting much attention owing to their infinite availability and absence of any harmful effect. Among all the available renewable energy sources, wind has the tremendous potential to fulfil the world's electricity demand. Due to its unique geographical location, India is one of the favourable destination for wind energy exploitation. The progress made in control strategies, semiconductor technologies, and robust wind turbine structure has helped in formulating efficient wind energy conversion system (WECS).

Wind turbine (WT) is the one of the most important part of entire system which is responsible for conversion of physical energy in the wind to mechanical energy. The efficacy of WECS is largely dependent upon appropriate design of wind turbine blade. Nowadays, small-scale wind turbines with the output power range from few watt to 50 kW are being employed at the off-grid location such as seashore and remote places. Small WT is very often located at the place or height where consistent wind flow may not be possible. Moreover, small WT has to operate under intermittent wind speed condition and low *Reynold number* by virtue of which it experiences low lift and high drag force. Therefore, careful design of wind turbine blade as per power requirement is very crucial in efficient energy extraction from the wind.

Many research works are available for the WT-blade design, however majority of them are dedicated to large-scale system and very few have paid attention to the small wind turbine [1–5]. In [6], methodology has been developed for the aero-structural and aero-dynamic design of sWT while taking the effect of several related parameters into consideration. In [7], a complete picture of modern state of art wind turbine blade design including, aerofoil selection, shape/quantity, and optimum angle of attack in the form of review has been presented. However, in none of these papers ([6] and [7]) effect of the *Reynold number* is considered, a significant factor that should be taken care while designing sWT. In [8], SG6043 airfoil has been taken for designing the 4 kW wind turbine. An algorithm has been developed in MATLAB to discover the parameter radius, chord length and blade twist. Based on the obtained parameter, performance analysis was carried out under changing wind speed and pitch angle. In [9], design and optimization of small HAWT have been done by using MATLAB programming based on blade element momentum theory [BEM]. Performance was investigated with multiple airfoils taken for analysis and comparison. However, both these papers [9, 10] propose blade design method which is based on simple blade element and momentum(BEM) theory and fails to produce desired result under stall developing region formed due to the wake above the airfoil. Wake is the phenomena produced due to rotation imparted by the blade to the air and separation of boundary layer from the blade surface at the high angle of attack(*AoA*). It is hence responsible for less energy extraction from the wind, thus must be incorporated while designing the blade.

After going through aforementioned articles, it can be inferred that research works are majorly devoted to design of large-scale wind turbine system, whereas small wind turbines which normally works in slightly different aerodynamic conditions have found little space so far. Primary aim of blade design for a WT blade is to find out its geometrical structure (chord, twist, and other related parameter) along the blade length for a specific airfoil to meet the power requirement and performance characteristics. BEM theory method has been used for blade design purpose; however, this method does not yield the expected result in high range of *angle of attack* when separation of air from boundary layer of blade starts and wake is developed.

This paper presents design method of a 4 kW wind turbine blade via a simple computational approach including effect of wake. Performance of the system is verified with both experimentally and theoretically result obtained in few past literatures. Though practical design of the optimal blade may fetch some constraints and unaddressed issues, this approach gives an insight of small wind turbine blade design approach with the extended version of BEM theory where effect of wake is also taken into account.

2 Airfoil and Its Operation

A WT (Fig. 1a) uses the aerodynamic force of the lift to rotate a shaft which in turn helps in the conversion of mechanical power to electricity by means of a generator. The development of an efficient WT depends upon selection and design of blades.

In order to obtain appropriate design and justified performance blade is divided into several parts (airfoil) of length (chord length, *c*) and width *dr* as shown in Fig. 1b. A detailed schematic of an airfoil showing its geometric parameters and forces acting on it is shown in Fig. 2. Wind flow causes pressure difference between upper (convex) and lower (concave) surface of the blade, due to which it experiences lift force in perpendicular direction and drag force in the in parallel direction of wind flow. Resultant of these two forces causes rotation in the wind turbine. Ratio of lift to drag should be should be maximum in order to capture the best possible power from the wind. It can be noted that value of lift: drag in any airfoil is maximum at a particular *angel of attack*. Design strategy should be such that this Aoa is maintained at this particular value throughout the blade. Small wind turbine usually works under low *angle of attack* and *Reynold number*. As mentioned earlier, a blade is divided into several small parts (airfoils) for design purpose. Hence, appropriate selection of airfoil according to aerodynamic conditions and power requirement is very crucial. Few numbers of terms are used to characterize the airfoil on the basis of which there are many types of airfoils. In this work, design of small wind turbine is proposed, where there is requirement of high lift: drag. Keeping view on this a comparison of

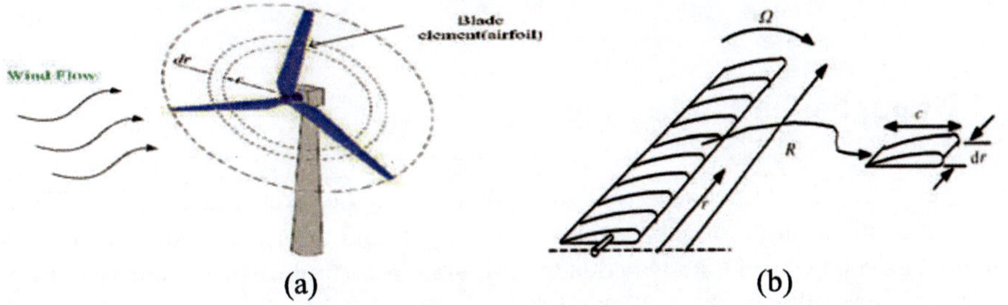

Fig. 1 **a** three-bladed wind turbine. **b** An airfoil out of the blade

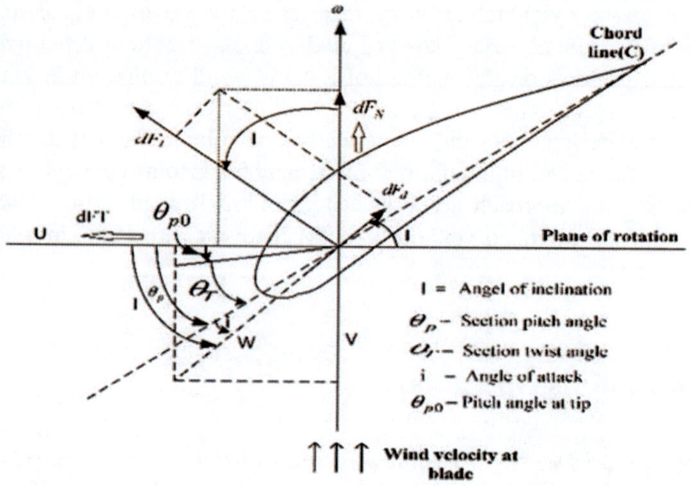

Fig. 2 Schematic of an airfoil with its geometrical parameters and forces acting on it

Fig. 3 Lift: drag of few airfoils with varying angle of attack

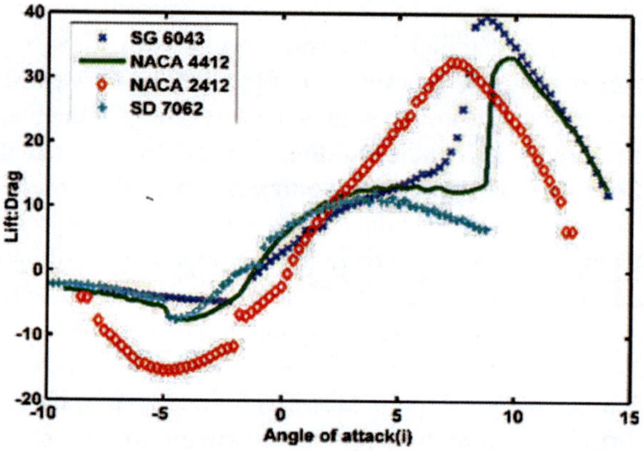

lift: drag of commonly used airfoils with varying angle of attack is shown in Fig. 3 [9]. It can be observed that SG6043 airfoils have got maximum lift: drag ratio in the low range of *angle of attack* under which a small WT normally operates [9].

3 Evaluation of Blade Shape

Few geometrical parameters of WT blade that need to be taken care are blade diameter, radial distribution of chord, twist, pitch angle, and tip-speed ratio. The airfoils should be designed in such a way that the lift: drag is maximum throughout the blade. This optimum value is obtained at a particular angle of incidence (angle of attack), which has to be taken from standard experimental data. To maintain same Aoa along the blade length, pitch angle (twist) should also changes, as angle of inclination varies along the blade length. To start with the process of obtaining the optimal blade shape, few mathematical formulation related to WT aerodynamics has to be introduced.

Axial induction factor (a) is defined as fractional decrease between wind velocity far from WT and that of passing through the blade.

$$a = \frac{V - V_1}{V} \tag{1}$$

where V is the velocity of wind coming from infinity, and V_1 is the velocity of wind passing through the blade. Some algebraic equations can be obtained from the Fig. 2.

$$\theta_T = \theta_p - \theta_{p0} \tag{2}$$

where θ_p, the angle between chord line and rotational plane of blade said to be sectional pitch angle, θ_{p0} is the pitch angle at the tip and θ_T is termed as blade tip. Angel of inclination (angle of relative wind, I) may be expressed as sum of angle of attack (i) and sectional pitch angle (θ_p).

$$I = i + \theta_p \tag{3}$$

Equation for 'angle of inclination' may be given as;

$$\tan I = \frac{V(1 - a)}{\Omega r(1 + a')} = \frac{1 - a}{(1 + a')\lambda_r} \tag{4}$$

ω is the angular velocity transmitted to wind flow due to wake and $a' = {}^{\omega}/_{2\Omega}$, is the angular induction factor. λ_r is the tip-speed ratio at distance r from the blade hub.

From the BEM theory, incremental force normal to the plane of rotation (dF_N) and force tangential to the circle of blade rotation (dF_T) can be obtained as.

$$dF_N = 0.5B\rho V_{rel}^2(C_L\cos I + C_D\sin I)cdr \tag{5}$$

$$dF_T = 0.5\rho V_{rel}^2(C_L\sin I - C_D\cos I)cdr \tag{6}$$

Similarly, tangential torque of 'B' bladed WT at the distance 'r' from the axis of rotation.

$$dQ = rBdF_T = 0.5B\rho V_{rel}^2(C_L\sin I - C_D\cos I)crdr \tag{7}$$

These relationships would be useful for designing the turbine blade of optimal shape and size. One of the important aspects of performance parameter is described by power coefficient, which is defined as:

$$C_p = \frac{P_{wt}}{P_{wind}} = \frac{\int_{r_h}^{R} \Omega dQ}{0.5\rho\pi R^2 V^3} \tag{8}$$

where P_{wt} is the power developed at WT rotor. P_{wind} is the power contained in the wind. r_h is the distance of hub from main shaft. From Eq. (5) expression for power coefficient can be obtained as, where P_{wt} is the power developed at WT blade. P_{wind} is the power contained in the flowing wind. r_h is the distance of hub from main shaft. From Eq. (7) expression for power coefficient can be obtained as,

$$C_p = \frac{8}{\lambda^2}\int_{\lambda_h}^{\lambda} (\sin I + \lambda_r \cos I)(\cos I - \lambda_r \sin I)\sin^2 I\left[1 - \left(\frac{C_D}{C_L}\right)\cot I\right]\lambda_r^2 d\lambda_r \tag{9}$$

λ_h is the value of tip-speed ratio at the rotor hub.

***A.** Blade shape for ideal WT blade (without wake rotation).*

For determining blade shape of the ideal rotor some assumption are to be made. Wake effect, drag forces, and losses between blades are neglected. Blade element theory and momentum theory have been combined (BEM theory) to obtain the expression for optimum chord and twist distribution [10].

$$\tan I = \frac{2}{3\lambda_r} \tag{10}$$

And chord length

$$c = \frac{8\pi r \sin I}{3C_L N_b \lambda_r} \tag{11}$$

These relationships can be used to find the optimum twist and chord variation of the blade along its length.

***B.** Blade shape for WT blade (with wake rotation).*

For the design of blade shape for ideal WT blade considering wake rotation (Fig. 4), effect of angular induction factor (a') has been introduced. An optimized blade shape can be obtained by partially differentiating the part of maximum power coefficient that is function of angle of inclination (I) and letting it to the equal of zero. Differentiation is done in the pursuit of getting maximum power coefficient with respect to I. From Eq. (9),

$$\frac{\partial}{\partial I}[(\sin I + \lambda_r \cos I)(\cos I - \lambda_r \sin I)\sin^2 I] = 0 \tag{12}$$

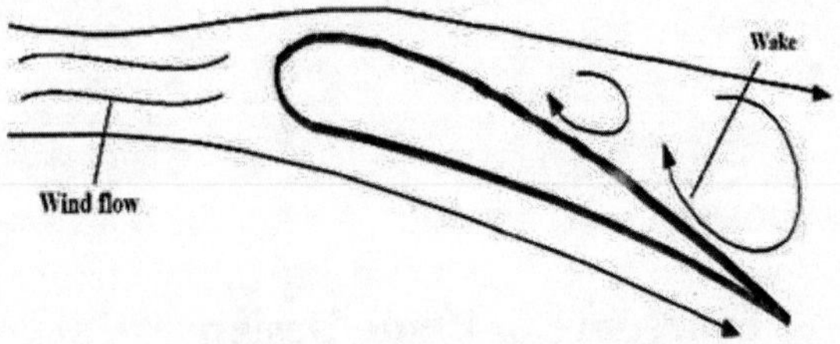

Fig. 4 Effect of wake onto the airfoil

$$\text{Hence, } \lambda_r = \frac{\sin I(2\cos I - 1)}{[(2\cos I + 1)(1 - \cos I)]} \tag{13}$$

After few exploration, section-wise shape along the blade in form of 'angle of inclination', I and chord length, c can be figured out as.

$$\left.\begin{aligned} I &= 0.66\tan^{-1}\left(\frac{1}{\lambda_r}\right) \\ c &= \frac{8\pi r}{BC_L}(1 - \cos I) \end{aligned}\right\} \tag{14}$$

Few other vital relationship may be obtained as:

$$a = a'\frac{\lambda_r}{\tan I} \tag{15}$$

$$a = \frac{1}{1 + \frac{4\sin^2 I}{\sigma C_L \cos I}} \tag{16}$$

$$\text{And,} \quad a' = \frac{1}{\frac{4\cos I}{\sigma' C_L} - 1} = \frac{1 - 3a}{4a - 1} \tag{17}$$

$$\text{Thrust coefficient } C_T = \frac{\sigma_k(C_L\cos I + C_D\sin I)(1 - a)^2}{\sin^2 I} \tag{18}$$

If calculated value of thrust C_T is below than 0.96, then for next iteration 'a' can be updated as:

$$a_{n+1} = \frac{1}{\frac{4P_n\sin^2 I_n}{\sigma C_{L_n}\cos I_n} + 1} \tag{19}$$

Else for, $C_T > 0.96$ $C_T > 0.96$

$$a_n = \left(\frac{1}{P_{k,n}}\right)\left[0.14 + \sqrt{0.02 - 0.65(0.89 - C_{T,k,n})}\right] \tag{20}$$

$$a'_{n+1} = \frac{1}{\frac{4P_{k,n}\cos I_{k,n}}{\sigma' C_{L,k,n}} - 1} \tag{21}$$

4 Procedure of Blade Design

Algorithm is proposed to design a 4 kW, 3-bladed wind turbine at a rated wind speed. In this work, SG6043 has been taken as an airfoil. A flow chart has been shown in the Fig. 5 to understand the whole design process in brief. An iterative procedure is followed to obtain the optimal blade shape and performance at specific tip-speed ratio (TSR). This design procedure is started by considering one particular value of TSR. Once obtained performance parameters for one value of TSR, procedure may be repeated to get the performance with its different values (Table 1).

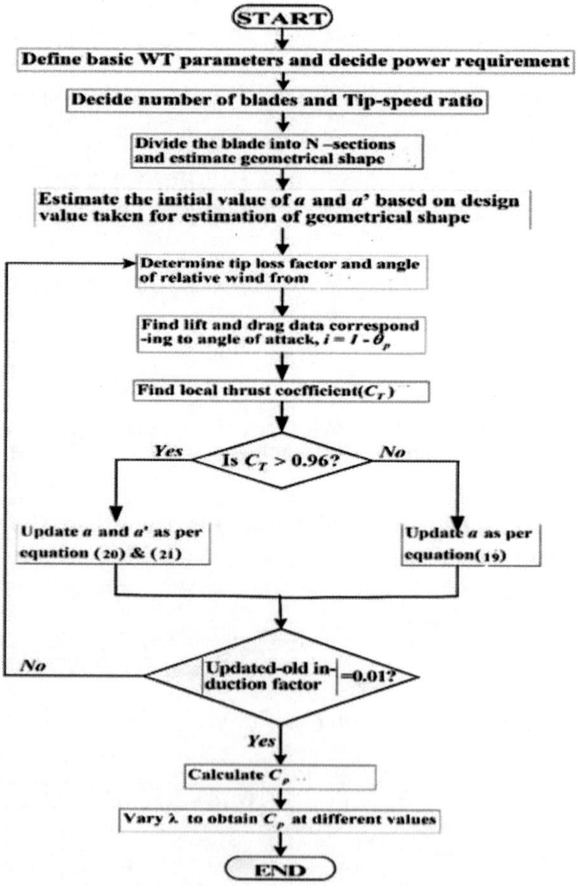

Fig. 5 Flow chart for proposed blade design

5 Results and Discussions

In this work, a simple design procedure for 4 kW wind turbine has been proposed. SG6043 airfoil has been taken for the design of blades. It is essential to find the value of *angle of attack* for the airfoil at which lift: drag ratio has the maximum value. In order to maintain the particular AoA at fixed value twist angle has to be varied as *inclination angle, I,* varies throughout the blade length. In the process of designing, optimum geometrical blade structure, i.e., variation of chord length, twist angle, etc., has been determined by considering both cases, with (parameters with subscript w in the Table 1) and without wake rotation as shown in Fig. 6. It can be observed that chord length decreases along the blade length from the blade hub to tip to ensure uniform thrust over the blade, whereas variation in twist angle is there to work entire blade under the AoA which provides maximum lift: drag (Fig. 6b).

After going through the iterative procedure performance of wind WT is obtained, which is curve between power coefficient (Cp) and tip-speed ratio. According to

Table 1 Variation of parameters along the blade length with and without considering wake rotation

r	I	θ_p	θ_T	C	I_W	$\theta_{p,w}$	$\theta_{T,w}$	C_w
0.1	60.00	51.25	54.56	0.45	45.80	38.80	40.40	0.31
0.3	29.59	20.84	24.15	0.25	27.05	20.05	21.63	0.27
0.5	19.24	10.49	13.79	0.17	18.13	11.13	12.71	0.21
0.7	13.93	5.18	8.47	0.12	13.43	6.43	8.03	0.17
0.9	10.67	1.92	5.22	0.09	10.61	3.62	5.22	0.13
1.1	8.78	0.03	3.33	0.08	8.77	1.76	3.33	0.11
1.3	7.46	−1.29	2.01	0.07	7.44	0.44	2.02	0.09
1.5	6.48	−2.27	1.03	0.06	6.47	−0.50	1.07	0.08
1.7	5.72	−3.03	0.29	0.05	5.71	−1.25	0.30	0.07

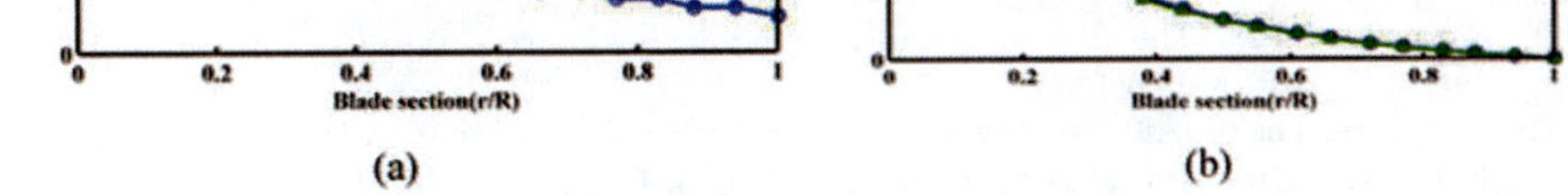

Fig. 6 Variation of Chord (**a**) and twist (**b**) along the blade length with and without considering wake rotation

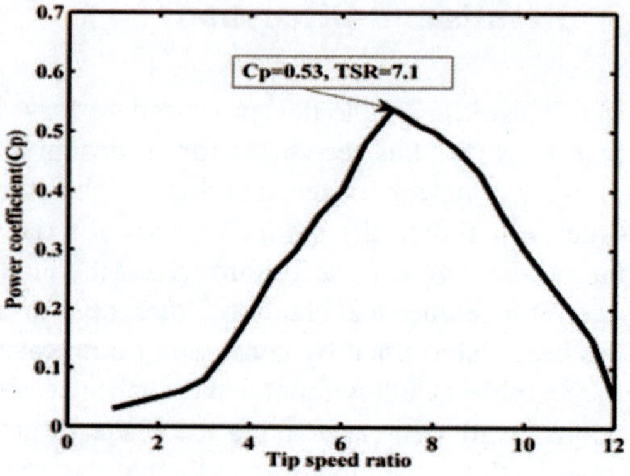

Fig. 7 Performance curve of wind turbine

Betz, maximum conversion efficiency or in other words, power coefficient can't be greater than 0.59. In the proposed work maximum value of Cp is found to be 0.53 at TSR = 7.1. This result nearly conforms to those presented in past literatures and experimental data [6–9].

6 Conclusions

Small wind turbine has not found much space in the research arena. This work presents a simple design procedure of 4 kW 3-bladed wind turbine. All the vital design parameters have been obtained and performance investigated by comparing results existing in past literatures. These geometrical design parameter of the blade can be obtained by combining forces from blade element and momentum theory. However, effect of rotating blade, which causes flow of opposite nature (wake) behind it especially in the range of high *AoA* and results in energy extraction lower than expectation has been neglected in this analysis. This lacunae in BEM theory has been eradicated in this work. This design plan can be useful in installing small wind turbine at the isolated areas where there is consistent wind flow and electric grid is far from reach.

References

1. Cheng M, Zhu Y The state of the art of wind energy conversion system and technologies. Energy Convers Manag 88:332–347
2. Devashish, Thakur AN (2019) Novel control strategy for standalone wind energy conversion system supplying power to isolated dc load. Majlesi J Electr Eng 13(1)
3. Wood D (2006) Small wind turbine analysis, design, and application. Springer
4. Devashish, Thakur AN (2017) A comprehensive review on wind energy systems for electric power generation: current situation and improved technologies to realize future development. Int J Renew Energy Res (IJRER) 7:1786–1805

5. Javed F, Javed S, Bilal T, Rastogi V (2016) Design of multiple airfoil HAWT blade using MATLAB programming. In: 5th international conference on renewable energy research and application, UK, 20–23 Nov 2016
6. Satpathy AS, Kastha D, Kishore NK (2016) Modeling of small wind turbine characteristics. In: IEEE International conference on Power System, New Delhi
7. Karthikeyan N, Murugavel KK, Kumar SA, Rajakumar S (2015) Review of aerodynamic developments on small horizontal axis wind turbine blade, vol 42, pp 801–822
8. Manwell JF, McGowan JG, Rogers AL (2002) Wind energy explained-theory, design and application. Wiley
9. Rehman S, Alam MM, Alhems LM, Rafique MM (2018) Horizotnal axis wind turbine blade design methodologies for efficiency enhancement- a review. Energies 11(3)
10. Jha D, Singh M, Thakur AN (2021) A novel computational approach for design and performance investigation of small wind turbine blade with extended BEM theory. Int J Energy Environ Eng

Numerical Investigation on the Thermal Performance of Hybrid Nano-Enhanced Phase Change Material in Heat Exchanger

R. Harish, Karthik Sekaran, Karan Das, and Neville Chrimson Noah

Abstract In the present study, a numerical investigation is performed to understand the flow and heat transfer characteristics of hybrid nano-enhanced phase change material in a heat exchanger. A heat exchanger of cylindrical cross-section is considered for the analysis. Paraffin wax (RT50) is considered as the phase change material (PCM) into which SWCNT-MgO hybrid spherical nanoparticles are dispersed. The parametric study is performed by varying the Nusselt number and heat transfer coefficient over a wide range of nano-particle volume fractions and different temperatures. The melting and heat transfer characteristics of the phase change material are investigated using ANSYS Fluent V.20.0. The problem is modeled as an unsteady, two-dimensional incompressible flow, considering the effects of melting and solidification. This study has identified that the melting rate of the PCM was significantly influenced by varying the particle volume fraction and temperature of the inner fluid. The transient melting behavior of the PCM is investigated by plotting contours of temperature distribution.

Keywords Heat exchanger · Phase change material · Hybrid nanoparticles · Single-walled carbon nanotubes · Latent heat thermal energy storage

1 Introduction

The modern age has seen an astronomical shift in the way energy is being supplied, with major companies in incessant pursuit of renewable sources of energy in order to meet the surging demands. The rift between energy supply and associated demand is currently a major challenge, one which can easily be solved by development of Thermal energy storage systems [1, 2]. Among these, latent heat energy storage using phase change materials (PCM) has been at the forefront of discussion in recent years [3]. These materials possess the ability to store and release a huge amount of latent heat during the phase transition period [4]. The flipside to PCM's however, lies in

R. Harish (✉) · K. Sekaran · K. Das · N. C. Noah
School of Mechanical Engineering, Vellore Institute of Technology, Chennai, Tamil Nadu 600127, India
e-mail: harish.r@vit.ac.in

A. Maurya et al. (eds.), *Recent Trends in Mechanical Engineering*, Lecture Notes in Mechanical Engineering, https://doi.org/10.1007/978-981-19-7709-1_67

their low thermal conductivity. There are multiple ways to get around this problem, including the use of fins in the geometry of the setup or the addition of nanoparticles in the PCM [5, 6], and sometimes even a combination of both. Nanoparticles are a brilliant solution to improving thermal conductivity [7, 8]. For instance, high heat generation owing to manufacturing, microelectronics, thermal plants, etc., is accompanied by a need for efficient heat dissipation technology. Conventional cooling techniques involve the use of oils, ethylene glycol, and water but these are hindered by their low thermal conductivity. Addition of nanoparticles to these coolants will help significantly increase the thermal conductivity of the system [9]. The use of "Hybrid" nanoparticles has also been studied in detail the last few years, whose properties find enormous potential [10, 11]. Different combinations of Nanoparticles such as Al_2O_3-Ag, SWCNT-MgO [12], Al_2O_3-SiO_2, Fe_2O_3-CNT, Al_2O_3-CNT, Al_2O_3-TiO_2, Ag-CNT, Cu–TiO_2 and MgO-MWCNT into the base fluids (Ethylene Glycol, Water, and a mixture of both) were extensively studied by researchers [13]. Du et. al [14] attempted to enhance the rate of heat transfer of latent heat storage unit using a coil heat exchanger by adding Copper nanoparticles to Paraffin. The results from their simulation revealed that 19.6% of the total melting time used up by the pure PCM, was saved owing to use of the nano-enhanced PCM. Furthermore, significant alleviation of non-uniformity in temperature was noted in the unit, because of the use of dispersed nanoparticles. Akhmetov et al. [15] used two integrated Latent Heat Thermal Energy Storage Systems, basing them on Paraffin Waxes PW-L and PW-H with varying phase change temperatures, and studied them numerically using COMSOL Multiphysics. On addition of aluminum oxide nanoparticles, they observed that when the nanoparticle comprises 4% by weight of PW-L mass, the thermal diffusivity could be increased up to 40% for PW-L and about 25% for PW-H. They also note that the addition of nanoparticles had no bearing on the latent heat and specific heat capacity of the PCM but did indeed enhance the heat transfer rate of the PCM. Shafee et al. [16] performed an unsteady simulation in which they charged Paraffin inside an enclosure equipped with a spiral pipe. They added nanoparticles to enhance the thermal performance. Their model involved increasing the inlet velocity, which led to a higher liquid fraction consequently leading to better performance. It was also noted that the inlet velocity had a diminished effect on melting as the time increased.

In this particular study, the enhancements provided by the addition of hybrid nano-particles of varying volume fractions are studied. Temperature variations of up to 127 °C (400 K) affecting heat transfer through NEPCM are studied. Such high-temperature studies are markedly absent from previously reviewed literature. Similarly, studying the effect of volume fractions of nanoparticles as high as 6% in the PCM is unique to our study. Among other hybrid nanoparticle studies, usage of single-walled carbon nanotubes in combination with any other metal oxide is quite rarely studied, something which our study delves into detail.

2 Methodology

Figure 1A shows the 2D geometry of the heat exchanger and Fig. 1b depicts the mesh used for the purpose of the simulation. A fluid having a temperature of 343 K is channeled through the inner cylinder while the nanofluid is present in the gap between the outer and inner cylinder. Initially, the nanofluid mixture, consisting of the PCM-paraffin wax and nanoparticles- Single-Walled Carbon Nano Tube (SWCNT)/MgO, exists in solid state but eventually melts due to transfer of heat from the hot fluid flowing through the inner cylinder. The walls of the enclosure are made of Aluminum and the outer wall is considered to be an adiabatic boundary.

The problem at hand requires the solution of all Navier Stokes equations due to advance of the melting front caused by the temperature differential in the material.

Continuity equation:

$$\frac{\partial \rho}{\partial t}+\frac{\partial(\rho u)}{\partial x}+\frac{\partial(\rho v)}{\partial y}=0 \tag{1}$$

Momentum equation:

$$\frac{\partial(\rho v)}{\partial t}+\frac{u\partial(\rho v)}{\partial x}+\frac{v\partial(\rho v)}{\partial y}=-\frac{\partial P}{\partial y}+\frac{\partial}{\partial x}\left(\frac{\mu\partial v}{\partial x}\right)+\frac{\partial}{\partial y}\left(\frac{\mu\partial v}{\partial y}\right) - Sv+\rho g\beta(T-T\mathrm{m}) \tag{2}$$

Energy equation:

$$\frac{\partial(\rho C_p T)}{\partial t}+\frac{u\partial(\rho C_p T)}{\partial x}+\frac{v\partial(\rho C_p T)}{\partial y}=\frac{\partial}{\partial x}\left(\frac{k\partial T}{\partial x}\right)+\frac{\partial}{\partial y}\left(\frac{k\partial T}{\partial y}\right) \tag{3}$$

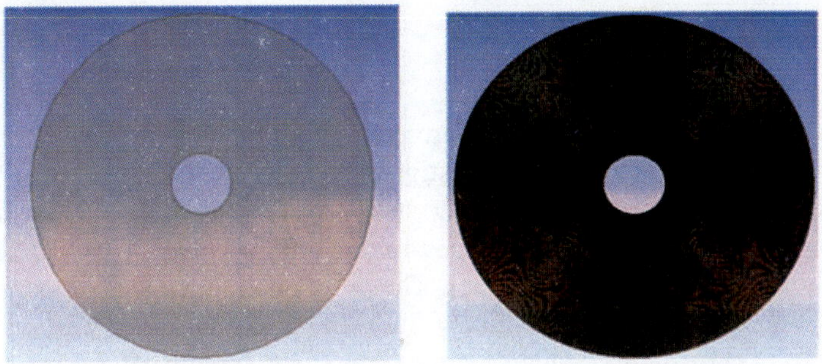

(a) (b)

Fig. 1 **a** Geometry of the heat exchanger and **b** Fine Mesh (70,000 cells)

Table 1 PCM and nanoparticle properties

Material	ρ (kg/m^3)	C_p (J/kgK)	L.H. (kJ/kg)	k (W/mK)	Φ
RT50	800	2000	168	0.2	–
SWCNT	1900	720	–	3000	0.04
MgO	3580	1030	–	60	0.04

The numerical solution to the governing equations given above was solved using the CFD package ANSYS Fluent R1 2020. The PRESTO scheme was applied to calculate the pressure and a second-order up-wind scheme was used to solve momentum and energy equations. Finally, the SIMPLE algorithm was used to solve the coupling between pressure and velocity. Under-relaxation factors of 1 for energy, 0.9 for liquid fraction, 0.75 for both momentum and pressure were assumed, in order to improve the stability of convergence [17].

Density is calculated using *boussinesq* approximation which offers results consistent with that expected of incompressible flows having constant or temperature-dependent quantities. In this approximation, the density of the material is considered to be a constant, and a constant thermal expansion coefficient is chosen, β = 0.0006. This model is suitable for natural convection models wherein the temperature differential is minimal

$$\text{i.e}\beta(T - T_o) \ll 1 \tag{4}$$

The density, thermal conductivity, latent heat, and heat capacity are calculated based on the expressions given by Esfahani et al. [18]. Table 1 cites the properties of the PCM and the nanoparticles used where ρ refers to the density of the pcm/nanofluid, Φ refers to the volume fraction of the nanoparticle(s), C_p refers to the specific heat, k refers to the thermal conductivity of the pcm/nanoparticle, L.H refers to the latent heat of the fluid.

3 Validation

The present two-phase model results are compared with the experimental findings of Nitsas and Koronaki [19]. They investigated the thermal characteristics of a cylindrical enclosure containing water as the heat transfer fluid and RT50 as the phase change material. The nanoparticles used in the simulation are Al_2O_3 /Cu and they are considered to be spherical particles. The outer and inner diameters of the model were 125 mm and 22 mm respectively. In both cases, the inner wall was given a temperature of 343 K (80 °C). As seen in Fig. 2, the results of the present model are consistent with that of the results of Nitsas' and Koronaki's model. The rising flume pattern is present in both cases and is indicative of heat transfer due to both conduction and convection.

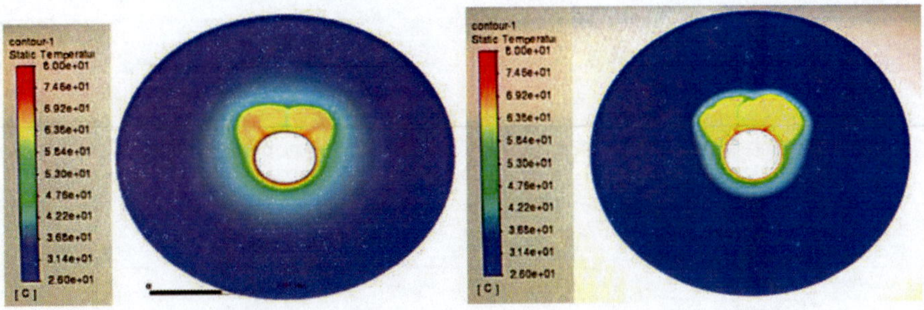

(a) (b)

Fig. 2 **a** Nitsas and Koronaki model. **b** Melting of RT50 (Present Study)

4 Results and Discussion

Figure 3 shows the temperature contours of the two-phase model embedded with SWCNT and MgO nanoparticles, i.e., hybrid nanofluid, at two different inner wall temperatures:70 °C and 60 °C. When compared to Fig. 2, the rising flume penetrates deeper into the cylindrical enclosure which can be attributed to the increased thermal conductivity of the mixture due to the addition of hybrid nanoparticles. Moreover, as the temperature difference between the inner wall and the outer wall increases, the shape of the flume tends to become more ovular as seen in Fig. 3a. Higher buoyancy forces create more intense velocity fields which in turn increases the natural circulation towards the outer wall.

Figure 4 depicts the variation of static temperature with vertical distance from the center. IWT refers to Inner Wall Temperature. As expected, the temperature gradually decreases the further we move away from the core. Initially, the temperature decreases

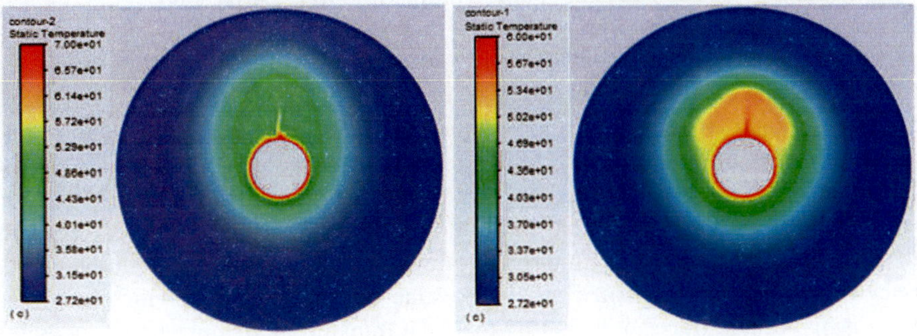

(a) (b)

Fig. 3 Temperature Contours at **a** 70 °C and **b** 60 °C

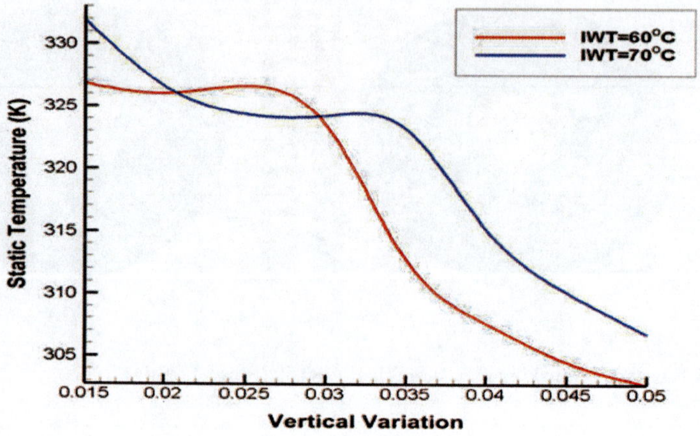

Fig. 4 Graph of static temperature versus vertical distance

Table 2 Nanofluid properties for various volume fractions

Φ (%)	ρ (kg/m^3)	C_p (J/kgK)	L.H. (kJ/kg)	k (W/mK)
2	877.17	1866.479	147.154	0.2248
4	953.44	1756.3634	129.911	0.2527
6	1028.84	1664.165	115.426	0.2834

at a slower rate but after a certain vertical distance, the temperature gradient rapidly increases. The graph was plotted at a constant horizontal distance, x = 0 mm.

Next, the inner fluid was modeled for flowing at different temperatures, namely 90°C (363 K), 100 °C (373 K), and 127 °C (400 K). Results for pure PCM and different volume fractions of nanoparticles in the PCM, namely 2%, 4%, and 6%, were obtained. For each corresponding volume fraction, the required properties are shown in Table 2.

From theoretical calculations (based on values taken off of Table 2), it is inferred that the thermal conductivity of the PCM increases by about 12.4% for 2% volume fraction addition of nanoparticles, by about 26.35% for 4% volume fraction and almost up to 42% for 6% volume fraction.

Nusselt number is a measure of how much heat is transferred via convection compared to that via conduction, in a fluid. A larger Nusselt number corresponds to more active convection, with turbulent flow typically in the 100–1000 range. As is evident from the graph in Fig. 5, the value of Nusselt number increases with increasing volume fraction indicating that heat transferred via convection increases when the concentration of nanoparticles is increased. For example, for a temperature difference of 60 K, it is inferred that the Nusselt number increases by about 5.43% for 2% volume fraction addition of nanoparticles, by about 10.22% for 4% volume fraction and almost up to 11.4% for 6% volume fraction.

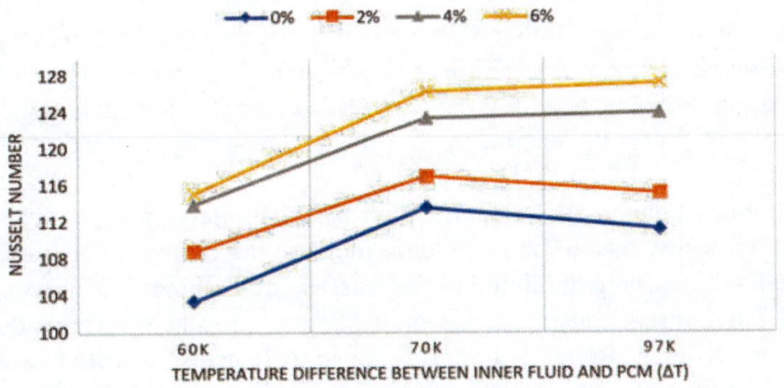

Fig. 5 Graph of Nusselt number versus temperature difference (ΔT)

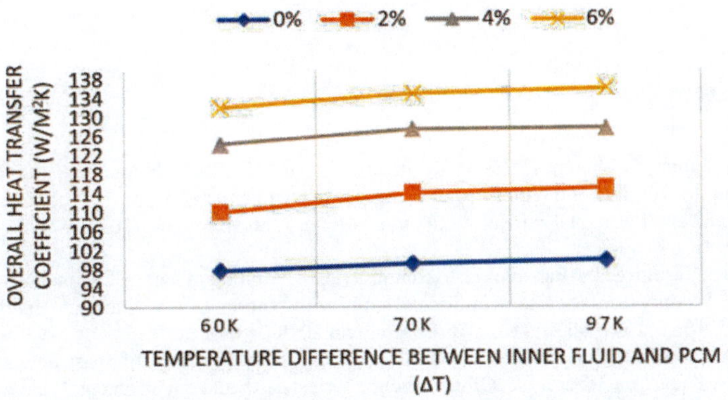

Fig. 6 Graph of heat transfer coefficient versus temperature difference (ΔT)

As seen in the graph in Fig. 6, heat transfer coefficient increases with increasing volume fraction indicating more heat gets transferred from water to the PCM. For example, for temperature difference of 60 K, it is inferred that the overall heat transfer coefficient increases by about 12.75% for 2% volume fraction addition of nanoparticles, by about 27.32% for 4% volume fraction, and almost up to 35.30% for 6% volume fraction.

5 Conclusion

The present numerical results are validated and are in excellent agreement with the benchmark results available in literature. Temperature was observed to decrease with increasing vertical distance from the inner core. The value of Nusselt number increases with increasing volume fraction indicating that heat transferred via convection increases when the concentration of nanoparticles is increased. Thermal conductivity enhancement of up to 42% is seen for 6% volume fraction addition of nanoparticles. Heat transfer coefficient increases with increasing volume fraction of nanoparticles. An increase of 35.30% is seen for 6% volume fraction for a temperature of 60 K. Addition of nanoparticles cannot increase indefinitely owing to the NEPCM's kinematic viscosity, which increases with increase in volume fraction of nanoparticles. Melt circulation is inhibited by an increase in the kinematic viscosity of the NEPCM. This consequently suspends natural convection and eventually, the transfer of heat to the NEPCM from the fluid. It is hoped that the results from this study help advance the usage of hybrid nanoparticles in combination with PCM's to enhance thermal conductivity, in particular, SWCNT's in combination with other metal oxide nanoparticles.

References

1. Rohit Kumar K, Sharan V, Harish R, Shyam Kumar MB (2020) Numerical study of thermal cooling for Lithium-ion battery pack using phase change material. In: IOP conference series: earth and environmental science, 2020, vol 573, no. 1. https://doi.org/10.1088/1755-1315/573/1/012018.
2. Ebadi S, Tasnim SH, Aliabadi AA, Mahmud S (2018) Melting of nano-PCM inside a cylindrical thermal energy storage system: numerical study with experimental verification. Energy Convers Manag 166. https://doi.org/10.1016/j.enconman.2018.04.016
3. Bahuguna A, Harish R, Shyam Kumar MB (2020) Thermal analysis of phase change materials inside a square enclosure. In: IOP conference series: earth and environmental science, vol 573, no. 1. https://doi.org/10.1088/1755-1315/573/1/012019
4. Leong KY, Abdul Rahman MR, Gurunathan BA (2019) Nano-enhanced phase change materials: a review of thermo-physical properties, applications and challenges. J Energy Storage 21. https://doi.org/10.1016/j.est.2018.11.008
5. Harish S, Harish R, Shyam Kumar MB (2020) Numerical simulation of melting behavior of nano-enhanced phase change material in differentially heated cavity. In: IOP conference series: earth and environmental science, vol 573, no 1. https://doi.org/10.1088/1755-1315/573/1/012028
6. Gorzin M, Hosseini MJ, Rahimi M, Bahrampoury R (2019) Nano-enhancement of phase change material in a shell and multi-PCM-tube heat exchanger. J Energy Storage 22. https://doi.org/10.1016/j.est.2018.12.023
7. Bala Akash M, Harish R, Shyam Kumar MB (2020) Thermal performance of nanoparticle enhanced phase change material in a concentric cylindrical enclosure. In: IOP conference series: earth and environmental science, vol 573, no 1. https://doi.org/10.1088/1755-1315/573/1/012016
8. Sheikholeslami M, Keshteli AN, Babazadeh H (2020) Nanoparticles favorable effects on performance of thermal storage units. J Mol Liq 300. https://doi.org/10.1016/j.molliq.2019.112329
9. Sajid MU, Ali HM (2018) Thermal conductivity of hybrid nanofluids: a critical review. Int J Heat Mass Transfer 126. https://doi.org/10.1016/j.ijheatmasstransfer.2018.05.021
10. Sundar LS, Irurueta GO, Venkata Ramana E, Singh MK, Sousa ACM (2016) Thermal conductivity and viscosity of hybrid nanfluids prepared with magnetic nanodiamond-cobalt oxide (ND-Co3O4) nanocomposite. Case Stud Therm Eng 7. https://doi.org/10.1016/j.csite.2016.03.001

11. Harish R, Sivakumar R (2021) Turbulent thermal convection of nanofluids in cubical enclosure using two-phase mixture model. Int J Mech Sci 190. https://doi.org/10.1016/j.ijmecsci.2020.106033
12. Hemmat Esfe M, Alirezaie A, Rejvani M (2017) An applicable study on the thermal conductivity of SWCNT-MgO hybrid nanofluid and price-performance analysis for energy management. Appl Therm Eng 111. https://doi.org/10.1016/j.applthermaleng.2016.09.091
13. Huminic G, Huminic A, Dumitrache F, Fleacă C, Morjan I (2020) Study of the thermal conductivity of hybrid nanofluids: recent research and experimental study. Powder Technol 367. https://doi.org/10.1016/j.powtec.2020.03.052
14. Du R, Li W, Xiong T, Yang X, Wang Y, Shah KW (2019) Numerical investigation on the melting of nanoparticle-enhanced PCM in latent heat energy storage unit with spiral coil heat exchanger. Build Simul 12(5). https://doi.org/10.1007/s12273-019-0527-3
15. Akhmetov B, Navarro ME, Seitov A, Kaltayev A, Bakenov Z, Ding Y (2019) Numerical study of integrated latent heat thermal energy storage devices using nanoparticle-enhanced phase change materials. Sol Energy 194. https://doi.org/10.1016/j.solener.2019.10.015
16. Shafee A, Jafaryar M, Alghamdi M, Tlili I (2020) Entropy generation for spiral heat exchanger with considering NEPCM charging process using hybrid nanomaterial. Eur Phys J Plus 135(3). https://doi.org/10.1140/epjp/s13360-020-00284-0
17. Mahood HB, Mahdi MS, Monjezi AA, Khadom AA, Campbell AN (2020) Numerical investigation on the effect of fin design on the melting of phase change material in a horizontal shell and tube thermal energy storage. J Energy Storage 29. https://doi.org/10.1016/j.est.2020.101331
18. Esfahani MR, Languri EM, Nunna MR (2016) Effect of particle size and viscosity on thermal conductivity enhancement of graphene oxide nanofluid. Int Commun Heat Mass Transf 76. https://doi.org/10.1016/j.icheatmasstransfer.2016.06.006
19. Nitsas MT, Koronaki IP (2020) Thermal analysis of pure and nanoparticle-enhanced PCM—application in concentric tube heat exchanger. Energies 13(15). https://doi.org/10.3390/en13153841

Numerical Simulation of Multiphase Flow and Heat Transfer Characteristics in Mixing Tank

P. Deepak, K. Reddy Rajesh, B. Veera Raghava, and R. Harish

Abstract In this project, the computational fluid dynamics approach is used to study the mixing in a stirred tank reactor. The main aim is to synthesize fine chemicals and pharmaceuticals involving multiple reactions which are critically dependent upon proper mixing and heat transfer in various zones of the tank domain and to increase the efficiency of the tank by providing visual imagery, simulations, and various data for support. To increase efficiency, there are many factors which are dimensions, number of rotors, rotor design, the type of fluids and materials and heat supply, etc. So, the design and scale-up are a great challenge for such reactors. Though there have been many measurement techniques developed over the years, they all have some limitations like a disturbance in the flow field and non-invasive techniques become inefficient in the ideal working environment. Through this paper, there is going to be a development and validation of the multiphase flow inside the mixing tank by varying impeller velocities. The eulerian method with k-ω turbulence is used along with the standard model in order to understand and eliminate losses.

1 Introduction

Mixing tanks, stirred tanks, agitators, and bio-reactors along with a variety of other mixing equipment are used in process industries, pharmaceutical industries, research centers for processing and manufacturing chemicals, and various pharmaceuticals. These devices can be designed to handle anything from minute amounts of ingredients, like microparticles, or to agitate and mix large drums, vats, and tanks. The type of chemical mixer necessary depends on numerous factors, such as the materials being mixed, the temperatures and consistencies involved, the amount of material being processed, and the desired end product.

There are various types of mixing equipment used for various processes and chemical reactions. Some include static mixers, portable mixers, handheld mixers,

P. Deepak · K. R. Rajesh · B. V. Raghava · R. Harish (✉)
School of Mechanical Engineering, Vellore Institute of Technology, Chennai, Tamil Nadu 600127, India
e-mail: harish.r@vit.ac.in

A. Maurya et al. (eds.), *Recent Trends in Mechanical Engineering*, Lecture Notes in Mechanical Engineering, https://doi.org/10.1007/978-981-19-7709-1_68

tank mixers, dispersion blades, etc. Also, various chemicals with different phase compositions are used to form the desired output. While mixing, there are certain parameters like the types of impeller blades, geometry, the rotation velocities, the type of fluids used, etc. which determine the quality and efficiency of mixing.

This paper focuses on a multiphase flow with Computational Fluid Dynamics Analysis. The two phases considered here are air and water at room temperature (300 k). The velocity, pressure, turbulence intensity, and volume fraction are examined w.r.t varying impeller velocities and addition of heat flux.

2 Literature Review

Pan Zhanga et al. [1] on performing the experimental and numerical tests on the mixing characteristics of cylindrical stirrer by the Three-dimensional time-dependent flow and mixing prediction suggesting that the cylindrical stirrer is more energy-efficient than Rushton and propeller turbines. Since it is difficult to examine the behavior of cultivated broth flow with experimental techniques, CFD analysis has been employed by Bahar Aslanbay Gulera et al. [2] have used Rushton turbine impeller in PBR and found the avg velocity, turbulence properties, and dead zone in-side PBR. Computational fluid dynamics analysis has been performed by Hoseini et al. [3] on u- and v-shape impellers and Rushton impellers for minimizing power consumption and reduced impeller equivalent stresses. The analysis showed that the power number is 21% and 18% for V- and U-shaped impellers when compared with Rushton impellers, whereas the maximum turbulent kinetic energy and maximum dissipation rate of the turbulent kinetic energy of U shape impeller are more than the V shape impeller and the lowest von mises stresses occur with V shape impeller. The CFD simulations by Patil et al. [4] with MRF performed on the fully baffled tank with Rushton impeller showed the optimized dimensions of the inner rotating fluid zone which plays an important role in accurately predicting the results on numerically based problems.

Sahu et al. [5] on studying the different designs of axial flow impellers by using the zonal model concept a new method has been proposed which is used to predict the mixing time of different impellers and the values of turbulent energy dissipation rate. Diana Carolina Hernandez-Jaramillo et al. [6] used CFD analysis with the two-phase model used to find the flow distribution in a crude oil container in which two impellers are side-mounted at 90° and 45° mounting angles. Masoud Rahimi et al. [7] found the effect of jet placing nearer to the impeller in a large crude oil tank on reducing the mixing time was investigated and showed that the homogenization time is lower when the angle between the jet and the impeller is 15° and 60°. Arjun Kumar Pukella et al. [8] using CFD analysis with the volume of fluid (VOF) approach and Lagrangian particle tracking designed a new baffle type called interface baffle for effective mixing with optimum baffling.

3 Mathematical Formulation

There are a wide range of mixing tanks with different capacities. To understand basic characteristics, a standard tank of 250 L capacity is modelled in Solid Works software.

Since the capacity is 250 L, the volume of tank is 0.25 m^3.

Considering the tank as a cylinder, the volume of cylinder V = πr2H;

$$\text{Volume of the Cylinder} = V = 0.25\ m^3 = \pi(D_t)2h/4 \tag{1}$$

The ratio between H:Dt is varying from 0.8 to 1.2

Our consideration ratio between H to D_t is 0.8

So, substitute H = 0.8 × Dt in Eq. (1).

Diameter of Tank, Dt ≈ 0.74 m, height of tank, H ≈ 0.59 m, diameter of Impeller, Da = 0.37 m, Impeller clearance, C = 0.123 m, width of the Impeller, W = 0.74 m, Length of the Impeller, L = 0.0925 m. The standard k-omega model is an empirical model based on model transport equation for the turbulence kinetic energy and the behavioral usage of model is for superior performance for wall-bounded boundary layer, free shear, and low Reynolds number flows compared to models from the k-eplison family. In this project, standard k-omega model has been used and the following governing equation is:

The Continuity Equation:

$$\frac{\partial \rho}{\partial t} + \nabla.(\rho \vec{v}) = 0 \tag{2}$$

The Momentum Equation:

$$\frac{\partial}{\partial t}(\rho \vec{v}) + \nabla.(\rho \vec{v}\,\vec{v}) = -\nabla p + \nabla.\left[\mu\left(\nabla\vec{v} + \nabla\vec{v}^{T}\right)\right] + \rho\vec{g} + \vec{F} \tag{3}$$

Energy Equation:

$$\rho c_P\left(\frac{\partial T}{\partial t} + u\frac{\partial T}{\partial x} + v\frac{\partial T}{\partial y} + w\frac{\partial T}{\partial z}\right) = k\left(\frac{\partial^2 T}{\partial x^2} + \frac{\partial^2 T}{\partial y^2} + \frac{\partial^2 T}{\partial z^2}\right) + \phi \tag{4}$$

Turbulence presence in the domain has been modeled using standard.

k-omega model.

$$\frac{\partial(\rho\kappa)}{\partial t} + \frac{\partial(\rho u_j \kappa)}{\partial x_j} = \rho p - \beta\rho\omega\kappa + \frac{\partial}{\partial x_j}\left[\left(\mu + \sigma_\kappa\frac{\rho\kappa}{\omega}\right)\frac{\partial\kappa}{\partial x_j}\right],\ \text{with } p = \tau_{ij}\frac{\partial u_i}{\partial x_j} \tag{5}$$

$$\partial\frac{(\rho\omega)}{\partial t} + \frac{\partial(\rho u_j\omega)}{\partial x_j} = \frac{\alpha\omega}{\kappa}P - \beta\rho\omega^2 + \frac{\partial}{\partial x_j}\left[\left(\mu + \sigma_\omega\frac{\alpha\kappa}{\omega}\right)\frac{\partial\omega}{\partial x_j}\right] + \frac{\rho\sigma_d}{\omega}\frac{\partial\kappa}{\partial x_j}\frac{\partial\omega}{\partial x_j} \tag{6}$$

4 Validation

The below schematic represents the mixing tank modeled using Solid works software. The tank consists of an impeller blade and an enclosure disk surrounding the blade for simulation purpose. The blade is rotated at an rpm of 50, 100, 150, 200, and 250 rpm. The different characteristics have been observed by varying the impeller velocity. Also, since the tank is closed, the fluids which are water and air in this case, have been placed carefully at top and bottom of the tank as primary and secondary phases with a contact region in the middle using Ansys Fluent Software and shown in Fig. 1.

4.1 *Principle of Heat Transfer Process*

The fluid in the stirred tank is heated by water vapor in the inner coil and the heat transfer process can be described as follows:

$$Q = K_o A_o \Delta T_{ln} \tag{7}$$

$$A_o = \pi d_o l \tag{8}$$

$$\Delta T_{ln} = \frac{(T_B - T_i) - (T_B - T_o)}{\ln \frac{(T_B - T_i)}{(T_B - T_o)}} \tag{9}$$

Here, A_o is the helical coil surface of heat transfer in formula (8) and ΔT_{ln} is the logarithmic mean temperature that was used for calculating the actual overall rate of heat transfer Q.

$$Q = Q_c + Q_s = \Delta E_m + \Delta E_w + Q_h \tag{10}$$

$$Q_c = M\left(C_p\right)_{cw}(T_o - T_i) \tag{11}$$

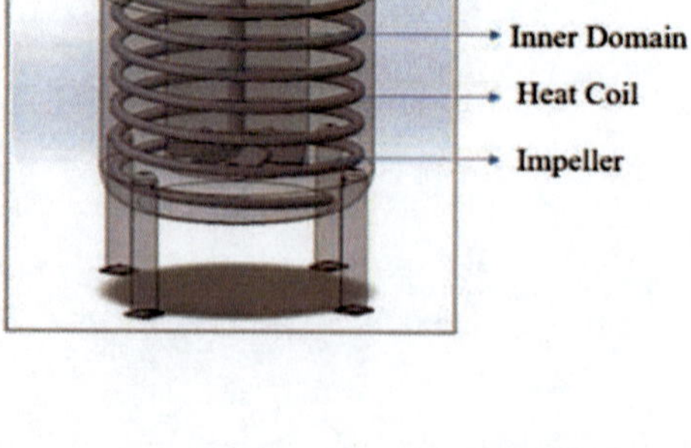

Fig. 1 Mixing Tank model

The actual overall rate of heat transfer Q in formula (10) originates from steam generators and work done by stirring. For low-viscosity fluid, the effect of the work Q_s by mechanical stirring is negligible. Thus, overall heat transfer rate Q is approximately equal to the heat flow Q_c in formula (11) generated by high-temperature steam. The method of calculation using the heat transfer coefficient hi of the inside coil wall deduces the heat transfer coefficient h_o of the outside coil wall in, and the Nusselt number Nu is obtained from h_o, d_o, and k_o.

$$Nu = \frac{h_o d_o}{k_o} \tag{12}$$

5 Results and Discussion

The results for 50, 150, and 200 rpm indicates the variation of pressure and velocity contours and are shown in Fig. 2.

From the above figures, it is clear that there is less turbulence due to low rpm of impeller. Hence most of the fluid phase does not get mixed properly and a transition or partially mixed layers can be observed in Fig. 3.

From the above figures, it is observed that at 150 rpm, a fully turbulent flow is generated and a swirl is created at the central axis, at the top of impeller blade. The pressure inside the tank is increased and high-pressure regions are created at the bottom ends of the tank. Also, the higher velocity regions are created just around the impeller and maximum velocity can be seen at the impeller shown in Fig. 4.

At this stage, a high turbulent flow can be observed. The high-pressure region is formed on top part of tank due to large swirl formation. Here, the fluids are thoroughly mixed without any interface regions and the variation of pressure, velocity and kinetic energy are tabulated in Table 1.

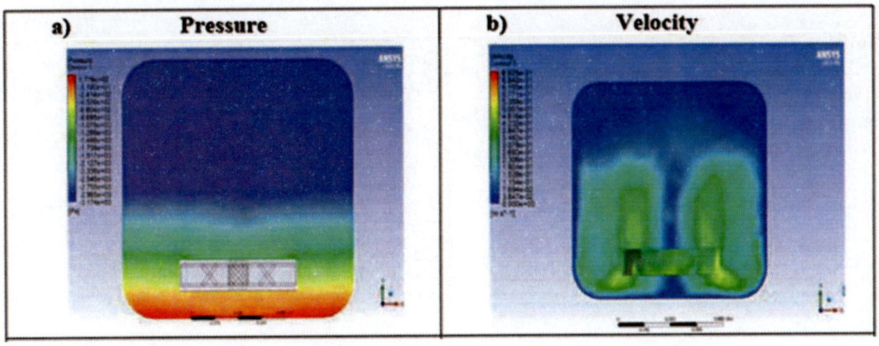

Fig. 2 Contours representing Pressure and Velocity at 50 rpm

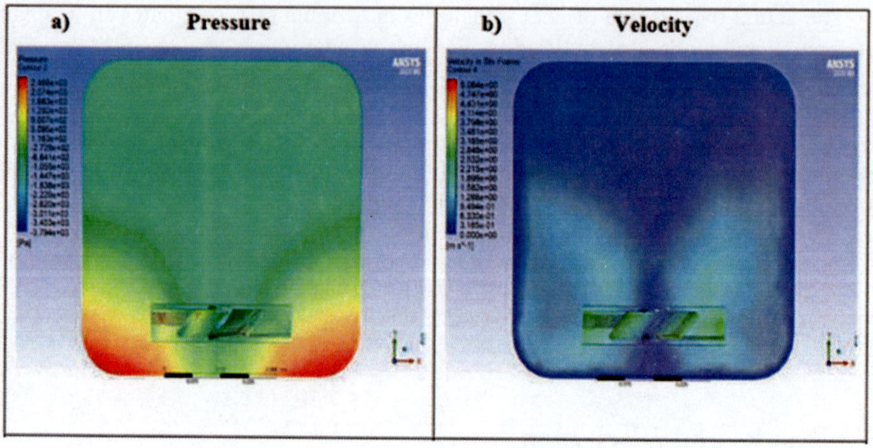

Fig. 3 Contours representing Pressure and Velocity at 150 rpm

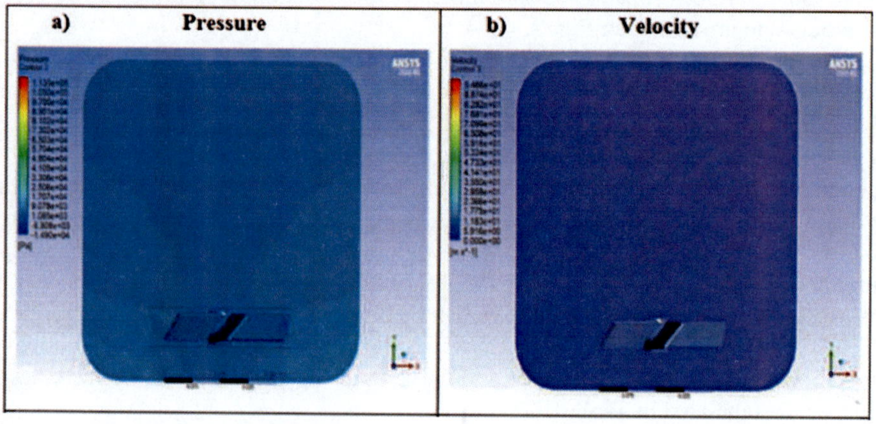

Fig. 4 Contours representing Pressure and Velocity at 250 rpm

Table 1 Pressure, Velocity, and Turbulent Kinetic Energy at different velocity of impeller

Velocity of impeller blade (rpm)	Pressure (Pa)	Velocity of fluid (m/s)	Turbulent kinetic energy (J/kg)
50	−707.9	6.585E-10	0.00065
100	−684.9	7.99E-10	0.0035
150	−603.8	9.60E-10	0.041
200	−451.9	2.75E-09	1.389
250	−263.4	2.00E-09	1.78

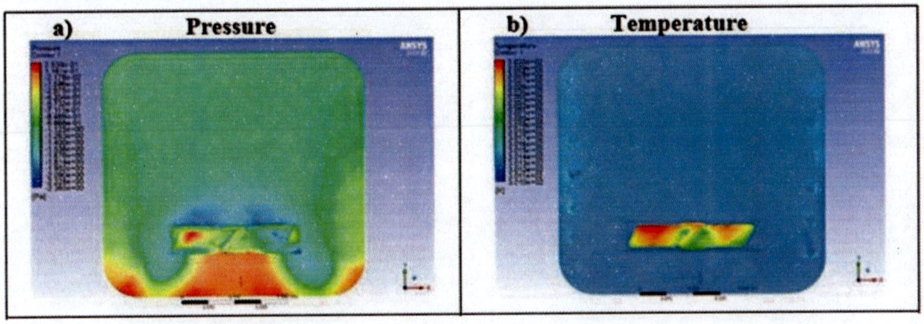

Fig. 5 Contours representing Pressure and Temperature at 5000 w/m^2

The results for 5000, 10,000, and 20,000 w/m^2 at 250 rpm are shown and pressure and temperature contours are shown in Fig. 5.

The above schematics represent different contours of pressure and temperature with a heat flux of 5000 W/m^2. Here too there are slight significant changes in pressure. The variations are minor and also the temperature is recorded as 301.3 k as shown in Fig. 6.

The above schematics represent different contours of pressure and temperature with a heat flux of 15,000 W/m^2. Here too there are significant changes in pressure. The variations are visible and also the temperature is recorded as 302.3 k and shown in Fig. 7.

Since the changes were very minimal, the heat flux was taken as 20,000 W/m^2, to observe the conditions. There was a significant increase of pressure. The temperature obtained is about 306.3 k and the variation of pressure and turbulent kinetic energy is shown in Fig. 8. Table 2 indicates the variation of parameters such as velocity, turbulent kinetic energy and temperature for various values of heat flux.

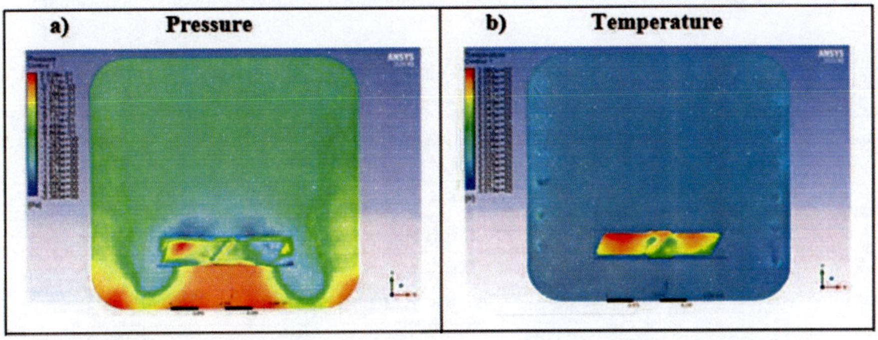

Fig. 6 Contours representing Pressure and Temperature at 10,000 w/m^2

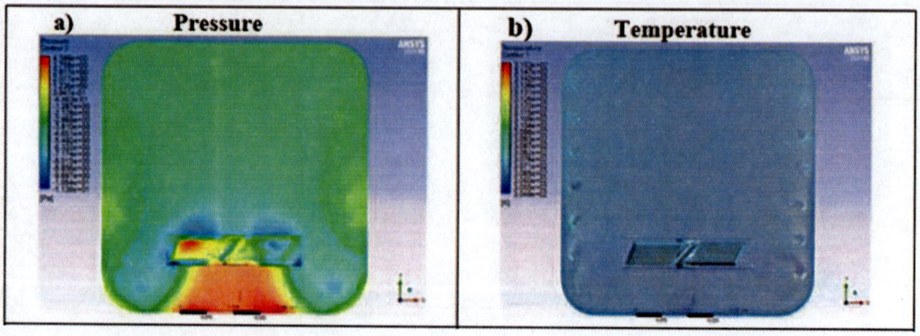

Fig. 7 Contours representing Pressure and Temperature at 20,000 w/m^2

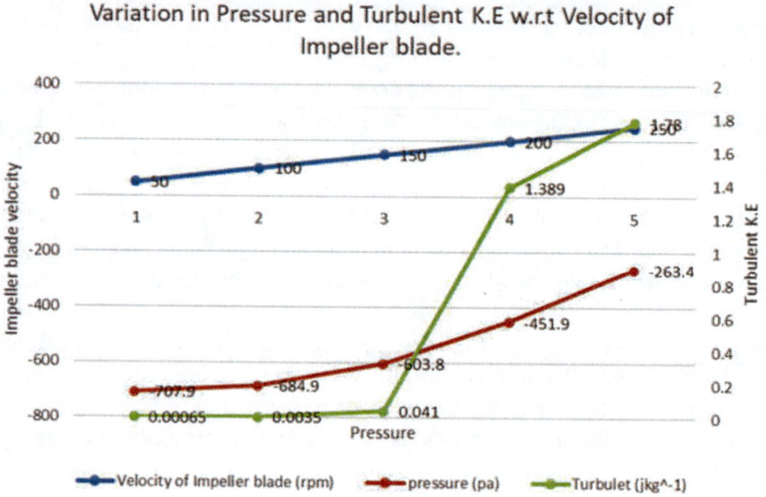

Fig. 8 A Graphical representation of pressure, Turbulent K.E w.r.t Velocity of Impeller blade

Table 2 Velocity, turbulent kinetic energy, and temperature at different heat flux

Heat flux (w/m^2)	Velocity (m/s)	Turbulent kinetic energy (J/kg)	Temperature (k)
500	124	16.4	300.12
1000	143	18.6	300.33
1500	166	22	300.4
2000	189.45	26.2	300.8
5000	205.7	31.1	301.3
10,000	221.64	33.4	302.3
20,000	241.32	36.3	306.3

6 Conclusion

The mixing tank with standard dimension has been designed using solid works as per the requirements and CFD is performed using Ansys by varying rpm and heat flux. The obtained graphs clearly show that on increasing the rpm of the impeller the mixing efficiency increases and the pressure distribution is even at 250 rpm. So, the 250 rpm is kept static and the heat flux is varied to observe the heat flux characteristics. Comparing the graphs obtained by varying the heat flux shows that, on conclusion of the heat flux, the mixing efficiency is increased. Various factors like impeller speed, size of the tank, material properties, the volume of the tank, etc. influence the mixing characteristics of the mixing tank among which heat transfer plays a critical role in pharmaceutical industries which has been studied.

References

1. Zhang P, Chen G, Duan J, Wang W (2018) Mixing characteristics in a vessel equipped with cylindrical stirrer. Results in Phys 10:699–705
2. Guler BA, Deniz I, Demirel Z, Oncel SS, Imamoglu E (2020) Computational fluid dynamics modelling of stirred tank photobioreactor for Haematococcus pluvialis production: hydrodynamics and mixing conditions. Algal Res 47:101854
3. Hoseini SS, Najafi G, Ghobadian B, Akbarzadeh AH (2020) Impeller shape-optimization of stirred-tank reactor: CFD and fluid structure interaction analyses. Chem Eng J 127497
4. Patil H, Patel AK, Pant HJ, Venu Vinod A (2018) CFD simulation model for mixing tank using multiple reference frame (MRF) impeller rotation. ISH J Hydraul Eng 1–10
5. Sahu AK, Kumar P, Patwardhan AW, Joshi JB (1999) CFD modelling and mixing in stirred tanks. Chem Eng Sci 54(13–14):2285–2293
6. Hernández-Jaramillo DC, Nieto- C, Álvarez-Sarmiento N, Barrero R, Novoa LÁ (2015) CFD simulation of mixing process in a large crude oil storage tank with side-entry impellers. Iteckne 12(1):25–32
7. Rahimi M, Parvareh A (2007) CFD study on mixing by coupled jet-impeller mixers in a large crude oil storage tank. Comput Chem Eng 31(7):737–744
8. Pukkella AK, Vysyaraju R, Tammishetti V, Rai B, Subramanian S (2019) Improved mixing of solid suspensions in stirred tanks with interface baffles: CFD simulation and experimental validation. Chem Eng J 358:621–633

CFD Analysis of an Automotive Turbocharger for Enhanced Engine Performance

Ashish Singh, Archit Sasane, Rohan Patney, and R. Harish

Abstract In recent years, developments of automobile downsizing promote the developers to enhance the performance of current turbocharging technology. Turbocharger has become one of the key components in the automotive industry as it helps to enhance the engine performance. Due to drawbacks of conventional radial turbine used in turbocharging techniques, preliminary design of axial turbine was proposed, in order to achieve highest performance of turbocharger axial turbine and therefore enhance the engine performance. In the present study, the optimal design is made based on the NACA profile blade of a single axial turbine for the turbocharger system on solidworks. A computational fluid dynamics (CFD) analysis is carried out, and the turbine design is modified based on the analysis results in order to attain optimum performance and minimal lag. The CFD analysis results of the velocity and pressure distributions identified the flow behaviour patterns such as flow separation, vortexes, and performance characteristics.

Keywords Axial turbine · Turbocharger · Computational Fluid Dynamics (CFD) · Volute · Compressor

1 Introduction

Turbocharger is a device which is the integral part of an internal combustion (IC) engine, which is widely used to enhance the engine performance. The compressor is driven by the impeller which is in turn driven by the exhaust gases. Presently, this technology is majorly utilized in automotive and aerospace industries. Factors like improved engine performance and fuel efficiency, government regulations, and engine downsizing have increased the need for turbocharging technologies. Computational fluid dynamics (CFD) may be a cost affected tool to supply detailed flow information inside the entire turbocharger. The research conducted by Chehhat et al. [1], Lintz [2] and Yang et al. [3] showed that volute significantly influences the overall

A. Singh · A. Sasane · R. Patney · R. Harish (✉)
School of Mechanical Engineering, Vellore Institute of Technology, Chennai, Tamil Nadu 600127, India
e-mail: harish.r@vit.ac.in

A. Maurya et al. (eds.), *Recent Trends in Mechanical Engineering*, Lecture Notes in Mechanical Engineering, https://doi.org/10.1007/978-981-19-7709-1_69

performance such as the operating range, the stability, and the location of the best efficiency point of the compressor. Not just the compressor impeller but the volute also affects the overall performance. So, it was necessary to include its effect in our study. Lintz et al. [2] went further and studied the significance of volute surface roughness and found that it had a significant impact on the automotive turbocharger performance. Improving the surface quality of the volute has potential benefits in turbine efficiency. In the research performed by Rinaldi [4] experiments and simulations were carried out to understand the performance of the CO_2 compressor and the CFD and experimental results were compared by addressing both the limitations of the adopted models, and those related to the experimental data. In this study [5], the researchers investigated different CFD modeling settings about interfaces and boundary conditions to successfully model compressible flow in various types of high-speed compressors. In the study carried out by Le. Sausse et al. [6], the compressor maps are obtained experimentally and then compared with CFD results. Several speed lines are calculated by varying input conditions. Galindo et al. [7, 8] carried out numerical investigations to understand the significance of geometry on the automotive turbocharger performance. Zheng et al. [9] studied and investigated the surge and stall conditions with a vaned diffuser by experiments. Jawad et al. [10] in their study proposed a preliminary design of axial turbines due to the drawbacks of conventional radial turbines used in turbochargers. A CFD analysis is carried out and is compared with the conventional radial turbine analysis results. Turbocharger's main function is to increase the efficiency of the engine [11]. In turbocharger the more the amount of air has been hit to an impeller and even by doing modification with the intake geometry the performance is enhanced significantly [12]. In the research paper done by Nicholas Anton et al. on axial turbochargers, they found that these types of turbochargers have less moment of inertia and also the design is wider than the conventional radial turbochargers and due to their more efficiency at higher flow rates which helps to get the most optimized solution [13–17]. In the research conducted on axial turbocharger by Serrano et al. [14] it is found that the tip leakage loss is directly proportional with the blade loading, whereas in case of radial turbocharger it is directly proportional to the rotational speed. Due to increase in chamber roughness the efficiency of the axial turbocharger is reduced nonlinearly [15]. Even though here we have seen the reasons for the use of axial turbochargers but in the study conducted by Jawad et al. [16] it has been found that the multi-stage axial turbochargers are more efficient and they increase the performance of the engine to a whole new level. In the present work, a turbocharger compressor along with the housing used in a DI diesel engine was designed using Solidworks. We have decided to select a backward curved Blade design for our compressor impeller. Backward curved blade has the highest efficiency. The benefit of using a backward curved impeller is that it does not have a stall point on its characteristic. This means that there is no point on the fan characteristic curve that should not be operated. A CFD flow analysis was done on the compressor blades at different blade speeds (rpm) and the results were analysed in order to give out maximum boost thereby enhancing the engine performance.

2 Methodology

In this paper, the work was divided into mainly three parts, the first was the making of a model which was done in the solidworks software. The mesh was generated on volume extraction of volute and Boolean subtraction on the blades collectively on Ansys Workbench. Here, fine mesh sizing was used which generated 1,429,503 cells in total and 420,115 nodes on both volute and blades. This model was imported to Ansys fluent where the simulations were calculated for a pressure-based solver, where the blades were set to a moving wall at the speed of 20,000 rpm and 40,000 rpm respectively in order to compare both simulations simultaneously. We have considered a low rpm range after studying surge and choke limits of various automobile turbochargers. These simulations were performed at three different inlet velocities, i.e. 10 m/s, 25 m/s, and 50 m/s. Figure 1 represents the geometrical configuration of the turbocharger compressor and the generated mesh using Ansys.

2.1 *Governing Equations*

$$\frac{\partial \rho}{\partial t} + \frac{\partial(\rho u)}{\partial x} + \frac{\partial(\rho u)}{\partial y} + \frac{\partial(\rho u)}{\partial z} = 0 \tag{1}$$

$$\frac{\partial(\rho u)}{\partial t} + \frac{u\partial(\rho u)}{\partial x} + \frac{v\partial(\rho u)}{\partial y} + \frac{w\partial(\rho u)}{\partial z} = -\frac{\partial p}{\partial x} + \frac{\partial}{\partial x}\left(\frac{\mu\partial u}{\partial x}\right) + \frac{\partial}{\partial y}\left(\frac{\mu\partial u}{\partial y}\right) + \frac{\partial}{\partial z}\left(\frac{\mu\partial u}{\partial z}\right) - S_x \tag{2}$$

$$\frac{\partial(\rho u)}{\partial t} + \frac{u\partial(\rho u)}{\partial x} + \frac{v\partial(\rho u)}{\partial y} + \frac{w\partial(\rho u)}{\partial z} = -\frac{\partial p}{\partial y} + \frac{\partial}{\partial x}\left(\frac{\mu\partial v}{\partial x}\right) + \frac{\partial}{\partial y}\left(\frac{\mu\partial v}{\partial y}\right) + \frac{\partial}{\partial z}\left(\frac{\mu\partial v}{\partial z}\right) - S_y \tag{3}$$

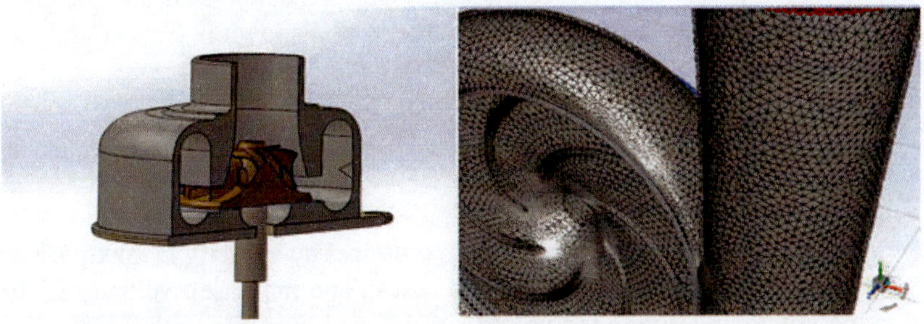

Fig. 1 Turbocharger compressor geometric model and generated mesh

$$\frac{\partial(\rho u)}{\partial t} + \frac{u\partial(\rho u)}{\partial x} + \frac{v\partial(\rho u)}{\partial y} + \frac{w\partial(\rho u)}{\partial z} = -\frac{\partial p}{\partial w} + \frac{\partial}{\partial x}\left(\frac{\mu \partial w}{\partial x}\right) + \frac{\partial}{\partial y}\left(\frac{\mu \partial w}{\partial y}\right) + \frac{\partial}{\partial z}\left(\frac{\mu \partial w}{\partial z}\right) - S_z \tag{4}$$

$$\frac{\partial k}{\partial t} + U_j \frac{\partial k}{\partial t} = P_k - \beta^* k\omega + \frac{\partial k}{\partial t}\left[(v + \sigma_k v_T)\frac{\partial k}{\partial t}\right] \tag{5}$$

Above mentioned are the few governing equations which are solved in each iteration to create the simulation. Equation (1) is the equation to solve continuity. Equations (2), (3) and (4) are the momentum equations on the respective axes, namely x, y and z direction. Finally, the Eq. (5) is the energy equation for the k-ω viscosity model equation.

2.2 Boundary Conditions

The pressure and temperature boundary conditions at the compressor inlet are 10 m/s specified with standard atmospheric conditions. The flow is axial to the impeller at 10 m/s, 25 m/s, and 50 m/s for different cases. At the outlet, static pressure is applied. Subsonic boundary was defined with the SST k-ω turbulence model with 5% turbulence intensity. The mesh motion method is adopted to rotate the impeller blades for a fixed impeller speed. The blades were defined as smooth surfaces moving at a specific speed and the compressor speed was set according to the load on the engine.

3 Results and Discussions

Simulation of the turbocharger was considered for two different impeller speeds (rpm): 20,000 and 40,000 and three different velocities: 10 m/s, 25 m/s, and 50 m/s. Velocity and pressure simulation results were obtained using CFD and the contours and graphs were analysed for both impeller speed and shown in Figs. 2 and 3.

3.1 Velocity Simulation

Following figure shows the velocity vector contour. The velocity is maximum at the compressor wheel blade endings in all the cases. The increased velocity is slowed down in the volute and due to the volute a swirl motion is generated. Figures 4 and 5 shows the air velocity contours at lower inlet speed of 10 m/s. Figures 6 and 7

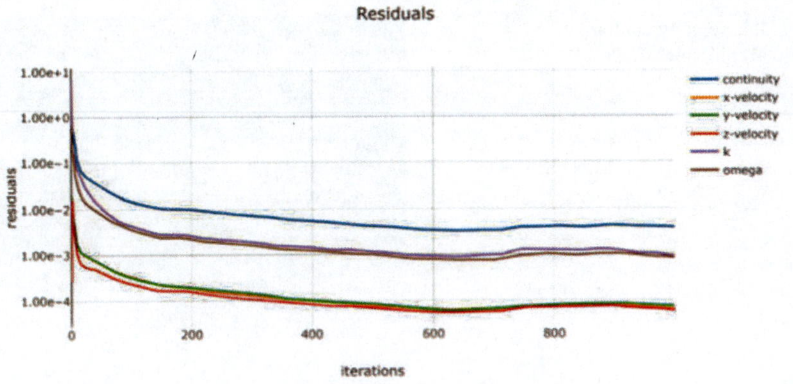

Fig. 2 Variation of residuals for an impeller speed of 20,000 rpm

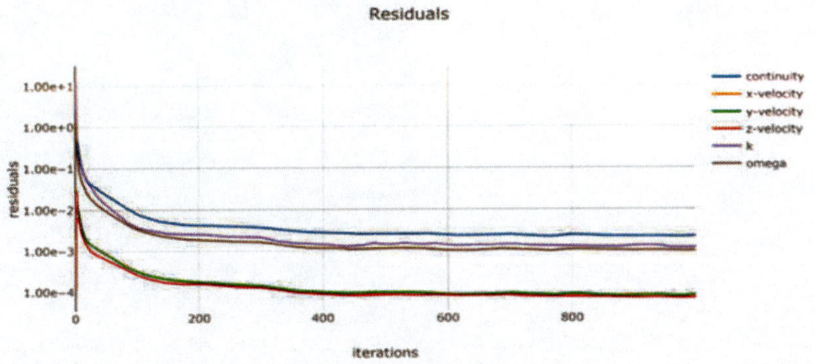

Fig. 3 Variation of residuals for an impeller speed of 40,000 rpm

indicates the air velocity distribution at transitional speeds of 25 m/s and 50 m/s. Figures 8 and 9 shows the velocity contours at a higher rotational speed of 40,000 rpm.

3.2 *Pressure Simulation*

The contours of the pressure distribution at the assembly of the compressor wheel and housing are investigated for different inlet speeds. The highest pressure is found in the downstream of the compressor wheel and the lowest upstream—suction opening of the compressor wheel. The pressure distribution looks symmetric and is similar for all the cases. Figures 10 and 11 shows the pressure distribution at a lower inlet speed of 10 m/s. Figures 12 and 13 indicates the pressure variations at transitional inlet speed of 25 m/s and 50 m/s respectively. Figures 14 and 15 shows the pressure contours at

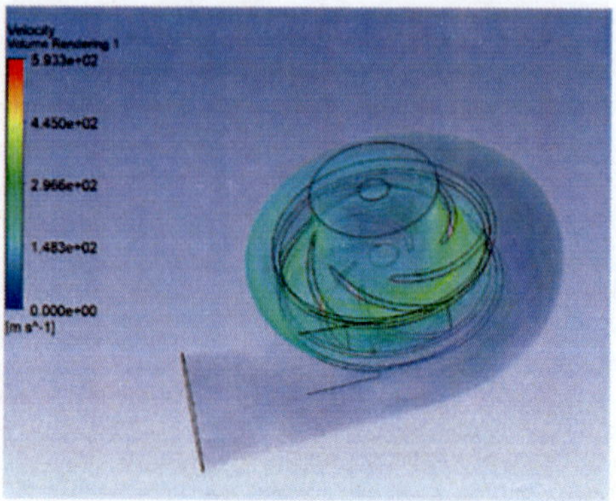

Fig. 4 Air velocity contours at 20,000 rpm, inlet speed 10 m/s

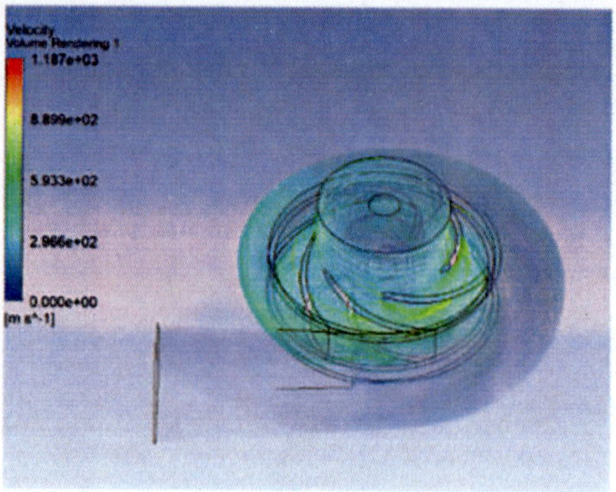

Fig. 5 Air velocity contours at 40,000 rpm, inlet speed 10 m/s

Fig. 6 Air velocity contours at 20,000 rpm, inlet speed 25 m/s

a higher rotational speed of 40,000 rpm. Figure 16 indicates the particle streamlines representing the development of swirl motion inside the turbocharger.

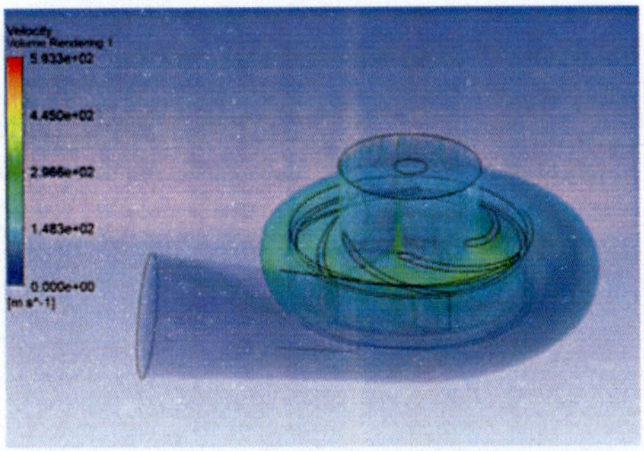

Fig. 7 Air velocity contours at 20,000 rpm, inlet 50 m/s

Fig. 8 Air velocity contours at 40,000 rpm, inlet speed 25 m/s

Fig. 9 Air velocity contours at 40,000 rpm, inlet 50 m/s

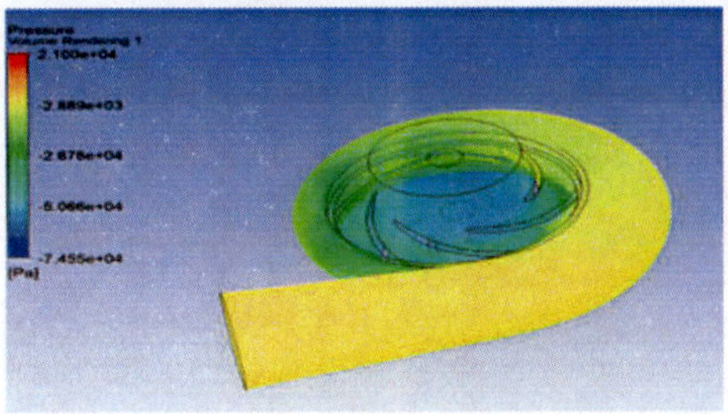

Fig. 10 Pressure contours at 20,000 rpm, inlet speed 10 m/s

Fig. 11 Pressure contours at 40,000 rpm, inlet speed 10 m/s

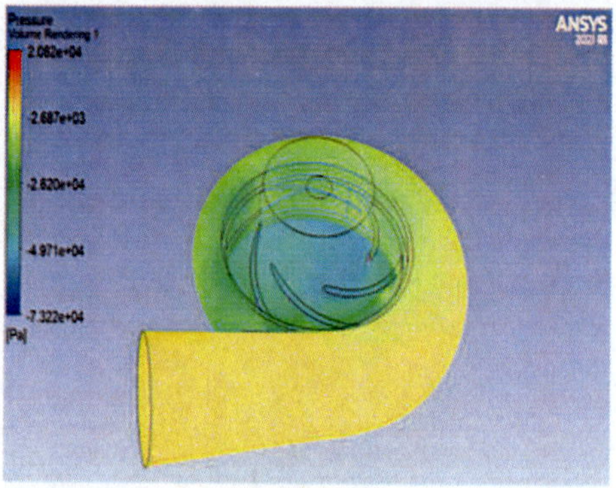

Fig. 12 Pressure contours at 20,000 rpm, inlet speed 25 m/s

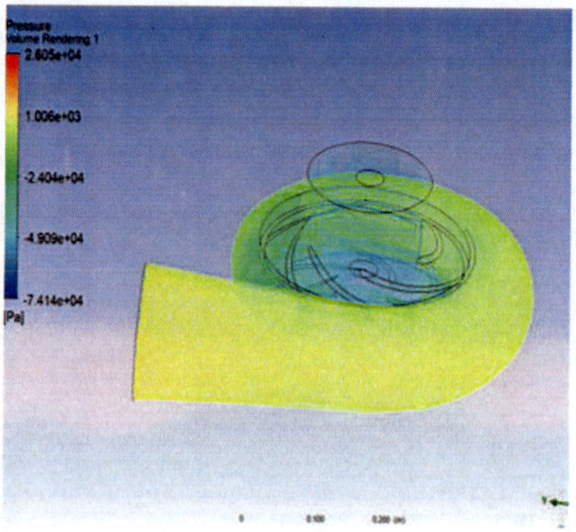

Fig. 13 Pressure contours at 20,000 rpm, inlet speed 50 m/s

4 Conclusions

This paper presents an effort to model and analyse the flow from inlet to the exit of an automobile turbocharger compressor with all the components in place. The flow was analysed at two different compressor blade speeds, i.e. 20,000 and 40,000 and three

Fig. 14 Pressure contours at 40,000 rpm, inlet speed 25 m/s

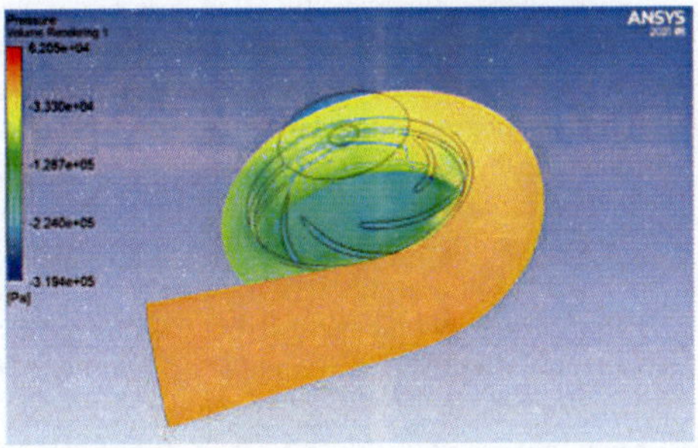

Fig. 15 Pressure contours at 40,000 rpm, inlet speed 50 m/s

Fig. 16 Particle streamline showing swirl motion at volute exit

different inlet velocities, i.e. 10 m/s, 25 m/s, and 50 m/s. Using the computational fluid dynamics (CFD) gave a better understanding of the behaviour of flow through a turbocharger compressor and how it impacts the turbocharger efficiency and how the turbocharger compressor performs in an automobile engine for enhancing its performance. The velocity simulation revealed that the velocity is maximum at the compressor blade endings and it is slowed down in the compressor volute and swirl motion is generated due to the volute. A pressure increase is seen at the compressor exit compared to the inlet. The highest pressure is found downstream the compressor wheel volute discharge. High pressure at the exit results in extra compressed air for the IC engine thereby increasing its power output.

References

1. Abdelmadjida C, Mohamedb S-A, Boussadc B (2013) CFD analysis of the volute geometry effect on the turbulent air flow through the turbocharger compressor, terragreen 13 international conference 2013
2. Lintz A (2017) The effect of volute surface roughness on the performance of automotive turbocharger turbines. In: Global power and propulsion forum, GPPF 2017
3. Yang M, Martinez-Botas R, Rajoo S, Yokoyama T, Ibaraki S (2015) An investigation of volute cross-sectional shape on turbocharger turbine under pulsating conditions in internal combustion engine. Energy Convers Manag
4. Rinaldi E, Pecnik R, Colonna P (2014) Computational fluid dynamics simulation of a supercritical CO2 compressor performance map. J Eng Gas Turbines Power
5. Schreiber J, Ottavy X, Ngo Boum G, Aubert S, Sicot F (2015) Numerical simulation of the flow field in a high speed multistage compressor- study of the time discretization sensitivity. In: Turbine technical conference and exposition GT
6. Le Sausse P, Fabrie P, Arnou D, Clunet F (2013) CFD comparison with centrifugal compressor measurements on a wide operating range. EDP Sci
7. Galindo J, Gil A, Navarro R, Tari D (2019), Analysis of the impact of the geometry on the performance of an automotive centrifugal compressor using CFD simulations. Appl Thermal Eng
8. Galindo J, Tiseira A, Navarro R, Tari D, Meano CM (2017) Effect of the inlet geometry on the performance, surge margin and noise emission of an automotive turbocharger compressor. Appl Thermal Eng
9. Zheng X, Sun Z, Kawakubo T, Tamaki H (2017) Experimental investigation of surge and stall in a turbocharger centrifugal compressor with vaned diffuser. Exp Therm Fluid Sci
10. Jawad LH, Razzaq HY, Hasan HM (2020), Aerodynamic design and performance investigation of an axial turbocharger turbine for automotive applications. Period Eng Nat Sci
11. Panayides C, Pesyridis A, Saravi SS (2019) Design of a sequential axial turbocharger for automotive application. Energies
12. Ramachandran D, Somashekarappa S, Mayandi B, Reddy Shanmugam R, Boolingam S, Ashok G (2018) A study on the influence of intake geometry on the turbocharger compressor performance, ASME
13. Anton N, Genrup M, Fredriksson C, Larsson PI, Christiansen-Erlandsson A (2018) Axial turbine design for a twin-turbine heavy-duty turbocharger concept, ASME
14. Serrano JR, Navarro R, García-Cuevas LM, Inhestern LB (2018) Turbocharger turbine rotor tip leakage loss and mass flow model valid up to extreme off-design condition with high blade to jet speed ratio, Elsevier
15. Wang X, Zhang X, Zuo Z, Zhu Y, Li W, Chen H, Ding Y (2021) Effect of chamber roughness and local smoothing on performance of a CAES axial turbine, Elseveir
16. Jawad LH, Hasan HM, Razzaq HY (2020) Performance evaluation of a multi-stage axial flow turbocharger turbine, solid state technology
17. Berchiolli M, Guarda G, Walsh G, Pesyridis A (2019) Turbocharger axial turbines for high transient response. Appl Sci

Grasshopper Optimization Algorithm and Its Application in Determination of Critical Failure Surface in Soil Slope Stability

Navneet Himanshu, Avijit Burman, Vinay Kumar, and Shiva Shankar Choudhary

Abstract The identification of critical failure surface (CFS) in soil slopes with lowest factor of safety, i.e., min (FOS) is a constrained global optimization problem. In the past decades, many metaheuristic optimization algorithms have been proposed to assess the CFS in slope stability analysis with their advantages and limitations. The task of evaluation of associated FOS is performed using limit equilibrium-based Morgenstern-Price method and is treated as the objective function of the optimization algorithm. This paper presents the results of recently proposed Grasshopper Optimization Algorithm (GOA) in soil slope stability analysis problems. Two soil slopes from earlier published literatures having different complexities have been analyzed using GOA. The observed results indicate that GOA technique used in the present study to minimize the objective function (i.e., the expression of factor of safety of the slope) has successfully achieved their goals. Statistical analysis of estimated min (FOS) obtained for different optimization parameters such as swarm size (N) and maximum number of iterations ($k_{\max}$) have been reported in detail. It has been observed that the value of standard deviation (SD) of evaluated min (FOS) based on ten runs for all combinations of N and $k_{\max}$ is quite low, i.e., in the order of E-03 or less for both soil slope problem, which for all practical geotechnical purposes may be deemed highly satisfactory. Furthermore, the best min (FOS), the worst min (FOS) and the mean min (FOS) are also compared for all combinations of N and $k_{\max}$ based on ten independent runs.

N. Himanshu
Department of Civil Engineering, Noida Institute of Engineering and Technology Uttar Pradesh, Noida, Uttar Pradesh 201306, India
e-mail: navneet.ce2014@nitp.ac.in

A. Burman · V. Kumar (✉) · S. S. Choudhary
Department of Civil Engineering, National Institute of Technology Patna, Patna 800005, India
e-mail: vinay1.ce16@nitp.ac.in

A. Burman
e-mail: avijit@nitp.ac.in

S. S. Choudhary
e-mail: shiva@nitp.ac.in

A. Maurya et al. (eds.), *Recent Trends in Mechanical Engineering*, Lecture Notes in Mechanical Engineering, https://doi.org/10.1007/978-981-19-7709-1_70

Keywords Grasshopper optimization algorithm · Metaheuristic · Swarm size · Critical failure surface · Factor of safety · Slope stability

1 Introduction

Stability analysis of soil slopes is key aspect in civil engineering projects, such as dams, embankments, and landfills. The analysis includes evaluation of minimum factor of safety (FOS) values and associated critical failure surface. In order to compute minimum factor of safety many researchers perform the stability analysis of soil slopes with different conventional optimization techniques with limit equilibrium techniques such as [1, 7, 13]. Addition to this, with the development of robust meta-heuristic techniques other researchers such as [2–6, 8–10]; utilizes Particle Swarm Optimization (PSO), Teaching–Learning-Based Optimization (TLBO), Multiverse Optimization (MVO), and Grey Wolf Optimization (GWO) and demonstrated their ability to locate critical failure surface efficiently. In this paper, recently developed Grasshopper Optimization Algorithm (GOA) are utilized to solve soil slope stability analysis problem. Grasshopper Optimization Algorithm (GOA) is nature-inspired algorithms which mimics the swarming behavior of grasshoppers searching for food source. The mathematical model simulating the swarming nature of grasshopper as an optimization algorithm was first modelled by [12]. Later, Saremi et al. (2017) discussed the selection of different parameter and demonstrated the efficiency of GOA in different unimodal, multimodal and composite benchmark problems. Also, Amiri et al. (2017) demonstrated the ability of GOA to solve the active power loss in Distributed Generation Placement Problem, and Hamour et al. (2018) used GOA to evaluate optimal switch combination for minimization of power loss in distribution network reconfiguration. This paper presents, two soil slopes from earlier published literatures having different complexities have been analyzed using GOA. The observed results indicate that GOA technique used in the present study to minimize the objective function (i.e., the expression of factor of safety of the slope) has successfully achieved their goals.

2 Methodology and Modeling

2.1 Grasshopper Optimization Algorithm

The Grasshopper Optimization Algorithm (GOA) mimics the behavior of grasshoppers in swarm seeking for food source. In nature, grasshoppers can found in both as individual as well as in swarm. The grasshopper swarm demonstrate unique characteristic that swarming behavior in both as nymph and adulthood.

The mathematical model simulating the swarming nature of grasshopper as an optimization algorithm was first modeled by [12]. Here, index i denotes the ith individual grasshopper in defined swarm (S) and k defines the current iteration step. At the beginning, all individual grasshoppers were randomly spread inside the search space A, since they don't know the position of food source. Mathematically swarm (S) is defined as:

$$S = \{G_1, G_2, \ldots\ldots, G_N\} \tag{1}$$

Individual grasshopper G_i (i.e., ith candidate solution) inside the defines swarm is identified as follows:

$$\vec{G}_i = (g_{i1}, g_{i2}, g_{i3}, \ldots\ldots\ldots, g_{im})^T \in \mathrm{A},\ i = 1,2,3,\ldots\ldots\ldots, N \tag{2}$$

Here, N is referred to as the number of individual bat grasshopper inside the swarm and m represents total number of variables. In this present study, individual grasshopper $\vec{G}_i$ refer to a potential failure surface out of N number of probable failure surfaces and m represent the control variable require to define a failure surface. Also, in GOA individual grasshoppers are characterized by their position vector $\left(\vec{Z}_i^k\right)$,

$$\vec{Z}_i^{k+1} = \vec{R}_i^k + \vec{M}_i^k + \vec{Y}_i^k \tag{3}$$

where $\vec{Z}_i^k$ defines the position of individual grasshopper, $\vec{R}_i^k$ represents the social interaction, $\vec{M}_i^k$ and $\vec{Y}_i^k$ represents the gravity force and wind advection.

In GOA, the social interaction $\vec{R}_i^k$ parameters modeled the interaction between individual grasshopper with all other grasshoppers in swarm. The movement of individual grasshoppers is based on the experience shared by others inside the swarm, and this guides the swarm toward the optimum goal (i.e., food position) mathematically,

$$\vec{R}_i^k = \sum_{\substack{j \neq i \\ j=1}}^{N} r(d_{ij})\,\hat{d}_{ij} \tag{4}$$

where r is a function to define the strength of social forces and mathematically,

$$r\left(z'\right) = f' e^{-\frac{z'}{l'}} - e^{-z'} \tag{5}$$

also, in Eq. (4) d_{ij} is the distance between two grasshoppers in GOA (i.e. ith and jth grasshoppers) and calculated as,

$$d_{ij} = \left|\vec{Z}_j^k - \vec{Z}_i^k\right| \tag{6}$$

similarly, the parameter $\hat{d}_{ij}$ in Eq. (4) represents unit vector between ith and jth grasshoppers and calculated as,

$$\hat{d}_{ij} = \frac{\vec{Z}_j^k - \vec{Z}_i^k}{d_{ij}} \tag{7}$$

In Eq. (5), the term f' represents the intensity of attraction and l' represents the attractive length scale.

The $\vec{M}_i^k$ component in the Eq. (3) represents the gravitational force acting on the individual grasshopper is calculated as follows:

$$\vec{M}_i^k = -g\hat{e}_g \tag{8}$$

where g is the gravitational constant and $\hat{e}_g$ represents a unit vector towards the center of earth. Similarly, the component $\vec{Y}_i^k$ in the Eq. (3) represents the wind drift acting on the grasshopper mathematically,

$$\vec{Y}_i^k = w'\hat{e}_w \tag{9}$$

where w' is the wind drift constant and $\hat{e}_w$ represents a unit vector towards direction of the wind.

Substituting Eqs. (4), (8), and (9) in Eq. (3), the following equation can be obtained:

$$\vec{Z}_i^{k+1} = \sum_{\substack{j \neq i \\ j=1}}^{N} r\left(\left|z_j^k - z_i^k\right|\right) \frac{z_j^k - z_i^k}{d_{ij}} - g\hat{e}_g + w'\hat{e}_w \tag{10}$$

The equation i.e., Eq. (10) cannot be utilized in the swarm simulation and optimization algorithm as it prevents the algorithm from exploring and exploitation the search space around the solution.

Another parameter c is introduced to original Eq. (10) modified version of this equation, useful for solving optimization problems, is presented as follows:

$$\vec{Z}_{im}^{k+1} = c^k \left(\sum_{\substack{j \neq i \\ j=1}}^{N} c^k \frac{z_{im}^{ub} - z_{im}^{lb}}{2} r\left(\left|z_{jm}^k - z_{im}^k\right|\right) \frac{z_{jm}^k - z_{im}^k}{d_{ij}} \right) + T_m^k \tag{11}$$

Here, index m represents control variables, ub and lb are upper bound and lower bound associated with each control variables. Also, c is a decreasing coefficient and T_m^k is the value of the mth control variable in the target position (best solution found so far).

It also evident that outer decreasing coefficient (c) contributes the search process around the current best position $\left(\vec{Z}_{sbest}^k\right)$ or current target position T^k as the iteration proceeds, while the inner decreasing coefficient (c) contributes to the decreases in the attraction/repulsion forces between the grasshoppers with increasing iteration counts.

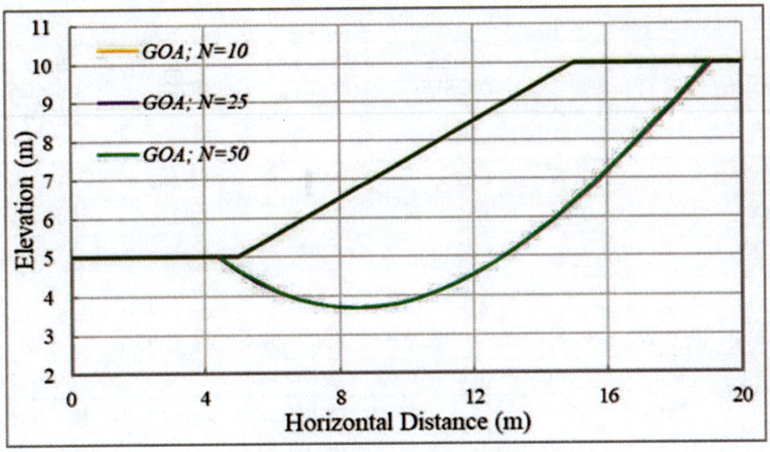

Fig. 1 Critical non-circular failure surfaces obtained by GOA for homogeneous soil slope

In order to maintain the balance between the ability of global and local exploration as well as ability of exploitation, the parameter c should be progressively decreased as iterations proceed forward. A linearly decreasing time-dependent coefficient (c) reduces the comfort zone proportional to the number of iterations 'k' and is mathematically expressed as:

$$c^k = c_{\max} - (c_{\max} - c_{\min})\frac{k}{k_{max}} \tag{12}$$

2.2 *Determination of FOS of the Slope*

The FOS of the slope is determined using [11] i.e. MPM method. It is a limit equilibrium-based method to estimate the FOS of the slope. The details of the algorithm used for estimation of FOS of the slope can be found in the works of [4, 14].

3 Result and Discussions

3.1 *Homogeneous Soil Slope [13]*

The analyzed soil slope incorporates a homogeneous soil slope. The considered slope problem has been investigated by various researchers [3, 7, 13] earlier. The

Table 1 Factor of safety values for homogeneous soil slope

Swarm size, max Iteration (N, $k_{\max}$)	Number of function evaluation (*NFE*)	Best min (FOS)F_b	Worst min (FOS)F_w	Mean min (FOS)$\overline{F}$	Standard deviation (SD)
(10, 200)	2000	1.3230	1.3273	1.3240	1.67E-03
(25, 200)	5000	1.3230	1.3240	1.3236	4.32E-04
(50, 200)	10,000	1.3230	1.3242	1.3234	5.60E-04

Table 2 Factor of safety values (previous studies) for homogeneous soil slope

Literatures	Limit equilibrium method	Optimization method	Minimum (FOS)
[13]	MPM	BFGS, DFP, powell method, simplex method	1.338–1.348
[7]	SM	MC technique	1.238
[3]	SM	PSO	1.3249–1.3285
		MPSO	1.3259–1.3273
[4]	BM	PSO	1.3136–1.3175
[6]	MPM	IW-PSO	1.3230–1.3273
		CS-PSO	1.3235–1.3273

geometric layout of the slope is shown in Fig. 1. The soil properties are as follows: unit weight $\gamma = 17.64\ \text{kN/m}^3$, effective cohesion $c' = 9.80\ \text{kN/m}^2$, and angle of internal friction $\varphi' = 10°$. In the present study, the search of minimum F value and associated CFS is performed with three different swarm size (N) 10, 25, and 50, respectively. Also, maximum number of iterations ($k_{\max}$) used during search process is 200 for each set of swarm size. The minimum F value obtained using GOA is tabulated in Table 1. It has been observed that the value of standard deviation (SD) of evaluated min (FOS) based on 10 runs for all combinations of N and $k_{\max}$ is quite low, i.e., in the order of E-03 or less for discussed homogeneous soil slope problem, which for all practical geotechnical purposes may be deemed highly satisfactory. However, the GOA work successfully reports minimum FOS values which are very close to the earlier reported global minimum FOS (refer Table 2). Small differences in minimum FOS in the present study, can be observed because of the differences in method employed to evaluate factor of safety as well as the discretization procedure of potential failure surface.

3.2 *Comparative Study of Convergence Curve for Soil Slopes*

Figure 2 demonstrates the convergence behavior of GOA during the search of minimum F for the homogeneous slope. The figure plots the Best (min. F) and

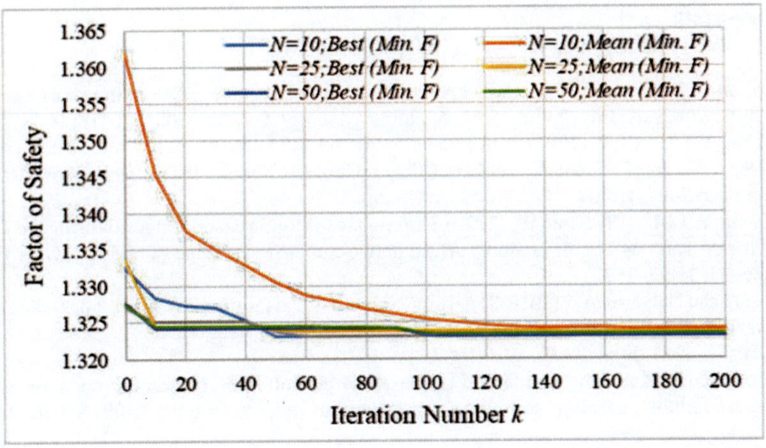

Fig. 2 Comparative Study of Convergence Curve for Homogeneous Soil Slope

Mean (min. F) of 10 runs with iteration count during the process of optimization for swarm size (N) equals to 10, 25, and 50 associated with termination criteria, i.e., k_{max} equals to 200 for the two-soil slope, respectively. It is found that GOA convergence before termination with little difference in Best (min. F) and Mean (min. F). These traits demonstrate that the results reported by GOA are converged global optimum/minimum for the soil slopes under consideration. Also, the above convergence characteristic demonstrates the efficiency of GOA in determination of critical failure for soil slope by showing very good reliability in the Best and Mean value (i.e., $\overline{F}$) of computed minimum F.

4 Conclusion

In the present work, Grasshopper Optimization Algorithm (GOA) has been used to search for the minimum Factor of Safety (FOS) of any slope and also corresponding critical failures surface has been located. The GOA has been found to be successful in achieving the global minimum F and locating the corresponding critical failure surface slope stability problems efficiently. Moreover, the Best (min. F) Mean (min. F) reflects convergence characteristics when the population size is equals to 10, 25 and 50.

References

1. Bolton H, Heymann G, Groenwold A (2003) Global search for critical failure surface in slope stability analysis. Eng Optim 35:51–65. https://doi.org/10.1080/0305215031000064749
2. Cheng Y, Li L, Chi S (2007) Performance studies on six heuristic global optimization methods in the location of critical slip surface. Comput Geotech 34:462–484. https://doi.org/10.1016/j.compgeo.2007.01.004
3. Cheng Y, Li L, Chi S, W W, (2007) Particle swarm optimization algorithm for the location of the critical non-circular failure surface in two-dimensional slope stability analysis. Comput Geotech 34:92–103
4. Himanshu N, Burman A (2019) Determination of critical failure surface of slopes using particle swarm optimization technique considering seepage and seismic loading. Geotech Geol Eng 37:1261–1281. https://doi.org/10.1007/s10706-018-0683-8
5. Himanshu N, Kumar V, Burman A et al (2020) Grey wolf optimization approach for searching critical failure surface in soil slopes. Eng Comput. https://doi.org/10.1007/s00366-019-00927-6
6. Himanshu N, Kumar V, Burman A, Gordan B (2020) Optimization of non-circular failure surface in slope based on particle swarm models. Innov Infrastruct Solut 5:9. https://doi.org/10.1007/s41062-019-0259-3
7. Malkawi AIH, Hassan WF, Sarma SK (2001) Global search method for locating general slip surface using Monte Carlo techniques. J Geotech Geoenviron Eng 127:688–698. https://doi.org/10.1061/(ASCE)1090-0241(2001)127:8(688)
8. Mishra M, Gunturi VR, Maity D (2019) Teaching–learning-based optimisation algorithm and its application in capturing critical slip surface in slope stability analysis. Soft Comput. https://doi.org/10.1007/s00500-019-04075-3
9. Mishra M, Gunturi VR, Maity D (2019) Multiverse Optimisation Algorithm for Capturing the Critical Slip Surface in Slope Stability Analysis. Geotech Geol Eng 2:. https://doi.org/10.1007/s10706-019-01037-2
10. Mishra M, Ramana V, Tiago G et al (2019) Slope stability analysis using recent metaheuristic techniques : a comprehensive survey. SN Appl Sci 1:1–17. https://doi.org/10.1007/s42452-019-1707-6
11. Morgenstern NR, Price VE (1965) The analysis of the stability of general slip surfaces. Géotechnique 15:79–93. https://doi.org/10.1680/geot.1965.15.1.79
12. Topaz CM, Bernoff AJ, Logan S, Toolson W (2008) A model for rolling swarms of lo-custs. Eur Phys J Spec Top 157:93–109
13. Yamagami T, Ueta Y (1988) Search for noncircular slip surfaces by the Morgenstern–Price method. In: Proceedings of 6th international conference on numerical methods in geomechanics. pp 1335–1340
14. Zhu DY, Lee CF, Qian QH, Chen GR (2005) A concise algorithm for computing the factor of safety using the Morgenstern Price method. Can Geotech J 42:272–278. https://doi.org/10.1139/t04-072

Prediction of End-Milling Optimal Parameters Using ANN-Based NSGA-II Model

H. Ramesh, S. Arockia Edwin Xavier, S. Muthu Pandi, S. Julius Fusic, and A. N. Subbiah

Abstract This paper investigates the influence of end-milling process parameters such as spindle speed (rpm), feed rate (mm/min), and depth of cut (mm) on the response parameters such as surface roughness (SR) in μm and material removal rate (MRR) in cu. inches/min. The economical production is obtained by optimizing the process parameters such as cutting speed, spindle speed, spindle power, feed rate, flow rate of coolant, and depth of cut, since simultaneous optimization of the selected parameter in the case of the end-milling process is difficult. Initially, process modeling of SR and MRR of aluminum workpiece using end-milling machine has been performed by artificial neural network (ANN). The data set for the model is generated using the full factorial method. In the next phase of this work, two genetic algorithm (GA)-based multi-objective algorithms are compared to find out the best trade-offs between the two conflicting output parameters SR and MRR. Finally, the post-optimality analysis was executed to figure out the affair between the optimal machining parameter and optimal response parameter of non-dominated sorting genetic algorithm-II (NSGA-II).

Keywords Artificial neural networks · Genetic algorithm · Material removal rate · Non-dominated sorting genetic algorithm · Surface roughness · Multi-objective optimized genetic algorithm

1 Introduction

In this modern age of global competition, product manufacturers are necessitated to create a way to increase the quality of the product and its productivity. The quality of the machined surface is assessed by the surface roughness factor which has a huge impact on the surface properties like fatigue, friction, and wear resistance.

H. Ramesh (✉) · S. M. Pandi · S. J. Fusic · A. N. Subbiah
Department of Mechatronics, Thiagarajar College of Engineering, Madurai, Tamil Nadu, India
e-mail: rameshh@tce.edu

S. A. E. Xavier
Department of E.E.E, Thiagarajar College of Engineering, Madurai, Tamil Nadu, India

A. Maurya et al. (eds.), *Recent Trends in Mechanical Engineering*, Lecture Notes in Mechanical Engineering, https://doi.org/10.1007/978-981-19-7709-1_71

The material removal rate gives the measurement of material removed from the given workpiece in the specified period. The efficiency of the production unit can be boosted by increasing the MRR, even a small amount of increase in MRR will result in a big gain.

Artificial neural networks (ANN) are computational algorithms inspired by the animal's nervous system. It is the machine learning algorithm whose model is inspired by the neurons of human beings. The brain of human beings consists of millions of neurons that send and receive the pattern of chemical and electrical signals. The finest solution is found by the optimization process. Generally, the optimization problems involve the minimization or maximization values or single objective or multi-objective problems. Multi-objective optimization is a problem that associates with more than one objective. It is the stochastic optimization technique where the concept of dominance is used to find the optimal Pareto front for the specific problems. By using the dominance relationship at each iteration, the calculated objective function value is compared with the other solutions, and the better solution is chosen for the management of the genetic population. NSGA-II has mostly used the multi-objective optimized genetic algorithm (MOGA) which uses the concept of elitism and dominance [1, 2] to choose the population for the next generation.

Hayajneh et al. developed a series of experiments to characterize the factors affecting surface roughness for the end-milling process. Feed rate, spindle speed, and depth of cut were considered as the machining parameter, and the machining was done on the aluminum workpiece to find out the surface roughness. The regression analysis was done using Statistica 6.0, a statistical software package. The created model behaves well with the deviation of 12% between the measured values and the predicted value of surface roughness (Hayajneh 2007) [3]. Dhabale and Jatti [4] created a regression model to optimize the material removal rate using the genetic algorithm. Experiments were performed on the NC machine with the machining parameters of the depth of cut, spindle speed, and feed rate on aluminum alloy as the workpiece material. The comparison was done between the experimental values and predicted values, and an error of 3.35% was obtained with a maximum material removal rate of 6021.411 mm^3/min. The relationship between the process and response parameters is found by using the Minitab 16 statistical software by employing the regression analysis [4]. Sahu and Nayak [5] proposed an ANN model to create a process modeling of material removal rate and tool wear rate. The experiments were done using the EDM machine with workpiece material of A2 steel. Pulse on time, discharge current, duty cycle and discharge voltage were considered as the process parameter with MRR and tool wear rate (TWR) as the response parameter. The response parameter was conflicting in nature; thus, the best trade-ups between these two parameters are founded by implementing the NSGA-II optimization algorithm. At the higher values of discharge current and duty cycle, the maximum MMR was obtained. Again, for the minimum value of TWR, the lower value of discharge current and the duty cycle is used with the mid-range of pulse on time [5]. Shaik Dawood and Saravana Kumar [6] had conducted an experiment on Inconel 718 material with process parameters such as depth of cut, cutting speed, and feed rate on the response parameters of MRR and SR. The machining process is done in a

CNC machine. A genetic algorithm is used as the technique for optimizing the input parameters. The regression equations are obtained from Minitab software which is used in the genetic algorithm to find out the optimum combinations of process parameter and their corresponding response parameters [6]. Azlan Mohd Zain et al. (2012) developed an ANN model integrated with GA for obtaining the minimum value of machining performance parameter. The operations are carried out using the end-milling machine with input parameters such as speed, feed and radial rake angle, and output parameter of surface roughness. From the proposed approach, the minimum surface roughness obtained is 0.139 μm with optimal machining parameters of feed = 167.029 m/min, speed = 0.025 mm/tooth, and rake angle = 14.769°. The ANN-GA method is compared with the experimental, ANN, regression, and response surface results [7]. Chaki (2015) proposed an integrated ANN-NSGA-II model to optimize the surface roughness and MRR during Nd: YAG laser cutting process of aluminum alloy. The data set for the integrated model is created for the independent parameters such as pulse width, pulse energy, and cutting speed using the full factorial method. The objective function for the optimization algorithm is developed from the ANN function which is trained with the Bayesian regularization algorithm. The NSGA-II algorithm generates the Pareto optimal fronts with a complete set of optimal solutions for response parameters. The ANN with 3–6-2 network architecture is considered as best ANN model. The mean percentage of error less than 1% is obtained from the ANN model. The maximum MRR of 57.494 mg/min is obtained at input parameters of speed = 1.8 mm/s, pulse energy = 6.8 J, and pulse width = 0.5 ms. The minimum SR of 6.23 μm is obtained for the following condition of cutting speed = 1.2 mm/s, pulse energy = 4.5 J, and pulse width = 0.5 ms [8]. Sahu and Nayak [9] conducted optimization experiments using genetic algorithms for the multiple response parameters like MRR, TWR, and OC of the EDM process. The composite objective function is created as the single objective function and given to the genetic algorithm with boundary conditions. From the genetic algorithm technique, the optimum values of MRR are 11.0658 mm^3/min, TWR of 0.055988 mm^3/min, and OC of 0.1934 mm that are obtained [9].

From the literature reviews, it is accepted that artificial neural networks are used effectively for developing the model using the machining input and output parameters. For obtaining the high-performance ANN model, the selection of process parameters should be done in an orderly manner. Also, the study of literature exposes that multi-objective evolutionary algorithms are used efficiently for finding the conflicting response parameters in the electric discharge machining process. It is found that a single objective optimization problem [10–12] is used mostly in the end-milling process. So here, in this work, the multi-objective optimization technique is implemented to find the optimal conflicting response parameters using an elitist non-dominated sorting genetic algorithm (NSGA) in the end-milling process. Here NSGA-II and multi-objective optimization using genetic algorithms (GA-MOOP) are compared, and the better one is validated with the experimental values.

Table 1 Combinations of parameters and their levels

S.no	Parameters	Levels
1	Spindle speed (rpm)	750,1000,1250 and 1500
2	Feed rate (mm/min)	150,225,300,375,450,525 and 600
3	Depth of cut (mm)	0.25,0.75 and 1.25

2 Experimental Setup and Materials

The machining process parameter and response parameters data of the End-milling process used here for modeling is referred from (Hayajneh 2007) [3]. The experiment was done on an end-milling machine with 3/4 inch four flute high-speed steel cutters. The experiment was performed on the aluminum workpiece. The cutting parameters and their levels are given in Table 1. The surface roughness Ra is measured in micrometers using a stylus-type profilometer. Another response parameter, i.e., the material removal rate, is measured using the formula.

Material removal rate = RDOC × ADOC × FEED RATE in cubic inches per minute where RDOC: radial depth of cut in inches, ADOC: axial depth of cut in inches, and FEED RATE in inches/min (Table 1).

3 Research Methodology

The improved ANN model can be obtained by using the full factorial design of experiments to achieve the learning algorithms, ANN architectures, and the number of hidden neurons. Table 4 gives the details of ANN architecture, training algorithms, and the number of hidden layers used in this work (Table 2).

Table 2 ANN process parameter and their levels

Process parameter	Levels			
	1	2	3	4
Neural architecture	Multi-layer perceptron (MLP)	Cascade feed forward network (GFNN)	–	–
Training algorithm	Levenberg–Marquardt	Bayesian regularization	Scaled conjugate gradient	–
No. of hidden neurons	4	8	12	16

4 Full Factorial Design Analysis

Figure 1 gives the flowchart of the methodology used in this work. With the help of full factorial design analysis, the design factors with discrete levels take on all possible sequence with all factors which results in the sequence of factors with four replicates to be performed in the ANN network. ANN is trained and tested according to the sequence obtained from full factorial design analysis. By using the Minitab 18 software, the main effect plot for the training MSE, testing MSE, training R (correlation coefficient), and testing R was generated. Generally, the training data set is used to fit the model, while the testing data set is used to evaluate the model. So here in this proposed model, the evaluation of the outstanding ANN network is selected based on the testing data set. Hence from the main effect plot of testing MSE and testing R, cascade forward network architecture with the scaled conjugate algorithm is selected. The number of hidden neurons is selected as four. Figures 2 and 3 shows the main effect plot for testing MSE and R.

Objective Function

In this proposed approach, two functions are selected as the objective functions: surface roughness Ra and material removal rate MRR. The NSGA-II algorithm was implemented for multi-objective optimization problem with three decision variables and two response variables.

The objectives functions are as follows:

Minimize: Ra (S, F, D).

Maximize: -MRR (S, F, D).

Subjected to:

$750 \leq \mathrm{S} \leq 1500.$

$150 \leq \mathrm{F} \leq 600.$

$0.25 \leq \mathrm{D} \leq 1.25$

where S (spindle speed), F (feed rate), and D (depth of cut) are the input parameters and MRR and Ra are the response parameters.

Regression analysis was carried out in Minitab 18 environment where the relationship between the process parameters and response parameters can be found. They are linearly related to each other. From the regression analysis, the model equations were generated to predict the surface roughness and MRR, respectively.

$$
\begin{aligned}
f1 = {} & 0.27 + 0.000317 * x(1) + 0.01239 * x(2) + 1.00 * x(3) \\
& - 0.000005 * x(1) * x(2) - 0.00024 * x(1) * x(3) - 0.00633 * x(2) * x(3) \\
& + 0.000003 * x(1) * x(2) * x(3)
\end{aligned}
$$

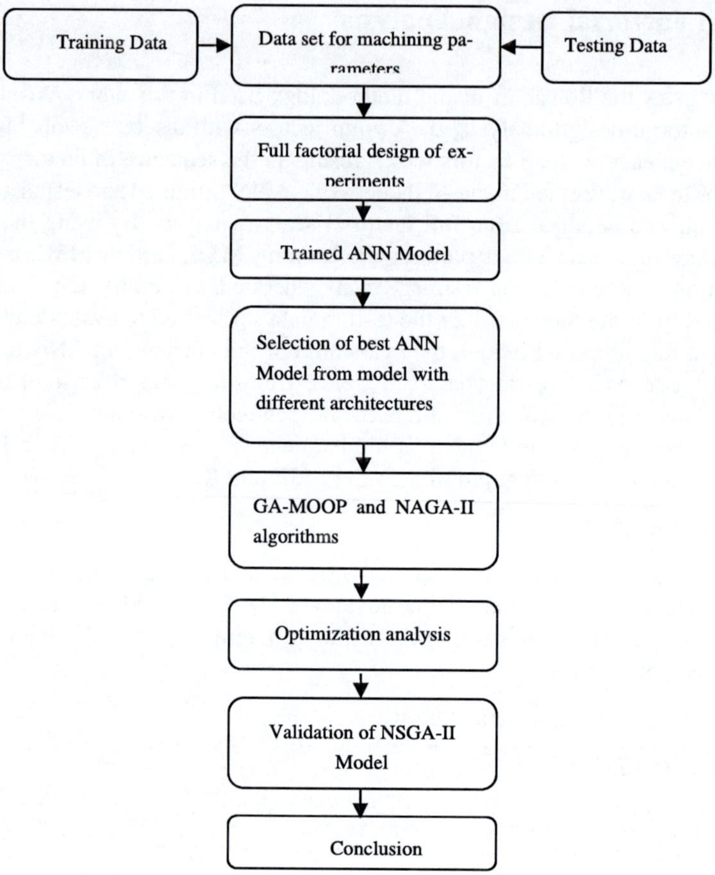

Fig. 1 Flowchart of methodology

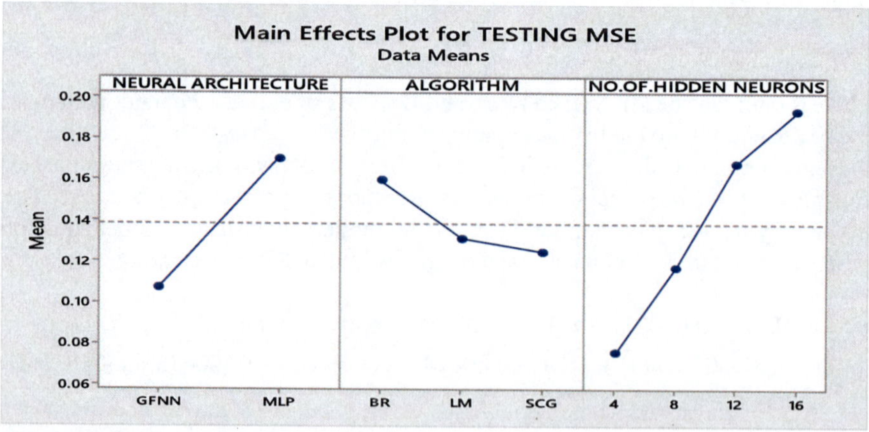

Fig. 2 Main effect plot for testing MSE

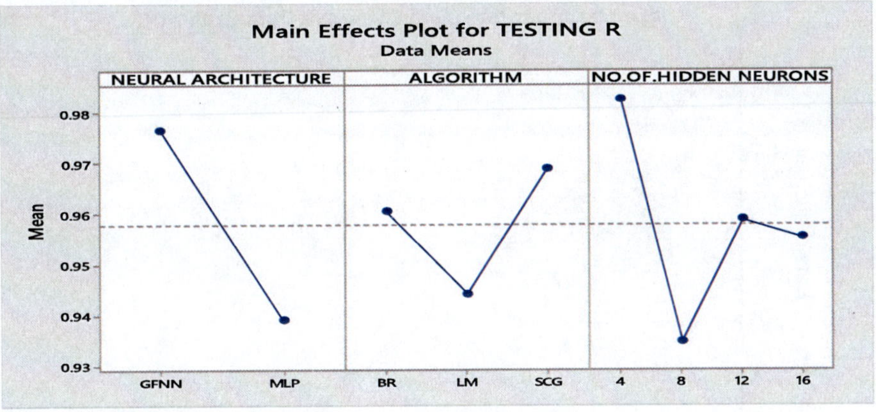

Fig. 3 Main effect plot for testing R

$$f2 = -(-0.00131 + 0.000001 * x(1) + 0.000003 * x(2) + 0.00352 * x(3) - 0.0000 * x(1) * x(2) - 0.000003 * x(1) * x(3) + 0.001150 * x(2) * x(3) + 0.000000 * x(1) * x(2) * x(3))$$

where

$f1, f2$—objective functions of surface roughness and MRR

$x(1)$—spindle speed

$x(2)$—feed Rate

$x(3)$—depth of cut.

5 Results and Discussion

The Pareto optimal vectors are obtained from the multi-objective optimization using the genetic algorithm and NSGA-II, respectively. On comparing the Pareto front of these two algorithms, the NSGA-II has a better spread of solutions and convergence near the true Pareto front than the multi-objective optimization using the genetic algorithm. Hence, the NSGA-II algorithm is selected because of its wide range selection of the optimal set of process parameters according to the requirement. The post-optimal analysis is done to check whether the optimal set of the solution generated by the NSGA-II is overwhelmed by a set of machining parameters obtained from the optimization algorithm (Figs. 4, 5 and Table 3).

Table 3 shows the comparative results of multi-objective optimization using a genetic algorithm and NSGA-II. From the table, it is clear that the NSGA-II algorithm has a wide range of response parameters, namely surface roughness and material

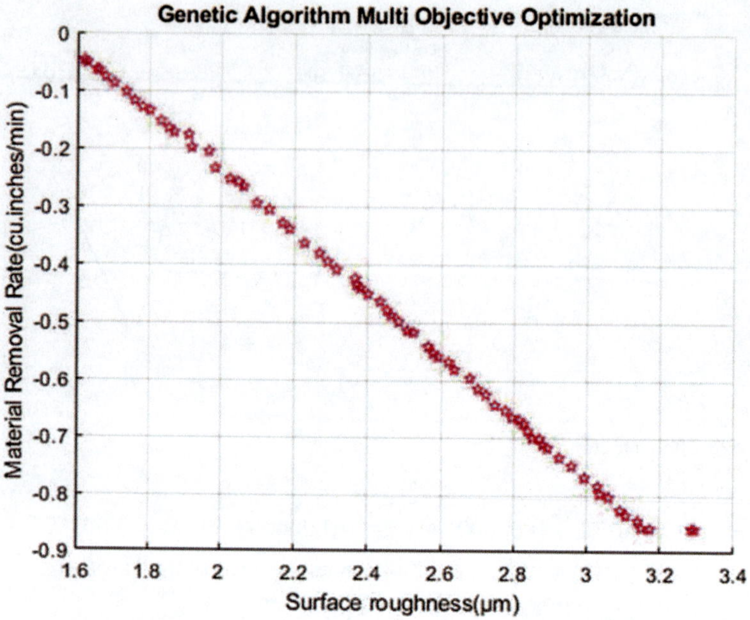

Fig. 4 Pareto optimal front obtained from GA-multi-objective optimization

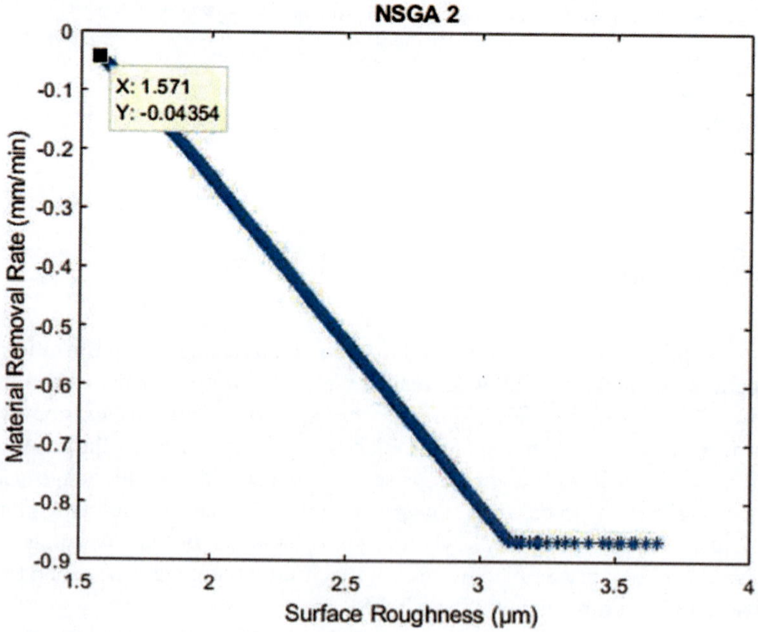

Fig. 5 Pareto optimal front obtained from NSGA-II

Table 3 Comparison of GA-MOOP versus NSGA-II

Optimization algorithm	Max · MRR	Min · MRR	Max · SR	Min · SR
GA-Multi-objective	0.8589	0.04693	3.294	1.62
NSGA-II	0.865	0.0435	3.656	1.571

removal rate. The surface roughness values lie between 1.571 μm and 3.656 μm, and MRR ranges between 0.0435 and 0.865. Thus, NSGA-II has a better spread of solutions and convergence near the true Pareto front than the multi-objective optimization using the genetic algorithm.

Figure 6a, b, and c shows the effect of optimal input parameters on optimized surface roughness.

From the above graph, the individual effect of input parameters with surface roughness is studied in detail, and it can be noted that a good surface finish is obtained at higher spindle speed (S), low feed rate (F), and low depth of cut (D).

The individual effect of input parameters with the material removal rate is shown in Fig. 7a, b and c. Thus, from the graph it can be observed that a higher material removal rate is obtained at a higher feed rate (F), higher depth of cut (D), and lower spindle rate (S).

6 Validation of Results

The optimal solution obtained from the NSGA_II algorithm is validated with the experimental data. The confirmation of the optimal results is done by randomly selecting two optimal sets of solutions containing the three input parameters along with two response parameters obtained by the optimization of NSGA-II. These corresponding input parameters spindle speed, feed rate, and depth of cut are given as the input to the experimental setup. The experiment is performed by the end-milling VMC machine with aluminum workpiece. The surface roughness of the machined workpiece is measured by the stylus-type surface roughness measurement meter, and the material removal rate is calculated by using the formula (Figs. 8 and 9). The comparison is done for the experimental and NSGA-II, and the result are tabulated in Tables 4 and 5.

7 Conclusions

In this study, the elementary objective is to establish an ANN model and to optimize the two conflicting output parameters, namely surface roughness and material removal rate of the end-milling process using the multi-objective optimization algorithms. With the help of full factorial design of experiments, the cascade forward

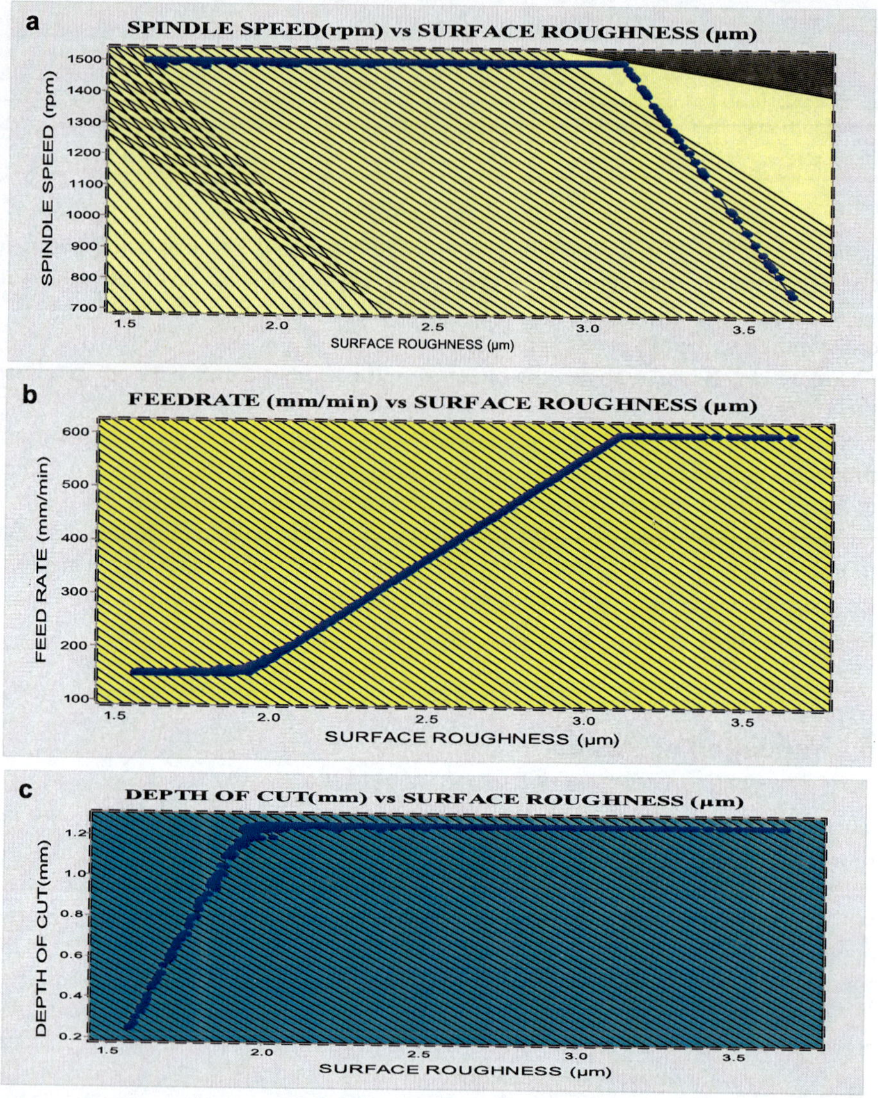

Fig. 6 **a** Effect of spindle speed on optimized surface roughness. **b** Effect of feed rate on optimized surface roughness. **c** Effect of depth of cut on optimized surface roughness

network architecture is selected with four hidden neurons trained by the scaled conjugate gradient algorithm. The multi-objective optimization using a genetic algorithm is compared with NSGA-II. The optimal solutions set generated by the NSGA-II is found to be unparalleled when compared to the multi-objective optimization using a genetic algorithm for the machining parameters of the end-milling process.

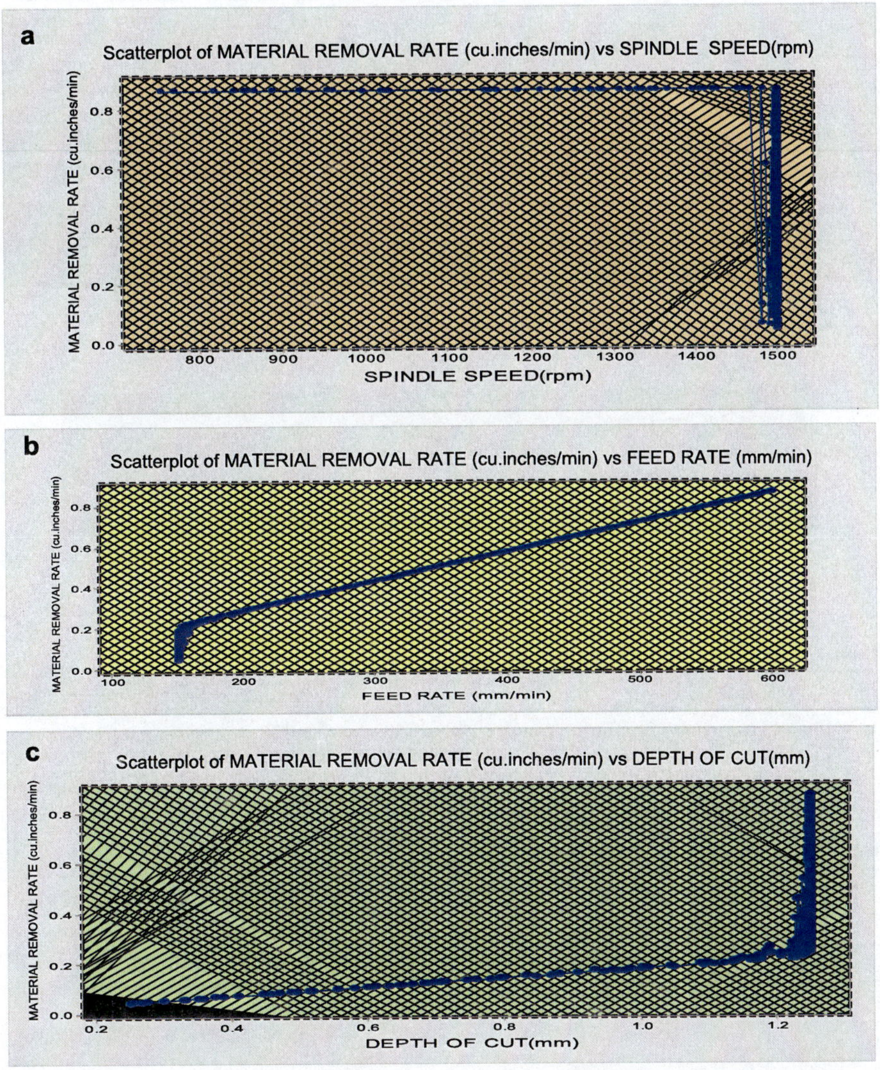

Fig. 7 **a** Effect of spindle speed on optimized material removal rate. **b** Effect of feed rate on optimized material removal rate. **c** Effect of depth of cut on optimized material removal rate

Fig. 8 Experimental setup using LMW JV 55—VMC

Fig. 9 Aluminum workpiece used for milling process

Table 4 Confirmation test for surface roughness

Optimal values	Predicted	Actual	Error (%)
S = 1492.44 rpm, F = 254.071 mm/min and D = 1.23 mm	3.206	3.550	9.69
S = 1499.98 rpm, F = 150.53 mm/min and D = 1.10 mm	2.884	2.657	8.58

Table 5 Confirmation test for material removal rate

Optimal values	Predicted	Actual	Error (%)
S = 1492.44 rpm, F = 254.071 mm/min and D = 1.23 mm	0.3614	0.3656	1.15%
S = 1499.98 rpm, F = 150.53 mm/min and D = 1.10 mm	0.1910	0.1935	1.29%

References

1. Deb K, Pratap A, Agarwal S, Meyarivan TA (2002) Fast and elitist multiobjective genetic algorithm: NSGA-II. IEEE Trans Evol Comput 6(2)
2. Yusoff Y, Ngadiman MS, Zain AM (2011) Overview of NSGA-II for optimizing machining process parameters. Procedia Eng 15:3978–3983
3. Hayajneh MT, Tahat MS, Bluhm J (2007) A study of the effects of machining parameters on the surface roughness in the end-milling process. Jordan J Mech Ind Eng 1:1–5
4. Dhabale R, Jatti VS (2015) Optimization of material removal rate of AlMg1SiCu in turning operation using genetic algorithm. Trans Appl Theor Mach 10:95–101
5. Sahu SN, Nayak NC (2018) Multi-objective optimization of EDM process using ANN integrated with NSGA-II algorithm. Int J Manuf Technol Manag 32:4–5
6. Shaik Dawood AK, Saravana Kumar K (2012) Optimization of CNC turning process parameters on INCONEL 718 using genetic algorithm. IRACST Eng Sci Technol Int J (ESTIJ). 2(4). ISSN: 2250-3498

7. Zain AM, Haron H, Sharif S (2012) Integrated ANN–GA for estimating the minimum value for machining performance. Int J Prod Res 50(1):191–213
8. Chaki S, Bathe RN, Ghosal S, Padmanabham G (2018) Multi-objective optimization of pulsed Nd: YAG laser cutting process using integrated ANN–NSGAII model. J Intell Manuf 29:175–190
9. Sahu SN, Nayak NC (2015) Multi-objective optimization of EDM process with performance appraisal of GA based algorithms in neural network environment. J Prod Eng 18
10. D'addona DM, Teti R (2013) Genetic algorithm-based optimization of cutting parameters in turning processes. In: Forty sixth CIRP conference on manufacturing systems 2013, Procedia CIRP 7, pp 323–328
11. Palanisamy P, Rajendran I, Shanmugasundaram S (2007) Optimization of machining parameters using genetic algorithm and experimental validation for end-milling operations. Int J Adv Manuf Technol 32:644–655
12. Petkovic D, Radovanovic M (2011) Using genetic algorithm for optimization of turning machining processes. J Eng Stud Res 17:10.29081
13. Fonseca CM, Fleming PJ (1993) Genetic algorithms for multi-objective optimization: formulation, discussion and generalization. In: Proceedings of the fifth international conference on genetic algorithms, pp 416–423
14. Fonseca CM, Fleming PJ (1998) Multi objective optimization and multiple constraint handling with evolutionary algorithms–part II: application example. IEEE Trans Syst Man Cybern Part A: Syst Hum 38–47
15. Sardinas RQ, Santana MR, Brindis EA (2006) Genetic algorithm-based multi-objective optimization of cutting parameters in turning processes. Published in Eng Appl Artif Intell 19:127–133
16. Ma G, Zhang F, Chiong R, Weise T, Michalewicz Z (2012) Genetic algorithms for manufacturing process planning. In: Variants of evolutionary algorithms for real-world applications. Springer, Berlin. https://doi.org/10.1007/978-3-642-23424,2012
17. Jameel A, Minhat M, Nizam M (2013) Using genetic algorithm to optimize machining parameters in turning operation: a review. Int. J. Sci. Res. Publ. 3(5)

Parametric Study of FDM Manufactured Parts to Analyze Dimensional Accuracy

Kailash Chaudhary, Naveen Kumar Suniya, Rahul Kumar Rakecha, and Bhuvnesh Dave

Abstract In the present scenario, fused deposition modeling (FDM) process is one of the most widely used processes in additive manufacturing (AM). The dimensional accuracy of parts manufactured using the FDM process depends on many factors. This paper aims to investigate the most influencing factors which affect the dimension accuracy. Taguchi L9 Orthogonal Array (OA) design is used to perform experiments. Selected input parameters for experiment are layer thickness, infill style, and infill density. ASTM D638 Type IV specimens of ABS material have been printed on the Accucrafti250 + 3D printer as per design of experiment (DOE). Analysis of measured data, S/N ratio, and ANOVA is established to understand the most influencing input parameter for dimensional accuracy.

Keywords FDM · 3D Printing · Dimensional accuracy · Design of experiment · S/N ratio · ANOVA

1 Introduction

In additive manufacturing (AM) technology, building of a product is done by adding material layer upon layer. AM technology was earlier used for rapid prototyping (RP) only but nowadays it is used for rapid manufacturing also. Fused Deposition Modeling (FDM) is one of the most emerging techniques used in RP and AM technologies due to its ability to manufacture complex geometry [1]. In this, material filament is unwound from a coil and supplies to an extrusion nozzle. Heated nozzle melts the material and it has a mechanism which allows the flow of the melted material continuously. The nozzle is mounted on such mechanical structure, so it can move in both horizontal and vertical directions. As the nozzle is moved over the table in the required geometry, it deposits a thin bead of extruded material to form

K. Chaudhary (✉) · R. K. Rakecha · B. Dave
Department of Mechanical Engineering, MBM University Jodhpur, Jodhpur, India
e-mail: kailash.me@mbm.ac.in

N. K. Suniya
Department of Production and Industrial Engineering, MBM University Jodhpur, Jodhpur, India

A. Maurya et al. (eds.), *Recent Trends in Mechanical Engineering*, Lecture Notes in Mechanical Engineering, https://doi.org/10.1007/978-981-19-7709-1_72

each layer. The material hardens immediately after being squirted from the nozzle and bonds to layer below. The entire system is contained within a chamber which is held at a temperature just below the melting point of the material. The temperature inside the chamber is maintained lower than the melting temperature of the filament. Therefore, the deposited filament rapidly solidifies when adhered to the previous layer thus involving highly transient heat transfer process [1].

The basic principle of AM is creating a 3D computer-aided design (CAD) model and fabricating directly without the need of process planning. But it has some drawbacks related to surface roughness, dimensional accuracy, strength, time and cost [2]. Dimensional accuracy is the most critical among these drawbacks. The factors that affect dimension accuracy are layer thickness, air gap, build orientation, raster width, orientation, print speed, and extrusion temperature [3]. Tiwari and Kumar [4] investigated the effect of orientation and support on dimensional accuracy. A rectangle block of PLA material was printed for which dimensions were measured and compared in Rapid-I Vision Measuring System. Experiment concluded that the error increases as the inclination angle of the slot with the base decreases and the effectiveness of support is maximum for horizontal printing and almost negligible for the vertical printing. Akande [5] performed experiment with cuboid model and selected process parameter layer thickness, speed multiplier, infill density at low level and high level on PLA material. The measurement was done with Mitutoyo digital caliper with 0.01 mm least count and compared with standard values. It was found that with high layer thickness, low speed of deposition, and low infill density, better dimension accuracy can be achieved. Wankhede et al. [6] performed experiment using Taguchi L8 Design of Experiment (DOE) to investigate for the most influential parameters. For the experiment, layer thickness, infill density, and support style were selected as the input parameters. The layer thickness was found most significant input variable.

The critical study of literature reveals that the various factors affecting the dimensional accuracy of FDM manufactured parts are layer thickness, infill density, infill style, bed temperature, speed, nozzle diameter, filament diameter, raster angle, orientation, air gap, and extrusion width. These factors directly or with combination of other affect the dimensional accuracy. Few among these factors are more influential than others. Literature survey indicates that many researchers have attempted to improve the dimensional accuracy by managing individual or interaction of process parameter.

This paper aims to investigate for most influencing process parameter among layer thickness, infill style and infill density with respect to the dimension accuracy of FDM manufactured parts. Taguchi L9 orthogonal array (OA) design of experiment is used to reduce number of specimen, time, and cost to find the better combination of these parameters. The measurements were taken using calibrated Vernier Caliper of 0.02 mm least count. S/N ratio and ANOVA will be establishing to find rank and percentage of contribution, respectively.

Table 1 Input parameters and their levels

Input parameters	Level 1	Level 2	Level 3
Layer thickness (mm)	0.20	0.25	0.30
Infill style	Straight	Octagonal	Rounded
Infill density (%)	0.25	0.33	0.50

2 Experiment Details

2.1 Design of Experiment

Taguchi method is used for reducing the number of printed specimen, i.e., reducing the variations in process through robust design of experiments. Objective of the design is to produce higher dimension accuracy of product. Dimension accuracy of product does not depend only on single parameter, but on many parameters. This research work is an attempt to find the optimum combination of all parameters. For analysis layer thickness (mm), infill style, and infill density (%) are chosen as input parameters (Table 1) which have three levels each and other parameters are constant during printing. In Taguchi L9, orthogonal array specifies the method of managing the least number of the experiment which could present with the full information of all the factors that affect the performance parameter [7]. The crux of each orthogonal arrays method lies in choosing the level combination of input design variables for each experiment.

2.2 CAD Design and 3D Printing

The dumbbell-shaped American Society for Testing and Materials (ASTM) D638 Type IV specimens is selected for this experiment. Firstly, a 3D CAD model was developed for ASTM D638 Type IV specimen in AUTODESK FUSION 360, and file was exported in STL format. Levels of the parameters were selected as low, medium, and high for DOE as per range available in slicer software "KISSlicer" for slicing of the specimen (Table 2). STL file was imported and corresponding extruder path was generated as G-Codes on which extruder moves. Acrylonitrile Butadiene Styrene (ABS) material is selected for experiment as it has excellent mechanical properties and interlayer adhesion along with reliable bed adhesion properties. Melting temperature of ABS is 220-245^{0}C [8]. For printing of specimens, Accucrafti250 + 3D printer manufactured by Divide by Zero was used. It has single extruder and build envelope area of 300 mm × 250 mm × 200 mm.

Vernier Calliper of 0.02 mm least count was used to measure Length Overall (LO), Width Overall (WO), and Thickness (T) of printed specimens (Fig. 1c). As per build orientation, here LO, WO, and T represent x, y, and z axis (Fig. 1b), respectively.

Table 2 Taguchi L9 orthogonal array design

Experiment no	Layer thickness (mm)	Infill style	Infill density (mm)
1	0.20	Straight	0.25
2	0.20	Octagonal	0.33
3	0.20	Rounded	0.50
4	0.25	Straight	0.33
5	0.25	Octagonal	0.50
6	0.25	Rounded	0.25
7	0.30	Straight	0.50
8	0.30	Octagonal	0.25
9	0.30	Rounded	0.33

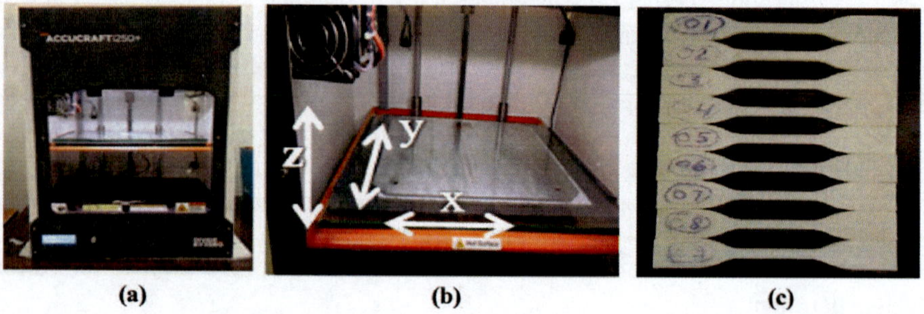

Fig. 1 **a** outside view of Accucrafti250 + 3D Printer. **b** Inside view Accucrafti250 + 3D Printer Chamber. **c** Printed specimens

3 Results and Discussion

To investigate the effects of change in process parameters on the dimensional accuracy of the manufactured parts, the measurements of all selected parameters were taken thrice and the average of these values was considered. Table 3 represents response table of Taguchi L9 orthogonal array.

3.1 Analysis of Length Overall (LO)

For response of Signal to Noise (S/N) Ratio for length overall (LO) presented in Table 4, the rank of infill density was found one, i.e., it is the highest influencing factor, affecting dimensional accuracy in length overall, i.e., x-axis. The next influencing factor is the layer thickness while the infill style is least influencing. From Analysis of Variance (ANOVA) for S/N ratios presented in Table 5, the contribution of infill density was found highest (37.56%) among three process parameters (Fig. 2a).

Table 3 Response table of Taguchi L9 orthogonal array

S. no	Layer thickness (mm)	Infill style	Infill density (%)	LO (mm)	ΔLO (mm)	WO (mm)	ΔWO (mm)	T (mm)	ΔT (mm)
1	0.20	Straight	0.25	114.420	0.580	19.027	0.027	4.133	0.133
2	0.20	Octagonal	0.33	114.320	0.680	18.940	0.060	4.140	0.140
3	0.20	Rounded	0.50	114.273	0.727	18.953	0.047	4.140	0.140
4	0.25	Straight	0.33	114.173	0.827	19.007	0.007	4.220	0.220
5	0.25	Octagonal	0.50	114.387	0.613	18.873	0.127	4.147	0.147
6	0.25	Rounded	0.25	114.607	0.393	18.840	0.160	4.153	0.153
7	0.30	Straight	0.50	114.280	0.720	18.967	0.033	4.047	0.047
8	0.30	Octagonal	0.25	114.247	0.753	18.933	0.067	4.107	0.107
9	0.30	Rounded	0.33	114.220	0.780	18.920	0.080	4.087	0.087

Table 4 Response table of length overall (LO) for SN ratio (smaller is better)

Level	Layer thickness	Infill style	Infill density
1	3.618	3.079	5.099
2	4.668	3.352	2.387
3	2.491	4.345	3.291
Delta	2.178	1.266	2.712
Rank	2	3	1

Table 5 ANOVA for length overall (LO) for S/N ratio

Source	DF	Seq SS	Adj SS	Adj MS	F-Value	P-Value	Contribution (%)
Layer thickness	2	7.115	7.115	3.558	0.77	0.565	23.36
Infill style	2	2.664	2.664	1.332	0.29	0.776	8.75
Infill density	2	11.439	11.439	5.719	1.24	0.447	37.56
Error	2	9.243	9.243	4.621			
Total	8	30.461					

3.2 *Analysis of Width Overall (WO)*

For response of S/N ratio for width overall(WO) presented in Table 6, the rank of infill style was found one, i.e., it is the highest influencing factor, affecting dimensional accuracy in width overall, i.e., y-axis. The next influencing factor is the infill density while the layer thickness is least influencing.

From ANOVA for S/N ratios presented in Table 7, the contribution of infill density was found highest (63.62%) among three process parameters (Fig. 2b).

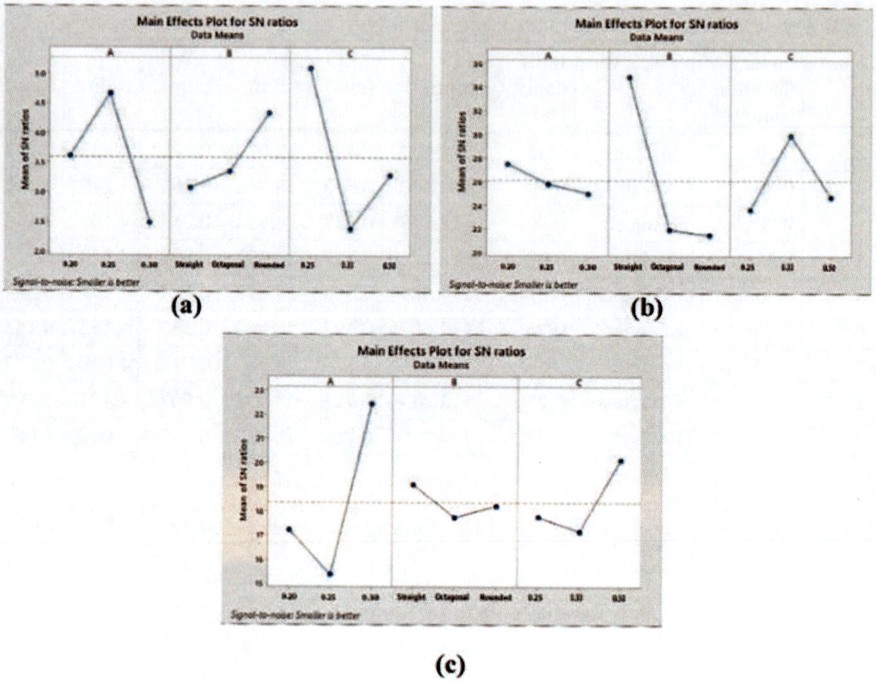

Fig. 2 Main Effects plot for **a** Length Overall (LO); **b** Width Overall (WO); **c** Thickness (T)

Table 6 Response table of width overall (WO) for S/N ratio (smaller is better)

Level	Layer thickness	Infill style	Infill density
1	27.51	34.85	23.64
2	25.80	21.97	29.97
3	25.00	21.49	24.70
Delta	2.51	13.36	6.33
Rank	3	1	2

Table 7 ANOVA for width overall (WO) for S/N ratio

Source	DF	Seq SS	Adj SS	Adj MS	F-value	P-value	Contribution (%)
Layer thickness	2	9.888	9.888	4.944	0.08	0.923	1.83
Infill style	2	344.509	344.509	172.255	2.91	0.256	63.62
Infill density	2	68.839	68.839	34.420	0.58	0.632	12.71
Error	2	118.245	118.245	59.122			
Total	8	541.482					

Table 8 Response table of thickness (T) for S/N ratio (smaller is better)

Level	Layer thickness	Infill style	Infill density
1	17.22	19.09	17.74
2	15.37	17.73	17.16
3	22.43	18.20	20.12
Delta	7.06	1.36	2.97
Rank	1	3	2

Table 9 ANOVA for Thickness (T) for S/N ratio

Source	DF	Seq SS	Adj SS	Adj MS	F	P	Contribution (%)
Layer Thickness	2	80.506	80.506	40.253	4.52	0.181	69.40
Infill Style	2	2.864	2.864	1.432	0.16	0.861	2.67
Infill Density	2	14.810	14.810	7.405	0.83	0.546	12.77
Error	2	17.815	17.815	8.907			
Total	8	115.995					

3.3 *Analysis of Thickness (T)*

For response of S/N Ratio for thickness (T) presented in Table 8, the rank of layer thickness was found one, i.e., it is the highest influencing factor, affecting dimensional accuracy in thickness, i.e., z-axis. The next influencing factor is the infill density while the infill style is least influencing. From ANOVA for S/N ratios presented in Table 9, the contribution of infill density was found highest (69.40%) among three process parameters (Fig. 2c).

4 Conclusion

The aim of research was to investigate the most influencing process parameter among layer thickness; infill style and infill density with respect to the dimension accuracy of FDM manufactured parts. Measured data was analyzed by S/N ratio and ANOVA. Based on this analyzed data, it has been analyzed that:

1. Infill density is most influential factor among all three process parameters that affects the dimensional accuracy in x-axis (LO) from Table 4 and contributes maximum in the variation of dimensional accuracy in x-axis (LO) from Table 5. Thus, infill density should be selected optimum to obtain higher dimensional accuracy in x-axis.
2. Infill style is most influential factor among all three process parameters that affects the dimensional accuracy in y-axis (WO) from Table 6 and contributes maximum in the variation of dimensional accuracy in y-axis (WO) from Table

7. Thus Infill style should be selected optimum to obtain higher dimensional accuracy in y-axis.
3. Layer thickness is most influential factor among all three process parameters that affects the dimensional accuracy in z-axis (T) from Table 8 and contributes maximum in the variation of dimensional accuracy in z-axis (T) from Table 9. Thus layer thickness should be chosen optimum to obtain higher dimensional accuracy in z-axis.

Thus to achieve the accuracy for the dimensions of a product, the optimum values of the process parameters should be chosen.

References

1. Noorani R (2006) Rapid prototyping: principles and applications. Wiley
2. Chaudhari M, Jogi BF, Pawade RS (2018) Comparative study of part characteristics built using additive manufacturing (fdm). Procedia Manuf 20:73–78
3. Dey A, Yodo N (2019) A systematic survey of FDM process parameter optimization and their influence on part characteristics. J Manuf Mater Process 3(3):64
4. Tiwari K, Kumar S (2018) Analysis of the factors affecting the dimensional accuracy of 3D printed products. Mater Today: Proc 5(9):18674–18680
5. Akande SO (2015) Dimensional accuracy and surface finish optimization of fused deposition modelling parts using desirability function analysis. Int J Eng Res Technol 4(4):196–202
6. Wankhede V, Jagetiya D, Joshi A, Chaudhari R (2020) Experimental investigation of FDM process parameters using Taguchi analysis. Mater Today: Proc 27:2117–2120
7. Roy RK (2001) Design of experiments using the Taguchi approach: 16 steps to product and process improvement. Wiley
8. Acrylonitrile Butadiene Styrene (ABS) material technical data by Farnell. http://www.farnell.com/datasheets/2310520.pdf

Application of Thermal Analysis to Study the Effect of Inoculation on the Solidification of Ductile Cast Iron

Bahubali B. Sangame, Y. Prasannatha Reddy, and Vasudev D. Shinde

Abstract Ductile cast iron solidification has a vital role in the production of sound-quality castings. Due to complex solidification of ductile cast irons, thermal analysis is used to predict molten metal quality by analyzing the pattern for the cooling curve. Present work investigates the effect of inoculation on solidification of ductile cast iron through thermal analysis technique. The influence of addition of Ca, Ce-FeSi inoculant, on the solidification of ductile cast iron was studied. The data of eutectic undercooling, recalescence, and solidus temperature of uninoculated and inoculated melts was gathered from thermal analysis curves. Comparison of solidification parameters and microstructure of uninoculated and inoculated melt samples is done in this study. The inoculated melt samples have shown increased minimum and maximum eutectic temperatures and reduced recalescence than uninoculated ones.

Keywords Ductile cast iron · Inoculation · Thermal analysis · Undercooling · Recalescence

1 Introduction

Ductile iron remains one of the materials of choice in the automotive and structural sectors due to its exceptional casting ability, outstanding damping capability, and a great blend of strength and ductility [1]. Considering the growing need for

B. B. Sangame (✉)
Mechanical Engineering Department, Sinhgad College of Engineering Vadgaon, Pune, Maharashtra 411041, India
e-mail: bahubali2010@gmail.com

B. B. Sangame · Y. P. Reddy
Mechanical Engineering Department, Sharad Institute of Technology College of Engineering, Yadrav, Maharashtra 416115, India
e-mail: ypreddy.scoe@sinhgad.edu

V. D. Shinde
Mechanical Engineering Department, DKTE Society's Textile and Engineering Institute, Ichalkaranji, Maharashtra 416115, India
e-mail: vdshinde@dkte.ac.in

A. Maurya et al. (eds.), *Recent Trends in Mechanical Engineering*, Lecture Notes in Mechanical Engineering, https://doi.org/10.1007/978-981-19-7709-1_73

high-quality casting taking into account two main demands in the industrial sector; metallurgical efficiency and economic factors, the quality of casting should be guaranteed. The microstructure of ductile cast iron plays a crucial role in the production of quality castings [2]. Control over internal microstructure mechanisms is important for the improvement of ductile cast iron properties and characteristics [3]. One of the important research exercises and the focus of scientific literature is the control casting microstructure [4]. The most frequent issue for all manufacturers of ductile cast iron castings is how they can be sure of the nucleation quality to obtain a good casting. In order to try to resolve this problem, thermal analysis is a useful method [5] and, on the other hand, may help determine the inoculation needs for the melt [6, 7]. Chemical composition, magnesium treatment, inoculation procedure, and melting background are some factors that alter the shape and nature of the cooling curve [8–10]. Inoculant is added to molten metal immediately after magnesium treatment to promote the formation of nucleation sites in heterogeneous graphite formation [11, 12]. Inoculants are FeSi-based alloys that contain one or perhaps more active key elements in the graphite production process, such as Ca, Ba, Al, Ce, Sr, and Zr. By interacting with other elements in the iron melt, such as Mg, Mn, S, O, N, Al, and Si, these elements enhance the potential of the existing compounds in the iron melt to nucleate nodular graphite [13–15]. With a detailed understanding of the nature of the cooling curve, foundries can get some important key points on the cooling curves. These key points help to predict the chemical composition and metallurgical formation of treated ductile cast irons, ensuring a closely monitored process of solidification. The current study concentrates on the comparison of thermal analysis parameters of uninoculated ad inoculated ductile cast iron melts, to analyze the influence of inoculation on the solidification of ductile cast iron.

2 Experimental Procedure

The raw material was melted into a 150 kg medium-frequency induction furnace. The charge includes 30% of pig iron, 25–30% of cold-rolled steel scrap and 30 to 40% of foundry runner riser return. The foundry uses a 5% magnesium-based ferrosilicon alloy to treat ductile iron with the use of a 1.8wt % addition rate. At a temperature of at 1450 °C, the molten metal has been tapped in a pre-heated ladle of 10–15 mm ferro-silicon-magnesium alloy at its bottom surrounded by stainless steel scrap. After magnesium treatment, the melt was immediately transferred to 20 kg ladles, where the inoculant of size 2–4 mm was added to the flow of molten metal. The experiment was conducted where metal was inoculated with 0.3 wt% of Ca, Ce-FeSi inoculant containing [73.2%Si, 1.03% Ca, 1.67% Ce, and 0.94% Al, Fe Bal]. For spectroscopic analysis, metal samples were collected during the pouring to determine the end chemistry of the casting. The standard cups having thermocouples with a cooling modulus of 0.75 cm were connected to data logger DAQ-3005 (MCC-USA) along with Dasylab 12.0 software for recording of time temperature data (Fig. 1). This data was further used for plotting of cooling (solidification) curves of melt.

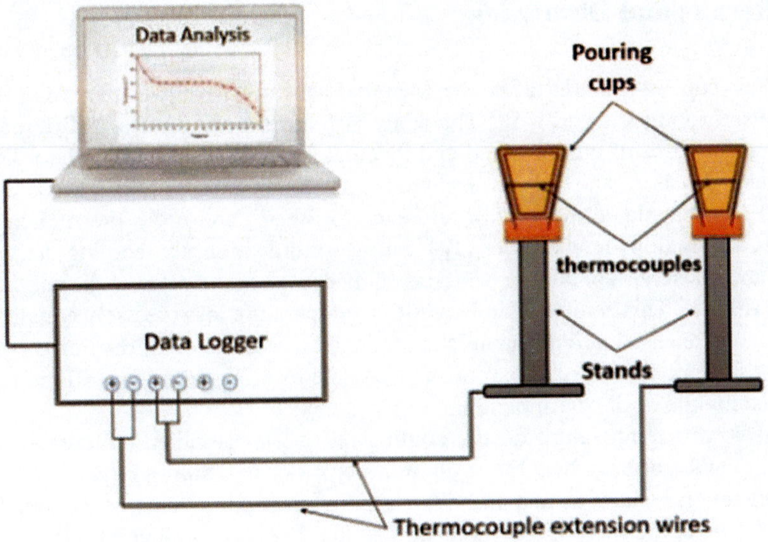

Fig. 1 Data acquisition system with pouring cups

Table 1 shows the chemical analysis of the inoculated and un-inoculated ductile cast iron melts. The calculated values of carbon equivalent (CE), temperature of eutectic graphite (TEG), and temperature of eutectic carbide (TEC) of the system are functions of the amount of silicon in the melt. Equations 1, 2, and 3 which are typically used in this study [10], demonstrates the capacity of elements to extend the eutectic interval (TEG-TEC) by increasing TEG and decreasing TEC values, respectively.

$$CE = \%C + \frac{1}{3}(Si + P) \tag{1}$$

$$TEG = 1149.1(^{\circ}C) + 4.7Si\% + 2.7Cu\% - 4Mn\% - 44P\% \\ + 1Ni\% - 10.5Cr \tag{2}$$

$$TEC = 1142.6(^{\circ}C) - 11.6Si\% - 1.4Cu\% - 0.75Mn\% \\ - 46.2P\% - 1.1Ni\% + 5.9Cr \tag{3}$$

Table 1 Chemical composition of melt after magnesium and inoculation treatment

Melt treatment	%	%	%	%	%	%	%	%	%	CE
	C	Si	Mn	P	S	Cu	Al	Cr	Mg	
Uninoculated	3.6	1.8	0.18	0.034	0.01	0.28	0.02	0.02	0.005	4.22
Inoculated	3.6	1.96	0.16	0.021	0.02	0.28	0.03	0.022	0.025	4.29

3 Results and Discussion

The first cup was poured after the Mg treatment to investigate the solidification response of uninoculated melt. The analysis focuses comparison of different critical temperatures of uninoculated and inoculated ductile iron melts. The following thermal analysis parameters are discussed to understand the solidification of un inoculated and inoculated melts with reference to Fig. 2a and b. As the molten metal attains the liquidus temperature (T_L) during solidification, the cooling curve indicates an almost horizontal line, which is equivalent to a zero point in the cooling rate curve (Fig. 2). This zero point indicates that temperature loss precisely equalizes the heat that is released into the sample at that time. The duration of the horizontal line depends on the time it requires for the austenite to grow from the walls of the cup into the middle of the thermocouple.

The lowest temperature on the cooling curve T_{EU} is called as temperature of eutectic undercooling. It is the minimum point on the cooling curve from which the temperature starts to rise and attains the highest point. This point is called as temperature of the eutectic recalescence T_{ER}. It is the maximum point on the cooling curve from which temperature starts to decrease during this period the precipitation of graphite and austenite takes place. Solidus temperature T_{ES} represents the complete solidification of cup casting. This is the temperature at which solidification ends. Inoculated ductile cast iron melts have higher T_{ES} temperature as compared to uninoculated Fig. 2a. Recalescence temperature(Δ_{TR}) is the difference between the highest (T_{ER}) and the lowest (T_{EU}) temperatures of the eutectic reaction. Recalescence represents the quantity of graphite and austenite that develop during the eutectic solidification and therefore its value influences the shrinkage cavities prediction [7]. The eutectic interval range (ΔTs) is the difference between the graphite eutectic temperature (TEG) and the carbide eutectic temperature (TEC) [11]. The difference between the graphite eutectic temperature and the lowest eutectic temperature represents the degree of maximum undercooling ($\Delta T_m = TEG - T_{EU}$). Uninoculated ductile

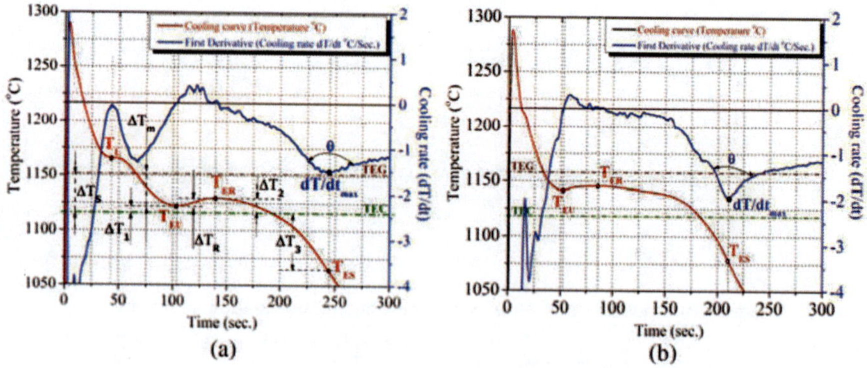

Fig. 2 **a** cooling curves of uninoculated, **b** cooling curve of inoculated melt

iron melts have larger value of ΔT_m than uninoculated one. The difference between minimum eutectic temperature (T_{EU}) and carbide eutectic temperature (TEC) is ΔT_1. At this point, the eutectic reaction starts. It represents the undercooling at the start of eutectic reaction. To avoid the carbides, ΔT_1 must be greater than zero [8]. The difference between maximum eutectic temperature (T_{ER}) and carbide eutectic temperature (TEC) is ΔT_2. At this point, the eutectic reaction ends. It represents the undercooling at the end of eutectic reaction. Undercooling (ΔT_3) at the end of solidification is the difference between carbide eutectic temperature (TEC) and end of solidification temperature (T_{ES}). To avoid the carbides at the center of casting the value of ΔT_3 must be small [8, 16]. The depth of derivative curve at solidus temperature is referred as maximum cooling rate (dT/dt_{max}). Angle (θ) formed by the cooling rate curve at the solidus represents the amount of graphite precipitated during last part of solidification. It indicates the thermal conductivity of the cast iron at the end of solidification, which indicates the possibility of formation of shrinkage and porosity [5, 11].

Table 4 represents the effect of inoculation on various representative temperatures. A very noticeable effect of inoculation is the increase of the minimum (T_{EU}) and maximum eutectic temperature (T_{ER}) Fig. 3a. At the cooling rate (derivative) curve, a zero point of the lowest eutectic temperature is seen, which results in the heat produced by the recalescence of the specific heat and the latent heat Fig. 3b. The eutectic reaction occurs at this zero point, and the recalescence heat induces the temperature to rise to maximum temperatures (T_{ER}). The first derivative curve then reaches to a second zero point. Due to the development of inherent heat called latent heat, the highest eutectic temperature (T_{ER}) is reached. The recalescence ($\Delta T_R = T_{ER}$-T_{EU}) represents the proportions of austenite and graphite precipitating during the first part of the eutectic solidification process.

Table 5 represents comparison of eutectic undercooling temperatures of inoculated and uninoculated ductile iron melts. The higher values of (ΔT_1 and ΔT_2) at the beginning of eutectic reaction and undercooling (ΔT_3) at the end of solidification of inoculated melts represent undercooling level is significantly reduced Fig. 3. Too high recalescence will increase the nucleation of graphite during the first part of eutectic reaction, and for the later part during the end of solidification, there will be insufficient amount of carbon to precipitate as graphite [11]. Lower recalescence levels are associated with high inoculant efficiency and a reduced risk of macro shrinkage and porosity. If the recalescence degree is high, it could mean an undesirable low nodule count and early graphite expansion, which raises the risk of wall expansion effects and primary shrinkage.

Table 4 Representative eutectic temperatures of uninoculated and inoculated melts

Melt treatment	TEG °C	TEC °C	T_{EU}°C	T_{ER} °C	T_{ES} °C	dT/dt_{max} °C/sec	Θ
Uninoculated	1155.9	1119.7	1120	1130	1070	−1.5	110°
Inoculated	1157.5	1118.5	1141.1	1145.2	1079	−1.98	77°

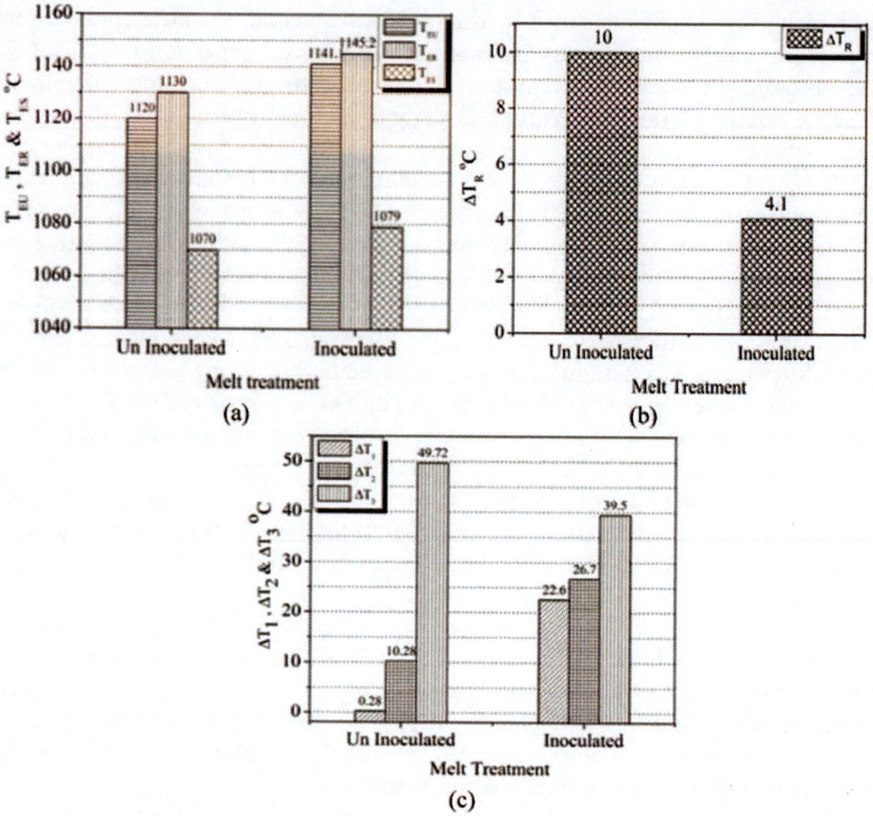

Fig. 3 Comparison of the eutectic temperatures, **a** Minimum (T_{EU}), maximum (T_{ER}) eutectic temperatures and solidus temperature (T_{ES}), **b** Recalescence $\Delta T_R = T_{ER} - T_{EU}$, **c** Eutectic undercooling ΔT_1, ΔT_2 and ΔT_3

Table 5 Undercooling temperatures of uninoculated and inoculated melts

Melt treatment	$\Delta TS = TEG - TEC$	$\Delta T_R = T_{ER} - T_{EU}$	$\Delta T_1 = T_{EU} - TEC$	$\Delta T_2 = T_{ER} - TEC$	$\Delta T_3 = TEC - T_{ES}$
	°C	°C	°C	°C	°C
Uninoculated	36.18	10	0.28	10.28	49.72
Inoculated	38.99	4.1	22.60	26.70	39.5

The inoculant addition increases the number of nodules for a given composition as shown in Fig. 4. Therefore, it indicates that inoculation provides more number of nucleation cites for the formation of nodular graphite during the solidification. Current experiments demonstrate not only the significant role of inoculation in the quality of ductile iron casting but also the effectiveness of thermal analyses as a simple solution for predicting the properties of ductile cast iron before pouring.

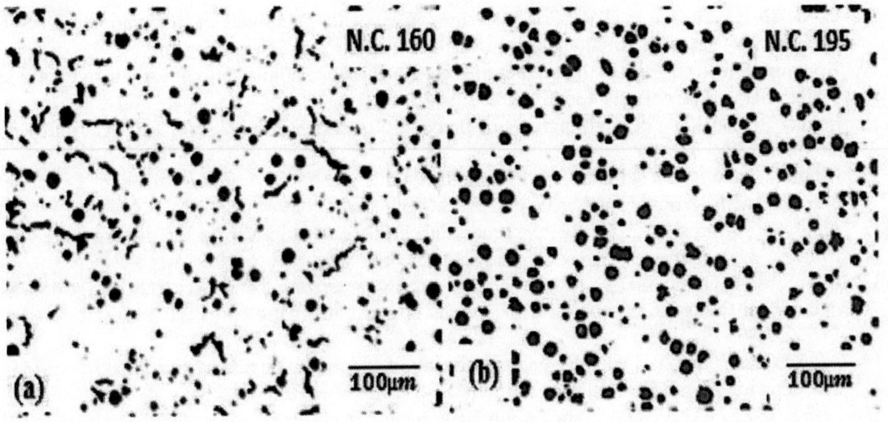

Fig. 4 Microstructure samples of **a** un-inoculated **b** inoculated melts

4 Conclusions

The following findings have been noted in the experiments conducted:

1. The cooling curve and its derivatives, related critical temperatures can be used to predict the solidification behavior of the melt.
2. Inoculation reduces the recalescence and under-cooling rate, thus enhancing overall ductile iron properties.
3. Inoculated melts are having higher values of undercooling ΔT_1 and ΔT_2 at the beginning of eutectic reaction which indicated the metal is more resistant to chill.
4. Inoculated ductile iron melts have higher nodule count than un inoculated one.

References

1. Karsay S (1985) Ductile iron production practices. In: American foundrymens society, pp 101–107
2. Skaland T (2003) Ductile iron shrinkage control through graphite nucleation and growth. Int J Cast Met Res 16:11–16
3. Seidu SO, Riposan I (2009) Graphite characteristics in preconditioned and inoculated ductile Irons. Metall Int 19(2):75–78
4. Vitor A, Carlos A, Silva R (2018) Maximization and control of nodular iron melt's self-feeding characteristics to minimize shrinkage. Mater Sci Forum 925:147–154
5. Stefanescu DM (2015) Thermal analysis theory and applications in metal casting. Int J Metal 09:07–22
6. Soivio K, Elmquist L (2013) Influence of inoculation on shrinkage defects in spheroidal graphite cast iron. Int J Cast Metals Res 26:221–227
7. Bhat M Afzal Khan D, Singh K (2018) Effect of inoculation and nodularization treatment temperature on recalescence and eutectic undercooling temperature in Spheroidal graphite (SG) cast iron. Int J Cast Metals Res 1–8

8. Cojocaru A, Riposan I (2019) Solidification influence in the control of inoculation effects in ductile cast irons by thermal analysis. J Therm Anal Calorim 26:1–13
9. Boonmee S et al (2019) On the prediction of residual magnesium and nodularity in ductile iron by thermal analysis. Trans Tech Publ 287:24–29
10. Seidua SO (2013) Thermal analysis of preconditioned ductile cast iron. Int J Current Eng Technol 3:813–818
11. Kanno T, Yuki I (2017) Prediction of graphite nodule count and shrinkage tendency In ductile cast iron, with 1 cup thermal analysis. Int J Metal Cast 11(1):94–100
12. Singh KK, Sethy SK (2013) Recalescence and shrinkage relationship in ductile iron: study on front engine support casting by thermal analysis. Int J Cast Met Res 26(3):168–175
13. Regordosa A, Llorca-Isern N (2016) Microscopic characterization of different shrinkage defects in ductile irons and their relation with composition and inoculation process. Int J Metalcasting 11:778–789
14. NG KL, Sasaki H (2017)Heterogeneous nucleation of graphite on rare earth compounds during solidification of cast iron. ISIJ Int Adv Publ J-STAGE
15. Fras E, Górny M (2001) An inoculation phenomenon in cast iron. Arch Metall Mater 767–777
16. Stefan E, Riposan I, Mihai Chisamera M (2019) Application of thermal analysis in solidification pattern control of La-inoculated grey cast irons. J Therm Anal Calorim 138(4):2491–2503

Design of a Cost-Effective Floating Waste Cleaning Robot

Mahedy Hasan, Sakib Asrar, Tanvirul Azim, Syeda Prioty Sultana, Rashed Shelim, and Riasat Khan

Abstract Plastic garbage in reservoirs causes significant harm to water quality, aquatic life, and the entire ecosystem. This paper presents a low-cost water waste cleaning robot to collect floating waste in ponds and lakes, composed of commonly available low-cost materials requiring little human labor. This study aims to develop a robot that can collect floating trash in place of humans and evaluate the performance of the proposed system. This automatic system is constructed of floatable material and will float on the water to gather waste materials. A simple smartphone application is used to control the robot's cage-like framework, resulting in an extremely user-friendly interface. The waste trapped inside will have to be manually taken out of the bot before a second launch. Successful experiments have been made to collect different types of plastic waste in a small waterbody. The robot's operating range and battery life are measured to ensure an efficient cleaning process in terms of time. Furthermore, the operator may adjust the robot's speed to make movement simple and precise.

Keywords Eco-friendly robot · Floating waste cleaner · Remote-controlled bot

1 Introduction

Bangladesh is known as the country of rivers with numerous lakes and tributaries that connect with the big rivers and eventually with the Bay of Bengal. Currently, a vast majority of the waterbodies are polluted with different floating plastic materials [1]. Plastic is already a danger for the world; plastic and non-recyclable waste materials in the water double the trouble. Most of the water sources of Bangladesh are rampantly polluted by different human and industrial contaminants [2]. To overcome this situation, an automated management system with modern robots can be employed to make this city's water as pure and clean as before industrialization [8].

M. Hasan (✉) · S. Asrar · T. Azim · S. P. Sultana · R. Shelim · R. Khan
North South University, Dhaka, Bangladesh
e-mail: mahedy.hasan@northsouth.edu

A. Maurya et al. (eds.), *Recent Trends in Mechanical Engineering*, Lecture Notes in Mechanical Engineering, https://doi.org/10.1007/978-981-19-7709-1_74

Many researchers, scientists, and inventors have been working on methods to automate the cleaning process of the waterbodies [9]. Many of them have been successful but at the expense of resource availability, economic sustainability, or inefficiency. In the subsequent paragraphs, some of the related works on waterbodies' floating waste collection devices and their various features are described. For instance, Ruangpayoongsak et al. aimed to build an uncrewed water vehicle designed to scoop the waste out of the water in vast quantities [3]. Their robot can collect round-shaped slippery plastic water bottles by using its five scoopers. As the side scoopers drop in and out of the water, it creates massive waves. Due to its colossal size, the device is only suitable for large waterbodies such as rivers and similar waterbodies. It would be creating a drag force [10] that would make the waste drift further away in smaller water bodies. In [4], Shridhar et al. proposed a smaller device composed of GI metal sheet, which has four principles of working. The first method of operation seemed very ineffective as the system would be entirely stationary. Another method of operation made the system quite complicated and, thus, expensive. Their device can collect a lot of waste without human intervention. The proposed system uses an Atmel AT89S52 microcontroller and various sensors to detect waste, and only a few types of waste can be collected with this device. A Netherlands-based drone technology company, RanMarine Technology, has built an autonomous commercial bot titled "WasteShark," that cleans the harbor area off the plastic waste [5]. Their device can investigate water quality by measuring water depth, chemical composition, and salinity. It is an aquadrone of 72 kg weight that removes plastics and other floating debris from the water surface. It is relatively rigid and can handle small waves in the harbor areas. This drone would be an excellent gadget for cleaning the floating plastic and other waste materials from the waterbodies, but it sells at a high price of USD 17,000. In a recent work [6], the authors designed a robot to collect plastic wastes from the water's surface. The device will monitor the water quality by measuring pH, turbidity, and temperature using Arduino Nano and NodeMCU. Though the paper showed some data on the water monitoring system, it did not indicate any results of waste collection rate, range, and cost of the device, specifications of the battery, etc. Recently, A. Akib and his team built a prototype of an autonomous robotic system that cleans floating waste materials from small waterbodies [7]. The proposed system can move in four directions by connecting to two propellers and DC motors of 300 RPM. An Arduino Uno microcontroller has been used to control the DC motors, overload detection, and communicate the proposed system with the Bluetooth mobile application controlling device. The 2.3 kg weighing robot has an estimated 4 h battery life and costs approximately USD 160. The authors showed some results on the device's power consumption in static water and analytical propeller speed with duty cycle.

In this work, we have built an efficient and automatic device to collect floating waste materials from ponds and lakes. The proposed bot is made of Styrofoam that will hover on the water to trap waste materials and clean the surface water. Arduino Uno microcontroller has been used to send the signals to the motor driver that control the orientation and rotating speed of the two individual motors used to change the total system's direction and speed. As a simple smartphone app controls the device, we believe the mass people will face little to no difficulty operating the bot. It is an unparalleled and incomparable device of its kind, which will be helpful for any still waterbody's authority responsible for its cleanliness.

2 Proposed System

The proposed floating waste cleaning system has a straightforward robotic system design that is very user-friendly and can be controlled with minimal instructions and prior experiences. All of its individual components are inexpensive, readily available in the market, and built with standard hardware tools available at home. The robotic system consists of an Arduino Uno microcontroller, a Bluetooth module HC05, two motor drivers BTS7960, and two 775 DC motors. A regular smartphone with our custom application is used as the controller of the bot. First, the Bluetooth module HC05 receives the input signals from the smartphone. Next, the Arduino Uno is used to send the signals to the H bridge motor driver boards that control the orientation and rotating speed of the two individual motors used to change the bot's direction and speed. Figure 1 shows the working procedures of the proposed floating waste collecting system.

In this work, two 12V 775 DC motors have been used as the main driving motor of this bot, which runs at a moderate speed of 7,000 RPM. The open-source micro-controller Arduino Uno and the BTS7960 for motor driver applications, which uses an H bridge module, are used in this research. It can withstand a maximum current of 43A. The specific motor drive module is selected because, for underwater motor propulsion, the current drawn often exceeds 20A due to high resistivity. The two motors are connected to two motor drivers to operate within the safety limit. We utilized the HC05 Bluetooth module, which has an adequate operational range of up to 10 m for connectivity. The total system has been powered through two 2200 mAh lithium-ion batteries, which consist of 3-cells and have a low voltage of 11.10 V DC with a discharge rate of 25 °C. We have used four plastic propellers (two on each motor) to drive the bot. Each propellor is composed of three blades with a diameter of 40 mm. In the next section, the required hardware and software tools to design the proposed system have been described in detail.

3 Required Tools

3.1 Hardware Tools

The main body of the proposed structure is constructed of plastic wood composite materials on the top and polystyrene Styrofoam on both sides attached to it at an angle of 150°. The 1" Styrofoam is cut into 1 square foot slice and then glued together to double the thickness. Styrofoam was used in this work because it is incredibly

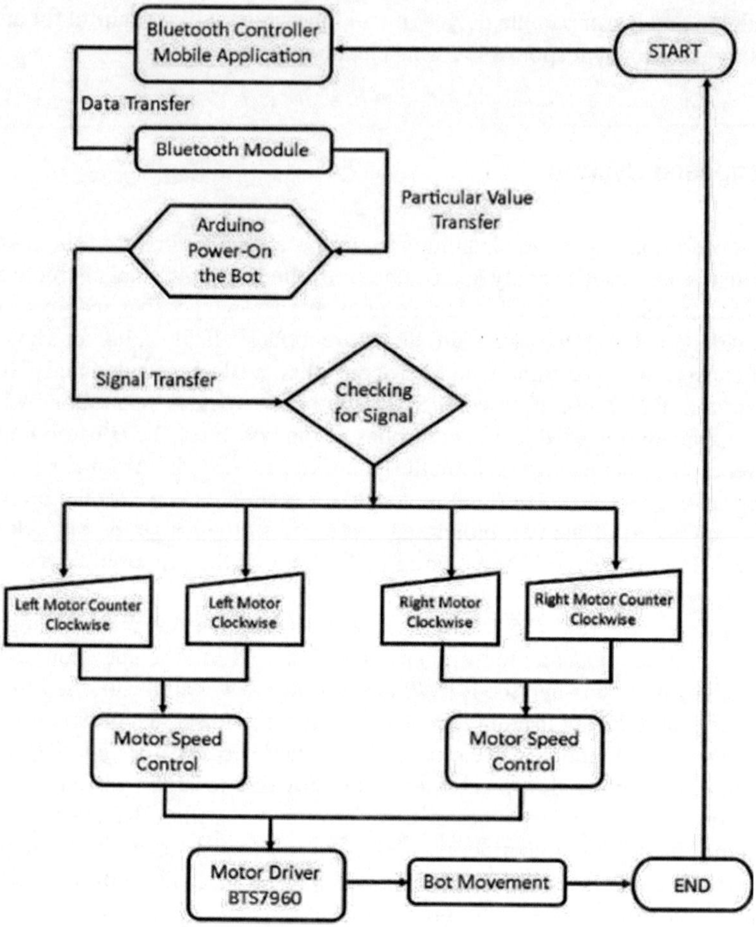

Fig. 1 Working flowchart representing the proposed robotic system

lightweight, cheap and provides much higher floatable property. Plastic wood was used because it is sturdy, which makes it suitable for holding electrical components. After drying, the Styrofoam slices were then attached with glue and skewers to the plastic wood at the angle, as mentioned above. Skewers were inserted to reinforce the joint between the foam and the plastic wood. Finally, the bot was kept in a safe place to dry.

After the chassis is ready, it was then laminated by wrapping it up with duct tape layers. This process makes the bot collision-proof and sturdy. This was done because the polystyrene foam parts could crumble and disintegrate over repeated use of the bot in a submerged position. The duct tape wrapping reduces contact between water and the foam. This process also reduces the chance of mold and bacteria growth in the foam. A mesh net is attached at the rear part and is extended to cover the bot's floor. The mesh net is one of the most crucial parts of the bot. It allows the water to flow out through the bot but traps the floating waste effectively. Two powerful 12V 775 DC motors were attached firmly with the help of zip-ties on either side of the bot. At the end of the motor shafts, duo-propellers were installed. The motors with the propellers provide the driving force for the bot as it helps it sail through the water and take suitable turns. The propellers help to make the bot turn by rotating in clockwise and anti-clockwise directions.

The control box is also made of plastic wood materials, and it is placed on the top of the bot. Another conical-shaped box is used (also out of plastic wood), purposed to contain disinfectants or water treatment chemicals in it. It is also attached with L clamps. It is shaped in a conical manner for the smooth disposal of chemicals. It can carry dry chemicals that are dropped remotely to purify the water. Figure 2 demonstrates the final designed physical structure of the proposed system.

Figure 3 shows the duo-propellers, which displaces more water per unit time to help faster propelling of the device. The high torque motors enabled this set-up as it draws a lot of current in the circuit whenever the motors feel resistance underwater. Finally, the bot weighs about 2.5 kg (equivalent to 5.5 pounds), including all the components and balancing weights. The effective range of the Bluetooth module is about 10 m.

This part (Fig. 4) explains how the electrical components of the bot are designed to make its remote and smooth operations. Arduino Uno microcontroller is used to establish a remote connection, and an HC05 Bluetooth module is employed to form a wireless connection with a smartphone. The proposed device has a built-in Bluetooth module that connects to the hotspot that a smartphone can create. Figure 4 illustrates the functional block diagram of the hardware interfacing of the proposed system. The batteries are connected with the BTS7960 motor driver. The motor driver sends the power to the motors attached to the side of the bot.

Fig. 2 Front view of the bot

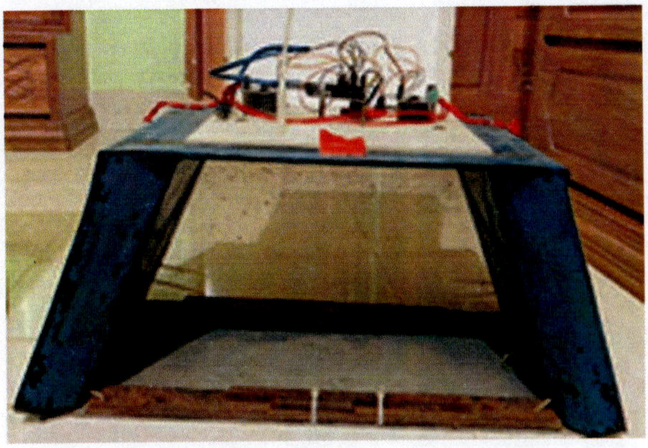

Fig. 3 Dual propellers attached to the bot

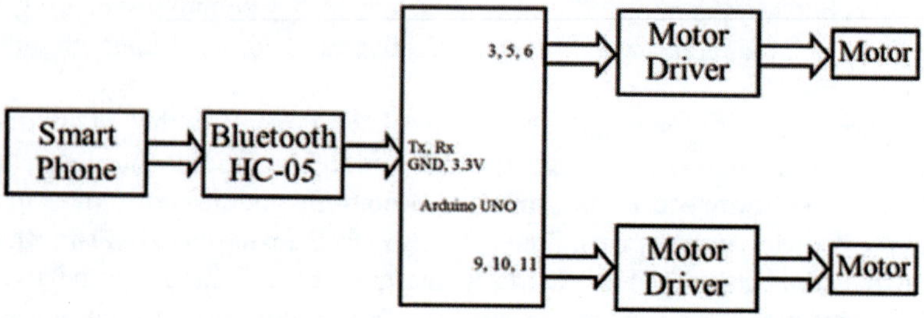

Fig. 4 Functional block diagram of the hardware interfacing of the proposed system

3.2 *Software Tools*

In this paper, we have used the open-source software tool, Arduino IDE, for Arduino Uno programming and the servo motor code uploading. The bot relies on a simple mechanism of motor control for steering. The direction and speed of rotation can be controlled independently. Thus, to make a right turn, the motor on the right is switched off while the only motor on the left operates at maximum speed and vice versa. We used the pulse-width-modulation technique to enable this operation. An app named "DC Motor" (available for Android) was used to pair the Bluetooth receiver with the phone's Bluetooth. The interface of the app is also straightforward to use.

3.2.1 Analysis

In this section, experimental results are presented for the successful test of the proposed floating waster-cleaning robot. The environment of the experiment has been done on a local pond located in Narayanganj city of Bangladesh. The dimensions of the pond were approximately 21.6 × 18.0 × 1.37 m. The water was stagnant for the majority of the time except for occasional waves caused by the wind. Initially, the bot was stable and balanced on the water surfaces during the first few tests, as demonstrated in Fig. 5. The motors were operating at an optimal voltage of 18–20V, moving the bot swiftly forward, backward, and sideways with the help of the controller. The bot had a battery life of approximately 1 h under continuous operation, which has two 2200 mAh lithium-ion batteries. In this period, the bot was launched nine times by three different male operators between the ages of 23 and 26 and collected approximately a total of 2.5 kg of plastic waste. Its operational range was 10 m, which is a standard Bluetooth controller range with no interfering body. On average, the bot collected about 130–140 g of floating wastes per minute. Table 1 contains nine sets of three operators' data, each conducting three tests of the waste collecting device. It can be observed that the average waste collection rate is 0.133 kg/minute. The factors causing slightly diverse results for the different operators were a piece of wood that weighed 200 g, how densely the waste materials were scattered, and how the entire bot was affected by wind at different points of the day.

Finally, five types of wastes were collected from the performed experiments of the proposed device. Most of the collected wastes were plastic wastes, such as bottles, one-time-use plates, and plastic packets. Some of them were high-density polymer waste, e.g., empty cans of sprays, food containers, and shopping bags. There were also little amounts of leaves and wooden debris. After one hour of rigorous operation, the battery completely drained, but there was no noticeable damage to the bot. The electrical component box at the top was utterly safe from coming in contact with water. The bot remained balanced and did not submerge beyond the desired level even when it was filled with waste materials.

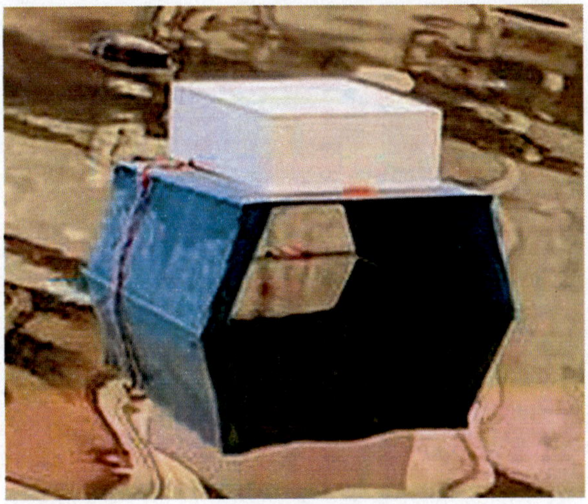

Fig. 5 Partially submerged view of the bot

Table 1 Waste collection rate representation

Operators	Waste collected (kg)	Time taken (min)	Average waste collection rate (kg/min)
Operator 1	0.423	2.33	0.132
	0.188	1.33	
	0.225	2.67	
Operator 2	0.237	1.50	0.137
	0.270	2.17	
	0.310	2.27	
Operator 3	0.325	1.93	0.129
	0.355	2.42	
	0.220	2.62	

Table 2 Comparative analysis of present work with the reported work

Work	Operating microcontroller	Experimental results	Approximated cost
[3]	Arduino Uno	Waste collection rate	Not reported
[4]	Atmel AT89S52	Not reported	USD 250
[5]	Arduino Uno	Not reported	USD 160
[8]	Arduino Nano	Not reported	Not reported
This work	Arduino Uno	Waste collection rate	USD 100

Table 2 contains a comparative analysis of the present work with similar systems which tried to implement similar functions of floating water waste collection. Our proposed device accomplishes the task of floating water garbage collection with minimal cost and operational complexity compared to other systems.

4 Conclusion

In this paper, a cost-effective robot has been implemented to collect the floating wastes on the water surfaces. Experimental results show that the designed system stably floats on the water surface, works perfectly, and can accumulate floating waste with ease. All the controls involved in this system are straightforward and can be operated by anyone with just a smartphone in hand. The total system is exceptionally inexpensive (USD 100). In the future, the proposed system can be entirely automatically operated to collect waste by exchanging data among other bots connected under one central server, and consequently, Raspberry Pi or Jetson Nano can be utilized. The Bluetooth module can be replaced with a Wi-Fi or RF module to increase the device's operating range.

References

1. Jambeck JR, Geyer R, Wilcox C, Siegler TR, Perryman M, Andrady A, Narayan R, Law KL (2015) Plastic waste inputs from land into the ocean. Science 347:768–771
2. Hasan MK, Shahriar A, Jim KU (2019) Water pollution in Bangladesh and its impact on public health. Heliyon 5(8)
3. Ruangpayoongsak N, Sumroengrit J, Leanglum M (2017) A floating waste scooper robot on water surface. In: International conference on control, automation and systems, pp 1543–1548. Jeju, Korea
4. Shridhar K, Vijay B, Siddarth K (2016) Aqua drone remote controlled unmanned river cleaning BOT. In: BVB college of engineering and technology, Hubballi
5. Hardiman SR (2021) Waste shark. 2016–17, project reference: RanMarine technology. https://www.ranmarine.io/. Accessed 15 July 2021
6. Turesinin M, Kabir AMH, Mollah T, Sarwar S, Hosain MS (2020) Aquatic Iguana: a floating waste collecting robot with IoT based water monitoring system. In: International conference on electrical engineering, computer sciences and informatics, Yogyakarta, Indonesia, pp 21–25
7. Akib A et al (2019) Unmanned floating waste collecting robot. In: IEEE region 10 conference (TENCON), Kochi, India, pp 2645–2650
8. Sumroengrit J, Ruangpayoongsak N (2017) Economic floating waste detection for surface cleaning robots. In: MATEC web of conferences, vol 95
9. Ichimura T, Nakajima S (2016) Development of an autonomous beach cleaning robot "Hirottaro". In: International conference on mechatronics and automation, Harbin, China, pp 868–872
10. Kolev NI (2005) Drag forces. In: Multiphase flow dynamics, vol 2. Springer, Berlin

Computational Fluid Dynamics (CFD) Analysis of Pesticide Flow-Repellent Helmet for Farmers

Chetan Bankar and Vipin B. Gawande

Abstract According to the Ministry of Chemicals and Fertilizers, per hectare consumption of pesticides in India shows a significant increase year after year. India is the fourth-largest global producer of pesticides. Pesticides are spread in the farms by the farmers/operators without using safety equipment. There is no safety equipment available in the market, which could be used as a flow repellent. So, there is a need to develop a cost-effective design of pesticide flow-repellent helmet for farmers. An innovative and cost-effective design of the helmet is created, and the flow analysis is carried out using ANSYS-CFD software. The fan in the helmet is used to create turbulence which has a sufficient velocity to repel the incoming toxic pesticide flow. The flow simulation data from CFD is used to predict the flow distribution in the helmet. CFD results are analyzed, and the fan position is adjusted to obtain the required flow in the helmet. Based on the flow distribution, a prototype model is made to check the flow distribution in actual working conditions. The design is validated by carrying out several laboratory experiments. Solar energy panels are used to charge the batteries to run the fan.

Keywords Pesticide flow repellent · Helmet · CFD · Solar energy · Personal protective equipment · Flow simulation

1 Introduction

In India, pesticides are extensively used in agriculture to control harmful pests and prevent crop yield losses. According to the Ministry of Chemicals and Fertilizers, per hectare consumption of pesticides in India shows a significant increase year after year. India is the fourth-largest global producer of pesticides. Figure 1 shows the consumption of chemical pesticides in various states/UTs during 2014–15 to 2019–20, as per data taken from the Ministry of Agriculture and Farmers Welfare. From

C. Bankar (✉) · V. B. Gawande
Department of Mechanical Engineering, VPKBIET, Baramati, India
e-mail: chetan.bankar.mech.2017@vpkbiet.org

A. Maurya et al. (eds.), *Recent Trends in Mechanical Engineering*, Lecture Notes in Mechanical Engineering, https://doi.org/10.1007/978-981-19-7709-1_75

the figure, it is noted that there is a significant increase in pesticide consumption in India every year.

Farmers usually face major exposure from direct spray during the pesticide spraying operations. This kind of exposure is often underestimated. In India, farmers rarely use safety equipment during pesticide spraying, and the process of spraying is manual as shown in Fig. 2.

Minnikanti Venkata Satya Sai et al. [1] evaluated from their survey that skin problems and neurological system disturbances were the most common pesticide-related health symptoms. They also emphasized pesticide safety education regarding the usage of personal protective devices, personal hygiene, and sanitation practices during and after the application of pesticides. It has been observed in various cases that prolonged exposure to pesticides can lead to cardiopulmonary disorders, neurological and hematological symptoms, and skin disease [2]. Pesticides entering the body through inhalation can cause serious damage to the nose, throat, and lung tissues [3].

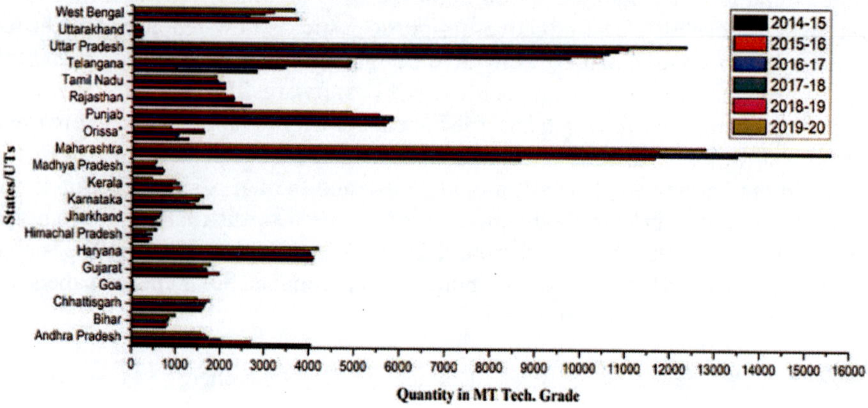

Fig. 1 Year-wise consumption of chemical pesticides in states/UTs of India [1]

Fig. 2 Conventional method of pesticides spraying [3]

The health disparities resulting from pesticide exposure in agricultural communities are addressed by Arcury et al. (2001), [4]. Chitra et al. [5] along with her co-authors revealed that 68.6% of farmers sprayed pesticides themselves and thus are directly exposed to toxic pesticide inhalation. They also found in their survey that about 88% of the farmers used no protection while handling pesticides. They also suggested that there is a need to spread awareness about the use of protective equipment while handling pesticides. Sarkar et al. [6] tried to identify the health risks associated with modern agricultural practices. They have observed significant health effects due to the massive use of pesticides.

As we have seen, many researchers suggested that the use of various types of personal protective equipment (PPE) can be used to prevent the adverse effect of pesticides during the spraying operation. The main reasons for not using personal protective equipment (PPE) like a respirator, glasses/goggles, gloves, hat, protective boots, etc., by the farmers are this personal protective equipment are too expensive and are not affordable by most of the farmers, and another reason is that use of this equipment in the local hot and humid climate reduces comfort while working. This also reduces the efficiency of working of the farmer.

As stated earlier, the greatest potential for poisoning via respiratory exposure is with vapors and extremely fine particles of the spray solution and is rarely prevented by using any of the PPE. There is no affordable and efficient safety equipment available in the market, which could be used as a flow repellent to prevent the inhalation of toxic pesticide particles. The innovative design of the helmet will solve this problem and uses the principle of fluid dynamics for flow distribution.

So, the main objectives of the proposed project are,

1. To design a pesticide flow-repellent helmet design using CAD software.
2. To conduct the flow analysis in the helmet using numerical tool ANSYS-CFD
3. To develop a prototype of helmet and selection of parts to make it cost-effective.
4. Laboratory testing to test the performance of pesticide flow-repellent helmet.

2 Methodology and CFD Set up

As per the literature review, it has been rectified that the major health hazards that occurred to the farmers are due to inhalation of the toxic pesticides. The farmers inhaled the toxic particles of pesticides due to the unavailability of suitable personal protective equipment (PPE). Instead, they used ordinary cloth, as shown in Fig. 3, to resist the entry of these particles in the body. The cloth used for this purpose hardly absorbed the particles on its surface and is not effective at all. A flow mechanism is required to repel the incoming pesticide flow. An innovative helmet design proposed in this project will come out as suitable protective equipment for spraying operation.

Fig. 3 Use of cloth as protective equipment during pesticide spraying operation

2.1 3D Computational CAD Model

A three-dimensional CAD model is created in ANSYS Workbench as shown in Fig. 4. A 3D printed helmet design uses the principle of fluid mechanics to repel the pesticide particles and prevent their inhalation by farmers. The helmet has a plastic shield with a glass placed in front of the human face for clear visibility. A fan is inserted in the helmet above the head of the farmer and is operated by the chargeable batteries. The fan is used to create turbulence in the helmet and increases the velocity of incoming fresh air in the helmet which repels the air mixed with pesticides' particles. Since the spraying operation is carried out during the daytime, the solar panel is fitted on the helmet which is used for charging the batteries used for running the fans in the helmet. Wind sensors are provided so that the farmers are informed about the wind direction and pesticide spraying must be carried out in that direction only. Solar panel and wind sensors are considered for the actual prototype model and are omitted from the computational domain of the helmet.

2.2 Mesh Generation

The computational domain along with a body, fan, and a human face is considered as a computational domain. Mesh is generated in the ANSYS-Mesh module and is an unstructured type and is shown in Fig. 4b. The mesh is made fine near the fan surface to capture the flow pattern generated by a rotating fan. A mesh independence test is carried out to check the correctness of the mesh used for the analysis.

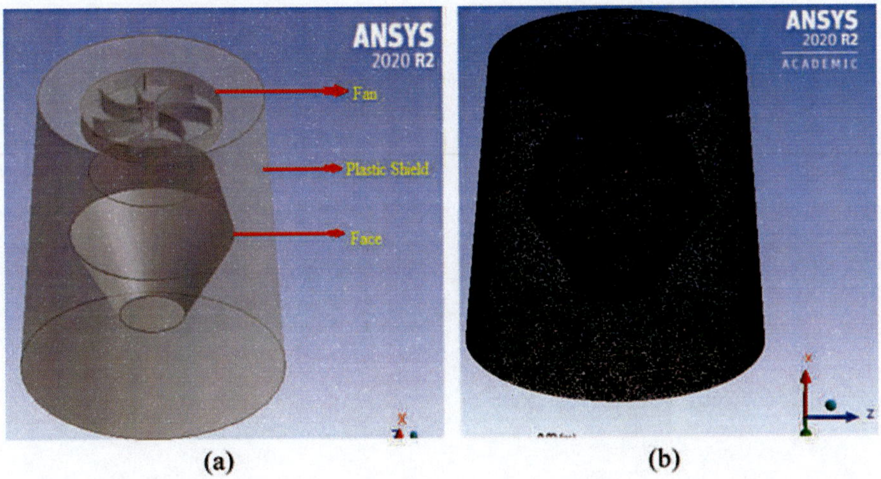

Fig. 4 **a** Three-dimensional CAD model of pesticide flow-repellent helmet. **b** Unstructured mesh generation on computational domain

2.3 CFD Solver Set up

The flow distribution in the helmet plays a vital role to repel the incoming toxic pesticide particles during spraying. The fan is provided on the top of the helmet and is useful for creating flow distribution in the helmet. The fan location in the helmet is decided by the flow distribution analysis by providing r.p.m to the fan in ANSYS solver-FLUENT. For present analysis, fan is given r.p.m of 1000. The inlet and outlet boundary conditions are maintained as pressure inlet and pressure outlet, respectively. k-ε turbulence model is selected for analysis [7, 8].

3 Results and Discussion

As mentioned earlier, the main reason for causing a health hazard to the operator is the inhalation of pesticide particles during the spraying operation. The fan, installed in the helmet, takes air from the top of the helmet, which comes from the filter so that the fine droplets of pesticide, if any, get absorbed. The vortex created due to rotation of the fan is shown in Fig. 5. Rotation of the fan increases the velocity of the incoming air, and hence, turbulence is generated. Turbulence is the main phenomenon due to which high-velocity air flows through a narrow space between the plastic shield and the operator's face. The high-velocity flow is sufficient to repel the incoming pesticide flow from the bottom part of the helmet. The flow distribution over the operator's face and in the entire helmet is shown in Fig. 6.

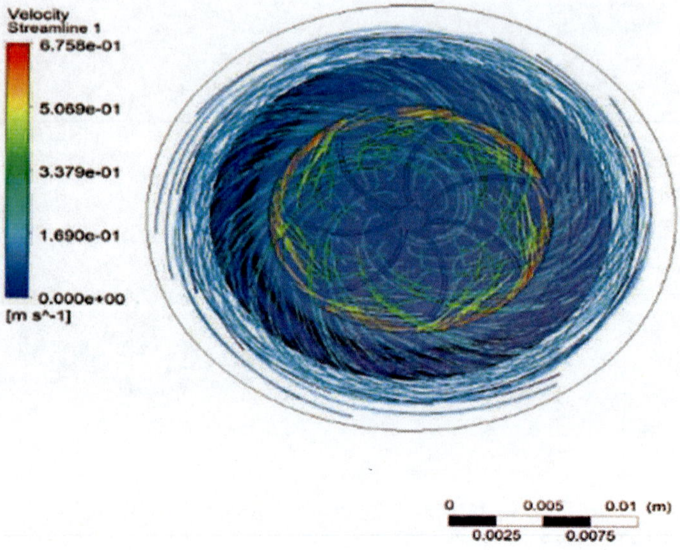

Fig. 5 Vortex generation in the helmet

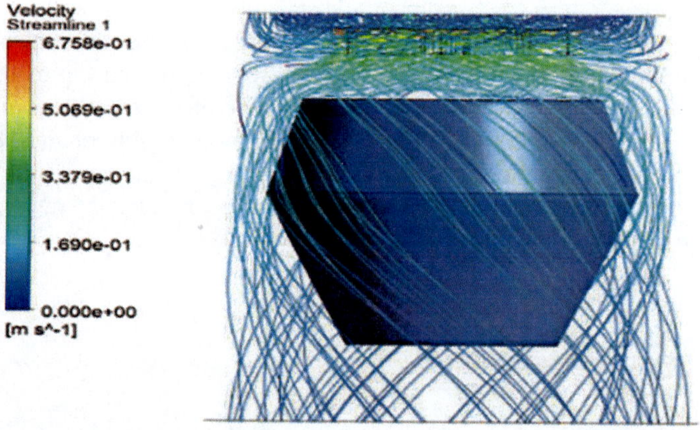

Fig. 6 Flow distribution over the operator's face

Figure 7a and b shows contour plots for velocity and turbulent kinetic energy, respectively. Turbulence is generated due to increase in turbulent kinetic energy. The turbulent kinetic energy depends upon the r.p.m of the fan. As the speed of the fan increases, turbulent kinetic energy also increases. The turbulence thus generated due to this energy is responsible for repelling the pesticide inflow into the helmet.

To make the helmet cost-effective, a plastic shield is used. The rechargeable batteries are used to operate the fan, and a solar panel is mounted on the helmet to charge the batteries. The flow distribution obtained from the CFD analysis helps to

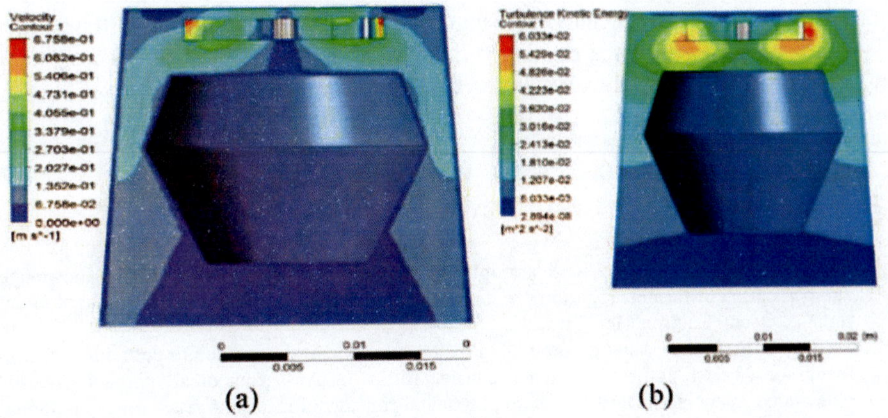

Fig. 7 **a** Velocity contour plot. **b** Turbulence kinetic energy

Fig. 8 Prototype model prepared based on CFD analysis result and its testing

locate the position of the fan in the helmet to get maximum flow effect in the helmet. These results are used to build a prototype of the helmet as shown in Fig. 8. The prototype model is tested using an artificial spray with actual operating conditions and using an operator. The helmet design is validated based on the operator's feedback.

4 Conclusion

Pesticide spraying causes various health problems to farmers/operator. Though the various types of personal protective equipment are available in the market, due to high cost, farmers/operators do not prefer to purchase these equipment. Also, there is a lack of health education among farmers/operators related to health hazards caused due to the inhalation of pesticides during the pesticide mixing and spraying process. This project proposed an innovative design of personal protective equipment, in the form of a helmet. The helmet is designed in such a way that it creates turbulence inside the helmet which causes high-velocity flow. This flow passes over the farmer's/operator's helmet and repels the incoming pesticide flow. The design is cost-effective, and the fan will be operated by rechargeable batteries. This innovative design of the helmet will save many lives in the coming future. Also, this could be used for any other working area where workers are exposed to toxic inhalation.

References

1. Sai MV, Revati GD, Ramya R, Swaroop AM, Maheswari E, Kumar MM (2019) Knowledge and perception of farmers regarding pesticide usage in a rural farming village, Southern India. Indian J Occup Environ Med 23(1):32
2. Davies JE, Freed VH, Whittemore FW (1990) An agromedical approach to pesticide management: some health and environmental considerations. In: An agromedical approach to pesticide management: some health and environmental considerations, US Agency for International Development (AID); Consortium for International Crop Protection; University of Miami. School of Medicine
3. Sarwar M (2015) The dangers of pesticides associated with public health and preventing of the risks. Int J Bioinform Biomed Eng 1(2):130–136
4. Arcury TA, Quandt SA, Dearry A (2001) Farmworker pesticide exposure and community-based participatory research: rationale and practical applications. Environ Health Perspect 109(3):429–434
5. Chitra GA, Muraleedharan VR, Swaminathan T, Veeraraghavan D (2006) Use of pesticides and its impact on health of farmers in South India. Int J Occup Environ Health 12(3):228–233
6. Sarkar A, Aronson KJ, Patil S, Hugar LB (2012) Emerging health risks associated with modern agriculture practices: a comprehensive study in India. Environ Res 115:37–50
7. Bichitkar SM, Buddiyal PP, Chavan SS, Kulkarni AA, Gawande VB (2020) CFD analysis of wind turbine with different flange angles. In: Advances in applied mechanical engineering. Springer, Singapore, pp 199–204
8. Gawande VB, Dhoble AS, Zodpe DB, Chamoli S (2016) Experimental and CFD investigation of convection heat transfer in solar air heater with reverse L-shaped ribs. Sol Energy 131:275–295

Adoption of Green Practices by Indian SMEs

Trilok Pratap Singh and Utkal Khandelwal

Abstract The objective of the study was to address, in particular, the top management position of the MNCs and the need to gain a competitive advantage could be included in environmental policy decision-making processes. 119 Indian SMEs have recently been investigated in this study. Based on a green approach and a non-green approach, the companies surveyed were divided into two categories. To define the variables that distinguish between these companies, single variance analysis and a progressively discriminatory analysis were used. The results show that the comprehensive preparation of green practices is a critical factor in the differentiation between two business groups by combining the importance of formulating and implementing green practices.

Keywords SMEs · Green practices · Environmental concern · Manufacturing processes · Globalization · Sustainable development

1 Introduction

In India, economic growth has accelerated gradually in the last half-decade and remained most critically. Globalization and the liberalization of foreign trade in goods and services have stimulated local technological and economic growth and increased employment. Today's globalization also increases the chances of all types of SMEs and enhances the environmental performance of the manufacturer by placing the customer's emphasis on environmental improvement [1]. In the present scenario, consumers are looking for green solutions towards the ecological challenges of the world (e.g. global warming, environmental degradation, deforestation, air, and water pollution, and resource depletion). The motive is to integrate their actions that are

T. P. Singh
Department of Management, Madhav Institute of Technology and Science, Gwalior, Madhya Pradesh, India

U. Khandelwal (✉)
Institute of Business Management, GLA University, Uttar Pradesh, Mathura, India
e-mail: utkal.khandelwal@gla.ac.in

A. Maurya et al. (eds.), *Recent Trends in Mechanical Engineering*, Lecture Notes in Mechanical Engineering, https://doi.org/10.1007/978-981-19-7709-1_76

good for the country [2]. As a result, an increasing number of SMEs is being held responsible for the environment. SMEs will develop new business approaches, such as the adoption of green practices, to meet these environmental requirements in line with these renewable options [3].

Typically, manufacturers categorized green practice into four categories: pollution mitigation solutions based on compliance, cost-effective, pollution-reduction approaches, and value-added quest strategies [4]. A leading research organization looked at how businesses deal with the environment and how green initiatives influence business decisions [5]. This study aimed to tackle, in particular, the MNC's top management role, and the need to gain a competitive advantage that could be included in decision-making processes and adoption of green practices.

2 Literature Review

Green practices are the common mantra of various environmental groups around the world for the protection and conservation of today's resources [6]. Awareness of sustainable development and greater focus in the recent years has led to initial efforts and best practices to identify eco-friendly assignment rules [7]. Green practices among business operators are widely documented and diversified; they are often implemented by modifying business policies, revamping manufacturing processes and technological innovations in manufacturing, health, and safety risks, and improving overall job security [8]. In these contexts, it is normal for corporations to identify and apply support strategies and increase resources in the implementation of green initiatives [9]. All these concepts are linked to developments that seek to significantly reduce or eliminate pollution from sources at the end of production processes through major processing and product adjustments [10].

A research study suggested a framework for the selection of green practices through the analysis and identification of drivers for the innovation of 'green' products to address the environmental aspects of the organization's strategy [11]. According to the company's product development or operation in green initiatives, the kind of green practices under consideration has a different impact on the results of the SMEs. Consumers found that their product reactions are more advantageous [12]. Continued industrialization and ever-increasing consumerism have led to a situation, in which SME's activities have begun to have an adverse environmental impact [13]. Developed countries have shown little respect for environmental concerns in their pursuit of industrial development and progress towards their current economic role. However, as at present, some industries, governments, and even customers are becoming more aware of how the atmosphere can be conserved and emissions reduced. On the other hand, companies in developing countries are not so receptive to the cause of environmental conservation, and there is a greater need for green practices to be implemented [14].

3 Hypothesis Development

In the physics world, businesses are constantly designing and implementing strategies to address specific issues like 'greenness'. Indian SMEs also communicate these strategies through management practices such as financial transactions and acquisitions [15]. On the basis of extensive literature review, the following hypotheses are developed:

- H_1: Indian SMEs are more likely to develop and implement green practices that are more responsive to environmental change.
- H2: Indian SMEs, which are more likely to adopt green practices, are less informed about environmental concerns.
- H3: Indian SMEs are the most innovative and physically aware of their environment and are more likely to pursue green practices.
- H4: Indian SMEs whose management and employees are more willing to adopt new environmental concerns in their business strategies.
- H5: Indian companies are more likely to perform well for environmental efficiency.
- H6: Green practices are more likely to be adopted by Indian SMEs run by younger and more qualified managers.
- H7: Green practices are more likely to be adopted for older and bigger Indian businesses.

4 Research Methodology

The survey was conducted on Indian manufacturing SMEs through online survey. Based on a recent literature analysis, a detailed questionnaire has been developed. Many of the concerns have been answered in this questionnaire, but efforts have been made to match SMEs with green approaches to those that have not (i.e. the same problems). Survey is comprised of the variables: developing and adopting green practices, green practice included the involvement of supervisors, business model changes in the SMEs, green practices organization and application, looking for a return on money, characteristics of respondents and SMEs. A random sample of 702 SMEs was obtained from an Indian SME database. The report sent an email to all organizations calling for the questionnaire to be completed. There are 60 out of 702 emails not available and there is a net sample of 642 companies. One hundred thirty-eight questionnaires were received at a response rate of 21.49% following an email follow-up. This was comparable to other online studies reported. It was possible to use 119 of the 138 replies received. In this analysis, the dependent variable is the 'green practices', and each business is classified in one of two categories: (1) a non-green company whose owners and managers have shown that they have not implemented a green strategy; and (2) a SMEs with a green strategy owner/manager who has tacitly or explicitly stated that it has a green approach policy to adopt it.

The proposed two-stage classification system was identified as consisting of 36 non-green firms and 83 green firms. The participating SMEs chosen for analysis represent a wide array of industries. Most of the companies were SMEs. Thirteen percent of SMEs reported sales below twenty-five lakhs, 39.3% of SMEs reported sales between twenty-five lakhs and five crores, and 47.3% of SMEs reported sales more than five corers per year. In many companies, there were fewer than 50 employees (45.9%). For more than ten years, the study had a large number of SMEs (81.6%). Only 4.7% of companies have less than five years of industry experience. The overwhelming majority (83.8%) were 45 years of age or older and 65.1% graduated from or after school.

5 Findings

A two-stage data analysis identified various elements that have an impact on the implementation of green practices. A one-way ANOVA will be used to analyse data at the first level. It demonstrates the organization's green strategy.

The results of Table 1 support H1 which suggests that SMEs with more excellent knowledge of environmental change are more likely to develop and implement green practices. Green SMEs are also more likely than non-green SMEs to feel that it is vital to develop and implement green practices. These results show that this reflects management practices as an organization formulates and implements green strategies; these include decisions on general management of resources, activities in finance and procurement, marketing strategies, and policies on growth. Environmental studies lead to the development of specific organizational capacity [16]. The outcome also supports H2, as given in Table 1. Mean values suggest that SMEs are more likely to implement an understanding of environmental management, discussion, engagement, and formal planning. Green SMEs are more vulnerable than non-green companies to discuss, engage, and organize green practices.

The results of Table 2 support H3, whereby it means SMEs with a better physical understanding of their natural environment is more likely to adopt green practices. Green SMEs tend to be greener and more impactful than non-green SMEs on existing business models. Yes, their top managers decided to change their business plan to justify 'going green' instead of small and medium-sized non-green businesses. However, there have been significant changes in the new business model between the two business classes.

The first three hypotheses are distinguished by their close links with the management itself. The development of a green solution, therefore, has very close ties with the context and thinking of the owner/manager. However, the findings of Table 1 support part of H4, suggesting that organizations with a higher probability of changing managers and workers are more vulnerable to new environmental problems. The study found that the two groups of companies had significant gaps in employee resistance and the timing needed to implement a green strategy and green practices. However, they state that, by applying new green principles, they have not

Table 1 Univariate test of significance

Variables	Means		p*
	Non-green SMEs	Green SMEs	
Developing and adopting green practices			
Pursuing an aggressive green practices	2.5	3.4391	0.000
Sectors affected by green practices	2.5071	5.1429	0.000
Developing and implementing green practices	2.4286	3.6735	0.000
Green practice included the involvement of supervisors			
Awareness of green practices at management level	3.0357	4.102	0.000
Discussion of management on green practices	2.3929	3.4898	0.000
Management commitment to develop green practices	2.6071	3.7551	0.000
Systematic implementation of a green practices	2.2143	3.5918	0.000
Timeframe for the implementation of green practices	3.1071	3.2449	0.000
Changes in the SMEs business model			
Greening impact on current business model	1.8571	2.9796	0.000
Management role in handling the changing business model	2.9286	3.4286	0.009
The level of business support for greening	3.1071	3.8163	0.000
Possible new business model	2.6071	3.1429	0.3
Organizing and applying green practices			
Changes to the business structure	2.5357	2.6327	0.000
Management style changes	2.3929	2.6531	0.000
Resistance of employees	2.4643	1.9796	0.014
Resistance of management	2.25	2.0204	0.000
Time horizon needed to accept the green practices	2.7857	3.2653	0.0204
Looking for a ROI			
Financial performance versus environment	2.7857	3.2449	0.039
Scope of financial rewards for green practices	3.1429	3.4286	0.000
Sufficient financial incentive to develop green products	3.0357	3.449	0.042
Extension of the financial reward of the eco-organization	2.8929	3	0.000
Extension of firm's green technology investment	3.0714	3.5306	0.002
Environmental policies rely on investment returns	2.8929	3.1429	0.000
Receiving any green practices award or recognition	1.1429	1.102	0.000
Participation in the setting of environmental criteria	1.2143	1.0408	0.000
Obtain government funding for green practices	1.2143	1.1633	0.000
Green practices impact on future exports	3.1071	3.5306	0.003

Note * The p-values are shown only that exceed 0.05

Table 2 Classification matrix

Group membership	Predicted group membership					
	Non-green firms	%	Green firms	%	Total	%
Non-green firms	40	71.4	16	28.6	46	100
Green firms	12	12.2	86	87.6	73	100
Predicted total	52	100	102	100	119	100

Notes Correct classification = 81.8%; the average Morrison proportional chance criterion is 54.43%

changed or modified their management style or green practices when referring to SMEs. Our findings show that practical standards need not change an organization or a company structure dramatically.

Discriminant coefficients (their ratios are the same) have statistical significance. Variables with high coefficients are therefore known to contribute more generally to the discriminant function. The value of the differentiating coefficients, independently of the symbol, indicates the relative power of each variable to distinguish between different classes (Table 3).

The first step involves the formal preparation of green practice policies. The findings show that the strategic planning of green and non-green organizations is very distinct. Green SMEs are more eager to develop and implement a green strategy than they are at the formal level. Overall, systematic management planning and the commitment to green development and implementation strategy tend to be critical factors that distinguish the two groups of SMEs. Green SMEs tend to plan regularly, understanding more than non-green companies the value of implementing green practices. Although the conduct of Indian SMEs enterprises is not significantly different from that of other countries, it is not meant to be a general reflection of small and medium-sized enterprises around the world. For this purpose, SMEs in different parts of the world will adopt various activities. In spite of this restriction, the above conclusions should be considered.

Table 3 Standardized canonical discriminant coefficients

Variables	Coefficient
Importance of developing and implementing green practices	0.617
Systematic implementation of a green practices	0.536
Wilki's λ	0.517
X^2	48.842
DF	2
Sig	0.000

6 Conclusions and Implications

The green practices have been hypothesized to Indian SMEs by seven constructs: the development and implementation of the green practice, the extent to which they participate in the management of the green practices, the change in the corporate model, the development of the green strategy, economic performance, and the outlook for management. Besides, we assumed that the decision of small- and medium-sized enterprises and the characteristics of the respondent to adopt green practices could have an impact on them. This research found that the implementation of a green policy by SMEs was mainly based on its management plan and on its understanding of the value of creating and implementing green methods for it. However, the remaining five schemes were not relevant to Indian enterprises implementing a green strategy. There are two interpretations of the phenomenon. First, in terms of green practice, Indian culture and societies are far more advanced than the other economies. There might have been a strong focus on medium-sized enterprises. These results should, therefore, be viewed with such limitations, including because, as opposed to only 43 non-green companies, 73 of the 119 SMEs regard themselves as sustainable. In this report, companies from a wide range of Indian industries have been chosen to demonstrate very different green practices for various industries and countries.

References

1. Chassé S, Courrent JM (2018) Linking owner–managers' personal sustainability behaviors and corporate practices in SMEs: the moderating roles of perceived advantages and environmental hostility. Bus Ethics: Eur Rev 27(2):127–143
2. Adebanjo D, Teh PL, Ahmed PK (2016) The impact of external pressure and sustainable management practices on manufacturing performance and environmental outcomes. Int J Oper Prod Manag 36(9):995–1013
3. Collett N, Pandit NR, Saarikko J (2014) Success and failure in turnaround attempts. An analysis of SMEs within the Finnish Restructuring of Enterprises Act. Entrep Reg Dev 26(1–2):123–141
4. Dhillon MK, Bentley Y, Bukoye T (2016) Investigating employee attitudes towards adopting green supply chain practices in Indian SMEs-using qualitative methods: literature review and research methodology. In: European conference on research methodology for business and management studies. Academic Conferences International Limited, p 367
5. Li Y, Ye F, Dai J, Zhao X, Sheu C (2019) The adoption of green practices by Chinese firms: assessing the determinants and effects of top management championship. Int J Oper Prod Manag 39(4):550–572
6. Bansal P, Roth K (2000) Why companies go green: a model of ecological responsiveness. Acad Manag J 43(4):717–736
7. Baregheh A, Rowley J, Sambrook S, Davies D (2012) Innovation in food sector SMEs. J Small Bus Enterp Dev 19(2):300–321
8. Chen TB, Chai LT (2010) Attitude towards the environment and green products: consumers' perspective. Manag Sci Eng 4(2):27–39
9. Govindan K, Kaliyan M, Kannan D, Haq AN (2014) Barriers analysis for green supply chain management implementation in Indian industries using analytic hierarchy process. Int J Prod Econ 147:555–568

10. Klassen RD, McLaughlin CP (1996) The impact of environmental management on firm performance. Manag Sci 42(8):1199–1214
11. Russo MV, Fouts PA (1997) A resource-based perspective on corporate environmental performance and profitability. Acad Manag J 40(3):534–559
12. Todd PR, Javalgi RRG (2007) Internationalization of SMEs in India: fostering entrepreneurship by leveraging information technology. Int J Emerg Mark 2(2):166–180
13. Manian K, Ashwin JN (2014) Present green marketing-brief reference to Indian scenario. Int J Sci Res Manag 2(3):51–57
14. Noci G, Verganti R (1999) Managing 'green' product innovation in small firms. R&d Manag 29(1):3–15
15. Dowell GW, Muthulingam S (2017) Will firms go green if it pays? The impact of disruption, cost, and external factors on the adoption of environmental initiatives. Strateg Manag J 38(6):1287–1304
16. Soda S, Sachdeva A, Garg RK (2015) GSCM: practices, trends and prospects in Indian context. J Manuf Technol Manag 26(6):889–910

Real-Time Welding Defect Detection and Classification Using Artificial Intelligence and Its Implementation in Manufacturing Plants

Anurag Kumar Singh, Tanya Maurya, Pankaj Kumar Sudarshi, and Richa Pandey

Abstract Weld joint flaws can lead to part and assembly rejection, costly repairs, a significant reduction in performance under normal operating conditions, and, in the worst-case scenario, serious property, and life loss. In reality, flawless welding is nearly impossible, and in most situations, providing the relevant service functions is insufficient. However, early detection and segregation are still preferable. To test the quality of welded joints, several methodologies have been fabricated time and again. Non-destructive technology has essentially replaced the destructive methods of testing today, as this form of testing allows continuous and in-service inspection and is the basis for quality inspection in modern world. In our work, we use combination of digital image processing techniques and a well-established deep learning model for real-time detection of surface welding defects along with displaying the severity of each defect after their classifications. Moreover, the paper also shows how can this technology be used in an actual large-scale manufacturing plant in order to test the quality of welded joints. The objective is achieved using a hardware and software system that consists of a low-cost vision system for acquiring the image of a job, computer vision libraries, and a deep learning model for statistical analysis for quality monitoring.

Keywords Defect detection · Quality inspection · Artificial intelligence · Deep learning · Digital image processing

A. K. Singh (✉) · T. Maurya · P. K. Sudarshi · R. Pandey
BIT Mesra, Ranchi, India
e-mail: be10354.17@bitmesra.ac.in

T. Maurya
e-mail: be10068.17@bitmesra.ac.in

P. K. Sudarshi
e-mail: be10582.17@bitmesra.ac.in

R. Pandey
e-mail: richarp@redeiffmail.com

A. Maurya et al. (eds.), *Recent Trends in Mechanical Engineering*, Lecture Notes in Mechanical Engineering, https://doi.org/10.1007/978-981-19-7709-1_77

1 Introduction

Welding defects are anomalies or inconsistencies in the weld surface that occur in joint pieces. Surface defects in joints may lead to component rejection, assembly failure, costly repairs, significant performance reductions, and, in the worst-case scenario, catastrophic failures resulting in loss of life and property [1]. Owing to the inherent weaknesses in welding technology and the properties of metals, defects are still present in the welding. Welded joints are often the site of crack initiation due to inherent metallurgical geometrical defects, as well as a lack of homogeneity in mechanical properties and residual stresses. At the same time, early detection and segregation is always preferred to mishaps. Using machine learning algorithms, we can easily detect faults in welding by image acquisition and its processing. At the same time, measure the severity of each fault precisely. It was found, using deep learning algorithm, it makes the process much more efficient. Resulting in an accuracy of 93.3% by the end of the work. In this work, we have tried to show our approach for defect detection using artificial intelligence on friction stir welded joint surface defects, but it is to be noted that this approach can be applied to any metal surface defect detection irrespective of type and technology involved.

The Welding Institute (TWI) invented friction stir welding (FSW) in 1991, and it is widely recognized as the most significant advancement in metal joining processes. Thanks to its ability to be fully automated and other special characteristics, this welding method is used by a wide range of industries.

Pre-heat, initial deformation, extrusion, forging, and cool-down metallurgical phases are all steps in the FSW process [2].

Despite its many advantages, the FSW process is still prone to defect formation due to the lack of potential for imbalances between the different processing zones [3]. The following are the most common surface defects that occur during the welding process:

Flash Defects: During welding, the material is subjected to extremely hot processing conditions as the tool pin rotates at a very high rpm. Excessive heat is produced as a result, thermally softening the material near the tool-shoulder boundary and expelling a large amount of material in the form of surface flash.

Voids: Voids are irregularly sized defects that can be oriented at any angle.

Cracks: Cracks are smaller than voids and have a rougher texture.

Line Defect: A series of unwanted small nuggets shaped along a line is known as a line defect (Fig. 1).

2 Survey of Defect Detection

Internal technology for detecting defects primarily involves the identification of internal bugs, hole detection, and detection of cracks. Currently, many approaches, including deep learning, magnetic powder, eddy current testing, ultrasonic testing,

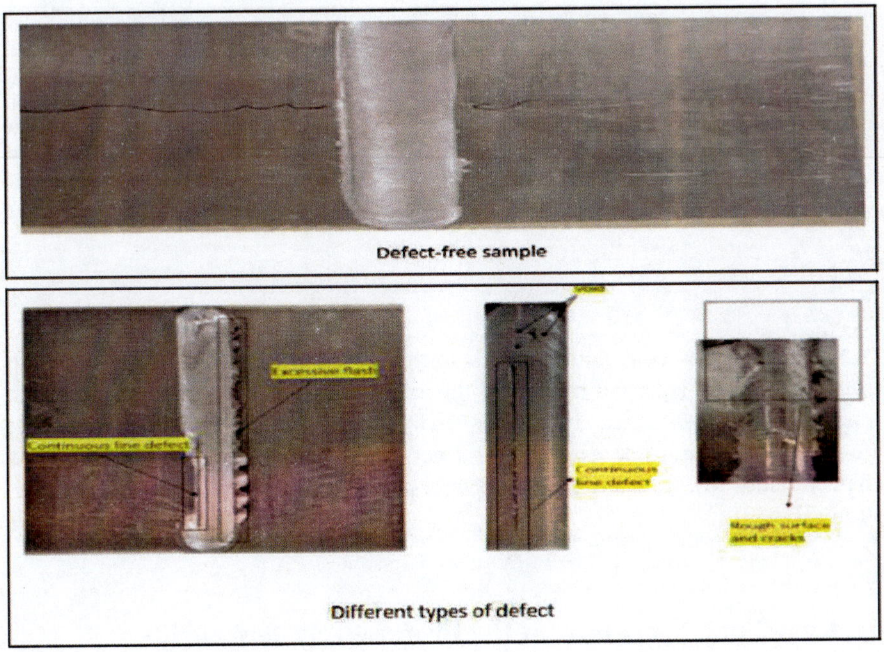

Fig. 1 Types of samples

and machine vision detection techniques, are used to detect product quality. Detection of machine vision primarily consists of image acquisition and detection and classification of defects. Machine vision is widely used due to its fast, reliable, non-destructive, and low-cost characteristics. Machine vision recognizes objects based primarily on the objects color, texture, and geometric characteristics [4]. The efficiency of the image processing algorithm, in turn, directly affects the identification and classification of defect accuracy and error detection rate. Regression models have also been used quite often in this domain for predicting important characteristics like predicting UTS of weld at different parameters [5].

3 Convolution Neural Network

A type of deep neural network widely used in image classification is the convolution neural network. Convolution layers add an action of convolution to the input that transfers the outcome to the next layer [6]. To train a neural network with a billion characteristics, the computational and memory requirements are prohibitive. The use of convolution layers over fully linked layers has two key benefits: parameter sharing and relation sparsity [7].

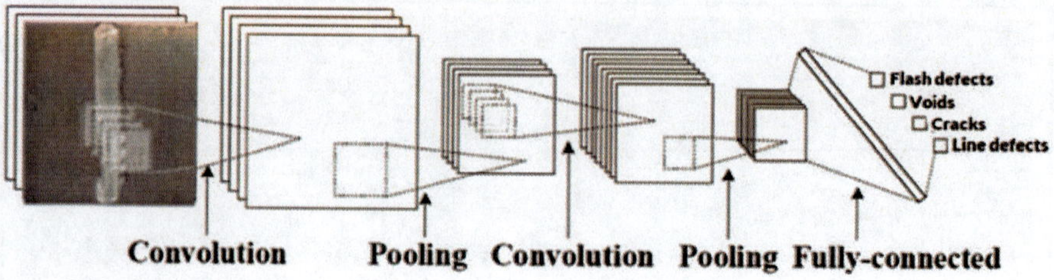

Fig. 2 CNN model

Neural network convolution looks for patterns in a picture. With a smaller matrix, the image is transformed, and this convolution finds patterns in the image. Lines/corners/edges etc. can be established by the first few layers, and these patterns are passed down into the deeper neural network layers to identify more complex characteristics [8]. In identifying objects in pictures, this property of CNNs is very strong (Fig. 2).

4 Real-Time Workflow of the Proposed System Inside a Plant

A smart camera vision system is installed along the production line mounted in an overhead position. Image acquisition of the surface of the part or assembly is carried out by the camera vision system. The image would be sent to a server for image pre-processing and then the trained model which is contained in the Server will accept this image as an input. These pre-processed images will be treated as the input data for the model. These scans will be fed through our neural network which is written using deep learning libraries like Tensor Flow and Keras. The results Generated will be analyzed and will be displayed to the operator through a graphic user interface. The user can decide whether to accept or reject the parts based on the severity of defect as well as can separate the defected parts based on its type of defect. The decision can be conveyed back to the system through the GUI. However, on requirement basis, this segment can also be automated by providing acceptable limits and other essentials. These generated results scan also be accessed by other users if the entire system is connected through a network. The whole setup has huge scope of customization depending on the requirements and pocket (Fig. 3).

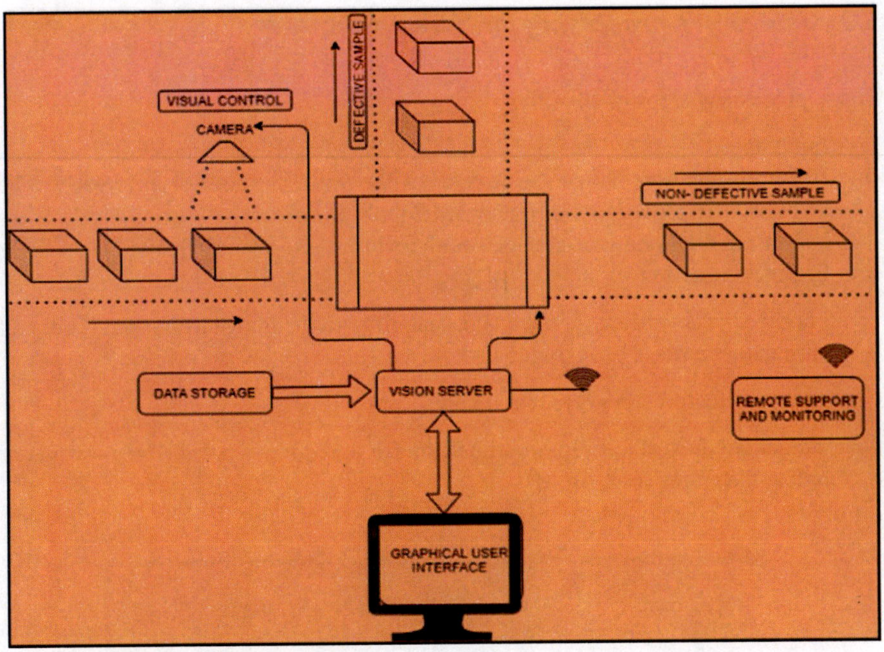

Fig. 3 Work flow in the system

5 Model Architecture Outline

5.1 Dataset

Training was conducted on a regular laptop with an NVIDIA GTX 850 M GPU card and 4 GB memory. The welded parts used in the study were separated as defective and non-defective through visual inspection, and a total of 1516 images were obtained from different friction stir welded joint parts through web scrapping. These images were split into training and testing sets and prepared for entry into the CNN model of deep learning. Of all the images, 1213 (80%) were used in training sets and 303 (20%) were reserved for testing.

5.2 Image Pre-processing and Preparation

Image processing steps involved:

(a) Input data
(b) Noise elimination: Since the texture of the weld surface varies greatly. The Contrast Limited Adaptive Histogram Equalization technique boosts an image's contrast and emphasizes cracks, rough textures, and voids [9] (Fig. 4).
(c) Feature extraction

Canny edge detection: Apply a Gaussian filter to the image to smooth it out and eliminate the noise [10] (Fig. 5).

Comparing features in grayscale images

- homogeneity in surface: Hough transform for homogeneity and edges—voids or rough at the edges (Fig. 6)

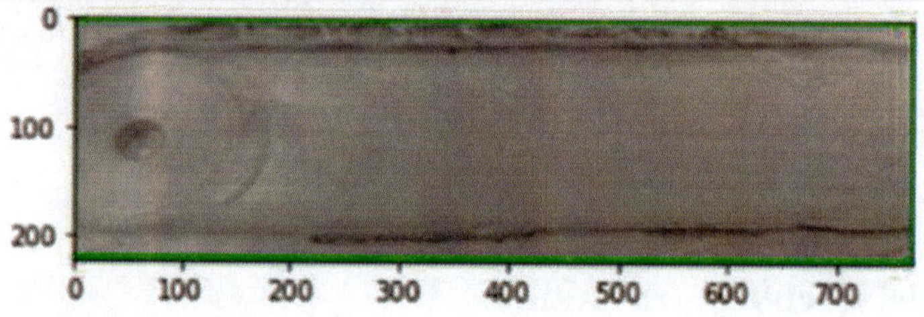

Fig. 4 Noise plotting

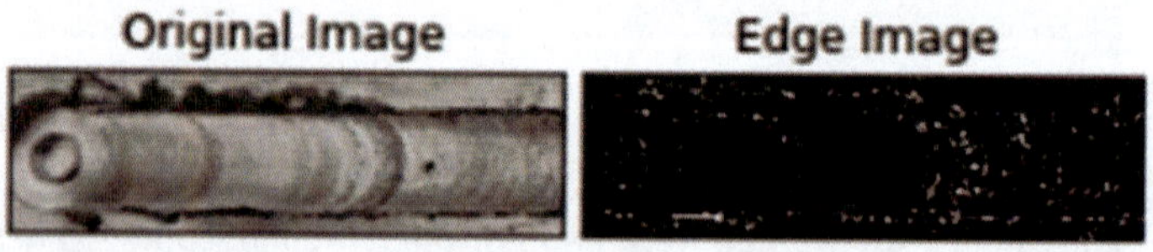

Fig. 5 Original image versus edge image

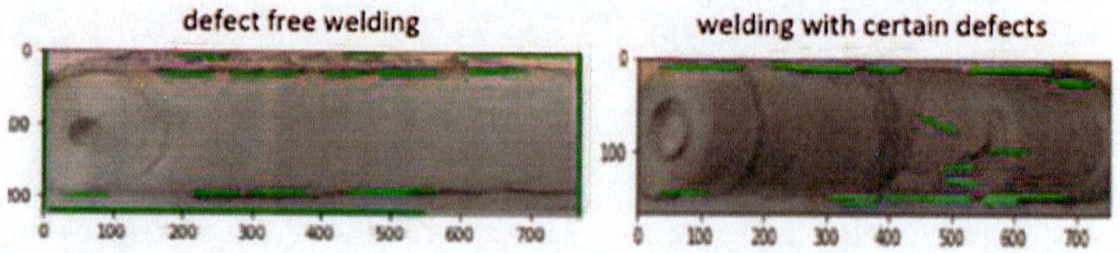

Fig. 6 Defect plot

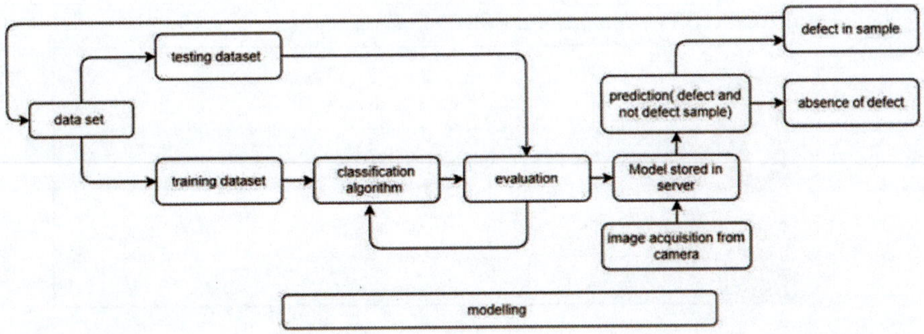

Fig. 7 Modelling

5.3 Modeling (Fig. 7)

6 Results and Discussion

The defect detection process is divided into two stages, first includes defect or no defect and other phase includes classification of types of defect present in the sample. Types of defects include voids, cracks, line, flash, and rough surface. Firstly, we are training our data containing defect and non-defect samples, using classification algorithm we are predicting our model with real-time images and our model is predicting if sample has defect/not defect. In the second part, we are training the model for classification of different types of defect present in defected section. By training different data set, we will be able to test with new sample. Training accuracy is over 97% and testing accuracy is more than 93%. Loss value is close to 0 in both the training and the testing set. Our model is predicting the percentage of resemblance to the training set and then calculating percentage as shown (Figs. 8 and 9).

7 Conclusion

Images of friction stir welded aluminum sheets were captured by a Logitech C270HD webcam for training a CNN deep learning model. The main conclusions drawn from the present study are:

1. Pre-processing libraries like Open CV in Python were successfully used for eliminating noise, extracting features, and comparing features in grayscale images.
2. CNN was successfully implemented for classifying the pre-processed sample images into different categories of defects like line, crack, flash, and void defects and an accuracy close to 90% was achieved.

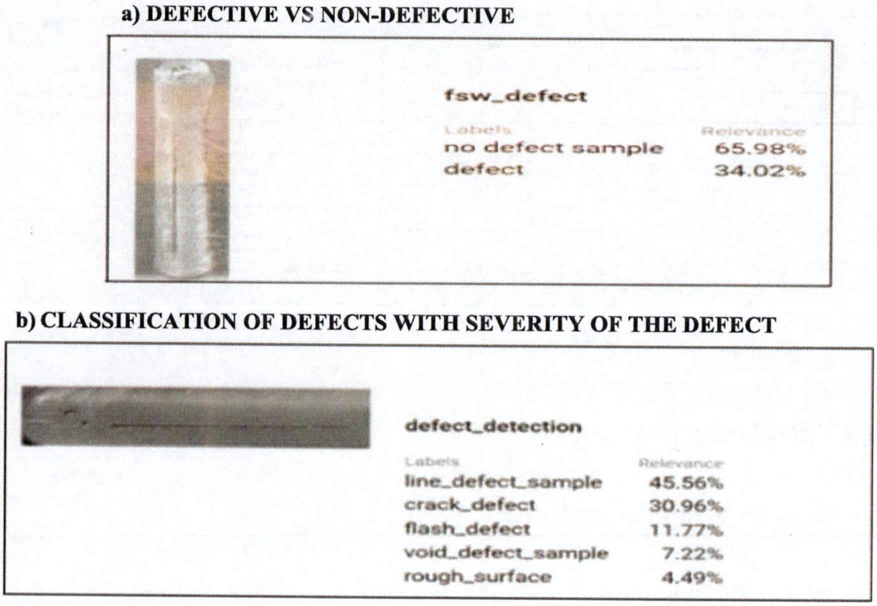

Fig. 8 Intensity of defect in the sample

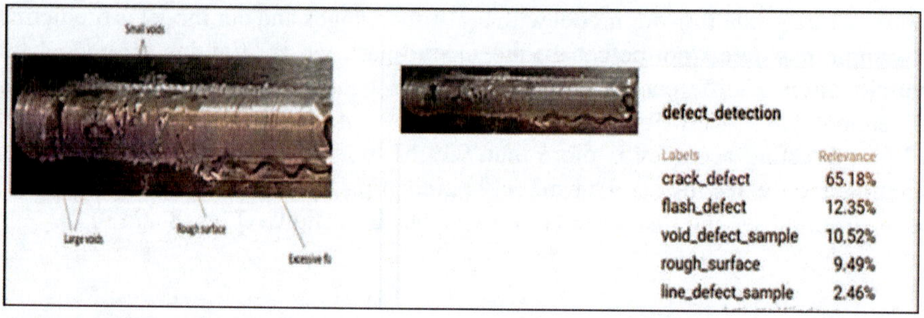

Fig. 9 Classification of defects

3. A real-time workflow of the proposed system is shown inside an actual manufacturing plant that comprises of a visual control unit for acquiring images, a vision server that contains the classifier software, and a graphical user interface that helps a quality monitoring engineer to read the results and input the final decisions.
4. The novelty lies in the proposed automated hardware setup that reduces labor in a quality inspection plant, improves accuracy, saves time, and has scope of giving a remote access by connecting the system to a cloud enabling usage of IOT further.

References

1. Sun J (2019) An effective method of weld defect detection and classification based on machine vision. IEEE
2. Sinha P (2006) Condition monitoring of first mode of metal transfer. Springer-Verlag London Limited 2006
3. Rajashekar R (2013) Digital image processing of friction stir weld bead surface. J Mech Civ Eng (IOSR-JMCE)
4. Fang X (2020) Research progress of automated visual surface defect detection for industrial metal planar materials. Sensors
5. Engineering with computers, pp 1–24. Analysis and characterization of weld quality during butt welding through friction stir welding. IJEMS 26(5&6):298–310
6. Yang J (2020) Using deep learning to detect defects in manufacturing. Materials
7. Hossain MA (2019) Classification of image using convolutional neural network (CNN). Glob J Comput Sci Technol: D
8. Prabhu (2018) Understanding of CNN
9. Chu H-H (2015) A vision-based system for post-welding quality measurement. Springer-Verlag London 2016, pp 2–4
10. Kumar G (2014) A detailed review of feature extraction in image processing systems. IEEE

Studies of Friction for Different Forging Lubricants Using Ring Compression Test

Manoj Kumar and Mathala Prithvi Raj

Abstract In closed die forging the interfacial frictional condition between the dies and workpiece influences the energy requirement, defects formation, and load on the dies. Forging lubricants are used at the interface to reduce the friction. Breakdown of lubricant results in excessive die wear and die damage. Hence, it is important to select correct forging lubricant which minimizes friction. So, it is important to evaluate the friction coefficient of the lubricant. In this paper, ring compression test was done to find the coefficient of friction at high temperature using two different lubricants. Standard calibration curve was drawn by using DEFORM software. Standard mild steel rings were compressed to different reductions at forging temperature in hydraulic press. To evaluate the friction coefficient, percentage change in internal diameter of the rings was plotted against the percentage reduction on the standard calibration curve. It was found that the DELTA FORGE gives minimum friction coefficient.

Keywords Forging lubricant · Die wear · Ring compression test · Calibration curves · Closed die forging · Friction coefficient

1 Introduction

In hot forging process after forging of every component, the dies are lubricated by spraying to maintain the die surface temperature below 450 °C [1]. In hot forging, dies are subjected to cyclic thermal stress which results in softening of the dies. Lubricating the dies will help in maintaining the die temperature below 450 °C. The interfacial condition between the dies and workpiece also affects the quality of the forged part as well as the load requirement in the forging [2]. The quality of the

M. Kumar (✉)
Department of Foundry Forge Technology, National Institute of Advanced Manufacturing Technology, Ranchi, Jharkhand 834003, India
e-mail: mknifft@gmail.com

M. P. Raj
Department of Mechanical Engineering, Anil Neerukonda Institute of Technology and Sciences, Visakhapatnam 531162, India

A. Maurya et al. (eds.), *Recent Trends in Mechanical Engineering*, Lecture Notes in Mechanical Engineering, https://doi.org/10.1007/978-981-19-7709-1_78

interaction surface depends on the type of machining used for manufacturing of dies and type of lubrication used while forging. In most of the cases, either CNC milling or EDM is used for machining the dies, so the quality of the surface obtained is similar. Hence the concentration can be drawn towards the type of lubricant used. If an improper lubricant is used, it seizes before the completion of forging operation and as a result the friction coefficient between the dies and workpiece increases. This increase in friction leads to increase in the load requirement, increase in die wear as well as may lead to defect formation in the component. Hence selection of lubricant is one of the most important aspects in forging.

The selection of lubricant can be done based on the operating temperature, availability, coefficient of friction (COF), and less fume formation [3, 4]. Among these parameters, COF is the most effecting parameter which decides the performance of a lubricant. There are many methods for evaluating the coefficient of friction of lubricants like ring compression test, spike test, double cup extrusion test, etc. The ring compression test is very simple test and is most widely used in industry for evaluation of coefficient of friction of lubricant and is mostly used by many researchers [5–7]. In ring compression test, standard rings of dimension 6:3:1 or 6:3:2 are made and deformed axially to different heights. Then the percentage change in the internal diameter can be calculated and can be plotted on standard calibration curves. The standard calibration curves can either be obtained by literature developed by Male and Cockcroft or can be made by simulation in finite element-based software. FE-based simulations are extensively used by many researcher [8–10] for development of friction models and calibration curves. In ideal condition, i.e., when interfacial friction is zero, the ring would deform in the same way as a solid cylinder. In which, each element in the ring moves outward at a rate proportional to its distance from center. As a result, the internal diameter would increase without any bulging. If the friction at the interface is small but finite, then internal diameter would increase with bulging. If friction at interface is more than a critical value, then the internal diameter would decrease i.e., the hole of the ring will shrink. Thus, the change in internal diameter of the deformed ring can be used as a measure of the interfacial friction. The correlation between reduction in height of the ring and reduction of internal diameter is given by friction calibration curve.

2 Experimental Procedure

2.1 Ring Compression Test

A mild steel specimen of standard dimensions in the ratio of 6:3:2 (Outer diameter: Inner diameter: Height) is considered for ring compression test. The dimensions of the specimen are—outer diameter $D_0 = 40$ mm, inner diameter $d_0 = 20$ mm and the height $H_0 = 13.33$ mm. To obtain the standard ring dimensions a solid mild steel rod of 40 mm diameter is taken and then it is machined to the required dimensions.

Table 1 Chemical composition of ring specimen and dies

	C	Mn	S	P	Cr	Mo	V	Ni	Si	Fe
AISI 1015	0.15	0.4	0.05	0.04	–	–	–	–	–	Bal
H13	0.4	0.3	0.03	0.03	5.2	1.25	1.2	0.3	0.8	Bal

Initially, the samples are cut to a height of 14 mm and then they are ground and polished to 13.33 mm using surface grinder and double disc polishing machine to get approximately same roughness value. The dies used are made of H13 hot worked tool steel. The chemical composition of ring specimen and dies are shown in the Table 1. After the preparation of rings, the rings are heated to a temperature of 1200 °C in a resistance-heating furnace and soaked for 30 min. The dies are preheated to 150 °C. Then, the rings are deformed in a 150 ton hydraulic press. The rings are deformed by 20, 30, 40, and 50% of their original height. Stoppers are used to maintain the upset height. Before upsetting both dies and the rings are coated with the lubricants. Before application of the lubricants, the dies are cleaned with acetone to remove the traces of the previous forging operation. Two different types of lubricants, namely, GRAPHOL GFL and DELTA FORGE, graphite-based lubricants, supplied by Anand Engg. Pvt. Ltd and compared with the dry state i.e., without applying any lubricant. It was required to know the coefficient of friction of the lubricants so that better lubricant (having minimum coefficient of friction) can be used for die forging of steel. Sample ring and experimental setup are shown in Fig. 1.

(b)

(a)

Fig. 1 **a** Ring sample, **b** 150 ton hydraulic press

After the completion of forging the inner diameter and height of the rings are measured using a vernier calipers. Then the percentage reduction in internal diameter is calculated using the following equation.

$$\%\text{ Change in Inner diameter } \Delta d = \frac{d_f - d_0}{d_0} \times 100 \quad (1)$$

where d_f = final internal diameter, d_0 = initial internal diameter.

Then the friction coefficient is calculated by using the formula [11].

$$\mu = 0.055 \exp\left(\frac{\Delta d}{\exp(0.044\Delta h + 1.06)}\right) \quad (2)$$

where Δd = % Change in Inner diameter, Δh = % change in height

After ring compression test the reduction in internal diameter is plotted on the standard calibration curves obtained from simulation. Standard calibration curves are obtained by simulating the rings to various deformation under various friction conditions in a finite element-based software DEFORM V12.0.1. Then the coefficient of friction of the two lubricants and in dry condition is estimated by using standard calibration curves and analytical approach.

2.2 *Finite Element Modeling*

To obtain the calibration curves the entire process is simulated in a commercially available FE-based software DEFORM V12.0.1 [12]. Rings with same dimensions as mentioned in the Sect. 2.1 are considered. Initial ring temperature is selected as 1200 °C and the die temperature is maintained at 150 °C. The ring material is considered as fully plastic and the dies as rigid material. The number of elements of the ring are considered as 20,000 for ring sample and 10,000 for dies, and the actual numbers of elements are 17,790 and 9324 for ring and dies, respectively. For obtaining the calibration curves various friction coefficients (0.08, 0.05, 0.03, 0.3, 0.4, and 0.5) are selected. For each friction coefficient, simulations are carried out by deforming the rings by 20, 30, 40, and 50% of the actual height. The ram speed is set at 1 mm/s. After simulation, the percentage change in the internal diameter is calculated in each case and the corresponding graph is plotted. The cross-sectional view of the ring with meshing and die set assembly is shown in the Fig. 2.

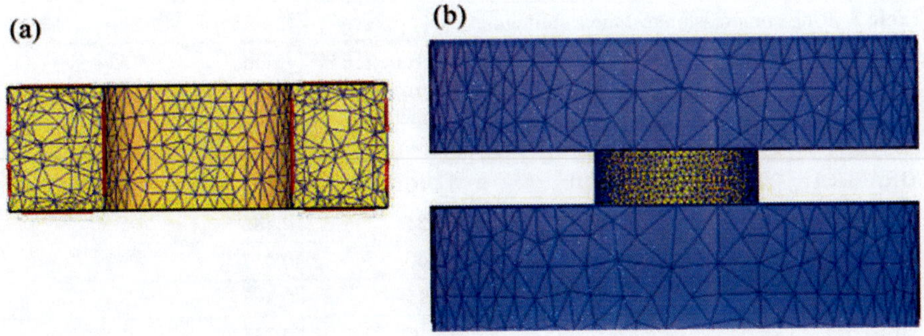

Fig. 2 **a** Cross-sectional view of ring specimen, **b** die set assembly

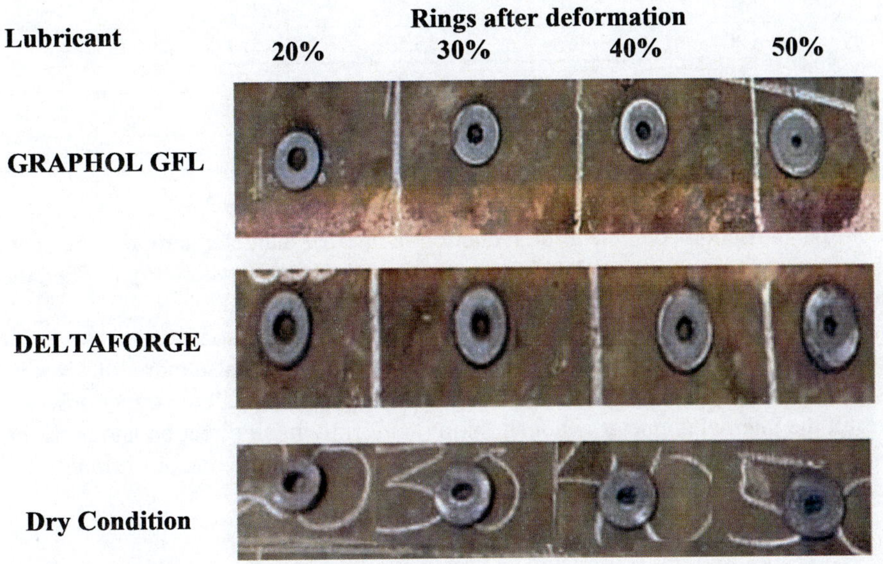

Fig. 3 Rings after deformation

3 Results and Discussion

3.1 Deformation of Rings

The rings after deformation are shown in Fig. 3. Three measurements are taken from each percentage reduction of the ring and the average internal diameter is reported in Table 2. The measurements corresponding to various deformation with the two lubricants and with dry condition are shown in Table 2.

It can be observed that for different lubricants internal diameter dimensions are different for same deformation. From the table, it can be observed that with an

Table 2 Ring compression test measurements

Lubricant	% deformation	Internal diameter after deformation (mm)	% reduction in internal diameter	Friction coefficient (μ)	Average (μ)
GRAPHOL GFL	20	17.18	14.10	0.417	0.423
	30	15.29	23.55	0.486	
	40	13.12	34.40	0.427	
	50	10.17	49.15	0.363	
DELTA FORGE	20	17.98	10.10	0.235	0.307
	30	15.68	21.60	0.406	
	40	14.32	28.40	0.299	
	50	11.35	43.25	0.289	
Dry condition	20	16.9	15.50	0.510	0.518
	30	14.96	25.20	0.567	
	40	12.05	39.75	0.588	
	50	9.56	52.20	0.408	

increase in friction coefficient the reduction in internal diameter is more. The coefficient of the friction is calculated for respective lubricants based on Eq. 2. Then the average of these friction coefficients is finally calculated.

As the friction increases the restriction for the metal flow also increases as a result the internal diameter shrinks more, and the barreling effect is also more. In dry state since there is no lubricant, the material flow is not smooth in all the directions, as a result the internal diameter is not shrinking uniformly in all direction and is taking an elliptical shape. From this, it can be said that there is more non uniform internal barreling taking place in dry condition.

3.2 *Calibration Curves*

The standard calibration curves are obtained by simulation in FE-based software DEFORM V12.0.1 [12]. The simulation results with friction coefficient minimum 0.03 and maximum 0.5 are shown in Fig. 4. The standard calibration curves are shown in Fig. 5. The results obtained by measuring the change in internal diameter with corresponding percentage reduction are superimposed on the standard calibration curves and shown in Fig. 5. By comparing the standard calibration curve and experimental values we can see that the coefficient of friction of GRAPHOL GFL lies between 0.4 and 0.5. Similarly, for DELTA FORGE lubricant, the friction coefficient varies around 0.3. In case of dry condition, the coefficient of friction is maximum and lying between 0.5 and 0.6. It can be observed that both analytical analysis and graphical approach gives approximately same results.

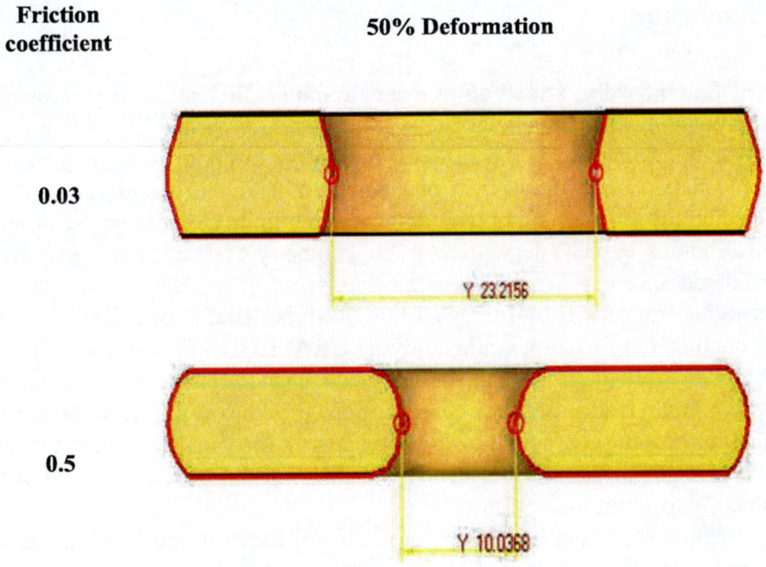

Fig. 4 Simulation results with interface friction 0.03 and 0.5

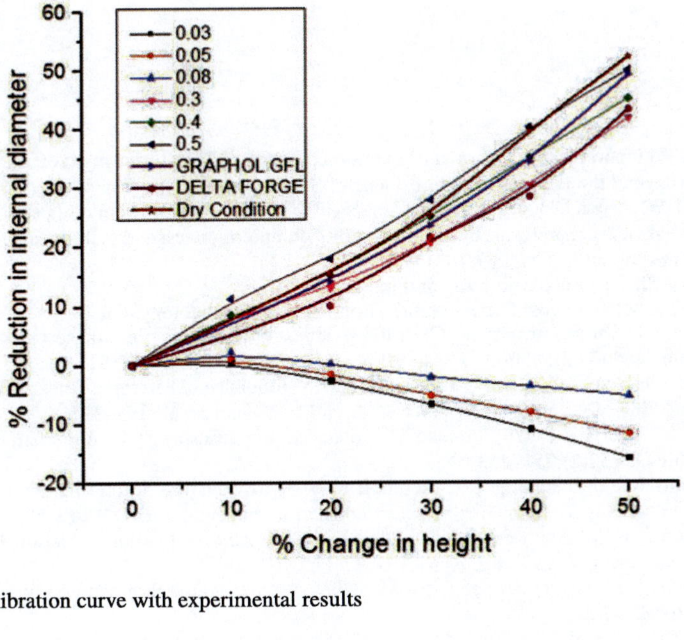

Fig. 5 Calibration curve with experimental results

4 Conclusion

- In the present work, a method to evaluate the coefficient of friction using ring compression test is explained.
- Ring compression test is the simplest method to evaluate coefficient of friction and is independent of flow stress of the material and load calculation.
- The calibration curves are obtained by carrying finite element-based simulation, the calibration curve is dependent on the geometry of the ring, and here we have considered 6:3:2.
- The coefficient of friction for DELTA FORGE is found to be 0.3.
- The coefficient of friction while applying GRAPHOL GFL is found to be 0.4.
- In case of closed die forging the main objective is to fill the die cavity completely with minimum load and energy consumption. To achieve this one of the conditions is to minimize the coefficient of friction between the workpiece and die material. Hence, out of these two lubricants, DELTA FORGE will give better result in closed die forging.
- The coefficient of friction in dry state (when there is no lubricant) is maximum due to formation of scales which are in direct contact with rings and dies. As a result, maximum die wear will occur in this condition.

References

1. Yang L, Shivpuri R (2007) A water evaporation based model for lubricant dryoff on die surfaces heated beyond the Leidenfrost point. J Manuf Sci Eng 129(4):717–725
2. Wolla DW, Joseph Davidson M, Khanra AK (2016) Evaluation of friction of powder metallurgical Al–4 wt% Cu preforms by employing ring compression test and FEM in hot compression test. Trans Indian Inst Met 69:1031–1041
3. (1993) ASM hand book, forming and forging, vol 14
4. (1992) ASM hand book, friction, lubrication, and wear technology, vol 18
5. Robinson T, Ou H, Armstrong CG (2004) Study on ring compression test using physical modelling and FE simulation. J Mater Process Technol 153–154:54–59
6. Bizet L, Charleux L, Balland P, Tabourot L (2017) Influence of heterogeneities introduced into the modelling of a ring compression test. Arch Civ Mech Eng 17:365–374
7. Reddy TY, Reid SR (1979) On obtaining material properties from the ring compression test. Nucl Eng Des 52(2):257–263
8. Horwatitsch D, Merstallinger A, Steinhoff K (2012) Evaluation of the ring compression test for the parameter determination of extended friction models. In: ICAA13 Pittsburgh 2012
9. Danckert J (1988) Analysis of the ring test method for the evaluation of frictional stresses in bulk metal forming processes. Ann CIRP 37(1):217–220
10. (1999) On the measurement of friction coefficient utilizing the ring compression test. Tribol Int 32(6):327–335
11. Prodhana A, Reddy SRS, Prasad SN (2000) Studies on solidification, aging and forgeability of modified stainless steels. J Mater Process Technol 103:324–331
12. (2020) DEFORM V12.0.1, user's manual

Neural Network-Based Flow Curve Modeling of High-Nitrogen Austenitic Stainless Steel

Abhishek Kumar Kumre, Ashvin Shrivastava, Mayank Mishra, and Matruprasad Rout

Abstract The constitutive relationship in terms of stress–strain behavior at high temperatures is important to study and understand the flow characteristics of the material. Artificial intelligent techniques like neural networks (NN) can be used to predict the flow behavior of the material. In this work, flow stress modeling of a high-nitrogen austenitic stainless steel has been done using two different models, viz., artificial neural network (ANN) and fuzzy neural network (FNN). The models have been developed on MATLAB and trained by the experimentally obtained flow curve data. Flow curves at three different temperatures and three different strain rates have been considered for the present work. Both the models were analyzed and compared based on their accuracy in prediction and their ability to interpolate and extrapolate the flow stress values for temperature, strain rate, and strain. The results revealed that for the same amount of experimental data, the FNN can predict the flow stress at intermediate values of temperature. Also, the FNN model predicted more accurately at the interpolated strain rate than ANN. However, the FNN model is unable to extrapolate the flow stress values beyond the trained value of strain. At extrapolated strain rate and temperature, both the models are unable to predict the expected result.

Keywords Neural network · Fuzzy · Flow stress · Steel · High-temperature

1 Introduction

Prediction of the load requirement for a specific hot working process is quite dependent on the material behavior at elevated temperatures [1]. So, to accurately predict the load and hence, deciding the capability of the machine to be used for the said hot working process, knowledge about the stress–strain behavior of the material at working temperature, strain, and strain rate ranges is essential. However, determining these behaviors through experimental studies is time-consuming as well as

A. K. Kumre · A. Shrivastava · M. Mishra · M. Rout (✉)
Department of Production Engineering, National Institute of Technology Tiruchirappalli, Tiruchirappalli 620015, Tamil Nadu, India
e-mail: matruprasadrout87@gmail.com

A. Maurya et al. (eds.), *Recent Trends in Mechanical Engineering*, Lecture Notes in Mechanical Engineering, https://doi.org/10.1007/978-981-19-7709-1_79

expensive. Hence, various empirical models were introduced to predict the constitutive behavior of the material [2]. However, these models have limitations as the material behavior is nonlinear and a single equation may not be able to completely describe the deformation behavior of the material [3]. Data-driven machine learning models like neural networks (NN) are being used as an alternative tool to predict material behavior efficiently [3–6]. These machine learning models do not require the knowledge of physics-based law for the material behavior and are purely based on the learning of mapping function between the input and output data. Hence, the accuracy of the prediction depends on the learning of the model. In the present work, two machine learning approaches, viz., artificial neural network (ANN) and fuzzy neural network (FNN) have been considered to predict the flow behavior of 304LN austenitic stainless steel.

2 Material and Models

2.1 *Material and Experiment*

Hot rolled and solution-annealed 304LN austenitic stainless steel (C-0.035, N-0.09, Cr-18.29, Ni-8.04 wt%) has been considered for the present work. Axisymmetric compression tests, along the normal direction of the as-received material, were carried out on cylindrical samples. The tests were performed at constant strain rates of 0.01, 0.1, and 1 s^{-1}, and at temperatures of 900, 1000, and 1100 °C [7].

2.2 *Artificial Neural Network (ANN) Model*

The ANN was used to train a learning algorithm for the prediction of flow stress of 304LN austenitic stainless steel, which is a function of strain, strain rate, and temperature. This model works on the principle of the human biological learning model. It takes the input and the corresponding output to learn the relation between them. This learning is inspired by the biological neural network where the algorithm tries to process the information like a human brain. It consists of artificial neurons which act as processing elements that are interconnected and arranged in three different layers, viz., input, output, and hidden layer as illustrated in Fig. 1. Every layer of the ANN architecture consists of several interconnected artificial neurons. An artificial neuron produces an output by receiving the input parameters from the input data or from the other nodes of a hidden layer that comes earlier. The input data which is used to train the ANN model is divided into two subsets. The first one is the training set, which is used to train the algorithm for the mapping function between the input parameter and the corresponding output. It is used to optimize the parameters of the ANN model during training. The other subset is the validating set which is not used

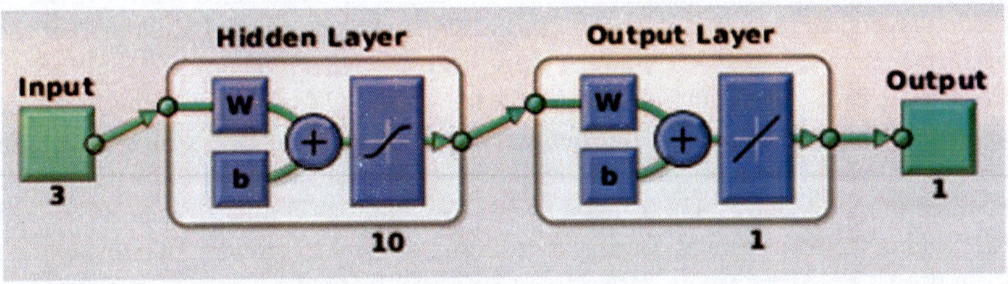

Fig. 1 ANN model architecture

to train the network but to validate the accuracy of the pre-trained ANN model [3, 4, 8].

2.3 *Fuzzy Neural Network (FNN) Model*

The FNN is a learning machine that learns from the input parameters for finding out the parameters of the fuzzy system. It utilizes the technique of approximation from the neural network. A fuzzy system-based neuro-fuzzy model uses a neural network approach that uses data inputs to train the model. This approach drives some changes in the fuzzy system locally by considering local data. This could be represented as a set of fuzzy rules, during the learning process. Thus, for initializing the system there is no need of knowing the fuzzy rule beforehand. Insurance of the semantic properties of the underlying fuzzy system is very vital under the learning process. An n-dimensional unknown function is approximately produced by a neuro-fuzzy system during the learning process that is partially represented by training examples. Therefore, fuzzy rules might be depicted as an indefinite group of the training data. There are three special layers viz. input variables, fuzzy rule, and output variables. These three layers together represent a whole neuro-fuzzy system [5, 6, 9, 10].

3 Results and Discussion

3.1 *ANN Model*

The ANN model was trained with the flow curve data of 304LN austenitic stainless steel obtained from the compression test. The purpose of the training was to make the model learn the mapping function that will be able to predict flow stress values at different inputs of temperature, strain rates, and strain values. The development of the ANN model can be described in three steps: (i) extraction of the data from the flow curves obtained from the experiment; (ii) setting hyper-parameters, number of hidden layers, and number of neurons in the hidden layers and (iii) evaluation of the performance of the final ANN model. MATLAB GRABIT, the built-in application, was used to extract the data from the flow curves obtained from experiments and used to train the mapping function at different temperatures and different strain rates. In the present work, the number of hidden layers considered is one and the number of neurons in the hidden layers is taken as 10. Around 30,000 rows of data were used for training the ANN model. During the training of the ANN model, two sets of data were used, i.e., 70% training data set and 30% validation data set.

The process of 'trial and error' was used for identifying the best architecture of the ANN model, for the material flow curve data, based on the hyper-parameter, iteratively. The best architecture was chosen from several tested architectures. Similarly, several tests were conducted on each architecture till the best possible case was found. Every architecture was trained for 1000 epochs. The best architecture was chosen based on the average number of hidden layers (HL), minimum mean squared error (MSE) achieved, equivalent flow stress, number of neurons (N), training time, and standard deviation. Training time for the different cases greatly varied as some of the networks stopped at early stages, whereas a few were continued till the entire epoch limit. Also, the training time increased with the increase in the complexity of the network. After several tests, the best architecture was found in terms of both training time and accuracy for the ANN with one hidden layer and 10 neurons. The best MSE and training time achieved were 23.6 and one second, respectively. Figure 2 shows the flow curves obtained from the best-trained ANN. A good prediction by the model can be seen. The achieved MSE score is equivalent to the average error in predicted flow stress and the computing time was around one second for the prediction of flow stress values from the pre-trained ANN model. For every input condition, the ANN was observed to adapt to the different flow curve shapes rather than predicting a generalized shape.

3.2 FNN Model

As mentioned earlier, the FNN works on the principle of data-driven learning algorithm which utilizes the known input and output parameters for the training purpose and to predict comparatively accurate results for unknown input variables. The structure of the FNN model which was trained to predict the flow stress of 304LN austenitic stainless steel is shown in Fig. 3a. The FNN structure consists of six layers. The first layer is the input layer, from which the input values move forward to the next layer. The transportation of input values to the next immediate layer occurs through the existing node in the layer. Layer two consists of several membership functions and fuzzy sets for each input value. The node present in layer three depicts one fuzzy rule. In this layer calculation of degrees of rules occur. Each node in the fourth and fifth layers depicts the multiplication and addition, respectively. The sixth layer is the output layer. In the present work, strain rate, temperature, and strain are taken as input parameters to train the FNN to model the flow curve. The development of

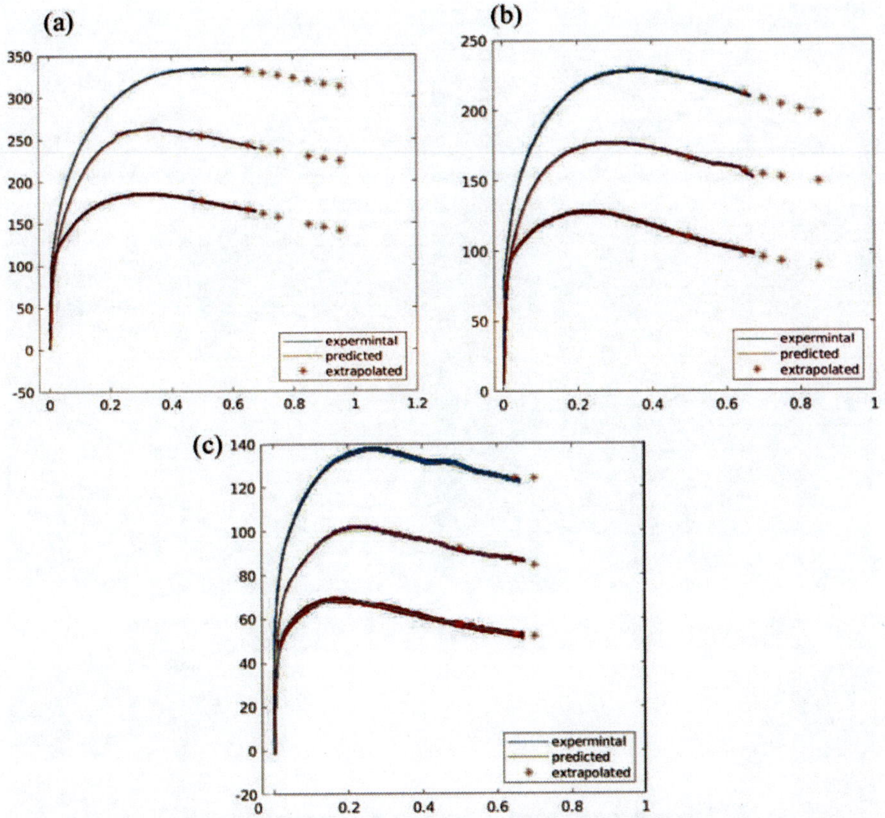

Fig. 2 Flow curve predicted by ANN with some extrapolated values of strain **a** 900 °C, **b** 1000 °C, and **c** 1100 °C

FNN was carried out using three main steps: (i) extraction of the data from the experimental flow curve, with approximate zero deviation, and dividing the whole data set into two different sets for training and testing of the model, (ii) loading up data in Neuro-fuzzy designer and setting up the membership functions to make optimal FNN model architecture, and (iii) evaluation of the performance of the final FNN model. The same set of data, extracted from the experimental flow curves (used in ANN model), was used for the FNN model and the two subsets of data used were 60% training data set and 40% validation data set.

The neuro-fuzzy designer, i.e., the built-in application of MATLAB was used to develop the architecture and two sets of data were uploaded for training and testing of the network. The initial fuzzy inference system (FIS) was generated using grid partitioning, and a number of the input membership function is added. The training of FIS was done by setting up the optimum method and the combination of minimum square regression and backpropagation was used to tune the parameters to get the optimum method. The training error plot, obtained after the training, is shown in

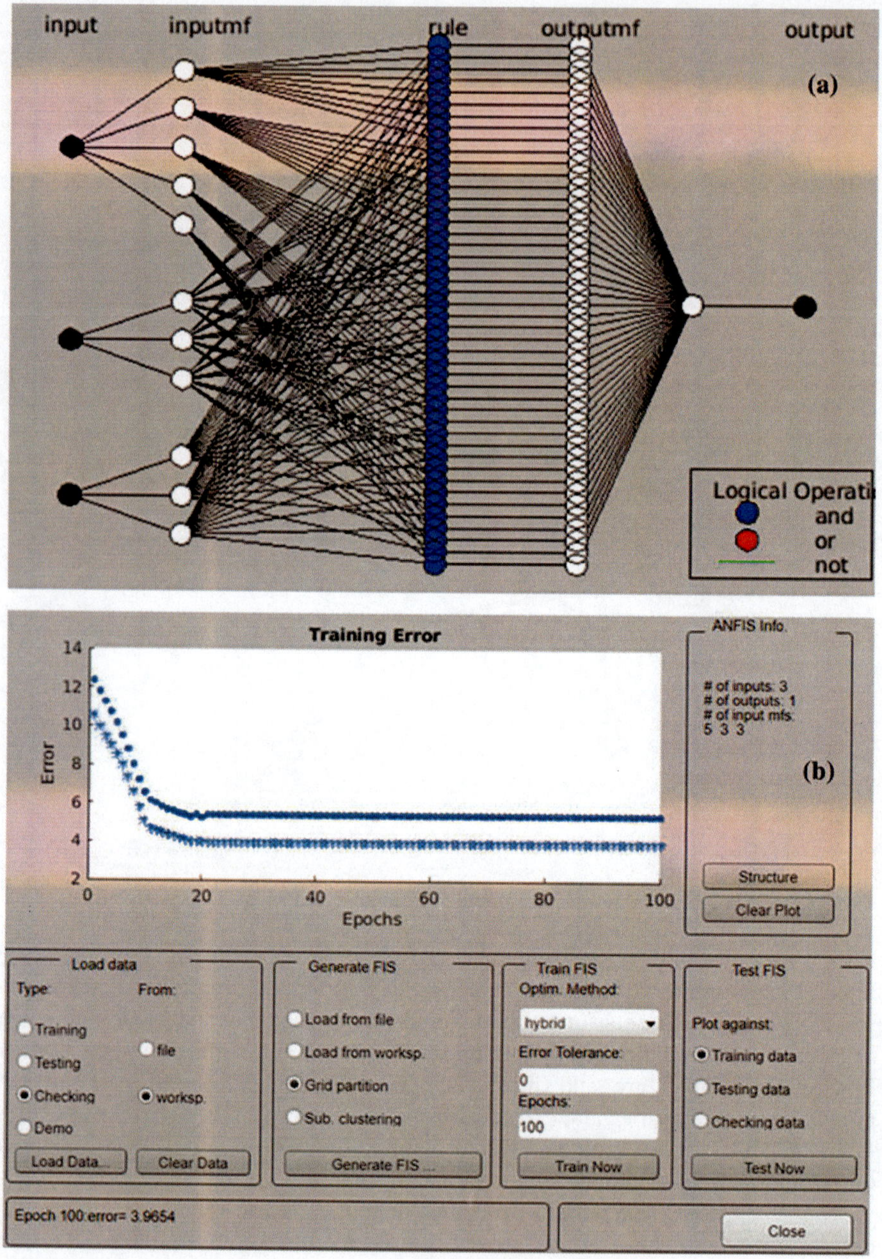

Fig. 3 **a** Architecture diagram and **b** FIS diagram for FNN model

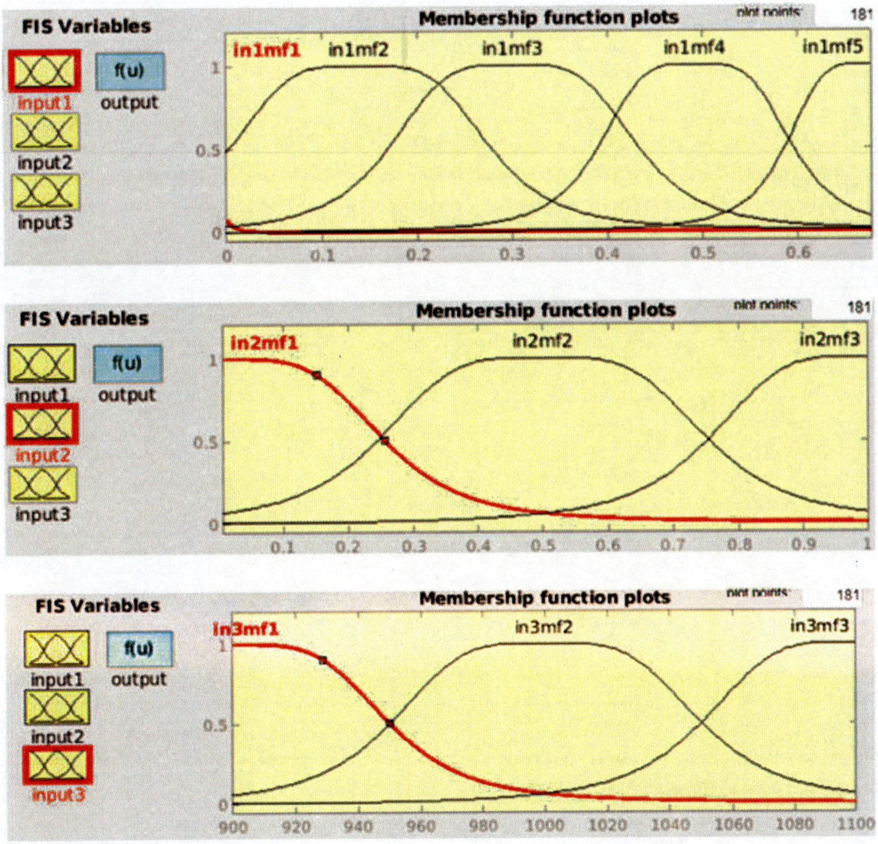

Fig. 4 Member function diagrams

Fig. 3b. The testing data set, which is different from the training data set, is used to validate the FIS. The process of trial and error is used for the recognition of the best membership function sets of FNN. Several tests with different sets of membership functions (Fig. 4) were carried out and the network with the best-achieved MSE 4.029 was chosen for predicting flow stress. The predicted flow curves are shown in Fig. 5.

3.3 Interpolation and Extrapolation by the FNN Model

An investigation on the FNN model's ability to predict for interpolated and extrapolated input parameters is carried out. Figure 6a–c shows the predicted flow curves at intermediated temperatures (continuous lines) for strain rates of 0.01, 0.1, and 1 s^{-1}. Similarly, Fig. 7a–c represents the predicted flow curves at intermediate strain rates

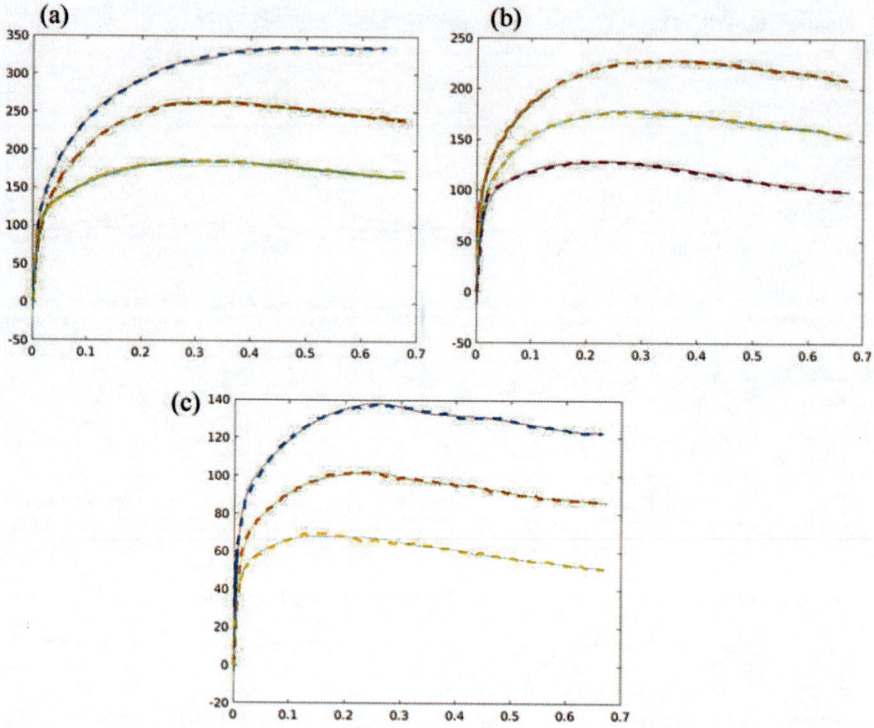

Fig. 5 Predicated flow curves by the FNN model **a** 900 °C, **b** 1000 °C, and **c** 1100 °C (strain rate value decreases in the downward direction)

between 0.01 and 0.1 s^{-1} at temperatures 900, 1000, and 1100 °C. These plots show the ability of the FNN model to predict the flow stress at intermediate values of input parameters. However, the predicted flow stress values between the strain rates of 0.1 and 1 s^{-1} were not within the acceptable values and hence not shown in the figure. Similarly, the predicted flow stress at the extrapolated values of input parameters does not give expected results, which shows the limitation of the present model. Further work is necessary to improve the model for better prediction at extrapolated values of input parameters. Also, the ability of the model is constrained by the availability of the experimental data (three temperatures and three strain rates).

4 Conclusion

In the present work, two different models, viz., ANN and FNN were used to train and predict the flow stress at temperatures of 900, 1000, and 1100 °C and strain rates of 0.01, 0.1, and 1 s^{-1}, using the MATLAB toolbox. Both the models were analyzed based on their accuracy of prediction, time taken, and the ability to predict

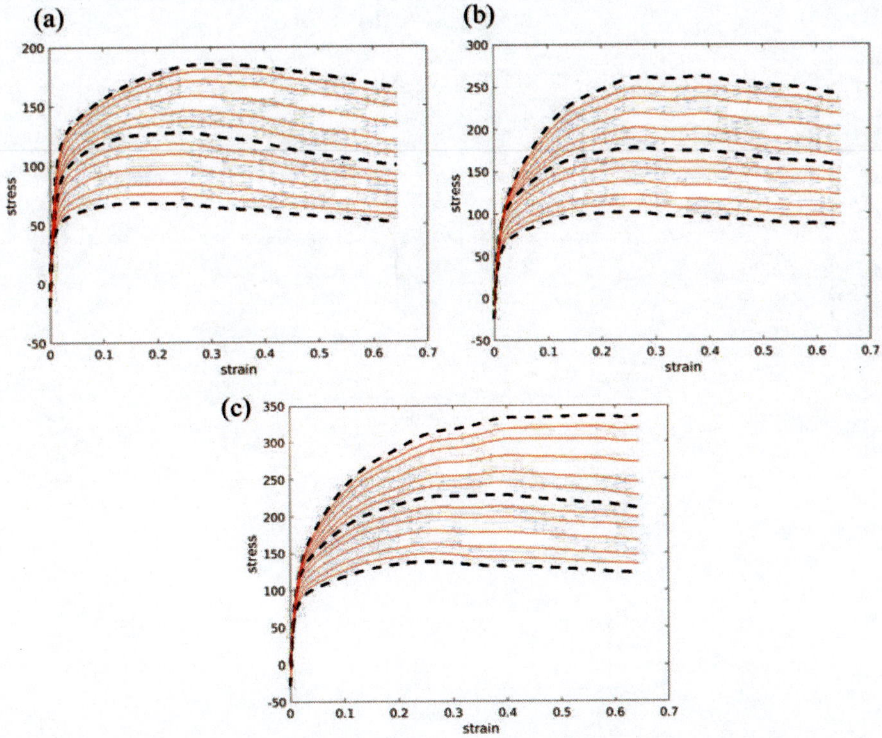

Fig. 6 Predicted flow curves at intermediate temperatures for strain rate **a** 0.01, **b** 0.1, and **c** 1 s^{-1} (temperature value increases in the downward direction, dashed line: experimental flow curve, continuous line: predicted flow curve)

flow stress values at interpolated and extrapolated values of input data. The work can be summarized below.

- During the training of both the models, the time taken to find out the minimum achieved MSE value is more as many training iterations were required.
- For the same experimental data set, the FNN model made predictions more accurately in comparison to the ANN model. Also, in terms of computing time, the FNN was more efficient.
- For the interpolated and extrapolated values of the input data, FNN was able to produce comparatively better results.
- The ability of the models to predict the flow stress values at the interpolated and extrapolated values of input parameters is restricted by the quantity of the input data set.

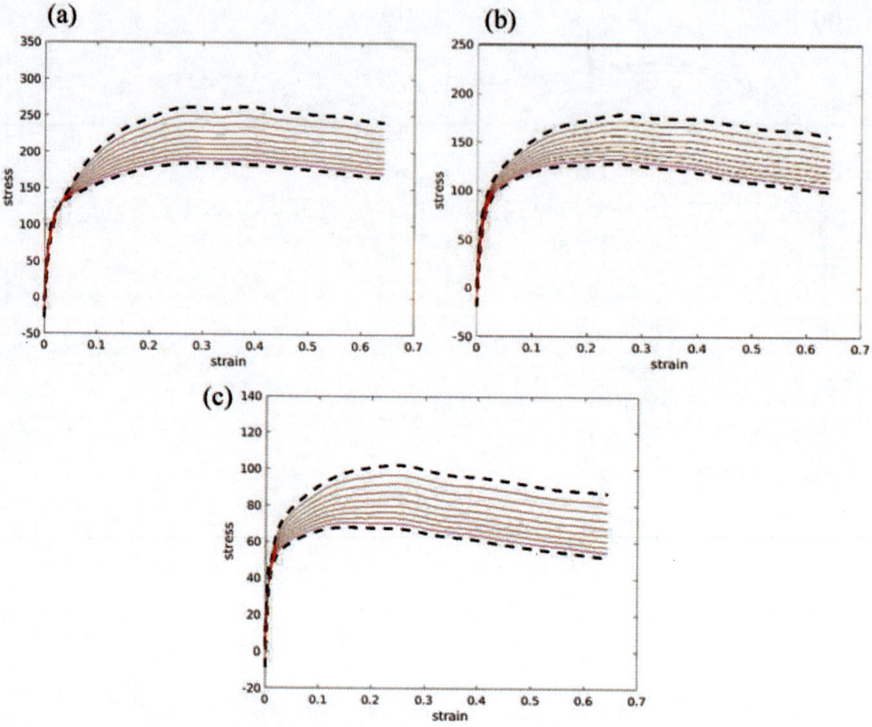

Fig. 7 Predicted flow curves at intermediate strain rates between 0.01 and 0.1 s^{-1} for temperatures **a** 900 °C, **b** 1000 °C, and **c** 1100 °C (strain rate value decreases in the downward direction, dashed line: experimental flow curve, continuous line: predicted flow curve)

References

1. Rout M, Pal SK, Singh SB (2016) Finite element simulation of a cross rolling process. J Manuf Process 24:283–292. https://doi.org/10.1016/j.jmapro.2016.09.012
2. Lin YC, Chen X-M (2011) A critical review of experimental results and constitutive descriptions for metals and alloys in hot working. Mater Des 32:1733–1759. https://doi.org/10.1016/j.matdes.2010.11.048
3. Mandal S, Sivaprasad PV, Venugopal S, Murthy KPN (2006) Constitutive flow behaviour of austenitic stainless steels under hot deformation: artificial neural network modelling to understand, evaluate and predict. Model Simul Mater Sci Eng 14:1053–1070. https://doi.org/10.1088/0965-0393/14/6/012
4. Kumar S, Aashranth B, Davinci MA, Samantaray D, Borah U, Bhaduri AK (2018) Assessing constitutive models for prediction of high-temperature flow behavior with a perspective of alloy development. J Mater Eng Perform 27:2024–2037. https://doi.org/10.1007/s11665-018-3237-6
5. Jia W, Zeng W, Han Y, Liu J, Zhou Y, Wang Q (2011) Prediction of flow stress in isothermal compression of Ti60 alloy using an adaptive network-based fuzzy inference system. Mater Des 32:4676–4683. https://doi.org/10.1016/j.matdes.2011.06.053

6. Liu YG, Luo J, Li MQ (2012) The fuzzy neural network model of flow stress in the isothermal compression of 300M steel. Mater Des 41:83–88. https://doi.org/10.1016/j.matdes.2012.04.043
7. Rout M, Biswas S, Ranjan R, Pal SK, Singh SB (2018) Deformation behavior and evolution of microstructure and texture during hot compression of AISI 304LN stainless steel. Metall Mater Trans A 49:864–880. https://doi.org/10.1007/s11661-017-4447-5
8. An Z, Li J, Feng Y, Liu X, Du Y, Ma F, Wang Z (2015) Modeling constitutive relationship of Ti-555211 alloy by artificial neural network during high-temperature deformation. Xiyou Jinshu Cailiao Yu Gongcheng/Rare Met Mater Eng 44:62–66. https://doi.org/10.1016/s1875-5372(15)30013-8
9. Tang B, Tang B, Li J, Zhang F, Yang G (2013) Modeling the high temperature deformation constitutive relationship of TC4-DT alloy based on fuzzy-neural network. Xiyou Jinshu Cailiao Yu Gongcheng/Rare Met Mater Eng 42:1347–1351. https://doi.org/10.1016/s1875-5372(13)60083-1
10. Han Y, Zeng W, Zhao Y, Zhang X, Sun Y, Ma X (2010) Modeling of constitutive relationship of Ti-25V-15Cr-0.2Si alloy during hot deformation process by fuzzy-neural network. Mater Des 31:4380–4385. https://doi.org/10.1016/j.matdes.2010.03.047

Design and Development of Personal Air Vehicles: A Review

Purvika Mittal and Mohammad Irfan Alam

Abstract The world is currently facing two significant problems associated with urban travel. The first problem is to reduce carbon footprints because of the high volume of vehicles running on conventional fuels. The next issue is the rise in traffic congestion, especially in mega-cities of the world. Clean and green energy is a way of the future. Therefore, interest in electric cars' design and development has increased significantly in the last few years. However, they are much more practical in solving the first problem. There is an urgent need to reduce travel time by avoiding traffic congestion. Therefore, the world needs a clean, fast, and reliable method of urban transpiration. Personal air vehicles, with vertical takeoff and landing (VTOL) capability and batteries as an energy source, can address both concerns. It can also play a vital role in connecting the people living in remote and inaccessible areas to its nearest cities. Several startups and aerospace giants are under the exploration of feasible solutions. Ehang 184, Lilium, Kitty Hawk Cora, Joby S4, Vahana, and Volocpter-2X are potential personal air vehicles in advanced design and testing phases. This paper presents a crisp review of the design and development of personal air vehicles worldwide. There are some critical challenges in getting a feasible design of these systems. Moreover, necessary infrastructure development to support the required ground handling of these systems that are an integral part of urban travel mode needs to be well optimized. Therefore, the paper also focuses on the system's design, development, and implementation challenges.

Keywords Flying taxi · Electric aircraft · Urban air mobility

1 Introduction

The pollution of the world has increased to an alarming level. The rise in vehicle emissions is causing severe respiratory diseases and climate change, rather disturbing the entire life balance and ecosystem of the earth. Fresh air, which is our prime

P. Mittal · M. I. Alam (✉)
Department of Aerospace Engineering, SMEC, VIT Bhopal University, Sehore 466114, MP, India
e-mail: mohammad.alam@vitbhopal.ac.in

A. Maurya et al. (eds.), *Recent Trends in Mechanical Engineering*, Lecture Notes in Mechanical Engineering, https://doi.org/10.1007/978-981-19-7709-1_80

requirement, is becoming impossible to get, especially in megacities. One of the significant contributors to the urban-pollution is the high volume of vehicles running on conventional fuel. The rise in electric cars' design and development and their rapid adoption is a sign of relief. However, they cannot help in reducing a load of vehicles on urban highways. The high volume of cars and other vehicles causes frequent traffic jams, resulting in reduced overall speed. Therefore, instead of saving our productive time and energy, which is the primary objective of transport, we lose them. Therefore, a vehicle is needed for uninterrupted movement between two locations.

Personal Air Vehicle (PAV, Urban-Air-Mobility Aircraft, or simply "flying taxi") can solve the above problems. A fully electric air vehicle with vertical takeoff and landing (VTOL) capability cannot only deal pertaining issues in moving from one location to another but could be vital in reducing air pollution and greenhouse gases significantly [1, 2]. A personal air vehicle under design and development by Airbus is shown in Fig. 1.

Exploration and feasibility analysis of all-electric versions are not limited to personal air vehicles and general aviation aircraft. It also includes large commercial aircraft. Fuel consumption is the major component of airline cost. However, a significant efficiency improvement in jet engine performance has not been seen over the decades, primarily because of saturation and other limitations. All-electric commercial aircraft are much more challenging to achieve with the present limits. Hybrid and more electric commercial aircraft are feasible. Nevertheless, with the rapid increase in all-electric airplanes' interest, consequently, a large volume of research articles has emerged. This paper presents a crisp review of the design and development efforts worldwide related to all-electric vertical takeoff and landing (eVTOL) aircraft. There are several challenges to achieving them. Therefore, it also briefly discusses challenges associated with its design, implementation, and operation in the context of using them as an urban air mobility (UAM) aircraft.

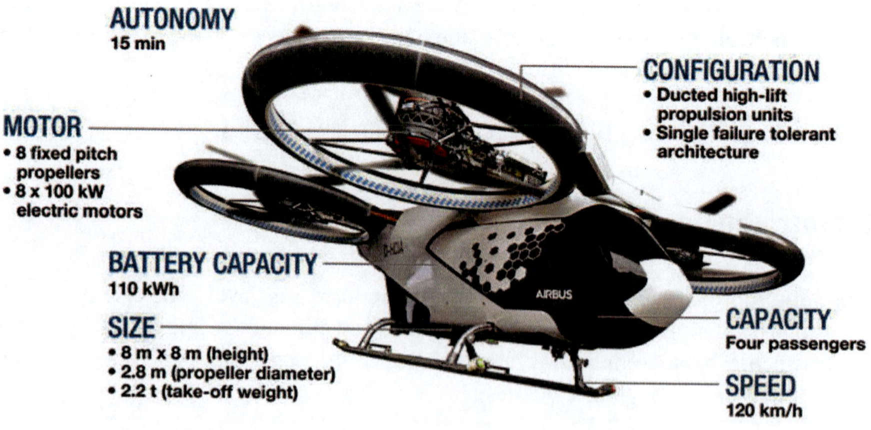

Fig. 1 A personal air vehicle CityAirbus by Airbus [3]

Table 1 Worldwide eVTOL UAM aircrafts under design and testing (multicopter)

 (*Source* ehang.com)	a. EHang-216 By Guangzhou EHang Intelligent Technology Co. Ltd • Origin: China • Passengers: 02 • Range: 35 km • Cruise speed: 100 km/h • Max. speed: 130 km/h
 (*Source* volocopter.com)	b. Volocopter 2X By Volocopter GmbH • Origin: Germany • Passengers: 01 • Range: 27 km • Cruise speed: 100 km/h • Max. speed: NA

1.1 Worldwide Design and Development Efforts

This section describes the design and development effort going over the globe. Several small companies and startups, including tech giants viz., Boeing and Airbus are currently exploring the area. Some of them are the advanced stage of design and testing. With Uber's proposal, the world is witnessing more and more interest to be the first in that race. Design requirements drive innovations. The design requirements for the models under development vary from one pax (e.g., Ehang-184, Volocopter-2X) to 6 pax (e.g., Lilium Jet). However, the models can be broadly be classified into two categories. The first categories are those with less range and passenger. They are simple in design and inherit the configuration of drones. They can be placed in multi-copter configurations, as listed in Table 1.

Another category is winged body models inspired from aircrafts as listed in Table 2. They offer higher payload capacity and longer range. These aircrafts use thrust vectoring for VTOL capability.

2 PAVs as UAM: State-of-the-Art

The landscape of UAM covers various aspects and disciplines. Several top universities are carrying their research in multiple fronts of PAV design. Straubinger et al. [4] have presented an overview of various aspects and findings of key research on UAM going worldwide. Patterson et al. [5] have described a three-pronged approach to study the UAM mission. With a large amount of design data, conventional aircraft has been standardized.

Table 2 Worldwide eVTOL UAM aircrafts under design and testing (thrust vectoring)

 (*Source* lilium.com)	a. Lilium Jet By Lilium GmbH • Origin: Germany • Passengers: 06 • Range: 300 km • Cruise speed: 300 km/h • Max. speed: 400 km/h
 (*Source* kittyhawk.aero)	b. Cora by Wisk By Boeing Company and Kitty Hawk Corporation • Origin: United States • Passengers: 02 • Range: 100 km • Cruise speed: 180 km/h
 (*Source* jobyaviation.com)	c. Joby S4 2.0 By Joby Aviation • Origin: United States • Passengers: 05 • Range: 241 km • Cruise speed: 322 km/h • Max. speed: NA
 (*Source* hyundai.com)	d. Hyundai Air-Taxi By Uber and Hyundai Motor Company • Origin: South Korea • Passengers: 05 • Range: 100 km • Cruise speed: 290 km/h
 (*Source* bartini.aero)	e. Bartini By Bartini Aero • Origin: Russia • Passengers: 04 • Range: 150 km • Cruise speed: 300 km/h • Max. speed: NA

The design and sizing of electrical aircraft, especially with VTOL capability, vary because of being a new area, thus lacking enough data. Kadhiresan and Duffy [6] have carried out conceptual design and mission analysis. Several potential concept configurations for the application of UAM are being explored [7, 8]. Finger et al. [9] have carried our initial sizing different VTOL configurations with hybrid electric systems. Since most companies follow their in-house developed method for design and sizing, their experimental models vary pretty significantly. Bachhini and Cestino [10] have carried out a comparative study of three different configurations, viz., Ehang 184, Kitty Hawk, and Lilium Jet. It can be concluded that multirotor formats are slow and less efficient in cruise. However, they offer a simple design and better VTOL capabilities.

In a conventional aircraft, fuel is a dynamic mass. Therefore, considering various segments of mission profile, fuel consumption can be estimated. In electric aircraft, power estimation is required to calculate the battery weight and motor weight. A detailed study on the effects of design range and battery technology on sizing and performance of eVTOL has been presented by Warren et al. [11]. Barra et al. [12] have described a power estimation method for eVTOL. The design of PAVs is multidisciplinary. For overall improvement, system and subsystem level analysis and mutual improvements in all the disciplines are essential. Goetzendorf-Grabowski and Figat [13] have carried an aerodynamic analysis of tandem-wing PAVs. Niklaß et al. [14] have presented a cross-discipline study and carried collaborative modeling and design approach.

Apart from design, UAM is a new dimension of mass mobility; implementation and operation are also crucial. Romli et al. [15] have carried out a public survey to understand Malaysia's feasibility from the above perspectives. UAM is meant to provide a door-to-door pick-and-drop facility. However, there are several constraints involved in the concept. For mass mobility, seamless integration of the system with another mode of travel is vital. Vascik et al. [16] have carried out the study to understand the underlying constraints in providing notional door-to-door service by UAM.

In a nutshell, the design and development of PAVs and using them for UAM poses several challenges on multiple fronts. Those challenges are discussed in the next section.

3 Design and Operational Challenges

With the incredible progress of electric vehicle technologies over the last few years, the all-electric plane does not seem impossible or science fiction anymore. The high popularity of electric cars and overwhelming customers' responses within a short span gives an immense hope, not just from the designers' or investors' perspective but also from the end-user's perspective. However, all-electric planes have several additional and severe challenges before becoming a reality and viable to the majority

[17, 18]. These challenges can be classified into three main categories, viz., design, implementation, and operational.

The first and the foremost is the design challenge. The design of conventional aircraft is standardized. The aircraft shapes and configurations have also been aggressively optimized. With sufficient data and empirical relations, a designer can quickly get reasonably good estimates of system and subsystem weight for given design requirements. In electric aircraft, neither design has not been standardized due to lack of sufficient data, nor are optimal configurations guaranteed. Designing such systems is susceptible to the technology and varies significantly with mission profiles [5]. With the extremely low energy density of battery technology to in contrast to aviation fuels, full-scale all-electric commercial aircraft seems impossible to achieve soon. Although Hydrogen Fuel Cell and other non-conventional approaches are being considered to address these issues [19]. Another challenge is to design a system within the acceptable noise level to operate from city center.

The second category of the challenge is related to its operational viability. Operating and certification requirements have not been defined clearly. The Federal Aviation Administration (FAA) collaborates with the National Aeronautics and Space Administration (NASA) and recently developed and shared the UAM Concept of Operations (ConOps) version 1.0 [20]. Similarly, planning and ground infrastructure support, e.g., vertiports, and other requirements, are crucial to these systems [21, 22]. Moreover, economic viability is also vital for wide adoption after its design and development [23]. Figure 2 highlights the criticality of the economic aspect in wide adoption of these system, especially for commercial level.

The third challenge is in its implementation, especially for UAM application. Traffic management and route planning is also a key area to focus [24]. Moreover, being a new mode of travel, people might feel psychological barriers before they

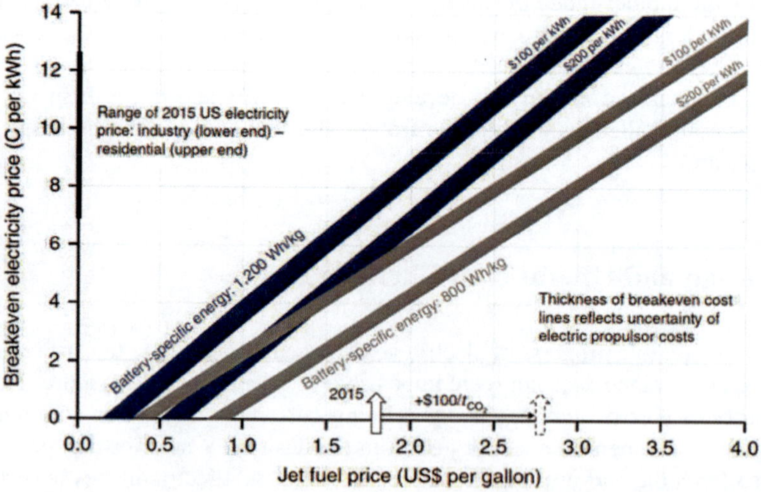

Fig. 2 Break-even electricity price for a first-generation all-electric aircraft [23]

accept it widely. With making these systems autonomous to fly, ensuring safety and privacy can be challenging. Similarly, environmental damage due to its infrastructure development, noise pollution, and visual disruption must be considered in advance.

4 Conclusions

The world is witnessing an unprecedented challenge in controlling city pollution and managing the city's free flow of traffic. Personal air vehicles have the potential to solve the problems. Therefore, several companies are involved in design and development of these vehicles. Some of the developed designs are in the advanced stage of testing and certification. Although due to a viable model to use them as a personal vehicle, several design and operational challenges are ahead. The low energy content is acting as a bottleneck. Therefore, more electric and hybrid propulsion is also gaining popularity, especially for commercial aircraft. Undoubtedly, several challenges exist in achieving a fully implemented all-electric version; however, people are working tirelessly on multiple fronts to achieve the common goal. Consequently, a large volume of research has started appearing in the area, which is a hope of a feasible solution and emphasizes the significance and research scope in the area.

References

1. Moore M (2010) NASA puffin electric tailsitter VTOL concept. In: 10th AIAA aviation technology, integration, and operations (ATIO) conference, September, p 9345
2. André N, Hajek M (2019) Robust environmental life cycle assessment of electric VTOL concepts for urban air mobility. In: AIAA aviation 2019 forum, p 3473
3. Urban air mobility—the sky is yours, 27 Nov 2018. https://www.airbus.com/newsroom/stories/urban-air-mobility-the-sky-is-yours.html. Accessed 15 Mar 2021
4. Straubinger A, Rothfeld R, Shamiyeh M, Büchter K-D, Kaiser J, Plötner KO (2020) An overview of current research and developments in urban air mobility—setting the scene for UAM introduction. J Air Transp Manag 87:101852
5. Patterson MD, Antcliff KR, Kohlman LW (2018) A proposed approach to studying urban air mobility missions including an initial exploration of mission requirements
6. Kadhiresan AR, Duffy MJ (2019) Conceptual design and mission analysis for eVTOL urban air mobility flight vehicle configurations. In: AIAA aviation 2019 forum, p 2873
7. Johnson W, Silva C, Solis E (2018) Concept vehicles for VTOL air taxi operations
8. Silva C, Johnson WR, Solis E, Patterson MD, Antcliff KR (2018) VTOL urban air mobility concept vehicles for technology development. In: 2018 aviation technology, integration, and operations conference, p 3847
9. Finger DF, Götten F, Braun C, Bil C (2018) Initial sizing for a family of hybrid-electric VTOL general aviation aircraft
10. Bacchini A, Cestino E (2019) Electric VTOL configurations comparison. Aerospace 6:26
11. Warren M, Garbo A, Kotwicz Herniczek MT, Hamilton T, German B (2019) Effects of range requirements and battery technology on electric VTOL sizing and operational performance. In: AIAA SciTech 2019 forum, p 0527

12. Barra F, Scanavino M, Guglieri G (2020) A methodology for multirotor aircraft power budget analysis. Aircr Eng Aerosp Technol
13. Goetzendorf-Grabowski T, Figat M (2017) Aerodynamic and stability analysis of personal vehicle in tandem-wing configuration. Proc Inst Mech Eng, Part G: J Aerosp Eng 231(11):2146–2162
14. Niklaß M, Dzikus N, Swaid M, Berling J, Lührs B, Lau A, Terekhov I, Gollnick V (2020) A collaborative approach for an integrated modeling of urban air transportation systems. Aerospace 7(5):50
15. Romli FI, Rashid H, Harmin MY (2018) Market potential for personal air vehicle (PAV) concept: a Malaysian case study. Int J Pure Appl Math 119(15):3755–3760
16. Vascik PD, Hansman RJ, Dunn NS (2018) Analysis of urban air mobility operational constraints. J Air Transp 26(4):133–146
17. Postorino MN, Sarné GML (2020) Reinventing mobility paradigms: flying car scenarios and challenges for urban mobility. Sustainability 12:3581
18. Pan G, Alouini MS (2021) Flying car transportation system: advances, techniques, and challenges. IEEE Access 9:24586–24603
19. Ng W, Patil M, Datta A (2021) Hydrogen fuel cell and battery hybrid architecture for range extension of electric VTOL (eVTOL) aircraft. J Am Helicopter Soc 66(1):1–13
20. Urban-air mobility: ConOps v.01. https://nari.arc.nasa.gov/sites/default/files/attachments/UAM_ConOps_v1.0.pdf. Accessed 16 Mar 2021
21. Ploetner KO, Al Haddad C, Antoniou C et al (2020) Long-term application potential of urban air mobility complementing public transport: an upper Bavaria example. CEAS Aeronaut J 11:991–1007
22. Urban air mobility (UAM) market study, NASA. https://www.nasa.gov/sites/default/files/atoms/files/uam-market-study-executive-summary-v2.pdf
23. Schäfer AW, Barrett SR, Doyme K, Dray LM, Gnadt AR, Self R, O'Sullivan A, Synodinos AP, Torija AJ (2019) Technological, economic and environmental prospects of all-electric aircraft. Nat Energy 4(2):160–166
24. Kleinbekman IC, Mitici MA, Wei P (2018) eVTOL arrival sequencing and scheduling for on-demand urban air mobility. In: 2018 IEEE/AIAA 37th digital avionics systems conference (DASC), September, pp 1–7

A Comparative Study of Standard Profiles of High-Altitude Airships Based on Initial Sizing

Aaftab Khan and Mohammad Irfan Alam

Abstract Seamless connectivity has become a necessity in a new era of technology. Yet, a sizable population does not have access to the internet and experiencing a cavity life. Therefore, several efforts are underway to provide uninterrupted communication to the people and incredibly remote locations; Project Loon is a classic example of similar attempts. High-altitude airships can provide an aerial platform that can be used to mount communication gears. These systems use buoyancy to remain airborne for a long time, thus providing an efficient solution to the problem. They are powered by solar energy harnessed through solar panels attached to the upper portion of the envelope conformal to the shape. The shape of the envelope is the largest contributor to the system's structural weight. Moreover, the profile also drives the drag coefficient and net energy generation from the solar panel. Therefore, selecting the optimal shape of the envelope play a vital role in the design and operation of such systems. The researcher worldwide explored several profiles suitable for high-altitude airships, viz., GNVR, NPL, Wang, etc. Based on the given design requirements and deployment location, the selection of an optimal shape may vary. This paper presents a comparative study of the standard profiles based on the initial sizing. The effect of operational parameters on the final results is discussed in detail. The results can help the designer understand the significance of selecting the correct shape for the given design requirements.

Keywords Stratospheric airship · HAPs · Solar-powered airship · Initial sizing

1 Background and Introduction

Incredible progress in Internet technology has watered in innovating several tech products and IoT devices that have become part of our daily life over the last decade. It has become difficult to imagine a life without all those smart devices connected using seamless connectivity. The entire world is moving from 3 to 4G, and now 5G

A. Khan · M. I. Alam (✉)
Department of Aerospace Engineering, SMEC, VIT Bhopal University, Sehore 466114, MP, India
e-mail: mohammad.alam@vitbhopal.ac.in

A. Maurya et al. (eds.), *Recent Trends in Mechanical Engineering*, Lecture Notes in Mechanical Engineering, https://doi.org/10.1007/978-981-19-7709-1_81

with notable pace. However, despite these rapid developments in providing high-speed Internet connectivity to the world, many are still deprived of reliable Internet. Various techniques are being explored to overcome this barrier and fulfill the need for connectivity to every human, especially those living in remote areas [1]. High-Altitude Airships have the potential to solve the above problems [2]. They are platforms that use buoyancy to remain deployed at a designated altitude, typically 20 km. These platforms will hold necessary equipment and gears to provide high-speed Internet connectivity, observation, monitoring, etc. High-altitude airships (shown in Fig. 1) are based on solar energy and work in a combination of regenerated fuel cells (RFC) for the day-night power supply to their mounted devices propulsion systems.

Airship envelopes are the most prominent and heaviest part of these systems. They hold buoyant gases and provide a top area to mount solar panels. However, they also draw a significant amount of power to overcome the drag encountered due to the large size. Several studies have been carried get the optimized profile of these systems. They can be categorized under conventional, a 2D body of revolution, and non-conventional, non-body of revolutions. Among traditional shapes, NPL (a low-drag airship profile suggested by National Physical Laboratory) [4], GNVR (in the memory of Prof. G. N. V. Rao) [5], and Wang [6] are the most common. Researchers have carried out initial sizing and design activity for those specific shapes. The effect of varying sizing parameters was also studied on a single profile [7]. Some non-bodies of revolution are also being explored [8]. However, a detailed comparison of the standard airship profiles based on initial sizing has not been carried out. A comparative study to size an airship for a given set of requirements can provide insight into a suitable profile for a given set of design requirements. Therefore, this paper presents a comparative study of output parameters based on initial sizing for defined input parameters. An in-house code was developed in Python to carry out the analysis. The study is relevant in getting the idea before carrying any field trial and prototype development. The following section describes the mathematical modeling and sizing approach adopted in the present study.

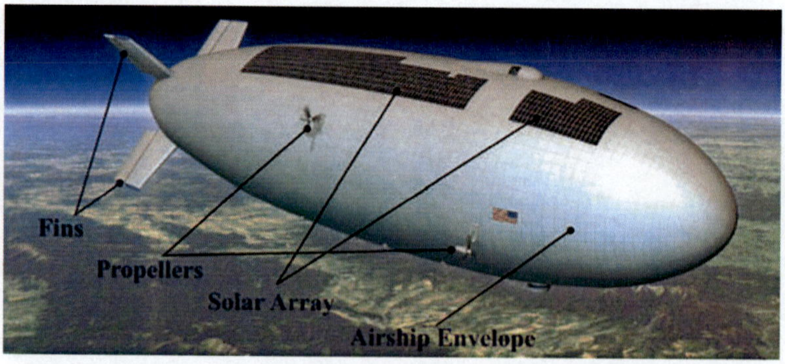

Fig. 1 A conceptual representation of high-altitude airship by NASA [3]

2 Mathematical Modeling and Sizing Approach

The design and sizing of high-altitude airships involve various disciplines [7, 8]. In most cases [8, 9], there are two main constraints required to meet. The first constraint is to meet the system's total power demand, and the second is to hold the necessary payload at the designated altitude. Researchers have proposed several sizing methodologies with different levels of complexity and keeping the above constraints in focus [8–10]. In this study, a simple physics-based modular method developed by Alam and Pant [9], as illustrated in Fig. 2 is used to carry the analyses. It is an iterative approach to satisfy the designed requirements by varying the envelope's size and shape.

This study compares three different airships' envelope profiles, viz., NPL [2], GNVR [3], and WANG [4]. Mathematical representations of these profiles have been described in Table 1. NPL, WANG, and GNVR have their fineness ratios (ratio of length to its diameter) of 4, 3.87, and 3.05, respectively [11, 12]. Table 2 presents key equations used in the initial sizing of high-altitude airships.

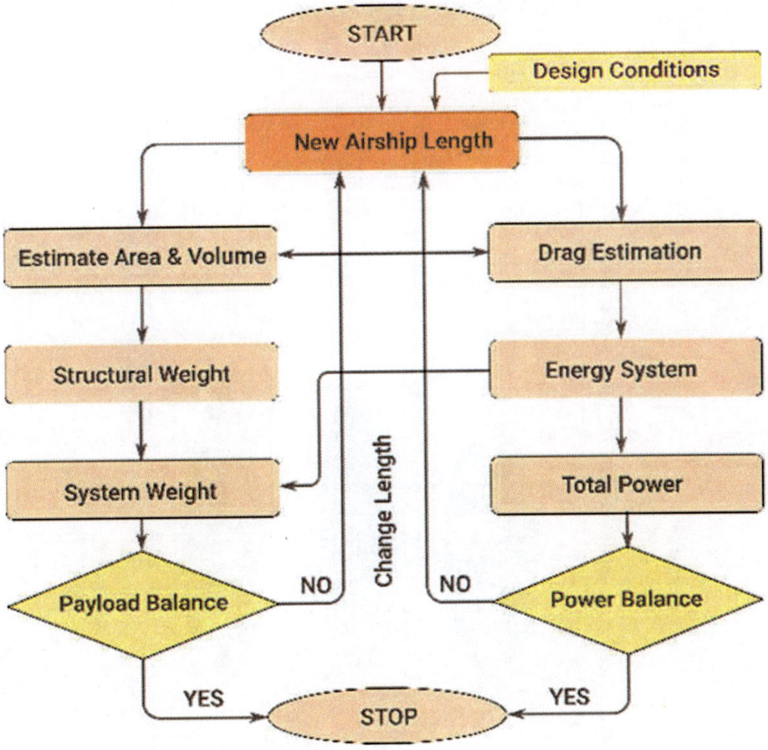

Fig. 2 Initial sizing methodology [9]

Table 1 Mathematical representation of GNVR, NPL, and WANG

GNVR	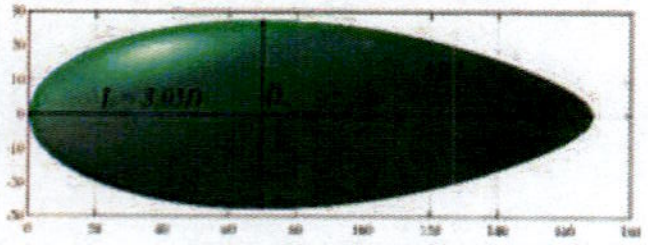	$Y = \begin{cases} 0.5D\sqrt{1-\left(\frac{X-1.25D}{1.25D}\right)^2} & 0.0 < X \leq 1.25D \\ \sqrt{16D^2-(X-1.25D)^2}-3.5D & 1.25D < X \leq 2.875D \\ \sqrt{0.1373D(1.8D-(X-1.25D))} & 2.875D < X \leq 3.05D \end{cases}$
NPL	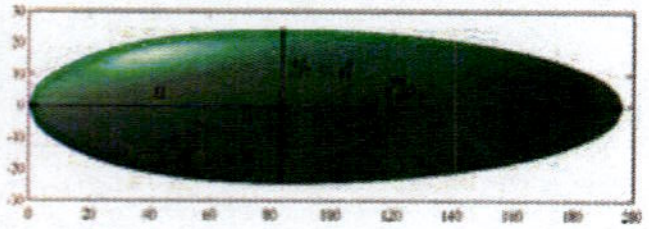	$y = \begin{cases} \pm b\sqrt{1-\frac{(x-a)^2}{a^2}} \text{ when } x \leq a \\ \pm b\sqrt{1-\frac{(x-a)^2}{2a^2}} \text{ when } x > a \end{cases}$
Wang	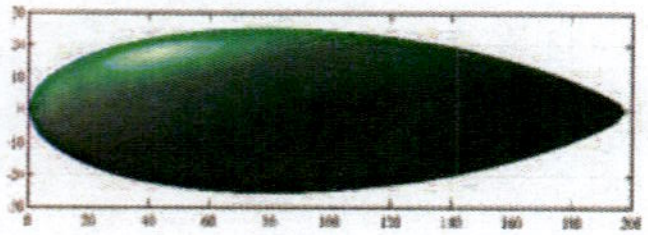	$y = \frac{\sqrt{a(l-x)\left(bx-l\sqrt{c}+\sqrt{cl^2-dlx}\right)}}{8}$

Table 2 Key equations used in sizing

Volume	$V_{\text{env}} = \int_0^l \pi y^2 dx$	$l =$ length
Surface area	$A_{\text{env}} = \int_{\xi=0}^{\xi=2\pi} \int_{x=0}^{x=l} y(x) \left[1 + \left(\frac{dy}{dx}\right)^2\right]^{1/2} d\xi dx$	
Buoyancy	$B_F = \rho_a g V_{\text{env}}$	$\rho_a =$ air density
Drag coefficient	$C_{DV,\text{env}} = \frac{0.172 f^{\frac{1}{3}} + 0.252 f^{-1.2} + 1.032 f^{-2.7}}{Re^{\frac{1}{6}}}$	
Drag	$D_{\text{total}} = \frac{1}{2} \rho_a v_{\text{opr}}^2 C_{DV,\text{ total}} V_{\text{env}}^{2/3}$	$v_{\text{opr}} =$ relative velocity
Structural weight	$W_{\text{st}} = 1.25(\rho_{\text{env}} \cdot A_{\text{env}})$	$\rho_{\text{env}} =$ envelope density
Weight of solar array	$W_{\text{sc}} = S.R_{\text{sc}} \cdot \rho_{\text{sc}} / \eta_{\text{sc}}$	$\rho_{\text{sc}} =$ solar array density
Propulsion weight	$W_{\text{ps}} = P_{\text{ps}} \cdot / \rho_{\text{ps}}$	$\rho_{\text{ps}} =$ power density
RFC weight	$W_{\text{RF}} = E_{\text{RF}} / (\eta_{\text{RF}} \cdot \rho_{\text{RF}})$	$\rho_{\text{RF}} =$ storage density

In calculating the structural weight of the airship, to accommodate the weight of fins, lightning protection unit, etc. 25% of the envelope weight is added to the total envelope weight of an airship for simplicity. For the airship to be stationary at the intended location during the operational period, a thrust is produced by propellers to overcome the drag due to wind speed. This thrust was used to calculate the power requirement from propulsion system. At the point of power meet, it has to be checked that whether the power requirements for propulsion as well as power for payload are fulfilled by power output from direct solar cells and/or RFC storage unit or not.

Table 3 enlists the input parameters. These constants may change with time due to research and development. Although the change of these parameters will influence output parameters, the sizing procedure will remain the same. Therefore, a new set of results can easily be obtained with a new set of constants.

3 Results and Discussion

Results presented in this section are preliminary and are based on the sizing approach illustrated in Fig. 2. Sensitivity analyses have been carried out to understand the effect of its design requirements on the output parameters. Figure 3 presents the required length of the airship at the various payload. It has been observed that GNVR-based airship can counter the load with a smaller envelope. Its payload requirement directly drives the volume of an airship. GNVR having a minimum fineness ratio can full the payload demand with a smaller-sized envelope. Figure 4 presents the analysis of power required versus needed solar array. High power demand can be obtained from larger solar panels, adding extra weight to the system, larger volume of the envelope, and so on. NPL drag coefficient is highest among all three, resulting in higher drag, thus needs a larger area of the solar array.

Table 3 Input parameters

Inputs	Baseline value
Payload mass (kg)	1000
Payload power (kW)	1
Floating height (m)	20,000
Mission speed (m/s)	20
Average irradiance $\left(W/m^2\right)$	480
Discharge time (h)	14
Off standard temperature (K)	+20
Geometric coverage ratio	0.25
Envelope mass/area ratio $\left(g/m^2\right)$	200
Solar cells mass area ratio $\left(g/m^2\right)$	250
Solar cells efficiency (%)	8
RFC storage energy/mass ratio (Wh/kg)	250
RFC efficiency (%)	60
Propeller power/mass ratio (W/kg)	75
Propeller efficiency (%)	85
Gear box transition efficiency (%)	80
Motor efficiency (%)	80

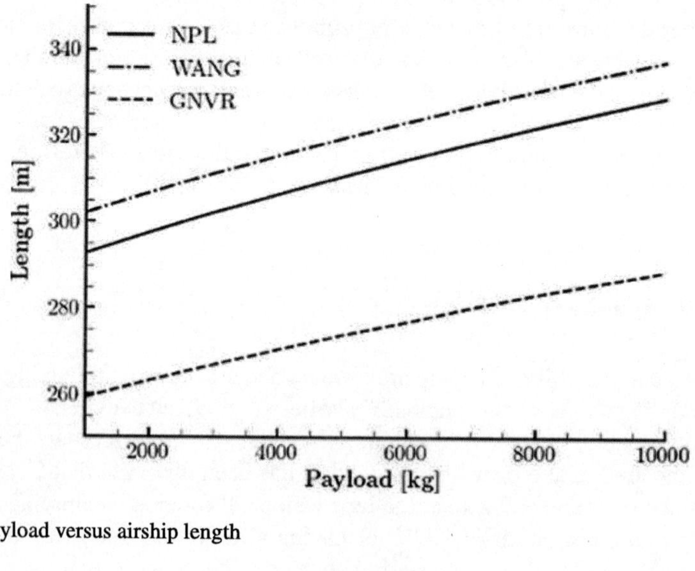

Fig. 3 Payload versus airship length

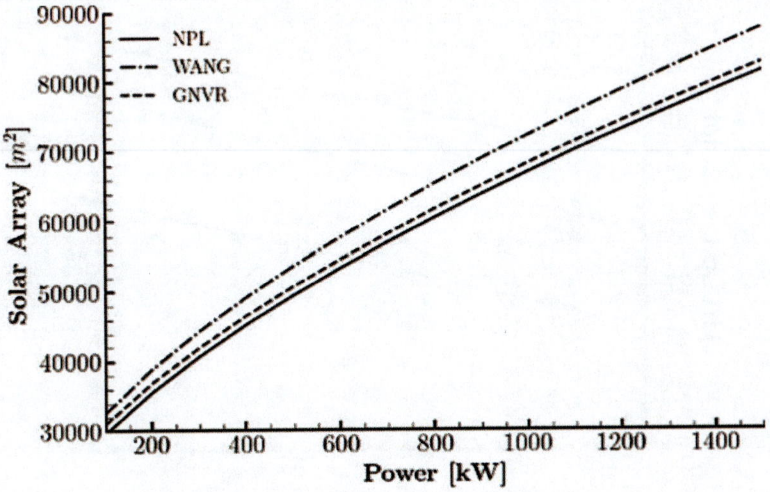

Fig. 4 Power versus solar array

The drag coefficient obtained of NPL is most significant while smallest in the case of GNVR. Reynold's number affect their drag coefficients. Therefore, a comparative study has also been carried out to understand its significance. Figures 5 and 6 present the results of increasing drag coefficients from baseline values to increasing them by 100% and their effect on airship power requirement and their length, respectively. Higher drag means more power needed by the system to counter, thus increased solar array.

4 Conclusions

High-altitude airships have the potential to provide a buoyancy platform at 17–25 km. Airship envelopes are the primary subsystem and the main contributor to the drag force. Though NPL, WANG, and GNVR are the main envelope profiles considered standard in airship design optimization, their modeling and shape vary significantly. In this study, a comparative analysis based on initial sizing is carried out. For the same payload requirement, GNVR offers 5–6% smaller envelopes with respect to others. For high power requirements, NPL needs a smaller solar array. WANG provides better aerodynamics, thus the minuscule increase in power demand with the increment of the drag coefficient. However, the observations in the study are based on the more straightforward approach of design and sizing. A detailed investigation with high-fidelity models is needed to get concrete conclusions.

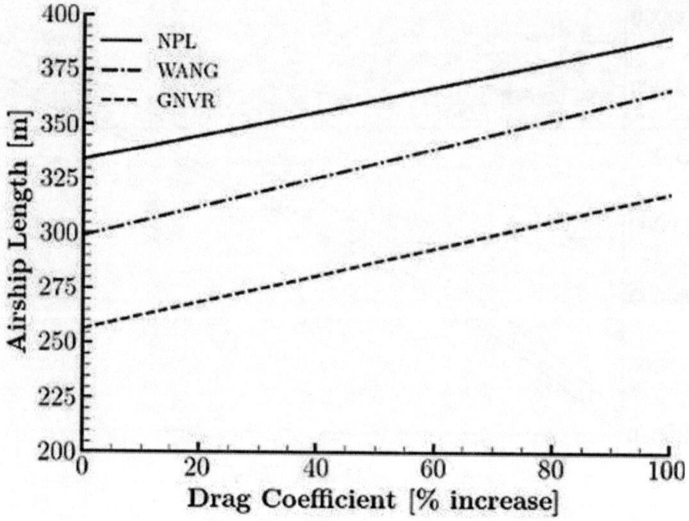

Fig. 5 Drag coeff. versus airship length

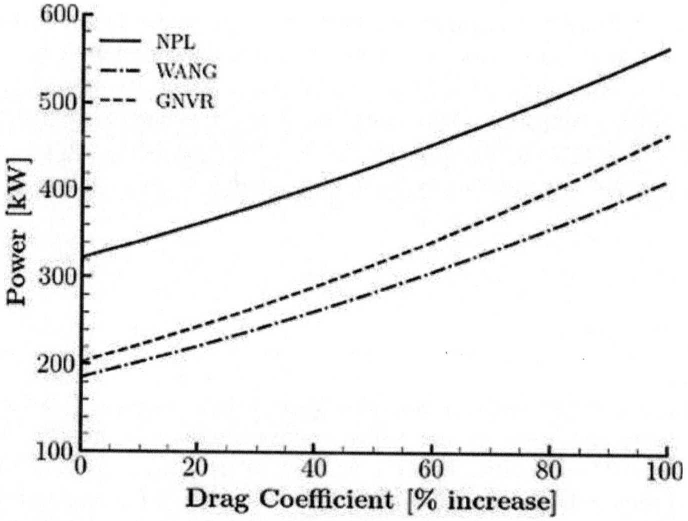

Fig. 6 Drag coeff. versus power required

References

1. Project Loon by Google. Loon LLC, Alphabet. https://loon.com/. Accessed 15 Mar 2021
2. Colozza A, Dolce J (2003) Initial feasibility assessment of a high-altitude long endurance airship. Tech. Rep. NASA/CR2003-212724, NASA
3. NASA seeks comments on possible airship challenge, 25 Nov 2014. https://www.nasa.gov/jpl/nasa-seeks-comments-on-possible-airship-challenge/. Accessed 15 Mar 2021
4. Cheeseman I (2012) Airship technology, ch. 2: aerodynamics. Cambridge University Press, pp 29–30
5. Ram CV, Pant RS (2010) Multidisciplinary shape optimization of aerostat envelopes. J Aircr 47(3):1073–1076
6. Wang QB, Chen JA, Fu GY, Duan DP (2009) An approach for shape optimization of stratosphere airships based on multidisciplinary design optimization. J Zhejiang Univ-SCI A 10(11):1609–1616
7. Chen Q, Zhu M, Sun KW (2011) Analysis to effects on conceptual parameters of stratospheric airship with specified factors. J Comput 6(5):1055–1062
8. Yang Y, Xu X, Zhang B, Zheng W, Wang Y (2020) Bionic design for the aerodynamic shape of a stratospheric airship. Aerosp Sci Technol 98:105664
9. Alam MI, Pant RS (2018) Multi-objective multidisciplinary design analyses and optimization of high-altitude airships. Aerosp Sci Technol 78:248–259
10. Liang H, Zhu M, Guo X, Zheng Z (2012) Conceptual design optimization of high-altitude airship in concurrent subspace optimization. In: 50th AIAA aerospace sciences meeting including the new horizons forum and aerospace exposition, January, p 1180
11. Alam MI (2018) Multi-objective multidisciplinary design optimization of high-altitude airships. PhD thesis
12. Shi YIN, Ming ZHU, Liang H (2019) Multi-disciplinary design optimization with variable complexity modeling for a stratosphere airship. Chin J Aeronaut 32(5):1244–1255
13. Wang XL, Shan XX (2006) Shape optimization of stratosphere airship. J Aircr 43(1):283–286
14. Alam MI, Pant RS (2017) Surrogate based shape optimization of airship envelopes. In: 24th AIAA aerodynamic decelerator systems technology conference, p 3393

Network Topology Model for Wear Behavior Prediction of Ti6Al4V Clad Magnesium Substrates

Ganesa Balamurugan Kannan, S. Revathi, K. Rajkumar, and M. Duraiselvam

Abstract In this work, Ti6Al4V alloy was clad on commercially pure magnesium substrate by laser cladding. The laser cladding parameters were laser scan speed and powder feed rate. Wear behaviors of the clad substrates were evaluated by a dry sliding wear testing on Ti-6Al-4V counter material by pin-on-disc method. The wear testing parameters like applied load and sliding velocity were varied by keeping sliding distance as constant. The experimental data were used to develop an optimized neural network model. Network models predicted that the clad deposition affects the wear behavior. The combination of lower laser scan and highest powder feed yields higher deposition compared to higher laser scan and lowest powder feed combination. The higher material deposition improves the wear resistance of the substrates. The experimental data confirms the network models prediction by showing the lower wear rate at 200 mm/min–10 g/min laser parametric condition which has higher clad deposition than lower wear rate at 300 mm/min–5 g/min condition with lower materials deposition. Overall, laser cladding of Ti6Al4V on commercially pure magnesium improves the wear resistance by ~75%.

Keywords Laser cladding · Ti6Al4V alloy · Pure magnesium · Clad deposition · Wear rate · Optimally pruned network topology

G. B. Kannan
IFET College of Engineering (Autonomous), Villupuram, India

S. Revathi
National Institute of Technology Puducherry, Karaikal, India

K. Rajkumar (✉) · M. Duraiselvam
National Institute of Technology, Tiruchirappalli, India
e-mail: rajkumar9033@gmail.com

A. Maurya et al. (eds.), *Recent Trends in Mechanical Engineering*, Lecture Notes in Mechanical Engineering, https://doi.org/10.1007/978-981-19-7709-1_82

1 Introduction

Magnesium has wide involvement in automobile, aeronautical, and aerospace applications because of its low density. Recently, it has contribution in bioimplants application due to its biocompatibility [1–7]. However, their inferior corrosion and wear resistances have restricted their wide usage in industries [8, 9]. Researchers are attempting various approaches to reduce the wear losses of them. Laser cladding is one of the surface coating technique shows higher metallurgical bonding and defects than other conventional methods [10, 11]. The selection of appropriate parameters of desired clad under operating conditions is an important attribute in achieving better clad behavior. This is influenced by cladding parameters. The identification of appropriate parameters is a crucial activity and it is commonly achieved with sophisticated practical trials. As an alternate approach, the benefits of the artificial intelligence were used to achieve similar goals by numerous researchers to correlate parameters in cladding by soft computing methodologies. Thus, as an indication of expansion of the current technique, in this study artificial feed-forward neural network topology based on the Levenberg–Marquardt backpropagation technique is used to estimate the wear behavior.

1.1 Levenberg–Marquardt Backpropagation Algorithm for Multilayer Feedforward Network

The backpropagation algorithm has been widely used as a learning algorithm in feed-forward multilayer neural networks. The first step of this supervised learning process is to initialize each node with random weights. The network essentially adjusts its weight based on the weight error correct rule. It has two stages: a forward pass and a backward pass [12, 13]. The forward pass involves information flow forward through the network in the natural sense of nodes' input–output relation. During the backward pass, the networks actual output is compared with the target output and the error estimates are computed for the output units and propagating this back from the output nodes to the hidden units. Thereby weights connected to the output unit are adjusted in order to reduce those errors. The implementation of backpropagation algorithm updates its weights incrementally after seeing each input output pair. This slow convergence problem is overcome by the Levenberg–Marquardt algorithm which is fast and has stable convergence. It inherits speed from Newton method, but it also has the convergence capability of steepest descent method. Firstly, the gradient is computed as $g = J^T e$ where J is the Jacobian matrix that contains first derivatives of the network errors with respect to the weights and biases, and e is a vector of network errors. Computing the newton like update

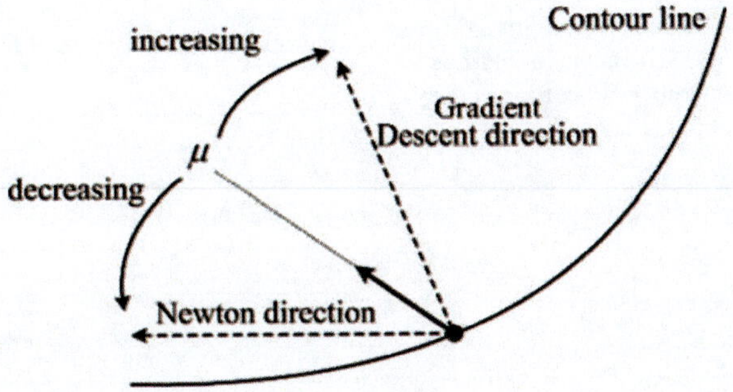

Fig. 1 Backpropagation learning algorithm based on Levenberg–Marquardt algorithm

$$X_{k+1} = X_k - \left[J^T J + \mu I\right]^{-1} J^T e \quad (1)$$

Here μ is the scalar damping parameter.

When,

$\mu = 0$; Newton's method
$\mu \gg$; Gradient descent.

And it is shown in Fig. 1.

Thus, in this training algorithm, initially when the error is greater, gradient descent is followed until μ becomes smaller, and it switches to work on newton method which is faster and more accurate with minimum error.

2 Experimental Procedure

Commercially pure magnesium was used as a substrate material. The surfaces were polished with metallographic sandpapers and washed with alcohol. The Ti6Al4V powders with the average particle size of 60 μm were used as the cladding material. The YB:YAG disk laser (Solid state laser), 4 kW capacity was used. The work parameters values are shown in Table 1. The Ti6Al4V powders were coaxially fed into the molten pool of the substrate material, and the Argon was supplied in the rate of 10 L/min to protect the molten pool. Metallographic characterizations were carried out by cross-sectioned the laser clad specimens by SEM. Microhardness measurements were taken on the coating, interface and the substrate with the test load of 0.5 kg. The pin-on disk equipment was used to conduct wear test on the unclad magnesium and clad magnesium. Wear testing parameters are shown in Fig. 2. The counterpart disk is a heat-treated Ti6Al4V plate. The volume loss and the wear rate were calculated by the Eqs. 2 and 3, respectively.

$$\text{Volume Loss } (V_L) = \frac{\text{Mass loss}}{\rho} \quad (2)$$

Table 1 Mean Square Error (MSE) values for various number of neurons in the hidden layer

Number of neurons in the hidden layer	MSE
05	1.5648×10^{-6}
10	2.0652×10^{-7}
15	0.6548×10^{-8}
20	3.8654×10^{-10}
25	1.0246×10^{-12}
30	0.9865×10^{-7}
35	3.6548×10^{-6}

$$\text{Wear Rate } (W_R) = \frac{\text{Volume Loss}}{\text{Sliding Distance}} \left(\text{mm}^3/\text{m}\right) \quad (3)$$

where ρ represents the density of the material.

3 Results and Discussion

Wear rate results showed that at constant applied loading of 30 and 50 N, wear rates are minimum at 2 m/s sliding velocity and higher at 1 m/s sliding velocity as evident from Fig. 2a, b. Wear rate results showed that laser cladding of Ti6Al4V on pure magnesium substrate significantly improves the wear resistance by ~75%. Samples clad at 200 m/s–10 g/min combination shows the higher wear resistance than other combinations in both constant loading conditions (Fig. 2). Similarly, lower wear resistance observed in 300 m/s–5 g/min combination in both 30 and 50 N test loading conditions (Fig. 2). When varying the loading condition by keeping the sliding velocities constant, higher wear rates are observed in 50 N loading condition than 30 N loading condition. Wear rate interpretations revealed that higher wear rates were observed in the samples clad at 300 mm/min scan speed. Likewise, samples clad at feed rate of 5 g/min showed the higher wear rates.

The input parameters of the neural network were laser scan speed, powder feed rate and applied load, and sliding velocity of wear testing conditions. The different state of these lasers clad and wear testing parameters were engaged for experimentation. In this process of learning and adapting, the samples used for training is 75% and for testing and validation purpose is 25%. The Levenberg–Marquardt (TRAINLM) training approach resulted in better performance, and hence it is utilized to adjust the weight coefficients for this neural network predictor. During the training phase, the weight adjustments are made in accordance with the error value accumulated at the output layer. In this work, input and output layer have linear transfer function, whereas activation function of sigmoid function (TANSIG) is used for hidden layer. TRAINLM is the network training function adopted to update the neuron weights and bias values according to Levenberg–Marquardt optimization approach.

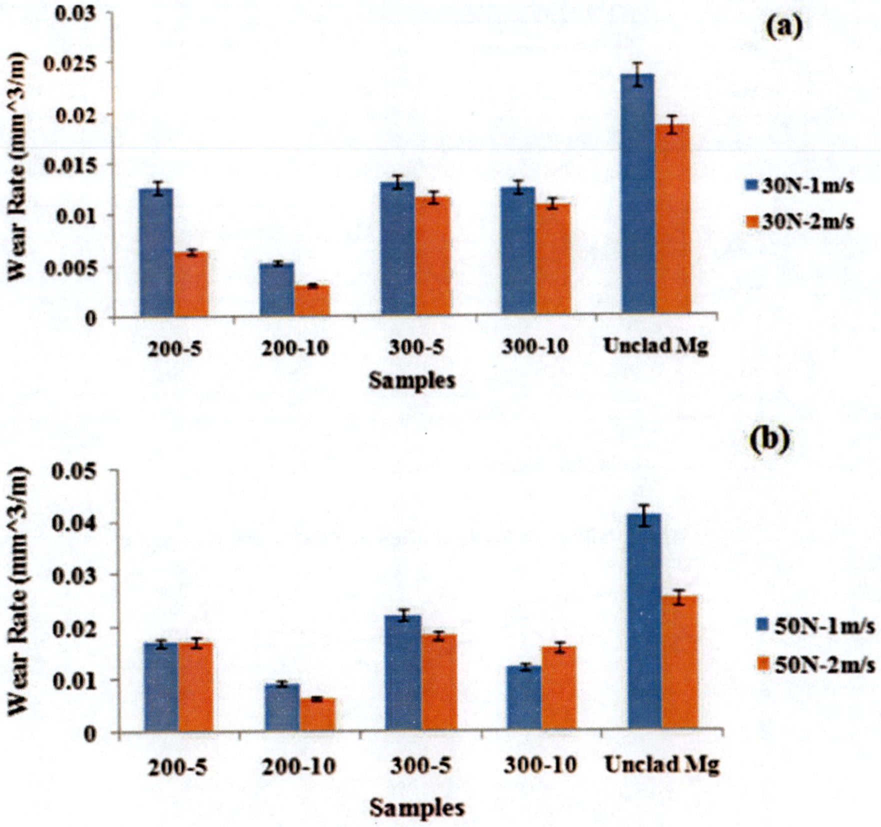

Fig. 2 Wear rate of clad and un-clad substrates tested under **a** 30 N, **b** 50 N

LEARNGDM is manifested as the adaption function based on the weight changes to set the network learning parameters, and here the values for learning rate are 0.02 and 0.9 for momentum constant. From the observations of Table 1, the hidden layer neurons is varied from 5 to 50 and it gives the better performance 25 hidden neurons. This feed-forward back propagation structure with 25 neurons in the hidden layer exhibited minimum mean square error (MSE) value of 1.0246×10^{-12}. And further increase in hidden layer neurons leads to increased MSE value.

The training phase of the neural network for wear and its performance is shown in Figs. 3 and 4, respectively. To foster generalization, the validation datasets were applied to curtail the training process at suitable stage. Here, the network performance is accessed by using the mean squared difference between output and target. The lower value of mean squared error indicates the supercilious network performance and the MSE is zero that there has been no prediction error in the neural network algorithm.

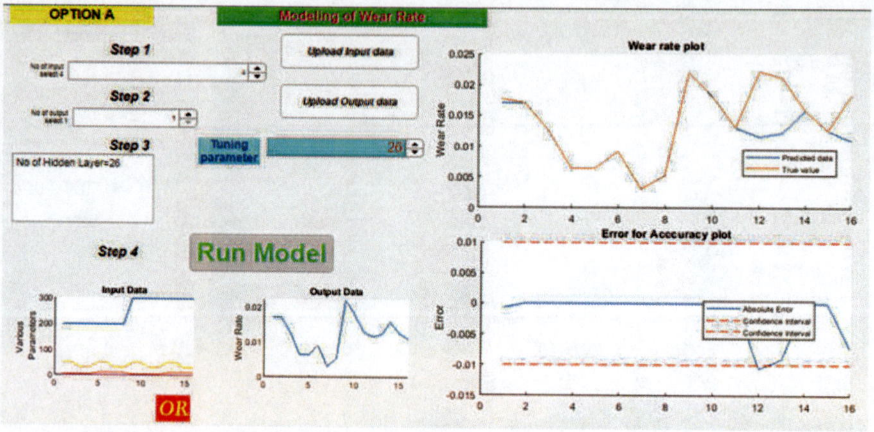

Fig. 3 Neural network training for prediction of wear

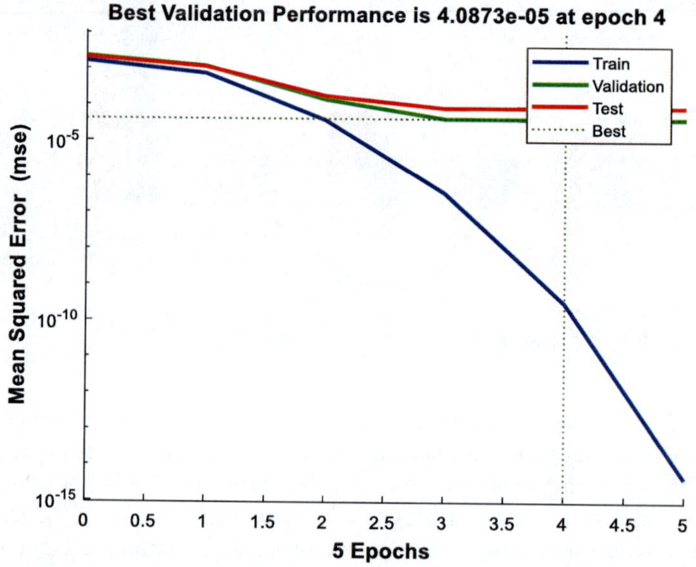

Fig. 4 Neural network training performance

The model was trained and tested to estimate the wear rate of cladding process. A total of 60 data points were used to train and develop the ANN model which was trained up to 5 epochs. It is observed from Fig. 5 the predicted values follow the same pattern of the experimental values under different loading condition. Further, errors between target values and predicted values after training are presented in Fig. 6 as error histogram. These result values indicated that the model could predict the wear behavior of the laser cladded material effectively. Artificial neural network had the average training MSE of 0.00235 and testing MSE 0.00654. This error produced by this estimator is very minimum.

Micro hardness assessment was carried out to illustrate the further insights of the prediction. The materials deposition and micro hardness have affected the wear rate of the clad substrates. Figure 7 shows the SEM images cross-sectional macro structures of clad substrates revealed the materials deposition nature. Higher powder feed rate and lower scan speed laser parameter combination yields a high compact cladding. Likewise, lower powder feed rate and higher scan speed yields low compact cladding.

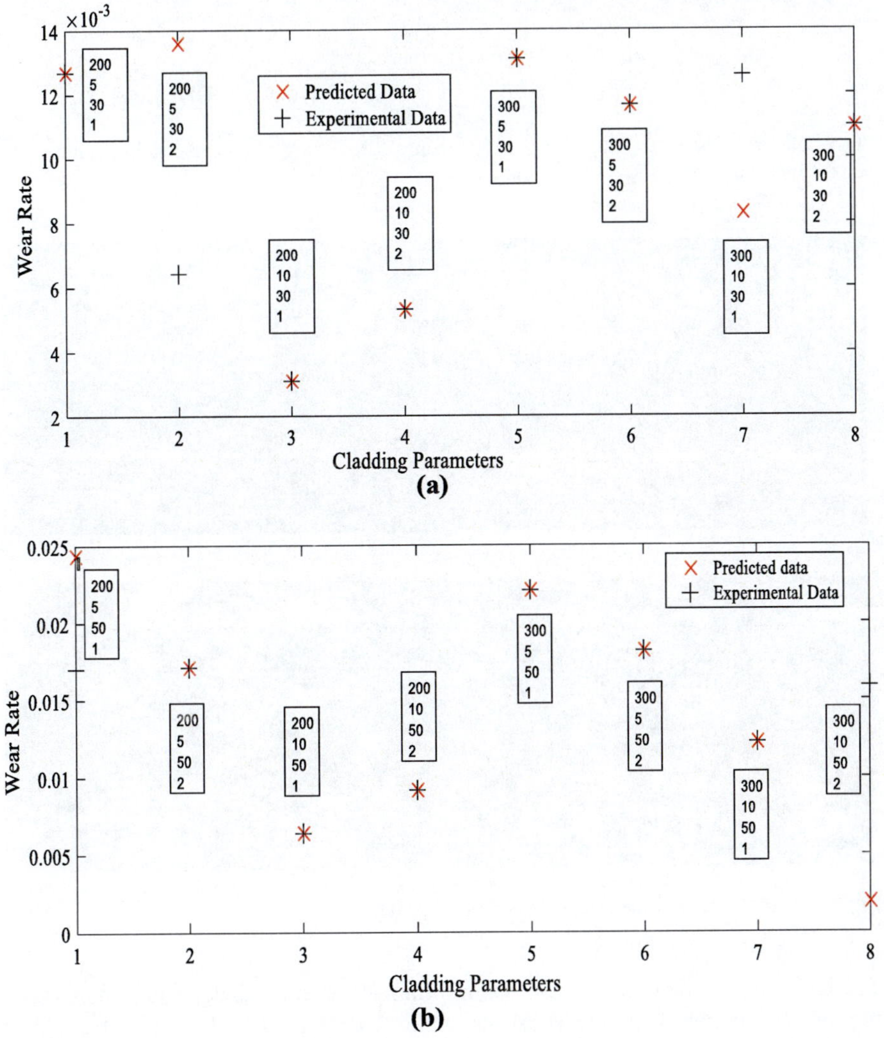

Fig. 5 Prediction of wear rate using neural network under different cladding and loading conditions **a** 30 N–1 m/s and 2 m/s, **b** 50 N–1 m/s and 2 m/s

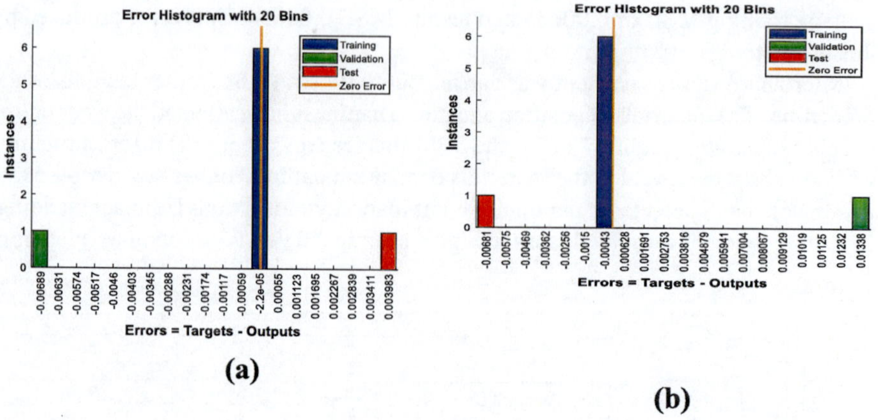

Fig. 6 Neural network training error histogram for different cladding and loading conditions **a** 30 N–1 m/s and 2 m/s, **b** 50 N–1 m/s and 2 m/s

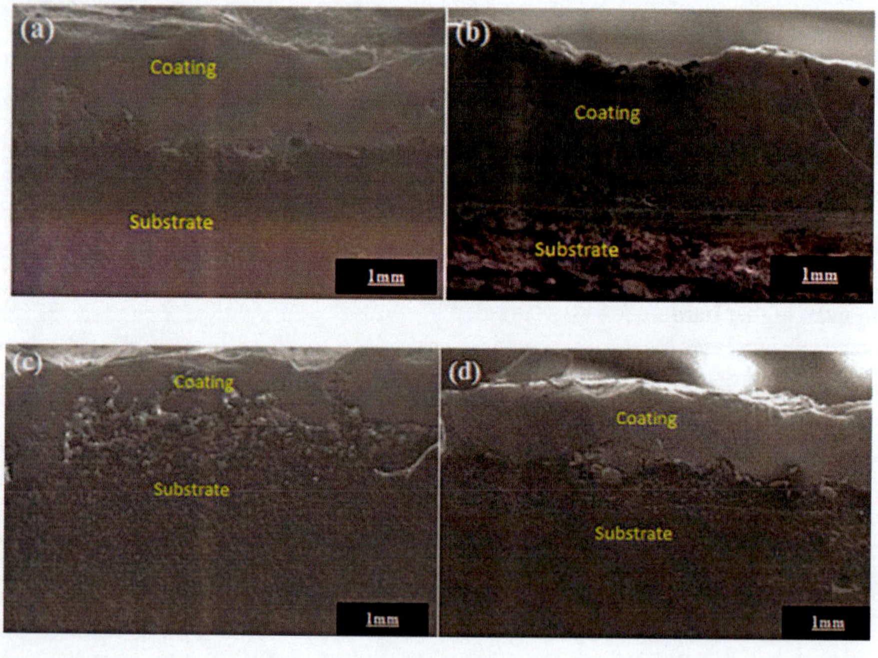

Fig. 7 Microstructure of the clad and un-clad substrates

The laser parametric combination 200 mm/min–10 g/min has deposited more quantity of clad material on the substrate which resulted in high compact cladding (Fig. 7b). Even though the scan speed is less in 200 mm/min–5 g/min combination, the compactness of the clad layer is less (Fig. 7a). This is due to minimum quantity of

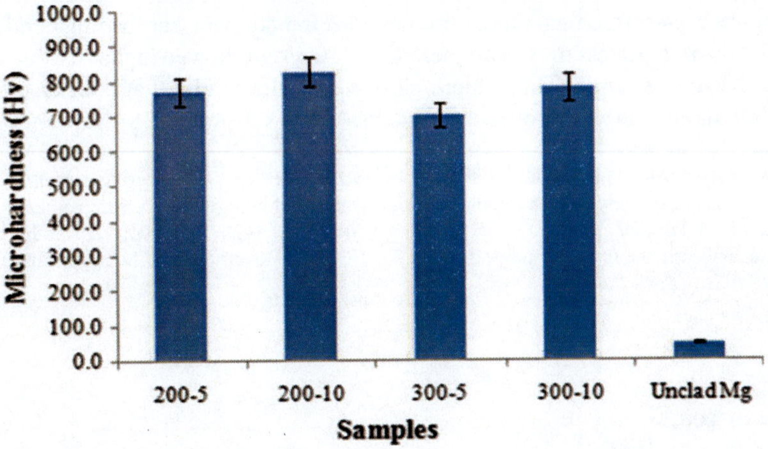

Fig. 8 Microhardness of the clad and un-clad substrates

materials deposition due to lower powder feed rate. The 300 mm/min-5 g/min combination has deposited minimum quantity of clad material due to increased scan speed and lower powder feed rate which resulted in low compact cladding layer (Fig. 7c). However, even at higher scan speed, due to higher feed rate of 10 g/min, a satisfactory compact clad layer has been formed on the substrate (Fig. 7d). More consolidation of clad materials on the substrate improves the hardness of the cladding. Figure 4 shows the micro hardness of the clad and un-clad magnesium substrates. Micro hardness results (Fig. 8) revealed that laser cladding of Ti6Al4V significantly improves the hardness of the substrate by 94%. Micro hardness of the samples clad at 10 g/min showed higher hardness than samples clad at 5 g/min. Higher consolidation of clad materials increases the intensity of the these intermetallic and subsequently increase the micro hardness. Increase in hardness retards the materials loss due to wear. The clad samples with higher hardness showed lower CoF advocates this concept.

4 Conclusion

This paper reports on the wear rate behavior of Ti6Al4V clad magnesium substrates for bio implants application. Materials deposition of the clad samples varies with laser parameters like scan speed and powder feed rate. Wear rate characteristics under different loading conditions were studied, and based on the experimental data, a regularized ANN model was developed to predict the wear resistance. The wear behavior analyzes under experimental and neural topology-based prediction method indicated that higher wear rates were observed in the samples clad at 300 mm/min scan speed. Likewise, samples clad at lower powder feed rate of 5 g/min showed the higher wear rates. Samples clad at 200 m/s–10 g/min combination shows the higher wear resistance than other combinations in both constant loading conditions. Further, micro hardness of the samples clad at 10 g/min showed higher hardness than samples clad at 5 g/min. Thus, micro hardness results exhibit that laser cladding of Ti6Al4V significantly improves the hardness of the substrate by 94%.

Acknowledgements This project is funded by the Science and Engineering Research Board (SERB), a statutory body of Department of Science and Technology (DST), Government of India under National Post-Doctoral Fellowship scheme (File No: PDF/2017/000412). The authors gratefully acknowledge the financial support by the DST-SERB, Government of India for this research work.

References

1. Pommiers S, Frayret J, Castetbon A, Potin-Gautier M (2014) Alternative conversion coatings to chromate for the protection of magnesium alloys. Corros Sci 84:135–146.3
2. Gray JE, Luan B (2002) Protective coatings on magnesium and its alloys—a critical review. J Alloy Compd 336:88–113
3. Yong Z, Zhu J, Qiu C, Liu Y (2008) Molybdate/phosphate composite conversion coating on magnesium alloy surface for corrosion protection. Appl Surf Sci 255:1672–1680
4. Hornberger H, Virtanen S, Boccaccini AR (2012) Biomedical coatings on magnesium alloys—a review. Acta Biomater 8:2442–2455
5. Zhong C, Liu F, Wu Y, Le J, Liu L, He M, Zhu J, Hu W (2012) Protective diffusion coatings on magnesium alloys: a review of recent developments. J Alloy Compd 520:11–21
6. Song W, Martin HJ, Hicks A, Seely D, Walton CA, Lawrimore WB II et al (2014) Corrosion behaviour of extruded AM30 magnesium alloy under salt-spray and immersion environments. Corros Sci 78:353–368
7. Asmussen RM, Jakupi P, Danaie M, Botton GA, Shoesmith DW (2013) Tracking the corrosion of magnesium sand cast AM50 alloy in chloride environments. Corros Sci 75:114–122
8. Yu L, Cao J, Cheng YL (2015) An improvement of the wear and corrosion resistances of AZ31 magnesium alloy by plasma electrolytic oxidation in asilicate–hexametaphosphate electrolyte with the suspension of SiC nanoparticles. Surf Coat Technol 276:266–278
9. Hussein RO, Northwood DO, Su JF, Nie X (2013) A study of the interactive effects of hybrid current modes on the tribological properties of a PEO (plasma electrolytic oxidation) coated AM60B Mg-alloy. Surf Coat Technol 215:421–430
10. Riquelme A, Rodrigo P, Escalera-Rodríguez MD, Rams J (2016) Analysis and optimization of process parameters in Al–SiCp laser cladding. Opt Lasers Eng 78:165–173
11. Zhong M, Liu W (2010) Laser surface cladding: the state of the art and challenges. Proc Inst Mech Eng C: J Mech Eng Sci 224(5):1041–1060
12. Jones SP, Jansen R, Fusaro RL (1997) Preliminary investigation of neural network techniques to predict tribological properties. Tribol Trans 40:312–320
13. Çetinel H, Öztürk H, Çelik E, Karlık B (2006) Artificial neural network-based prediction technique for wear loss quantities in Mo coatings. Wear 261(10):1064–1068

Effect of Forging on Mechanical and Tribological Properties of Aluminium Alloy Composites: A Review

Luckshya Kem, Lakshya Tyagi, Kalpana Gupta, and Shailesh Mani Pandey

Abstract Aluminium is one of the most abundant materials available on the earth surface. Aluminium alloys have vast applications in various sectors like manufacturing, automobile, defence, and aircrafts parts, etc. Aluminium alloys are divided into cast alloy and wrought alloys. Forging is a type of manufacturing process which is done to provide specific shapes to a raw metal, and it is basically classified on the basis of the temperature as hot forging and cold forging. This study reviews the effect of forging on mechanical and tribological properties of aluminium and its alloys and from all mechanical and tribological properties; hardness, tensile strength, and wear are taken into consideration in this study.

Keywords Aluminium · Tribological · Hardness · Hot forging · Cold forging

1 Introduction

Aluminium is most abundant element present in the earth crust about 8.2% of the earth crust comprises of aluminium, but it has poor mechanical and tribological properties that's why pure aluminium cannot be used in industries, i.e., steel and its alloys are used, but it is three times heavier than aluminium but from post twentieth century there is drastic increase in the use of aluminium alloys [1] due to their low cost, high wear resistance and high power-to-weight ratio [2] that's why aluminium alloys always remains in the centre of research work [3] and also used in manufacturing of various components like brake drums, engine piston, and much more [4]. Due to low thermal expansion and conductivity, Al composites are used as connecting rods, automotive drives and cylinder liners [5, 6]. Automobile industries always in search of a material which is light weight and have good set of mechanical and tribological

L. Kem · L. Tyagi · K. Gupta
Department of Mechanical Engineering, Delhi Technological University, New Delhi 110042, India

S. M. Pandey (✉)
National Institute of Technology, Patna, India
e-mail: smp.me@nip.ac.in

A. Maurya et al. (eds.), *Recent Trends in Mechanical Engineering*, Lecture Notes in Mechanical Engineering, https://doi.org/10.1007/978-981-19-7709-1_83

properties; this was fulfilled by aluminium alloys [7]. Al Alloys can be reinforced with many reinforcements like: SiC, B_4C, Al_2O_3, TiC, MgO, TiO_2, etc. [8, 9]. Al alloys can be shaped by using various metal forming processes like casting, extrusion, spraying, and forging [10, 11]. In context of forging, aluminium alloys can be forged into various types of shapes with very broad range [12]. Forging is the process of shaping metal by use of compressive forces. Forging process is classified in three main categories, i.e., hot forging, warm forging, and cold forging [13]. There are two type of working conditions for forging process; it can be hot working condition and cold working condition [14]. Hot working condition is when the temperature of forging is above the recrystallization temperature and cold working condition is when the forging temperature is below the recrystallization temperature [15]. Hot forging and warm forging both done in hot working conditions and cold forging is done is cold working condition. The forging process is economically attractive when a large number of parts must be produced. Though forging process gives superior quality product compared to other manufacturing processes [16]. Components of automobiles like crankshaft, gears, yoke, piston, connecting rod, shaft, etc., are manufactured by forging process [17].

2 Literature Review

2.1 Hardness

Keshavamurthy et al. [18] utilised two samples of Al2024 Alloy and Al2024—5% TiB_2 each. 1st samples of both the material are casted and 2nd is forged. They concluded that (Fig. 1) with the addition of reinforcement the microhardness enhanced. The observed microhardness of forged sample was more than that of casted samples. Pradeep Kumar et al. [22] compared the effect of casting and forging, on microhardness of Al6061 reinforced with TiB_2 particles. Microhardness was more for the forged alloy and microhardness of the material for particular process increased with the increasing content of TiB_2. Pattnaik et al. [23] studied the effect of forging on hardness of aluminium alloy reinforced with 10 wt% SiC reinforcement. It was observed that hardness values decreased and flowability increased. Senapati et al. [24] investigated the effect of hot forging on hardness of aluminium alloy and it was observed that hardness was comparatively improved after forging. Narayan and Rajeshkannan [26] observed the hardness value of hot forged aluminium alloy which was later cooled by direct sinter forged cooling. This cooling was done by two techniques, i.e. water cooled technique and furnace cooled technique. It was found that the hardness value of specimen cooled by water cooled technique was higher when compared with the specimen cooled by furnace cooled technique. It was found that hardness values were higher in water cooled technique. Bharathesh et al. [28] analysed an effect of reinforcement on harness of Al Alloy and hot forged Al-Alloy reinforced with TiO_2. It was observed that as the wt% of reinforcement increases

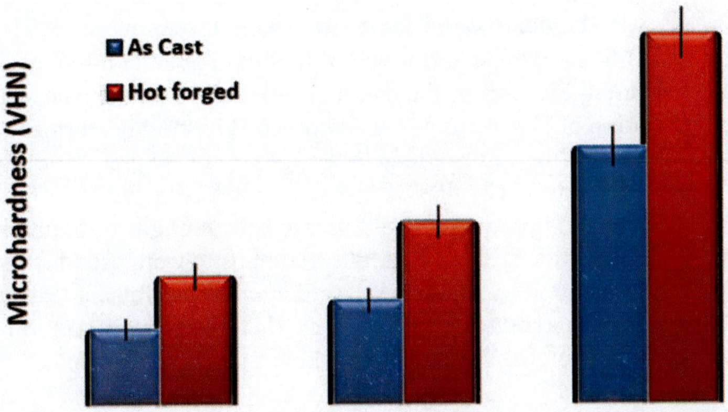

Fig. 1 Effect of TiB_2 addition as reinforcement on microhardness values

the hardness value also increases; another result was that forging of one sample improved its hardness on comparison with unforged Al-Alloy. Ramesh and Keshavamurthy [29] studied the effect of forging process on the hardness and tensile strength of aluminium alloy. Ni–P coated silicon nitride-reinforced Al6061 composite were used for the process, and this was manufactured by the use of liquid metallurgy. Vickers microhardness tester was used for hardness test, and it was found that as the concentration of silicon nitride increased the increment in hardness and tensile strength seen.

2.2 *Tensile Strength*

Adeosun et al. [19] analysed the strength of forged AA1200 aluminium alloy which was reinforced with steel particles. Separate samples were made with different sizes (106, 181, 256, 362, and 512 μm) of steel particles. Stir casting method used in the development of samples after that samples were homogenized, then forged and subsequently annealed. It was found that tensile strength for a sample with particle size 106 μm is maximum which was around 280 MPa. Pattnaik et al. [23] studied the effect of forging on tensile properties of aluminium alloy reinforced with 10 wt% SiC reinforcement. It was observed that tensile strength increased due to forging. Naser et al. [25] analysed a tensile strength of AA-6082 which was forged multiple times upto 5 times, and it was observed that tensile strength increases after first pass of forging then after that there is slight change on each pass of forging. Narayan and Rajeshkannan [26] observed the tensile strength value of hot forged aluminium alloy which was later cooled by direct sinter forged cooling. This cooling was done by two techniques i.e. water cooled technique and furnace cooled technique. It was found that the hardness value of specimen cooled by water cooled technique was higher when compared with the specimen cooled by furnace cooled technique. It was found that hardness values were higher in water cooled technique. Ozdemir et al. [27] reinforced aluminium with silicon carbide and forged it to investigate the effect of forging on tensile strength of Al–SiC composite and it was observed that tensile strength was improved by addition of SiC, but only till addition of 17 wt% SiC, further addition of reinforcement deceases the value of tensile strength. With the use of forging process, tensile strength got increased by 40%. Ramesh and Keshavamurthy [29] studied the effect of forging process on the hardness and tensile strength of aluminium alloy. Ni–P coated silicon nitride reinforced Al6061 composite were used for the process, and this was manufactured by the use of liquid metallurgy. Tensile tests were done as per ASTM A370 standards, and it was found that as the concentration of silicon nitride increased the increment in hardness and tensile strength seen.

2.3 Wear

Keshavamurthy et al. [18] analysed the wear of casted and forged aluminium alloy and aluminium composite having TiB_2 as reinforcement. In both the cases wear loss of forged material was less and also the wear loss decreased with the addition of reinforcement. Krushna et al. [20] reinforced aluminium with agro and industrial, the composite was forged and cast, wear behaviour of both the processed material was analysed. Wear loss was less for the forged material compared to that of casted material. Girish et al. [21] analysed wear behaviour of hot forged AA6061-SiC aluminium alloy, and two observations were made from the study was that as the weight percentage of silicon carbide increased the value of wear resistance decreases, and due to hot forging, wear rate got decreased. Pradeep Kumar et al. [22] studied an influence of hot forging on wear properties of Al6061 reinforced with TiB_2 particles and wear properties of casted and forged alloy were compared. It was found that wear rate decreases on increasing the wt% of reinforcement, and wear rate is comparatively low in the case of forged alloy. Senapati et al. [24] investigate an effect of hot forging on wear rate of aluminium alloy and found that wear rate comparatively decreases after forging. Bharathesh et al. [28] analysed an effect of increasing reinforcement content in hot forged aluminium alloy reinforced with TiO_2. It was observed that in hot forged condition as the wt% of reinforcement increases the value of wear rate decreases. Feyzullahoğlu et al. [30] observed the effect of hot forging on wear resistance of two aluminium alloys, i.e. Al–SiC and Al–Pb. It was found that the wear resistance for Al–Pb alloy was better than the wear resistance of Al–SiC alloy.

3 Conclusion

From Literature Review, we can conclude that:

1. Addition of reinforcement like SiC, TiB_2, Al_2O_3, etc., in aluminium alloy tends to improve the value of hardness and with the application of forging on reinforced aluminium the value of hardness increased even further. So, in terms of hardness forging shows positive effect.
2. In terms of tensile properties, forging method shows some unexpected results and it is observed that when forging is done multiple times and wt% of reinforcement increases in aluminium alloy, value of tensile strength increases to certain limit but after that limit addition of reinforcement and doing forging after that gives negative effect and value of tensile strength tends to decrease.
3. Ductility experienced a reduction when the reinforcements are added to the alloy and even forging contributed to decrease in ductility.
4. Positive effect of adding reinforcement and doing forging can be seen clearly on wear also, that by addition of reinforcement and further application of forging tend to decrease the value of wear rate means material loss is reduced by application of forging.

4 Summary

Researcher name	Material studied	Properties studied		
		Hardness	Tensile	Wear
Keshavamurthy et al. [18]	Al2024, Al2024/TiB_2 both in cast as well as forged condition were studied	Forged sample showed the better hardness compared to the cast material. Material having TiB_2 had better properties compared to non-reinforced samples	Nil	Wear properties of forged material was better and also reinforced samples showed better wear resistance
Adeosun et al. [19]	AA1200 reinforced with steel particles, the composite produced was annealed and forged	Nil	Annealing improved the tensile strength and elongation of the forged material	Nil

(continued)

(continued)

Researcher name	Material studied	Properties studied		
		Hardness	Tensile	Wear
Krushna et al. [20]	AA6061 reinforced with rice husk and fly ash respectively. The composite were subjected to forging after fabrication	Nil	Nil	Wear resistance of forged composite was more than cast. Reinforcing decreased the wear rate of the material and least wear rate was recorded for AA6061/FA composite
Girish et al. [21]	SiC reinforced AA6061 fabricated via stir casting technique. The fabricated samples were then forged	Hardness value of composite was found to increase with the increasing content of SiC particle and also it was more for forged material compared to unforged material	Nil	Wear rate of unforged composite were more than that of forged composite. Wear resistance followed a direct relation with that of reinforcement content
Pradeep Kumar et al. [22]	Al60161-TiB_2 synthesized in in-situ reaction fabricated using stir casting technique. The composite was hot forged	Recorded microhardness was more for forged composite and it increased with the increasing content of reinforcement	Introduction of reinforcement enhanced the tensile strength and it was more for secondary processed material	Coefficient of friction lowered and wear resistance improved with the formation of TiB_2 particles. Compared to cast materials, forged ally and composites showed enhanced tribological properties

(continued)

(continued)

Researcher name	Material studied	Properties studied		
		Hardness	Tensile	Wear
Pattnaik et al. [23]	AA2014 matrix reinforced with SiC (10 wt%) prepared through stir casting was used. The as-cast billet was hammer forged	Highest hardness was recorded at the center position of the forged billet which is attributed to localized deformation and clustering of reinforcement	Elongation to failure and tensile both improved on forging for the produced composite	Nil
Senapati et al. [24]	LM6 alloy reinforced with rice husk and fly ash respectively is studied. The composite is produced via stir casting and later it is hot forged	Hardness increased after forging for all materials and the maximum hardness was recorded for fly ash reinforced LM6 composite	Nil	Wear rate decreased significantly after forging. This might be attributed to increased hardness of the materials
Naser et al. [25]	AA6082 was the material of focus in the study	After first pass hardness of the material increased significantly and on increasing the number of pass the hardness increased slightly with every pass	Tensile strength increased and elongation decreased significantly after first pass and with the following pass strength increased slightly	Nil
Narayan and Rajeshkannan [26]	Aluminium composite reinforced with Al_4TiC, Al_4Fe_3C, Al_4Mo_2C, Al_4WC prepared using die set assembly	Hardness of carbide reinforced aluminium composite was higher for water cooled compared to furnace cooled for direct sinter-forged cooling	Ultimate tensile strength improved after reheat process. UTS and hardness almost followed similar kind of trend	Nil

(continued)

(continued)

Researcher name	Material studied	Properties studied		
		Hardness	Tensile	Wear
Ozdemir et al. [27]	Al–SiC alloy reinforced with particulate SiC produced via die casting are used. They are studied in as cast and hot forged conditions	Nil	Tensile and yield strength increased with the addition of reinforcement till 17%, later addition contributed to decrease both the strengths. Forging process contributed to enhance the tensile and yield strength. Ductility was observed to follow the inverse trend from that of tensile strength	Nil
Bharathesh et al. [28]	Liquid metallurgy route fabricated aluminium composite reinforced with Ni–P coated TiO_2 particles. The composite were subjected to hot forging, later to heat treatment and then quenching in ice media	With the increasing content of TiO_2 microhardness of the hot forged composite enhanced. Maximum hardness resulted after heat treatment and quenching, for a particular material	Nil	Coefficient of friction decreased with the increasing content of reinforcement for all tested materials. It also decreased with the increasing ageing duration

(continued)

(continued)

Researcher name	Material studied	Properties studied		
		Hardness	Tensile	Wear
Ramesh and Keshavamurthy [29]	Liquid metallurgy route fabricated Al6061 composite reinforced with Ni–P coated Si_3N_4, subjected to hot forging was studied	Microhardness increased for both as cast and hot forged composite, when compared that with the matrix alloy. It was highest for the forged sample containing maximum amount of reinforcement	With the increasing content of reinforcement ductility decreased and tensile strength increased. Ductility was more for the hot forged as compared to as cast samples	Nil
Feyzullahoğlu et al. [30]	Al–Si and Al–Pb composite subjected to forging are studied	Head treatment and forging enhanced the hardness of the composite studied	Nil	Wear properties were degraded with forging as compared to cast alloys. The wear resistance of Al–Pb composite were superior of the Al–Si composites

References

1. Shriwas AK, Kale VC (2016) Impact of aluminum alloys and microstructures on engineering properties—review. IOSR J Mech Civ Eng (IOSR-JMCE) 13(3):16–22
2. Stojanovic B, Bukvic M, Epler I (2018) Application of aluminum and aluminum alloys in engineering. Appl Eng Lett: J Eng Appl Sci 3(2):52–62. https://doi.org/10.18485/aeletters.2018.3.2.2
3. Dasgupta R (2012) Aluminium alloy-based metal matrix composites: a potential material for wear resistant applications. ISRN Metall 2012:1–14. https://doi.org/10.5402/2012/594573
4. Singh S, Singh R, Singh Gill S (2014) Development of aluminium MMC with hybrid reinforcement—a review. Mater Sci Forum 808:109–119. https://doi.org/10.4028/www.scientific.net/msf.808.109
5. Khan KB, Kutty TRG, Surappa MK (2006) Hot hardness and indentation creep study on Al–5% Mg alloy matrix–B_4C particle reinforced composites. Mater Sci Eng: A 427(1–2):76–82. https://doi.org/10.1016/j.msea.2006.04.015

6. Auradi V, Rajesh GL, Kori SA (2014) Processing of B_4C particulate reinforced 6061 Aluminum matrix composites by melt stirring involving two-step addition. Procedia Mater Sci 6:1068–1076. https://doi.org/10.1016/j.mspro.2014.07.177
7. Chaudhary A, Dev AK, Goel A, Butola R, Ranganath MS (2018) The mechanical properties of different alloys in friction stir processing: a review. Mater Today: Proc 5(2):5553–5562. https://doi.org/10.1016/j.matpr.2017.12.146
8. Butola R, Tyagi L, Kem L, Ranganath M, Murtaza Q (2020) Mechanical and wear properties of aluminium alloy composites: a review. In: Lecture notes on multidisciplinary industrial engineering, pp. 369–391. https://doi.org/10.1007/978-981-15-4619-8_28
9. Butola R, Singari RM, Bandhu A, Walia RS (2017) Characteristics and properties of different reinforcements in hybrid aluminium composites: a review. IJAPIE 10
10. Shin CS, Huang JC (2010) Effect of temper, specimen orientation and test temperature on the tensile and fatigue properties of SiC particles reinforced PM 6061 Al alloy. Int J Fatigue 32(10):1573–1581. https://doi.org/10.1016/j.ijfatigue.2010.02.015
11. Corrochano J, Walker JC, Lieblich M, Ibáñez J, Rainforth WM (2011) Dry sliding wear behaviour of powder metallurgy Al–Mg–Si alloy-$MoSi_2$ composites and the relationship with the microstructure. Wear 270(9–10):658–665. https://doi.org/10.1016/j.wear.2011.01.029
12. Kuhlman G (2005) Forging of aluminum alloys. https://materialsdata.nist.gov/bitstream/handle/11115/223/Forging%20of%20Aluminum%20Alloys.pdf?sequence=1&isAllowed=y. Accessed 22 Oct 2020
13. Bharti S (2017) Advancement in forging process: a review. Int J Sci Res (IJSR) 6(12):465–468. https://doi.org/10.21275/art20178736
14. Forcellese A, Gabrielli F (2000) Warm forging of aluminium alloys: a new approach for time compression of the forging sequence. Int J Mach Tools Manuf 40(9):1285–1297. https://doi.org/10.1016/s0890-6955(99)00127-3
15. Rathi MG, Jakhade NA (2014) An overview of forging processes with their defects. Int J Sci Res Publ 4(6)
16. Patel B, Thakkar H, Mehta S (2014) Review of analysis on forging defects for quality improvement in forging industries. J Emerg Technol Innov Res (JETIR) 1(7)
17. Nagarale S, Patil A, Dhake R, Gambhire G (2020) Productivity improvement in forging industry using industrial engineering techniques
18. Keshavamurthy R, Ahmed S, Laxman A, Anil Kumar N, Shashidhara M, Reddy Y (2014) Tribological properties of hot forged Al2024-TiB2 in-situ composite. Int J Adv Mater Manuf Charact 4(2):87–92. https://doi.org/10.11127/ijammc.2014.08.02
19. Adeosun S, Oyetunji A, Akpan E (2015) Strength and ductility of forged 1200 aluminum alloy reinforced with steel particles. Niger J Technol 34(4):710. https://doi.org/10.4314/njt.v34i4.7
20. Krushna M, Shekhar P, Kumar S (2019) Effect of hot forging on high temperature tribological properties of aluminium composite reinforced with agro and industrial waste. Int J Eng Adv Technol 8(6):1607–1612. https://doi.org/10.35940/ijeat.f8204.088619
21. Girish B, Shivakumar B, Hanamantraygouda M, Satish B (2020) Wear behaviour of hot forged SiC reinforced aluminium 6061 composite materials. Aust J Mech Eng 1–8. https://doi.org/10.1080/14484846.2020.1714353
22. Pradeep Kumar G, Keshavamurthy R, Kuppahalli P, Kumari P (2016) Influence of hot forging on tribological behavior of Al6061-TiB_2 in-situ composites. IOP Conf Ser: Mater Sci Eng 149:012087. https://doi.org/10.1088/1757-899x/149/1/012087
23. Pattnaik A, Das S, Jha B (2019) Effect of forging on microstructure, mechanical properties and acoustic emission characteristics of Al alloy (2014): 10 wt.% SiCp composite. J Mater Eng Perform 28(5):2779–2787. https://doi.org/10.1007/s11665-019-04094-z
24. Senapati A, Panda S, Mohanta G (2018) Preparation of papers for "the effect of hot forging on mechanical and tribological properties of Al alloy based MMC. 9(4)
25. Naser T, Bobor K, Krállics G (2014) Tensile behavior of multiple forged 6082 Al alloy. Period Polytech Mech Eng 58(2):113–117. https://doi.org/10.3311/ppme.7275
26. Narayan S, Rajeshkannan A (2017) Hardness, tensile and impact behaviour of hot forged aluminium metal matrix composites. J Mater Res Technol 6(3):213–219. https://doi.org/10.1016/j.jmrt.2016.09.006

27. Özdemir İ, Cöcen Ü, Önel K (2000) The effect of forging on the properties of particulate-SiC-reinforced aluminium-alloy composites. Compos Sci Technol 60(3):411–419. https://doi.org/10.1016/s0266-3538(99)00140-2
28. Bharathesh T, Ramesh C, Verma S, Keshavamurthy R (2013) Influence of heat treatment on tribological properties of hot forged Al6061-TiO_2 composites. Int J Emerg Technol Adv Eng 3(6)
29. Ramesh C, Keshavamurthy R (2012) Influence of forging on mechanical properties of Ni–P coated Si_3N_4 reinforced Al6061 composites. Mater Sci Eng: A 551:59–66. https://doi.org/10.1016/j.msea.2012.04.081
30. Feyzullahoğlu E, Ertürk A, Güven E (2013) Influence of forging and heat treatment on wear properties of Al–Si and Al–Pb bearing alloys in oil lubricated conditions. Trans Nonferrous Metals Soc China 23(12):3575–3583. https://doi.org/10.1016/s1003-6326(13)62903-9

A New Journey of Hard Turning with Coated Carbide Insert: A Review

Anupam Alok, Amit Kumar, Shailesh Mani Pandey, Ajit Kumar Pandey, and Manas Das

Abstract Machining with coated carbide tool of hard steel is one of the challenging jobs for any manufacturing industries. The hard turning process is one of the best options than other manufacturing process. The main issue in the hard turning process is the selection of the cutting insert with less cost. Coated carbide insert is the best alternative to the costly cutting insert. This present paper focused on the review on the machining of hard material with coated carbide insert and tries to empathize the crucial dispute associated with this process. Further, this paper focused the previous work on the effect of white layer in the hard machining. Also, the previous work on force modeling in the field of hard turning is highlighted.

Keywords Hard turning · Coated carbide insert · White layer · Force simulation

1 Introduction

Processes like metal cutting, metal forming, metal joining, and metal casting are the pillars of any manufacturing industries. The metal cutting operation is one of the maximum essential production strategies for any industry. The foremost precept of a metallic cutting procedure is the removal of metal through shearing motion because of the relative movement among the workpiece and a tool.

A. Alok · A. Kumar · S. M. Pandey (✉) · A. K. Pandey
National Institute of Technology, Patna 800005, Bihar, India
e-mail: smp.me@nitp.ac.in

A. Alok
e-mail: anupam.me@nitp.ac.in

A. Kumar
e-mail: amit@nitp.ac.in

A. K. Pandey
e-mail: ajit.me@nitp.ac.in

M. Das
Indian Institute of Technology, Guwahati 781039, Assam, India
e-mail: manasdas@iitg.ac.in

A. Maurya et al. (eds.), *Recent Trends in Mechanical Engineering*, Lecture Notes in Mechanical Engineering, https://doi.org/10.1007/978-981-19-7709-1_84

Manufacturing is continuously changing in recent times due to positive key elements like aid drawback, opposition in the global market, and an increase in expectation of clients. The importance of health and environmental factors related to the use of metal cutting fluids is growing. Simultaneously, the demands for dry and near dry machining using wear resistant tool materials are also increasing. Enhancement of surface quality, tight tolerance, and high accuracy in manufacturing are few of the key drivers to the progress of cutting technology. Byrne et al. [9] observed that these factors have a direct impact on cutting tool material, its geometry, cutting parameters, and workpiece material. Machining is one of the most important manufacturing processes in any industry. Machining in broad term encompasses several manufacturing processes like turning, milling, drilling, finishing, grinding, lapping, honing, and so on. Machining processes are widely adopted because of high dimensional tolerances and surface quality that can be attained.

1.1 Hard Turning

Hard tuning is accomplished for material of hardness greater than 45 HRC using an assortment of slanted inserts. Although, for achieving better surface finish at high feed, grinding is a good option, however, hard turning can also produce good surface finish at significantly higher material removal rate (MRR). Worldwide, enormous pressure is there on the manufacturer due to the demand for a better quality of the machined part at a lower price. According to the market demand, everyone looks for the cost-effective manufacturing process with higher product quality. In the automotive industries, generally very hard materials are required. Currently, the predominant method for finishing these parts like gears, shafts, bearings, and pinions is grinding because of its features like easy to operate and control. However, the grinding process has some serious drawback like it takes longer time to finish due to its fixed shape of a bonded wheel. Also, it has an issue with the cost increment due to the use of coolant. The shaped wheel is used as a solution to the first problem. However, it is very expensive. Furthermore, the use of coolant has its own environmental hazards associated with its use and disposal. These issues gave rise to the efforts to develop a cost-effective and environmentally safe process to finish hardened parts to a similar level as of grinding.

1.2 Cutting Tool Material

Cemented carbide is the most widely used insert material. Machining operations which require complex cutting tools like reamers, drills, broaches, tap, milling cutter, etc., use high-speed steel (HSS). HSS is used for 40% of total tool material, and rest 60% considers other materials. Powder metallurgy is the process behind the making of cemented carbides using a certain particular type of carbides like tantalum carbide, titanium carbide, tungsten carbide, and niobium carbide. The percentage of hard particles lies somewhere between 60 and 90%. The International Organization for Standardization (ISO) uses the symbol M (yellow color, alloyed tungsten carbide grades) for machining nickel-based super alloy, steel and ductile cast iron, P (blue color, highly alloyed tungsten carbide grades) for machining steel, and K (red color, straight tungsten carbide) for machining gray cast iron, nonmetallic material, and nonferrous alloys. A number is assigned to each grade in the group to denote its position from maximum hardness to maximum toughness. The rating for M grade is from 10 to 40, for K grade 1 to 40, and P grade 1 to 50. Cobalt enrichment is one of the fields of development in carbide inserts. A higher concentration of cobalt binder improves toughness of outer layer while maintaining core material hard. Collectively cemented carbides, ceramic plus metals are called as 'cermet' comprising hard particles based on titanium carbonatite, titanium carbide, and titanium nitride rather than tungsten carbide (WC). Cermet has higher resistance to abrasive wear, higher chemical stability and hot hardness, and minimal tendency for oxidation wear than cemented carbide. However, it has lesser toughness and strength than cemented carbide. The vast range of cermet's application takes care a certain range of cutting speed and feed.

Ceramic tool materials are used in milling and turning of super alloys, cast iron, and finishing of hard materials. Ceramic tool materials can be further classified on the basis of matrix materials, i.e., Al_2O_3 and Si_3N_4. Ceramics have properties like inertness to chemicals, high hardness, and high resistance to wear. Ceramics minimize wear by chemical erosion and adhesion. High strength is one of the characteristics of these materials that avoid plastic deformation during machining.

PCBN and CBN are harder than ceramics or cemented carbides and have exceptionally high hot hardness. CBN is widely used for machining of ferrous materials when cutting temperature is relatively high. The main chemical and physical properties of CBN are stability at elevated temperature, chemical inertness, and hot hardness. At 750 °C temperature, the hardness of CBN is still equivalent to the of tungsten carbide and oxide ceramics at normal temperature. According to several studies, polycrystalline diamond (PCD) is considered as the hardest material and has a higher resistance to abrasive wear as compared to any other insert materials. However, the main problem with PCD inserts is its high affinity to ferrous materials.

1.3 Cutting Tool Coatings

The demand for economic and environmental-friendly manufacturing processes in the field of sustainable manufacturing is increasing rapidly.

Hence, commercialization in the field of coating of tool materials is also increased. It is considered as one of the best solutions to address the environmental issues faced in wet machining. Figure 1 shows the growth of production in the German machine industry, where, during 3rd millennium's first decade, there has been a remarkable growth in the surface treatment industry.

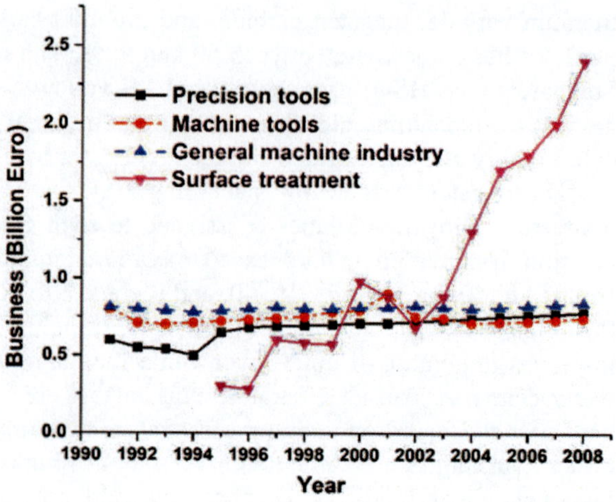

Fig. 1 Growth of production in the German machine industry [12]

Coating technologies for cutting tools have undergone rapid development in the past few years as shown in Fig. 2. In 1980, only TiN coating was used, and until 1988, only CrN and TiCN coatings were used. Fourteen standard coatings were available in 2000. Nowadays, many tool materials are coated with differently structured soft/hard layers of coating to increase gains in tool life and to increase its productivity by applying higher cutting speeds.

Presently, there are 70 different types of coatings with different chemical compositions which are available in the market. The coating can alter the performance of cutting process by increasing resistance to few particular types of tool wear, changing heat generation, friction, or heat flow which is basically the reason behind such fast growth and popularity of coating technology and the coated tools in metal

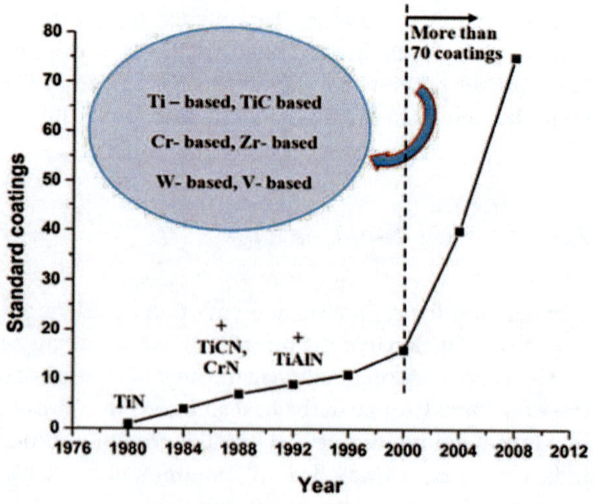

Fig. 2 Standard coatings developed in last three decades [12]

cutting industries. As coating reduces the frictional coefficient, the chip thickness ratio decreases. Therefore, a reduction in the cutting force is observed. Materials used in coating have extremely low solubility in iron at the machining temperature and therefore protect the tool from heat-affected wears such as diffusion, chemical, and oxidation wear.

1.4 White Layer Formation

One of the key elements of hard cutting is the formation of a white layer that requires special attention. The three approaches are aimed at creating a white layer as discussed below.

- The generation of homogeneous structure or very fine grain structure due to plastic flow mechanism.
- The white layer formation is due to the quick heating and cooling mechanism during hard turning.
- A surface reaction such as carburizing and ploughing is another probable reason for the formation of white layer.

The cooling rate also plays a major role in the white layer formation. Because of austenite having higher density than ferrite, the transformation temperature will reduce with specific cutting pressure. The generation of heat caused by high strain rate reduces the necessary stress for deformation. The hardness and the transformation temperature are significantly affected by the quench rate in case of low carbon steel.

2 Literature Review

In the present age of economic growth, the need for industries is to produce high-quality machined parts and components. With the help of a proper selection of machining process and cutting parameters, the required aim can be achieved. In this field, hard turning is one of the suitable machining processes. Hard turning is a machining process which has its own consequences to be considered. It requires an adamant and cost-effective machine tool which produces better surface finish, or at least, comparable to that of grinding.

2.1 Forces, Surface Roughness, and Tool Life

Figure 3 shows some important factors which are associated with machining performance [17]. The three directional forces available in this section are discussed in the available literature, surface quality, and tool life, in which frequently used

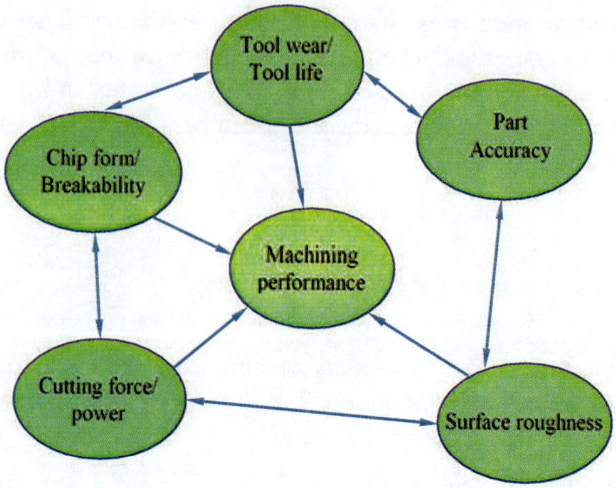

Fig. 3 Representation of the factors influencing machining performance [17]

coated carbide tools are displayed. Cutting forces give a better understanding of the machining process because they are directly related to the cutting conditions and practice conditions during machining.

Softening the workpiece at a high cutting temperature reduces the cutting forces. It has been observed that the radial force is very sensitive to the visibility of the working part: negative rake angle and wear achievement development. By increasing the force, one can understand the relationship of the cutting force with the feed, the depth of the cut, and the radius of the nose. In machining, the cost-effectiveness of a process is a critical factor and therefore requires a basic understanding of the cutting conditions and process variables. Nguyen [32] "analyzed the performance of rotatory tool which is self-propelled, and also, energy-competent optimization of the parameters were done. He focused on the enhancement of machining rate and improvement in energy saving and surface roughness. He found that rotatory tool is very effective in hard machining". Mia and Dhar [23] "performed experiment for the prediction of surface roughness with the help of artificial neural network (ANN)-based surface roughness model. Also, they have analyzed the benefit of high-pressure coolant (HCP) over dry turning. They found that ANN is very accurate for the prediction with a very less % age error. Also, they conclude that HCP helps in improvement in surface roughness". Mia and Dhar [25] "optimized the surface roughness and temperature values with the help of Taguchi technique for three different hard steels (40, 48, and 56 HRC). They were used coated carbide insert for all the experiments and also compared high-pressure coolant jet over dry machining. They found that high-pressure coolant gives a very positive result for the reduction in the surface roughness and cutting temperature". Mia et al. [26] "performed experiment for AISI 1060 steel with multi-coated carbide insert. They have used different cooling systems for better result. Further, they have optimized parameter by using Taguchi L8 orthogonal array".

Alok and Das [2] performed cost-effective hard turning for hard material AISI 52100 (55 HRC) with HSN^2-coated carbide insert. They found that coated carbide insert can successfully machine hard steel. Mia et al. [27] "applied smart manufacturing, i.e., evolutionary algorithms for the optimization of cutting parameters during machining of AISI 1060 hard steel with coated (TiCN/Al_2O_3/TiN) carbide insert". Panda et al. [36] "studied flank wear during hard turning of AISI 52100 steel by using both multi-coated carbide and mixed ceramic insert at different cutting conditions. They found that the value of flank wear for both the tools is within acceptable limit of 0.3 mm". Kumar et al. [18] "observed the performance of both uncoated and Al_2O_3-coated carbide inserts for hard turning of AISI D2 steel having hardness 55 HRC. They found that the uncoated tool completely failed during machining". Sampath Kumar et al. [45] "analyzed machining performance while turning Inconel 825 alloy with TiAlN, AlCrN, TiAlN/AlCrN-coated carbide insert and compared with uncoated carbide insert. They found that the performance of TiAlN/AlCrN-coated insert is better in terms of forces, surface roughness, and the tool wear". Sahoo and Sahoo [40] "studied surface roughness, chip morphology, cutting force, and flank wear during hard turning of AISI 4340 steel with uncoated as well as two different multilayer coated carbide tools (TiN/Al_2O_3/TiN/TiCN and TiN/TiCN/ZrCN/Al_2O_3). They found higher tool life of multilayer TiN/Al_2O_3/TiN/TiCN-coated carbide tools as compared to TiN/TiCN/ZrCN/Al_2O_3-coated carbide and uncoated inserts. Also, the cutting temperature in case of multilayer coated carbide inserts was relatively lower. Further, they developed a regression model which can predict surface roughness of cold worked high chromium tool steel (22 HRC) with coated carbide tools during turning operation. From ANOVA, they concluded that surface roughness is mostly affected by feed followed by cutting speed and depth of cut". Mia et al. [22] "analyzed the effect of minimum quantity lubrication during machining of 40 HRC hard material (AISI 1060 steel) with coated carbide insert. Also, they used Taguchi-based signal-to-noise (S/N) ratio for the optimization of surface roughness, tool wear, and material removal rate (MRR)". Mia and Dhar [24] "developed predictive models for temperature (workpiece–tool interface) with the help of response surface methodology (RSM) and artificial neural network (ANN) during hard turning of AISI 1060 steels with coated carbide insert. They found that ANN is more accurate during the present analysis".

Fang et al. [15] "compared and analyzed the effect of both sharp and round-edged tools in orthogonal cutting. From their analysis, it was inferred that cutting force variation is more stable in case of sharp tool than the round edge tool". Li et al. [19] "studied the progression of tool failure during milling of Ti-6Al-4V at high speed with multilayer CVD-coated carbide tool. An increase in cutting force and cutting temperature was observed as tool wear progresses". Asiltürk and Akkuş [6] "did optimization of machining parameters, using Taguchi method to reduce surface roughness during turning process of hardened AISI 4140 steel with TiC and Al_2O_3-coated carbide inserts". They summarized that the factors affecting surface roughness are feed rate, cutting [41, 42, 46] speed, and depth of cut. Different researchers analyzed how the characteristics of TiN coatings deposited on tungsten carbide tools are being influenced when coating thickness is varied between 1.8 and 6.8 μm during machining of carbon steels. The best turning performance was found in case of TiN coating with thickness of 3.5 μm on carbide inserts. They came up with the conclusion that the reason behind the coating failure in thicker coating is because of higher level of compressive stresses due to the chipping off of the cutting edge in coating material. Cakir et al. [10] "investigated how different coating layers and various cutting parameters affect output responses such as workpiece surface roughness. For this analysis, they used multi-layer CVD-coated Al_2O_3/TiN/TiCN and single-layered PVD-coated TiAlN while turning AISI P20 tool steel having hardness of 52–55 HRC. They found that the most influencing parameter is feed followed by cutting speed for surface roughness. However, depth of cut was found as insignificant factor". The better surface finish was observed with PVD-coated single-layer TiAlN inserts by some group of researchers [14, 16, 39] "while machining titanium and nickel-based alloys. Few researchers also observed that tool life is improved in case of multi-layer CVD coatings" [47]. However, in another analysis by some group of researchers [13, 31], it was found that thin layer of TiAlN/TiN or TiN/(Ti,Al,Si)N/TiN-coated tools having coating thickness less than 4 μm performs almost same or far better than cemented carbide inserts having thick layer (greater than 10 μm) of TiCN/Al_2O_3/TiN coating. During hard turning, Aouici et al. [4] "examined the effect of different parameters (workpiece hardness, feed, cutting speed, and depth of cut) on surface roughness and cutting force. AISI H11 steel was hardened to 40, 45, and 50 HRC and further turned using CBN inserts. Further, a regression model for surface roughness was developed using the RSM technique. Their experimental results indicated that the most important parameters that affected cutting force were depth of cut and workpiece hardness". However, surface roughness was greatly influenced by feed and workpiece hardness. Pal et al. [35] "concluded that tool wear detection must be precise as machining with a dull tool influences the machining performance, and at the same time, economy of the machining process is greatly influenced by replacing the tool before its actual tool life. Also, an optimum cutting condition was obtained by progressive tool wear information". Mia et al. [28] "did experiment for hard steel (AISI 1060 steel) with uncoated carbide tool. They have used time-control MQL and analyzed its effect on responses. They found that MQL with one second time gap gives best result". Mia et al. [29] "did the expedient for the optimization of parameter which affects the tool–chip interaction. For the optimization, they have used Taguchi (gray) technique. They performed all the experiment in both, i.e., dry and MQL conditions. They found that cutting speed is the most dominant factor during the all experiments".

2.2 *White Layer*

The white layer is the most important feature of hard turning needs special attention. Research is going on regarding the formation of white layer for different machining processes. Aramcharoen et al. [5] "analyzed the formation of white layer during hard turning of H13 tool steel (57 HRC) with two different types of coated carbide inserts, i.e., CrTiAlN and CrTiAlN/MoST. They found that at lower speed, the value of white layer thickness is 2.0 and 1.7 μm for uncoated and CrTiAlN-coated carbide inserts, respectively. Also, for CrTiAlN/MoST-coated carbide insert, an intermittent thickness of white layer as compared to above came out at lower speed. At higher cutting speed keeping other parameters constant, no white layer is detected". Chou and Evans [11] "analyzed white layer depth during turning of AISI 52100. They found that white layer depth initially increases with cutting speed, and at a critical cutting speed, the formation of white layer is reduced or saturated". Bosheh and Mativenga [8] "analyzed the formation of white layer at different cutting speeds. They also observed the effect of white layer on the workpiece hardness. They concluded that the depth of white layer decreases at higher cutting speed". Ramesh et al. [38] "observed the formation of white layer during hard turning of AISI 52100 hard steel. They found that at low-to-medium cutting speed, the generation of white layer is due to the grain refinement which leads to plastic deformation". Shi et al. [44] "analyzed that the white layer thickness for machining of AISI 5200 hard steel with ceramics as well as CBNcutting inserts for both dry and wet conditions at different cutting speeds and feeds. Also, the nose radius and rake angle are varied during experimentation. They found that white layer thickness increases with increased cutting speed, feed, and nose radius. Also, white layer thickness decreases for wet condition". Bartarya and Choudhury [7] "observed that the white layer produced on workpiece subsurface during hard turning of H13 steel with CBNtool was harder than the bulk material".

Alok and Das [3] "analyzed white layer thickness for hard steel AISI 52100 at high speed. They found that the thickness of white layer decreases at high cutting speed because of reduction in temperature of workpiece. Figure 4 shows the trend of white layer thickness with respect to cutting speed".

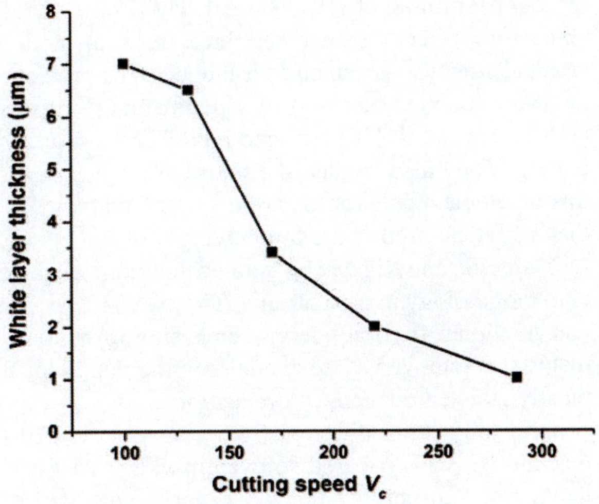

Fig. 4 Variation of white layer thickness at different cutting speeds [3]

2.3 Force Modeling

Lotfi et al. [20] "conducted 3D FEM simulation of tool wear during ultrasonic-assisted rotary turning of AISI 4140 steel, and the results are compared with conventional turning process. Johnson–Cook material's model is used for defining material properties. They investigated heat generation, surface roughness, and cutting force for vibratory-rotary motion. They found that this method is very useful and the calculated value of tool wear and heat generation is low".

Pervaiz et al. [37] "performed 2D orthogonal FEM-based simulation for machining Ti6Al4V for predicting forces, temperature, and Von–Misses stress. They used two different material models like power law (PL) and Johnson–Cook (J–C) material model. They found that PL model provided segmented chip (realistic saw tooth) formation".

Akbar et al. [1] "validated experimental results with the simulated one for forces and chip thickness while turning AISI/SAE 4140 steel using carbide inserts. For simulation, they used a 2D orthogonal model and fully coupled thermomechanical FEM model. They studied heat partition between tool and chip. They concluded that heat partition plays a great role in determining stresses, chip morphology, chip–tool interface temperature, and tool–chip contact length".

Xie et al. [48] "used ABAQUS® FEM code with CPE4RT element to predict tool wear. They used three steps during simulation, in first two steps, the tool was considered as rigid, and in third step, they considered it as deformable".

Shi and Liu [43] "compared different material models, i.e., Johnson–Cook, Litonski–Batra, Bodner–Partom, and power law models using ABAQUS 6.2 for orthogonal turning of HY-100 steel. They found that Litonski–Batra model predicts some forces better than other three material models. The chip morphology and residual stresses vary in all four models. The prediction of chip thickness and shear angle are found to be closer to experimental results for all four models".

Miguélez et al. [30] "studied both FEM and analytical modeling for orthogonal cutting. They also studied the both Lagrangian and ALE approaches. They found that no element deletion criterion is required for ALE formulation. They concluded that analytical models are time saving".

Mabrouki and Rigal [21] "studied thermomechanical analysis for the prediction of chip morphology in hard turning. They studied varying thermal contact conductance and fractional friction energy conversion to heat. In this study, they showed that material is removed by adiabatic shearing due to localized deformation. The chip is mostly affected due to heat evacuation".

Özel [33] "used different friction models in FEM formulation to analyze hard machining. A FEM-based software package, Deform®-2D, was used for the simulation while machining LCFCS material with P20 grade carbide tool. Lagrangian method with implicit integration technique was used during simulation. He found that the chip–tool interface friction models have significant effect on the chip morphology, as well as on forces and stress distribution".

Özel and Zeren [34] "did dynamic–explicit–arbitrary Lagrangian (ALE) formulation based for FEM analysis of high-speed machining of AISI 1045, 4340 steels, and Ti6Al4V with round edge cutting tools. From simulation, they found a very high deformation zone near tool nose area. They predicted very high and localized temperature at the tool–chip interface due to the application of proper friction model".

3 Conclusions

The following facts were observed after reviewing the works discussed above.

On the basis of the above review, it is concluded that the radial force is sensitive force in all three direction forces in the hard turning process. It is dominated on other forces during hard turning.

- There is a huge research gap in the field of coated tool after the year of 2000. Up to year 2000, researchers claimed that TiAlN is the best coating material for machining.
- Very less number of previous work is available for hard turning with coated carbide insert having coating thickness more than 4–5 μm.
- Very less number of previous work is available for turning of hard material having hardness more than 50 HRC with coated carbide insert.
- From above literature, TiAlN coating material gives best result in all available coating materials in the field of hard turning.
- From literature, it is concluded that the white layer is a drawback and can be controlled by the proper selection of the cutting parameters.
- Past research showed that some software package is available like ABAQUS® and deforms 3D or 2D for the force simulation in the field of hard turning.

References

1. Akbar F, Mativenga PT, Sheikh MA (2010) An experimental and coupled thermo-mechanical finite element study of heat partition effects in machining. Int J Adv Manuf Technol 46(5–8):491–507. https://doi.org/10.1007/s00170-009-2117-5
2. Alok A, Das M (2018) Cost-effective way of hard turning with newly developed HSN^2-coated tool. Mater Manuf Process 33(9). https://doi.org/10.1080/10426914.2017.1388521
3. Alok A, Das M (2019) White layer analysis of hard turned AISI 52100 steel with the fresh tip of newly developed HSN^2 coated insert. J Manuf Process 46:16–25. https://doi.org/10.1016/J.JMAPRO.2019.08.016
4. Aouici H, Yallese MA, Chaoui K, Mabrouki T, Rigal J-F (2012) Analysis of surface roughness and cutting force components in hard turning with CBN tool: prediction model and cutting conditions optimization. Measurement 45(3):344–353. https://doi.org/10.1016/J.MEASUREMENT.2011.11.011
5. Aramcharoen A, Mativenga PT, Manufacturing and Laser Processing Group (2008) White layer formation and hardening effects in hard turning of H13 tool steel with CrTiAlN and

CrTiAlN/MoST-coated carbide tools. Int J Adv Manuf Technol 36(7–8):650–657. https://doi.org/10.1007/s00170-006-0899-2
6. Asiltürk İ, Akkuş H (2011) Determining the effect of cutting parameters on surface roughness in hard turning using the Taguchi method. Measurement 44:1697–1704. https://doi.org/10.1016/j.measurement.2011.07.003
7. Bartarya G, Choudhury SK (2016) Effect of tool wear on white layer thickness and subsurface hardness on hard turned EN31 steel. Int J Mach Mach Mater 18(5/6):483. https://doi.org/10.1504/ijmmm.2016.078993
8. Bosheh SS, Mativenga PT (2006) White layer formation in hard turning of H13 tool steel at high cutting speeds using CBN tooling. Int J Mach Tools Manuf 46(2):225–233. https://doi.org/10.1016/J.IJMACHTOOLS.2005.04.009
9. Byrne G, Dornfeld D, Inasaki I, Ketteler G, König W, Teti R (1995) Tool condition monitoring (TCM)—the status of research and industrial application. CIRP Ann 44(2):541–567. https://doi.org/10.1016/S0007-8506(07)60503-4
10. Cakir MC, Ensarioglu C, Demirayak I (2009) Mathematical modeling of surface roughness for evaluating the effects of cutting parameters and coating material. J Mater Process Technol 209(1):102–109. https://doi.org/10.1016/J.JMATPROTEC.2008.01.050
11. Chou YK, Evans CJ (1999) White layers and thermal modeling of hard turned surfaces. Int J Mach Tools Manuf 39(12):1863–1881. https://doi.org/10.1016/S0890-6955(99)00036-X
12. Cselle T, Coddet O, Galamand C, Holubar P, Jilek M, Jilek J, Morstein M et al (2009) Triple-Coatings—new generation of PVD-coatings for cutting tools. J Mach Manuf 49(E1):19–25
13. Dobrzański LA, Gołombek K (2005) Structure and properties of the cutting tools made from cemented carbides and cermets with the TiN + mono-, gradient- or multi(Ti, Al, Si)N + TiN nanocrystalline coatings. J Mater Process Technol 164–165:805–815. https://doi.org/10.1016/J.JMATPROTEC.2005.02.072
14. Ezugwu EO, Okeke CI (2002) Behavior of coated carbide tools in high speed machining of a nickel base alloy. Tribol Trans 45(1):122–126. https://doi.org/10.1080/10402000208982530
15. Fang N, Pai PS, Mosquea S (2011) A comparative study of sharp and round-edge tools in machining with built-up edge formation: cutting forces, cutting vibrations, and neural network modeling. Int J Adv Manuf Technol 53(9–12):899–910. https://doi.org/10.1007/s00170-010-2887-9
16. Ibrahim GA, Che Haron CH, Ghani JA (2009) The effect of dry machining on surface integrity of titanium alloy Ti-6Al-4V ELI. J Appl Sci 9(1):121–127. https://doi.org/10.3923/jas.2009.121.127
17. Jawahir IS (1988) The chip control factor in machinability assessments: recent trends. J Mech Work Technol 17:213–224. https://doi.org/10.1016/0378-3804(88)90023-X
18. Kumar R, Sahoo AK, Mishra PC, Das RK (2018) Comparative investigation towards machinability improvement in hard turning using coated and uncoated carbide inserts: part I experimental investigation. Adv Manuf 6(1):52–70. https://doi.org/10.1007/s40436-018-0215-z
19. Li A, Zhao J, Luo H, Pei Z, Wang Z (2012) Progressive tool failure in high-speed dry milling of Ti-6Al-4V alloy with coated carbide tools. Int J Adv Manuf Technol 58(5–8):465–478. https://doi.org/10.1007/s00170-011-3408-1
20. Lotfi M, Amini S, Aghaei M (2018) 3D FEM simulation of tool wear in ultrasonic assisted rotary turning. Ultrasonics 88:106–114. https://doi.org/10.1016/j.ultras.2018.03.013
21. Mabrouki T, Rigal J-F (2006) A contribution to a qualitative understanding of thermo-mechanical effects during chip formation in hard turning. J Mater Process Technol 176(1–3):214–221. https://doi.org/10.1016/j.jmatprotec.2006.03.159
22. Mia M, Dey PR, Hossain MS, Arafat MT, Asaduzzaman M, Shoriat Ullah M, Tareq Zobaer SM (2018) Taguchi S/N based optimization of machining parameters for surface roughness, tool wear and material removal rate in hard turning under MQL cutting condition. Meas: J Int Meas Confed 122:380–391. https://doi.org/10.1016/j.measurement.2018.02.016
23. Mia M, Dhar NR (2016) Prediction of surface roughness in hard turning under high pressure coolant using artificial neural network. Measurement 92:464–474. https://doi.org/10.1016/j.measurement.2016.06.048

24. Mia M, Dhar NR (2016) Response surface and neural network based predictive models of cutting temperature in hard turning. J Adv Res 7(6):1035–1044. https://doi.org/10.1016/j.jare.2016.05.004
25. Mia M, Dhar NR (2017) Optimization of surface roughness and cutting temperature in high-pressure coolant-assisted hard turning using Taguchi method. Int J Adv Manuf Technol 88(1–4):739–753. https://doi.org/10.1007/s00170-016-8810-2
26. Mia M, Gupta MK, Singh G, Królczyk G, Pimenov DY (2018) An approach to cleaner production for machining hardened steel using different cooling-lubrication conditions. J Clean Prod 187:1069–1081. https://doi.org/10.1016/j.jclepro.2018.03.279
27. Mia M, Królczyk G, Maruda R, Wojciechowski S (2019) Intelligent optimization of hard-turning parameters using evolutionary algorithms for smart manufacturing. Materials 16(6). https://doi.org/10.3390/ma12060879
28. Mia M, Razi MH, Ahmad I, Mostafa R, Rahman SMS, Ahmed DH, Dhar NR et al (2017) Effect of time-controlled MQL pulsing on surface roughness in hard turning by statistical analysis and artificial neural network. Int J Adv Manuf Technol 91(9–12):3211–3223. https://doi.org/10.1007/s00170-016-9978-1
29. Mia M, Rifat A, Tanvir MF, Gupta MK, Hossain MJ, Goswami A (2018) Multi-objective optimization of chip-tool interaction parameters using Grey-Taguchi method in MQL-assisted turning. Measurement 129:156–166. https://doi.org/10.1016/J.MEASUREMENT.2018.07.014
30. Miguélez H, Zaera R, Rusinek A, Moufki A, Molinari A (2006) Numerical modelling of orthogonal cutting: influence of cutting conditions and separation criterion. J Phys IV (Proc) 134:417–422. https://doi.org/10.1051/jp4:2006134064
31. Moreno LH, Ciacedo JC, Martinez F, Bejarano G, Battaille TS, Prieto P (2010) Wear evaluation of WC inserts coated with TiN/TiAlN multinanolayers. J Braz Soc Mech Sci Eng 32(2):114–118. https://doi.org/10.1590/S1678-58782010000200003
32. Nguyen TT (2020) An energy-efficient optimization of the hard turning using rotary tool. Neural Comput Appl 1–24. https://doi.org/10.1007/s00521-020-05149-2
33. Özel T (2006) The influence of friction models on finite element simulations of machining. Int J Mach Tools Manuf 46(5):518–530. https://doi.org/10.1016/J.IJMACHTOOLS.2005.07.001
34. Özel T, Zeren E (2005) Finite element modeling of stresses induced by high speed machining with round edge cutting tools. Manufacturing Manuf Eng Mater Handl, Parts A and B 2005:1279–1287. https://doi.org/10.1115/IMECE2005-81046
35. Pal S, Heyns PS, Freyer BH, Theron NJ, Pal SK (2011) Tool wear monitoring and selection of optimum cutting conditions with progressive tool wear effect and input uncertainties. J Intell Manuf 22(4):491–504. https://doi.org/10.1007/s10845-009-0310-x
36. Panda A, Sahoo AK, Rout AK, Kumar R, Das RK (2018) Investigation of flank wear in hard turning of AISI 52100 grade steel using multilayer coated carbide and mixed ceramic inserts. Procedia Manuf 20:365–371. https://doi.org/10.1016/J.PROMFG.2018.02.053
37. Pervaiz S, Anwar S, Kannan S, Almarfadi A (2018) Exploring the influence of constitutive models and associated parameters for the orthogonal machining of Ti6Al4V. IOP Conf Ser: Mater Sci Eng 346(1):012058. https://doi.org/10.1088/1757-899X/346/1/012058
38. Ramesh A, Melkote SN, Allard LF, Riester L, Watkins TR (2005) Analysis of white layers formed in hard turning of AISI 52100 steel. Mater Sci Eng, A 390(1–2):88–97. https://doi.org/10.1016/J.MSEA.2004.08.052
39. Ramesh S, Karunamoorthy L, Senthilkumar VS, Palanikumar K (2009) Experimental study on machining of titanium alloy (Ti64) by CVD and PVD coated carbide inserts. Int J Manuf Technol Manag 17(4):373. https://doi.org/10.1504/IJMTM.2009.023954
40. Sahoo AK, Sahoo B (2012) Experimental investigations on machinability aspects in finish hard turning of AISI 4340 steel using uncoated and multilayer coated carbide inserts. Measurement 45(8):2153–2165. https://doi.org/10.1016/J.MEASUREMENT.2012.05.015
41. Sargade VG, Gangopadhyay S, Paul S, Chattopadhyay AK (2011) Effect of coating thickness on the characteristics and dry machining performance of TiN film deposited on cemented carbide inserts using CFUBMS. Mater Manuf Process 26(8):1028–1033. https://doi.org/10.1080/10426914.2010.526978

42. Saxena A, Singh N, Kumar D, Gupta P (2017) Effect of ceramic reinforcement on the properties of metal matrix nanocomposites. Mater Today: Proc 4(4):5561–5570. https://doi.org/10.1016/J.MATPR.2017.06.013
43. Shi J, Liu CR (2004) The influence of material models on finite element simulation of machining. J Manuf Sci Eng 126(4):849. https://doi.org/10.1115/1.1813473
44. Shi J, Wang J-Y, Liu CR (2006) Modelling white layer thickness based on the cutting parameters of hard machining. Proc Inst Mech Eng, Part B: J Eng Manuf 220(2):119–128. https://doi.org/10.1243/095440505X32977
45. Sampath Kumar T, Ramanujam R, Vignesh M, Tamiloli N, Sharma N, Srivastava S, Patel A (2018) Comparative evaluation of performances of TiAlN, AlCrN, TiAlN/AlCrN coated carbide cutting tools and uncoated carbide cutting tools on turning Inconel 825 alloy using Grey Relational Analysis. Sens Actuators A: Phys 279:331–342. https://doi.org/10.1016/J.SNA.2018.06.041
46. Tuffy K, Byrne G, Dowling D (2004) Determination of the optimum TiN coating thickness on WC inserts for machining carbon steels. J Mater Process Technol 155–156:1861–1866. https://doi.org/10.1016/J.JMATPROTEC.2004.04.277
47. Ulutan D, Ozel T (2011) Machining induced surface integrity in titanium and nickel alloys: a review. Int J Mach Tools Manuf 51:250–280. https://doi.org/10.1016/j.ijmachtools.2010.11.003
48. Xie L-J, Schmidt J, Schmidt C, Biesinger F (2005) 2D FEM estimate of tool wear in turning operation. Wear 258(10):1479–1490. https://doi.org/10.1016/J.WEAR.2004.11.004

A Critical Review of Thermal-Barrier Coatings and Critical Examination on Post Heat Treatment

Shubhangi Chourasia, Ankit Tyagi, Shailesh Mani Pandey, and Qasim Murtaza

Abstract In the present scenario, industries are more concerning about pollution-free environment, global warming, and endeavor to produce goods which causes less hazardous to the environment. Thermal-barrier coatings (TBCs) are a crucial method to protect the metallic parts of components against high temperatures of more than 1000 °C. Through the Thermal Barrier Coating provide extended life and excellent performance to components in the domain of aviation, automobiles, marines, and power generation by reducing the temperature of components by providing various thermal layers. These review works are divided into two parts: the first part is based on a critical review of thermal-barrier coatings materials used, coating process, the framework of TBC, challenges faced in TBCs, and detailed study of thermally grown oxides, and the second part is based on the effect of post-treatment on TBC and high-velocity oxygen fuel (HVOF) coatings.

Keywords Thermal-barrier coatings · Challenges · Post heat treatment

1 Introduction

The demand for non-renewable sources like crude petroleum oil is increasing rapidly since it has been discovered. The usage of crude petroleum will surely rise year after year. It was observed from the last conducted many surveys that the total annual consumption of petroleum in India has increased from 10.33 million tons in the year of 2007–2008 to 23.76 million tons in the year of 2016–2017 with a compound annual increment rate of 8.69 per year. At the same time, the use of diesel oil has

S. Chourasia · A. Tyagi (✉)
Department of Mechanical Engineering, SGT University, Haryana, India
e-mail: tyagiankit10506@yahoo.com

S. M. Pandey (✉)
Department of Mechanical Engineering, National Institute of Technology, Patna, India
e-mail: smp.me@nitp.ac.in

S. Chourasia · A. Tyagi · Q. Murtaza
Department of Mechanical Engineering, Delhi Technological University, Delhi, India

A. Maurya et al. (eds.), *Recent Trends in Mechanical Engineering*, Lecture Notes in Mechanical Engineering, https://doi.org/10.1007/978-981-19-7709-1_85

raised from 47.67 million tons to 76.03 million tons in the same respective years with a growth of 4.78 every successive year [1]. That is why it is the need for time to improve the efficiency of IC engines to support the environment, to reduce the emission of greenhouse gases as well as economic benefits [2]. Thermal-barrier coatings are made up of refractory oxide ceramics which are applied to the surface of hot metallic parts of engines to protect them from heat, wear, hot corrosion, etc. It has been found from many conducted studies that the efficiency of engines depends upon the peak cylinder pressure; as it increases, efficiency also increases with an increase in combustion temperature [2] which causes the reason defects in material such as hot corrosion, microstructure damage, creep [3]. For protecting the components of the engine, TBCs can be employed with a thickness of 120–400 μm approximately which tends to protect the metallic components from high temperature and can be used above the melting point. Besides this, TBC also protects the hot metallic parts of engines from hot corrosion, hazardous thermal stress, wear as well as also assists the engine components from temperature gradient which are continuously passed to an adverse environment. Additionally, spark-ignition engines can also attain excellent thermal efficiency by use of thermal-barrier coatings for components of the ignition system and also provide insulation to the braking system of automobiles during high temperature [4]. For achieving effective TBCs, the common necessities to provide good results in a high-temperature environment are: first, thermal coefficient of expansion should meet to the substrate; second, lower thermal conductivity; third, high melting point; fourth, low value of sintering; fifth, no change of phases during working at ambient temperature as well as at operating temperature; and sixth one, chemical inertness with excellent connection to the substrate materials [5, 6]. Over the past several decades decades, the continuous growth of insulating coatings is increased to resist the mechanical moving part components from degradation under the adverse working conditions. TBCs work as insulating layer for critical operating components in the various fields of gas turbines, IC engines, aerospace, etc. It facilitates to defend the components from wear, erosion, oxidation, chemical ablation, and corrosion (Fig. 1).

2 Thermal-Barrier Coating

Ceramic layers can be used as a coating stratum for insulation purpose of IC engine components which are called thermal-barrier coatings. The thermal-barrier coatings are made up of ceramic refractory oxides of a thickness (120–400 μm) which are coated over the alloy structure that passes through high temperature and can also use ahead of melting temperature. As well as, the thermal stress, rate of wear, rate of erosion, and hot corrosion could easily be prevented by the aid of thermal-barrier coatings and easy to consider temperature rise to the parts of components in IC engines, turbine blades, piston rings, etc., which usually face to the tough working environment [7–9]. The various processes included in coating with the features are shown in Table 1 [10–12]. The design consideration factors for making the TBCs are

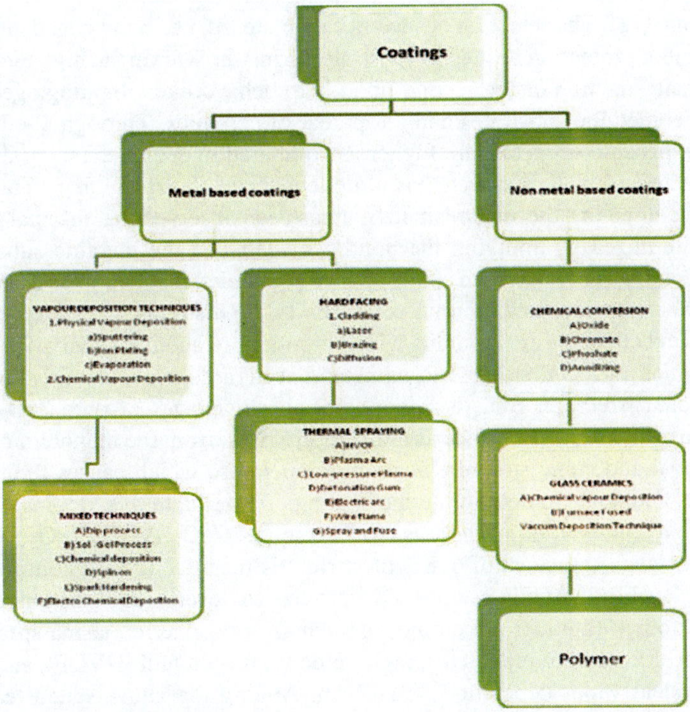

Fig. 1 Types of coating process

Table 1 Coating salient features and processes [10–12]

S. No	Coating thickness	Material used	Features	Coating methods
1	500–5000 mm	Ni-based alloy	Wear resistance	Weld overlay
2	40–3000 mm	Ceramic and metallic alloys	Wear and corrosion resistance	Thermal spray
3	1–50 micro mm	SiC	Wear resistance	CVD
4	10–100 mm	Chrome	Wear resistance	Hard chromium plate
5	1–5 micro mm	Ti(C,N)	Wear resistance	PVD

important for enabling the protection of materials from high temperature. Therefore, during the design of TBC, three layers are coated, where the first layer is formed over the surface of the substrate which comes in direct contact to exposure of high temperature, i.e., thermally grown oxide (TGO), second formed layer is an insulating ceramic layer whose value of thermal conductivity is approximately 0.8–1.2 W/m·K which depends on the deposition of coating techniques, and last coating layer is the bond coating which consists of metallic layers normally consisting of Co matrix or MCrAlY M-γNi acts as an umbrella to protect from oxidation on to the substrate and has a tendency to improve the adhesion which occurs between the substrate structure and ceramic [13]. The selection of second coat material, i.e., bond coat depends upon the amount of protection needed to substrate during the working at high temperature.

The main function of the second bond is to behave like a binding agent among the various metallic substrates and a top coat of ceramics. Through coefficients of thermal expansion of metals are higher as compared to ceramics materials, for this reason, the metal-made substrate has more tendencies to expand than ceramic coatings and resulted in a thermal mismatch. It has been observed that thermal mismatch can be minimized by applying the bond coat between the metallic substrate and ceramic coatings [14]. Herman et al. [8] have concluded that ZrO_2 has a linear value of thermal expansion, low thermal conductivity, high strength, fracture, toughness as well as excellent elastic modulus which is similar to another super alloy elements like Ni which makes it suitable to withstand at high degree of temperature due to its tetragonal structure. But, there are some disadvantages of pure ZrO_2 coating; phase transformation takes place as temperature rises, from the monoclinic structure to tetragonal and cubic structure at room temperature which causes the formation of stresses and tends to deteriorate the coating. To facilitate this, doping technique, ZrO_2 with a combination of various oxides such as CeO_2, MgO, Y_2O_3, or CaO, is used. The standard topcoat of 6–8% of yttrium is manufactured at commercial and industrial level for partially established zirconia. To develop the excellent quality of TBCs, various techniques are employed such as atmospheric plasma spray (APS), vapor deposition technique (VDP), high-velocity oxygen fuel (HVOF), and electron beam physical vapor deposition (EB-PVD). Among the above-stated techniques, APS and EB-PVD are widely used techniques for the fabrication of TBC, which are based on the requirements as well as applications. On other hand, APS techniques are usually used for the coating purpose due to their excellent capability of coating robust designs and larger shapes at a faster rate as well as at a reasonable cost as compared to EB-PVD [15, 16]. Also, in the newest studies, it has been examined that APS is modified by two distinguished names such as solution precursor plasma spraying (SPPS) and suspension plasma spraying (SPS) that have brought huge evolution in the development of an area of coatings due to its better strain tolerance capacity and low thermal conductivity than old plasma spray coatings [17]. The use of ceramic materials for thermal-barrier coatings plays a very important role in various metallic components such as turbines and piston rings to save the super alloy from damage at high temperatures [18, 19]. TBCs should have necessary properties for obtaining better results such as high melting point, low thermal conductivity, and good damage tolerance [20]. Normally, the material used for making TBCs should have possessed the low phonon thermal conductivity, i.e., <2 W/m·K in the free environment of electrons participation during thermal conductivity [21]. Nowadays, partially stabilized yttria zirconia is widely used for making TBCs [22–24]. But, partially stabilized yttria zirconia (YSZ) is not suitable for high temperature due to loss of phase stability and tolerance at thermal exposure [29]. Thus, the material with low thermal conductivity and stable phase at a high degree of hotness must be preferred for making TBCs [25]. Considering the present scenario, metal matrix composite coatings are also in high demand due to their various applications in advanced engineering as well as their phenomenal mechanical properties [26]. Nowadays, researchers are rigorously working on aluminum-based metal matrix coatings such as TiC, Al_2O_3, B_4C, WC, SiC, bamboo leaf ash, bagasse ash, rice husk, red mud which are being used for developing the aluminum-based metal matrix with their excellent and improved properties [24, 27, 28]. It has been studied by Bhushan et al. [29] wear and surface finish behavior by reinforcing Silicon carbide particles with Al 7075 and observed increase in tool wear and decrement in surface finish. From another study, it has been coming onto the light, Pitchayyapillai et al. [30] strength and wear-resisting properties of materials can be enhanced to increase the contents of SiC particles into the alloy of aluminum, with the increase of machining cost, while at nominal cost with improved wear and strength properties are obtained from a hybrid composite of Al-SiC-Gr. Adeosun et al. [31] have developed aluminum-based composite (SiC and bamboo leaves) coating by employing a stir casting process which has showed improved behavior of corrosion resistance with a decrease in density. Another researcher Balasivanandha et al. [32] have developed a tailor-made hybrid coating of Al 2219/B_4C/MoS_2 and examined that microhardness of substrate has increased after coating with an increase in wear resistance behavior. The hybrid coating of Al 2219/B_4C/MoS_2 has also shown good thermal stability. Hybrid composites, i.e., Al/Zn/Al_2O_3 made by use of stir casting process is very much used in precision applications as well as in thermal packaging. The coating of carbon onto the substrate material is influenced by several deposition parameters such as arc current, additives used, spray distance, layer thickness.

3 Heat Treatment

The muffle furnace-based heat treatment process is generally cheaper and easily available as compared to post spray heat treatment process. HVOF is deposition technique in which about approximately 12 mm thickness layer of coatings is deposited over the substrate by using mixture of oxygen and liquid fuel which fed into combustion chamber and ejected by converging and diverging nozzle at high velocity and low temperature on to substrate. It has to be seen that when HVOF coatings are treated at constant temperature, residual stress is generated which considerably affects the coatings. These changes occur due to the development of secondary stress in the HVOF-sprayed coated layer for the period of cooling at the end of heat treatment process and easiness of stress at high temperature [33–36]. Figure 2 shows the features and advantages of post heat-treated HVOF coatings.

Effect of Post-treatment on HVOF Coatings

It has been investigated after conducting numerous experiments that while HVOF-coated parts are passed through to cavitations and subjected to compressive loadings, it revealed many defects on the surface of substrates such as a large amount of porosity, unmelted microparticles, and poor adhesion properties concerning the substrate during the spraying process, which limits the appropriation of HVOF process. It has been observed by many studies that adhesion properties of HVOF coatings on to the surface of the substrate can be improved by following post-treatment

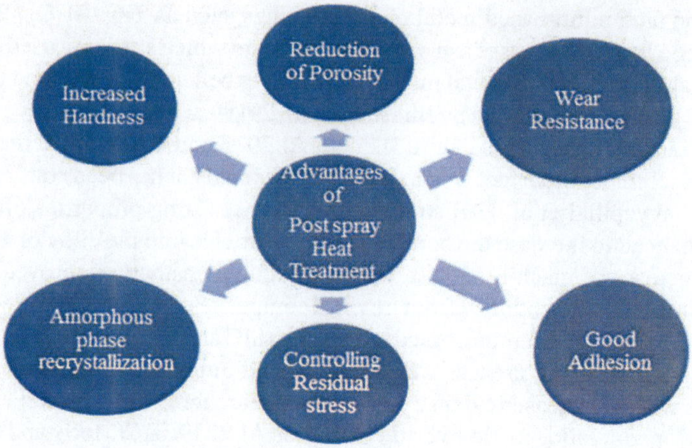

Fig. 2 Advantages of post spray heat treatment [33–36]

to facilitate filter or refine the layers [37, 38]. When HVOF-coated part components are passed through the heat treatment process at a constant temperature or reversible process, residual stresses are produced and could affect the pre-coated coatings. These changes can occur because HVOF works at high temperature, and at high temperature, stresses become relaxed and secondary stress is formed in the coating layers during the cooling process or the end-to-end heat-treated process.

Different Post-treatment Process

The aim of the current work is too detailed and discusses the variety of techniques and mechanisms which are used for making composite coating to reduce friction among mechanical components as well as for enhancing corrosion-resisting properties and wear-resisting properties on the dissimilar substrates to provide a carbon-based coating of carbon. Also, we will discuss in the current work the future opportunity for the world to accepting different and new innovative techniques of deposition to improved tribological properties.

Drawbacks of Post-treatments

Following treatment of thermally sprayed coating has been noticed in sprayed layers such as cracking appears in coated layers, removal of coating from the surface, oxide phase formations, and nonhomogeneous coating composite is noted. Hirata et al. [35] have observed abrasive wear behavior of tailor-made composite coatings of Cr_3C_2–25 wt% NiCr, before and after heat treatment at the furnace for 1 h at various temperatures, i.e., 723, 115 and 10333 K and observed that at the temperature of 10,333 K, metal oxides are formed which improve the hardness of coated coatings. The main challenge is to control the structural decohesion which leads to a bad effect on abrasive wear resistance due to performing the heat-treating process in the oxidizing atmosphere as compared to Cr_3C_2–25 wt% NiCr coatings.

4 Conclusions

The use of the HVOF method has increased among the industries due to its endless advantages more than the thermal spray technologies for deposition purposes of different materials like carbide cermets. It has been observed that development of TBCs and post-treatment of HVOF-coated substrates are generally changed and improved the microstructure of coatings, as well as stress, and also removed from it after the heat treatment process of as-sprayed coatings. It has been studied that surfaces treated by post-treatment provide a hard and dense coating which increases and improves the adhesion tribological property of the substrate or coatings. The continuous growth in the innovation of HVOF technologies and the appropriate choice of selection of post-treatment methods will no doubt have open the door of opportunities for a new market.

References

1. Energy Statistics 2018, Central Statistics Office, Ministry of Statistics and Programme Implementation, Govt. of India
2. Miguel M, Westmoreland B, Jones G, Phillips F, Eberle D, Tussing M, Yeomans N SAE Int J Engines 4(3):2736–2755
3. Kumar V, Kandasubramanian B (2016) Particuology 27:1–28
4. Thorne S, Mahalle AM (2019) Thermal barrier coating materials for SI engine. J Mater Res Technol 8:1532–1537
5. Tyagi A, Pandey SM, Walia RS, Murtaza Q (2019) Characterization and parametric optimization of tribological properties of Mo blend composite coating. Mater Res Express 6:086428
6. Cao XQ, Vassen R, Stoever D (2004) Ceramic materials for thermal barrier coatings. J Eur Ceram Soc 24:1–10. https://doi.org/10.1016/S0955-2219(03)00129-8
7. Schulz U, Leyens C, Fritscher K, Peters M, Saruhan-Brings B, Lavigne O, Dorvaux JM, Poulain M, Mévrel R, Caliez M (2003) Some recent trends in research and technology of advanced thermal barrier coatings. Aerosp Sci Technol 7:73–80
8. Herman H, Sampath S, McCune R (2000) Thermal spray: current status and future trends. MRS Bull 25:17–25
9. Perepezko JH (2009) The hotter the engine, the better. Science 326:1068–1069
10. Pawlowski L (2008) The science and engineering of thermal spray coatings. Wiley, New Jersey, US
11. Ren X, Wang F, Wang X (2005) High-temperature oxidation and hot corrosion behaviors of the NiCr–CrAl coating on a nickel-based superalloy. Surf Coat Technol 198(1–3):425–431
12. Bertrand G, Mahdjoub H, Meunier C (2000) A study of the corrosion behaviour and protective quality of sputtered chromium nitride coatings. Surf Coat Technol 126(3):199–209
13. Metting CJ, Bunn JK, Underwood E, Smoak S, Hattrick-Simpers J (2013) Combinatorial approach to turbine bond coat discovery. ACS Comb Sci 15:419–424
14. Bhushan B, Balkishan K (1991) Handbook of tribology: materials, coatings, and surface treatments. Krieger Publishing Company, Florida, USA
15. Han Z, Yin W, Zhang J, Jiang J, Niu S, Ren L (2013) J Bionic Eng 10(4):479–487
16. Włodarczyk-Fligier A, Dobrzannski LA, Adamiak M (2008) JAMME 2:147–150
17. Dyuti S, Mridha S, Shaha SK (2011) Adv Mater Res 264–265:1427–1432

18. Tyagi A, Koli Y, Sharma D (2018) Fabrication & mechanical testing of AA6082/Si3N4 composites. Int J Sci Res Sci Technol 4(2):579–582. ISSN: 2395-6011 (Print), Online ISSN: 2395-602X
19. Gell M, Wang J, Kumar R, Roth J, Jiang C, Jordan EH (2018) Higher temperature thermal barrier coatings with the combined use of yttrium aluminum garnet and the solution precursor plasma spray process. J Therm Spray Technol 27:543–555
20. Ghasemi R, Vakilifard H (2017) Plasma-sprayed nanostructured YSZ thermal barrier coatings: thermal insulation capability and adhesion strength. Ceram Int 43:8556–8563. https://doi.org/10.1016/j.ceramint.2017.03.074
21. Rai AK, Schmitt MP, Dorfman MR, Zhu D, Wolfe DE (2018) Comparison of single-phase and two-phase composite thermal barrier coatings with equal total rare earth content. J Therm Spray Technol 27:556–565. https://doi.org/10.1007/s11666-018-0713-3
22. Li F, Zhou L, Liu J-X, Liang Y, Zhang G-J (2019) High-entropy pyrochlores with low thermal conductivity for thermal barrier coating materials. J Adv Ceram 1–7
23. Thermal-barrier coatings for more efficient gas-turbine engines | MRS Bulletin | Cambridge Core (n.d.) https://www.cambridge.org/core/journals/mrs-bulletin/article/thermalbarrier-coatings-for-more-efficient-gasturbine-engines/72FDF8FDDCD21040144D75579E3C1358. Accessed 14 July 2019
24. Padture NP (2016) Advanced structural ceramics in aerospace propulsion. Nat Mater 15:804–809
25. Feng J, Ren X, Wang X, Zhou R, Pan W (2012) Thermal conductivity of ytterbiastabilized zirconia. Scr Mater 66:41–44
26. Secura S (1985) Optimization of the NiCrAl-Y/ZrO-Y_2O_3 thermal barrier system
27. Yan C, Lifeng W, Jianyue R (2008) CJA 21 578. https://www.sciencedirect.com/science/article/pii/S1000936108601776
28. Christy TV, Murugan N, Kumar S (2010) JMMCE 9:57
29. Bhushan RK, Kumar S, Das S (2013) Int J Adv Manuf Technol 65:611
30. Pitchayyapillai G, Seenikannan P, Balasundar P, Narayanasamy P (2017) Trans Nonferrous Met Soc China 27:2137
31. Adeosun SO, Osoba LO, Taiwo OO (2014) Int J Chem, Mol, Nucl, Mater Metall Eng 8:731. https://publications.waset.org/10000007/characteristics-of-aluminum-hybrid-composites
32. Balasivanandha SP, Karunamoorthy L, Kathiresan S, Mohan B (2006) J Mater Process Technol 171:268
33. Ghadami F, Sohi MH, Ghadami S (2015) Effect of bond coat and post-heat treatment on the adhesion of air plasma sprayed Wc-Co coatings. Surf Coat Technol 261:289–294
34. Ghadami F, Ghadami S, Abdollah-por H (2013) Structural and oxidation behavior of atmospheric heat treated plasma sprayed WC-Co coatings. Vaccum 94:64–68
35. Hirata Y, Kato T, Choi J (2015) DLC coating on a trench-shaped target by bipolar PBII. Int J Refract Met Hard Mater 49:392–399
36. Kang S, Lim H-P, Lee K (2015) Effects of TiCN interlayer on bonding characteristics and mechanical properties of DLC-coated Ti-6Al-4V ELI alloy. J Refract Met Hard Mater 53:13–16
37. Rübig B, Heim D, Forsich C, Dipolt C, Mueller T, Gebeshuber A, Kullmer R, Holecek R, Lugmair C, Krawinkler M, Strobl V (2017) Tribological behavior of thick DLC coatings under lubricated conditions. Surf Coatings Technol 314:13–17
38. Stewart DA, Shipway PH, McCartney DG (2009) Influence of heat treatment on the abrasive wear behavior of HVOF coatings. J Mater Process Technol 209(7):3172–3181

A Critical Review on Design and Examination of High-Temperature Thermal Spray Carbon-Based Composite Coatings at High Temperature

Shubhangi Chourasia, Ankit Tyagi, Shailesh Mani Pandey, Qasim Murtaza, and Kalpana Gupta

Abstract In metalworking industries, power generation, and aerospace industries, friction and wear are discerned as serious problems in mechanical moving machines that work in high-temperature environments. Wear and friction directly relate to vastly the manufacturing capital cost in the diligence. This paper furnishes a critical review on the effect of high temperature on high-velocity oxygen fuel (HVOF)-coated coatings including various designs' approach in the environment of a high wear resistance, a low friction, eco-friendly atmosphere, for high hardness and residual stress. And lastly, applications and future trends of high-temperature coating materials are introduced. These developments in the field of coatings are generously offered to deepen the understanding of high-temperature tribological behavior as well as expanding toward the various practical applications.

Keywords High temperature · HVOF · Design · Applications

1 Introduction

Wear is the accumulative result of loss of materials from the moving surface of components due to relative motion among the components between the medium of solid, gaseous, and liquid bodies. The problem related to wear and frictions is identified as a huge problem in moving mechanical bodies at operating conditions. Distinctive examples can be seen in the field of manufacturing industries, power generation,

S. Chourasia · A. Tyagi (✉)
Department of Mechanical Engineering, SGT University, Haryana, India
e-mail: tyagiankit10506@yahoo.com

S. M. Pandey (✉)
Department of Mechanical Engineering, National Institute of Technology, Patna, India
e-mail: smp.me@nitp.ac.in

S. Chourasia · A. Tyagi · Q. Murtaza · K. Gupta
Department of Mechanical Engineering, Delhi Technological University, Delhi, India

S. M. Pandey
Department of Mechanical Engineering, Patna University, Patna, India

A. Maurya et al. (eds.), *Recent Trends in Mechanical Engineering*, Lecture Notes in Mechanical Engineering, https://doi.org/10.1007/978-981-19-7709-1_86

aerospace, etc. Wear and friction highly contribute to almost 90% of manufacturing costs in manufacturing industries. The effect of wear on capital cost is higher climacteric than friction as it could head to sudden breakdowns and ruinous stoppage of machine and results to high cost of capital. According to the survey, 23% of overall energy consumption costs occurred due to wear and friction. This cost can be saved by applications of new materials, new methods, surface morphology modification, and better use of lubrication for minimizing the wear and friction, and approximately 40% of energy could be saved [1]. To reduce the friction at high temperature, i.e., 350 °C mechanically operated moving parts, the use of solid lubricant is the only way to reduce the friction, because, at a temperature of more than 350 °C, liquid lubricants stated to degrades, due to solid lubricating, high performance, better efficiency, and durability at high temperature [2–4].

This paper includes:

(1) Design approach for a high wear resistance, low friction, eco-friendly atmosphere, for high hardness and residual stress at high temperature,
(2) Applications of high-temperature friction-reducing lubricants,
(3) Trends of future materials at high temperature.

2 Design of High-Temperature Thermal Spray Carbon-Based Composite Coatings

2.1 Design for Low Friction

The overall energy efficiency of the system is surely improved by lowering the coefficient of friction. Normally, moving components are required a low coefficient of friction at high temperatures to keep low consumption of energy in form of money, efficiency, and power to maintain mobility of components such as pistons, sliding bearing, and gears. The low friction of coefficient can be obtained in lubricating boundary when physical and chemical absorption layer of lubricant with low shear stress wraps on the surface of contact, i.e., based on the basic theory of solid lubrication and classical theory [5–9]. It has to be noted that at the high temperature, friction between the components is getting reduced due to the formation of a solid-lubricating layer between the surface of components or on the worn surface of components which efficiently lowers the rate of friction. The perfect design for obtaining the low coefficient of friction can be attained when elastically rigid and tough substrate can withstand at normal load and carries a small area of contact, and at the same time, lubricating film on the surface of substrate gives shear which tends to lower the adhesive strength. This is the ideal concept of design for attaining a low value of coefficient friction.

2.2 Design for Wear Resistance at High Temperature

At a high degree of temperature, the solid-lubricating film on the worn components and materials is capable of providing high precision, accuracy along with long life, and high wear resistance rate. Corrosive wear, adhesive wear, fatigue wear, and abrasive wear are the terms that come under the principle of wear mechanism at high temperature. The solid-lubricating layers are formed over the surface of the worn surface of components which provide the improved abrasive and adhesive wear resistance properties to substrates. The low value of adhesive wear is obtained when the solid-lubricating layer gives minimum shear film and tends to reduce the straight contact with the sliding or rubbing parts. Also, due to the mechanical properties of the material such as ductility, hardness, toughness, strength, etc. are competent to keep away plowing the asperities of one surface to another one and also save softer material from the cutting action of harder material over its surface as well as expansion of crack, and fatigue over the surface of worn parts and substrate. At high temperatures, oxidation wear is responsible for dominating corrosive wear in a general manner. On one side, it prevents material solid film of lubrication from oxidation at high temperature and on the other side protects from glaze film. This glaze film is depending on tested parameters, tested conditions, oxygen pressure, and composition of trio-pair during friction measuring processes. The glaze film is composed of lubricating film which consists of solid lubricant particles formed during the reaction of tribo-oxidation and employed for reducing the wear. Wear is the collective result of the failure of materials or metals from the exterior surface due to the relative motion among the various parts of the components. Surface wear does not occur alone; it depends upon the many mechanical factors such as environmental conditions, load frequency, thermal properties, loads, force, intensity, and many more [10].

2.3 Design for Hardness Residual Stress at High Temperature

Guo et al. [12] evaluated the result of two self-developed coatings which were based on a study of tribological behavior of composition of composite coatings at high temperature. The composite coatings of NiCrBSi/WC-Ni and NiCrBSi are deposited onto the substrate of stainless steel with the use of laser cladding coating technique. At 500 C, NiCrBSi showed a high value of hardness due to the development of hard precipitates, i.e., 5 times more than steel substrate, whereas at the same temperature, NiCrBSi/WC-Ni has shown more hardness than NiCrBSi as a result of the presence of hard phase of WC/W2C. Zikin et al. [13] studied the wear and erosion behavior of coating at the ranges of temperature, i.e., 300, 550, and 700 °C. Researchers compose a tailor-made coating of NiCrBSi by bombarding Cr_3C_2-Ni, WC, and TiC-NiMo with the help of plasma arc cladding technique and observed that at 700 °C, hardness of

WC matrix has started to decrease due to high oxidation of fracture of WC. Liu et al. [14] investigated the tribological behavior of NiCrBSi-based coatings with the help of plasma-sprayed techniques on the cylinder liners at different temperatures by performing various heat treatment processes. Muffle furnace is used for performing heat treatment process at a temperature of 700, 500, and 300 °C, and researcher noted that microhardness has increased after heat treatment of specimens because of precipitation takes place in the phase of CrB and Ni_3B. Winkelmann et al. [15] studied the tribological behavior, i.e., wear and hardness of plasma-sprayed coating of NiCrBSi-60%WC at different temperature values, i.e., 750, 500, 650, 600, and 700 °C. Researchers have used the Vickers hardness test measuring rig for measurement of hot hardness and observed that hardness of NiCrBSi-60%WC coating is decreased continually at a temperature of 800 °C. Fernandez et al. [16] studied the behavior of NiCrBSi-based coating onto a substrate of AISI 1045 steel deposited by laser clad technique with changing the percentage of WC from 0 to 50%. The wear test and hardness test have performed to knowing the tribological behavior of coatings, and it was found that due to the enhancement of WC, the hardness of coating has increased because of assisting of WC reinforced in quick cooling which resulted in precipitation of borides and fine and hard chromium bromide from the matrix. Yeh et al. [17] have worked on the coating of NiAl on to substrate of 304 stainless steel using HVOF spray technique where researcher studied about different tribological properties of materials such as microstructure, porosity, and hardness of the coating and examined that due to increase in the percentage of oxygen content, there is an increase in hardness with a decrease of porosity, whereas at 20 °C, value of Young's modulus has observed and it has kept decreasing with the temperature rise.

Das et al. [18] observed the tribological behavior of 10 wt% diamond bronze-based coating on the substrate of steel bearing and compared with self-made tailor coating of composite coating of bronze with the pure bronze coatings. It has been reported that manifestation of diamond coating on bronze has improved the microhardness and elastic modulus by around 35% in pure bronze and 15% in composite bronze coatings which results in demonstrated microstructure integrity with the low level of porosity and defects on inert lamella. Hong et al. [19] have examined the cavitations' erosion responses of HVOF-sprayed coatings on WC-10Ni and WC-$20Cr_3C_2$–7Ni by performing systematic investigation of different rate of flow velocities onto substrates. It has to be found that both the coatings are showing dense and good bonded behavior onto the substrate of steel, whereas WC-10Ni has shown the highest porosity and modulus of elasticity but lowest hardness to WC-$20Cr_3C_2$–7Ni coating due to combination of WC, W2C, and W phases in composition of WC-10Ni coatings. The lower slope of cavitations behaviour has been observed in the coatings of WC-10Ni with lower values of H/E, H3/E2, and η as compared to WC-$20Cr_3C_2$–7Ni at each flow of velocity with a great decrease in volume than WC-10Ni coatings, i.e., critical velocity rate of flow for WC-$20Cr_3C_2$–7Ni coating was in region is 33.5–41.9 $m \cdot s^{-1}$. Researchers have fabricated composite coatings of WC-10Co4Cr for the substrate of 304 stainless steel by using HVOF and HVGOF deposition techniques where they found that the behavior of splat formation and cavitations when coatings are dipped in the solution of 3.5 wt% of sodium chloride solution for fulfilling testing purposes. And, it has been observed that disk-shaped splats were formed with a low rate of decarburization with low porosity, low hardness [20–22], fracture, corrosion, and slurry erosion resistance on the coated substrate, due to spraying of WC-10Co4Cr particles at low thermal temperature with high velocity as compared to HVGOF system. Higher material loss has been examined in HVOGF during the erosion testing in 3.5 wt% of sodium chloride solution, and it led to the higher rate of cavitations and small size of particles in freshwater.

2.4 *Residual Stress at High Temperature*

Around the world, because of limited availability of raw materials, high pollution rate, authoritarian rule and regulation toward the environment, and worry to a reduction in capital cost with a saving of energy which made great area of interest among the researchers, that is why the requirement of energy-saving technology and economical manufacturing processes are in demand. Thus, due to high technical applications of welding, coating and additive manufacturing technologies in numerous fields are in high-tech production lines for the production of intricate components which are more in the spotlight than ever. For achieving the required mechanical properties of materials, metallic component has to pass through the diverse heat treatment processes. Due to variable heat input during heat treating processes leads to the cause of residual stresses.

3 Benefits and Application of High-Temperature Thermal Spray Coating

The latest advancements in technology have enabled the industries to develop and use more efficient and effective machines and equipment. These modern machines and equipment are designed to perform different functions with less human intervention, leading to increased production and improved quality [23]. Thermal spray coating offers several important mechanical, tribological and metallurgical properties that includes

Enhanced Wear Resistance: High temperature thermal spray coatings offer excellent resistance against wear and tear. They protect components from damage caused by friction, abrasion, and erosion, which increases their service life.

Improved Corrosion Resistance: Thermal spray coatings are highly resistant to corrosion, making them ideal for use in aggressive environments such as marine, chemical, and oil and gas industries.

High Hardness: Thermal spray coatings are known for their high hardness, which makes them capable of withstanding high levels of stress and pressure. This makes them an excellent choice for applications that require strength and durability.

Reduced Friction: High-temperature thermal spray coatings have low coefficients of friction, which makes them ideal for use in applications that require smooth and low-friction surfaces. This helps to improve energy efficiency and reduces wear and tear on the components.

Thermal Barrier Protection: High-temperature thermal spray coatings provide protection against high temperatures and thermal shock. The coatings act as thermal barriers that protect components from extreme heat, thereby preventing stress cracking and premature failure.

Versatility: High-temperature thermal spray coatings can be applied to a wide variety of materials, including metals, ceramics, and polymers. This makes them an ideal solution for a range of applications, including aerospace, automotive, and industrial machinery.

Cost-Effective: High-temperature thermal spray coatings are a cost-effective solution for protection against wear, corrosion, and thermal damage. The coatings offer excellent performance at a fraction of the cost of replacing or repairing components.

Applications:

Aerospace and Defense Industry: The thermal spray carbon-based composite coatings are widely used in the aerospace and defense industry due to their high-temperature, wear, and corrosion resistance properties.

Automotive Industry: The coatings are used in the automotive industry to protect critical components from wear, corrosion, and chemical attack.

Oil and Gas Industry: The coatings are used in the oil and gas industry to protect critical components from high-temperature corrosion and wear.

Electronics Industry: The coatings are used in the electronics industry to protect critical components from wear and chemical attack.

Medical Industry: The coatings are used in the medical industry to protect critical components such as implants from wear and corrosion.

4 Future Perspectives, Challenges, and the Conclusion

Wear and friction directly relate to vastly the manufacturing capital cost in the diligence. This paper furnishes a critical review on the effect of high temperature on HVOF-coated coatings including various designs' approach in the environment of a high wear resistance, a low friction, eco-friendly atmosphere, for high hardness and residual stress. And lastly, applications and future trends of high-temperature coating materials are introduced. These developments in the field of coatings are generously offered to deepen the understanding of high-temperature tribological behavior as well as expanding toward the various practical applications. Modern industries are demanding novel and advanced materials which can fulfill all the requirements like wear and corrosion resistant, improved hardness, durable and high temperature resistant etc. With the high presence of developed technology in the area of computer technology and information technologies, 3D-based technology and material genome structure can fetch a good opportunity for high-temperature tribological.

Even if, researches are working out to develop high wear resistance coatings (solid lubricants) or which can withstand at high temperature and provide excellent tribological properties which are still a huge challenge among the scholars as well as modern industries. Currently, tribological studies based on high temperature are still lacking in the field of solid lubrication as compared to liquid lubrication. To fulfill this gap in the area of solid lubricant, solid lubricant (powder form) to be designed based on generalized creation of some parameters, i.e., coefficient of friction should be less than 0.2 (<0.2), whereas the value of wear rate should be 10^{-6} $mm^3/N{\cdot}m$ for a wide range of temperatures, i.e., high to low or vice versa. To grab the above-stated aim, there is a need to require the given studies in a systemic manner [23].

References

1. Holmerg K, Erdemir A (2017) Influence of tribology on global energy consumption, costs and emissions. Friction 5(3):263–264
2. Peterson MB, Murray SF, Florekk JJ (1959) Consideration of lubricants for temperatures above 1000F. Tribol Trans 2:225–234
3. Sliney HE (1982) Solid lubricant materials for high-temperature—a review. Tribol Int 15:303–315
4. Allam IM (1991) Solid lubricant for applications at elevated temperatures—a review. J Mater Sci 26:3977–3984
5. Bowden FP, Tabor D (1986) The friction and lubrication of solids. Clarendon Press, Oxford
6. Scharf TW, Prasad SV (2013) Solid lubricants: a review. J Mater Sci 48:511–531
7. Charf TW, Prasad SV, Dugger MT, Kotula PG, Goeke RS, Grubbs RK (2006) Growth, structure, and tribological behaviour of atomic layer-deposited tungsten disulphide solid lubricant coatings with applications to MEMS. Acta Mater 54:4731–4743
8. Jungk JM, Michale JR, Prasad SV (2008) The role of substrate plasticity on the tribological behaviour of diamond-like nano composite coatings. Acta Mater 56:1956–1966
9. Scharf TW, Michael JR, Prasad SV (2008) The role of substrate plasticity on the tribological behaviour of diamond-like nano composite coatings. Acta Mater 56:1956–1966
10. Davis JR (2001) USA, ASM Int 1–10
11. Janka I, Norpoth J, Trache R, Berger L-M (2016) Influence of heat treatment on the abrasive wear resistance of a Cr3C2NuCr coating deposited by an ethane-fuelled HVOF spray process. Surf Coat Technol 291:444–451
12. Guo C, Zhou J, Chen J, Zhao J, Zhou H (2011) High temperature wear resistance of laser cladding NiCrBSi and NiCrBSi/WC-Ni composite coatings. Wear 270(7–8):492–498
13. Zikin M, Antonov M, Hussainova I, Katona L, Gavrilovic A (2013) High temperature wear of cermet particle reinforced NiCrBSi hardfacings. Tribol Int 68:45–55
14. Liu L, Xu H, Xiao J, Wei X, Zhang G, Zhang C (2017) Effect of heat treatment on structure and property evolutions of atmospheric plasma sprayed NiCrBSi coatings. Surf Coat Technol 325:548–554
15. Winkelmann H, Badisch E, Varga M, Danninger H (2010) Wear mechanisms at high temperatures. Part 3: changes of the wear mechanism in the continuous impact abrasion test with increasing testing temperature. Tribol Lett 37:419–429
16. Fernandez MR, Garcia A, Cuetos JM, Gonzalez R, Noriega A, Cadenas M (2015) Effect of actual WC content on the reciprocating wear of a laser cladding NiCrBSi alloy reinforced with WC. Wear 324–325:80–89

17. Yeh C-H et al (2020) Temperature dependent elastic modulus of HVOF sprayed Ni-5%Al on 304 stainless steel using nondestructive laser ultrasound technique
18. Bandyopadhyay PP, Das S, Madhusudan S, Chattopadhyay AB (2019) Tribological behaviour of HVOF sprayed diamond reinforced bronze coatings. Diam Relat Mater 93:16–25
19. Hong S et al (2020) Effect of flow velocity on cavitation erosion behavior of HVOF sprayed WC10Ni and WC-20Cr3C2–7Ni coatings. Int J Refract Met Hard Mater 92:105330
20. Tyagi A, Pandey SM, Murtaza Q, Walia RS, Tyagi M (2020) Tribological behavior of carbon coating for piston ring applications using Taguchi approach. Mater Today Proc 25:759–764
21. Tyagi A, Pandey SM, Walia RS, Murtaza Q Characterization and parametric optimization of % change in residual stress of Mo composite coating using Taguchi approach. Mater Res Express (accepted)
22. Tyagi A, Walia RS, Murtaza Q Tribological behavior of HVOF carbon coating for wear resistance applications. Mater Res Express 6(12)
23. Shengyu Z, Jun C, Zhuhui Q, Jun Y (2019) High temperature solid-lubricating materials: a review. Tribol Int

The Role of Blockchain Technology: COVID-19 Pandemic Point of View

Prasun Sarote, Om Ji Shukla, and Shailesh Mani Pandey

Abstract Blockchain technology is a distributed ledger technology. This technology is based on creating a peer-to-peer network (P2P). As natural disasters and global pandemics expose severe functional vulnerabilities in these industries, key features backed by BCT such as decentralization, consensus mechanism, immutability, and smart contracts offer data integrity, traceability, and structured authorization through proof of work (POW), tokens, distributed ledgers, and encryption and can embolden to route some crucial problems—poor network management, loop in chains, frauds, misinformed structure, and biased authorization. BCT-supported start-ups with promising growth reports are examples to support the argument of blockchain establishing a sustainable economy that is both effective and resilient. Thus, the paper provides insights to the role of blockchain technology with respect to COVID-19 pandemic point of view.

Keywords Blockchain technology · COVID-19 · Immutability · Smart contracts

1 Introduction

Blockchain technology was originally illustrated in 1991 by a group of researchers Scott Stornetta and Stuart Haber who published and documented a theoretical aspect of blockchain titled 'How to Time-Stamp a Digital Document' which was designed to time-stamp digital documents to avoid backdating or tempering of information contained in them [1]. However, it went by mostly unused until it was 2008 to create a digital cryptocurrency bitcoin. Bitcoin, often labeled as cryptocurrency, is a virtual encrypted (coded) currency that moderates generation, verification, and independent transfer of funds [2]. Further, next-generation blockchain attributed to 'ethereum' was implemented in the year 2015. Ethereum is an open-source computing operating system that caters access to a financial service to everyone over a distributed platform and an internet connection [3]. Ethereum attracted a greater network of

P. Sarote (✉) · O. J. Shukla · S. M. Pandey
Department of Mechanical Engineering, National Institute of Technology Patna, Patna, India
e-mail: prasunsarote@gmail.com

A. Maurya et al. (eds.), *Recent Trends in Mechanical Engineering*, Lecture Notes in Mechanical Engineering, https://doi.org/10.1007/978-981-19-7709-1_87

users as it established a unique digital function called 'smart contracts' [4]. These predetermined conditions are referred to as 'consensus mechanism' (a system of arrangements to execute the required blockchain operations). Ethereum technology solved the problem of confined banking to provide an accessible banking service to everyone easily. Legitimate transaction statements are then added to the chain through a process called 'mining' for viewing of the records whenever necessary done by miners. Business-to-consumer (B2C) commerce transactions on the blockchain are relatively low, whereas business-to-business (B2B) commerce that includes major enterprises needs to process large volumes of transactions within a very secured platform. Hence, in 2017, 'hyperledgers' were developed to create a permissioned blockchain. Hyperledger is not a cryptocurrency, rather a collaborative project constituting some framework and tools to channel the development of a private blockchain between the selected enterprises in business [5]. The reason for the interest in blockchain is its central attributes that provide security, anonymity, and data integrity without any third-party organization in control of the transactions, and therefore, it creates interesting research areas, especially from the perspective of technical challenges and limitations. The meaning and definitions of block, chain, and blockchain are as follows:

- Block: A block is designed to store data, i.e., information in them. A hash is a cryptographic code associated with a certain block that is further in the identification of it.
- Chain: A continuous system of blocks references to one another creates a chain of these blocks.
- Blockchain: Blockchain technology is a distributed ledger technology. This technology is based on creating a peer-to-peer network (P2P). This is a distinctive from a client–server network system as data are stored and shared upon each party, technically referred to as nodes distributed a particular network [6]. Therefore, it is a decentralized form of networking platform that guarantees equal authority to all participating nodes in a transparent manner.

A pandemic refers to a global spread of an infectious disease, i.e., an epidemic that transmits through human interaction at a large scale. A global pandemic is bound to affect livelihood, financial loss, and worklessness creating a vacuum in the economy, e.g., COVID-19. Hence, blockchain can offer effective measures and insights to overcome threats posed by COVID-19 pandemic through appropriate implementation and supervision.

2 Role of BCT in Various Application Areas

2.1 Healthcare

The pandemic in many ways uncovers vulnerable steep cases that could be solved through the implementation of blockchain technology. Most pharma supply chains are under-utilized and ill-equipped; therefore, they are deeply impacted whenever encountered with a crisis. Decentralization of authority in BCT can help prevent loss of information along the chain [8]. Blockchain adoption can connect medical suppliers over a platform to carry out the logistics of supplying important gears, PPE, medicines, etc., in a simple manner. This procedure encourages reliable just-in-time supply chain solutions to avoid backorders and financial loss promoting a transparent and resilient chain with records and contracts stored electronically. Data aggregation for research and innovations in the field of healthcare is crucial as it delivers a base for further developments (modern equipment, vaccines, research study, etc.) [9]. As poor management and compilation of medical data harness are an organization's credibility and notoriety, implementation of blockchain can contribute to build a research network that can enable organizations and researchers to share data of new advancements such as vaccines in healthcare and overcome major health challenges. A pandemic is widely caused due to the spread of an epidemic, transmitted by human contact and interaction at a large scale (globally).

To address the above-said problem, contact tracing at an individual level, especially near one's surroundings, helps curbs major escalation inactive cases. BCT constitutes of hash cryptography to privatize the identity of the infected individual. Many mobile applications work on activated Bluetooth low energy (BLE) technology which regularly notifies and monitors the registered member on their proximity and possible interaction with the infected entity as well as their health status. Smart contracts can also facilitate and evaluate efficacy of drugs and reports. Immutability can securely store health records and clinical reports. Considering these crucial features of the technology, it can be established that it addresses the need for proper management of medical records which is governed by a robust clinician credentialing management structure. Blockchain technology adoption will result registration of medical professionals at better speed.

2.2 Food Industry

The agricultural sector is deeply affected during the span of a pandemic as the food value chain is adversely adjourned. Small-scale farmers and informal workers are more susceptible to great loss as these groups clock challenges selling and buying inputs due to poor access to the market. Since blockchain offers transparency and digital contracts, long paper-based agreements can be replaced and stored digitally; data can be exchanged without arbitrating authority of both buyers and sellers with easy transactions. The lasting effect of such a crisis endorses chronic undernourishment and hunger escalation around the globe and leads to acute food insecurity. Blockchain technology can easily bridge the gap between philanthropic institutions and local vendors and simplify the process of securely acquiring food to aid these poor sections of the society. The IBM Food Trust partnership employs blockchain to certify the quality of the food items and security in every part of the food supply chain. Appropriate consolidation of business model, governance structures, standards, and inter-operability with BCT in the food industry will promise optimal security and sanitary outcomes, especially, during a pandemic.

2.3 Education

A crisis is capable of halting educational establishments rupturing the process of imparting knowledge through institutes as these organizations become a hotspot for speedy transmission of an infectious disease. For example, at the helm of the COVID-19 pandemic, several countries imposed lockdown in most of their states as well as all educational institutes (school, university) to suppress the rapid spread of the virus. Blockchain provides real-time service monitoring including verifiable digital identity registries for service agents and students and claims to reduce program administration costs [11]. Blockchain can also be used to reinforce electronically automated standard procedures for educational portals and their installation for online content sharing that makes tracking the learning process easier and implementable. Furthermore, smart contracts and token ensure the fast and efficient payment gateways making the system more approachable, attractive, and reliable. Thus, the above arguments and examples present practical insights to the fact that BCT can make education accessible, secure, and engaging, even during a pandemic.

2.4 Labor Migration

Migrant laborers face abounding challenges due to several social, cultural, and economic factors concorded with migration. Interstate as well as international migration of laborers mirrors serious obstacles that call for urgent attention to management and policy changes for the improved predicament of these laborers [12, 13]. Blockchain can anchor and transact a huge number of assets that are valuable through its distributed ledger which is networked publicly. This channels frictionless storage and transfer of important documents such as passport, driving license, certificate. BCT meets the need for compilation documentation of these laborers which cannot be faked as they are time-stamped and public. Besides, digitalization combats theft and loss these records. Smart contracts can record original proof of employable contacts between the two parties. It ensures a well-built and engaged network between employers and stakeholders that will enable accessible formal labor markets and combat work abuse.

Blockchain can supervise sincere governance via the decentralization of the main body. It can play a pivotal role in assembling and strengthening the database on exodus and migration households to define and curate essential data such as volume and characteristics of these workers and their movement. Subsequently, camps, vehicles shelter and food, and medical arrangements can be made. Registration to social schemes granted by the government can be easily conferred by the use of blockchain technology.

2.5 *Logistics*

Our supply chain systems have been exercising lean managerial processes over the past years, which is a perk, but not good enough to overthrow, dissenting impacts of a pandemic. BCT can be a suitable technology to tackle this problem, as a logistic scale is scattered geographically and involves the need for authentication, coordination, traceability, and budget trading specifically where participants are either known or plentiful [14]. BCT provides four key features that can enhance integration and coordination among the members of a supply chain [15]: (1) transparency, (2) validation, (3) automation, and (4) tokenization. BCT allows fetching of real-time information on goods (location, etc.) along the chain. This information can be easily viewed by the permissioned members on the network. In a normal scenario, such visibility can help the supply network optimize efficiency with resiliency. Smart contracts enable a supply chain to operate on pre-specified rules automatically [16]. Therefore, BCT can easily detect the pace, position, and payments of goods along the chain. It incorporates intelligent practices that enable unique identification for all products and locations.

2.6 *Insurances*

Blockchain [7, 10] is often used as a tool prepared for the investigation by insurers. A comprehensive study through the history of pandemics linked to the insurance industry displays its profound impact on it. Blockchain solutions enhance client commitment through prominent visibility and automated smart contracts. This includes binary, easily verified information which facilitates the use of smart contracts on the top of blockchain [17]. With the removal of third-party intermediaries through blockchain, it can set up a system at a low cost that facilitates accelerated settlement structure. This can also speed up loan-processing time granted to enterprises affected due to the pandemic. Blockchain implementation saves time and prevents delay in processing of contracts. The benefits of such a system include faster processing time, lower cost, reduced operational risk, and more rapid settlement for all stakeholders involved.

2.7 Medicalchain

It provides control to patients of their medical data and secure management of health records and authority to share the most comprehensive version of their record with organizations in their medical network [18]. Implementation of blockchain technology enables the company to drive cooperation and efficiency in every aspect of healthcare. Electronic health records (EHRs) give authority patients to share their health records with other users with a property of limited time gateway ensuring and improving data security.

2.8 Agridigital

Agridigital ensures safe and transparent grain supply chains and distribution [19]. They manage farmers, bulk handlers, and traders through BCT. They make a grain supply chain easier by supervising the harvest, deliveries, inventory, and finance through blockchain technology. The company makes accessibility of important records and reports (smart contracts, inventory list, consignments, shipping, and delivery prices with accurate invoices) easier with a single live view and real-time notifications on digital platforms used by the receiver party through blockchain. This also makes communications of bodies simple so that a loop does not exist. The payments are completely online and affordable.

2.9 APPII

It employs BCT to verify the credentials of prospective students and professors [20]. Their users are allowed to create a profile that enables them to fill their academic CV along with educational history and transcripts. Once verified, the user's information is locked into its blockchain. The company has teamed up with the Open University for creating and developing an accreditation platform that manages entries of students through generating immutable academic records.

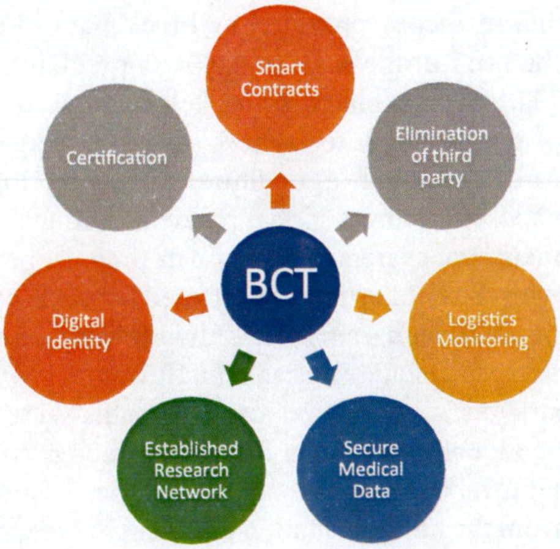

Fig. 1 Application of blockchain technology across sectors of the economy

2.10 Handshake

It is a start-up powered by professional coders that develops an open interface via blockchain to ensure safe working conditions for migrant workers worldwide [21]. They generate smart contracts for laborers and the employer that cannot be tampered. Their software also ensures staffing availability and probability while keeping a track of contracts and thus people to record a sudden exodus as well.

2.11 Provenance

A dynamic start-up that believes every product has a story. It brings the supply chain to shoppers with technology powered by blockchain and open data. They align stakeholders through their platform that gathers key product information on transparency strategy and increased engagement [22]. Their software supports and presents information to businesses about products and their supply chains and verified data to support them, i.e., origin, journey, and impact of their products. They offer programs such as transparency strategy, configuration, and publishing. Their programs facilitate marketing teams to access assured information on the engagement of customers with the products (Fig. 1).

3 Conclusion

Blockchain technology is a distributed ledger technology. This technology is based on creating a peer-to-peer network (P2P). As a global pandemic strangles the economy affecting lives, jobs, and financial losses, the urgency of transparency, immutability, and resiliency provided by blockchain in times of crisis caused by a pandemic is the need of the hour to tackle unpredicted economic as well as managerial issues. The paper explains some key features of BCT and links it with various sectors of the economy to address several managerial voids in the sectors such as healthcare, education, agriculture, insurance, logistics, and migrant labor management exposed during a global pandemic. It analyzes the problems of poor data management, transparency, and just-in-time frameworks and renders solution for the same with the help of features offered by the blockchain. It also derives possible solution of establishing a trusted network between the involved parties of several organizations that include research platforms and contact tracing, inter-operable food industry structures, automated standard paths with smart contacts, digitalization of records and identity proofs, transparency along the supply chain, and immutability, and the removal of third-party intermediaries is some of the many significant key, takeaway from the implementation of blockchain technology among the mentioned industries. Furthermore, the paper forwards examples of real companies that practice blockchain technology and address these crucial problems into practice. The paper is concluded with a firm argument that BCT has the potential to elucidate pandemic's threats to benefit the affected economies in future.

References

1. Haber S, Stornetta WS (1990) How to time-stamp a digital document. In: Conference on the theory and application of cryptography, August. Springer, Berlin, Heidelberg, pp 437–455
2. Nakamoto S, Bitcoin A (2008) A peer-to-peer electronic cash system. Bitcoin. https://bitcoin.org/bitcoin.pdf
3. Namasudra S, Deka GC, Johri P, Hosseinpour M, Gandomi AH (2020) The revolution of blockchain: state-of-the-art and research challenges. Arch Comput Methods Eng
4. Wood G (2014) Ethereum: a secure decentralised generalised transaction ledger. Ethereum Proj Yellow Pap 151:1–32
5. Paul MS (2018) Hyperledger—chapter 2|hyperledger frameworks & modules. The Startup, pp 1–12
6. Swan M (2015) Blockchain: blueprint for a new economy. O'Reilly Media, Inc.
7. Chang J, Han F (2016) Blockchain: from digital currency to credit society
8. Mackey TK, Kuo TT, Gummadi B, Clauson KA, Church G, Grishin D, Obbad K, Barkovich R, Palombini M (2019) 'Fit-for-purpose?'—challenges and opportunities for applications of blockchain technology in the future of healthcare. BMC Med 17(1):1–17
9. Fusco A, Dicuonzo G, Dell'Atti V, Tatullo M (2020) Blockchain in healthcare: insights on COVID-19. Int J Environ Res Public Health 17(19):7167
10. Prokofieva M, Miah SJ (2019) Blockchain in healthcare. Australas J Inf Syst 23
11. Wendelboe AM, Amanda Miller JD, Drevets D, Salinas L, Miller EJ, Jackson D, Chou A, Jill Raines JD, Public Health Working Group (2020) Tabletop exercise to prepare institutions of higher education for an outbreak of COVID-19. J Emerg Manag 18(2):1–20
12. Khanna A (2020) Impact of migration of labour force due to global COVID-19 pandemic with reference to India. J Health Manag 22(2):181–191
13. Blossey G, Eisenhardt J, Hahn G (2019) Blockchain technology in supply chain management: an application perspective. In: Proceedings of the 52nd Hawaii international conference on system sciences, January

14. Babich V, Hilary G (2018) What operations management researchers should know about blockchain technology. Georgetown Working Paper
15. Diedrich H (2016) Ethereum: blockchains, digital assets, smart contracts, decentralized autonomous organizations. Wildfire Publishing
16. Sayegh K, Desoky M (2018) Blockchain application in insurance and reinsurance. No. October, 2019
17. https://medicalchain.com/en/
18. https://www.agridigital.io/223218347281742314-438942f-hidden/blockchain
19. Employee background checks and CV verification underpinned by blockchain technology. https://appii.io/
20. Decentralized naming and certificate authority. https://handshake.org/
21. Blockchain: the solution for transparency in product supply chains. https://www.provenance.org/whitepaper

Experimental Investigations of Butanol as a Diesel Engine Fuel Blends

Ashish Kumar Singh, Harveer Singh Pali, Shailesh Mani Pandey, and Ashish Karnwal

Abstract Diesel engine development has been accelerated for performance improvement and pollution reduction as per current stringent emission norms. In this study, recognition of homogenous butanol blends in various volume ratios has been done to obtain effects of combustion, performance, and emission behavior in a diesel engine. At overall operating situations, the results show that increasing the butanol concentration in the mixed fuels boosted the brake thermal efficiency within considerable limits up to 7%. Also, high latent heat of alcohol results reduced in-cylinder temperature causing better combustion efficiency and lower unburnt hydrocarbon emissions (20 ppm downward) and higher value of NO_x at peak load. This study endorses replacement of diesel with the ability of alcoholic fuel making butanol to be a viable solution for CI engines.

Keywords Butanol · Combustion · Emissions · Higher alcohol · Performance

1 Introduction

Numerous studies are going on CI engine as they are most competent prime mover which governs the economy. Owing to its reliability, simple construction, protracted life span, and greater thermal efficiency, it is used widely in industrial, agricultural, power production, and transportation applications [1, 2]. Although having many benefits, it is also a major contributor of substantial pollution creating havoc to public health issues. One of the major issues is environmental deterioration and tighter government control of exhaust emissions. Various organizations are focusing their efforts on reduction of impact of NO_x and particulate matter emission on human

A. K. Singh · H. S. Pali
National Institute of Technology Srinagar, Srinagar, J&K, India

S. M. Pandey (✉)
National Institute of Technology Patna, Patna, Bihar, India
e-mail: smp.me@nitp.ac.in

A. Karnwal
Krishna Institute of Engineering and Technology, Ghaziabad, U.P., India

A. Maurya et al. (eds.), *Recent Trends in Mechanical Engineering*, Lecture Notes in Mechanical Engineering, https://doi.org/10.1007/978-981-19-7709-1_88

health. Abundant efforts are made by altering fuel to reduce emissions into developing in-cylinder formation and after treatment. It is challenging to concurrently decrease emissions using conventional method because of existence of trade-off curve between particulate matter and oxides of nitrogen [3–5]. Widespread concern about pollution as well as fast depletion, unequal distribution, and ever-increasing price of petroleum fuels stimulated a pursuit to substitute orthodox fuel by an unconventional fuel. To resolve this issue, a viable option is to supply abundant oxygen which increases combustion and reduce emissions by utilizing oxygenated fuel [6].

A synthetic fuel with oxygen bonding in fuel structure is popular oxygenated fuel. The oxygen bond in fuel delivers chemical energy, resulting in minimal efficiency loss during burning. To significantly reduce emissions, use of standalone oxygenated fuel or as an additive has been optimized potentially. Alcohols and ethers are both oxygenated fuel. An approach proposed by Deep et al. [1] employed lower alcohols as alternative fuel, but their miscibility in diesel is an issue. Furthermore, this work is also limited by its consideration of increased brake-specific fuel consumption (BSFC) and diminished engine power. Heating value and cetane number attained for butanol are significantly higher in comparison to ethanol; as well as, volatility and hydrophilicity are also reduced. Moreover, lesser polarity of butanol than shorter alcohol makes it a viable option to ethanol as a fuel in CI engine compatible with diesel fuel. To gather more information in this sector combustion, performance and emissions are explored for butanol in blended form with diesel [7]. Usually, a 4 carbon alcohol (butanol) made from fossil fuel is commonly used as transportation fuel. Biobutanol can be made by fermentation of biomass with feedstocks, viz. maize, sugar beets. High oxygen content, flammability temperature, and heat of vaporization of butanol create better effect on performance and emissions of diesel engine [8, 9].

Several experts have undertaken tests on diesel engines that are fueled with diesel mixed fuels. Izzudin et al. [10] compared butanol and pentanol blends in unmodified CI engine and resulted minute loss in engine power, with better brake thermal efficiency on a successful run up to 30% butanol blends and 25% pentanol blends. Result also quoted better BSFC for butanol blends than pentanol. With this alcohol mixed fuel, number [11] showed lower value of particulate matter and lesser soot production due to availability of oxygen. A slight increase in HC and NO_x is discovered with modest reduction in CO. In comparison to the diesel fuel used as a reference, Singh et al. [12] examined the influence of oxygen on cold start combustion uncertainties and found it as greater effect than cetane number and enthalpy of vaporization under warm temperatures. Lamani et al. [13] conducted experiments in CRDI compression ignition engines with butanol–diesel blends and showed optimum trade-off of NO_x and smoke at higher injection pressure than 120 MPa. Through numerous unconventional fuels, Kumar and Kumar [14] investigated injection timing effect on engine performance. In relevance to diesel fuel, result reveals rise in exhaust temperature with reduced emissions and increased brake thermal efficiency (BTE) while increasing injection timing, whereas delayed injection timing emits more HC and CO, but lesser NO_x. With increasing oxygenate content in blends, increase in brake thermal efficiency is documented by Praveen et al. [15] and Karabektas and Hosoz

[16], resulted that ethanol addition to diesel increases hydrocarbon, CO_2, and NO_x and decreases concentrate of particulate emission.

Tsang et al. [17] reported high in-cylinder pressure and maximum heat release closer to TDC with rise in amount of ethanol in blend. Result also showed similarity of CO and HC emission for 30% ethanol to baseline of diesel, but substantially higher with 40 and 50% ethanol with slight reduction in NO_x. Datta and Mandal [18] observed better engine power for DE10 at 3000 rpm with an advancement of 45°CA in injection angle. Furthermore, at 1400 rpm with 25°CA, maximum torque has been achieved, and also with 35°CA injection advance, least BSFC has been reported for DE20 blend at 1200 rpm. Li et al. [19] resulted a significant quicker combustion of oxygenated fuel than mineral diesel. This is due to reduced volatility of oxygenated fuel which evaporates faster, and proper atomization of fuel is achieved causing better exhaust emissions.

Combustion characteristics of oxygenated fuel in diesel engine are studied by Yesilyurt et al. [20], and considerable improvement of emissions and thermal efficiency has been reported. Banapurmath and Tewari [21] shown enhanced combustion properties and combustion duration with addition of ethanol. Increased oxygen mass percentage in fuel reveals considerable reduction in emissions like smoke, CO, and HC at higher engine loads. But greater NO_x emission is also documented, despite smoke-NO_x trade-off. Emiroğlu and Mehmet [22] studied the effects of air–fuel ratio and engine speed on CO and HC emission with higher ethanol mass percentage and found lower NO_x levels with greater HC emissions for richer combustible mixes when using ethanol blended fuels. Hansdah and Murugan [23] used fumigation approach fueled with bioethanol in single-cylinder diesel engine. Fumigation of bioethanol had 2–3°CA ignition delay than diesel at all flow rates at part load. When compared to diesel, lower smoke and NO_x were determined in the bioethanol fumigation. Mahla et al. [24] conducted an experiment to see effect of NO_x, PM, and smoke emissions using diesel–ethanol–biodiesel mixture. The observed result showed higher HC emission and same level of CO emissions as diesel fuel and quoted DBE10 as best alternative fuel.

Literature review revealed that combustion, performance, and emissions characteristics have shown diverse behavior at dissimilar blends and various loading conditions. Butanol has good miscibility due to its chemical structure compared to lower alcohols. Also, better physicochemical properties of butanol make it applicable to investigate various butanol blends with varying percentage of butanol (v/v) up to 20% in mineral diesel at varying engine loads.

2 Materials and Method

For the present work, biobutanol is procured from authorized vendor at 99.9%. An unmodified naturally aspirated single-cylinder four-stroke water-cooled diesel engine unit was fed with butanol–diesel blends in current experiment. The engine schematic diagram is shown in Fig. 1. A hydraulic dynamometer is loaded to engine.

Flow rate of fuel on volumetric basis is measured basis using burette and timer. A constant speed of 1500 rpm is maintained to obtain a steady-state condition. To measure emission characteristics, an AVL di-gas analyzer and AVL smoke meter are used. Current experiments are carried using unmodified engine having hemispherical shaped combustion chamber with an injection pressure of 205 bar and injection timing of 23°bTDC. The obtained results are examined finally. Properties of utilized butanol diesel fuel are shown in Table 1, whereas specification of engine is shown in Table 2.

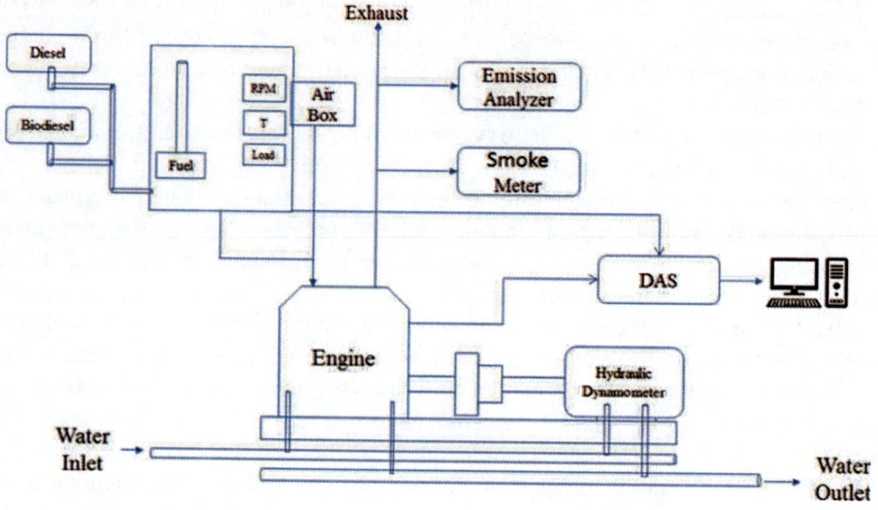

Fig. 1 Engine schematic

Table 1 Properties of the fuels used

S. No.	Property	Diesel	Butanol	DBOH 5	DBOH 10	DBOH 15	DBOH 20
1	Density (kg/m^3)	850	809.8	847.6	845.2	841.9	840.6
2	Boiling point (°C)	266	117.6	258.6	251.2	243.8	236.4
3	Flash point (°C)	85	37	80	76	72	65
4	Kinematic viscosity at 40 °C (cSt)	3.05	2.22	3	2.95	2.89	2.85
5	Autoignition temperature (°C)	316	343	317	318	320	321
6	Heating value (kJ/kg)	43,000	37,000	42,700	42,400	42,100	41,800
7	Surface tension at 20 °C (N/m)	0.023	0.024	0.023	0.023	0.023	0.023
8	Latent heat of vaporization (kJ/kg)	250	550	265	280	295	310
9	Cetane number	47	17	–	–	–	–

Table 2 Engine specification

S. No.	Parameters	Specification
1	Engine type	Kirloskar TV1
2	Injection pressure	205 bar
3	Injection timing	23° bTDC
4	Power	5.2 kW at 1500 rpm
5	Bore	87.5 mm
6	Stroke	110 mm
7	Compression ratio	17.5:1

3 Result and Discussion

In this section, the combustion, performance, and emissions characteristics of various concentration blends of diesel and butanol are explained.

Brake Thermal Efficiency: The variation of BTE for butanol blends is shown in the graph. All blends show increase in BTE with surge in load. Improved combustion due to better atomization happens with addition of butanol as it reduces interfacial tension between interacting immiscible fluids. In diesel–butanol blends, the presence of oxygen leads to improved combustion and better brake thermal efficiency to 30.09% compared to 28.13% of diesel. Higher latent heat of butanol causes cooling effect, which reduces exhaust temperature. Furthermore, lower cetane number and lower flame temperature of butanol cause longer ignition delay and lesser in-cylinder pressure, respectively (Fig. 2).

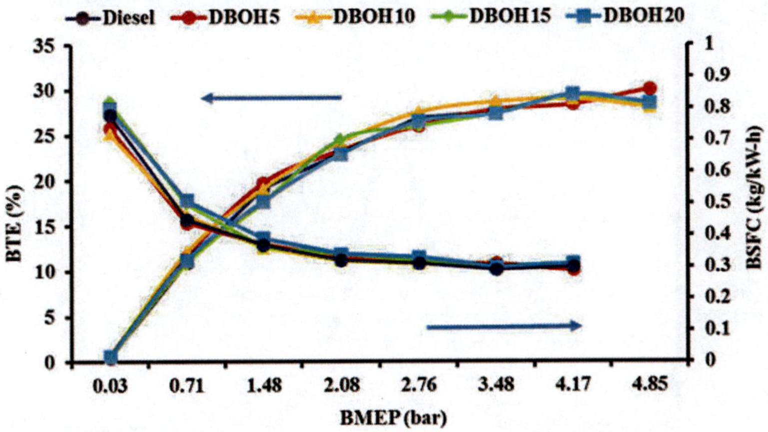

Fig. 2 Variation of BTE and BSFC with BMEP

3.1 Combustion Characteristics

Peak Pressure: The graphs suggest that brake power of diesel–butanol blends is greater than that of conventional diesel. Engine's pressure angle diagram is majorly affected by the autoignition temperature of fuel. Combustions of diesel confirm that formation of peak pressure and high latent heat of vaporization of butanol depresses due to continued heat absorption. Higher autoignition of butanol causes delayed combustion, which depicts sudden pressure and temperature rise. Shorten combustion time and slow ignition time result in fast combustion due to stabilization of mixture (Fig. 3).

Combustion Duration and Ignition Delay: The graph portraits the relationship of fuel blends and combustion time for diesel–butanol blends. With rise in load and butanol percentage, combustion duration also increases. The tendency is explained with higher autoignition temperature and latent heat of evaporation than diesel (Figs. 4 and 5).

Heat Release Rate: At full-load conditions near to TDC, heat release rate change with crank angle for diesel–butanol blends is shown in figure. Quicker heat release rate resulted in increase in combustion for all fuel blends and quicker laminar flame speed due to oxygen content of butanol. Heat release curve displays a tine dip during ignition delay phase due to heat loss in cylinder during vaporization of fuel. Longer ignition delay with diesel–butanol blends resulted in higher heat generation during premixed combustion due to low cetane value and higher latent heat of vaporization (Fig. 6).

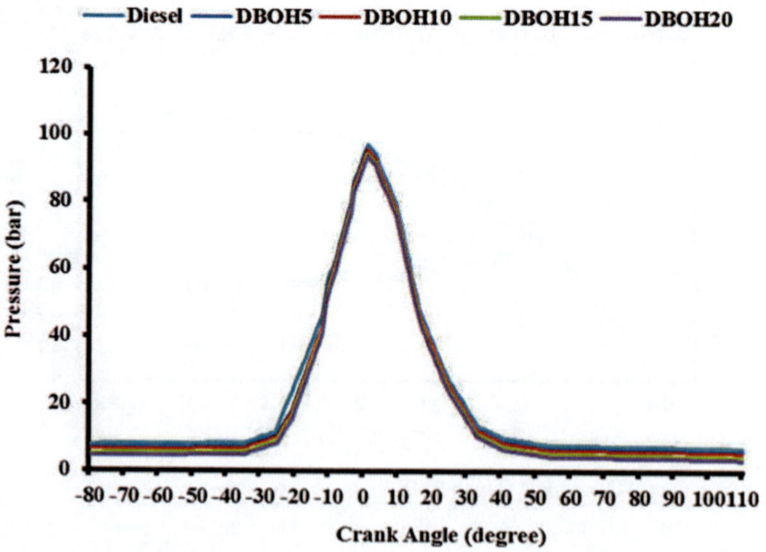

Fig. 3 Pressure crank angle diagram

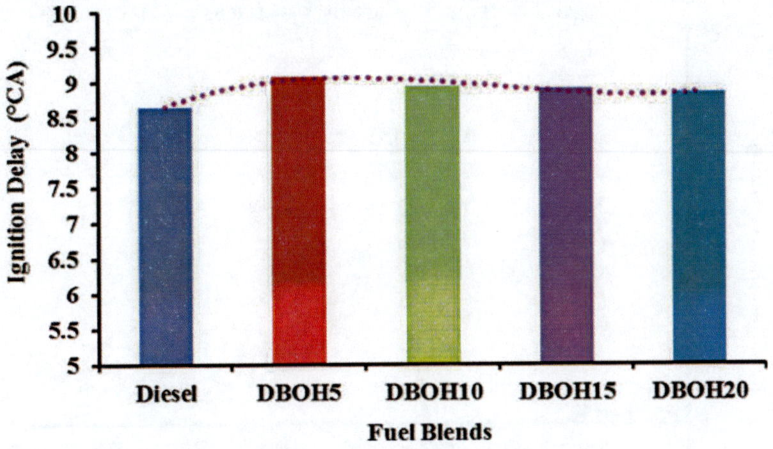

Fig. 4 Ignition delay variation for different fuel blends

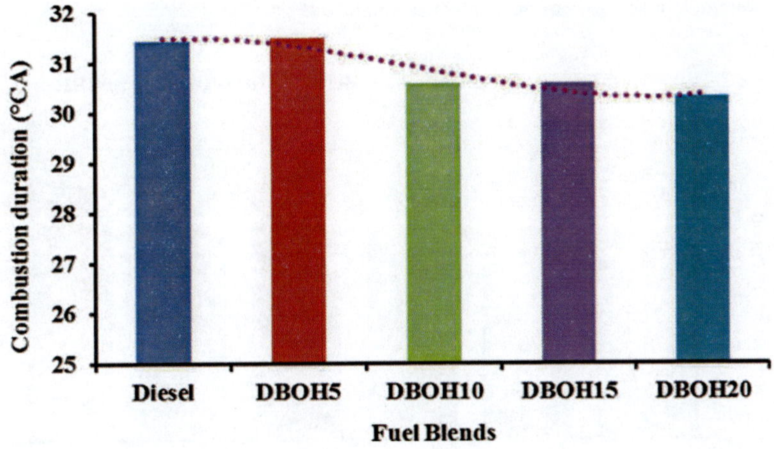

Fig. 5 Combustion duration variations for different blends

3.2 Emissions Characteristics

HC Emission: Unburnt and partially burnt exhaust are constituted of HC. Variation of total hydrocarbon with load for various diesel–butanol blends is showed in the graph. Increased quenching effect owing to low cetane numbers for blends resulted to rise in HC emissions. There is no discernible difference in HC emission for blends and diesel at higher load (Fig. 7).

CO Emission: The graphs depict the CO variation with load for diesel–butanol blends. Higher CO emissions are found at lower load, owing to higher latent heat of evaporation of butanol blends than diesel. As a result of insufficient vaporization,

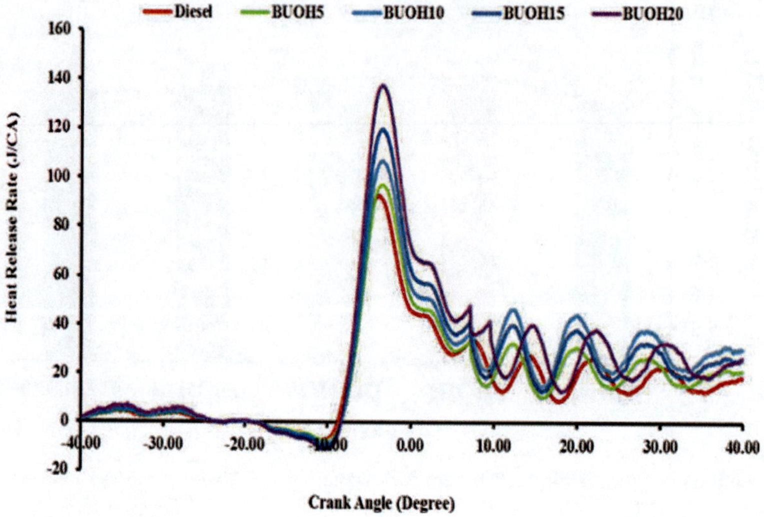

Fig. 6 Variation of heat release rate with crank angle

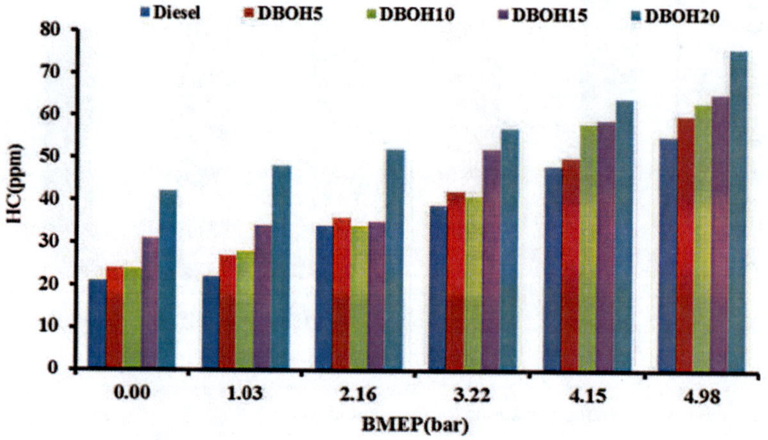

Fig. 7 Variation of HC with load

a little time to burn the fuel causes significant growth in CO emissions which is observed at lower load. But, at higher load, ample time causes better mixing, and inherent oxygen in butanol blends leads to full combustion and lower CO emissions (Fig. 8).

NO_x Emission: Formation of NO_x is determined by the flame temperature. High latent heat and low calorific value of butanol give comparable value of NO_x to diesel at low load, whereas at high load, higher oxygen availability and combustion

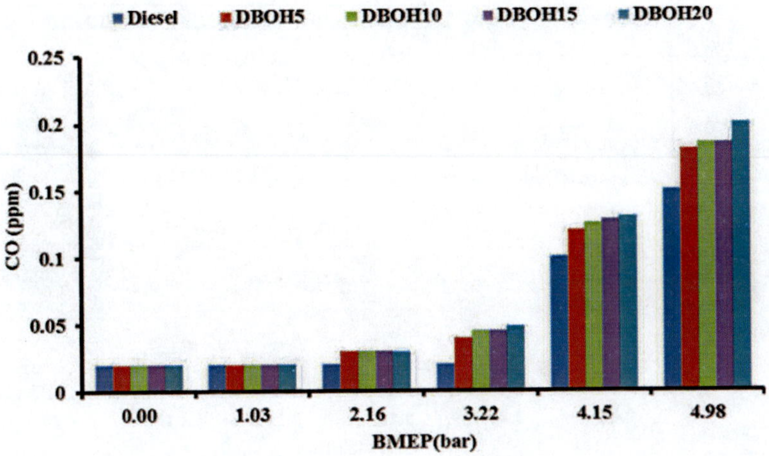

Fig. 8 Variation of carbon monoxide with load

temperature due to increased volume of injected fuel increase slightly greater NO_x. These are depicted in given graph (Fig. 9).

Smoke: Smoke opacity for diesel–butanol blend for engine blend is shown in figure. At low load, short combustion cycle at high speed causes incorrect mixing due to use of rich fuel resulting in higher production of smoke opacity than diesel. At high load, rise in flame temperature results in lower smoke opacity for diesel–butanol blends. Furthermore, at higher load, butanol's strong volatility has a significant impact on decrease in smoke opacity (Fig. 10).

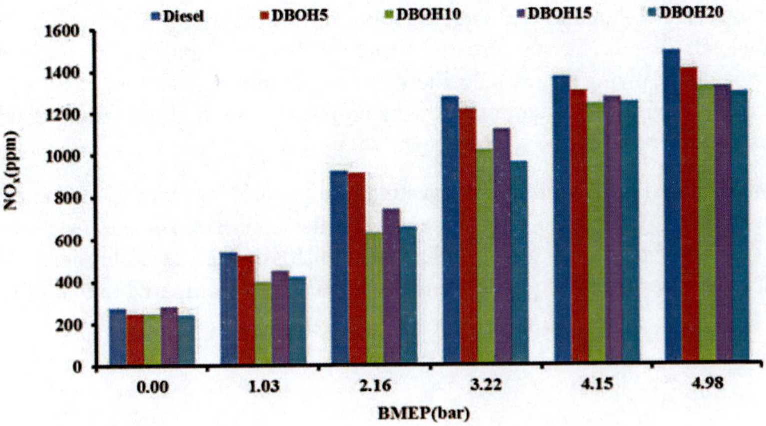

Fig. 9 Variation of NO_x

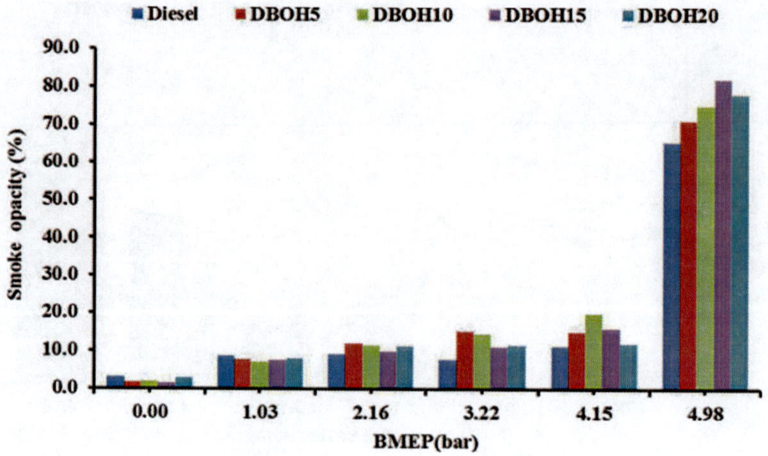

Fig. 10 Smoke opacity variation with load

4 Conclusion

The conventional CI engine is employed for diesel–butanol blends run at similar operating temperature. The following findings were obtained under different loading conditions of experimental research for the diesel–butanol blends run on diesel engine performance.

- Ignition delay, combustion duration, peak pressure, and heat release rate increases with increase in butanol.
- With rising blend ratios of alcohol up to 20%, the BTE revealed an increasing trend in diesel–butanol fuel injected in diesel engine by 7%.
- However, blends having greater value of butanol tend to decrease engine performance owing to low calorific value of butanol.
- Increasing butanol content also lowers emissions, but higher loading conditions increase them, so did NO_x emissions.

Finally, based on experimental investigation, butanol is emerged as viable choice to lower fossil fuel use in CI engine paving the approach for energy security of country in the face of rising cost and reliance of fossil fuel. Overall, diesel–butanol blended fuels perform adequate on engine performance compared to that of mineral fuel.

References

1. Deep A et al (2014) Assessment of the performance and emission characteristics of 1-octanol/diesel fuel blends in a water cooled compression ignition engine. SAE Tech Pap 2014-Octob. https://doi.org/10.4271/2014-01-2830
2. Ecklund EE, Bechtold RL, Timbario TJ, McCallum PW (1984) State-of-the-art report on the use of alcohols in diesel engines. SAE Tech Pap. https://doi.org/10.4271/840118
3. Bharathiraja M, Venkatachalam R, Murugesan A, Tiruvenkadam N (2017) Experimental investigation of a novel alcohol fumigation in a single-cylinder constant speed diesel engine. Int J Ambient Energy 38(8):794–802. https://doi.org/10.1080/01430750.2016.1222950
4. Hoseinpour M, Sadrnia H, Ghobadian B, Tabasizadeh M (2018) Effects of gasoline fumigation on exhaust emission and performance characteristics of a diesel engine with mechanically-controlled fuel injection pump. Environ Prog Sustain Energy 37(5):1845–1852. https://doi.org/10.1002/ep.12852
5. Altun Ş, Öner C, Yaşar F, Adin H (2011) Effect of n-butanol blending with a blend of diesel and biodiesel on performance and exhaust emissions of a diesel engine. Ind Eng Chem Res 50(15):9425–9430. https://doi.org/10.1021/ie201023f
6. Kumar N, Bansal S, Vibhanshu V, Singh A (2013) Utilization of blends of Jatropha Oil and N-butanol in a naturally aspirated compression ignition engine. SAE Tech Pap 11. https://doi.org/10.4271/2013-01-2684
7. Kumar N, Bansal S, Pali HS (2015) Blending of higher alcohols with vegetable oil based fuels for use in compression ignition engine. SAE Tech Pap 2015-April. https://doi.org/10.4271/2015-01-0958
8. Mittal N, Athony RL, Bansal R, Ramesh Kumar C (2013) Study of performance and emission characteristics of a partially coated LHR SI engine blended with n-butanol and gasoline. Alexandria Eng J 52(3):285–293. https://doi.org/10.1016/j.aej.2013.06.005
9. Kumar N, Pali HS (2016) Effects of n-butanol blending with Jatropha methyl esters on compression ignition engine. Arab J Sci Eng 41(11):4327–4336. https://doi.org/10.1007/s13369-016-2127-1
10. Izzudin I et al (2020) Experimental studies of single cylinder engine run on diesel-biodiesel-butanol blends. IOP Conf Ser Mater Sci Eng 863(1). https://doi.org/10.1088/1757-899X/863/1/012060
11. No SY (2020) Utilization of pentanol as biofuels in compression ignition engines. Front Mech Eng 6:1–19. https://doi.org/10.3389/fmech.2020.00015
12. Singh SB, Dhar A, Agarwal AK (2015) Technical feasibility study of butanol-gasoline blends for powering medium-duty transportation spark ignition engine. Renew Energy 76:706–716. https://doi.org/10.1016/j.renene.2014.11.095
13. Lamani VT, Yadav AK, Gottekere KN (2017) Performance, emission, and combustion characteristics of twin-cylinder common rail diesel engine fuelled with butanol-diesel blends. Environ Sci Pollut Res 24(29):23351–23362. https://doi.org/10.1007/s11356-017-9956-7
14. Kumar P, Kumar N (2018) Study of ignition delay period of n-butanol blends with JOME and diesel under static loading conditions. Energy Sources, Part A Recover Util Environ Eff 40(14):1729–1736. https://doi.org/10.1080/15567036.2018.1486904
15. Praveen A, Lakshmi Narayana Rao G, Balakrishna B (2018) Performance and emission characteristics of a diesel engine using Calophyllum Inophyllum biodiesel blends with TiO_2 nanoadditives and EGR. Egypt J Pet 27(4):731–738. https://doi.org/10.1016/j.ejpe.2017.10.008
16. Karabektas M, Hosoz M (2009) Performance and emission characteristics of a diesel engine using isobutanol-diesel fuel blends. Renew Energy 34(6):1554–1559. https://doi.org/10.1016/j.renene.2008.11.003
17. Tsang KS, Zhang ZH, Cheung CS, Chan TL (2010) Reducing emissions of a diesel engine using fumigation ethanol and a diesel oxidation catalyst. Energy Fuels 24(11):6156–6165. https://doi.org/10.1021/ef100899z

18. Datta A, Mandal BK (2016) Impact of alcohol addition to diesel on the performance combustion and emissions of a compression ignition engine. Appl Therm Eng 98:670–682. https://doi.org/10.1016/j.applthermaleng.2015.12.047
19. Li L, Wang J, Wang Z, Xiao J (2015) Combustion and emission characteristics of diesel engine fueled with diesel/biodiesel/pentanol fuel blends. Fuel 156:211–218. https://doi.org/10.1016/j.fuel.2015.04.048
20. Yesilyurt MK, Yilbasi Z, Aydin M (2020) The performance, emissions, and combustion characteristics of an unmodified diesel engine running on the ternary blends of pentanol/safflower oil biodiesel/diesel fuel, vol 140, no 6. Springer International Publishing
21. Banapurmath NR, Tewari PG (2010) Performance, combustion, and emissions characteristics of a single-cylinder compression ignition engine operated on ethanol-biodiesel blended fuels. Proc Inst Mech Eng Part A J Power Energy 224(4):533–543. https://doi.org/10.1243/09576509JPE850
22. Emiroğlu AO, Mehmet Ş (2018) Combustion, performance and emission characteristics of various alcohol blends in a single cylinder diesel engine. 212:34–40. https://doi.org/10.1016/j.fuel.2017.10.016
23. Hansdah D, Murugan S (2014) Bioethanol fumigation in a di diesel engine. Fuel 130:324–333. https://doi.org/10.1016/j.fuel.2014.04.047
24. Mahla SK, Goga G, Cho HM, Dhir A, Chauhan BS (2020) Separate effect of biodiesel, n-butanol, and biogas on performance and emission characteristics of diesel engine: a review. Biomass Convers Biorefinery. https://doi.org/10.1007/s13399-020-01056-7

A Detailed Review of Friction Stir Processing

Sachendra, Shailesh Mani Pandey, Satyajeet Kumar, Shailesh Kumar Singh, and Kuldeep Singh

Abstract Friction stir processing (FSP) is a surface modification technique which has widely grabbed the attention by significantly contributing in the development of functionally graded materials with improved surface structure and enhanced mechanical properties. By using FSP, the modified alloys and composites have ultrafine-grained structure with surface hardened with or without using reinforcement particles. In this paper, the review of latest achievement in FSP on aluminum alloys, with surface treatment and with particle reinforcement, is done. It can be seen that by varying the various process parameters along with particle reinforcement, the mechanical properties such as tensile strength, hardness, wear resistance, corrosion resistance, impact toughness have improved significantly.

Keywords Friction stir processing · Al alloys · Mechanical properties · Microstructure

1 Introduction

Aluminum alloys and composites have a wide application in the field of industrial and aerospace multifunctional products and components. These are aircraft and rocket parts, wheel disk, automotive components, heat exchangers, aluminum pipelines, air conditioners, fames and bases of many others. The main attribute of such products and components is their long serving life with high operational achievement quality, which mainly depends upon the selection of right material to furnish the desired

Sachendra (✉) · S. M. Pandey (✉) · S. Kumar
Department of Mechanical Engineering, NIT, Patna, India
e-mail: sachendra43@gmail.com

S. M. Pandey
e-mail: smp.me@nitp.ac.in

S. K. Singh
Department of Mechanical Engineering, IIT-ISM, Dhanbad, India

K. Singh
Department of Mechanical Engineering, KIET Group of Institutions, Ghaziabad, India

A. Maurya et al. (eds.), *Recent Trends in Mechanical Engineering*, Lecture Notes in Mechanical Engineering, https://doi.org/10.1007/978-981-19-7709-1_90

properties. The desired features of an alloy depend upon the standard method of processing which impart the required mechanical properties to the product. It has been well established that properties of a material can be enhanced by surface treatment or by reinforcing them with other elements in form of particles, powder or fiber form of different sizes and varying chemical composition. In the past few years, much focus has been given to techniques like cladding, spraying [1–3], laser remelting [4–6], ion implantation [7–9] and others for surface property enhancement. But, the mentioned methods have limitations like non-uniform distribution and agglomeration of second phase particles, surface reactions due to high processing temperature, leading to growth of unwanted phases and so on. Friction stir processing technique is a good replacement to avoid above-mentioned causes, because of its operational attributes. This technology is an adaptation of the principle of friction stir welding (FSW), a solid-state joining process originally developed at the Welding Institute in the UK. It allows one to transform and refine the structure of cast components and thus improve their properties and performance [10–12].

2 Friction Stir Processing

2.1 Principles and Process

The basic principle of FSP lies in the fact that this process has been widely used to modify the morphological changes and better control over layered surfaces of metallic components. This process additionally facilitates to modify the substrate surface at selective locations for enhancement of mechanical properties like toughness. The process came into existence on the concept of well-known solid-state joining process, i.e., friction stir welding (FSW). The only difference between the two is that the process is used to enhance the surface topography rather than to join the metals [13]. The friction-heated and plasticized metal is subjected to severe plastic deformation by stirring, which results in obtaining a homogeneous recrystallized fine-grained microstructure. The principal diagram of the FSP/FSW process is shown in Fig. 1 [14]. The process can be summarized into three major steps [13–15].

Step I: Friction between piece non-consumable rotating tool with pin and shoulder and work is used to produce the required heating and deformation. Rotation of tool provides uninterrupted heating of workpiece which plasticizes the metal and then transfers the metal from the front leading face of the pin to its trailing edge.
Step II: Revolving tool is used to produce the stirring motion and pushed laterally through the workpiece. A volume of processed material is produced by movement of materials from the front of the pin to the back of the pin.
Step III: Depth of penetration can be controlled by the tool shoulder and length of entry probe. When the tool shoulder moves on the metal surface, its rotation generates additional heat due to friction and plasticizes a larger cylindrical metal column around the inserted pin. The shoulder also imparts a forging force that initiates metal flow in upward direction caused by the tool pin.

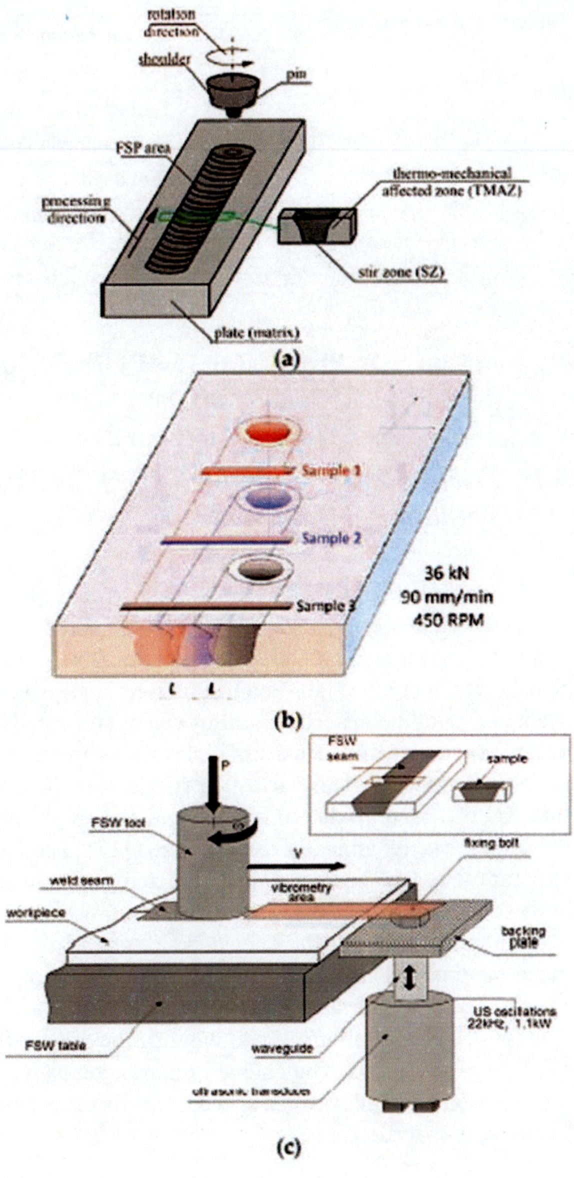

Fig. 1 Schematic diagram of FSP process: **a** and **b** multi-pass FSP process **c** FSP process with ultrasound assistance [14]

2.2 *Operation Parameters*

During the process, the material undergoes intense heat environment and severe plastic deformation, and hence, significant morphological changes can be observed.

The process is characterized by recrystallization and grain growth of additional phase particles. Various process parameters are tabulated in Table 1 [16].

Table 1 Process parameter

S. No	Process parameters
1	Tool rotational speed (rpm)
2	Tool travel speed (mm/min)
3	Tool inclination (degree)
4	Tool depth (mm)
5	Tool dimensions (shape, diameter, mandrel length (mm))
6	Diameter of backup rim (mm)

3 Friction Stir Processing (FSP) Technique in Aluminum Alloys

3.1 Friction Stir Processing (FSP) for Aluminum Alloys Without Reinforcement

Mechanical characteristics of crystalline structural materials are largely dependent on defects (porosity, shrinkage, cracks, etc.) and the coarse-grained structure of cast material. It is known that the strength of an alloy is enhanced by decreasing the grain size, according to the Hall–Petch equation. To turn a coarse-grained alloy into a fine-grained crystalline material, the alloy can be subjected to severe plastic deformation in order to produce a high density of dislocations and then to rearrange these dislocations to form a grain boundary array. In friction stir processing of structural alloys, the material is heated up due to friction and severe plastic deformation, and as a result, the stir zone is dynamically recrystallized [17]. The FSP forms a subsurface gradient structure with fine-equiaxed recrystallized grains of uniform size, due to which the alloy strength and hardness increase. According to the literature data, FSP was applied to aluminum. The FSP efficiency depends on the tool rotation rate, traverse speed and the number of passes and is determined for each type of alloy differently. Reduce the average grain size in aluminum alloys by 80–90%. Table 2 gives the experimental data for FSPed aluminum alloys, showing the effect of FSP on the grain structure and strength of the alloys. The choice of parameters such as the number of FSP passes, tool rotation rate and traverse speed depends on the aluminum alloy grade and leads to ambiguous results.

3.2 Friction Stir Processing (FSP) for Particle Reinforced Aluminum Alloys

The increasing curiosity toward particle-reinforced metallic material, particularly aluminum, has been seen in the past decade. These materials were commonly referred

Table 2 Experimental data on FSP of aluminum alloys

Material	Number of passes	Tool transverse speed (mm/min)	Tool rotation rate (rpm)	Average grain size of base alloy/average grain size after FSP (μm)	Mechanical properties	Reference number
6063	1		300	134/5.3	UTS: ↓ 6% Elong.: ↓ 42%	[18]
	2			134/8.6	UTS: ↓ 21% Elong.: ↓ 40%	
	1		500	134/5.5	UTS: no change Elong.: ↓ 28%	
	2			134/9.6	UTS: ↓ 10% Elong.: ↓ 29%	
	1		700	134/7.5	UTS:↑ 15% Elong.: ↓ 36%	
	2			134/9.7	UTS: ↑ 5% Elong.: ↓ 36%	
	1		1000	134/8	–	
	1		1200	134/7.8	–	
5086	1	30	1025	48/7	MH: ↑ 8.6% UTS: ↑ 3.8% Elong.: ↑ 30.7%	[19]
		80		48/10.5	MH: ↑ 8.6% UTS: ↑ 9.6% Elong.: ↑ 23%	
		150		48/3.8	MH: ↑ 10% UTS: ↑ 1.9% Elong.: ↑ 19.2%	

(continued)

Table 2 (continued)

Material	Number of passes	Tool transverse speed (mm/min)	Tool rotation rate (rpm)	Average grain size of base alloy/average grain size after FSP (μm)	Mechanical properties	Reference number
	12 (intermittent)	30		48/8	MH: ↑ 6.9% UTS: ↑ 5.7% Elong.: 40.3%	
		80		48/13.5	MH: ↑ 5.7% UTS: ↓ 19.2% Elong.: ↑ 19.2%	
		150		48/4	MH: ↑ 5.6% UTS: ↓ 3.8% Elong.: ↑ 15.3%	
	12 (continuous)	30		48/10.5	MH:↓ 4.3% UTS: ↑ 1.9% Elong.: ↑ 32.7%	
		80		48/15	MH: ↑ 1.4% UTS:↓ 30.8% Elong.: ↑ 3.8%	
		150		48/6	MH: ↑ 4.3% UTS: no change Elong.: ↑ 7.6%	
Al5052	1	80	1120	243/16.5	MH: ↑ 13.3%	[20]

(continued)

Table 2 (continued)

Material	Number of passes	Tool transverse speed (mm/min)	Tool rotation rate (rpm)	Average grain size of base alloy/average grain size after FSP (μm)	Mechanical properties	Reference number
AA5005-H34	1	127	490	–/10.7	MH: 42.6 HV UTS: 135.3 MPa Elong.: 34.4%	[21]
			970	–/18.5	MH: 38.9 HV UTS: 118.7 MPa Elong.: 37.3%	
			1200	–/20.4	MH: 37.9 HV UTS: 119.3 MPa Elong.: 41.4%	
AA1050	1	20	1600	42.85/10.58	MH: ↑ 47.6% CF: ↓ 13.8%	[22]
Al-12Si	1	28	1400	25/–	MH: ↑ 20.9% UTS: ↑ 15.1% Elong.: ↑ 3.7 times	[23]
A356	1	16	350	–/0.74	MH: 68 HV	[24]
	2			–/0.58	MH: 92 HV	
	3			–/0.45	MH: 113 HV	
	6			–/0.51	MH: 133 HV	

as hybrid composites or simply composites, metal matrix composites and others [20, 25–29]. This method is used for developing surface composite coatings with average thickness varying from 50 to 200 μm. Platelets, fiber or powder form of reinforcing additives is popularly used in specially milled grooves for the surface composites.

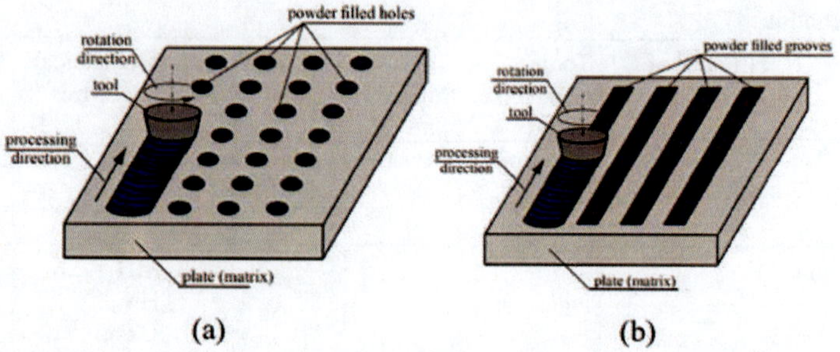

Fig. 2 Schematic of the FSP process with the reinforcement filling **a** into holes in the substrate and **b** into a groove

A typical subsurface structure of the stir zone with particle introduction is shown in Fig. 2.

Hard fine-grained particles can be admixed to the substrate during FSP by the following mechanism. Severe plastic deformation and dynamic recrystallization are obtained due to heat generation by friction of tool shoulder and pin in the metal matrix around the tool. It results in fine-equiaxed grain microstructure in the fabricating zone, with reinforcing particles located both inside the grains and grain boundaries [27, 28]. No interfacial reactions were observed within the matrix, when reinforcing particles were introduced. There exists a distinct boundary between the matrix and the reinforcement, as confirmed by the EDX plot done for AA6063 alloy FSPed with the addition of vanadium particles [28]. The performance characteristics and mechanical properties of FSP a reinforced material are also heavily impacted by the percentage of particle introduced. Also, multi-pass FSP creates a material with homogeneous particle distribution and enhanced tensile strength in comparison to single-pass material of same composition [20, 22, 30–32]. As found for cast aluminum alloy A356, addition of Ti_3AlC_2 in different proportions causes substantial elimination of coarse needle-like silicon particles and large primary aluminum dendrites and produces a uniform distribution of fine Si and Ti_3AlC_2 particles in the matrix. Microhardness and mechanical properties are also increased by two-fold [27]. The same above result of characteristic improvement was seen in AA7005 which is FSPed and particle reinforced by B_4C and TiB_2 in varying volume percentage [32]. Carbon-based materials like grapheme, carbon fibers, SWCNTs, MWCNTs, etc., are used as high-strength reinforcing agents for next-generation aerospace and automotive materials. Zhang et al. [31] demonstrated that the microstructure and mechanical properties of nanocomposites are closely related to the energy input. For an AA6061–graphene–TiB_2 hybrid nanocomposite synthesized by Nazari et al. [33], it was shown that the simultaneous addition of graphene and TiB_2 particles during FSP led to a significant grain structure refinement in the stir zone; the average grain size was reduced to <1 μm. Both grapheme and TiB_2 particles retained their structure while being high-speed stirred into the aluminum alloy matrix. The hardness of the aluminum alloy increased to ~102 HV, mainly at the cost of TiB_2 particles introduced together with graphene with an optimal hybrid ratio of 1 wt% graphene–20 wt% TiB_2 [33]. With the same ratio of components, the processed hybrid nanocomposites demonstrated the best combination of tensile properties, namely three times higher yield strength and ~70% higher ultimate tensile strength.

4 Applications of FSP for Aluminum Alloys

Friction stir processing (FSP) technique is used to convert a diversified microstructure to a more homogeneous, refined microstructure. There are variety of feasible methods available that can be imposed to alloys of different compositions, shapes and sizes. It is seen that the reprocessed areas have superior formability and strength as compared to the parent material, e.g., reprocessing in aluminum castings leads to elimination or fusion of voids, or extrusions can be enhanced in highly stressed areas. Along with superplastic forming, FSP offers the possibility to form intricate-shaped parts at elevated strain rates and in section thicknesses which were not possible by only conventional superplastic processing. Though FSP is a relatively novel processing technique, it has been broadly utilized to alter the microstructure of aluminum alloys for aerospace applications to achieve high-strength, fracture and fatigue-resistant aluminum alloys/composites. Due to variety of applications of friction stir processing, its operations have important implications for critical and high integrity components and for fabricated components of diesel engines.

5 Conclusions

This paper summarizes the latest progress in the study of friction stir processing of aluminum alloys. Intense plastic deformation and thermal effects during FSP cause the destruction of primary dendrites and second phase particles, grain refinement in the matrix, elimination of porosity, as well as the formation of a homogeneous fine-grained structure. It has emerged as a surface engineering technology that can locally eradicate defects of casting and produce refine microstructures, thereby enhancing ductility and strength, providing resistance to fatigue and corrosion, increasing formability and improvement in other properties. FSP can also produce fine-equiaxed microstructures throughout the thickness to set and provide super plasticity.

References

1. Bijalwan P, Pandey KK, Mukherjee B, Islam A, Pathak A, Dutta M, Keshri AK (2019) Tailoring the bimodal zone in plasma sprayed CNT reinforced YSZ coating and its impact on mechanical and tribological properties. Surf Coat Technol 377:124870
2. Gui M, Kang SB, Euh K (2005) Influence of spraying conditions on microstructures of Al-SiC p composites by plasma spraying. Metall Mater Trans A 36(9):2471–2480
3. Maharajan S, Ravindran D, Rajakarunakaran S, Khan MA (2020) Analysis of surface properties of tungsten carbide (WC) coating over austenitic stainless steel (SS316) using plasma spray process. Mater Today: Proc 27:2463–2468
4. Hu C, Baker TN (1997) A new aluminium silicon carbide formed in laser processing. J Mater Sci 32(19):5047–5051
5. Pantelis D, Tissandier A, Manolatos P, Ponthiaux P (1995) Formation of wear resistant Al–SiC surface composite by laser melt–particle injection process. Mater Sci Technol 11(3):299–303
6. Katipelli LR, Dahotre NB (2001) Mechanism of high temperature oxidation of laser surface engineered TiC/Al alloy 'composite' coating on 6061 aluminium alloy. Mater Sci Technol 17(9):1061–1068
7. Dong M, Cui X, Lu B, Jin G, Cai Z, Feng X, Liu Z, Wang H (2019) Effect of Ti+N and Zr+N ions implantation on mechanical and corrosion performance of carburized layer. Thin Solid Films 692:137597
8. Chen T, Castanon E, Gigax JG, Kim H, Balerio R, Fan J, Shao L et al (2019) Nitrogen ion implantation into pure iron for formation of surface nitride layer. Nucl Instrum Methods Phys Res, Sect B 451:10–13
9. Acciari HA, Palma DP, Codaro EN, Zhou Q, Wang J, Ling Y, Zhang Z et al (2019) Surface modifications by both anodic oxidation and ion beam implantation on electropolished titanium substrates. Appl Surf Sci 487:1111–1120
10. Ma ZY (2008) Friction stir processing technology: a review. Metall and Mater Trans A 39(3):642–658
11. Kumar RA, Kumar RA, Ahamed KA, Alstyn BD, Vignesh V (2019) Review of friction stir processing of aluminium alloys. Mater Today: Proc 16:1048–1054
12. Węglowski MS (2018) Friction stir processing—state of the art. Arch Civ Mech Eng 18(1):114–129
13. Mishra RS, Ma ZY (2005) Friction stir welding and processing. Mater Sci Eng: R Rep 50(1–2):1–78
14. Kalashnikov KN, Tarasov SY, Chumaevskii AV, Fortuna SV, Eliseev AA, Ivanov AN (2019) Towards aging in a multipass friction stir–processed AA2024. Int J Adv Manuf Technol 103(5–8):2121–2132
15. Li K, Liu X, Zhao Y (2019) Research status and prospect of friction stir processing technology. Coatings 9(2):129
16. Padhy GK, Wu CS, Gao S (2018) Friction stir based welding and processing technologies-processes, parameters, microstructures and applications: a review. J Mater Sci Technol 34(1):1–38
17. Luo PENG, McDonald DT, Xu W, Palanisamy S, Dargusch MS, Xia KENONG (2012) A modified Hall-Petch relationship in ultrafine-grained titanium recycled from chips by equal channel angular pressing. Scr Mater 66(10):785–788
18. Zhao H, Pan Q, Qin Q, Wu Y, Su X (2019) Effect of the processing parameters of friction stir processing on the microstructure and mechanical properties of 6063 aluminum alloy. Mater Sci Eng, A 751:70–79
19. Ramesh KN, Pradeep S, Pancholi V (2012) Multipass friction-stir processing and its effect on mechanical properties of aluminum alloy 5086. Metall Mater Trans A 43(11):4311–4319
20. Dolatkhah A, Golbabaei P, Givi MB, Molaiekiya F (2012) Investigating effects of process parameters on microstructural and mechanical properties of Al5052/SiC metal matrix composite fabricated via friction stir processing. Mater Des 37:458–464

21. Abrahams R, Mikhail J, Fasihi P (2019) Effect of friction stir process parameters on the mechanical properties of 5005–H34 and 7075–T651 aluminium alloys. Mater Sci Eng, A 751:363–373
22. Jain VKS, Varghese J, Muthukumaran S (2019) Effect of first and second passes on microstructure and wear properties of titanium dioxide-reinforced aluminum surface composite via friction stir processing. Arab J Sci Eng 44(2):949–957
23. Sun H, Yang S, Jin D (2018) Improvement of microstructure, mechanical properties and corrosion resistance of cast Al–12Si alloy by friction stir processing. Trans Indian Inst Met 71(4):985–991
24. Rao AG, Ravi KR, Ramakrishnarao B, Deshmukh VP, Sharma A, Prabhu N, Kashyap BP (2013) Recrystallization phenomena during friction stir processing of hypereutectic aluminum-silicon alloy. Metall Mater Trans A 44(3):1519–1529
25. Kumar H, Prasad R, Kumar P, Tewari SP, Singh JK (2020) Mechanical and tribological characterization of industrial wastes reinforced aluminum alloy composites fabricated via friction stir processing. J Alloy Compd 831:154832
26. Huang CW, Aoh JN (2018) Friction stir processing of copper-coated SiC particulate-reinforced aluminum matrix composite. Materials 11(4):599
27. Manochehrian A, Heidarpour A, Mazaheri Y, Ghasemi S (2019) On the surface reinforcing of A356 aluminum alloy by nanolayered Ti3AlC2 MAX phase via friction stir processing. Surf Coat Technol 377:124884
28. Abraham SJ, Dinaharan I, Selvam JDR, Akinlabi ET (2019) Microstructural characterization of vanadium particles reinforced AA6063 aluminum matrix composites via friction stir processing with improved tensile strength and appreciable ductility. Compos Commun 12:54–58
29. Dinaharan I, Sathiskumar R, Murugan N (2016) Effect of ceramic particulate type on microstructure and properties of copper matrix composites synthesized by friction stir processing. J Market Res 5(4):302–316
30. Bourkhani RD, Eivani AR, Nateghi HR (2019) Through-thickness inhomogeneity in microstructure and tensile properties and tribological performance of friction stir processed AA1050-Al2O3 nanocomposite. Compos Part B: Eng 174:107061
31. Zhang S, Chen G, Wei J, Liu Y, Xie R, Liu Q, Shi Q et al (2019) Effects of energy input during friction stir processing on microstructures and mechanical properties of aluminum/carbon nanotubes nanocomposites. J Alloy Compd 798:523–530
32. Pol N, Verma G, Pandey RP, Shanmugasundaram T (2019) Fabrication of AA7005/TiB2-B4C surface composite by friction stir processing: evaluation of ballistic behaviour. Def Technol 15(3):363–368
33. Nazari M, Eskandari H, Khodabakhshi F (2019) Production and characterization of an advanced AA6061-graphene-TiB2 hybrid surface nanocomposite by multi-pass friction stir processing. Surf Coat Technol 377:124914

Performance and Emissions Characteristics of Unmodified Diesel Engine Running on Waste Plastic Fuel, Diesel, and n-Butanol Blends

Parvesh Kumar, Harveer Singh Pali, Vikash Kumar, Sidharth, and Shailesh Mani Pandey

Abstract The disposal of single waste plastic is becoming from worst to disaster due to its continuous increasing demand in the market. The majority of single-used plastic ends to landfill after its useful life that leads to several problems including air and water pollution. In the present research, the single-used plastic is converted into liquid fuel by the pyrolysis process. The waste plastic fuel (WPF) is mixed with petroleum diesel in equal volumetric proportion to test it on the engine. To enhance the combustion characteristics of the fuel, oxygenated fuel is added to the blend, and the performance and emissions characteristics of the unmodified engine are tested. The better properties of WPF compared to diesel fuels improve the brake thermal efficiency of the engine by 1% for all loading conditions. However, a slight increase in emissions such as CO and NO_X is observed. Further improvement in the BTE is observed by adding a small proportion of n-butanol.

Keywords WPF · n-Butanol · Performance · Emissions characteristics · Single-use plastic

---

P. Kumar (✉)
Mechanical Engineering Department, Vaagdevi College of Engineering, Warangal, India
e-mail: pkkhatkar12@gmail.com

H. S. Pali
Department of Mechanical Engineering, NIT Srinagar, Srinagar, India
e-mail: hspali@nitsri.ac.in

V. Kumar
Department of Mechanical Engineering, MANIT Bhopal, Bhopal, India

Sidharth
Department of Mechanical and Automation Engineering, Maharaja Agrasen Institute of Technology, Delhi, India

S. M. Pandey (✉)
Department of Mechanical Engineering, NIT Patna, Patna, India
e-mail: smp.me@nitp.ac.in

P. Kumar · H. S. Pali · Sidharth
CASRAE, Delhi Technological University, Delhi, India

A. Maurya et al. (eds.), *Recent Trends in Mechanical Engineering*, Lecture Notes in Mechanical Engineering, https://doi.org/10.1007/978-981-19-7709-1_91

1 Introduction

Plastic and its products start gaining popularity right from its invention because of their unmatchable properties such as durability, lightweight, waterproof, higher load-carrying capacity. All these properties of plastic bring revolutionary changes to the packaging and food industries [1]. Further, plastic products are easy to mold; therefore, in recent years, the use of plastic is also increasing in the toy sector as well. As a result, the production of plastic and its products reached 335 million tons in 2016 [2] with an average increase of 10% annually from its commercialization [3]. As per the report, the per capita production of plastic waste in India is 4 kg and at 94th in the list which is headed by Australia and Singapore with per capita single-use plastic waste of 56 kg and 76 kg, respectively [4]. However, India is ranked third in overall single-use plastic waste production, whereas China is at the top followed by the USA. Furthermore, the production of plastic products is increasing continuously because of its high demand. But, plastic products have problems in their disposal technique after their useful life. The self-decay period of most of the plastic variants is around 500 years that makes it difficult to dispose of once used. As a result, only 9% of single-use plastic waste is recycled mechanically, 12% is cremated, and the rest ended with landfill and ocean fill [5]. The cremation, landfill, and ocean fill of single-use plastic waste lead to degradation of air and water quality that give birth to several airborne and waterborne diseases. Almost all varieties of plastic are a rich source of carbon and hydrogen with some traces of nitrogen, sulfur, and other compounds that make it a good source of energy [6]. Therefore, the conversion of single-use plastic waste into an energy-rich source helps to overcome environmental problems caused by plastic waste and provides a strategic alternative to the energy crisis. For this, pyrolysis is a more suitable and technically sound technique to convert single-use plastic waste to energy-rich end products like liquid fuel, syngas, and char. The end energy-rich products obtained after pyrolysis of plastic waste have very impressive properties, and all are user-friendly. The properties of liquid fuel obtained from pyrolysis are very comparable to petroleum diesel and are used in the unmodified diesel engine directly or as a blend with petroleum diesel. Also, the synthetic gas commonly known as syngas obtained is a rich mixture of methane, hydrogen, carbon monoxides, and carbon dioxide [7]. However, the properties of syngas are depended upon their composition. Further, the char obtained in the reactor after the pyrolysis process is richest in carbon content among all other products. The quality of end products obtained depends upon several factors including reactor type, reactor size, the catalyst used, heating process, heating rate, catalyst concentration, reacting environment, etc. Further, the quality of syngas obtained during gasification heavily depends upon the rate of steam or hot water supplied. Xue et al. [8] investigated the effect of operating temperature on the yield of liquid fuel obtained from high-density polyethylene in a continuous fluidized bed reactor. The authors took the temperature range between 525 and 675 °C and conclude that the maximum yield of 57.6 wt% can be obtained at 625 °C; however, further increase in temperature leads to the formation of light gases that decrease the yield of liquid fuel. Artetxe et al. [9] determine the effect of temperature and gas flow rate on the composition and product yield. Therefore, it can be concluded that the yield of the liquid fuel obtained during pyrolysis is dependent upon operating conditions and parameters; however, the quality of fuel obtained is nearly constant and very comparative to the petroleum diesel fuel. Conclusively, it can be said that the conversion of plastic waste into a liquid fuel contributes to the solution of air pollution and crude oil scarcity. However, long engine trials are not done with waste plastic fuel (WPF), so it is very difficult to project its effectiveness for the long run at present, but the experimental results obtained by the researchers are very positive and acceptable. Therefore, for the safer side, it is better to run the engine with the WPF and diesel blends. The blend varies according to the requirement and availability. To improve the combustion characteristics of the fuel, some oxygen rich additives such as biodiesel and higher alcohol can be used. The use of higher alcohols provide positive results while used with diesel fuel directly or in diesel–biodiesel blends. The higher alcohols such as butanol, octanol have better combustion properties such as cetane number, calorific value, pour point, and cloud point as compared to basic and lower carbon alcohol. All these properties are suitable for diesel engines this is the reason why higher carbon alcohols are gaining popularity over the lower carbon alcohols. The calorific value and cetane values of alcohols improve with an increase in carbon atoms, but the synthesis becomes difficult. Also, at present, only a limited type of alcohols is synthesized naturally, whereas most alcohols are synthesized through a chemical process. Therefore, it becomes important to select the right alcohol for the research. In the present work, the WPF and petroleum diesel are mixed in an equal volume percentage. The n-butanol is selected to improve the combustion characteristics of WPF and diesel blend. The composition of n-butanol in the blend is limited to 10% as a further increase in concentration decreases the calorific value of the blend. The tested fuels obtained are D-100, D-WPF, 95D-WPF05nB, 90D-WPF10nB, and 85D-WPF15nB. The prepared test fuels are tested on unmodified diesel engine to analyze the performance and emissions characteristics. The detailed nomenclature of the tested fuels is provided in the experimental section.

2 Experimental

2.1 Materials and Preparations

To prepare waste plastic fuel, the waste plastic products such as cold drinks and water bottles and food packaging papers are collected from the central canteen of the Vaagdevi College of Engineering, Warangal. The collected plastic waste is first cut into smaller pieces of 5–10 mm size. The cut pieces are thoroughly washed in the water and put into sunlight for drying. The initial process of preparation of the raw material takes 2–5 days depending on the availability of sunlight. It makes sure that all the moisture in the raw material is removed completely for a better pyrolysis process. The chemicals such as catalysts and alcohols are purchased from an institutional local vendor. The catalyst bed is synthesized in the Department of Mechanical engineering of Vaagdevi College of Engineering, Warangal.

2.2 *Pyrolysis Setup*

To perform the pyrolysis process, one setup is developed in the Department of Mechanical Engineering of VCE, Warangal. The setup consists of one pyrolysis reactor attached with a thermocouple and pressure gauge for continuous monitoring of temperature and pressure. The reactor is heated with the help of a coil-based electric heater purchased from the local market. One setup for catalyst bed is also developed in the department only. This catalyst bed can carry catalysts and is easily installed over the reactor head. The gases generated are passed through the catalyst bed where they cracked into simple molecules. The gases released during the pyrolysis process are a mixture of condensable and non-condensable gases. Therefore, one condenser is installed to the outlet passage of the reactor in such a way that the pyrolysis gases will pass through the condenser after pass through the catalyst bed. To maintain the controlled temperature in the condenser, one chiller is attached to the condenser. The chiller cools the hot fluid coming out from the condenser. In most cases, the fluid used in the condenser is water so as in the present work. The condensable gases lose the heat in the condenser and are converted into liquid fuel that can be collected at the bottom of the condenser, whereas the uncondensed gases remain in gaseous form even after losing heat. One entry to the reactor is provided for the inert gas supply, so to eliminate any traces of the oxygen gas in the setup, nitrogen is passed through the complete setup before each pyrolysis trail. The power to the electric heater is supplied through temperature controller and transformer arrangement and designed in such a way that the power supply can be changed as per required temperature and heating rate. The thermocouple attached to the reactor is connected to a temperature controller and indicator to maintain the desired temperature during the pyrolysis process. The schematic diagram of the pyrolysis experiment setup is shown in Fig. 1.

2.3 *Production of WPF*

First of all, the dried small pieces of plastic waste are put into the reactor. After that, to vanish any traces of the oxygen in the setup, the nitrogen gas is supplied at least for 10 min. All the valves are closed just after stopping the nitrogen supply. This process stops any combustion during the pyrolysis process. Now heating to the reactor is started with a rate of 10 °C/min. Only 80% of the reactor is filled with solid plastic waste pieces that become 30% after melting. The liquid plastic starts boiling after some time depending upon the variety of plastic waste used. Initially, the vapor

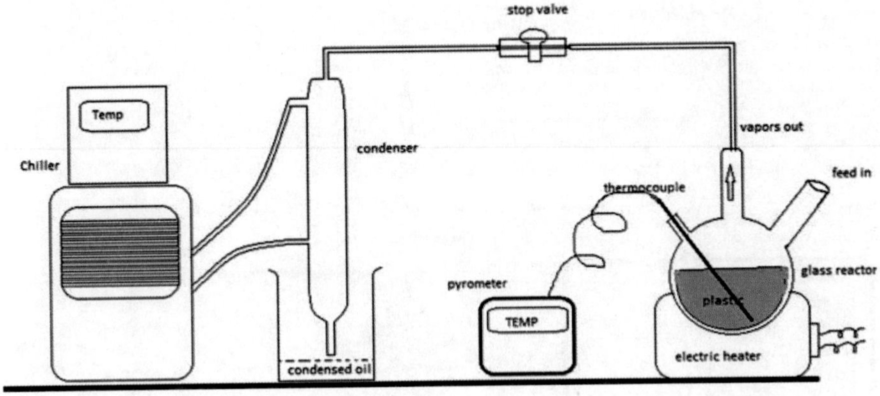

Fig. 1 Setup for pyrolysis and condensation

formed is rich in wax and uncondensed gases due to lower reaction temperature. Further, the continuous heating of the reactor increases the reaction temperature that improves the rate of conversion of the complex compound to the smaller and simpler compound. The reaction temperature maintained in the present research is 600–650 °C. The generated gases pass through the catalyst bed that further enhances the cracking rate of complex compounds. The simple compounds then pass through the series of the condenser where the condensable gases get condensed and converted into liquid, whereas the uncondensed gases leave the condenser in gaseous form only. The condensed liquid get collected from the bottom of the condenser for further analysis and usage, whereas the uncondensed gases which are very rich in carbon and hydrogen are collected in a cylinder and can be used for any heating purposes. Sometimes, the hybrid heaters are used that can use both electric powers when the supply of gaseous fuel is not available and gaseous fuels when it is available in enough quantity. The complete conversion of plastic waste into end products takes 4–10 h depending upon various factors. In the end, when no further release of gases is observed, it is a clear indication of completion of the pyrolysis process for that lot. The content of liquid fuel, gaseous fuel, and char obtained at the end of the pyrolysis process vary in the range of 40–70%, 20–50%, and 5–15%, respectively. In the present research, 58% of plastic waste is converted into liquid fuel, the char left in the reactor is 9 w/w% to solid plastic waste, and the remaining 33% is uncondensed gases.

2.4 Blend Preparation and Engine Setup

The properties of WPF obtained during the current trial are very much similar to the petroleum diesel fuel; therefore, the blend is prepared by taking both in equal percentage by volume, i.e., 50% of petroleum diesel and 50% of WPF. To enhance

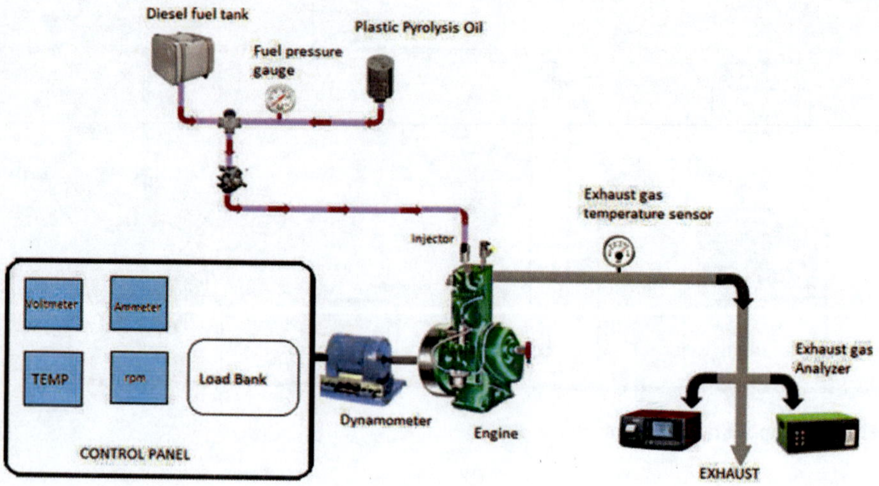

Fig. 2 Schematic diagram of the experimental test rig

the combustion characteristics of the fuel, isobutanol is mixed in the prepared blend in 5, 10, and 15% by volume. The physicochemical properties of all the tested fuels are measured as per the ASTM standard and presented in detail in the next section. The prepared blends are tested on the engine. The engine test rig used in the present research is the same (Kirloskar four-stroke, single-cylinder, vertical, air-cooled, manually cracked started, gravity fuel feeding system direct injection diesel engine coupled with an eddy current dynamometer) used in the last research. The schematic diagram of the test rig used in the present research is given in Fig. 2.

3 Results and Discussion

3.1 *Physicochemical Properties*

The physicochemical properties of all the tested fuels are measured as per ASTM standard, and the processes opt to measure the properties of all tested fuels are the same opt in our previous research [10]. The nomenclature and the results obtained for physicochemical properties of all the tested fuels are given in Tables 1 and 2, respectively.

Table 1 Nomenclature of all tested fuels

Notification	Nomenclature
D100	100% petroleum diesel
WPF	100% waste plastic fuel
D-WPF	50% petroleum diesel + 50% waste plastic fuel
nB	100% n-butanol
95D-WPF05nB	95% D-WPF + 05% n-butanol
90D-WPF10nB	90% D-WPF + 10% n-butanol
85D-WPF15nB	85% D-WPF + 15% n-butanol

Table 2 Physicochemical properties of all tested fuels

Property	Density	Viscosity	Calorific value	Cetane number	Flash point	Autoignition temperature
D100	823	2.6	43.3	45	53	230
WPF	816	2.8	45.2	51	42	195
D-WPF	820.5	2.7	44.65	50	43	202
nB	807	2.3	32.6	16	38	358
95D-WPF05nB	820	2.68	44.05	48	43	210
90D-WPF10nB	819	2.66	43.45	47	43	218
85D-WPF15nB	818	2.64	42.84	45	42	225

3.2 Performance Characteristics

The performance of all tested fuels is analyzed by testing them on an unmodified diesel engine. The parameters such as brake thermal efficiency (BTE), brake specific energy consumption (BSEC), and exhaust gas temperature (EGT) are analyzed in the current section. The effect of blending WPF with petroleum diesel is analyzed first, and after that, the effect of n-butanol on the performance of the blend is also analyzed. The results obtained are compared with conventional petroleum diesel results, taking it as baseline data. The comparative study of all tested fuels with baseline data for BTE–BSEC and EGT is shown in Figs. 3 and 4, respectively.

The blending of WPF with petroleum diesel in equal proportion improves the properties such as calorific value and cetane number of the diesel fuel blend. However, other properties remain very comparable to petroleum diesel. Therefore, the prepared blend can be used in the unmodified diesel engine. After testing the blend, it is analyzed that the improved properties of the blend lead to better combustion, and as a result, the improvement in the BTE can be observed for D-WPF as compared to conventional diesel fuel. Further, adding n-butanol in the D-WPF blend provides free oxygen to the fuel that improves the combustion characteristics of fuel. Hence, improvement in the BTE of the engine can be observed with an increase in the content of n-butanol in the blend. However, the calorific value and cetane number of n-butanol

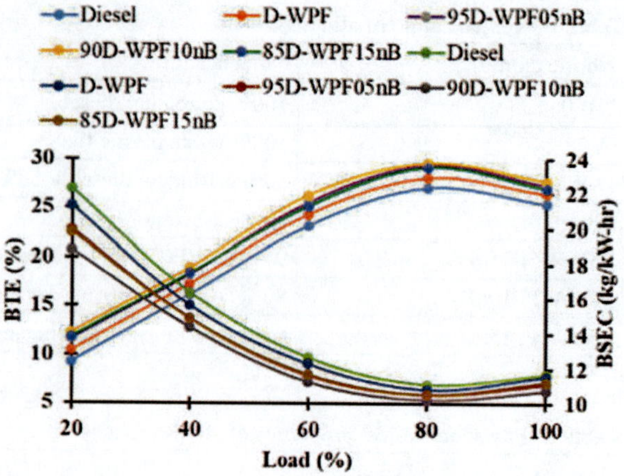

Fig. 3 Trends of BTE and BSEC for all test fuels

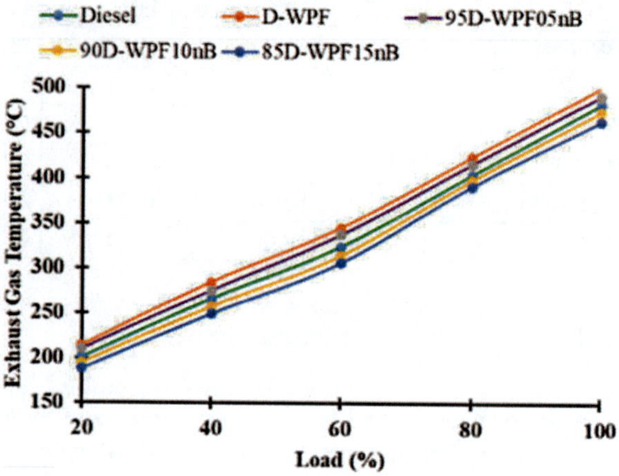

Fig. 4 Variation in EGT with respect to load

are less as compared to petroleum diesel and WPF; therefore, the adding of n-butanol in the blend beyond some limit put an adverse effect on the fuel properties. This can be observed with the 85D-WPF15nB blend result. The BTE of the engine increases with an increase in n-butanol content in the fuel up to 10%, and once the n-butanol content increase beyond 10%, the reduction in BTE of the engine is noticed. The BSEC is inversely proportional to the BTE; therefore, the trend followed by all the fuels in BSEC is opposite to the trend followed in BTE. However, the trends of EGT are slightly different from BTE and BSEC. The exhaust gas temperature for D-WPF is higher than conventional diesel fuel for all working loads. But, the addition of n-butanol provides some cooling effect to the fuel, and as a result, the increasing content of the n-butanol in the fuel decreases the EGT of the engine.

3.3 *Emissions Characteristics*

Emissions such as carbon monoxide (CO), unburnt hydrocarbons (HC), and oxides of nitrogen (NO_X) of unmodified diesel engines are analyzed in the present section for all the tested fuels. The combative results of all the emissions for all tested fuels are given in Figs. 5, 6, and 7, respectively. It is observed that the CO emissions of the D-WPF fuel are slightly higher when compared with conventional diesel fuel. The higher CO emission might be due to a slightly higher density of blend while compared with petroleum diesel. However, the addition of n-butanol provides extra oxygen in the combustion chamber, and as a result, the CO emissions decrease with an increase in the composition of n-butanol in the fuel. However, a higher percentage of n-butanol further increases the CO emissions, and the reason is lower combustion temperature in the combustion chamber. On the other hand, the HC emissions of all the tested fuels are observed less compared to baseline data. This is due to the higher cetane number of D-WPF and oxygenated fuels. The HC emissions of oxygenated fuels are found lesser as compared to the blend and the diesel fuel. Similar to the CO emissions, the increase in the composition of n-butanol beyond 10% starts increasing the HC emissions, and the reason is the same as that of CO emissions. The oxides of nitrogen are dominant in diesel engines while compared to other internal combustion engines. The main reason for the formation of NO_X is higher temperature and pressure along with the availability of free oxygen in the combustion chamber. All the conditions are more favorable in diesel engines compared to gasoline engines. It is observed that NO_X emissions of D-WPF are higher than conventional diesel fuel because of higher combustion temperature. However, the addition of n-butanol reduces the energy content of the fuel and also provides a cooling effect in the combustion chamber before and during combustion. As a result, the NO_X emissions reduce with increasing the n-butanol content in the blend. Conclusively, the highest NO_X emissions of 17.21 g/kWh are observed for D-WPF, whereas the lowest NO_X emissions of 3.4 g/kWh are observed for 85D-WPF15nB at 20% load and 100% load, respectively.

4 Conclusions

Some important conclusions drawn from the present research work are as follows:

- Most physicochemical properties of D-WPF are better or comparable to petroleum diesel. Therefore, WPF can be used as an energy source for conventional diesel engines.
- The addition of n-butanol improves the combustion characteristics of the fuel.
- The improvement of 1% of the BTE is observed for D-WPF compared to diesel fuel, and the BTE further increases by 1.3% with the addition of 10% n-butanol in the D-WPF.

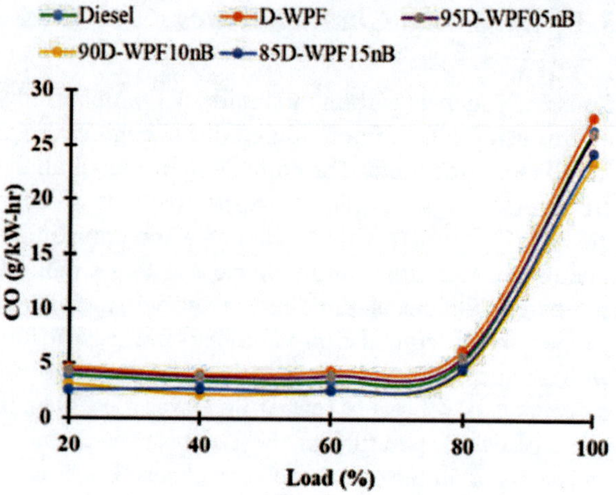

Fig. 5 CO emissions trend for all the tested fuels

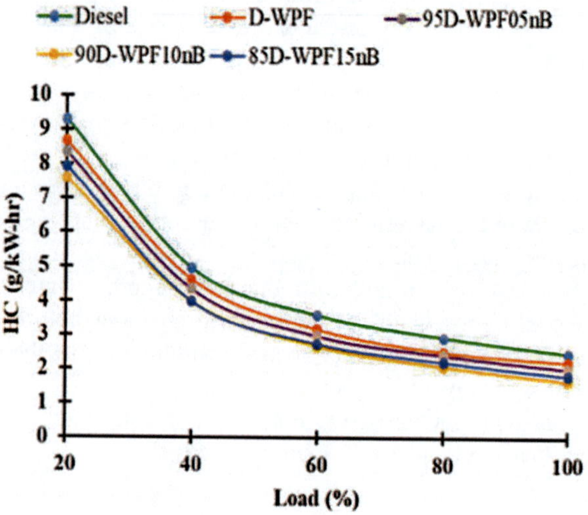

Fig. 6 HC emissions trend for all the tested fuels

- The EGT for D-WPF is decreased by 50 °C by adding 15% of n-butanol by volume.
- The emissions such as CO and NO_X are increased for the D-WPF compared to conventional diesel. However, with the addition of n-butanol, a reduction in all the emissions is observed.

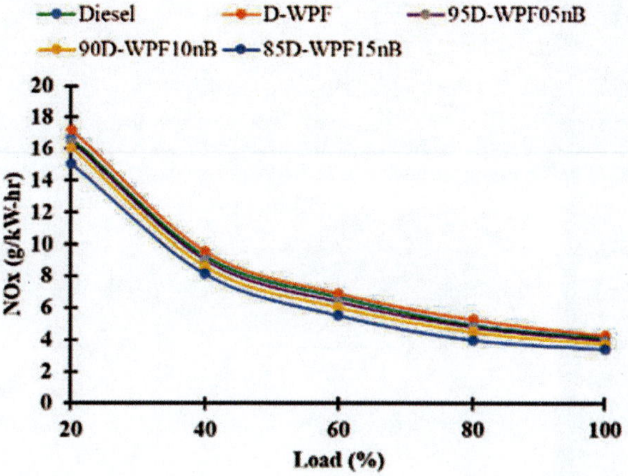

Fig. 7 NO_X emissions trend for all the tested fuels

References

1. Hammoodi SI, Almukhtar RS (2019) Thermal pyrolysis of municipal solid waste (MSW). IOP Conf Ser Mater Sci Eng 579
2. Quesada L et al (2019) Optimization of the pyrolysis process of a plastic waste to obtain a liquid fuel using different mathematical models. Energy Convers Manag 188:19–26
3. Damodharan D, Sathiyagnanam AP, Rana D, Rajesh Kumar B, Saravanan S (2017) Extraction and characterization of waste plastic oil (WPO) with the effect of n-butanol addition on the performance and emissions of a DI diesel engine fueled with WPO/diesel blends. Energy Convers Manag 131:117–126
4. Charles D, Kimman L, Saran N (2021) The plastic waste makers index. 86
5. Mangesh VL, Padmanabhan S, Tamizhdurai P, Ramesh A (2020) Experimental investigation to identify the type of waste plastic pyrolysis oil suitable for conversion to diesel engine fuel. J Clean Prod 246:119066
6. Tulashie SK, Boadu EK, Dapaah S (2019) Plastic waste to fuel via pyrolysis: a key way to solving the severe plastic waste problem in Ghana. Therm Sci Eng Prog 11:417–424
7. Saebea D, Ruengrit P, Arpornwichanop A, Patcharavorachot Y (2020) Gasification of plastic waste for synthesis gas production. Energy Rep 6:202–207
8. Xue Y, Zhou S, Brown RC, Kelkar A, Bai X (2015) Fast pyrolysis of biomass and waste plastic in a fluidized bed reactor. Fuel 156:40–46
9. Artetxe M et al (2015) Styrene recovery from polystyrene by flash pyrolysis in a conical spouted bed reactor. Waste Manag 45:126–133
10. Kumar P, Kumar N (2016) Experimental investigation of Jatropha oil methyl ester (JOME) as pilot fuel with CNG in a dual fuel engine. Biofuels 7:511–520